# Study on Developmental and Behavioral Theories of Honeybee Biology

# 蜜蜂发育及行为生物学理论研究

曾志将 等 著

Edited by Zeng Zhijiang *et al.*

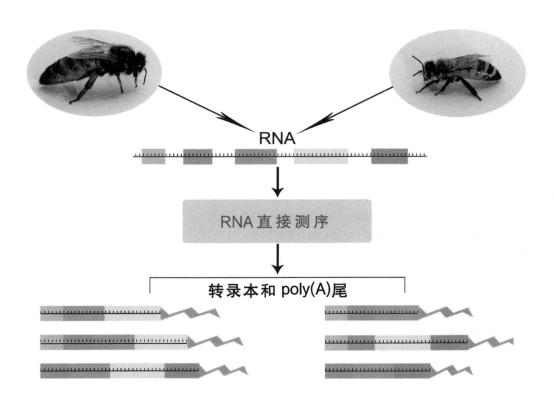

中国农业科学技术出版社

图书在版编目（CIP）数据

蜜蜂发育及行为生物学理论研究 = Study on Developmental and Behavioral Theories of Honeybee Biology/ 曾志将等著 . —— 北京：中国农业科学技术出版社，2023.9
ISBN 978-7-5116-6440-2

Ⅰ . ①蜜⋯ Ⅱ . ①曾⋯ Ⅲ . ①蜜蜂—生物学—研究 Ⅳ . ① S891

中国国家版本馆 CIP 数据核字（2023）第 176173 号

责任编辑　朱　绯
责任校对　马广洋
责任印制　姜义伟　王思文

| | |
|---|---|
| 出 版 者 | 中国农业科学技术出版社 |
| | 北京市中关村南大街 12 号　　邮编：100081 |
| 电　　话 | （010）82109707（编辑室）　（010）82109702（发行部） |
| | （010）82109709（读者服务部） |
| 网　　址 | https://castp.caas.cn |
| 经 销 者 | 各地新华书店 |
| 印 刷 者 | 北京建宏印刷有限公司 |
| 开　　本 | 210 mm × 285 mm　1/16 |
| 印　　张 | 35 |
| 字　　数 | 1100 千字 |
| 版　　次 | 2023 年 9 月第 1 版　2023 年 9 月第 1 次印刷 |
| 定　　价 | 300.00 元 |

◀ 版权所有·侵权必究 ▶

# 内容提要

本书是江西农业大学蜜蜂研究所曾志将教授团队在国家自然科学基金、国家蜂产业技术体系、江西省一流学科建设等项目连续资助下取得的研究成果。内容包括:"蜜蜂级型分化及其分子机制""优质蜂王培育及其分子机制""蜜蜂行为生物学及其分子机制""蜜蜂基因组生物学及CRISPR/Cas9技术"等蜜蜂发育及行为生物学主题。全书知识新颖、理论性强,可以使读者了解更多的蜜蜂生物学新知识,可供广大生物科技工作者及研究生参考使用。

# 序

蜜蜂生物学是养蜂学中历史最悠久学科。因为人类饲养蜜蜂，是从观察蜜蜂生物学习性开始，同时蜜蜂生物学也是养蜂学中发展最迅速和最热门的领域。

《蜜蜂发育及行为生物学理论研究》一书是江西农业大学蜜蜂研究所曾志将教授研究团队取得的最新研究成果。曾志将教授是中国养蜂学会副理事长、国家蜂产业技术体系岗位科学家、我国著名蜜蜂生物学家，在蜜蜂生物学研究方面颇有建树，在我国蜂学界享有崇高声誉。多年来，在国家自然科学基金、国家蜂产业技术体系、江西省一流学科建设等项目连续资助下，曾志将教授带领团队成员，围绕蜜蜂生物学这一主题进行了系统研究，并在 *Current Biology*、*Molecular Ecology* 等世界高水平期刊发表学术论文，取得了许多原创性研究成果。

曾志将教授也很热心全国蜜蜂生物学青年人才培养，2019年6月21—23日在江西农业大学举办了全国蜜蜂生物学青年英才培训班，拓宽了青年人才视野，同时提升了青年人才研究能力。

我相信《蜜蜂发育及行为生物学理论研究》的出版发行，会让读者了解到更多的蜜蜂发育及行为生物学新知识，该书具有很好的学术性和创新性，将有助于推动我国养蜂业高质量、可持续发展。

特作此序，以示祝贺。

中国养蜂学会理事长 吴杰 研究员

2023年5月

# 序

蜜蜂是自然生态系统中，与人类关系极为密切的昆虫。养蜂业作为畜牧业中的特色产业，在满足人民美好生活需要、促进农业绿色发展、保护生态环境、助力乡村振兴等方面都具有重要意义。

曾志将教授是我国著名的蜜蜂专家，他对蜜蜂事业的执着和热爱，给我留下了深刻印象。1987年，他从福建农学院养蜂专业毕业，被分配到江西农业大学从事"养蜂学"教学和研究工作，由于"养蜂学"在江西农业大学属于新开设课程，没有养蜂教学团队、养蜂教学蜂场、养蜂实验室和养蜂科研课题。面对困难，他知难而进，在学校支持下，从零起步，从无到有，用蜜蜂"采蜜"式的办法，一点一滴地积累，经过36年的努力，慢慢把"芝麻"变成了"西瓜"。目前，江西农业大学的养蜂教学与科研在全国养蜂行业中具有重要影响。

曾志将教授带领的蜜蜂研究团队瞄准国际前沿，在蜜蜂生物学研究领域取得了大量原创研究成果，公开发表了100多篇蜜蜂生物学研究论文，为我国蜜蜂科技创新作出重要贡献。

《蜜蜂发育及行为生物学理论研究》一书主要是曾志将教授团队2015—2023年有关蜜蜂生物学研究成果展示，内容丰富，创新性强，学术水平高，是一部不可多得蜜蜂生物学研究专著。

《蜜蜂发育及行为生物学理论研究》一书的出版必将有力推动蜜蜂生物学研究深入开展，进而为我国养蜂业健康稳定提供技术支撑。

国家蜂产业技术体系首席科学家<br>
中国农业科学院蜜蜂研究所所长　　研究员<br>
2023年5月

# 前　言

蜜蜂生物学是养蜂学领域中最基础的学科。人类饲养蜜蜂是从观察蜜蜂生物学特性逐渐开始的，并且随着对蜜蜂生物学特性的深入了解，蜜蜂饲养技术得到了不断提高。养蜂生产不但可为人类提供营养丰富的蜂产品，更通过蜜蜂授粉大幅度提高农作物产量和品质，同时对维持自然生态系统的生物多样性有重要意义。另外，蜜蜂是一种理想的社会型模式昆虫，蜜蜂生物学研究结果对整个社会生物学研究有重要的参考价值，因此，蜜蜂生物学研究长期以来一直是热门领域，每年有大量论文发表，文献浩如烟海。

2015年1月，本团队在科学出版社出版了专著《蜜蜂生物学理论中若干问题研究》，书中主要收录了2008—2014年本团队在国内外学术刊物公开发表的38篇有关蜜蜂生物学理论研究论文。

光阴似箭，日月如梭。《蜜蜂生物学理论中若干问题研究》出版已8年。在这8年里，在国家自然科学基金、国家蜂产业技术体系、江西省一流学科建设等项目连续资助下，本团队围绕蜜蜂发育及行为生物学中的"蜜蜂级型分化及其分子机制""优质蜂王培育及其分子机制""蜜蜂行为生物学及其分子机制""蜜蜂基因组生物学及CRISPR/Cas9技术"等问题进行了系统研究。非常欣慰的是，在大家的共同努力下，取得了可喜的研究成果，并获批成立了国内第一个专门的蜜蜂生物学省部级重点实验室"江西省蜜蜂生物学与饲养重点实验室"。

《蜜蜂发育及行为生物学理论研究》集结了本团队2015—2023年已公开发表的39篇有关蜜蜂发育及行为生物学理论研究论文，作为《蜜蜂生物学理论中若干问题研究》的延续，供大家参考使用。

《蜜蜂发育及行为生物学理论研究》收录的39篇文章公开发表在 Current Biology, Molecular Ecology, Insect Biochemistry and Molecular Biology, Insect Science, iScience, International Journal of Molecular Sciences, Journal of Comparative Physiology A, Science of Nature, G3-Genes Genomes Genetics, Frontiers in Genetics, Frontiers in Physiology, BMC Developmental Biology, Insectes, Scientific Reports, Apidologie, Journal of Economic Entomology, Archives of Environmental Contamination and Toxicology, Journal of Applied Entomology, Insectes Sociaux, Life, Journal of Asia-Pacific Entomology, Journal of Apicultural Science, African Entomology, Sociobiology，以及《中国农业科学》《昆虫学报》《江西农业大学学报》等国内外学术刊物上。其中多篇论文获得《中国科学报》、科学网、New Scientist、Science Daily、PNAS-Front Matter 等国内外权威科学媒体专题报道。

世界著名蜜蜂生物学家、1973年诺贝尔奖获得者卡尔·冯·弗里希（Karl Von Frisch）教授曾经说过，蜜蜂世界是一个"魔井"，当我们发现的新知识越多，越有更多奥秘等待我们去探索。

让我们共同努力，去探索蜜蜂世界中更多的奥秘！

江西农业大学蜜蜂研究所　曾志将<br>
2023年5月于南昌

# 目 录

## I 蜜蜂级型分化及其分子机制
## Honeybee Caste Differentiation and its Molecular Mechanism

1. Extent and Complexity of RNA Processing in Honey Bee Queen and Worker Caste Development ········ 003
2. The Diverging Epigenomic Landscapes of Honeybee Queens and Workers Revealed by Multiomic Sequencing ·········· 017
3. Phenotypic Dimorphism between Honeybee Queen and Worker is Regulated by Complicated Epigenetic Modifications ·········· 037
4. A Comparison of Honeybee (*Apis mellifera*) Queen, Worker and Drone Larvae by RNA-Seq ············· 050
5. DNA Methylation Comparison between 4-day-old Queen and Worker Larvae of Honey Bee ············ 062
6. H3K4me1 Modification Functions in Caste Differentiation in Honey Bees ·········· 073

## II 优质蜂王培育及其分子机制
## High Quality Queen Bee Breeding and its Molecular Mechanism

1. A Maternal Effect on Queen Production in Honey Bees ·········· 087
2. Making a Queen: an Epigenetic Analysis of the Robustness of the Honey Bee (*Apis mellifera*) Queen Developmental Pathway ·········· 102
3. Transgenerational Accumulation of Methylome Changes Discovered in Commercially Reared Honey Bee (*Apis mellifera*) Queens ·········· 114
4. Transcriptomic, Morphological and Developmental Comparison of Adult Honey Bee Queens Reared from Eggs or Worker Larvae of Differing Ages ·········· 131
5. Effects of Commercial Queen Rearing Methods on Queen Fecundity and Genome Methylation ············ 142
6. Honeybee (*Apis mellifera*) Maternal Effect Causes Alternation of DNA Methylation Regulating Queen Development ·········· 152

7. Transcriptome Comparison between Newly Emerged and Sexually Matured Bees of *Apis mellifera* ...... 162
8. Effects of Queen Cell Size and Caging Days of Mother Queen on Rearing Young Honey Bee Queens *Apis mellifera* L. ...... 171
9. Feeding Asian Honeybee Queens with European Honeybee Royal Jelly Alters Body Color and Expression of Related Coding and Non-coding RNAs ...... 180
10. 全基因组 DNA 甲基化揭示西方蜜蜂表观遗传印迹 ...... 199

# III 蜜蜂行为生物学及其分子机制
## Behavioral Biology of Honeybee and its Molecular Mechanism

1. RFID Monitoring Indicates Honeybees Work Harder before a Rainy Day ...... 211
2. Starving Honey Bee (*Apis mellifera*) Larvae Signal Pheromonally to Worker Bees ...... 216
3. The Involvement of a Floral Scent in Plant-honeybee Interaction ...... 231
4. The Capping Pheromones and Putative Biosynthetic Pathways in Worker and Drone Larvae of Honey Bees *Apis mellifera* ...... 247
5. Differential Protein Expression Analysis following Olfactory Learning in *Apis cerana* ...... 260
6. Deltamethrin Impairs Honeybees (*Apis mellifera*) Dancing Communication ...... 274
7. The Effects of Sublethal Doses of Imidacloprid and Deltamethrin on Honeybee Foraging Time and the Brain Transcriptome ...... 284
8. Honeybees (*Apis mellifera*) Modulate Dance Communication in Response to Pollution by Imidacloprid ...... 295
9. Absence of Nepotism in Waggle Communication of Honeybees (*Apis mellifera*) ...... 309
10. Influence of RNA Interference-mediated Reduction of *Or*11 on the Expression of Transcription Factor *Kr-h*1 in *Apis mellifera* Drones ...... 317
11. Cloning and Expression Pattern of Odorant Receptor 11 in Asian Honeybee Drones, *Apis cerana* (Hymenoptera, Apidae) ...... 331
12. Expression Patterns of Four Candidate Sex Pheromone Receptors in Honeybee Drones (*Apis mellifera*) ...... 344
13. Age Components of Queen Retinue Workers in Honey Bee Colony (*Apis mellifera*) ...... 352
14. Genotypic Variability of the Queen Retinue Workers in Honeybee Colony (*Apis mellifera*) ...... 356
15. 蜂王上颚腺信息素对雄蜂候选性信息素受体基因表达的影响 ...... 364
16. 蜜蜂蜂王与雄蜂幼虫饥饿信息素鉴定及其生物合成通路 ...... 378
17. 东方蜜蜂幼虫封盖信息素含量及生物合成通路 ...... 391

# IV 蜜蜂基因组生物学及 CRISPR/Cas9 技术
## Honeybee Genome Biology and CRISPR/Cas9

1. A Chromosome-scale Assembly of the Asian Honeybee *Apis cerana* Genome ················ 407
2. Signatures of Positive Selection in the Genome of *Apis mellifera carnica*: A subspecies of European Honeybees ················ 417
3. Histomorphological Study on Embryogenesis of the Honeybee *Apis cerana* ················ 431
4. Dynamic Transcriptome Landscape of Asian Domestic Honeybee (*Apis cerana*) Embryonic Development Revealed by High-quality RNA-seq ················ 443
5. High-efficiency CRISPR/Cas9-mediated Gene Editing in Honeybee (*Apis mellifera*) Embryos ··········· 462
6. A Comparison of RNA Interference via Injection and Feeding in Honey Bees ················ 477

附录 A  彩图 ················ 489

作者简介 ················ 548

# 蜜蜂级型分化及其分子机制

## Honeybee Caste Differentiation and its Molecular Mechanism

蜂王和工蜂都是由受精卵发育而成的雌性蜜蜂，遗传物质完全相同。蜂王和工蜂在幼虫阶段，由于发育过程中得到的食物和发育空间不同，其结果是两者不仅在外观形态上差异很大，而且在生殖能力、寿命、行为等方面也迥然不同。蜜蜂级型分化调控是一个非常复杂的过程，是很多因素共同调控的结果。蜜蜂级型分化调控机制可为优质蜂王培育提供新思路、新方法。

传统的研究认为级型分化过程是受大量基因的表达差异调控，然而一个基因通过可变剪切可以产生多个不同的转录本，而这些转录本可以翻译成不同的蛋白质，进而产生不同的生物学功能。利用 Direct RNA sequencing 技术分别对 2 日龄、4 日龄、6 日龄蜂王和工蜂测序发现：蜂王和工蜂在发育过程中存在十分复杂的 RNA 加工和修饰过程。首次发现蜜蜂级型分化过程受到了更为精确转录本表达差异调控，并证明这些差异转录本来自一个非常灵活多变且复杂可变剪切加工系统。同时首次发现跟随在每条转录本后面的 poly（A）尾可以通过负调控转录本表达，参与调控蜜蜂级型分化。

利用 Hi-C、ATAT-seq、ChIP-seq 和 RNA-seq 四组学技术，比较分析了蜂王和工蜂在发育早期（2 日龄幼虫）和发育关键期（4 日龄幼虫）的表观遗传调控分子机制。结果发现：2 日龄蜂王幼虫和工蜂幼虫染色体 Hi-C（染色体内 Cis 和染色体间 Trans）互作无显著差异，而 4 日龄蜂王幼虫在染色体内互作更强，工蜂幼虫则在染色体间互作更强。Hi-C、ATAC-seq、ChIP-seq 结果都与 RNA-seq 结果呈正相关，且 4 日龄蜂王幼虫比 2 日龄幼虫有更多的差异基因，这些基因主要与级型分化相关。蜜蜂级型分化是通过复杂多组学互作来调控的。在幼虫发育早期，表观遗传修饰较少，且较低程度地参与调控基因表达，因此，2 日龄幼虫发育具可塑性；但到幼虫发育关键期，表观遗传修饰变化加剧，且多种表观遗传修饰通过协同互作，特别是与级型分化关键基因受到表观修饰多重调控，因此，4 日龄幼虫发育则不可逆。首次提出表观遗传修饰能根据生物体发育状态，自动启动自身的表观遗传调控程序，进而调控生物体发育。研究结果解释了表观遗传修饰在调控蜜蜂表型可塑性分子机理。

利用 Hi-C、ChIP-seq 和 RNA-seq 三组学技术，比较分析了刚羽化出房蜂王和工蜂表观遗传调控机制。首次发现蜂王比工蜂染色体 Hi-C（包括染色体内 Cis 和染色体间 Trans）互作都更强；蜂王比工蜂有更多特异性 H3K27ac 修饰，而工蜂有更多 H3K4me1 修饰。蜂王与工蜂表观遗传差异相关基因主要集中在寿命、卵巢发育以及级型分化等通路。该研究不仅深入解析了蜂王与工蜂的表观遗传调控分子机制，而且研究结果对衰老生物学研究有重要参考价值。

利用 RNA-Seq 技术检测了意蜂三型蜂 2 日龄与 4 日龄幼虫的 mRNA，分析比较了三型蜂幼虫在发育过程中的基因表达差异。结果表明：各三型蜂幼虫组检测到的总基因个数没有差异，但各组之间存在大量差异表达基因；2 日龄蜂王幼虫与工蜂幼虫差异表达基因数为 121 个，但 4 日龄时，两者差异表达基因数增至 475 个。

利用 DNA 甲基化免疫共沉淀技术比较了 4 日龄蜂王幼虫和工蜂幼虫 DNA 甲基化模式。结果表明：与 4 日工蜂幼虫相比，蜂王幼虫的甲基化下调基因主要富集在生物学调控、免疫系统和代谢调控，且这些差异基因在调控雌性蜜蜂幼虫发育的 MAPK、Notch、Insulin/Wnt pathway 信号通路中发挥重要作用。

利用 ChIP-seq 和 RNA-seq 技术研究发现：ChIP-seq 与 RNA-seq 呈正相关。H3K4me1 修饰主要在蜂王幼虫内含子区域，而工蜂幼虫则主要在启动子区域。H3K4me1 修饰密切参与了工蜂幼虫发育调控。

# 1. Extent and Complexity of RNA Processing in Honey Bee Queen and Worker Caste Development

Xujiang He[1,2], Andrew B. Barron[3], Liu Yang[4], Hu Chen[4], Yuzhu He[1], Lizhen Zhang[1], Qiang Huang[1], Zilong Wang[1], Xiaobo Wu[1], Weiyu Yan[1], Zhijiang Zeng[1,2]*

1. Honeybee Research Institute, Jiangxi Agricultural University, Nanchang, Jiangxi 330045, P. R. of China
2. Jiangxi Province Honeybee Biology and Beekeeping, Nanchang, Jiangxi 330045, P. R. of China
3. Department of Biological Sciences, Macquarie University, North Ryde, NSW 2109, Australia
4. Wuhan Benagen Tech Solutions Company Limited, Wuhan, Hubei 430021, P. R. of China

**Abstract**

The distinct honey bee (*Apis mellifera*) worker and queen castes have become a model for the study of genomic mechanisms of phenotypic plasticity. Here we performed a Nanopore-based direct RNA sequencing with exceptionally long reads to compare the mRNA transcripts between queen and workers at three points during their larval development. We found thousands of significantly differentially expressed transcript isoforms (DEIs) between queen and worker larvae. These DEIs were formatted by a flexible splicing system. We showed that poly(A) tails participated in this caste differentiation by negatively regulating the expression of DEIs. Hundreds of isoforms uniquely expressed in either queens or workers during their larval development, and isoforms were expressed at different points in queen and worker larval development demonstrating a dynamic relationship between isoform expression and developmental mechanisms. These findings show the full complexity of RNA processing and transcript expression in honey bee phenotypic plasticity.

**Keywords:** Honey bees, phenotypic plasticity, direct RNA sequencing, transcriptome complexity, isoform expression

**Introduction**

Phenotypic plasticity is a phenomenon that the same genotype showed distinct phenotypes if raised in different conditions, which is the most interesting aspects of development (Agrawal, 2001; Borges, 2005). Phenotypic plasticity is particularly dramatic in eusocial insects, and the honey bee (*Apis mellifera*) has become a valuable system for exploring this phenomenon. Honey bee queens and workers both develop from fertilized (diploid) eggs, but differences in nutrition in their larval diets determines their diverging developmental

---

* Corresponding author: bees1965@sina.com (ZJZ).
注：此文发表在 *iScience*，2022，25(5)，104301。

trajectories that result in different phenotypes (Asencot & Lensky, 1984; Brouwers, 1984; Maleszka, 2008; Spannhoff *et al*, 2011; Mao *et al*, 2015). The queen is the colony's only fecund female with well-developed ovaries producing around 1500 eggs per day, whereas workers are facultatively sterile females that perform all the essential tasks in the colony such as brood care, foraging, defence, construction and cleaning (Winston, 1991).

Prior foundational studies have shown how the process of phenotypic plasticity is epigenetically determined (Chen *et al*, 2012; Foret *et al*, 2012; Cameron *et al*, 2013; Guo *et al*, 2016; He *et al*, 2017; Wojciechowski *et al*, 2018). Studies have compared the transcriptomes between honey bee queens and workers using short-length RNA sequencing (RNA-Seq) (Chen *et al*, 2012; Cameron *et al*, 2013; He *et al*, 2017) or cDNA microarrays (Yamazaki *et al*, 2006; Barchuk *et al*, 2007). Thousands of genes were differentially expressed between queens and workers during their development (Chen *et al*, 2012; Cameron *et al*, 2013; He *et al*, 2017). Consistently a few key signalling pathways have been implicated as important in queen/worker differentiation. These include target of rapamycin (TOR), the fork head box O (FoxO), Notch, Wnt, insulin/insulin-like signaling (IIS) pathways, mitogen-activated protein kinase (MAKP), Hippo, transforming growth factor beta (TGF-beta) (Patel *et al*, 2007; Mutti *et al*, 2011a; Duncan *et al*, 2016; Xiao *et al*, 2017; Yin *et al*, 2018; Wang *et al*, 2021); Key genes such as *Juvenile hormone esterase precursor (Jhe)*, *ecdysone receptor (Ecr)*, *Vitellogenin (Vg)*, *Hexamerin 70b (Hex70b) major royal jelly proteins (Mrjps)* also participate in the determination of queen-worker fate (Barchuk *et al*, 2007; Buttstedt *et al*, 2013; Mello *et al*, 2014). Moreover, epigeneic modifications such as DNA methylation, microRNA and alternative chromatin states have been shown to be involved in the regulation of honey bee queen-worker dimorphism (Foret *et al*, 2012; Ashby *et al*, 2016; Guo *et al*, 2016; Wojciechowski *et al*, 2018). The latest evidence showed that m6A methylation on RNA also participates in the genomic regulation leading to queen-worker dimorphism (Bataglia *et al*, 2021; Wang *et al*, 2021).

These studies have gone a long way to establish the honey bee as a model for genomic analyses of phenotypic plasticity, but thus far, we know rather little about how RNA processing might be involved in the process. RNA processing determines the mRNA coding sequence, and can lead to different isoforms of a gene. Distinct isoforms can perform different biological functions, and these can be important for animal phenotypic plasticity (Marden, 2008). Alternative transcripts produced from the same gene can differ in the position of the start site, the site of cleavage and polyadenylation, and the combination of exons spliced into the mature mRNA (Parker *et al*, 2020). Poly(A) tails are also very important for RNA processing. Length of the Poly(A) tail is correlated with translational efficiency and regulates transcript stability (Workman *et al*, 2019; Roach *et al*, 2020).

Alternative splicing (AS) of genes diversifies the transcriptome and increases protein coding capacity through the production of multiple distinct isoforms (Modrek & Lee, 2001; Reddy *et al*, 2013). AS is involved into many biological processes in plants and animals, and particularly responses and adaptation to changing environments (Modrek & Lee, 2001; Reddy *et al*, 2013). AS of pre-mRNA has been associated with phenotypic plasticity in insects such as the bumble bee *Bombus terrestris* and the pea aphid *Acyrthosiphon pisum* (Grantham & Brisson, 2018; Price *et al*, 2018). Different alternative transcripts are also involved in honey bee caste differentiation, and alternative splicing plays an important role in queen-worker differentiation (Aamodt, 2008; Mutti et al, 2011b). These studies have focused on the effects of AS on the regulation of gene expression. Thus far none have investigated how AS determines isoform expression genome wide in honey bees.

So far honey bee RNA-Seq studies have been based on short-length RNA sequencing, which is limited to detect full-length isoforms or alternative transcripts of a gene. The new direct RNA sequencing technology (DRS) based on the Oxford Nanopore technology (ONT) directly sequences the native full-length mRNA molecules, which offers detailed information on the mRNA and RNA modifications (Garalde et al, 2018; Harel et al, 2019; Workman et al, 2019; Parker et al, 2020; Roach et al, 2020; Zhang et al, 2020). It avoids biases from reverse transcription or amplification and yields full-length, strand-special mRNA (Garalde et al, 2018; Harel et al, 2019; Zhang et al, 2020). This method allows a genome wide investigation of RNA processing [e.g. alternative splicing and poly(A) tails].

Our objective in this study was to deploy this DRS technology to examine differences in RNA processing between the worker and queen honey bee castes. This allows us to investigate fundamental questions on caste-specific transcriptome patterns in honey bees: the variety of full-length isoforms and how they expressed differentially in queens and workers, and the extent and complexity of the RNA processing in honey bee development.

We sampled honey bee queens and workers at different points in their larval development to assess the process of developmental canalization into the different phenotypes. Our results reveal extremely prevalent and highly complex differences in RNA processing between workers and queens. These differences changed across larval development showing a complex dynamical interaction between the epigenomic programs and the developmental pathways.

## Results

### *Quality of direct RNA sequencing data*

Twelve libraries were generated and sequenced by ONT direct RNA sequencing. The number of clean reads of each sample was $5,416,798 \pm 1,510,975$ (Table I-1-1). The N50 read length was $1,297 \pm 30$ nucleotide (nt), and the read quality was $10.63 \pm 0.13$ (Table I-1-1) indicating a high integrity of the sequenced RNAs. $98.13\% \pm 0.79\%$ clean reads were successfully mapped to the *Apis mellifera* reference genome (*Amel HAv 3.1*) (Table I-1-1). Moreover, the total number of genes detected from DRS and RNA-seq was equal (9,592 vs 9,838), and over 88.8% of them were overlapped (Fig. S1). We also mapped the transcript isoforms to honeybee reference genes and showed a great quality of percentile of gene body (5'-3') in each DRS sample (Fig. S2). These results again confirmed a high and reliable integrity of the sequenced RNAs. The Pearson correlation coefficients between two biological replicates of each group were all over 0.85 (Fig. S3) and the sample cluster tree also showed that two biological replicates of each group were clustered together (Fig. S4), reflecting a good repeatability of biological replicates.

### *Differences of isoform and gene expression in queen-worker comparisons*

Our results showed genome wide differentially expressed isoforms in honey bee caste differentiation.

Significantly differentially expressed transcript isoforms (DEIs) and genes (DEGs) between queen and worker larvae were identified. We identified 662, 1855 and 1042 DEIs in 2d, 4d and 6d queen-worker larval comparisons respectively, which were notably more than the DEGs (281, 1369 and 645 respectively, see Fig. I-1-1A, refer to Appendix A, page 489. Detailed information see Table S1-6). To compare the differences between DEIs and DEGs, we firstly mapped DEIs to reference genes (DEIGs) and showed that only 271 (34.72%), 977 (53.45%) and 410 (38.32%) genes overlapped between DEIGs and DEGs in 2d, 4d and 6d queen-worker comparisons respectively (Fig. I-1-1 B, C, D, see page 489). This shows a considerable difference between DEIs and DEGs in queen-worker differentiation.

We identified DEIs (24, 54, 48 in 2d, 4d and 6d comparisons respectively) that are involved in several key KEGG signaling pathways for honey bee caste differentiation, such as mTOR, Notch, FoxO, Wnt, IIS, Hippo, TGF-beta and MAPK (Table S1-3). The number of DEGs enriched in these signal pathways were 7, 34, 33 in 2d, 4d and 6d respectively (Table S4-6). Other DEIs mapped to key genes such as *Jhe*, *Mrjps*, *Ecr*, *Vg* and *Hex70b* (Table S1-3). The number of isoforms with significantly different poly(A) length in 2d, 4d and 6d comparisons showed a negative trend as DEIs and DEGs, with more in queens (Fig. I-1-1A, detailed information see Table S7-9) and less in workers.

DEIs and DEGs were enriched in five key KEGG signaling pathways (mTOR, Notch, FoxO, Wnt, IIS) which have been previous shown to be participating in of queen-worker differentiation. DEIs mapped to 35 proteins in these five pathways, whereas DEG mapped onto just 17 proteins (Fig. I-1-2A, see page 490). In detail, 23 proteins were uniquely mapped by DEIs whereas only 5 were uniquely mapped by DEGs. 12 proteins mapped to both DEI and DEG (Fig. I-1-2A). We also provided detailed information about the expression of DEIs and DEGs enriched in these five key pathways (Fig. I-1-2B, see page 490). Clearly, more isoforms were significantly differentially expressed in these five key pathways compared to DEGs, and the differentially expressed isoforms of each DEIG are presented in Fig. I-1-2B.

These findings suggest that caste-specific isoform expression provided more accurate and detailed information of the real transcriptome differences between honey bee queen and worker castes than measurements of differences in gene expression.

### *Uniquely expressed isoforms between queens and workers*

We identified 187, 357 and 364 uniquely expressed isoforms in queen or worker larvae at 2d, 4d and 6d age, with more occurring in queens than workers (Fig. I-1-3A, see page 491). The number of uniquely expressed isoforms in queens reached a maximum at the 4d stage whereas in workers it reached a maximum in the 6d sample. This could suggest that the queen developmental pathway diverges more quickly and at an early larval stage while the developmental pathway of workers diverges slightly later.

Uniquely expressed isoforms were enriched into eight key KEGG signaling pathways such as mTOR and IIS for queen-worker differentiation as well as some key genes such as *DNA methyltransferase 3 (Dnmt3)* and *Ecr* (Fig. I-1-3, details see Table, S10-12). This indicated that there are isoforms which are likely to be involved in queen-worker differentiation that are uniquely expressed in queens or workers during their development.

### *DEIs and alternative splicing*

We identified 21,574 isoforms in each sample on average, and these isoforms could map to averagely 8509 honeybee reference genes (Table I-1-1). This suggested that the honey bee genome produces notably more isoforms than genes. DRS technology allows to detect the splicing events causing an isoform. We reconstructed the complex splicing events, as well as the DEIs resulted from the alternative splicing. We then define a DEI from a single gene, which is Differentially Spliced compared to the longest isoform of this gene as a DS-DEI. The majority of DEIs (73.26%, 67.98% and 63.53% of 2d, 4d and 6d DEIs) were DS-DEIs, and most of them (71.13%, 66.16% and 65.56% in 2d, 4d and 6d) were generated by at least two types of splicing events in a single isoform. A combination of RI, A5 and A3 in a single DEI was the most common type (Fig. I-1-4, see page 492). This indicates that the formation of transcripts in honey bee development is more complex than previous known.

**Table I-1-1  Overview of direct RNA sequencing**

| Samples | 2Q-1 | 2Q-2 | 2W-1 | 2W-2 | 4Q-1 | 4Q-2 | 4W-1 | 4W-2 | 6Q-1 | 6Q-2 | 6W-1 | 6W-2 | Average |
|---|---|---|---|---|---|---|---|---|---|---|---|---|---|
| Clean read numbers | 7,268,687 | 5,793,325 | 6,113,875 | 6,093,102 | 6,067,308 | 6,945,178 | 6,207,113 | 3,807,886 | 5,039,594 | 2,057,391 | 5,910,366 | 3,697,756 | 5,416,798 |
| Average length | 1061.94 | 997.4 | 1001.55 | 1022.06 | 1007.76 | 998.23 | 967.9 | 1027.53 | 1036.56 | 1069.02 | 1027.87 | 976.65 | 1,016 |
| N50 length | 1356 | 1275 | 1285 | 1301 | 1295 | 1277 | 1242 | 1315 | 1338 | 1283 | 1306 | 1286 | 1,296 |
| Max read length | 23283 | 16804 | 16550 | 17946 | 29881 | 19966 | 16391 | 19519 | 17266 | 17829 | 13661 | 15496 | 18,716 |
| Average read quality | 10.71 | 10.61 | 10.75 | 10.71 | 10.26 | 10.53 | 10.66 | 10.62 | 10.77 | 10.68 | 10.66 | 10.64 | 10.63 |
| Alignment % identity | 98.80 | 98.35 | 98.42 | 98.46 | 98.23 | 98.63 | 98.19 | 98.38 | 96.85 | 98.40 | 98.71 | 96.19 | 98.13 |
| Uni mapped genes | 8,781 | 8,690 | 8,658 | 8,658 | 8,431 | 8,656 | 8,411 | 8,234 | 8,666 | 7,741 | 8,539 | 8,644 | 8,509 |
| Isoforms | 24,058 | 22,978 | 22,853 | 22,939 | 21,746 | 23,018 | 20,498 | 19,790 | 22,365 | 16,487 | 20,885 | 20,714 | 21,527 |

Of these DEIs, 9.82% (65), 5.71% (106), 9.50% (99) of DEIs from the 2d, 4d and 6d comparisons were produced by significantly different AS events when comparing workers and queens (Fig. I-1-5, see page 493, Table S13-15). Similarly, many of them (46.15%, 42.45% and 45.45% in 2d, 4d and 6d comparisons respectively) contained at least two types of significantly different AS events (Fig. I-1-6, see page 494, Table S13-15). Consequently, a part of DEIs were significantly differentially spliced.

### *Correlation between poly(A) length and expression of DEIs and DEGs*

We showed a negative correlation between isoform expression and their poly(A) tail length (Fig. S7). Expression of DEIs (Fig. I-1-5) and DEGs (Fig. S8) was also negatively correlated with their poly(A) lengths. It is believed that poly(A) tails are involved in the regulation of isoform expression. More interestingly, the poly(A) lengths were positively correlated with $\log_2$ fold change values of DEIs, which was stronger than that of DEGs (Fig. I-1-5, D-F). This firstly showed that poly(A) lengths were more closely related to the biased expression of DEIs than DEGs.

### *Correlation of isoform expression and poly(A) lengths for two key genes*

Two key genes *Jhe* and *Ecr* which participate in queen-worker differentiation were selected as examples to show the correlation pattern of gene and isoform expression and its poly(A) length. *Jhe* had only one isoform, and was negatively correlated with its poly(A) lengths (Fig. I-1-6A). The *Ecr* gene has 5 isoforms and we detected 4 isoforms (Fig. I-1-6B). Queen and worker larvae showed a more detailed difference in the expression of these 4 isoforms (see the middle of Fig. I-1-6B) than gene expression (see the left of Fig. I-1-6B). We selected the most differentially expressed isoform (*Ecr.t4*) to measure the correlation between its expression and poly(A) length. Similarly, the expression of *Ecr.t4* isoform was also negatively correlated with their poly(A) lengths (Fig. I-1-6B).

We selected two isoforms of the *Ecr* gene (*Ecr.t1* and *Ecr.t2*) for TaqMan real-time PCR confirmatory experiment. The expression of these two isoforms in TaqMan real-time PCR (Fig. I-1-7) was mainly consistent

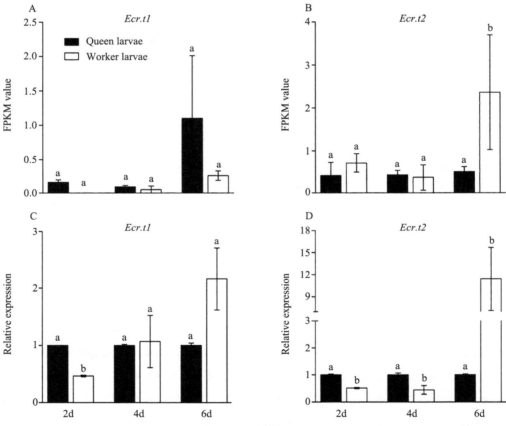

**Figure I-1-7** Verification of two isoforms of Ecr gene between 2d, 4d and 6d queen and worker larvae by TaqMan real-time PCR. See also Table S18. (A) The Ecr.t1 expression in three queen-worker comparisons (2d, 4d and 6d) from DRS. The expression of the isoform was the FPKM value. Data were presented as mean ± SE. Different letters in the top of bars indicate significant difference ($p < 0.05$ and $\log_2$ Fold change values > 1.5 using DESeq2) and same letter indicate no significant difference. Same in B. (B) The Ecr.t2 expression in three queen-worker comparisons (2d, 4d and 6d) from DRS. (C) The Ecr.t1 expression in three queen-worker comparisons (2d, 4d and 6d) from TaqMan real-time PCR. Data were presented as mean ± SE. Different letters in the top of bars indicate significant difference ($p < 0.05$, one-way ANOVA test) and same letter indicate no significant difference. Same in D. (D) The Ecr.t2 expression in three queen-worker comparisons (2d, 4d and 6d) from TaqMan real-time PCR.

with the DRS results, confirming a biased isoform expression in honeybee caste differentiation.

## Discussion

RNA processing act as a major driver of animal phenotypic plasticity (Gingeras, 2007), but the complexity and characteristics of RNA processing underlying this phenotypic variability is not well understood. The present study has shown the extent of RNA processing in honey bee queen-worker differentiation. Differences in isoform expression revealed the true extent of the differences in the transcriptome between queens and workers during their development, providing more detailed and accurate information on mRNA expression compared to measures of gene expression. We also showed the flexibility of alternative splicing in queen-worker differentiation. Most DEIs were formed by more than one type of alternative splicing. Poly(A) length negatively correlated with the expression of DEIs demonstrating another part of the genomic mechanisms of honey bee caste differentiation.

Thousands of DEIs were identified between queens and workers, which were notably more than DEGs (Fig. I-1-1A). Many genes were not differentially expressed, but different isoforms from these genes were significantly differentially expressed (Fig. I-1-1B-D, Table S1-3), demonstrating how DRS gives a very different perception of the transcriptome differences between developing worker and queen honey bees. DEIs enriched into more proteins of five key KEGG signaling pathways such as mTOR and IIS than DEGs (Fig. I-1-2). The mTOR signaling pathway is the central component of a conserved eukaryotic signaling pathway, regulating cell and organismal growth in response to nutrient status and has been shown to be important for determining the queen-worker developmental paths (Colombani *et al*, 2003; Oldham & Hafen, 2003; Patel *et al*, 2007). The IIS and Notch signaling pathway have been shown to be involved in the development of honey bee ovaries (Mutti *et al*, 2011a; Duncan *et al*, 2016). More DEIs were identified in these key pathways than DEGs, indicating that DEIs could provide more nuanced differences of mRNA expression between honey bee queens and workers during their development. Therefore, compared to DEGs that conceals a more nuanced picture of transcription differences, DEIs show the true nature of transcriptome differences between queens and workers. Consequently, DEIs identified in the present study revealed a more informative knowledge and precise difference in the transcriptome expression of honey bee caste differentiation compared to previous RNA-Seq results (Chen *et al*, 2012; Cameron *et al*, 2013; He *et al*, 2017; Yin *et al*, 2018).

Hundreds of uniquely expressed isoforms were identified in queens or workers. Many uniquely expressed isoforms were enriched in mTOR, IIS, Notch, FoxO etc eight key KEGG signaling pathways (Fig. I-1-3). Some uniquely expressed isoforms mapped to key genes such as DNA methyltransferase 3 (*Dnmt3*) and *Ecr* genes (Fig. I-1-3). *Dnmt3* is a key driver of epigenetic global reprogramming controlling honey bee queen-worker fate (Kucharski *et al*, 2008). Here we showed one isoform of *Dnmt3* (*Dnmt3.t3*) was uniquely expressed in 4d worker larvae, suggesting that this isoform of *Dnmt3* may play an important role in the worker developmental pathway at this critical stage. For the *Ecr* gene, 2d worker uniquely expressed *Ecr.t3* whereas 2d queen larvae uniquely expressed *Ecr.t4*. Different isoforms from the same gene were expressed in queens and workers, revealing the complexity of of RNA processing in the honey bee queen-worker differentiation.

We have documented many genes where, isoform expression was not consistently biased toward either queens or workers throughout all stages of larval development. In some cases, an isoform was uniquely expressed in either queens or workers in one larval stage, but at other developmental stages the isoform occurred in both workers and queens (Fig. I-1-3B). Isoforms like *LOC409856.t4* and *LOC727227.t8* uniquely expressed in queens at 2d and 4d stages but oppositely expressed in workers at 6d stage (Fig. I-1-3B). This illustrates the dynamic relationship between RNA processing and the developmental process. A simple model might imagine different isoforms of a gene specific to either the queen or worker developmental pathway. Instead we see isoforms featuring in both pathways, but at different stages.

The number of uniquely expressed isoforms also showed a different trend in queens and workers. It reached a maximum in queens at the 4d larval stage but it peaked later in workers (Fig. I-1-3A). These different trends correlate with the known faster development of queens (Winston, 1991) and suggest the queens developmental pathway differentiates at an earlier larvae stage than the worker developmental pathway. Queen larvae at early stage receive significantly more food than worker larvae, and the nutrient contents in royal jelly also differ from worker jelly in terms of sugar, vitamins, proteins, acids, microRNA (Asencot & Lensky, 1984; Brouwers, 1984; Maleszka, 2008; Mao *et al*, 2015). In Fig. I-1-3B, most isoforms enriched in mTOR signaling pathway which is a signaling pathway in response to nutrient status (Colombani *et al*, 2003; Oldham & Hafen, 2003) were uniquely expressed in queens at 2d stage and its number decreased at 4d and 6d stages, which supports our speculation.

We detected 21,528±2,040 isoforms in each sample which could map to 8,509±286 honeybee reference genes (Table I-1-1). This is consistent with a previous study that 60.3 % of honey bee multi-exon genes are alternatively spliced (Li *et al*, 2013). The majority of DEIs, contained AS events and over 65% of them had at least two types of AS patterns in a single isoform (Fig. I-1-4). Some DEIs contained more than five different AS types (Fig. I-1-4). A few DEIs were significantly differently spliced between queens and workers by multiple AS patterns (Fig. S3, Table S13-15). It means several distinct AS types together contributed to the construction of a single isoform (Fig. I-1-4), revealing a flexible system in honeybee caste differentiation.

Poly(A) tails are a regulator of translation and transcript stability (Lim *et al*, 2016; Tudek *et al*, 2018; Woo *et al*, 2018). Poly(A) tail length is not stable but dynamic and condition-dependent (Lim *et al*, 2016; Woo *et al*, 2018). This study showed that poly(A) tail lengths were strongly and negatively correlated with expression of isoforms in honey bees (Fig. I-1-5). The negative correlation between isoform expression and poly(A) length has been shown in the regulation of *Caenorhabditis elegans* development (Roach *et al*, 2020). Here we showed that poly(A) tails also participate in the regulation of isoform expression, and therefore contributes to complexity of RNA processing in honeybee queen-worker differentiation. Moreover, we examined the relationship between poly(A) length and the degree of biased expression of isoforms or genes in queen-worker differentiation. Interestingly, the $\log_2$ fold change values of DEIs were positively correlated with the poly(A) lengths and their correlation were stronger than that of DEGs (Fig. I-1-5D-F), suggesting that poly(A) tails play an important role in the RNA processing of honey bee caste differentiation. This also firstly showed that poly(A) tails had a stronger correlation with isoforms than genes in terms of insect phenotypic plasticity.

Consequently, poly(A) tails possibly contribute to the determination of honey bee queen-worker dimorphism via regulating the isoform expression.

Environmental stimulus altering patterns of gene expression has been considered as a primary molecular mechanism of phenotypic plasticity (Schlichting & Simth, 2002; Aubin-Horth & Renn, 2009). However, one gene can produce different transcripts by RNA processing, and different transcripts have distinct biological functions that are vital for phenotypic plasticity (Marden, 2008). Therefore, increasing evidence revealed that the complex RNA processing may be a principal contributor to the phenotypic complexity (Gommans et al, 2008; 2010). The present study supports this hypothesis and showed a genome-wide different isoform expression in queen-worker differentiation, which was far more complex than the gene expression (Fig. I-1-1 and Fig. I-1-2). Especially hundreds of isoforms were uniquely expressed in queens or workers (Fig. I-1-3). These uniquely expressed isoforms cannot be easily seem in gene expression, but may play an important role in the determination of honeybee caste differentiation. We also showed a flexible alternative splicing generating a variety of isoforms (Fig. I-1-4 and S6), and the polyadenylation [poly(A) tails] negatively regulating isoform expression (Fig. I-1-5 and Fig. I-1-6). All findings demonstrated a complexity of RNA processing in honeybee phenotypic plasticity.

In summary, RNA processing plays an important role in shaping honey bee phenotypic plasticity, but investigations of gene expression by short-length RNA-Seq fail to reveal the full complexities of RNA processing (Chen et al, 2012; Cameron et al, 2013; He et al, 2017; Yin et al, 2018). By using DRS, this study has shown the extent of differential isoform expression between workers and queens, flexibility of transcript splicing, and polyadenylation. It provides a more detailed understanding of the molecular mechanisms underlying the divergence of the queen and worker developmental paths. It also contributes to our understanding of the extent and complexity of RNA processing in animal phenotypic plasticity.

## Limitations of Study

Although this study showed an extensive and complex RNA processing in honey bee caste differentiation including biased isoform expression, flexible alternative splicing and polyadenylation, other RNA processing events like 5' capping are limited to be detected by the DRS technology but may also be involved into the honey bee caste differentiation. In addition, to extract enough amount of total RNA for DRS, we used a whole body of a 4d and 6d queen or worker larva as a sample and collected twenty 2d larvae as a mixed sample due to its small body size. This amount of RNA used in DRS was much lower than that in RNA-Seq, however, we note that the isoform expression of honey bee queen and worker larvae might differ in their distinct tissues.

## Acknowledgement

We thank Mr Yi Cheng Yang and Mrs Yan Niu for help on data analysis and Mrs Xin Zhang for help on project management. This work was supported by the National Natural Science Foundation of China (32172790, 32160815, 31872432), Major Discipline Academic and Technical Leaders Training Program of Jiangxi Province (20204BCJL23041) and the Earmarked Fund for China Agriculture Research System (CARS-

44-KXJ15).

## Authors' Contribution

Z.J.Z. and X.J.H. designed research; X.J.H. performed the research; X.J.H., A.B.B. provided guidance for data; X.J.H., L.Y., and H.C. analyzed data; Y.Z.H., L.Z.Z., Q.H., Z.L.W., X.B.W., and W.Y.Y. assisted in the experiments. X.J.H., A.B.B. and Z.J.Z. wrote the paper.

## Declaration of Interests

The authors declare no competing interest. All authors were involved in the preparation of the final manuscript.

STAR ★ Methods
KEY RESOURCES TABLE
RESOURCE AVAILABILITY
Lead contact
Materials availability
Data and code availability
EXPERIMENTAL MODEL AND SUBJECT DETAILS
    Insects
METHOD DETAILS
    Larva sampling
    RNA preparation
    Direct RNA sequencing
    Preprocessing and alignments
    DEG and DEI calculation
    KEGG pathway enrichment analysis
    Alternative splicing identification
    Poly(A) length estimation and correlation with expression of genes and isoforms
    Short-length RNA sequencing
    TaqMan real-time PCR
QUANTIFICATION AND STATISTICAL ANALYSIS
SUPPLEMENTAL INFORMATION
Supplemental Information can be found online.

## References

AAMODT, R.M. (2008) The caste- and age-specific expression signature of honeybee heat shock genes shows an alternative splicing-dependent regulation of Hsp90. *Mech Ageing Dev* 129: 632-637.

AGRAWAL, A.A. (2001) Phenotypic plasticity in the interactions and evolution of species. *Science* 294: 321-326.

ASENCOT, M., LENSKY, Y. (1984) Juvenile hormone induction of 'queenliness' on female honeybee (*Apis mellifera* L.) larvae reared on worker jelly and on stored royal jelly. *Comp Biochem Phys B* 78: 109-117.

ASHBY, R., FORÊT, S., SEARLE, I., MALESZKA, R. (2016) MicroRNAs in honey bee caste determination. *Sci Rep* 6: 18794.

AUBIN-HORTH, N., RENN, S.C.P. (2009) Genomic reaction norms: using integrative biology to understand molecular mechanisms of phenotypic plasticity. *Mol Ecol* 18: 3763-3780.

BATAGLIA, L., SIMES, Z.L.P., NUNES, F.M.F. (2021) Active genic machinery for epigenetic RNA modifications in bees. *Insect Mol Biol* 33: 566-579.

BARCHUK, A.R., CRISTINO AS, KUCHARSKI R, COSTA LF, SIMÕES ZL, MALESZKA R. (2007) Molecular determinants of caste differentiation in the highly eusocial honeybee *Apis mellifera*. *BMC Dev Biol* 7: 70.

BORGES, R.M. (2005) Do plants and animals differ in phenotypic plasticity? *J Biosciences* 30: 41-50.

BROUWERS, E.V.M. (1984) Glucose/fructose ratio in the food of honeybee larvae during caste differentiation. *J Apicult Res* 23: 94-101.

BUTTSTEDT, A., MORITZ, R.F., ERLER, S. (2013) More than royal food-Major royal jelly protein genes in sexuals and workers of the honeybee *Apis mellifera*. *Front Zool* 10: 72.

CAMERON, R.C., DUNCAN, E.J., DEARDEN, P.K. (2013) Biased gene expression in early honeybee larval development. *BMC Genomics* 14: 903.

CHEN, X., HU, Y., ZHENG, H., CAO, L., NIU, D., YU, D., SUN, Y., HU, S., HU, F. (2012) Transcriptome comparison between honey bee queen- and worker-destined larvae. *Insect Biochem Mol Biol* 42: 665-673.

COLOMBANI, J., RAISIN, S., PANTALACCI, S., RADIMERSKI, T., MONTAGNE, J., LÉOPOLD, P. (2003) A nutrient sensor mechanism controls Drosophila growth. *Cell* 114: 739-749.

COSTER, W.D., D'HERT, S., SCHULTZ, D.T., CRUTS, M., VAN BROECKHOVEN, C. (2018) NanoPack: visualizing and processing long-read sequencing data. *Bioinformatics* 15: 2666-2669.

DUNCAN, E.J., HYINK, O., DEARDEN, P.K. (2016) Notch signalling mediates reproductive constraint in the adult worker honeybee. *Nat Commun* 7: 12427.

FESENKO, I., KNYAZEV, A. (2020) Direct RNA sequencing dataset of SMG1 KO mutant Physcomitrella (Physcomitrium patens). *Data in Brief* 33: 106602.

FORET, S., KUCHARSKI, R., PELLEGRINI, M., FENG, S., JACOBSEN, S.E., ROBINSON, G.E., MALESZKA, R. (2012) DNA methylation dynamics, metabolic fluxes, gene splicing, and alternative phenotypes in honey bees. *P Nat Acad Sci USA* 109: 4968-4973.

GARALDE, D.R., SNELL, E.A., JACHIMOWICZ, D., SIPOS, B., LLOYD, J.H., BRUCE, M., PANTIC, N., ADMASSU, T., JAMES, P., WARLAND, A., *et al.* (2018) Highly parallel direct RNA sequencing on an array of nanopores. *Nat Methods* 3: 201-206.

GINGERAS, T.R. (2007) Origin of phenotypes: Genes and transcripts. *Genome Res* 17: 682-690.

GOMMANS, W.M., DUPUIS, D.E., MCCANE, J.E., TATALIAS, N.E., MAAS, S. (2008) Diversifying exon code through A-to-I RNA editing. In: Smith, H., editor. DNA RNA editing. Wiley & Sons, Inc.; Hoboken, NJ: pp. 3-30.

GOMMANS, W. M., MULLEN, S.P., MAAS, S. (2010) RNA editing: a driving force for adaptive evolution?. *Bioessays* 31: 1137-1145.

GRANTHAM, M., BRISSON, J. (2018) Extensive differential splicing underlies phenotypically plastic aphid morphs. *Molecul Biol Evol* 8: 1934-1946.

GUO, X., SU, S., GEIR, S., LI, W., LI, Z., ZHANG, S., CHEN, S., CHEN, R. (2016) Differential expression of miRNAs related to caste differentiation in the honey bee, *Apis mellifera*. *Apidologie* 47: 495-508.

HAREL, N., MEIR, M., GOPHNA, U., STERN, A. (2019) Direct sequencing of RNA with MinION Nanopore: detecting mutations based on associations. *Nucletic Acids Res* 47: 22,e148.

HE, X.J., JIANG, W.J., ZHOU, M., BARRON, A.B., ZENG, Z.J. 2019. A comparison of honeybee (*Apis mellifera*) queen, worker and drone larvae by RNA-Seq. *Insect Sci* 26: 499-509.

KUCHARSKI, R., MALESZKA, J., FORET, S., MALESZKA, R. (2008) Nutritional control of reproductive status in honeybees via DNA methylation. *Science* 319: 1827-1829.

LI, H., HANDSAKER, B., WYSOKER, A., FENNELL, T., RUAN, J., HOMER, N., MARTH, G., ABECASIS, G., DURBIN, R. (2009) The sequence alignment/map format and SAMtools. *Bioinformatics* 25: 2078-2079.

LI, Y., HONGMEI, L.B., PAUL, B., MARK, B., ROBINSON, G.E., MA, J. (2013) TrueSight: a new algorithm for splice junction detection using RNA-seq. *Nucleic Acids Res* 41: e51.

LI, H. (2018) Minimap2: versatile pairwise alignment for nucleotide sequences. *Bioinformatics* 18: 1-6.

LIM, J., LEE, M., SON, A., CHANG, H., KIM, V.N. (2016) mTAIL-seq reveals dynamic poly(A) tail regulation in oocyte-to-embryo development. *Gene Dev* 30: 1671-1682.

LOVE, M.I., HUBER, W., ANDERS, S. (2014) Moderated estimation of fold change and dispersion for RNA-seq data with DESeq2. *Genome Biol* 15: 550.

MALESZKA, R. (2008) Epigenetic integration of environmental and genomic signals in honey bees: the critical interplay of nutritional, brain and reproductive networks. *Epigenetics* 3: 188-192.

MARDEN, J. (2008) Quantitative and evolutionary biology of alternative splicing: how changing the mix of alternative transcripts affects phenotypic plasticity and reaction norms. *Heredity* 100: 111-120.

MAO, W., SCHULER, M.A., BERENBAUM, M.R. (2015) A dietary phytochemical alters caste-associated gene expression in honey bees. *Sci Adv* 1: e1500795.

MELLO, T.R.P., ALEIXO, A.C., PINHEIRO, D.G., NUNES, F.M.F., BITONDI, M.M.G., HARTFELDER, K., BARCHUK, A.R., SIMÕES, Z.L.P. (2014) Developmental regulation of ecdysone receptor (EcR) and EcR-controlled gene expression during pharate-adult development of honeybees (*Apis mellifera*). *Front Genet* 5: 445.

MODREK, B., LEE, C. (2001) A genomic view of alternative splicing. *Nat Genet* 30: 13-19.

MUTTI, N.S., DOLEZAL, A.G., WOLSCHIN, F., MUTTI, J.S., GILL, K.S., AMDAM, G.V. (2011a) IRS and TOR nutrient-signaling pathways actvia juvenile hormone to influence honey bee caste fate. *J Exp Biol* 214: 3977-3984.

MUTTI, N.S., YING, W., OSMAN, K., AMDAM, G.V., MORITZ, R. (2011b) Honey bee PTEN-description, developmental knockdown, and tissue-specific expression of splice-variants correlated with alternative social phenotypes. *PLoS One*, 7: e22195.

OLDHAM, S., HAFEN, E. (2003) Insulin/IGF and target of rapamycin signaling: a TOR de force in growth control. *Trends*

*Cell Biol* 13: 79-85.

PAN, Q.Z., WU, X.B., GUAN, C., ZENG, Z.J. (2013) A New method of queen rearing without grafting larvae. *American Bee Journal* 12: 1279-1280.

PATEL, A., FONDRK, M.K., KAFTANOGLU, O., EMORE, C., HUNT, G., FREDERICK, K., AMDAM, G.V. (2007) The making of a queen: TOR pathway is a key player in diphenic caste development. *PLoS One* 2: e509.

PATRO, R., DUGGAL, G., LOVE, M.I., IRIZARRY, R.A., KINGSFORD, C. (2017) Salmon provides fast and bias-aware quantification of transcript expression. *Nat Methods* 14: 417-419.

PARKER, M.T., KNOP, K., SHERWOOD, A.V., SCHURCH, N.J., MACKINNON, K., GOULD, P.D., HALL, A.J., BARTON, G.J., SIMPSON, G.G. (2020) Nanopore direct RNA sequencing maps the complexity of Arabidopsis mRNA processing and m6A modification. *eLife*, 9: e49658.

PERTEA, M., PERTEA, G.M., ANTONESCU, C.M., CHANG, T., MENDELL, J.T., SALZBERG, S.L. (2015) StringTie enables improved reconstruction of a transcriptome from RNA-seq reads. *Nat Biotechol* 33: 290-295.

PERTEA, G., PERTEA, M. (2020) GFF Utilities: GffRead and GffCompare [version 2; peer review: 3 approved]. *F1000 Research* 9: 304.

PRICE, J., HARRISON, M.C., HAMMOND, R.L., ADAMS, S., GUTIERREZ-MARCOS, J.F., MALLON, E.B. (2018) Alternative splicing associated with phenotypic plasticity in the bumble bee *Bombus terrestris*. *Mol Ecol* 27: 1036-1043.

REDDY, A., MARQUEZ, Y., KALYNA, M., BARTA, A. (2013) Complexity of the alternative splicing landscape in plants. *Plant Cell* 25: 3657-3683.

ROACH, N.P., SADOWSKI, N., ALESSI, A.F., TIMP, W., TAYLOR, J., KIM, J.K. (2020) The full-length transcriptome of C. elegans using direct RNA sequencing. *Genome Res* 30: 299-312.

SCHLICHTING, C.D., SMITH, H. (2002) Phenotypic plasticity: linking molecular mechanisms with evolutionary outcomes. *Evol Ecol* 16: 189-211.

SPANNHOFF, A., KIM, Y.K., RAYNAL, N.J.M, GHARIBYAN, V., SU, M.B., ZHOU, Y.Y., LI, J., CASTELLANO, S., SBARDELLA, G., ISSA, J.P.J. (2011) Histone deacetylase inhibitor activity in royal jelly might facilitate caste switching in bees. *EMBO reports* 12, 3: 238-243.

TANG, A.D., SOULETTE, C.M., BAREN, M., HART, K., HRABETA-ROBINSON, E., WU, C.J., BOOKS, A.N. (2020) Full-length transcript characterization of SF3B1 mutation in chronic lymphocytic leukemia reveals downregulation of retained introns. *Nat Commun* 11: 1438.

TRINCADO, J.L., ENTIZNE, J.C., HYSENAJ, G., SINGH, B., SKALIC, M., ELLIOTT, D.J., EYRAS, E. (2018) SUPPA2: fast, accurate, and uncertainty-aware differential splicing analysis across multiple conditions. *Genome Biol* 19: 40.

TUDEK, A., LLORET-LLINARES, M., JENSEN, T.H. (2018) The multitasking polyA tail: nuclear RNA maturation, degradation and export. *Phil Trans Biol Sci* 373: 20180169.

WANG, J.R., HOLT, J., MCMILLAN, L., JONES, C.D. (2018) FMLRC: Hybrid long read error correction using an FM-index. *BMC Bioinformatics* 19: 50.

WANG, L., WANG, S., LI, W. (2012) RSeQC: quality control of RNA-seq experiments. *Bioinformatics* 28: 2184-2185.

WANG, M., XIAO, Y., LI, Y., WANG, X., QI, S., WANG, Y., ZHAO, L., WANG, K., PENG, W., LUO, G., et al. (2021)

RNA m6A modification functions in larval development and caste differentiation in honeybee (*Apis mellifera*). *Cell Rep* 34: 108580.

WINSTON, M. (1991) The biology of the honey bee. Harvard university press, Cambridge, MA, USA.

WOJCIECHOWSKI, M., LOWE, R., MALESZKA, J., CONN, D., MALESZKA, R., HURD, P.J. (2018) Phenotypically distinct female castes in honey bees are defined by alternative chromatin states during larval development. *Genome Res* 28: 1532-1542.

WOO, Y.M., KWAK, Y., NAMKOONG, S., KRISTJÁNSDÓTTIR, K., LEE, S.H., LEE, J.H., KWAK, H. (2018) TED-Seq identifies the dynamics of Poly(A) length during ER stress. *Cell Rep* 24: 3630-3641.

WORKMAN, R.E., TANG, A.D., TANG, P.S., JAIN, M., TYSON, J.R., RAZAGHI, R., ZUZARTE, P.C., GILPATRICK, T., PAYNE, A., QUICK, J., *et al.* (2019) Nanopore native RNA sequencing of a human poly(A) transcriptome. *Nat Methods* 16: 1297-1305.

XIAO, C., MA, C., CHAO, C., QIAN, L., GUO, H. (2017) Integration of lncRNA-miRNA-mRNA reveals novel insights into oviposition regulation in honey bees. *PeerJ* 5: e3881.

XIE, C., MAO, X., HUANG, J., DING, Y., WU, J., DONG, S., KONG, L., GAO, G., LI, C., WEI, L. (2011) KOBAS 2.0: a web server for annotation and identification of enriched pathways and diseases. *Nucletic Acids Res* 39: W316-W322.

YAMAZAKI, Y., SHIRAI, K., PAUL, R.K., FUJIYUKI, T., WAKAMOTO, A., TAKEUCHI, H., KUBO, T. (2006) Differential expression of HR38 in the mushroom bodies of the honeybee brain depends on the caste and division of labor. *FEBS Letters*, 580: 2667-2670.

YIN, L., WANG, K., NIU, L., ZHANG, H., CHEN, Y., JI, T., CHEN, G. (2018) Uncovering the changing gene expression profile of honeybee (*Apis mellifera*) worker larvae Transplanted to queen cells. *Front Genet* 9: 416.

ZHANG, S., LI, R., ZHANG, L., CHEN, S., XIE, M., YANG, L., XIA, Y., FOYER, C.H., ZHAO, Z., LAM, H. (2020) New insights into *Arabidopsis* transcriptome complexity revealed by direct sequencing of native RNAs. *Nucletic Acids Res* 48: 7700-7711.

ZHANG, X., LI, G., JIANG, H., LI, L., MA, J., LI, H., CHEN, J. (2019) Full-length transcriptome analysis of *Litopenaeus vannamei* reveals transcript variants involved in the innate immune system. *Fish Shellfish Immun* 87: 346-359.

## 2. The Diverging Epigenomic Landscapes of Honeybee Queens and Workers Revealed by Multiomic Sequencing

Yong Zhang[a,b,†], Xujiang He[a,b,†], Andrew B. Barron[c], Zhen Li[a,b], Mengjie Jin[a,b], Zilong Wang[a,b], Qiang Huang[a,b], Lizhen Zhang[a,b], Xiaobo Wu[a,b], Weiyu Yan[a,b], Zhijiang Zeng[a,b,*]

a. Honeybee Research Institute, Jiangxi Agricultural University, Nanchang, Jiangxi 330045, P. R. of China

b. Jiangxi Province Honeybee Biology and Beekeeping Nanchang, Jiangxi 330045, P. R. of China

c. Department of Biological Sciences, Macquarie University, North Ryde, NSW 2109, Australia

**Abstract**

The role of the epigenome in phenotypic plasticity is unclear presently. Here we used a multiomics approach to explore the nature of the epigenome in developing honey bee (*Apis mellifera*) workers and queens. Our data clearly showed distinct queen and worker epigenomic landscapes during the developmental process. Differences in gene expression between workers and queens become more extensive and more layered during the process of development. Genes known to be important for caste differentiation were more likely to be regulated by multiple epigenomic systems than other differentially expressed genes. We confirmed the importance of two candidate genes for caste differentiation by using RNAi to manipulate the expression of two genes that differed in expression between workers and queens were regulated by multiple epigenomic systems. For both genes the RNAi manipulation resulted in a decrease in weight and fewer ovarioles of newly emerged queens compared to controls. Our data show that the distinct epigenomic landscapes of worker and queen bees differentiate during the course of larval development.

**Keywords:** honeybees, caste differentiation, development, epigenetic modifications, gene expression

## Introduction

Our growing appreciation of epigenomics has transformed our understanding of development and genome environment interactions. The epigenome describes the interacting mechanisms that collectively regulate the expression of the genome (Skvortsova *et al.*, 2018; Nacev *et al.*, 2020; Tsai and Cullen, 2020). These mechanisms include biochemical modifications of the genome (such as DNA methylation), modulation of the affinity of DNA to proteins, that provide structural support to the DNA molecule, and changes to the folding of the DNA molecule and/or the chromosome to alter the accessibility of sections of DNA to gene transcription

---

† These two authors contributed equally to this paper.
*Corresponding author: bees1965@sina.com (ZJZ).
注：此文发表在 *Insect Biochemistry and Molecular Biology* 2023, 115: 103929。

machinery (Robertson, 2005; Karlić et al., 2010; Krijger and De Laat, 2016; Verhoeven et al., 2016; Luo et al., 2022). Multiple different specific mechanisms exist in each of these classes. The complexity of the epigenome is such that metaphor of the "epigenomic landscape" has been used to capture the multidimensional structural complexity and diversity of the epigenome (Berdasco and Esteller, 2010; Rijnkels et al., 2010; Herman et al., 2014). The epigenomic landscape was first described by Conrad Waddington, who believed that during development, cells are affected by different epigenetic factors, resulting in different forms of cell development (Waddington and Kacser, 1957).

The significance of the epigenome is apparent in naturally occurring phenotypic plasticity. Here one genome can yield different phenotypes in different environments with no change to the DNA sequence of the genome (West-Eberhard, 1989; Whitman and Agrawal, 2009). The interactions between the genome and the environment, and the developmental pathways that lead to different phenotypic outcomes are all functions of the epigenome (Duncan et al., 2014; Schlichting and Wund, 2014).

Waddington famously considered the process of development as a progressive series of developmental decisions, which he conceptualized as forked paths in an imagined developmental landscape. Two identical entities (two single cells in an organism or two embryos) that took different paths at a fork early in development could find themselves channeled (or canalized) along progressively more divergent paths to ultimately develop into irreversibly distinct forms. In this way an undifferentiated cell could give rise to different tissues, or organisms could show developmental plasticity.

Waddington, and other researchers building on his frameworks, imagined that functional networks of genes could shape the bifurcating paths and valleys of the imagined developmental landscape (Waddington, 1942; Waddington and Kacser, 1957; Moris et al., 2016). Waddington's work preceded our contemporary understanding of epigenetic mechanisms. In the mid-1970s Riggs and Holliday both proposed that DNA methylation could affect gene expression, hence serve as an epigenetic regulatory mechanism. Further discoveries have added chromatin modifications, noncoding RNAs and miRNAs as other epigenetic modifications affecting gene expression and altering the operation of Waddington's functional gene networks. We might conceptualise the function of differential gene regulation in development in two different ways. Does differential gene regulation set the developmental landscape, establishing the valleys along which different developmental trajectories can flow? Or is differential gene regulation a response to moving through a developmental landscape, with different configurations of gene expression forming as diverging developmental paths canalize?

Honeybee (*Apis mellifera*) caste differentiation is a striking example of naturally occurring phenotypic plasticity, and it has become an important model for studies of genomics and epigenomics in phenotypic plasticity (Corona et al., 2016; Maleszka, 2018; Duncan et al., 2020). Young female larvae (less than 3.5 days old) can develop into two very different castes: queens and workers (Winston, 1991). The queen is large in size, with a long lifespan, large spermatheca and ovaries, and is usually the only female in the colony that can produce male and female offspring. The worker is small in size, with a short lifespan and usually infertile (Wilson, 1971; Michener and Michener, 1974). These different developmental outcomes come about because larvae are raised by workers in different developmental environments. These are different sized cells and different diets that differ in sugars, fatty acids, para-coumaric acid (*p*- coumaric acid) and specific royal jelly proteins (Kamakura, 2011; Shi *et al.*, 2011). We now understand that these nutritional differences cause many changes in the epigenetic regulation of several key genomic signaling pathways. These include mitogen-activated protein kinase signaling pathway (MAPK), target of rapamycin signaling pathway (TOR) and insulin receptor substrate signaling pathway (IRS) (Patel *et al.*, 2007; Mutti *et al.*, 2011). Developing queen and worker larvae are recognized to differ in extent and distribution of DNA methylation, histone modification, microRNA (miRNA) expression, $m^6A$ modification and poly (A) tails (Kucharski *et al.*, 2008; Shi *et al.*, 2011; Shi *et al.*, 2013; Wojciechowski *et al.*, 2018; Yi *et al.*, 2020; Wang *et al.*, 2021a; He *et al.*, 2022). Collectively these studies illustrate very different epigenomic profiles in developing workers and queens.

The queen and worker developmental pathways are flexible, to a degree. Transplanting young larvae from worker cells to queen cells can result in a normal queen phenotype, but transplanting larvae that are more than 3.5 days old from the worker rearing environment to the queen rearing environment result in inter-castes (a caste that is morphologically intermediate between the queen and workers), or workers rather than queens (Weaver, 1957; Weaver, 1966). There is a point (3.5 day) in developmental beyond which the worker developmental trajectory is irreversibly canalized and cannot be successfully redirected to yield the queen phenotype.

In this study we examined how the epigenomic profiles of the queen and worker change during honeybee during development. We used a multiomic approach to assess different aspects of the epigenome of workers and queens sampled before and after the 3.5 day developmental point at which the worker/queen trajectories are canalized. We compared gene expression levels with RNA sequencing. We compared histone modifications using chromatin immunoprecipitation with high-throughout sequencing (ChIP-seq). We compared which regions of the chromosome were more available for transcription using the assay for transposase accessible chromatin with high-throughout sequencing (ATAC-seq). Finally, we compared the three-dimensional chromosome structure of samples with the high-throughput chromosome conformation capture (Hi-C) assay. Hi-C can detect topologically associated domains (TAD) in the genome. TADs are influenced by the three-dimensional configuration of the chromosome and genes within TADs are more likely to interact with each other and to be co-regulated than genes at random (Sikorska and Sexton, 2020; Eres and Gilad, 2021). ATAC-seq can distinguish the compartments of the genome in which genes are more or less likely to be transcribed (Klemm *et al.*, 2019; Agbleke *et al.*, 2020). ChIP-seq assays histone modifications associated with different genes, which might influence the affinity of the DNA for the histone protein (Grant, 2001; Zentner and Henikoff, 2013). These three assays each give different perspectives on biochemical processes that regulate gene expression, whereas RNA-seq measures the outcome of that regulation: the expression of each gene.

Using these four methods we compared the epigenomic landscapes of queen and worker larvae that were two and four days old to measure developmental timepoints before and after the age at which a worker pathway is canalized. This approach allowed us to study how the distinct epigenomic profiles of queen and worker bees develop. We also selected two key genes which were significantly different in all four omics for RNA interference (RNAi) to verify the functions of these genes in honeybee caste differentiation.

## Material and Methods

### *Insects*

Honeybees (*Apis mellifera*) were from Jiangxi Agricultural University (28.46°N, 115.49°E), Nanchang, China, in 2020. Queens were restricted for 6h (8 am-2 pm) to a plastic frame designed by Pan *et al* (Pan *et al*., 2013) to lay eggs in worker cells. The queen laid her eggs on a removable plastic base, which could be transferred to a plastic queen cell without touching the egg itself. Half of the eggs were transferred to queen cells at 2 pm on the second day after laying, while the other half remained in the worker cells. All eggs (in both queen and worker cells) were cared for by workers. Eggs hatched on the third day after laying. To collect 2-day and 4-day old queen and worker larvae, we sampled larvae from both queen and worker cells in each of three colonies at 4 pm on the 5th and 7th day after laying. Larvae were picked with sterilized tweezers and rinsed in ddH$_2$O three times. Filter paper was used to drain the water from the larvae, and larvae were placed immediately in liquid nitrogen. Whole larvae were used for sequencing, as in previous studies (Wang *et al*., 2021a; He *et al*., 2022), since they were very small (2-day worker and queen larvae: 1.63 mg/larva; 4 day worker larvae: 11.60 mg/larva; 4 day queen larvae: 25.92 mg/larva) (Stabe, 1930; Liu *et al*., 2021; Wang *et al*., 2021a; He *et al*., 2022). Because larvae of different ages differed in mass, we sampled different numbers of 2-day and 4-day larvae so that each of our samples delivered similar amounts of DNA or RNA before sequencing. The reference genome used throughout this paper was Amel_HAv3.1 (GCF_003254395.2) downloaded from the NCBI. The detailed methods of high-throughput sequencings (Hi-C, ATAC-seq, ChIP-seq and RNA-seq) described below. All omic analyses had three biological replicates.

### *Hi-C*

In situ Hi-C was performed using larval samples, with minor modification to previously described methods (Rao *et al*., 2014). In 2-day queen (2Q) and worker (2W) larval groups each sample had 80 larvae, whereas in 4-day queen and worker larval ones each sample mixed 8 larvae. Hi-C was performed using a modification of the method described by Szabo *et al* (Szabo *et al*., 2018). Hi-C libraries were sequenced with paired-end, 150 bp reads on an Illumina Hiseq3000. Reads were filtered for contaminants, then were processed with Homer (version 4.11; http://homer.ucsd.edu/homer/) to produce genome-wide contact maps (Heinz *et al*., 2010). Principal component values were produced using the HiCExplorer, and A and B compartments were assigned to each 100-kb window according to the sign of the first component (PC1) values. We would reverse the sign of eigenvalues based on gene density content (Wolff *et al*., 2018). After correction, the positive value was the A compartment and the negative value was the B compartment. Lieberman *et al*. (2009) first found that A compartment is more open and has more transcriptional activation regions whereas B dcompartment has more chromatin inhibited regions (Lieberman-Aiden *et al*., 2009). TAD is a topological association domain and the frequency of gene interactions within a TAD is significantly higher than that between the adjacent two TADs, which establishes patterns of gene co-regulation (Burgess, 2022). According to the following hidden Markov model formula, our data were divided into thousands of windows on the chromosome with a length of

100 kb, and the directionality index value of each window was calculated. The size of TAD can be determined according to the directionality index value. A loop is a ring structure of chromosomal interaction, and the entrance of the ring is two regions of the chromosome that are close in space. The approach frequency here is multiple times that of other random approaches (Rao et al., 2014), so a peak can be formed. Based on this, we distinguished loops from the background (random connection) using the algorithm (HiCCUPS) (Durand et al., 2016), which is called call-peak and is often the action site of CCCTC-binding factor (CTCF), and also the action site of other regulatory elements (enhancers, etc.).

### *ATAC-seq*

ATAC-seq was performed using a modification of the method described by Corces et al (2017). Briefly, approximately 100 mg of larvae (40 larvae per sample in 2Q and 2W; 4 larvae per sample in 4Q and 4W) were ground and the nuclei were extracted, then the nuclei pellet was re-suspended in Tn5 transposase reaction mix. The transposition reaction was incubated at 37°C for 30 min. Equimolar Adapter1 and Adatper2 were added after transposition. PCR was then performed to amplify the library. ATAC libraries were sequenced on an Illumina Hiseq3000 and 150 bp paired-end reads were generated.

We employed Burrows Wheeler Aligner (BWA, version 0.7.12) to align the ATAC-seq reads to the honeybee genome as above. After mapping reads to the reference genome, we used the Model-based Analysis of ChIP-Seq (MACS) (version 2.1.0; https://github.com/taoliu/MACS/) to identify peaks with parameters as reported previously (Zhang et al., 2008; Corces et al., 2017). A P-adj enrichment threshold of 0.05 was used for all data sets. The analysis of different peaks (Fold change ≥ 2) is based on the folding enrichment of different experimental peaks. ChIPseeker (Yu et al., 2015) was used to identify the nearest Transcription Start Sites (TSS) of every peak and the distance distribution between peaks and TSS was shown. An analysis of the distribution of peak summits on different functional regions of the gene, such as 5'-untranslated region (5'UTR), 3'-untranslated region (3'UTR), distal intergenic, coding DNA sequence (CDS), was performed. Peak related genes were confirmed by PeakAnnotator (Zhu et al., 2010), and then KOBAS software (version 3.0) was used to test the statistical enrichment of peak related genes in KEGG pathways (Xie et al., 2011). HMMRATAC was used for range identification of single, double and triple nucleosomes in honeybees (Tarbell and Liu, 2019). We used MEME Suite to conduct sequence analysis based on Motif (Bailey et al., 2009). Motif is implemented by MEME using the search mode "anr". It is assumed that each peak sequence can contain any number of non-overlapping occurrences of each motif. The most significant motif is firstly detected, and its significance is evaluated by the e value, which is the estimated value of the number of expected memes under a given log-likelihood ratio.

### *ChIP-seq*

H3K27ac has been shown to be associated with caste differentiation of honeybees (Wojciechowski et al., 2018), therefore it was chosen in this study. Chromatin immunoprecipitation was performed as described by Wojciechowski et al (2018) with slight modifications. Approximately 800 mg of larvae (320 larvae per sample in 2Q and 2W; 32 larvae per sample in 4Q and 4W) were cross-linked for 10 min in 1% ChIP-seq-grade formaldehyde. The antibody used for immunoprecipitation was H3K27ac (ab4729, Abcam). The H3K27ac library was sequenced (150 bp paired-end reads) on an Illumina Hiseq3000 sequencer.

We employed Burrows Wheeler Aligner (BWA; version 0.7.12) to align the ChIP-Seq reads to the honeybee genome. After mapping reads to the reference genome, we used the MACS finding algorithm to identify regions of Immunoprecipitation (IP) enrichment over background (inputs were used as a control for this experiment). A p-*adj* enrichment threshold of 0.05 was used for all data sets. The analysis of different peaks (Fold change ≥ 2) was based on the folding enrichment of different experimental peaks. ChIPseeker (Yu *et al.*, 2015) was used to identify the nearest TSS of every peak and the distance distribution between peaks and TSS was shown. Besides, the distribution of peak summits on different function regions of the gene structure, such as 5' UTR, 3'UTR, distal intergenic, CDS, was performed. Peak related genes were confirmed by PeakAnnotator (Zhu *et al.*, 2010), and then KOBAS software was used to test the statistical enrichment of peak related genes in KEGG pathways.

*RNA-seq*

RNA-seq was performed as previously reported (He *et al.*, 2019). Total RNA was extracted from 100mg larvae (40 larvae per sample in 2Q and 2W; 4 larvae per sample in 4Q and 4W). Paired-end 150-cycle sequencing was performed on Illumina Hiseq3000 sequencers according to the manufacturer's directions.

We employed Hierarchical Indexing for Spliced Alignment of Transcripts (HISAT; version 2.0.5) (Kim *et al.*, 2015) to align the RNA-Seq reads to the reference genome. Expression levels were reported as Fragments Per Kilobase Million (FPKM) to normalize for the length of annotated transcripts and for the total number of reads aligned to the transcriptome. Analysis of differential expression was performed using the DESeq2 R package (1.32.0). The p-value was corrected for multiple comparisons with a false discovery rate (FDR< 0.05). Genes with P-*adj* ≤ 0.05 were defined as differentially expressed genes. By integrating the Hi-C, ATAC-seq, ChIP-seq and RNA-seq data, we selected 58 key DEGs that also differed in one or more epigenetic modifications. These 58 DEGs were chosen because they have been associated with caste differentiation in other studies (Table S8).

*Application of RNAi*

One-day-old worker larvae were fed with semi-artificial diet in a petri dish according to the previous method (Mao *et al.*, 2015) and were incubated at 34°C and 75% humidity. Artificially manufactured siRNA for 4-coumarate-CoA Ligase (*4CL*)

(F: GGUGAAAGAUAUGCUAAUATT; R: UAUUAGCAUAUCUUUCACCTT)

and L-lactate dehydrogenase (*LLD*)

(F: CGGUCGACAUUCUCACCUACG; R: UAGGUGAGAAUGUCGACCGGA)

were added to the semi-artificial diet, with a final concentration of 100 μg/mL. Similarly the NCsiRNA

(F: UUCUUCGAACGUGUCACGUTT; R: ACGUGACACGUUCGGAGAATT)

was added into semi-artificial diet and fed other larvae as a control group. Totally 118 honey bee larvae with 36 in each group (*4CL*-RNAi, *LLD*-RNAi and the control) were used, and the mortality rate was 29.67%. At least 15 biological replicates were successfully reared and sampled in each group. Some of the larvae fed with the above siRNA diets for two days and four days were taken for qRT-PCR validation to verify the effect of RNAi on the expression of the *4CL* gene and *LLD* gene. Other larvae were reared until emergence and these newly emerged bees were weighed using an analysis balance (accuracy: 0.1 mg, ME204, METTLER) and photographed under a microscope (6.5X, GL99TI, VISHENT). Samples then were used for ovariole counts.

### Gene expression differences were verified by qRT-PCR

2Q, 2W, 4Q and 4W larvae were sampled. In each case four larvae of one type were combined as one sample. We collected four samples for each type, with each sample coming from a different colony. Therefore, for this study we sampled a total of 16 of each larval type (2Q, 2W, 4Q and 4W) across four different colonies.

qRT-PCR was performed according to the method of Pan et al (Pan et al., 2022). Briefly, total RNAs were extracted by TransZol Up Plus RNA Kit (TransGen Biotech, China). Total RNAs were then transcribed into cDNA using a PrimeScript™ RT reagent Kit (Takara, Japan). GAPDH was used as a reference gene

(F: GCTGGTTTCATCGATGGTTT; R: ACGATTTCGACCACCGTAAC).

The primers were as follows: *4CL*

(F: CAAGTGGACCTTTCGTGGTT; R: CGTCAACATGACACCTTTCG)

and *LLD*

(F: ATTCGAGCGTGTGCATAGTG; R: GACCGGATTCGAGACGATAA).

The primer sequences were designed by Prime Primer 5.0. A 10 μL (5 μL of SYBR®Premix Ex Taq™ II, 3 μL of $H_2O$, 1 μL of cDNA, 0.4 μL each of forward and reverse primers and 0.2 μL of ROX) qRT-PCR reaction system was established. The PCR conditions were as follows: 95 °C, 5 min; 94 °C, 2 min; 40 cycles (95 °C, 10 s, Tm, 15 s, 72 °C, 15 s); 72 °C, 10 min. To establish the melting curve of the qRT-PCR product, the primers were heated slowly with a gradual increase of 1 °C every 5 s from 72 to 99 °C. The data were analyzed by $2^{-\Delta\Delta CT}$ (Livak and Schmittgen, 2001).

### Paraffin sectioning of the honeybee ovaries

The left ovaries of the newly emerged samples were dissected under a microscope. Ovaries were then fixed in 4% paraformaldehyde fix solution for 16 h and paraffin sections of the ovary were created using methods described by Yi et al (2020). The sections were photographed and ovarioles counted under the same microscope (40X) according to our previous methods (Yi et al., 2020).

### Quantification and statistical analysis

For the data analysis of qRT-PCR were calculated using $2^{-\Delta\Delta Ct}$ comparative Ct method (Livak and Schmittgen, 2001) and were transformed by taking their square root to be normally distributed. Data were analyzed by one-way $t$ test in SPSS (version 25.0). The p-value < 0.05 was considered as significantly different. The relative expression was evaluated as mean ± SEM. Other data analyses were also processed using SPSS (version 25.0) and the methods used are also written in the figure legend.

### Data and code availability

Raw sequencing reads for Hi-C ATAC-seq, ChIP-seq and RNA-seq are available at SRA accession PRJNA770835.

## Results

### *Data quality control*

This study performed four different sequencing techniques (Hi-C, ATAC-seq, ChIP-seq and RNA-Seq) to analyze 2Q, 2W, 4Q and 4W. The Q30 value of each sample in all four omics was all over 90 % (Table S1). The Pearson's correlations between the three biological replicates of four omics were all over 0.85 (see Fig. S1). These results indicated that the sequencing was acceptable and the biological replicates were reliable.

### *Comparison of Hi-C and RNA-seq analysis*

Chromosome interactions were similar between 2Q and 2W (Fig. I-2-1A, C, see page 495), but 4Q had significantly stronger cis interaction while 4W had significantly stronger trans interaction (Fig. I-2-1B, C). The 4Q/4W comparison also had more genes that switched between the A/B compartments (325) than the 2Q/2W comparison (247) (Table S2). Compared with 2Q/2W, the 4Q/4W comparison had more A/B switched regions and more genes within the A/B switched region (Fig. S2A). The 4Q/4W comparison had longer TAD boundaries than the 2Q/2W comparison (Fig. I-2-1F). These indicated that the differences of 3D genomic structure between honeybee queens and workers were different at the 2-day larval stage and the 4-day larval stage.

The A/B compartment differences between queens and workers were positively related to differences in gene expression at both 2 day and 4 day stages. Active compartments were strongly associated with up-regulated genes in queen larvae whereas inactive compartments were associated with down-regulated genes in queen larvae (Fig. I-2-1D, E). More significantly differentially expressed genes (DEGs) were located in the TAD boundaries of 4Q and 4W than 2Q and 2W, and more DEGs were located in the TAD boundaries of 4W compared to 4Q (Fig. I-2-1G). Many of these DEGs were enriched in Kyoto Encyclopedia of Genes and Genomes (KEGG) pathways involved in honeybee caste differentiation, including insect hormone biosynthesis, TOR signaling and MAPK signaling (Fig. S2E, S2F).

The number of 4Q unique loops increased compared to 2Q, whereas this was decreased in 4W (Fig. I-2-1H). There were more DEGs located in unique loops of the 4Q/4W comparison than the 2 day comparison (Fig. I-2-1I).

### *Comparison of ATAC-seq and RNA-seq analyses*

ATAC-seq results showed that 1.02 Mb (0.46%), 1.52 Mb (0.69%), 8.65 Mb (3.92%) and 1.50 Mb (0.68%) of the honeybee genome was detected as unique open accessible chromatin in 2Q, 2W, 4Q and 4W respectively (Fig. I-2-2A, see page 496). 4Q had more open accessible regions in its genome compared to the other samples.

In 2Q/2W comparisons we identified 253 unique ATAC peaks from 2Q and 382 from 2W (Table S3, Fig. S3A). In 4Q/4W comparisons, more unique peaks from 4Q (4,618) were identified but only 448 from 4W (Table S4, Fig. S3A). The proportion of mononucleosome, di-nucleosomes and tri-nucleosomes in queens dramatically increased with age ($p = 8.11e-22$), whereas workers showed an opposite trend ($p = 1.25e-39$, see Fig. I-2-2B). When mapping these unique peaks to gene expression differences, considerably more DEGs which contain significantly different ATAC peaks were identified in the 4 day comparison compared to 2 day comparison (Fig. I-2-2C, S3B). There was a strong positive correlation between the differences of ATAC and biased gene expression in queen-worker comparisons (Fig. I-2-2D, S3C-S3F). 4Q had the largest number of DEGs containing unique ATAC peaks, and also the most DEGs associated with caste differentiation (Fig. I-2-2E). There were more DEGs containing unique ATAC peaks in 4Q or 4W compared to the 2 day samples. More DEGs containing significantly different ATAC peaks in the 4 day comparison were involved in honeybee caste differentiation than in the 2 day comparison (Fig. I-2-2F). DEGs which contain significantly different ATAC peaks from 4Q/4W were enriched in eight key pathways involved in honeybee caste differentiation, whereas DEGs which contain significantly different ATAC peaks from 2Q/2W were enriched in three key pathways only (Fig. S3G, S3H). The most significantly enriched transcription factor in all larval groups was activating transcription factor 1 (Atf1). This transcription factor was predicted to regulate 3729 target genes including the 4-coumarate-CoA Ligase (*4CL*, Loc726040) which was selected for the later RNAi experiment (Fig. S3I, Table S9). These findings suggest that chromatin accessibility is strongly related to the biased gene expression that is known to be causal of honeybee caste differentiation (Fig. I-2-2C and D, Fig. S3E and F).

*Comparison of ChIP-seq and RNA-seq analyses*

The number of unique ChIP peaks increased with age and were more abundant in queen than worker samples (Fig. S4A, Table S7). Similar to the ATAC results, the 4-day comparison had more DEGs containing significantly different ChIP peaks than the 2-day comparison (Fig. I-2-3A, B, see page 497, S4B). There was a strong positive correlation between H3K27ac modification and gene expression (Fig. I-2-3C, D, S4C-S4F). Compared with the 2 day comparison, the 4-day comparison had more genes containing unique peaks, and the genes associated with caste differentiation were also more abundant (Fig. I-2-3E). Compared with 2Q/2W, there were more DEGs containing significantly different peaks associated with caste differentiation in 4Q/4W (Fig. I-2-3F). Many DEGs containing significantly different ChIP peaks were enriched in caste-differentiation related pathways, with 8 pathways in 2Q/2W and 9 in 4Q/4W (Fig. S4G, S4H). These results suggest that H3K27ac also partly contributes to the honeybee caste differentiation and differences are more pronounced at the day 4 stage than day 2.

*Multiomics analysis of caste differentiation*

In the 2W/2Q comparisons, Hi-C, ATAC and H3K27ac histone modification all differed between honeybee queens and workers, and the numbers of genes related to the significant differences of Hi-C, ATAC and histone modification were 228, 344 and 4308 respectively. A very low proportion of these differences overlapped with the DEGs (Fig. I-2-4A, see page 498). In the 4W/4Q comparison more genes were related to the significant differences of Hi-C (813), ATAC (4096) and histone modification (7275), and a higher proportion of these genes overlapped with DEGs (Fig. I-2-4B). Among them, the 72 genes that were different in the four omics were mainly involved in the propanoate metabolism, pyruvate metabolism and insect hormone biosynthesis pathway and oxidoreductase activities (Fig. S7).

We selected 58 key DEGs that have previously been associated with honeybee caste differentiation (Table S8). More of these genes were differentially regulated by at least one epigenomic system in 4 day comparisons compared to 2 day comparisons (Fig. I-2-4C and D). These suggest that there is a small divergence in gene expression between queen and workers at an early developmental stage that is not widely reinforced by differential genomic regulation. At the 4 day larval stage the differences in genes expression between workers and queens were more profound, and many more of the differences are reinforced by at least one form of genomic regulation.

In the 2 day comparison, only a few of the 58 key DEGs were regulated by three different epigenetic modifications and most of them were influenced by one or two types of modifications (see the color-marked symbols in Fig. I-2-4E). By contrast, more key DEGs were regulated by three epigenomic modifications in the 4 day comparison, and more genes were differentially regulated by more than one genomic system (see the color-marked symbols in Fig. I-2-4E). These suggest that epigenomic control of honeybee caste differentiation involves complex multi-omic interactions that develop as the worker and queen phenotypic diverge.

***Gene expression differences associated with multiple epigenetic modifications***

In 2 day comparison, genes containing significant differences of three epigenetic modifications between queen and worker larvae rarely overlapped (Fig. S8A), but these notably increased in 4 day comparison (Fig. S8B). The proportion of DEGs that were associated with at least one other epigenomic difference was lower in 2 day comparison (51.42 %, Fig. I-2-4F) than the 4 day comparison (76.15 %, Fig. I-2-4G). Moreover, the proportion of DEGs associated with more than one epigenomic difference was also higher in the 4 day than 2 day comparisons (Fig. I-2-4F and G). These results show that epigenomic differences between queens and workers become more extensive, more layered and more complex at a later developmental stage compared to an earlier stage. This pattern was also seen in the 58 key DEGs that are associated with honeybee caste differentiation (Fig. I-2-4H and I), a higher proportion of these key DEGs were differentially regulated by at least 2 types of epigenetic modifications (Fig. I-2-4I) compared to total DEGs (Fig. I-2-4G), suggesting that the key genes involved into queen-worker dimorphism are likely regulated by more layers of epigenetic modification compared to other genes. The highest proportion was DEGs was associated with differences in chromatin accessibility and histone modification rather than 3D chromosome structure (Fig. I-2-4I), perhaps suggesting that gene expression is more directly regulated by chromatin accessibility and histone modification and less directly influenced by chromosome structure.

***Effects of a key candidate gene on caste differentiation***

We selected 6 key genes [vitellogenin (*VG*) (Zhang *et al.*, 2022), hexamerin 70a (*Hex70a*) (Cameron *et al.*, 2013), lethal(2)essential for life (*L(2)EFL*) (Garcia *et al.*, 2009), probable cytochrome P450 6a13 (*P450-6a13*) (Mao *et al.*, 2015), juvenile hormone acid O-methyltransferase (*JHAMT*) (Bomtorin *et al.*, 2014) and heat shock protein 90 (*HSP90*) (Evans and Wheeler, 2000)] that have been shown involved in caste differentiation, and 2 genes [*4CL* and L-lactate dehydrogenase (*LLD*, Loc411188)] which we suspected as key genes involved in caste differentiation also. The *4CL* is involved in *p*-coumaric acid synthesis in honeybee larval diets and may be involved in honeybee caste differentiation (Stuible *et al.*, 2000; Cukovica *et al.*, 2001; Mao *et al.*, 2015; Islam *et al.*, 2019). Our results showed a clear trend that all 8 genes had greater differences in epigenomic regulation in 4Q/4W comparison than the 2Q/2W comparison (Fig. I-2-5A, B see page 499, and

Fig. S6).

*4CL* and *LLD* differed in all four omics assays in our 4Q/4W comparison. We did an interaction analysis on the four-omic overlapped 72 DEGs from 4 day comparison, and two key DEGs (*4CL* and *LLD*) had the largest node in the gene interaction network. These two key genes linked to 36 genes which were significantly different in at least one omic, in which seven genes are known to be important for caste differentiation such as *HSP90* (Evans and Wheeler, 2000), *Loc408567* (Hasegawa *et al.*, 2009) and *Loc412541* (Li *et al.*, 2010) (Fig. I-2-5C). Reducing the expression of these two genes with RNAi (Fig. I-2-5D) resulted in a decrease in body weight (Fig. I-2-5E) and body size (Fig. I-2-5F) of newly-emerged queens compared to the control. Reduced expression of *4CL* resulted in fewer ovarioles in the adult queen ovary when compared to controls (Fig. I-2-5G; *t* test, T = 16.86, $P = 1.3e^{-5}$).

## Discussion

Here we used a multiomics approach to obtain different perspectives on the divergent epigenomic profiles of worker and queen bees. The methods we used (Hi-C, ATAC-seq, ChIP-Seq and RNA-Seq) explored different types of epigenomic differences between the worker and queen castes. Our results clearly showed that all these four omics were involved in honeybee caste differentiation. Thousands of DEGs were identified between worker and queen larvae (Fig. S9), which are similar to previous studies (He *et al.* 2019). We found differences in 3D chromosome structure and dramatically different proportions of accessible chromatin between queen and worker larvae (Fig. I-2-1 and I-2-2), including hundreds of differences in A/B compartments, TAD boundaries and loops between queen and worker larvae at both 2-day and 4-day larval stages. Moreover, many DEGs that were enriched for key pathways such as TOR, IRS, Notch, Hippo etc (Patel *et al.*, 2007; Mutti *et al.*, 2011; Chen *et al.*, 2017) (see Fig. S2C-S2F) were related to these distinct A/B switch areas, TAD boundaries and loops (Fig. I-2-1F-I, Fig. S2A). Therefore, our findings suggest that differences in chromosome structure are involved in honeybee caste differentiation, possibly via regulating gene expression, as has been suggested previously (Rodríguez-Carballo *et al.*, 2017; Wang *et al.*, 2018; Sikorska and Sexton, 2020; Eres and Gilad, 2021; Wang *et al.*, 2021b).Our ChIP-seq results revealed a clear difference in histone modification (H3K27ac) between workers and queens, which were consistent with Hurd's study that H3K27ac participates in the regulation of honeybee caste differentiation (Wojciechowski *et al.*, 2018). Our ATAC-seq results found differences of unique peaks and significantly different peaks between queen and workers (Fig. I-2-2A and C), which were associated with thousands of DEGs (Fig. I-2-2E and F). The ratio of mononucleosome, di-nucleosomes and tri-nucleosomes in queens dramatically increased with age while decreased in workers (Fig. I-2-2B), indicating a different transcription activity between queen and workers during their larval development.

In our 4Q sample we observed a notable increase in unique accessible chromatin regions compared to 4W, 2Q and 2W (Fig. I-2-2A). The 4Q samples were fed with royal jelly and were undergoing extremely rapid growth and development (Winston, 1991). The large-scale open accessible chromatin in 4Q could indicate increased transcriptional processing to support their active metabolism. In support of this conclusion more DEGs were up-regulated in 4Q (1355) than 4W (951), (Table S7). These results indicate that the development of queen-worker dimorphism involves distinct patterns of accessible chromatin. In addition, we found that ATF1 was the most abundant transcription factor in all samples. ATF1 is an important transcription factor involved in many biological processes such as the production of adenosine triphosphate, mitochondrial respiration, DNA synthesis and repair, and general cell growth (Bleckmann *et al.*, 2002; Ghosh *et al.*, 2002; Jin and O'Neill, 2010; Jin and O'Neill, 2014; Nickels *et al.*, 2022). Therefore, further investigations should explore the possible biological functions of ATF1 in honeybee caste differentiation.

All three epigenetic modifications we explored (histone modification, chromatin accessibility and chromosome structure) had positive correlations with gene expression (Fig. I-2-1D, E, Fig. I-2-2C, Fig. I-2-3C, D). Chromosome structure was positively correlated with chromatin accessibility and histone modification (Fig. S5), emphasizing how these epigenetic modifications can interact with each other to regulated gene expression (Ikegami *et al.*, 2009; Wang *et al.*, 2009; Bannister and Kouzarides, 2011; Buitrago *et al.*, 2021; Lu *et al.*, 2021). Our data suggest that gene expression differences underlying caste differentiation are regulated by an interacting system of epigenetic modifications rather than dominated by a single epigenetic modification (Fig. I-2-4) revealing very different epigenomic landscapes for developing worker and queen bees which develop during the course of larval development.

Moreover, the differences in chromosome structure, chromatin accessibility and histone modification between workers and queens at the early developmental stage (see Fig. I-2-1E-I, Fig. I-2-2A-F, Fig. I-2-3A, E, F), were not very pronounced and they had little association with DEGs (Fig. I-2-4A) or key DEGs known to be involved in the caste differentiation (Fig. I-2-4C). By contrast, at the later day 4 developmental stage differences in all three types of epigenetic modifications increased and were more highly overlapped with DEGs (Fig. I-2-4 and 5). This pattern was even more pronounced for our 58 key DEGs that have previously been associated with caste differentiation (Fig. I-2-5E and F). The implication is the development of a layering of epigenomic control of gene expression differences during larval development as the phenotypes diverge. This might serve to stabilise critical functional differences in gene expression between the worker and queen phenotypes.

Stable differences in gene expression depend on both the nature of environmental change and the regulatory epigenomic landscape (Herman *et al.*, 2014). The 4 day sample point was beyond the point at which a worker larva can be successfully transformed into adult queens by feeding them with more nutritional diets, suggesting the worker developmental trajectory is canalized at this point (Weaver, 1957; Weaver, 1966). In the 4 day worker/queen larval comparison a higher proportion of the 58 key DEGs related to caste differentiation differed in two or three different epigenetic modifications (Fig. I-2-5F). Moreover, 8 selected key genes also had more differences in genomic regulation in 4Q/4W comparisons compared to 2Q/2W comparisons (Fig. I-2-5A and 5B, Fig. S6). From this we infer that gene expression differences that are critical to the formation of the distinct worker and queen phenotypes become stabilised in development through regulation by multiple degenerate epigenomic systems.

To test our inference that genes of key importance for queen/worker differences are subject to the most extensive epigenomic regulation we selected the candidate DEG *LLD* and *4CL* for RNA interference (RNAi) verification. Both genes have been causally linked to caste differentiation and both genes differed between 4W and 4Q in all four omics assays (Fig. I-2-5B). Previous studies have shown that *4CL* is involved in *p*-coumaric acid synthesis, and this acid which exists in honeybee larval diets plays an important role in honeybee caste differentiation (Stuible *et al*., 2000; Cukovica *et al*., 2001; Mao *et al*., 2015; Islam *et al*., 2019). Here RNAi knock down of *4CL* expression resulted in newly emerged queens of lower weight, smaller body size and fewer ovariole numbers than controls (Fig. I-2-5C-F). Knocking down of expression of *LLD* resulted in lower weight of newly emerged queens but did not change ovariole counts (Fig. I-2-5C-F). In the interaction network, *4CL* and *LLD* were associated with 36 genes, seven of which have been linked to caste differentiation (Fig. I-2-5C). Therefore, RNA interference of *4CL* may also have an indirect effect on these caste differentiation related genes. In addition, *ATF1* was also predicted to regulate the expression of *4CL* (Fig. S3I, Table S9), again reflecting that this key gene is essential for caste differentiation and can be regulated by various epigenetic modifications. Consequently, these findings confirm the importance of expression of these two genes for caste differentiation, and it is notable that they were also subject to the most extensive epigenomic differences between workers and queens.

Honeybee phenotypic plasticity and caste differentiation is an outcome of millions of years of evolution (Seeley, 1989). This is achieved through different epigenomic regulation of the bee genome during development. Our results, together with previous studies, demonstrate that various epigenetic modifications are all involved in bee phenotypic plasticity (Kucharski *et al*., 2008; Shi *et al*., 2011; Guo *et al*., 2013; Shi *et al*., 2013; Shi *et al*., 2015; Guo *et al*., 2016; Wojciechowski *et al*., 2018; Wang *et al*., 2021b). The key insights from this present study are that different kinds of epigenomic modification work together to establish the different worker and queen phenotypes. The very different epigenomic profiles of the worker and queen are not programs. They are not pre-established to control the running of different developmental paths. Instead the epigenomic profiles diverge during development as the worker and queen phenotypes diverge. In this example, phenotypic plasticity epigenomic mechanisms are possibly part of the developmental autoshaping process, divering as workers and queens take progressively divergent developmental pathways through Waddington's developmental landscape. Epigenetic modifications also play an important role in the determination of caste differentiation in ants, wasps and other social insects (Weiner *et al*., 2013; Simola *et al*., 2016; Marshall *et al*., 2019). As a classical social insect, our findings in honeybees serve as a model of epigenetic landscape in the caste system for other Hymenoptera species. Consequently, our study has a potential contribution not only to the field of honey bee caste development, but also to the area of other social insects' phenotypic plasticity driven by environmental factors.

## Author Contributions

Z.J.Z., and X.J.H., designed the research; Y.Z. and M.J.J. performed experiments; Z.J.Z., A.B.B., X.J.H, Q.H., X.B.W., L.Z.Z., Z.L.W. and W.Y.Y. provided guidance for data; Y.Z., Z. L., and X.J.H. analyzed data; Y.Z., X.J.H., A.B.B. and Z.J.Z wrote the paper. The authors declare no competing interest. All authors were involved in the preparation of the final manuscript.

## Funding

This work was supported by the National Natural Science Foundation of China (32172790, 32160815, 31872432), Major Discipline Academic and Technical Leaders Training Program of Jiangxi Province (20204BCJL23041) and the Earmarked Fund for China Agriculture Research System (CARS-44-KXJ15).

## Supplementary Information

Supplementary data to this article can be found online at github. https://github.com/Catsheet/Supplementary.

## Data Availability

Raw sequencing reads for Hi-C ATAC-seq, ChIP-seq and RNA-seq are available at SRA accession PRJNA770835.

### *Ethics approval and consent to participate*

Ethics approval was not applicable.

### *Consent for publication*

Not applicable.

### *Competing interests*

The authors declare that they have no competing interests.

## Acknowledgement

We thank Mr. Xiaofeng He, Pengfei Yu and Han Li for help in data analysis.

## Appendix Supplementary data

Supplementary data to this article can be found online at https://doi.org/10.1016/j.ibmb.2023.103929.

## References

AGBLEKE, A.A., AMITAI A., BUENROSTRO J.D., CHAKRABARTI A., CHU L., HANSEN A.S., KOENIG K.M., LABADE A.S., LIU S., NOZAKI T., 2020. Advances in chromatin and chromosome research: perspectives from multiple fields. *Mol. Cell* 79, 881-901.

BAILEY, T.L., BODEN M., BUSKE F.A., FRITH M., GRANT C.E., CLEMENTI L., REN J., LI W.W., NOBLE W.S., 2009. MEME SUITE: tools for motif discovery and searching. *Nucleic Acids Res*. 37, W202-W208.

BANNISTER, A.J., KOUZARIDES T., 2011. Regulation of chromatin by histone modifications. *Cell Res*. 21, 381-395.

BERDASCO, M., ESTELLER M., 2010. Aberrant epigenetic landscape in cancer: how cellular identity goes awry. *Dev. Cell* 19, 698-711.

BLECKMANN, S.C., BLENDY J.A., RUDOLPH D., MONAGHAN A.P., SCHMID W., SCHÜTZ G.N., 2002. Activating transcription factor 1 and CREB are important for cell survival during early mouse development. *Molecular and Cellular Biology* 22, 1919-1925.

BOMTORIN, A.D., MACKERT A., ROSA G.C.C., MODA L.M., MARTINS J.R., BITONDI M.M.G., HARTFELDER K., SIMÕES Z.L.P., 2014. Juvenile hormone biosynthesis gene expression in the corpora allata of honey bee (*Apis mellifera* L.) female castes. *PLoS One* 9, e86923.

BUITRAGO, D., LABRADOR M., ARCON J.P., LEMA R., FLORES O., ESTEVE-CODINA A., BLANC J., VILLEGAS N., BELLIDO D., GUT M., 2021. Impact of DNA methylation on 3D genome structure. *Nat. Commun.* 12, 1-17.

BURGESS, D.J., 2022. A TAD refined for gene regulation. *Nat. Rev. Genet.*, 1-1.

CAMERON, R.C., DUNCAN E.J., DEARDEN P.K., 2013. Biased gene expression in early honeybee larval development. *BMC Genomics* 14, 1-12.

CHEN, X., MA C., CHEN C., LU Q., SHI W., LIU Z., WANG H., GUO H., 2017. Integration of lncRNA-miRNA-mRNA reveals novel insights into oviposition regulation in honey bees. *PeerJ* 5, e3881.

CORCES, M.R., TREVINO A.E., HAMILTON E.G., GREENSIDE P.G., SINNOTT-ARMSTRONG N.A., VESUNA S., SATPATHY A.T., RUBIN A.J., MONTINE K.S., WU B., 2017. An improved ATAC-seq protocol reduces background and enables interrogation of frozen tissues. *Nat. Meth.* 14, 959-962.

CORONA, M., LIBBRECHT R., WHEELER D.E., 2016. Molecular mechanisms of phenotypic plasticity in social insects. *Curr. Opin. Insect Sci.* 13, 55-60.

CUKOVICA, D., EHLTING J., ZIFFLE J.A.V., DOUGLAS C.J., 2001. Structure and evolution of 4-coumarate: coenzyme A ligase (4CL) gene families. *Biol. Chem.* 382, 645-654.

DUNCAN, E.J., GLUCKMAN P.D., DEARDEN P.K., 2014. Epigenetics, plasticity, and evolution: How do we link epigenetic change to phenotype? Journal of Experimental Zoology Part B: *Molecular and Developmental Evolution* 322, 208-220.

DUNCAN, E.J., LEASK M.P., DEARDEN P.K., 2020. Genome architecture facilitates phenotypic plasticity in the honeybee (*Apis mellifera*). *Mol. Biol. Evol.* 37, 1964-1978.

DURAND, N.C., SHAMIM M.S., MACHOL I., RAO S.S., HUNTLEY M.H., LANDER E.S., AIDEN E.L., 2016. Juicer provides a one-click system for analyzing loop-resolution Hi-C experiments. *Cell Systems* 3, 95-98.

ERES, I.E., GILAD Y., 2021. A TAD skeptic: is 3D genome topology conserved? *Trends Genet.* 37, 216-223.

EVANS, J.D., WHEELER D.E., 2000. Expression profiles during honeybee caste determination. *Genome Biol.* 2, 1-6.

GARCIA, L., SARAIVA GARCIA C.H., CALÁBRIA L.K., COSTA NUNES DA CRUZ G., SÁNCHEZ PUENTES A., BÁO S.N., FONTES W., RICART C.A., SALMEN ESPINDOLA F., VALLE DE SOUSA M., 2009. Proteomic analysis of honey bee brain upon ontogenetic and behavioral development. *J. Proteome Res.* 8, 1464-1473.

GHOSH, S.K., GADIPARTHI L., ZENG Z.-Z., BHANOORI M., TELLEZ C., BAR-ELI M., RAO G.N., 2002. ATF-1 mediates protease-activated receptor-1 but not receptor tyrosine kinase-induced DNA synthesis in vascular smooth muscle cells. *J. Biol. Chem.* 277, 21325-21331.

GRANT, P.A., 2001. A tale of histone modifications. *Genome Biol.* 2, 1-6.

GUO, X., SU S., GEIR S., LI W., LI Z., ZHANG S., CHEN S., CHEN R., 2016. Differential expression of miRNAs related to caste differentiation in the honey bee, *Apis mellifera. Apidologie* 47, 495-508.

GUO, X., SU S., SKOGERBOE G., DAI S., LI W., LI Z., LIU F., NI R., GUO Y., CHEN S., 2013. Recipe for a busy bee: microRNAs in honey bee caste determination. *PLoS One* 8, e81661.

HASEGAWA, M., ASANUMA S., FUJIYUKI T., KIYA T., SASAKI T., ENDO D., MORIOKA M., KUBO T., 2009. Differential gene expression in the mandibular glands of queen and worker honeybees, *Apis mellifera L.*: implications for caste-selective aldehyde and fatty acid metabolism. *Insect Biochem. Mol. Biol.* 39, 661-667.

HE, X.J., BARRON A.B., YANG L., CHEN H., HE Y.Z., ZHANG L.Z., HUANG Q., WANG Z.L., WU X.B., YAN W.Y., 2022. Extent and complexity of RNA processing in honey bee queen and worker caste development. *Iscience* 25, 104301.

HE, X.J., JIANG W.J., ZHOU M., BARRON A.B., ZENG Z.J., 2019. A comparison of honeybee (*Apis mellifera*) queen, worker and drone larvae by RNA-Seq. *Insect Sci.* 26, 499-509.

HEINZ, S., BENNER C., SPANN N., BERTOLINO E., LIN Y.C., LASLO P., CHENG J.X., MURRE C., SINGH H., GLASS C.K., 2010. Simple combinations of lineage-determining transcription factors prime cis-regulatory elements required for macrophage and B cell identities. *Mol. Cell* 38, 576-589.

HERMAN, J.J., SPENCER H.G., DONOHUE K., SULTAN S.E., 2014. How stable 'should' epigenetic modifications be? Insights from adaptive plasticity and bet hedging. *Evolution* 68, 632-643.

IKEGAMI, K., OHGANE J., TANAKA S., YAGI S., SHIOTA K., 2009. Interplay between DNA methylation, histone modification and chromatin remodeling in stem cells and during development. *Int. J. Dev. Biol.* 53, 203-214.

ISLAM, M.T., LEE B.R., LEE H., JUNG W.J., BAE D.W., KIM T.H., 2019. *p*-Coumaric acid induces jasmonic acid-mediated phenolic accumulation and resistance to black rot disease in Brassica napus. *Physiol. Mol. Plant Pathol.* 106, 270-275.

JIN, X., O'NEILL C., 2010. The presence and activation of two essential transcription factors (cAMP response element-binding protein and cAMP-dependent transcription factor ATF1) in the two-cell mouse embryo. *Biol. Reprod.* 82, 459-468.

JIN, X., O'NEILL C., 2014. The regulation of the expression and activation of the essential ATF1 transcription factor in the mouse preimplantation embryo. *Reproduction* 148, 147-157.

KAMAKURA, M., 2011. Royalactin induces queen differentiation in honeybees. *Nature* 473, 478-483.

KARLIĆ, R., CHUNG H.-R., LASSERRE J., VLAHOVIČEK K., VINGRON M., 2010. Histone modification levels are predictive for gene expression. *PANS* 107, 2926-2931.

KIM, D., LANGMEAD B., SALZBERG S.L., 2015. HISAT: a fast spliced aligner with low memory requirements. *Nat. Methods* 12, 357-360.

KLEMM, S.L., SHIPONY Z., GREENLEAF W.J., 2019. Chromatin accessibility and the regulatory epigenome. *Nat. Rev. Genet.* 20, 207-220.

KRIJGER, P.H.L., DE LAAT W., 2016. Regulation of disease-associated gene expression in the 3D genome. *Nat. Rev. Mol. Cell Biol.* 17, 771-782.

KUCHARSKI, R., MALESZKA J., FORET S., MALESZKA R., 2008. Nutritional control of reproductive status in honeybees via DNA methylation. *Science* 319, 1827-1830.

LI, J., WU J., BEGNA RUNDASSA D., SONG F., ZHENG A., FANG Y., 2010. Differential protein expression in honeybee (Apis mellifera L.) larvae: underlying caste differentiation. *PLoS One* 5, e13455.

LIEBERMAN-AIDEN, E., VAN BERKUM N.L., WILLIAMS L., IMAKAEV M., RAGOCZY T., TELLING A., AMIT I., LAJOIE B.R., SABO P.J., DORSCHNER M.O., 2009. Comprehensive mapping of long-range interactions reveals folding principles of the human genome. *Science* 326, 289-293.

LIU, Y.B., YI Y., ABDELMAWLA A., ZHENG Y.L., ZENG Z.J., HE X.J., 2021. Female developmental environment delays development of male honeybee (*Apis mellifera*). *BMC Genomics* 22, 1-12.

LIVAK, K.J., SCHMITTGEN T.D., 2001. Analysis of relative gene expression data using real-time quantitative PCR and the 2-$\Delta\Delta$CT method. *Methods* 25, 402-408.

LU, J., HUANG Y., ZHANG X., XU Y., NIE S., 2021. Noncoding RNAs involved in DNA methylation and histone methylation, and acetylation in diabetic vascular complications. *Pharmacol. Res.* 170, 105520.

LUO, L.H., GRIBSKOV M., WANG S.F., 2022. Bibliometric review of ATAC-Seq and its application in gene expression. *Brief. Bioinform.*

MALESZKA, R., 2018. Beyond Royalactin and a master inducer explanation of phenotypic plasticity in honey bees. *Communications Biology* 1, 1-7.

MAO, W., SCHULER M.A., BERENBAUM M.R., 2015. A dietary phytochemical alters caste-associated gene expression in honey bees. *Sci. Adv.* 1, e1500795.

MARSHALL, H., LONSDALE Z.N., MALLON E.B., 2019. Methylation and gene expression differences between reproductive and sterile bumblebee workers. *Evolution Letters* 3, 485-499.

MICHENER, C.D., MICHENER C.D. (1974). *The social behavior of the bees: a comparative study*. (Harvard University Press).

MORIS, N., PINA C., ARIAS A.M., 2016. Transition states and cell fate decisions in epigenetic landscapes. *Nat. Rev. Genet.* 17, 693-703.

MUTTI, N.S., DOLEZAL A.G., WOLSCHIN F., MUTTI J.S., GILL K.S., AMDAM G.V., 2011. IRS and TOR nutrient-signaling pathways act via juvenile hormone to influence honey bee caste fate. *J. Exp. Biol.* 214, 3977-3984.

NACEV, B.A., JONES K.B., INTLEKOFER A.M., YU J.S., ALLIS C.D., TAP W.D., LADANYI M., NIELSEN T.O., 2020. The epigenomics of sarcoma. *Nat. Rev. Cancer* 20, 608-623.

NICKELS, J.F., DELLA-ROSA M.E., GOYENECHE I.M., CHARLTON S.J., SNEPPEN K., THON G., 2022. The transcription factor Atf1 lowers the transition barrier for nucleosome-mediated establishment of heterochromatin. *Cell Reports* 39, 110828.

PAN, L.X., HU W.W., CHENG F.P., HU X.F., WANG Z.L., 2022. Transcriptome analysis reveals differentially expressed genes between the ovary and testis of the honey bee Apis mellifera. *Apidologie* 53, 1-12.

PAN, Q.-Z., WU X.-B., GUAN C., ZENG Z.-J., 2013. A new method of queen rearing without grafting larvae. *Am. Bee J.* 153, 1279-1280.

PATEL, A., FONDRK M.K., KAFTANOGLU O., EMORE C., HUNT G., FREDERICK K., AMDAM G.V., 2007. The making of a queen: TOR pathway is a key player in diphenic caste development. *PLoS ONE* 2, e509.

RAO, S.S., HUNTLEY M.H., DURAND N.C., STAMENOVA E.K., BOCHKOV I.D., ROBINSON J.T., SANBORN A.L., MACHOL I., OMER A.D., LANDER E.S., 2014. A 3D map of the human genome at kilobase resolution reveals principles of chromatin looping. *Cell* 159, 1665-1680.

RIJNKELS, M., KABOTYANSKI E., MONTAZER-TORBATI M.B., BEAUVAIS C.H., VASSETZKY Y., ROSEN J.M., DEVINOY E., 2010. The epigenetic landscape of mammary gland development and functional differentiation. *J. Mammary Gland Biol. Neoplasia* 15, 85-100.

ROBERTSON, K.D., 2005. DNA methylation and human disease. *Nat. Rev. Genet.* 6, 597-610.

RODRÍGUEZ-CARBALLO, E., LOPEZ-DELISLE L., ZHAN Y., FABRE P.J., BECCARI L., EL-IDRISSI I., HUYNH T.H.N., OZADAM H., DEKKER J., DUBOULE D., 2017. The HoxD cluster is a dynamic and resilient TAD boundary controlling the segregation of antagonistic regulatory landscapes. *Genes Dev.* 31, 2264-2281.

SCHLICHTING, C.D., WUND M.A., 2014. Phenotypic plasticity and epigenetic marking: an assessment of evidence for genetic accommodation. *Evolution* 68, 656-672.

SEELEY, T.D., 1989. The honey bee colony as a superorganism. *Am. Sci.* 77, 546-553.

SHI, Y.Y., HUANG Z.Y., ZENG Z.J., WANG Z.L., WU X.B., YAN W.Y., 2011. Diet and cell size both affect queen-worker differentiation through DNA methylation in honey bees (Apis mellifera, Apidae). *PLoS One* 6, e18808.

SHI, Y.Y., YAN W.Y., HUANG Z.Y., WANG Z.L., WU X.B., ZENG Z.J., 2013. Genomewide analysis indicates that queen larvae have lower methylation levels in the honey bee (Apis mellifera). *Naturwissenschaften* 100, 193-197.

SHI, Y.Y., ZHENG H.J., PAN Q.Z., WANG Z.L., ZENG Z.J., 2015. Differentially expressed microRNAs between queen and worker larvae of the honey bee (Apis mellifera). *Apidologie* 46, 35-45.

SIKORSKA, N., SEXTON T., 2020. Defining functionally relevant spatial chromatin domains: it is a TAD complicated. *J. Mol. Biol.* 432, 653-664.

SIMOLA, D.F., GRAHAM R.J., BRADY C.M., ENZMANN B.L., DESPLAN C., RAY A., ZWIEBEL L.J., BONASIO R., REINBERG D., LIEBIG J., 2016. Epigenetic (re) programming of caste-specific behavior in the ant Camponotus floridanus. *Science* 351, aac6633.

SKVORTSOVA, K., IOVINO N., BOGDANOVIĆ O., 2018. Functions and mechanisms of epigenetic inheritance in animals. *Nat. Rev. Mol. Cell Biol.* 19, 774-790.

STABE, H.A., 1930. The Rate of Growth of Worker, Drone and Queen Larvae of the Honeybee, *Apis Mellifera* Linn. *J. Econ. Entomol.* 23, 447-453.

STUIBLE, H.-P., BÜTTNER D., EHLTING J., HAHLBROCK K., KOMBRINK E., 2000. Mutational analysis of 4-coumarate: CoA ligase identifies functionally important amino acids and verifies its close relationship to other adenylate-forming enzymes. *FEBS Lett.* 467, 117-122.

SZABO, Q., JOST D., CHANG J.-M., CATTONI D.I., PAPADOPOULOS G.L., BONEV B., SEXTON T., GURGO J., JACQUIER C., NOLLMANN M., 2018. TADs are 3D structural units of higher-order chromosome organization in Drosophila. *Sci. Adv.* 4, eaar8082.

TARBELL, E.D., LIU T., 2019. HMMRATAC: a Hidden Markov ModeleR for ATAC-seq. *Nucleic Acids Res.* 47, e91-e91.

TSAI, K., CULLEN B.R., 2020. Epigenetic and epitranscriptomic regulation of viral replication. *Nat. Rev. Microbiol.* 18, 559-570.

VERHOEVEN, K.J., VONHOLDT B.M., SORK V.L., 2016. Epigenetics in ecology and evolution: what we know and what we need to know. *Mol. Ecol.* 25, 1631-1638.

WADDINGTON, C., KACSER H., 1957. The strategy of the genes. A discussion of some aspects of theoretical biology. With an appendix by H. Kacser. Strateg. genes. A Discuss. some Asp. Theor. *Biol. With an Append.* by H. Kacser.

WADDINGTON, C.H., 1942. Canalization of development and the inheritance of acquired characters. *Nature* 150, 563-565.

WANG, L., WUERFFEL R., FELDMAN S., KHAMLICHI A.A., KENTER A.L., 2009. S region sequence, RNA polymerase II, and histone modifications create chromatin accessibility during class switch recombination. *J. Exp. Med.* 206, 1817-1830.

WANG, M., XIAO Y., LI Y., WANG X., QI S., WANG Y., ZHAO L., WANG K., PENG W., LUO G.-Z., 2021a. RNA m6A modification functions in larval development and caste differentiation in honeybee (*Apis mellifera*). *Cell Reports* 34, 108580.

WANG, Q., SUN Q., CZAJKOWSKY D.M., SHAO Z., 2018. Sub-kb Hi-C in D. melanogaster reveals conserved characteristics of TADs between insect and mammalian cells. *Nat. Commun.* 9, 1-8.

WANG, Y.L., LIU Y.Q., XU Q., XU Y., CAO K., DENG N., WANG R.M., ZHANG X.Y., ZHENG R.Q., LI G.L., 2021b. TAD boundary and strength prediction by integrating sequence and epigenetic profile information. *Brief. Bioinform.* 22, bbab139.

WEAVER, N., 1957. Effects of larval age on dimorphic differentiation of the female honey bee. *Ann. Entomol. Soc. Am.* 50, 283-294.

WEAVER, N., 1966. Physiology of caste determination. *Annu. Rev. Entomol.* 11, 79-102.

WEINER, S.A., GALBRAITH D.A., ADAMS D.C., VALENZUELA N., NOLL F.B., GROZINGER C.M., TOTH A.L., 2013. A survey of DNA methylation across social insect species, life stages, and castes reveals abundant and caste-associated methylation in a primitively social wasp. *Sci. Nat.* 100, 795-799.

WEST-EBERHARD, M.J., 1989. Phenotypic plasticity and the origins of diversity. *Annu. Rev. Ecol. Syst.* 20, 249-278.

WHITMAN, D.W., AGRAWAL A.A., 2009. What is phenotypic plasticity and why is it important. Phenotypic plasticity of insects: Mechanisms and consequences, 1-63.

WILSON, E.O., 1971. The insect societies. Harvard University Press, Cambridge.

WINSTON, M.L., 1991. The biology of the honey bee. Harvard University Press, Cambridge.

WOJCIECHOWSKI, M., LOWE R., MALESZKA J., CONN D., MALESZKA R., HURD P.J., 2018. Phenotypically distinct female castes in honey bees are defined by alternative chromatin states during larval development. *Genome Res.* 28, 1532-1542.

WOLFF, J., BHARDWAJ V., NOTHJUNGE S., RICHARD G., RENSCHLER G., GILSBACH R., MANKE T., BACKOFEN R., RAMÍREZ F., GRÜNING B.A., 2018. Galaxy HiCExplorer: a web server for reproducible Hi-C data analysis, quality control and visualization. *Nucleic Acids Res.* 46, W11-W16.

XIE, C., MAO X., HUANG J., DING Y., WU J., DONG S., KONG L., GAO G., LI C.-Y., WEI L., 2011. KOBAS 2.0: a web server for annotation and identification of enriched pathways and diseases. *Nucleic Acids Res.* 39, W316-W322.

YI, Y., HE X.J., BARRON A.B., LIU Y.B., WANG Z.L., YAN W.Y., ZENG Z.J., 2020. Transgenerational accumulation of methylome changes discovered in commercially reared honey bee (*Apis mellifera*) queens. *Insect Biochem. Mol. Biol.* 127, 103476.

YU, G., WANG L.G., HE Q.-Y., 2015. ChIPseeker: an R/Bioconductor package for ChIP peak annotation, comparison and visualization. *Bioinformatics* 31, 2382-2383.

ZENTNER, G.E., HENIKOFF S., 2013. Regulation of nucleosome dynamics by histone modifications. *Nat. Struct. Mol. Biol.* 20, 259-266.

ZHANG, W., WANG L., ZHAO Y., WANG Y., CHEN C., HU Y., ZHU Y., SUN H., CHENG Y., SUN Q., 2022. Single-cell transcriptomic analysis of honeybee brains identifies vitellogenin as caste differentiation-related factor. *iScience* 25, 104643.

ZHANG, Y., LIU T., MEYER C.A., EECKHOUTE J., JOHNSON D.S., BERNSTEIN B.E., NUSBAUM C., MYERS R.M., BROWN M., LI W., 2008. Model-based analysis of ChIP-Seq (MACS). *Genome Biol.* 9, 1-9.

ZHU, L.J., GAZIN C., LAWSON N.D., LIN S.M., LAPOINTE D.S., GREEN M.R., 2010. ChIPpeakAnno: a Bioconductor package to annotate ChIP-seq and ChIP-chip data. *BMC Bioinformatics* 11, 1-10.

# 3. Phenotypic Dimorphism between Honeybee Queen and Worker is Regulated by Complicated Epigenetic Modifications

Mengjie Jin[1,2][†], Zilong Wang[1,2][†], Zhihao Wu[1,2], Xujiang He[1,2], Yong Zhang[1,2], Qiang Huang[1,2], Lizhen Zhang[1,2], Xiaobo Wu[1,2], Weiyu Yan[1,2], Zhijiang Zeng[1,2]*

1. Honeybee Research Institute, Jiangxi Agricultural University, Nanchang, Jiangxi 330045, P. R. of China
2. Jiangxi Province Honeybee Biology and Beekeeping Nanchang, Jiangxi 330045, P. R. of China

**Abstract**

Phenotypic dimorphism between queens and workers is an important biological characteristic of honeybees that has been the subject of intensive research. The enormous differences in morphology, lifespan, physiology and behavior between queens and workers are caused by a complicated set of factors. Epigenetic modifications are considered to play an important role in this process. In this study, we analyzed the differences in chromosome interactions and H3K27ac and H3K4me1 modifications between the queens and workers using High-through chromosome conformation capture (Hi-C) and Chromatin immunoprecipitation followed by sequencing (ChIP-seq) technologies. We found that the queens contain more chromosome interactions and more unique H3K27ac modifications than workers, in contrast, workers have more H3K4me1 modifications than queens. Moreover, we identified *Map3k15* as a potential caste gene in queen-worker differentiation. Our results suggest that chromosomal conformation and H3K27ac and H3K4me1 modifications are involved in regulating queen-worker differentiation, which reveals that the queen-worker phenotypic dimorphism is regulated by multiple epigenetic modifications.

**Keywords:** chromosome interaction, histone modification, multi-omic, Hi-C, ChIP-Seq

## Introduction

Honeybees are eusocial insects that have a division of labor and an advanced information exchange system.[1-3] They are economically valuable pollinators that are essential for the ability of many crops and wild plants to produce seed. In addition, they can maintain an ecological balance through pollination.[4,5]

The queens and workers develop from a similar genetic background but show enormous differences in morphological characters, lifespan, reproductive ability, and behavior. The queen has an approximately 10-fold longer lifespan compared with the workers and is dedicated to reproducing offspring.[6] Usually, the queen

---

† These two authors contributed equally to this paper.
*Corresponding author: bees1965@sina.com (ZJZ).
注：此文发表在 *iScience* 2023,26(4),106308。

has 150-180 ovarian tubes per ovary and can lay up to 2,000 eggs per day.[7,8] In contrast, the workers are usually sterile; with 3-26 ovarian tubes per ovary.[7] Moreover, both queen and larval pheromones can inhibit the development of worker ovaries.[9,10] The workers have a behavioral maturity period where they perform different tasks as they age. They serve as housekeepers when they are young and conduct such tasks as keeping eggs warm, cleaning the hive, and feeding larvae among others and eventually change to foraging behavior.[11,12]

The study of queen-worker differentiation has been the focus of concern on honeybees over a long period of time. The mechanism of caste differentiation is still not totally understood despite the large number of studies that have been conducted. The existing findings suggest that many factors contribute to the differentiation of castes in honeybees, including the size of the larval developmental space (worker cell and queen cell),[13,14] food quality and quantity,[15-18] hormones,[6,14,19,20] differences in gene expression,[21-26] DNA methylation,[20,23,27-29] histone modifications,[30-32] microRNAs,[33-35] and poly(A) tail.[36] To date, many signaling pathways have been reported to be involved in honeybee caste differentiation, including FoxO, mTOR, MAPK, Hippo, Hedgehog, Wnt, TGF-beta, Toll and Imd, longevity regulation (multiple species), dorsoventral axis formation and insect hormone biosynthetic signaling pathways.[23,37-40]

Epigenetic modifications play an important role in the queen-worker differentiation of honeybees, and DNA methylation is crucial. In addition, histone modifications also have influence on caste differentiation. There are four types of core histone modifications (H2A, H2B, H3 and H4), and the regulation of gene expression by H3 has been more widely studied.[41,42] It has been shown that H3K4me3, H3K27ac, and H3K36me3 differ extensively in the whole genome between worker and queen larvae and that H3K27ac can influence the caste differentiation of honeybees.[43]

In addition, we hypothesize that chromosomal interactions have an impact on the queen-worker differentiation of honeybees. Studies have shown that chromosomal interactions can dynamically regulate gene expression.[44-46] In higher eukaryotes, chromosomes are folded in three-dimensional (3-D) structure within the nucleus, and extensive interactions exist within the same chromosome and between different chromosomes.[47-49] Recent advances in the genome-wide localization of chromatin interactions, such as Hi-C,[44,46] have facilitated the identification of important 3-D genomic features, such as genome-wide chromatin loops, topologically associated domains (TADs), and A/B compartments.[44,46,50,51] Considering that chromosomal interactions extensively regulate individual development by influencing gene expression, we hypothesize that differences in chromosomal interaction are one of the important reasons for queen-worker differentiation.

Here, we studied the epigenetic regulatory mechanisms behind the phenotypic differences between queens and workers using multi-omics data, such as (Hi-C, ChIP-Seq, and RNA-Seq). We systematically investigated the differences in patterns of chromosome interaction and histone modification (H3K27ac and H3K4me1) between the two castes; and identified the A/B compartments, loops, TADs, and H3K27ac and H3K4me1 modifications related to queen-worker differentiation. To our knowledge, this is the first report of comprehensive insights into the epigenetic regulation mechanisms of queen-worker differentiation.

# Results

## Data quality control

Three omics, including Hi-C-seq, Chip-seq and RNA-Seq, were performed on newly emerged queens and workers of western honeybee (*Apis melfliera*) to study the epigenetic mechanism of honeybee caste differentiation. In Hi-C-seq, a total of 180.13 G and 180.45 G clean reads were obtained from queens and workers, respectively, with an average of 600.97 M clean reads per sample, and the Q30 of each sample was higher than 89.42% (Table S1). In Chip-seq, the clean reads of each sample were between 2.75-and 4.28 G, and the Q30 of each sample was higher than 86.83% (Table S2). In RNA-seq, the clean reads of each sample were from 5.97 G to 7.59 G, and the Q30 of each sample was higher than 92.71% (Table S3). These results indicated that the sequencing quality of these three omics were high enough and reliable.

## The queens contain more chromosome interactions than the workers

There is a significant difference in chromosome interaction between the queens and workers (Fig. I-3-1A, B, see page 500). Compared with the workers, the queens have a significantly larger number of *cis*- and *trans*- interactions (Fig. I-3-1B). A total of 2,165 bins were identified in both queens and workers, and 78 bins had the A/B compartment switched in the queen vs worker comparison (Table S4). We analyzed the changes in expression-fold changes (queen/worker) of the genes associated with the switch of A and B compartments. We found that in the queen vs worker comparison, the B-to-A related genes were transcriptionally activated overall in the queen (the mean of log2[queen/worker] >0), while the A-to-B related genes were transcriptionally repressed overall in the queen (the mean of log2[queen/worker] <0) (Fig. I-3-1C). The loop numbers in queen and worker were also comparable, and most of them were unique to each other (Fig. I-3-1D). We found that there were 243 differentially expressed genes (DEGs) in the queen loops by co-analyzing the loop data with the RNA-seq data (Fig. I-3-1E), and 73 DEGs unique to the queen loops were enriched in multiple signaling pathways (Fig. S1A), of them, the insect hormone biosynthesis and apoptosis signaling pathways could be related to the growth and development of the queens. In addition, there were 78 unique DEGs in the worker loop, which are enriched in pathways associated with growth and development, immune function lifespan and caste differentiation in honeybees (Fig. I-3-1E, S1B). In addition, the queen and worker have approximately similar amounts of TADs (373 vs 385) (Table S5). Of the TAD coverage regions in the genome, 69.36% are common to both queens and workers; 15.11% are unique to the queens, and 15.53% are unique to the workers (Fig. I-3-1F). Moreover, 311 and 333 genes were identified in the queen unique and worker unique regions, respectively (Fig. I-3-1G).

## The queens contain more unique H3K27ac modifications than the workers

The distribution of H3K27ac between queens and workers is shown in Fig. I-3-2A(see page 501). We found that H3K27ac is enriched around the transcriptional start sites (TSS) of the genes in both the queen and worker castes (Fig. I-3-2B), which is similar to previous studies of honeybee larvae.[43] There were 3,601 differential peaks between the queens and workers, and more peaks were up-regulated in the queen, which were distributed in the intron and promoter regions (Fig. I-3-2C). The genes associated with these differential peaks were enriched in the pathways associated with growth and development, ovarian development, and lifespan regulation in honeybees (Fig. I-3-2D, Table S6). Compared with the workers, the queens have exceptionally more unique H3K27ac peaks that are primarily distributed in the intron and promoter regions (Fig. I-3-2E, Table S7). The queen unique peak-related genes were enriched in the pathways related to differences between the queens and workers in body size, longevity, immunity, and ovarian development, such as the Hippo, Wnt, MAPK, Hippo, FoxO, TGF-beta, Notch, mTOR, Dorso-ventral axis formation and longevity regulating signaling pathways (Fig. S2A). In contrast, the worker unique H3K27ac peaks were few and enriched in pathways related to metabolism (Fig. S2B). Moreover, silencing the expression of the histone acetylase gene *p300* in honeybee larvae by RNAi resulted in a significant reduction in body weight, body length and the content of H3K27ac of the newly emerged bees in the RNAi group compared with the control group (Fig. S3). In conclusion, we hypothesize that H3K27ac histone modification is important in shaping the differentiation between the queens and workers.

### *The workers have more H3K4me1 modifications than the queens*

The distribution of H3K4me1 between the queens and workers is shown in Fig. I-3-3A, see page 502. Compared with that of the queens, the H3K4me1 of the workers were more widely distributed throughout the genome. In addition, the workers have more up-regulated peaks (Fig. I-3-3A, B, see page 502), and the genes associated with these differential peaks are enriched in pathways related to the formation of differences between the queens and workers. (Fig. I-3-3C, Table S8). There were 2,908 and 1,979 unique peaks in the queens and workers, respectively, which were primarily enriched in intron and promoter regions (Fig. I-3-3D). The queen unique peak-related genes were enriched in the Wnt, MAPK, Hippo, TGF-beta, FoxO, Dorso-ventral axis formation, Notch, Hedgehog, mTOR, and Phototransduction signaling pathways (Fig. S4A). These pathways are associated with body size, ovary development, lifespan, regulation and caste differentiation in honeybees. Pathway enrichment of the worker unique peak-associated genes exhibited results similar to those of the queens (Fig. S4B). We analyzed the correlation between the levels of methylation of the peaks located in the promoter region and the expression levels of the corresponding genes and found a significant negative correlation between them (Fig. I-3-3E). Moreover, silencing the expression of the histone methylase gene *setd1* led to a significant increase in body length and reduction in H3K4me1 content of the newly emerged bees in the RNAi group compared with those in the control group (Fig. S5). In conclusion, we hypothesize that H3K4me1 histone modification is important in shaping the differences between the queens and workers.

### *Association analysis of Hi-C, H3K27ac and H3K4me1 with DNA methylation*

In this study, 839 A/B compartment switch-related genes, 1,868 differential H3K27ac peak-related genes and 5383 differentially H3K4me1 peak-related genes were identified (Table S9, S10, S11). The distributions of CpG Observe/Expect (O/E) in the genomic regions of these genes showed a bimodal pattern (Fig. I-3-4A, B, C, see page 503), which is consistent with the findings of previous studies.[52,53]

We compared these genes with the 381 differentially methylated genes (DMG) reported by Lyko et al.,[52] and found that 25 A/B compartment switch-related genes and 72 differential H3K27ac peak-related genes and 223 differential H3K4me1 peak-related genes were overlapped between these two studies (Fig. I-3-4D, E, F,

Table S12, S13, S14).

### *Map3k15 is a potential caste gene in queen-worker differentiation*

There were 26 differential genes between the queens and workers identified in the integrated analysis of the Hi-C, H3K27ac and RNA-seq, and 74 differential genes in the integrated analysis of the Hi-C, H3K4me1 and RNA-seq (Fig. I-3-5, see page 504, S6). Four genes of these differential genes were chosen to verify their function in queen-worker differentiation using RNAi (Table S15). After knocking down the expression of *Map3k15* gene (loc408533) by RNAi (Fig. I-3-6A, see page 505), the newly emerged bees in the RNAi group had significantly reduced body weight and body length compared with the control group although other worker specific traits such as typical pollen baskets were not observed (Fig. I-3-6B, C). They were obviously smaller in size (Fig. I-3-6D), but the other three genes had no significant effect on body weight and body length of the newly emerged bees after RNAi (Fig. S7). These results suggest that *Map3k15* has a significant effect on the queen-worker differentiation and is a potential caste gene in honeybees.

## Discussion

The dimorphism of queens and workers caused by external environmental factors is a typical epigenetic model.[15] To explore the genetic mechanism behind this phenomenon, previous studies have focused on the differences in transcriptome, proteome, and DNA methylation[54] between the queens and workers during the larval stage. In contrast, only a few studies to date have been conducted to resolve the molecular mechanisms that underlie the differences between adult queens and workers, and the results showed that there were significant differences in DNA methylation between them[55]. In this study, epigenetic differences between the newly emerged queens and workers were compared for the first time using Hi-C, ChIP-Seq, and RNA-Seq technologies. A combined multi-omics analysis was used to reveal differences between them, thus, providing a complex model of genome-wide epigenetic regulation in queens and workers. We found that there were significant differences in chromosomal interactions and H3K27ac and H3K4me1 modifications between the queens and workers. This suggests that developmental differentiation between these two female castes is regulated by multiple epigenetic modifications.

Hi-C sequencing showed that the queens contain more chromosome interactions compared with those of the workers, and A/B switches can significantly affect gene expression. An increasing number of studies have shown that the intra- and inter-chromosomal interactions are relatively common events in regulating gene expression. For example, heart failure is associated with the reduced stability of chromatin interactions around disease-causing genes.[56] Our results suggest that chromatin conformation is involved in queen-worker caste differentiation by regulating the expression of related signaling pathway genes.

In this study, we found that the queen and worker each contained a large number of unique chromatin loops. Typically, chromatin loops enable two regions of the chromosome that are far apart to interact with each other.[46,57] Studies on chromatin structure have shown that regulatory elements, including enhancers, promoters, and insulators, can often form DNA loops to regulate the expression of related genes.[58-60] Our results suggest that these queen- and worker-unique loops could contain a large number of enhancers, promoters, or insulators to regulate transcription of genes located in the loops.

We found that the queens have many more unique H3K27ac peaks compared with those of the workers. In contrast, a previous study at the larval stages showed that the worker larvae had more specific peaks compared with those of the queen larvae. This could suggest that H3K27ac modification is dynamic over the time course of honeybee development. The H3K27ac modification is a robust mark of active enhancers and promoters that are strongly correlated with gene expression and transcription factor binding.[61] We found that the unique peaks of the queens were primarily distributed in the intron and promoter regions, suggesting that H3K27ac modifications in the promoter region of these related genes that are unique to the queens lead to the development of queen-specific phenotype.

We found that H3K4me1 modifications at the genome-wide level were higher in the workers than in the queens overall and had more up-regulated peaks in the workers. Moreover, the differential H3K4me1 peaks between the queens and workers were primarily distributed in promoter and intron regions. H3K4me1 is found at both transcriptional active promoters and distal regulatory elements, such as enhancers, and the H3K4me1 modification of promoters is often associated with the conditional repression of inducible genes.[62,63] Similarly, we found a negative correlation between methylation of the H3K4me1 promoter and gene expression. Our results suggest that the queen-worker differential H3K4me1 peaks could result in worker phenotypes by repressing gene transcription in the workers.

Among the genes that showed difference in all three omics, we used RNAi to confirm that the *Map3k15* (*ask3*) gene of the MAPK signaling pathway has an obvious effect on caste differentiation between the queens and workers. *Map3k15* is a member of the apoptosis signal-regulating kinases and plays an indispensable role in the signal transduction pathway implicated in cell death triggered by various types of cellular stresses, as well as in tumor initiation and progression[64,65]. Our RNAi results suggest that *Map3k15* could regulate cell proliferation and differentiation in honeybees. Moreover, we found that the differentially regulated H3K4me1 peaks of *Map3k15* between the queens and workers was located in the intron region of this gene (Fig. I-3-6E), suggesting that H3K4me1 modifications could act on the introns of *Map3k15* to regulate its differential expression between the queens and workers, which, in turn leads to caste differentiation.

In conclusion, we found significant differences in chromatin interactions and the modifications of H3K27ac and H3K4me1 between the queens and workers, and these epigenetic modifications could be important causes of phenotypic dimorphism between the queens and workers. Furthermore, we found that the *Map3k15* gene, which showed differences in three omics, is a potential caste gene in queen-worker differentiation. These results suggest that phenotypic differentiation between the queens and workers is a complex process that is regulated by multiple epigenetic modifications.

## Limitations of Study

This study found extensive caste differences in chromosome interaction and H3K27ac and H3K4me1 modifications between queens and workers by Hi-C and ChIP-seq technologies. However, the queen-worker differences are exhibited in many tissues/organs, our study just considered these epigenetic differences at the whole body level while not at the tissue/organ level. In further studies, we need to explore how epigenetic modifications affect the developmental differentiation of specific tissues/organs between queens and workers.

STAR★Methods

KEY RESOURCES TABLE

CONTACT FOR REAGENT

Lead contact

Materials availability

Data and code availability

EXPERIMENTAL MODEL AND SUBJECT DETAILS

    Insects

METHOD DETAILS

    Rearing of queens and workers

    RNA-seq

    ChIP-seq

    Hi-C

    GO and KEGG analysis

    RNAi

    qRT-PCR

    Elisa

QUANTIFICATION AND STATISTICAL ANALYSIS

SUPPLEMENTAL INFORMATION

Supplemental information can be found online at https://doi.org/10.1016/j.isci.2023.106308.

## Acknowledgement

We thank Ms. Zhen Xiu Zeng for her help on the experimental honeybee colony and Novogene Corporation (Beijing, China) for the sequencing service. This work was supported by the National Natural Science Foundation of China (32172790, 31702193), Jiangxi provincial academic and technical leader project (20204BCJL23041) and the Earmarked Fund for China Agriculture Research System (CARS-44-KXJ15).

## Author Contributions

Z.J.Z. designed the research. M.J.J., Z.L.W., X.J.H., Y.Z. performed the research. M.J.J. conducted most experiments. M.J.J., Z.L.W., X.J.H., Y.Z., and Q.H. analyzed the data. M.J.J., Z.L.W. performed validation experiments. Z.L.W., L.Z.Z. collected experimental data, X.B.W. and W.Y.Y. conceived of the study, designed the study, coordinated the study and helped draft the manuscript. Z.L.W., M.J.J., L.C., P. J.H., and Z.J.Z., wrote and revised the manuscript.

## Declaration of Interests

The authors declare that they have no competing interest. All authors were involved in the preparation of the final manuscript.

## References

1. ROBINSON, G.E. (2002). Genomics and integrative analyses of division of labor in honeybee colonies. *The American Naturalist 160*, S160-S172. https://doi.org/10.1086/342901.

2. ZHANG, H., ZENG, Z.J., YAN, W.Y., WU, XB., AND ZHANG, Y.L. (2010). Effects of three aliphatic esters of brood pheromone on development and foraging behavior of Apis cerana cerana workers. *Acta Entomol Sinica. 53*, 55-60https://doi.org/10.1016/S1002-0721(10)60377-8.

3. I'ANSON PRICE, R., DULEX, N., VIAL, N., VINCENT, C., AND GRÜTER, C. (2019). Honeybees forage more successfully without the "dance language" in challenging environments. *Science Advances 5*, eaat0450. https://doi.org/10.1126/sciadv.aat0450.

4. KEARNS, C.A., AND INOUYE, D.W. (1997). Pollinators, flowering plants, and conservation biology. *Bioscience 47*, 297-307. https://doi.org/10.2307/1313191.

5. DELAPLANE, K.S., MAYER, D.R., and MAYER, D.F. (2000). Crop pollination by bees (Cabi). https://doi.org/10.1079/9780851994482.0000.

6. PAGE JR, R.E., and PENG, C.Y.-S. (2001). Aging and development in social insects with emphasis on the honey bee, Apis mellifera L. *Experimental gerontology 36*, 695-711. https://doi.org/10.1016/S0531-5565(00)00236-9.

7. SAKAGAMI, S.F., AND AKAHIRA, Y. (1958). Comparison of ovarian size and number of ovarioles between the workers of japanese and european honeybees: Studies on the Japanese honeybee, Apis indica cerana Fabricius. I. 昆虫 *26*, 103-109. http://dl.ndl.go.jp/info:ndljp/pid/10649688.

8. GROZINGER, C.M., FAN, Y., HOOVER, S.E., and WINSTON, M.L. (2007). Genome-wide analysis reveals differences in brain gene expression patterns associated with caste and reproductive status in honey bees (Apis mellifera). *Molecular Ecology 16*, 4837-4848. https://doi.org/10.1111/j.1365-294X.2007.03545.x.

9. MOHAMMEDI, A., PARIS, A., CRAUSER, D., and LE CONTE, Y. (1998). Effect of aliphatic esters on ovary development of queenless bees (Apis mellifera L.). *Naturwissenschaften 85*, 455-458.

10. HOOVER, S.E., KEELING, C.I., WINSTON, M.L., and SLESSOR, K.N. (2003). The effect of queen pheromones on worker honey bee ovary development. *Naturwissenschaften 90*, 477-480. https://doi.org/10.1007/s00114-003-0462-z.

11. FAHRBACH, S.E., and ROBINSON, G.E. (1995). Behavioral development in the honey bee: toward the study of learning under natural conditions. *Learning & Memory 2*, 199-224. https://doi.org/10.1101/lm.2.5.199.

12. ROBINSON, G.E., FAHRBACH, S.E., and WINSTON, M.L. (1997). Insect societies and the molecular biology of social behavior. *Bioessays 19*, 1099-1108. https://doi.org/10.1002/bies.950191209.

13. DE WILDE, J., and BEETSMA, J. (1982). The physiology of caste development in social insects. *Advances in Insect Physiology 16*, 167-246. https://doi.org/10.1016/S0065-2806(08)60154-X.

14. USHERWOOD, P.N.R. (1985). Comprehensive insect physiology, biochemistry and pharmacology. In Insect control,

GA Kerkut and LI Gilbert, ed. (Oxford University Press), pp. 849-863.

15. HAYDAK, M.H. (1970). Honey bee nutrition. *Annual review of entomology 15*, 143-156.

16. TAO, T., SU, S.K., CHEN, S, L., and DU, H.H. (2008). Biological functions of the royal jelly proteins. *Chinese Bulletin of Entomology*. https://doi.org/10.3724/SP.J.1141.2008.00459.

17. KAMAKURA, M. (2011). Royalactin induces queen differentiation in honeybees. *Nature 473*, 478-483. https://doi.org/10.1038/nature10093.

18. MAO, W., SCHULER, M.A., and BERENBAUM, M.R. (2015). A dietary phytochemical alters caste-associated gene expression in honey bees. *Science Advances 1*, e1500795. https://doi.org/10.1126/sciadv.1500795.

19. RACHINSKY, A., STRAMBI, C., STRAMBI, A., and HARTFELDER, K. (1990). Caste and metamorphosis: hemolymph titers of juvenile hormone and ecdysteroids in last instar honeybee larvae. *General and Comparative Endocrinology 79*, 31-38. https://doi.org/10.1016/0016-6480(90)90085-Z.

20. MUTTI, N.S., DOLEZAL, A.G., WOLSCHIN, F., MUTTI, J.S., GILL, K.S., and AMDAM, G.V. (2011). IRS and TOR nutrient-signaling pathways act via juvenile hormone to influence honey bee caste fate. *Journal of Experimental Biology 214*, 3977-3984. https://doi.org/10.1242/jeb.061499.

21. EVANS, J.D., and WHEELER, D.E. (1999). Differential gene expression between developing queens and workers in the honey bee, Apis mellifera. *Proceedings of the National Academy of Sciences 96*, 5575-5580. https://doi.org/10.1073/pnas.96.10.5575.

22. SHILO, B.-Z. (2005). Regulating the dynamics of EGF receptor signaling in space and time. *Development, 132*, 4017-4027. https://doi.org/10.1242/dev.02006.

23. BARCHUK, A.R., CRISTINO, A.S., KUCHARSKI, R., COSTA, L.F., SIMÕES, Z.L., and MALESZKA, R. (2007). Molecular determinants of caste differentiation in the highly eusocial honeybee *Apis mellifera*. *BMC Developmental Biology 7*, 1-19. https://doi.org/10.1186/1471-213X-7-70.

24. YAMANAKA, N., and O'CONNOR, M.B. (2011). Apiology: royal secrets in the queen's fat body. *Current Biology 21*, R510-R512. https://doi.org/10.1016/j.cub.2011.05.037.

25. CAMERON, R.C., DUNCAN, E.J., and DEARDEN, P.K. (2013). Biased gene expression in early honeybee larval development. *BMC Genomics 14*, 1-12. https://doi.org/10.1186/1471-2164-14-903.

26. SHAO, X.L., HE, S.Y., ZHUANG, X.Y., FAN, Y., LI, Y.H., and YAO, Y.G. (2014). mRNA expression and DNA methylation in three key genes involved in caste differentiation in female honeybees (*Apis mellifera*). *Dongwuxue. Yanjiu. 35*, 92-98. https://doi.org/10.11813/j.issn.0254-5853.2014.2.092.

27. KUCHARSKI, R., MALESZKA, J., FORET, S., and MALESZKA, R. (2008). Nutritional control of reproductive status in honeybees via DNA methylation. *Science 319*, 1827-1830. https://doi.org/10.1126/science.1153069.

28. SHI, Y.Y., HUANG, Z.Y., ZENG, Z.J., WANG, Z.L., WU, X.B., and YAN, W.Y. (2011). Diet and cell size both affect queen-worker differentiation through DNA methylation in honey bees (*Apis mellifera, Apidae*). *PLoS One 6*, e18808. https://doi.org/10.1371/journal.pone.0018808.

29. FORET, S., KUCHARSKI, R., PELLEGRINI, M., FENG, S., JACOBSEN, S.E., ROBINSON, G.E., and MALESZKA, R. (2012). DNA methylation dynamics, metabolic fluxes, gene splicing, and alternative phenotypes in honey bees. *Proceedings of the National Academy of Sciences 109*, 4968-4973. https://doi.org/10.1073/pnas.1202392109.

30. BARKER, S.A., FOSTER, A.B., LAMB, D.C., and JACKMAN, L.M. (1959). Biological Origin and Configuration of 10-Hydroxy-Δ2-decenoic acid. *Nature 184*, 634-634. https://doi.org/10.1038/184634a0.

31. WELLEN, K.E., HATZIVASSILIOU, G., SACHDEVA, U.M., BUI, T.V., CROSS, J.R., and THOMPSON, C.B. (2009). ATP-citrate lyase links cellular metabolism to histone acetylation. *Science 324*, 1076-1080. https://doi.org/10.1126/science.1164097.

32. SPANNHOFF, A., KIM, Y.K., RAYNAL, N.J.-M., GHARIBYAN, V., SU, M.-B., ZHOU, Y.-Y., LI, J., CASTELLANO, S., SBARDELLA, G., and ISSA, J.-P.J. (2011). Histone deacetylase inhibitor activity in royal jelly might facilitate caste switching in bees. *EMBO Reports 12*, 238-243. https://doi.org/10.1038/embor.2011.9.

33. GUO, X., SU, S., SKOGERBOE, G., DAI, S., LI, W., LI, Z., LIU, F., NI, R., GUO, Y., and CHEN, S. (2013). Recipe for a busy bee: microRNAs in honey bee caste determination. *PLoS One 8*, e81661. https://doi.org/10.1371/journal.pone.0081661.

34. SHI, Y.Y., ZHENG, H.J., PAN, Q.Z., WANG, Z.L., and ZENG, Z.J. (2015). Differentially expressed microRNAs between queen and worker larvae of the honey bee (*Apis mellifera*). *Apidologie 46*, 35-45. https://doi.org/10.1007/s13592-014-0299-9.

35. ZHU, K., LIU, M., FU, Z., ZHOU, Z., KONG, Y., LIANG, H., LIN, Z., LUO, J., ZHENG, H., and WAN, P. (2017). Plant microRNAs in larval food regulate honeybee caste development. *PLoS Genetics 13*, e1006946. https://doi.org/10.1371/journal.pgen.1006946.

36. HE, X.J., BARRON, A.B., YANG, L., CHEN, H., HE, Y.Z., ZHANG, L.Z., HUANG, Q., WANG, Z.L., WU, X.B., and YAN, W.Y. (2022). Extent and complexity of RNA processing in honey bee queen and worker caste development. *Iscience 25*, 104301. https://doi.org/10.1016/j.isci.2022.104301.

37. WHEELER, D.E., BUCK, N.A., and EVANS, J.D. (2014). Expression of insulin/insulin-like signalling and TOR pathway genes in honey bee caste determination. *Insect Molecular Biology 23*, 113-121. https://doi.org/10.1111/imb.12065.

38. ASHBY, R., FORÊT, S., SEARLE, I., and MALESZKA, R. (2016). MicroRNAs in honey bee caste determination. *Scientific Reports 6*, 1-15. https://doi.org/10.1038/srep18794.

39. DUNCAN, E.J., HYINK, O., and DEARDEN, P.K. (2016). Notch signalling mediates reproductive constraint in the adult worker honeybee. *Nature Communications 7*, 1-10. https://doi.org/10.1038/ncomms12427.

40. FERNANDEZ-NICOLAS, A., and BELLES, X. (2016). CREB-binding protein contributes to the regulation of endocrine and developmental pathways in insect hemimetabolan pre-metamorphosis. *Biochimica et Biophysica Acta (BBA)-General Subjects 1860*, 508-515. https://doi.org/10.1016/j.bbagen.2015.12.008.

41. GRUNSTEIN, M. (1997). Histone acetylation in chromatin structure and transcription. *Nature 389*, 349-352. https://doi.org/10.1038/38664.

42. LACHNER, M., and JENUWEIN, T. (2002). The many faces of histone lysine methylation. *Current Opinion in Cell Biology 14*, 286-298. https://doi.org/10.1016/S0955-0674(02)00335-6.

43. WOJCIECHOWSKI, M., LOWE, R., MALESZKA, J., CONN, D., MALESZKA, R., and HURD, P.J. (2018). Phenotypically distinct female castes in honey bees are defined by alternative chromatin states during larval development. *Genome Research 28*, 1532-1542. https://doi.org/10.1101/gr.236497.118.

44. LIEBERMAN-AIDEN, E., VAN BERKUM, N.L., WILLIAMS, L., IMAKAEV, M., RAGOCZY, T., TELLING, A., AMIT, I., LAJOIE, B.R., SABO, P.J., and DORSCHNER, M.O. (2009). Comprehensive mapping of long-range interactions reveals folding principles of the human genome. *Science 326*, 289-293. https://doi.org/10.1126/science.1181369.

45. JIN, F., LI, Y., DIXON, J.R., SELVARAJ, S., YE, Z., LEE, A.Y., YEN, C.-A., SCHMITT, A.D., ESPINOZA, C.A., and REN, B. (2013). A high-resolution map of the three-dimensional chromatin interactome in human cells. *Nature 503*, 290-294. https://doi.org/10.1038/nature12644.

46. RAO, S.S., HUNTLEY, M.H., DURAND, N.C., STAMENOVA, E.K., BOCHKOV, I.D., ROBINSON, J.T., SANBORN, A.L., MACHOL, I., OMER, A.D., and LANDER, E.S. (2014). A 3D map of the human genome at kilobase resolution reveals principles of chromatin looping. *Cell 159*, 1665-1680. https://doi.org/10.1016/j.cell.2014.11.021.

47. BICKMORE, W.A., and VAN STEENSEL, B. (2013). Genome architecture: domain organization of interphase chromosomes. *Cell 152*, 1270-1284. https://doi.org/10.1016/j.cell.2013.02.001.

48. BONEV, B., and CAVALLI, G. (2016). Organization and function of the 3D genome. *Nature Reviews Genetics 17*, 661-678. https://doi.org/10.1038/nrg.2016.112.

49. ROWLEY, M.J., and CORCES, V.G. (2018). Organizational principles of 3D genome architecture. *Nature Reviews Genetics 19*, 789-800. https://doi.org/10.1038/s41576-018-0060-8.

50. DIXON, J.R., SELVARAJ, S., YUE, F., KIM, A., LI, Y., SHEN, Y., HU, M., LIU, J.S., and REN, B. (2012). Topological domains in mammalian genomes identified by analysis of chromatin interactions. *Nature 485*, 376-380. https://doi.org/10.1038/nature11082.

51. SEXTON, T., YAFFE, E., KENIGSBERG, E., BANTIGNIES, F., LEBLANC, B., HOICHMAN, M., PARRINELLO, H., TANAY, A., and CAVALLI, G. (2012). Three-dimensional folding and functional organization principles of the Drosophila genome. *Cell 148*, 458-472. https://doi.org/10.1016/j.cell.2012.01.010.

52. LYKO, F., FORET, S., KUCHARSKI, R., WOLF, S., FALCKENHAYN, C., and MALESZKA, R. (2010). The honey bee epigenomes: differential methylation of brain DNA in queens and workers. *PLoS Biology 8*, e1000506. https://doi.org/10.1371/annotation/2db9ee19-faa4-43f2-af7a-c8aeacca8037.

53. ZENG, J., and YI, S.V. (2010). DNA methylation and genome evolution in honeybee: gene length, expression, functional enrichment covary with the evolutionary signature of DNA methylation. *Genome Biology and Evolution 2*, 770-780. https://doi.org/10.1093/gbe/evq060.

54. FORET, S., KUCHARSKI, R., PITTELKOW, Y., LOCKETT, G.A., and MALESZKA, R. (2009). Epigenetic regulation of the honey bee transcriptome: unravelling the nature of methylated genes. *BMC Genomics 10*, 1-11. https://doi.org/10.1186/1471-2164-10-472.

55. ELANGO, N., HUNT, B.G., GOODISMAN, M.A., and YI, S.V. (2009). DNA methylation is widespread and associated with differential gene expression in castes of the honeybee, *Apis mellifera*. *Proceedings of the National Academy of Sciences 106*, 11206-11211. https://doi.org/10.1073/pnas.0900301106.

56. ROSA-GARRIDO, M., CHAPSKI, D.J., SCHMITT, A.D., KIMBALL, T.H., KARBASSI, E., MONTE, E., BALDERAS, E., PELLEGRINI, M., SHIH, T.-T., and SOEHALIM, E. (2017). High-resolution mapping of chromatin conformation in cardiac myocytes reveals structural remodeling of the epigenome in heart failure. *Circulation 136*,

1613-1625. https://doi.org/10.1161/CIRCULATIONAHA.117.029430.

57. DUGGAL, G., WANG, H., and KINGSFORD, C. (2014). Higher-order chromatin domains link eQTLs with the expression of far-away genes. *Nucleic Acids Research 42*, 87-96. https://doi.org/10.1093/nar/gkt857.

58. DOYLE, B., FUDENBERG, G., IMAKAEV, M., and MIRNY, L.A. (2014). Chromatin loops as allosteric modulators of enhancer-promoter interactions. *PLoS Computational Biology 10*, e1003867. https://doi.org/10.1371/journal.pcbi.1003867.

59. PRIEST, D.G., KUMAR, S., YAN, Y., DUNLAP, D.D., DODD, I.B., and SHEARWIN, K.E. (2014). Quantitation of interactions between two DNA loops demonstrates loop domain insulation in E. coli cells. *Proceedings of the National Academy of Sciences 111*, E4449-E4457. https://doi.org/10.1073/pnas.1410764111.

60. GRUBERT, F., SRIVAS, R., SPACEK, D.V., KASOWSKI, M., RUIZ-VELASCO, M., SINNOTT-ARMSTRONG, N., GREENSIDE, P., NARASIMHA, A., LIU, Q., and GELLER, B. (2020). Landscape of cohesin-mediated chromatin loops in the human genome. *Nature 583*, 737-743. https://doi.org/10.1038/s41586-020-2151-x.

61. CREYGHTON, M.P., CHENG, A.W., WELSTEAD, G.G., KOOISTRA, T., CAREY, B.W., STEINE, E.J., HANNA, J., LODATO, M.A., FRAMPTON, G.M., and SHARP, P.A. (2010). Histone H3K27ac separates active from poised enhancers and predicts developmental state. *Proceedings of the National Academy of Sciences 107*, 21931-21936. https://doi.org/10.1073/pnas.1016071107.

62. BARSKI, A., CUDDAPAH, S., CUI, K., ROH, T.Y., SCHONES, D.E., WANG, Z., WEI, G., CHEPELEV, I., and ZHAO, K. (2007). High-resolution profiling of histone methylations in the human genome. *Cell 129*, 823-837. https://doi.org/10.1016/j.cell.2007.05.009.

63. CHENG, J., BLUM, R., BOWMAN, C., HU, D., SHILATIFARD, A., SHEN, S., and DYNLACHT, B.D. (2014). A role for H3K4 monomethylation in gene repression and partitioning of chromatin readers. *Molecular cell 53*, 979-992. https://doi.org/10.1016/j.molcel.2014.02.032.

64. KAJI, T., YOSHIDA, S., KAWAI, K., FUCHIGAMI, Y., WATANABE, W., KUBODERA, H., and KISHIMOTO, T. (2010). ASK3, a novel member of the apoptosis signal-regulating kinase family, is essential for stress-induced cell death in HeLa cells. *Biochemical and Biophysical Research Communications 395*, 213-218. https://doi.org/10.1016/j.bbrc.2010.03.164.

65. MORISHITA, K., WATANABE, K., and ICHIJO, H. (2019). Cell volume regulation in cancer cell migration driven by osmotic water flow. *Cancer Science 110*, 2337-2347. https://doi.org/10.1111/cas.14079.

66. KIM, D., PAGGI, J.M., PARK, C., BENNETT, C., and SALZBERG, S.L. (2019). Graph-based genome alignment and genotyping with HISAT2 and HISAT-genotype. *Nat. Biotechnol. 37*, 907-915. https://doi.org/10.1038/s41587-019-0201-4.

67. LIAO, Y., SMYTH, G.K., and SHI, W. (2014). featureCounts: an efficient general purpose program for assigning sequence reads to genomic features. *Bioinformatics 30*, 923-930. https://doi.org/10.1093/bioinformatics/btt656.

68. LOVE, M.I., HUBER, W., and ANDERS, S. (2014). Moderated estimation of fold change and dispersion for RNA-seq data with DESeq2. *Genome Biol. 15*, 550. https://doi.org/10.1186/s13059-014-0550-8.

69. ROBINSON, M.D., MCCARTHY, D.J., and SMYTH, G.K. (2010). edgeR: a Bioconductor package for differential expression analysis of digital gene expression data. *Bioinformatics 26*, 139-140. https://doi.org/10.1093/

bioinformatics/btp616.

70. YU, G., WANG, L G., HAN, Y., and HE, Q Y. (2012). clusterProfiler: an R package for comparing biological themes among gene clusters. *Omi. a J. Integr. Biol. 16*, 284-287. http://doi.org/10.1089/omi.2011.0118.

71. CHEN, S., ZHOU, Y., CHEN, Y., and GU, J. (2018). fastp: an ultra-fast all-in-one FASTQ preprocessor. *Bioinformatics 34*, i884-i890. https://doi.org/10.1089/omi.2011.0118.

72. ANDREWS, S. (2010). FastQC: a quality control tool for high throughput sequence data.

73. LI, H., and DURBIN, R. (2009). Fast and accurate short read alignment with Burrows-Wheeler transform. *Bioinformatics 25*, 1754-1760. https://doi.org/10.1093/bioinformatics/btp324.

74. RAMÍREZ, F., DÜNDAR, F., DIEHL, S., GRÜNING, B.A., and MANKE, T. (2014). deepTools: a flexible platform for exploring deep-sequencing data. *Nucleic Acids Res. 42*, W187-W191. https://doi.org/10.1093/nar/gku365.

75. ZHANG, Y., LIU, T., MEYER, C.A., EECKHOUTE, J., JOHNSON, D.S., BERNSTEIN, B.E., NUSBAUM, C., MYERS, R.M., BROWN, M., and LI, W. (2008). Model-based analysis of ChIP-Seq (MACS). *Genome biology 9*, 1-9. https://doi.org/10.1186/gb-2008-9-9-r137.

76. YOUNG, M.D., WAKEFIELD, M.J., SMYTH, G.K., and OSHLACK, A. (2010). Gene ontology analysis for RNA-seq: accounting for selection bias. *Genome Biol. 11*, 1-12. https://doi.org/10.1186/gb-2010-11-2-r14.

77. LANGMEAD, B., and SALZBERG, S.L. (2012). Fast gapped-read alignment with Bowtie 2. *Nat. Methods 9*, 357-359. https://doi.org/10.1038/nmeth.1923.

78. SERVANT, N., VAROQUAUX, N., LAJOIE, B.R., VIARA, E., CHEN, C.J., VERT, J.P., HEARD, E., DEKKER, J., and BARILLOT, E. (2015). HiC-Pro: an optimized and flexible pipeline for Hi-C data processing. *Genome Biol. 16*, 1-11. https://doi.org/10.1186/s13059-015-0831-x.

79. HEINZ, S., BENNER, C., SPANN, N., BERTOLINO, E., LIN, Y.C., LASLO, P., CHENG, J.X., MURRE, C., SINGH, H., and GLASS, C.K. (2010). Simple combinations of lineage-determining transcription factors prime cis-regulatory elements required for macrophage and B cell identities. *Molecular Cell 38*, 576-589. https://doi.org/10.1016/j.molcel.2010.05.004.

80. XIE, C., MAO, X., HUANG, J., DING, Y., WU, J., DONG, S., KONG, L., GAO, G., LI, C.Y., and WEI, L. (2011). Kobas 2.0: a web server for annotation and identification of enriched pathways and diseases. *Nucletic Acids Res. 39*, W316-W322. https://doi. org/10.1093/nar/gkr483.

81. PAN, Q.Z., WU, X.B., GUAN, C., and ZENG, Z.J. (2013). A new method of queen rearing without grafting larvae. *American Bee Journal 153*, 1279-1280.

82. WOLFF, J., BHARDWAJ, V., NOTHJUNGE, S., RICHARD, G., RENSCHLER, G., GILSBACH, R., MANKE, T., BACKOFEN, R., RAMÍREZ, F., and GRÜNING, B.A. (2018). Galaxy HiCExplorer: a web server for reproducible Hi-C data analysis, quality control and visualization. *Nucleic Acids Research 46*, W11-W16. https://doi.org/10.1093/nar/gky504.

# 4. A Comparison of Honeybee (*Apis mellifera*) Queen, Worker and Drone Larvae by RNA-Seq

Xujiang He[1], Wujun Jiang[1], Mi Zhou[2], Andrew B. Barron[3], Zhijiang Zeng[1*]

1. Honeybee Research Institute, Jiangxi Agricultural University, Nanchang, 330045, China
2. Biomarker Technologies Co., Ltd. Beijing, 101300, China
3. Department of Biological Sciences, Macquarie University, North Ryde, NSW 2109, Australia

**Abstract**

Honeybees (*Apis mellifera*) have haplodiploid sex determination: males develop from unfertilized eggs and females develop from fertilized ones. The differences in larval food also determine the development of females. Here we compared the total somatic gene expression profiles of 2-day and 4-day-old drone, queen and worker larvae by RNA-Seq. The results from a co-expression network analysis on all expressed genes showed that 2-day-old drone and worker larvae were closer in gene expression profiles than 2-day-old queen larvae. This indicated that for young larvae (2-day-old) environmental factors such as larval diet have a greater effect on gene expression profiles than ploidy or sex determination. Drones had the most distinct gene expression profiles at 4-day larval stage, suggesting that haploidy, or sex dramatically affects the gene expression of honeybee larvae. Drone larvae showed fewer differences in gene expression profiles at the 2-day and 4-day time points than the worker and queen larval comparisons (598 against 1190 and 1181), suggesting a different pattern of gene expression regulation during the larval development of haploid males compared to diploid females. This study indicates that early in development the queen caste has the most distinct gene expression profile, perhaps reflecting the very rapid growth and morphological specialization of this caste compared to workers and drones. Later in development the haploid male drones have the most distinct gene expression profile, perhaps reflecting the influence of ploidy or sex determination on gene expression.

**Keywords:** Honeybees, haploid and diploid, larval development, environmental factors, gene expression, caste differentiation

## Introduction

In the animal kingdom, approximatively 20% of species are haplodiploid animals such that haploid eggs develop into males and diploid eggs develop into females (Beye 2004). Honeybees (*Apis mellifera*), an eusocial

* Corresponding author: bees1965@sina.com (ZJZ).
注：此文发表在 *Insect Science* 2019 年第 2 期。

insect that reproduces by arrhenotokous parthenogenesis (Trivers & Hare, 1976), are a good model for studies of development of haplodiploid organisms. Queens and workers are females that develop from fertilized eggs, whereas drones are males from unfertilized eggs (Ratnieks & Keller, 1998). These three castes dramatically differ in gender, morphology, physiology and behaviour. Drones are highly specialised for mating, with larger compound eyes to detect virgin queens in flight, more wing sensilla, smaller mandibles, a large endophallus, no sting and hypopharyngeal glands; Queens are specialised for egg laying with large ovaries (Snodgrass, 1925). Several studies have investigated how the honeybee haplodiopoid genome is involved in sex determination and developmental regulation of honeybee males and females. The drone developmental pathway is considered to be primarily controlled by a genetic mechanism related to the ploidy of the embryo. Heterozygosity at the hypervariable complementary sex determiner gene (*csd*) locus determines that the embryo develops into a queen and worker, whereas homozygosity at the locus or (more commonly) one single copy determines drone development (Beye, 2004; Gempe et al., 2009). Pires *et al.* (2016) compared gene expression between honeybee male and female embryos and observed certain mRNAs and miRNAs were expressed differently during haploid and diploid embryogenesis. Vleurinck *et al.* (2016) compared gene expression in brains of honeybee males and females and reported that both sex and caste signals are involved in the gene regulation in male and female brains. Similar results have been demonstrated in the fire ant (*Solenopsis invicta*): the gene expression profiles in haploid males are very different from those of dioploid females and even diploid males at three developmental time points (Nipitwattanaphon et al., 2014).

The effects of animal haploid genome on developmental regulation are far more complex as they have endopolyploidy. Diploidy is found in the muscles of male bumblebees (*Bombus terrestris* L.) (Aron et al., 2005). In honeybees, Woyke and Paluch (1985) showed that haploid males reach endopolyploidy levels during their development and the endoreduplication is also observed in particularly active tissues, such as ventriculus, malpoghian tubules, fat body and silk gland. More interestingly, the endopolyploidy levels in drone larvae are much higher than in embryos and adults (Woyke & Paluch, 1985). However, the molecular mechanisms of regulating the larval development of drone, queen and worker remain unclear.

The honeybee queen and worker developmental pathways are controlled primarily by an environmental mechanism. Queens and workers are both females, but while queens have hundreds of ovarioles and hypertrophied ovaries, most workers normally have inactive ovaries and an underdeveloped spermatheca so that they cannot mate. Both workers and queens develop from diploid eggs, but queen larvae are supplied with an abundance of royal jelly over their whole larval stage, whereas workers are fed with worker jelly in the first three days and then are fed with a yellowish, pollen-containing food (Haydak, 1970). The royal jelly for queen larvae is dramatically different from jellies fed to either worker or drone larvae in terms of minerals and vitamin, sugar content, juvenile hormone and major royal jelly protein content (Haydak, 1970; Asencot & Lensky, 1984; Brouwers, 1984; Kamakura, 2011). This determines their caste differentiation.

Thousands of genes are differentially expressed between queen and worker larvae, including the signalling molecules vitellogenin, juvenile hormone and the mTOR pathway, which are involved in the caste differentiation process (Hepperle & Hartfelder, 2001; Guidugli *et al.*, 2005; Barchuk *et al.*, 2007; Patel *et al.*,

2007; Chen *et al.*, 2012). However, thus far no study has compared gene expression in developing honeybee male and female larvae. The objective of this study was to systematically compare gene expression in workers, queens and drones at two points in larval development by RNA-Seq. The larval stage is crucial for the development of honeybees, as the polyploidization is higher at this point than at any other point during their lifespan (Woyke & Paluch, 1985). This would allow a genomic comparison of how and when the sex and caste differentiation pathways of honeybees diverge, and also a comparison of the gene expression differences arising from genetic and environmentally regulated developmental pathways.

## Materials and methods

### *Insects*

Six hives of the standard Chinese commercial strain of western honeybee (*Apis mellifera*) with a mature egg-laying queen and eight frames were located at the Honeybee Research Institute of Jiangxi Agricultural University (28.46° N, 115.49 ° E).

### *RNA-Seq analysis of queen, worker and drone larvae*

For RNA-Seq, the mated queen was controlled on an empty worker and a drone frame to lay diploid and haploid eggs for 6 hours. Subsequently, 2- and 4-day of worker and drone larvae (6 biological replicates from 6 colonies) were sampled using a bee grafting pen to lift the larvae from the wax cells in which they were developing. Larvae were immediately flash-frozen in liquid nitrogen. For 2-day-old larval samples (36- 42 h after hatching), approximately 30 larvae were collected for each RNA sequencing sample, while for 4-day old larval samples (84 - 90 h after hatching) 6 were collected. Four-day-old larvae are much larger than 2 d larvae and hence fewer larvae were needed for adequate RNA yield (6 μg). For the queen larvae samples, a new queen rearing methodology developed by Pan *et al.* (2013) was employed to control the queen to lay eggs for 6 h. Afterwards, eggs were immediately removed from queen cells and returned back to their natal colonies. 2- and 4-day queen larvae samples were then collected from queen cells at 2- and 4-day time points. Each larval group has 6 biological replicates from three different honeybee colonies (each colony provided 2 biological replicates for each group, therefore totally 36 samples were collected).

Total RNA of each sample was extracted from honeybee larvae according to the standard protocol of the TRIzol Reagent (Life technologies, California, USA). RNA integrity and concentration were checked using an Agilent 2100 Bioanalyzer (Agilent Technologies, Inc., Santa Clara, CA, USA).

mRNA was isolated from total RNA using a NEBNext Poly(A) mRNA Magnetic Isolation Module (NEB, E7490). A cDNA paired-end library was constructed following the manufacturer's instructions for the NEBNext Ultra RNA Library Prep Kit（NEB, E7530）and the NEBNext Multiplex Oligos（NEB, E7500）from Illumina. In brief: enriched mRNA was fragmented into approximately 200nt RNA inserts, which were used as templates to synthesize the cDNA. End-repair/dA-tail and adaptor ligation were then performed on the double-stranded cDNA. Suitable fragments were isolated by AgencourtAMPure XP beads (Beckman Coulter, Inc.), and enriched by PCR amplification. Finally, the constructed cDNA libraries of the honeybee were sequenced on a flow cell using an IlluminaHiSeq™ 2500 sequencing platform.

Low quality reads, such as adaptor-only reads or reads with > 5% unknown nucleotides were filtered from subsequent analyses. Reads with a sequencing error rate less than 1% (Q20 >98%) were retained. These remaining clean reads were mapped to the *Apis mellifera* official genes (OGSv3.2) using Tophat2 software (Kim *et al.*, 2013). The aligned records from the aligners in BAM/SAM format were further examined to remove potential duplicate molecules. Gene expression levels were estimated using FPKM values (fragments per kilobase of exon per million fragments mapped) by the Cufflinks software (Trapnell *et al.*, 2010).

*Co-expression network analysis*

To explore the correlations of gene expression among queen, worker and drone larvae, the read count of each gene from 6 larval groups were used for weighted correlation network analysis (WGCNA) in R package ("Number of total genes detected in all samples") according to the method developed by Langfelder and Horvath (2008), resulting in a gene clustering tree for each larval group shown in Fig. I-4-1.

*Identification of differentially expressed genes*

DESeq and Q-value were employed and used to evaluate differential gene expression among queen, worker and drone larvae by estimating the count data from high-throughput sequencing assays and testing for differential expression based on a model using the negative binomial distribution (Anders & Huber, 2010). The false discovery rate (FDR) control method was used to identify the threshold of the *P*-value in multiple tests in order to compute the significance of the differences by using the read count of each gene. Here, only genes with an absolute value of $\log_2$ ratio ≥1 and FDR significance score <0.01 were used for subsequent analysis. Gene abundance differences between sample groups were calculated based on the ratio of the FPKM values which were used for presenting the gene expression of each gene.

Sequences differentially expressed between sample groups were identified by comparison against various protein database by BLASTX, including the National Center for Biotechnology Information (NCBI) non-redundant protein (Nr) database, and Swiss-Prot database with a cut-off E-value of $10^{-5}$. Furthermore, genes were searched against the NCBI non-redundant nucleotide sequence (Nt) database using BLAST by a cut-off *E*-value of $10^{-5}$. Genes were retrieved based on the best BLAST hit (highest score) along with their protein functional annotation. All significantly differentially expressed genes (DEGs) were mapped to terms in the GO database. The GO enrichment analysis of functional significance used a hypergeometric test (*p*-value < 0.05 indicates the significance) (He *et al.*, 2014; Qin *et al.*, 2014) to identify significantly enriched GO terms in DEGs compared to the complete genome. DEGs from queen vs drone larvae and worker vs drone larvae comparisons were selected for the GO enrichment analysis. DEGs between 2- and 4-day larvae in three larval castes were also employed in the GO enrichment analysis.

DEGs were mapped to the KEGG protein database (KEGG database: http://www.genome.jp/kegg/kegg1.html) by BLAST (*E*-value < $1e^{-5}$). KOBAS 2.0 software was used to test the statistical enrichment of DEGs in KEGG pathways using a hypergeometric test (*Q*-value < 0.05) (Xie *et al.*, 2011).

*Statistics of raw data and Saturation analysis of sequencing*

In RNA-Seq, 6 libraries were generated from our experimental groups, and summaries of RNA sequencing analyses are shown in Table S1. In each library, more than 97% clean reads were unique mapped reads of which more than 86% reads were paired reads. Very few clean reads (<2.3%) were multiple mapped reads. Each library had a sufficient coverage of the expected number of distinct genes (stabilized at 3M reads). The Pearson correlation coefficient among three biological replicates of each experimental group were all > 0.80 (Table S2), which is a conventionally accepted threshold for valid replicates (Trapnell *et al.*, 2010) indicating that there was acceptable sequencing quality and repeatability among the biological replicates of each group.

## Results

### *Number of total genes detected in all samples*

We compared gene expression measured with RNA-seq in queen, worker and drone larvae sampled when 2- and 4-days old. There was no significant difference ($p>0.05$, ANOVA followed by Fisher's PLSD test) among the 6 larval groups in terms of total gene number detected in the RNA-Seq (Fig. I-4-2A). Most genes were expressed in all these three castes, very few (<1.6%) were uniquely expressed in only one group (Fig. I-4-2B).

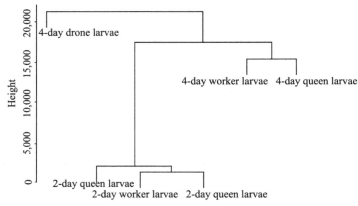

**Figure I-4-1** The gene clustering tree of 2- and 4-day drone, worker and queen larvae. Read counts of all genes from six larval groups were analyzed by a weighted correlation network analysis (WGCNA) in R package (Number of total genes detected in all samples) for the correlations of whole gene expression among drone, worker and queen larvae. The y-axis value is euclidean distance.

### *Results of co-expression network analysis*

A weighted correlation network analysis (WGCNA, Fig. I-4-1) showed that all the 2-day-old samples formed quite a tight cluster. Within that cluster, 2-day-old worker and drone larvae were more similar than 2-day-old queen larvae. The 4-day-old larval samples were more divergent. 4-day drone larvae were the most distinct group whereas 4-day-old queen and worker larvae (the two female castes) remained quite closely clustered.

### *Differentially expressed genes among queen, worker and drone larvae*

When considering the numbers of DEGs between groups (Fig. I-4-3, Table S3-8), for both 2- and 4-day larvae the fewest number of DEGs were between worker and queen larvae (the two female castes), and the greatest number of DEGs were between queen and drone larvae (the fertile male and female sexes).

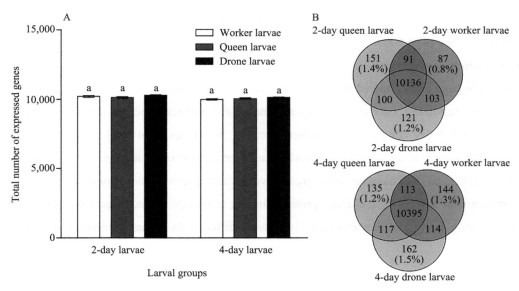

**Figure I-4-2** Total number of expressed genes in 6 larval groups (A) and venn diagram of expressed genes among 6 larval groups (B). Open, grey and black bars represent worker, queen and drone larvae groups respectively. Each group has 6 biological replicates. Same letters "a" on top of bars indicate no significant difference ($p<0.05$, ANOVA followed by Fisher's PLSD test in Statview 6.0).

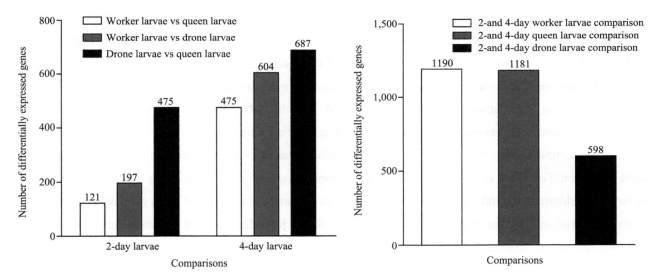

**Figure I-4-3** Number of DEGs among queen, worker and drone larvae at 2- and 4-day stages. Open, grey and black bars represent comparisons of worker larvae vs queen larvae, worker larvae vs drone larvae and queen larvae vs drone larvae respectively.

**Figure I-4-4** Number of DEGs between 2- and 4-day larvae in the same larval types. Open, grey and black bars represent comparisons of 2-day vs 4-day worker larvae, 2-day vs 4-day drone larvae and 2-day vs 4-day queen larvae respectively.

Further, comparisons among 4-day-old larvae had higher numbers of DEGs compared to 2-day-old larvae comparisons (Fig. I-4-3). While comparing the DEGs between 2- and 4-day larvae of the same larval type, queen and worker larvae showed more DEGs than drones (1181 and 1190 against 598, Fig. I-4-4, Table S9-11).

Interestingly, 33 genes of the DEGs between male and females were involved in hormone biosynthesis, oocyte maturation, venom, eye development, sex determination, wnt signaling pathway and notch signaling pathway (Fig. I-4-5, see page 506).

***Gene ontology enrichment analysis***

Fig. I-4-6(see page 507) showed that DEGs between 2-day female and male larvae were enriched in 11 categories such as membrane, catalytic activity and metabolic process, whereas there were more DEGs and categories (29) in the 4-day comparisons. Furthermore, Fig. I-4-7(see page 508) showed that categories and percentages of DEGs in 2- and 4-day queen, worker and drone larval comparisons were similar, though drone larvae had fewer DEGs in each category compared to worker and queen larvae. However, percentages of DEGs in four GO enrichment categories including protein binding transcription factor activity, translation regulator activity, antioxidant activity and morphogen activity were different in drones compared to female queen and workers.

## Discussion

Here we used genomic analyses to compare the divergence of the three honey bee castes during larval development: queen, workers and drones. Our results indicated that in early larval development (day 2) queens had the most distinct gene expression profiles when compared to workers and drones, but by day four of larval development male drones had the most distinct gene expression profiles when compared to the queen and worker castes (Fig. I-4-1). This suggests that developmental pathways segregate most strongly by sex and a genetically controlled developmental system, but these differences only manifest in the later stages of larval development. Previous studies showed that cell nuclei in 1st-instar male larvae is haploid but become polyploidy in later instar larvae (Risler, 1954; Woyke & Paluch, 1985). Perhaps the polyploidization plays a very important role in the regulation of gene expression in older male larvae.

Larvae sampled when two days old are quite similar in terms of gene expression profiles regardless of caste allocation. For 2-day-old larvae, queens were the most divergent group, suggesting that in the early stages of larval development the influence of the rearing environment and royal jelly diet has a greater impact on gene expression profiles than the sex determination system. Royal jelly dramatically differs in terms of minerals and vitamins, sugar content, juvenile hormones and major royal jelly protein content compared to worker and drone jelly (Haydak, 1970; Asencot & Lensky, 1984; Brouwers, 1984; Kamakura, 2011), and these nutritional differences can induce thousands of DEGs between queen and worker larvae (Chen *et al.*, 2012). Vleurinck *et al.* (2016) showed that more DEGs and differentially spliced genes (DSGs) were detected in worker and queen pupae brains than worker and drone brains. Our findings indicate that these nutritional differences have a greater impact on gene expression than haplodioploidy in early larval development.

Although the total number of expressed genes in the three castes was not significantly different (Fig. I-4-2), hundreds of genes were significantly differentially expressed among them (Fig. I-4-3). The drone larvae vs queen larvae comparison had the greatest number of DEGs, indicating that both genomic and nutritional differences expand the extent of developmental differentiation of drones and queens. When comparing DEGs between 2- and 4-day larvae of the same larval type, queen and worker larvae showed more DEGs than drones

(Fig. I-4-4), further emphasizing the clear difference between the male and female developmental trajectories. Comparisons among 4-day-old larvae had higher numbers of DEGs compared to 2-day-old larvae comparisons (Fig. I-4-3), and GO enrichment categories were higher in 4-day male and female larvae compared to 2-day comparisons, suggesting that the extent of differentiation among drone, queen and worker larvae increased with their age.

To explain the morphological, physiological and behavioural differences among female and male honeybees, we noted 33 interesting DEGs which are involved in hormone biosynthesis, oocyte maturation, venom, eye development, sex determination, wnt signaling pathway and notch signaling pathway (Fig. I-4-5). Both 2- and 4-day drone larvae differed from female honeybees in terms of the expression of genes involved in hormone biosynthesis, oocyte maturation and the mTOR pathway (Fig. I-4-5, Fig. S1-S3). Many previous studies have showed that genes involved in hormone biosynthesis such as *juvenile hormone esterase precursor, vitellogenin precursor* and *ecdysteroids* that are involved in the caste differentiation of the queen and workers (Wirtz & Beetsma, 1972; Rachinsky et al., 1990; Chen et al., 2012; Cameron et al., 2013). In this study, these hormone-related genes were also significantly differentially expressed between male drones and female queen and workers (Fig. I-4-5), suggesting that these genes may be involved in sex determination also. *Insulin-like receptor-like, Insulin-like peptide A chain, cytoplasmic polyadenylation element-binding protein 1* and *G2/mitotic-specific cyclin-B (LOC551860)* that are involved in oocyte maturation and the mTOR pathway (a pathway that determines ovary development in queen and workers. Wheeler et al., 2006; de Azevedo & Hartfelder, 2008), were differentially expressed among drone, queen and worker larvae (Fig. I-4-5). This may reflect an early genomic signal of the key reproductive differences between the three castes.

Honeybee drones have much larger compound eyes than workers and queens, and females produce venom but not in males (Trivers & Hare, 1976). Here we showed that 2-day worker and queen larvae had 1 and 2 DEGs involved in the development of eyes compared to 2-day drone larvae, and 1 DEG was found in 4-day worker larvae (Fig. I-4-5). Marco and Hartfelder (2016) reported that honeybee workers and drones have already started their eye development process from 3rd instar stage, and a few genes involved in this process are differentially expressed between the sexes. Roat and Cruz-Landim (2011) also showed the antennal lobes of workers, queens and drones are already different at larval stage and drones have bigger antennal lobes than that of worker and queens. Our results are consistent with their findings, reflecting that the differentiation of eye development between honeybee males and females is from a very early larval stage. Furthermore, male and female honeybee larvae were differed in the venom gene expression, with 4 and 3 DEGs in 2-day queen vs drone and worker vs drone comparisons respectively and 1 in 4-day comparisons (Fig. I-4-5). Though it is unclear when honeybee females start the developmental process of their venom glands, these significantly differentially expressed venom-gland related genes might play an important role in regulating in the development of venom glands in female castes. This still requires further investigation.

Honeybee sex is determined by the *csd* gene and interacted with *feminizer* (*fem*) and *doublesex* (*dsx*) genes (Charlesworth, 2003; Gempe et al., 2009). Here we did not find *fem* and *dsx* were differentially expressed between male and female larvae. Only the *csd* gene was downregulated in 2-day worker larvae compared to

2-day drone larvae (Fig. I-4-5). A previous study showed that *csd* and *fem* genes are required only to initiate sex-specific differentiation in early embryogenesis (Gempe *et al.*, 2009). Therefore, it still remains unclear how sex-determining genes play a role in larval development and needs further investigations.

Many genes from wnt and notch signalling pathways were significantly downregulated in female larvae compared to male larvae. In detail, 3 and 2 genes from wnt signalling pathway in 2- and 4-day queen larvae were downregulated compared drone larvae respectively, and 1 was downregulated in 4-day worker larvae compared to drone larvae (Fig. I-4-5 and Fig. S4). Three genes involved in notch signalling pathway were downregulated in 2-day queen and drone comparison, whereas 2 and 1 were in downregulated 2- and 4-day worker and drone comparisons (Fig. I-4-5 and Fig. S5). Wnt and notch signalling pathways are two conserved pathways playing an important role in embryogenesis, morphogenesis and imaginal disc development in insects and other animals (Dearden *et al.*, 2006; Bolos *et al.*, 2007; Komiya & Habas, 2008; Wilson *et al.*, 2011). Therefore these DEGs may play an important role in the body development and formation of morphological traits of male and female honeybees, which requires further investigations.

## Conclusion

Consequently, our data show the complexity of honeybee caste differentiation, and that clear differences between castes are established even very early in larval development, long before the formation of adult or reproductive structures. Honeybee ploidy, sex determination and environmental factors may all influence the larval gene expression profiles and developmental trajectory. Our findings contribute information on how genetic and environmental factors regulate male and female development in haplodiploid insects.

## Competing Interests

The authors declare that they have no competing interests.

## Authors' Contributions

ZJZ and XJH conceived and designed the experiments. XJH and WJJ performed the experiments. XJH, XCZ and ABB analyzed the data. XJH, ZJZ and ABB wrote the paper. All authors read and approved the final manuscript. We thank Dr. Lianne Meah for revising the paper.

## Acknowledgement

This work was supported by the Science and Technology Project of Colleges and Universities of Jiangxi Province (KJLD13028), the Earmarked Fund for China Agriculture Research System (CARS-44-KXJ15) and China Scholarship Council (No. 201408360073).

## Additional Information

RNA-Seq data of 2d honeybee worker larvae: NCBI SRA: SRS1249139;

RNA-Seq data of 2d honeybee queen larvae: NCBI SRA: SRS1263242;

RNA-Seq data of 2d honeybee drone larvae: NCBI SRA: SRS1263244;

RNA-Seq data of 4d honeybee worker larvae: NCBI SRA: SRS1263211;

RNA-Seq data of 4d honeybee queen larvae: NCBI SRA: SRS1263243;

RNA-Seq data of 4d honeybee drone larvae: NCBI SRA: SRS1263256.

## References

ANDERS, S., and HUBER, W. (2010) Differential expression analysis for sequence count data. *Genome Biology*, 11, R106.

ARON, S., DE MENTEN, L., VAN BOCKSTAELE, D.R., BLANK, S.M., ROISIN, Y. (2005) When hymenopteran males reinvented diploidy. *Current Biology*, 15, 824-827.

ASENCOT, M., and LENSKY, Y. (1984) Juvenile hormone induction of 'queenliness' on female honey bee (*Apis mellifera* L.) larvae reared on worker jelly and on stored royal jelly. *Comparative Biochemistry & Physiology Part B Comparative Biochemistry*, 1, 109-117.

BARCHUK, A.R., CRISTINO, A.S., KUCHARSKI, R., COSTA, L.F., SIMÕES, Z.L.P., MALESZKA, R.(2007) Molecular determinants of caste differentiation in the highly eusocial honeybee *Apis mellifera*. *BMC Developmental Biology*, 7, 70.

BEYE, M. (2004) The dice of fate: the csd gene and how its allelic composition regulates sexual development in the honey bee, *Apis mellifera*. *Bioessays*, 26, 1131-1139.

BOLOS, V., GREGO-BESSA, J., DE LA POMPA, J.L. (2007) Notch signaling in development and cancer. *Endocrine Reviews*, 28, 339-363.

BROUWERS, E. (1984) Glucose/fructose ratio in the food of honeybee larvae during caste differentiation. *Journal of Apicultural Research*, 2, 94-101.

CAMERON, R.C., DUNCAN, E.J., DEARDEN, P.K. (2013) Biased gene expression in early honeybee larval development. *BMC Genomics*, 14, 903.

CHARLESWORTH, B. (2003) Sex determination in the honeybee. *Cell*, 114, 397-398.

CHEN, X., HU, Y., ZHENG, H., CAO, L., NIU, D., YU, D., *et al*. (2012) Transcriptome comparison between honey bee queen- and worker-destined larvae. *Insect Biochemistry and Molecular Biology*, 42, 665-673.

DEARDEN, P.K., WILSON, M.J., SABLAN, L., OSBORNE, P.W., HAVLER, M., MCNAUGHTON, E., *et al*. (2006) Patterns of conservation and change in honey bee developmental genes. *Genome Research*, 16, 1376-1384.

DE AZEVEDO, S.V., and HARTFELDER, K. (2008) The insulin signaling pathway in honey bee (*Apis mellifera*) caste development - differential expression of insulin-like peptides and insulin receptors in queen and worker larvae. *Journal of Insect Physiology*, 54, 1064-1071.

GEMPE, T., HASSELMANN, M., SCHIOTT, M., HAUSE, G., OTTE, M., BEYE, M., *et al*. (2009) Sex determination in honeybees: two separate mechanisms induce and maintain the female pathway. *PLoS Biology*, 7, e1000222.

GUIDUGLI, K.R., PIULACHS, M.D., BELLES, X., BELLÉS, X., LOURENÇO, A.P., SIMÕES, Z.L.P. (2005) Vitellogenin expression in queen ovaries and in larvae of both sexes of *Apis mellifera*. *Archives of Insect Biochemistry and Physiology*, 59, 211-218.

HAYDAK, M. (1970) Honey bee nutrition. *Annual Review of Entomology*, 1, 143-156.

HEPPERLE, C. AND HARTFELDER, K. (2001) Differentially expressed regulatory genes in honey bee caste development. *Naturwissenschaften*, 88, 113-116.

HE, X.J., TIAN, L.Q., BARRON, A.B. GUAN, C., LIU, H., WU, X.B., et al. (2014) Behavior and molecular physiology of nurses of worker and queen larvae in honey bees (*Apis mellifera*). *Journal of Asia-Pacific Entomology*, 17, 911-916.

KAMAKURA, M. (2011) Royalactin induces queen differentiation in honeybees. Nature, 473, 478-483.

KIM, D., PERTEA, G., TRAPNELL, C., PIMENTEL, H., KELLEY, R., SALZBERG, S. (2013) TopHat2: accurate alignment of transcriptomes in the presence of insertions, deletions and gene fusions. *Genome Biology*, 14, R36.

KOMIYA, Y., and HABAS, R. (2008) Wnt signal transduction pathways. *Organogenesis*, 4, 68-75.

LANGFELDER, P., and HORVATH, S. (2008) WGCNA: an R package for weighted correlation network analysis. *BMC Bioinformatics*, 9, 559.

MARCO, A.D., and HARTFELDER, K. (2016) Toward an understanding of divergent compound eye development in drones and workers of the honeybee (*Apis mellifera* L.): a correlative analysis of morphology and gene expression. *Journal of Experimental Zoology Part B Molecular and Developmental Evolution*, 00B, 1-18.

NIPITWATTANAPHON, M., WANG, J., ROSS, K.G., RIBA-GROGNUZ, O., WURM, Y., KHUREWATHANAKUL, C., et al. (2014) Effects of ploidy and sex-locus genotype on gene expression patterns in the fire ant *Solenopsis invicta*. *Proceedings Biological Sciences*, 281, 20141776.

PAN, Q.Z., WU, X.B., GUAN, C., ZENG, Z.J. (2013) A new method of queen rearing without grafting larvae. *American Bee Journal*, 12, 1279-1280.

PATEL, A., FONDRK, M.K., KAFTANOGLU, O., EMORE, C., HUNT, G., FREDERICK, K., et al. (2007) The making of a queen: TOR pathway is a key player in diphenic caste development. *PLoS One*, 2, e509.

PIRES, C.V., FREITAS, F.C.D.P., CRISTINO, A.S., DEARDEN, P.K., SIMÕES, Z.L. (2016) Transcriptome analysis of honeybee (*Apis Mellifera*) haploid and diploid embryos reveals early zygotic transcription during cleavage. *PLoS One*, 11, e0146447.

QIN, Q.H., WANG, Z.L., TIAN, L.Q., GAN, H.Y., ZHANG, S.W., ZENG, Z.J. (2014). The integrative analysis of microRNA and mRNA expression in *Apis mellifera* following maze-based visual pattern learning. *Insect Science*, 21,5, 619-636.

RACHINSKY, A., STRAMBI, C., STRAMBI, A., HARTFELDER, K. (1990) Caste and metamorphosis: hemolymph titers of juvenile hormone and ecdysteroids in last instar honeybee larvae. *General & Comparative Endocrinology*, 79, 31-8.

RATNIEKS, F.L.W., and KELLER, L. (1998) Queen control of egg fertilization in the honey bee. *Behavioral Ecology & Sociobiology*, 44, 57-61.

RISLER, H. (1954) Die somatische polyploidie in der entwichlung der honigbiene (*Apia mellifera* L.) und die wiederherstellung der diploide bei den drohnen. *Zeitschrift Für Zellforschung Und Mikroskopische Anatomie*, 1, 1-78.

ROAT, T.C., and CRUZLANDIM, C.D. (2011) Differentiation of the honey bee (*Apis mellifera* l.) antennal lobes during metamorphosis: a comparative study among castes and sexes. *Animal Biology*, 61, 153-161.

SNODGRASS, R.E. (1925) Anatomy and physiology of the honeybee. McGraw-Hill book company, New York and London. pp: 248-263.

TRAPNELL, C., WILLIAMS, B.A., PERTEA, G., MORTAZAVI, A., KWAN, G., VAN BAREN, M.J., *et al.* (2010) Transcript assembly and quantification by RNA-Seq reveals unannotated transcripts and isoform switching during cell differentiation. *Nature Biotechnology*, 28, 511-515.

TRIVERS, R., and HARE, H. (1976) Haplodiploidy and the evolution of the social insects. *Science*, 4224, 249-263.

VLEURINCK, C., RAU, S., STURGILL, D., OLIVER, B., BEYE, M. (2016) Linking genes and brain development of honeybee workers: a whole-transcriptome approach. *PLoS One*, 11, e0157980.

WHEELER, D.E., BUCK, N., EVANS, J.D. (2006) Expression of insulin pathway genes during the period of caste determination in the honey bee, *Apis mellifera*. *Insect Molecular Biology*, 15, 597-602.

WILSON, M.J., ABBOTT, H., DEARDEN, P.K. (2011) The evolution of oocyte patterning in insects: multiple cell-signaling pathways are active during honeybee oogenesis and are likely to play a role in axis patterning. *Evolution & Development*, 13, 127-137.

WIRTZ, P., and BEETSMA, J. (1972) Induction of caste differentiation in the honeybee (*Apis mellifera*) by juvenile hormone. *Entomologia Experimentalis Et Applicata*, 4, 517-520.

WOYKE, J., and PALUCH, W. (1985) Changes in tissue polyploidzation during development of worker, queen haploid and diploid drone honeybees. *Journal of Apicultural Research*, 4, 214-224.

XIE, C., MAO, X., HUANG, J., DING, Y., DONG, S., KONG, L., *et al.* (2011) KOBAS 2.0: a web server for annotation and identification of enriched pathways and diseases. *Nucleic Acids Research*, 39(Web Server issue), W316-322.

# 5. DNA Methylation Comparison between 4-day-old Queen and Worker Larvae of Honey Bee

Yuanyuan Shi[1,2], Hao Liu[1,2], Yafeng Qiu[1], Zhiyong Ma[1], Zhijiang Zeng[2*]

1.Shanghai Veterinary Research Institute, Chinese Academy of Agricultural Sciences, Shanghai 200241, China;

2.Honeybee Research Institute, Jiangxi Agricultural University, Nanchang, Jiangxi 330045, China

**Abstract**

The honey bee is a social insect that is famous for queen-worker differentiation. Numerous studies indicate that queen larvae (QL) and worker larvae (WL) have different expressed genes and proteins. DNA methylation has been found to play an important role in regulating gene expression. To further explore the roles of the methylated genes in queen-worker differentiation, we analyzed DNA methylome profiles of 4-day-old QL and WL (*Apis mellifera*). The results demonstrated that there were 7.2 gigabases of sequence data from six methylated DNA immunoprecipitation libraries, and provided a genome-wide DNA methylation map as well as a gene expression map for 4-day-old QL and WL. The genome coverage of every sample was 4.79. According to CpG representation, all promoters in the *A. mellifera* genome were classified into high CpG promoters, intermediate CpG promoters and low CpG promoters. The methylated cytosines of larvae were enriched in introns, followed by coding sequence regions, 2 K downstream of genes, 5' untranslated regions (UTRs), 2 K upstream of genes, and 3' UTRs. Compared with 4-day-old WL, a number of genes in QL were down-methylated that were involved in biological regulation, immune system and metabolic regulation. In addition, these DMGs were involved in many signal pathways of caste differentiation such as Mitogen-activated protein kinase (MAPK), Notch, Insulin and Wnt signaling pathways.

**Keywords:** honey bee, queen larvae, worker larvae, caste differentiation, differentially methylated gene

## Introduction

Because of its precise labor division and high levels of social cohesion, the honey bee (*Apis mellifera*) is the most important model organism to study the animal behaviour (Seeley, 1989; Zeng *et al.*, 2009). An integrated honey bee colony is made up of three castes: one single reproductive queen, hundreds of haploid drones, and thousands of nearly sterile workers (Winston, 1987; Seeley, 1989). Despite having an identical genome, queen and workers exhibit striking differences in behaviour, morphology, longevity and reproduction

* Corresponding author: bees1965@sina.com (ZJZ).

注：此文发表在 *Journal of Asia-Pacific Entomology* 2017 年第 1 期。

(Weaver, 1957; Zeng et al., 2009).

The mechanism for queen-worker differentiation has been relatively well-studied (Weaver, 1966; Hartfelder et al., 1993; Kucharski et al., 2008; Kamakura, 2011; Leimar et al., 2012; Buttstedt et al., 2016) and the main factor regulating this phenomenon is known to the quality of the food fed to queen larvae (Winston, 1987; Buttstedt et al., 2016). Caste differentiation is more affected by sugars and the actual amount of food available to the developing larvae (Buttstedt et al., 2016).The major royal jelly proteins also play a role in reproductive maturation and brain development in queen larvae (Drapeau et al., 2006). Royalactin activates the ribosomal protein S6K kinase by activation of epidermal growth factor receptors in the fat body, resulting queen characteristics in the larvae (Kamakura, 2011). Queen larvae (QL) and worker larvae (WL) exhibit differential genes (Evans and Wheeler, 2001; Hepperle and Hartfelder, 2001; Barchuk et al., 2007; Chen et al., 2012; Cameron et al., 2013), proteins (Wu and Li, 2010), and microRNAs (Chen et al., 2010; Shi et al., 2015).

DNA methylation has been shown to be closely associated with queen-worker determination (Wang et al., 2006; Schaefer and Lyko, 2007; Kucharski et al., 2008; Elango et al., 2009; Foret et al., 2009; Lyko et al., 2010; Foret et al., 2012; Kucharski et al., 2016). A. mellifera become the first insect known to have three varieties of DNA methyltransferase (Wang et al., 2006). The reduction of DNA methyltransferase 3 expression may enhance queen production (Kucharski et al., 2008). Both diet and cell size can affect queen-worker differentiation through DNA methylation (Shi et al., 2011). Methylated genes in A. mellifera are divided into CpG low-content and CpG high-content genes, with caste specific genes tending to have high CpG content (Elango et al., 2009). The majority of methylated cytosines are located in the exon regions of the genes in brains of queens and workers (Lyko et al., 2010). There are 2,399 differentially methylated genes (DMGs) in brains of 4-day-old QL and WL, with 82 % of DMGs up-methylated in WL (Foret et al., 2012). In adult queens and workers, 561 DMGs are found, with 56 % of DMGs up-methylated in workers (Lyko et al., 2010).

In our previous study, we demonstrated that DNA methylation of QL and WL were less than twenty percent, and that differences were most apparent in 4-day-old larvae. In order to confirm our findings and to provide clearer information about the epigenetic factors affecting caste differentiation, we used three biological replicates to compare the DNA methylation of 4-day-old QL and WL by methylated DNA immunoprecipitation-sequencing analysis (MeDIP-seq). MeDIP-seq recently became an efficient method of analysis for large-scale DNA methylation, and has been used successfully in studies of plant genomics (Zhang et al., 2006) and pig muscle tissues (Li et al., 2012). 4-day-old QL and WL have differential expressed genes, proteins and microRNAs, however, there is little study on the differences in methylated genes between these two castes. In the present study, we sought to describe the distribution of methylation modifications in the A. mellifera genome, and to identify the DMGs that are involved in caste differentiation between 4-day-old QL and WL.

## Materials and Methods

### Experimental honey bee colonies

According to standard beekeeping techniques, the A. mellifera colonies were kept at Honey Bee Research

Institute, Jiangxi Agricultural University, Nanchang, China.

### Honey bee queen and worker larvae

The two castes were sampled from three *A. mellifera* colonies by having three biological replicates (10 larvae for each replicate) for 4-day-old QL and WL. All samples (QL1, QL2, QL3, WL1, WL2, and WL3) were flash frozen in liquid nitrogen until use.

### Methylated DNA immunoprecipitation-sequencing analysis

The whole larval DNA was isolated and purified by Universal Genomic DNA Extraction Kit (TaKaRa, DV811A), and then was sent to BGI (Beijing Genomics Institute at Shenzhen, China) for meDIP-seq analysis by using Illumina HiSeq™ 2000 (Illumina Inc, CA, USA). The detailed protocol was published in Li and Shi (Li *et al.*, 2012), these clean reads were deposited in the National Center for Biotechnology Information sequence read archive (SRR4897308 for QL1, QL2, and QL3, SRR4897290 for WL1, WL2, and WL3).

To compare DNA methylation rates among samples of 4-day-old larvae, read depth was normalized by averaging the number of reads in every group. A 1Mb sliding window was then used to smooth the distributions. The CpGo/e ratio, density of SNPs, genes, repeats and CpG islands were all calculated using a 1Mb sliding window. Measurement of DNA methylation levels along chromosomes revealed that in every chromosome there was a different pattern of DNA methylation in 4-day-old QL compared with 4-day-old WL (Fig. S1). According to CpG representation (Weber *et al.*, 2007; Li *et al.*, 2012), the promoters in the *A. mellifera* genome were also classified into high CpG (HCPs), intermediate CpG (ICPs), and low CpG promoters (LCPs) (Fig.S2).

### Gene ontology and Kyoto Encyclopedia of genes and genomes annotation

The reference genome of *A. mellifera*, together with gene information, was downloaded from the NCBI database (ftp://ftp.ncbi.nih.gov/genomes/*Apis_mellifera*/). The information about Gene Ontology (GO) and Kyoto Encyclopedia of Genes and Genomes (KEGG) terms was downloaded from the UniProtKB-GOA database. Both GO and KEGG with $p<0.05$ were considered as significant enriched.

### Real-time quantitative PCR

According to manufacturer's instructions, total RNA was extracted by using TRIzol reagent (Invitrogen, Carlsbad, CA, USA). 200 ng of RNA was reverse-transcribed to cDNA with Prime Script™ RT Master Mix kit (TaKaRa). Real-time quantitative PCR was performed on Step One real-time PCR system using the Real Time SYBR master mix kit (TaKaRa). Gene-specific primers were listed in Table I-5-1.

### Data analyses

The nucleotides of every sample library were 1.2 Gb, and the genome size of *A. mellifera* was 250.287 Mb (ftp://ftp.ncbi.nih.gov/genomes/*Apis_mellifera*/). The following formula was used to calculate the genome coverage of every sample library:

$$\text{The genome coverage} = (\text{nucleotides per library}) / (\text{genome size}).$$

Model-based analysis of chip sequencing (MACS) was used to scan the methylated levels in *A. mellifera* genome. The genes with DNA methylation went through GO analysis. The number of up- and down-methylated genes was the result of comparisons between QL and WL. The following formulae were used to calculate fold-changes and *P*-values from the normalized gene methylation of QL and WL samples:

The normalized gene methylation of QL = clean reads of the methylated gene in QL / total clean reads in QL *1000000,

The normalized gene methylation of WL = clean reads of the methylated gene in WL / total clean reads in WL *1000000,

Fold-change=log$_2$ (the normalized gene methylation of QL / the normalized gene methylation of WL) (Guan *et al.*, 2013),

$$p(x|y)=(\frac{N_2}{N_1})^y \frac{(x+y)!}{x!y!(1+\frac{N_2}{N_1})^{(x+y+1)}} \quad \begin{aligned} C(y \leq y_{min}|x) &= \sum_{y=0}^{y \leq y_{min}} p(y|x) \\ D(y \geq y_{max}|x) &= \sum_{y \geq y_{max}}^{\infty} p(y|x) \end{aligned}$$

where $x$ and $y$ indicate the mapped clean reads number for the same methylated gene in QL and WL library, $N_1$ and $N_2$ represent the total reads number for the two libraries respectively. In this study, absolute log$_2$ (fold change) > 1 and $P < 0.05$ were considered to represent significant differences in methylated gene (Guan *et al.*, 2013).

The real-time PCR results were examined by analysis of variance (ANOVA) using StatView (v 5.01, SAS Institute, Gary, NC, USA). Multiple comparisons of the mean values were performed using Fisher's Protected Least Significant Difference only after ANOVA showed a significant effect ($P < 0.05$).

## Results

### *Global mapping of DNA methylation in 4-day-old QL and WL*

We generated 7.2 gigabases (Gb) methylated DNA MeDIP-seq data from six samples, of which 5.8 Gb (80.5%) clean reads were aligned on the *A. mellifera* genome. The genome coverage of every sample was 4.79. After removing the ambiguously mapped reads and reads which may have come from duplicate clones, we used 5.1 Gb (70.8%) uniquely aligned non-duplicate reads in the following analysis (Table S1). To avoid false positives in enrichment, we required > 10 reads to determine a methylated CpG in a sample.

### *Differentially methylated genes in 4-day-old QL and WL*

Distributions of up- or down-methylated genes among gene functional elements in 4-day-old QL are presented in Fig.S3. The methylated cytosines of larvae were enriched in introns, followed by coding sequence (CDS) regions, 2 K downstream of genes, 5' untranslated regions (UTRs), 2 K upstream of genes, and 3' UTRs (Fig.S3). We found that > 7355 methylated genes were in the larval body (Table S2). Down-methylated genes outnumbered up-methylated genes in QL (Table S2). The index number of the linear relationships about DNA methylation of the genes in QL and WL replicates were $0.88 < R^2 < 0.94$ and $0.84 < R^2 < 0.94$, respectively.

By use of GO analysis, we found that a large number of genes in QL were down-methylated involved in many processes including biological regulation, growth, immune system process, and metabolic regulation (Fig. I-5-1 and Table S3). After mapping all of the DMGs to terms in the KEGG database (Ogata *et al.*,

1999), we found that the DMGs participated in "Calcium signaling pathway", "Phosphatidylinositol signaling system", "Purine metabolism", "Long-term potentiation", "Regulation of actin cytoskeleton", "Notch signaling pathway", "Wnt signaling pathway", "Tight junction", "Taste transduction" and "Pyrimidine metabolism" (Table I-5-2).

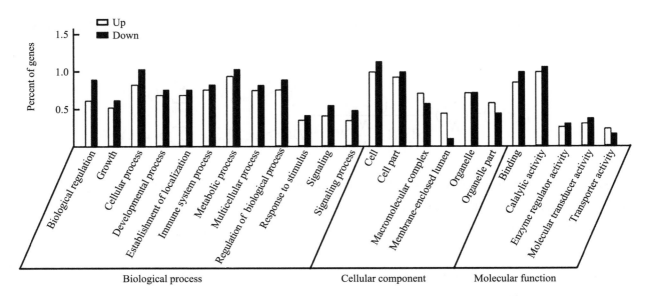

Figure I-5-1 Gene Ontology classification of down- and up-methylated genes in 4-day-old queen larvae, relative to worker larvae. The results are summarized in three main categories: biological process, cellular component and molecular function. The Y-axis indicates the precentage of genes in a category. The X-axis indicates category.

By use of the NCBI annotations, we found 39 DMGs of QL and WL that had been established in adult honey bees (Wheeler et al., 2006; Barchuk et al., 2007; Ikeda et al., 2011; Wolschin et al., 2011) were related to caste differentiation pathways, namely, the insulin (InR-2, Kinesin-3C, Sirt6-PA), mitogen-activated protein kinase (MAPK) (p38b, RL), Wnt (Hex110, Wnt1, Wnt4, Wnt6, Wnt7), and notch (LOC411086, LOC413289, LOC100578232, LOC410351, LOC100578826, LOC552725) signal pathway (Table I-5-3). In addition, we also found many differentially methylated miRNAs in QL and WL (Table I-5-3). Compared with WL, Dnmt1, Dnmt3, Wnt4, and Wnt6 were up-methylated, while Dynactin, Hex110, His3.3A-PA, mRpL45, Sir2, Sirt6-PA, Trap1-PA, Wnt1, Wnt7 were down-methylated in QL (Table I-5-3).

DNA methylation has negative correlations with gene expression (Li et al., 2012). In order to validate the sequencing results, LOC725827 and LOC412638 in the insulin signaling pathway, LOC411915 and LOC409523 in MAPK signaling pathway, LOC413502, LOC411919 in Wnt signaling, LOC411086, LOC413289, LOC100578232, LOC410351, LOC100578826, LOC552725 in Notch signaling pathway were selected for real-time quantitative PCR analysis (Fig. I-5-2). These results showed that the real-time PCR results of all these genes were consistent with the meDIP-seq data.

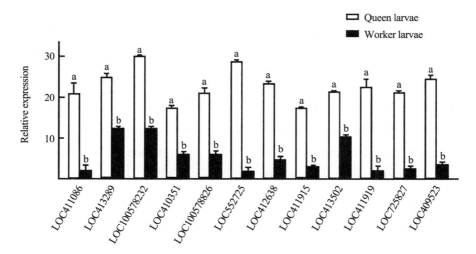

**Figure I-5-2** Verification of ten differentially methylated genes between 4-day-old queen larvae and worker larvae by RT-PCR. Different letters on top of bars indicate significant difference ($P<0.05$) with Fisher's Protected Least Significant Difference. Each bar corresponds to a single group represented as the mean ± S.E. of its biological replicates.

## Discussion

It is widely believed that only *A. mellifera* WL that are < 3 days old are able to become queens (Weaver, 1966), and our previous results supported to the notion that 3.5 day old larvae were at a critical juncture in caste development. By using digital gene expression (DGE) tag profile, many DEGs between 4-day-old QL and WL were identified (Barchuck *et al.*, 2007; Chen *et al.* 2012). In this study, we used three biological replicates to compare the DNA methylation of 4-day-old QL and WL by MeDIP-seq in order to confirm our previous findings that those results were based on a sample size of one, and to provide clearer information about the epigenetic factors affecting the developing larvae. We found that every chromosome had different DNA methylation in QL compared to WL (Fig. S1). All promoters in the *A. mellifera* genome were classified into HCPs, ICPs and LCPs (Fig. S2).

Several studies have proved that in *A. mellifera* methylated genes are predominantly clustered in CDS regions (Wang *et al.*, 2006; Kucharski *et al.*, 2008; Lyko *et al.*, 2010; Foret *et al.*, 2012) and exons (Lyko *et al.*, 2010; Cingolani *et al.*, 2013). In this study, DNA methylation occurred preferentially in introns, CDS regions and promoter regions, consistent with our previous research (Fig. S3). Comparing with previously published larval head and adult brain methylation patterns, we found that the number of methylated genes was highest in larval body: > 7355 methylated genes (Table S2) were in the larval body, 6086 methylated genes in the larval head, and 5854 in an adult brain (Lyko *et al.*, 2010; Foret *et al.*, 2012). These marked differences might be due to the different levels of growth and functioning between larval tissues and the adult brain. For the whole larval body, higher numbers of tissues were involved in the methylation analysis compared to the adult brain. In addition, larval tissues are closely related to protein turnover and high growth, whereas brain cells are postmitotic.

Contrary to our findings in 4-day-old WL, our results indicated that some genes in QL showed a decrease in DNA methylation. The genes Hex110, His3.3A-PA, mRpL45, Sirt6-PA, Trap1-PA (Table I-5-3) have been shown previously to be methylated in adult honey bees (Wheeler *et al.*, 2006; Barchuk *et al.*, 2007; Wolschin *et al.*, 2011), while our study indicated that these genes were also methylated at the larval stage. It is well known that individual development involves many signaling pathways, such as insulin (Wolschin *et al.*, 2011), MAPK (Kamakura, 2011), Target of Rapamycin (TOR) (Patel *et al.*, 2007), Notch and Wnt signaling pathways (Li *et al.*, 2011). In the present study, we found that InR-2 and Kinesin-3C in insulin signaling pathway, p38b and RL in MAPK signaling pathway, Wnt1, Wnt7 in Wnt signaling, LOC411086, LOC413289, LOC100578232, LOC410351, LOC100578826, LOC552725 in Notch signaling pathway (Table I-5-3) were down-methylated in QL compared to WL.

Caste differentiation is a significant and unique natural phenomenon. DNA methylation provides new ideas for the explanation of honey bee larval development (Kucharski *et al.*, 2008; Lyko *et al.*, 2010; Shi *et al.*, 2011; Foret *et al.*, 2012; Weiner and Toth, 2012) after its initial discovery in the system (Wang *et al.*, 2006). Gene expression regulation due to the nutrition intake might be the major contribution (Shi *et al.*, 2011; Leimar *et al.*, 2012). We measured, for the first time, the DNA methylation profiles of QL and WL in every chromosome. Further studies are needed to understand how DMGs are translated into physiological changes in the developing queens and workers of the honey bee.

## Conclusions

In a word, our results provide clearer information about the epigenetic factors affecting caste differentiation, and used three biological replicates to compare the DNA methylation profiles of 4-day-old QL and WL by MeDIP-seq. There is a different pattern of DNA methylation in every chromosome between 4-day-old QL and WL. This study suggests that differentially methylated genes in the larvae might also function in caste differentiation.

## Acknowledgements

We thank Doctor Xu-Jiang He for reviewing this manuscript. This work was supported by the National Natural Science Foundation of China (No.31502046) and the Earmarked Fund for the China Agriculture Research System (No.CARS-45-KXJ12).

Table I-5-1 Gene-specific primers used in real-time quantitative PCR

| Gene name | Gene description | Gene ID | Sense and antisense sequences (5′–3′) |
| --- | --- | --- | --- |
| LOC410122 | GAPDH | GB50902 | CTCAGGTTGTTGCCATTA/TTGCCTCTCGTTCACTAA |
| LOC411086 | Protein jagged-1 | GB48639 | GTGAATGAGTGCGTGTAA/ATTATGTCGTTGTCGTTGT |

(continued)

| Gene name | Gene description | Gene ID | Sense and antisense sequences (5'-3') |
|---|---|---|---|
| LOC413289 | Notch protein | GB49149 | TTCCAACTCCAATATCCAA/GTCGCAATGATGATTACC |
| LOC100578232 | Notch 2 like protein | GB44370 | AGAGGTAAGCGTAGCGATA/GTTCCTTCACAGCGACAT |
| LOC410351 | Neurogenic locus protein delta | GB45005 | CGCTAGTGTTGCCTATTAG/CTATCCTTCTCCGTTCTCT |
| LOC100578826 | Protein notch-like | GB41174 | CCTTGACTCTTCTGTAATCG/TCATCACTGCTGCTATCT |
| LOC552725 | N-acetylglucosamine-1-phosphotransferase | GB49324 | ACTTAGAACGAGACAACTT/GCATATCAGCAACAATCC |
| LOC412638 | Kinesin-3C | GB48375 | GTTATTGTTACTTGTGAATCT/ACCAGTTTGTCCATAAG |
| LOC411915 | p38b | GB43914 | TGACTGGTTATGTTGCTA/TAAGACTCGTGTTAGATGAT |
| LOC413502 | Wnt1 | GB45510 | AGAGTACGCTTTGCTAAA/GGATTACGAAGAACATCAC |
| LOC411919 | Wnt7 | GB44402 | CCTTCTTCTTCTCCTTCT/GTTCATTCGTTTCGTTTC |
| LOC725827 | InR-2 | GB55245 | GATAGATACGAGCAAGATT/GTTCATTCCTTCCTTCTC |
| LOC409523 | RL | GB51503 | GTATTACATAGAGACTTG/CCTTGGAATTAAGCATTA |

### Table I-5-2 KEGG analysis of the differentially methylated genes in 4-day-old queen larvae, relative to worker larvae

| Pathway | DMGs in pathway | All genes in pathway | P-value | Q-value | Pathway ID |
|---|---|---|---|---|---|
| Calcium signaling pathway | 21 (6.18%) | 193 (2.04%) | 5.74E-06 | 0.000816 | ko04020 |
| Phosphatidylinositol signaling system | 11 (3.24%) | 76 (0.8%) | 7.88E-05 | 0.005595 | ko04070 |
| Long-term potentiation | 10 (2.94%) | 70 (0.74%) | 0.000186 | 0.00879 | ko04720 |
| Purine metabolism | 16 (4.71%) | 171 (1.81%) | 0.000435 | 0.015046 | ko00230 |
| Regulation of actin cytoskeleton | 19 (5.59%) | 230 (2.44%) | 0.000631 | 0.015046 | ko04810 |
| Notch signaling pathway | 8 (2.35%) | 54 (0.57%) | 0.000636 | 0.015046 | ko04330 |
| Wnt signaling pathway | 9 (2.65%) | 70 (0.74%) | 0.000856 | 0.017362 | ko04310 |
| Tight junction | 13 (3.82%) | 135 (1.43%) | 0.001131 | 0.020069 | ko04530 |
| Taste transduction | 9 (2.65%) | 74 (0.78%) | 0.001282 | 0.020233 | ko04742 |
| Pyrimidine metabolism | 10 (2.94%) | 98 (1.04%) | 0.002695 | 0.038273 | ko00240 |

Table I-5-3 Differentially methylated genes in 4-day-old queen larvae, relative to worker larvae

| Gene | Gene annotation/Pathway | DNA methylation (up/down) | | | Gene | Gene annotation/Pathway | DNA methylation (up/down) | | |
|---|---|---|---|---|---|---|---|---|---|
| | | 4d QL1 | 4d QL2 | 4d QL3 | | | 4d QL1 | 4d QL2 | 4d QL3 |
| 5-HT1 | G protein coupled receptor | Down | Down | Down | Let-7 | mircoRNA | Down | Down | Down |
| 5-HT2beta | | Down | Down | Down | Mir184 | | Down | Down | Down |
| AChE-1 | | Down | Down | Down | mir71 | | Up | Down | Down |
| DopR2 | | Down | Down | Down | mir927 | | Down | Down | Down |
| nAChRa1 | | Down | Down | Down | LOC411086 | Notch signaling pathway | Down | Down | Down |
| nAChRa4 | | Down | Down | Down | LOC413289 | | Down | Down | Down |
| nAChRa6 | | Down | Down | Down | LOC100578232 | | Down | Down | Down |
| Atg13 | Adenosine Triphosphate (ATP) | Down | Down | Up | LOC410351 | | Down | Down | Down |
| Atg2 | | Down | Up | Down | LOC100578826 | | Down | Down | Down |
| ATP7 | | Down | Down | Down | LOC552725 | | Down | Down | Down |
| Dnmt1 | DNA methyltransferases | Up | Up | Up | p38b | MAPK signaling pathway | Down | Down | Down |
| Dnmt3 | | Up | Up | Up | RL | | Down | Down | Down |
| Dynactin | Calmodulin-binding protein | Down | Down | Down | RpL4 | Tim 44-like domain | Down | Down | Down |
| HDAC1 | Histone deacetylation | Down | Down | Down | Rpn5 | Code ribosomal protein | Down | Down | Down |
| HDAC6 | | Down | Down | Down | Rpn6 | | Down | Down | Down |
| Hex110 | Wnt signaling pathway | Down | Down | Down | Sir2 | Sir2 family | Down | Down | Down |
| His3.3A-PA | Histone deacetylation | Down | Down | Down | Sirt6-PA | Insulin signaling pathway | Down | Down | Down |
| InR-2 | Insulin signaling pathway | Down | Down | Down | Trip1-PA | TRAF zinc finger | Down | Down | Down |
| Kinesin-3A | Calmodulin-binding protein | Down | Down | Down | Wnt1 | Wnt signaling pathway | Down | Down | Down |
| Kinesin-3C | Insulin signaling pathway | Down | Down | Down | Wnt4 | | Up | Up | Up |
| mRpL45 | Tim 44-like domain | Down | Down | Down | Wnt6 | | Up | Up | Up |
| mRpS22 | | Down | Down | Down | Wnt7 | | Down | Down | Down |

## References

BARCHUK, A.R., CRISTINO, A.S., KUCHARSKI, R., COSTA, L.F., SIMÕES, Z.L., MALESZKA, R., 2007. Molecular determinants of caste differentiation in the highly eusocial honey bee *Apis mellifera*. BMC Dev. Biol. 7, 70.

BUTTSTEDT, A., IHLING, C.H., PIETZSCH, M., MORITZ, R.F., 2016. Royalactin is not a royal making of a queen. *Nature* 537 (7621), E10-12.

CAMERON, R.C., DUNCAN, E.J., DEARDEN, P.K., 2013. Biased gene expression in early honeybee larval development. *BMC Genomics* 14, 903.

CHEN, X., YU, X., CAI, Y., ZHENG, H., YU, D., LIU, G., ZHOU, Q., HU, S., HU, F., 2010. Next-generation small RNA sequencing for microRNAs profiling in the honey bee *Apis mellifera. Insect Molecular Biology* 19 (6), 799-805.

CHEN, X., HU, Y., ZHENG, H., CAO, L., NIU, D., YU, D., SUN, Y., HU, S., HU, F., 2012. Transcriptome comparison between honey bee queen- and worker-destined larvae. *Insect Biochem. Mol. Biol.* 42 (9), 665-673.

CINGOLANI, P., CAO, X., KHETANI, R.S., CHEN, C.C., COON, M., SAMMAK, A., BOLLIG-FISCHER, A., LAND, S., HUANG, Y., HUDSON, M.E., GARFINKEL, M.D., ZHONG, S., ROBINSON, G.E., RUDEN, D.M., 2013. Intronic non-CG DNA hydroxymethylation and alternative mRNA splicing in honey bees. *BMC Genomics* 14, 666.

GUAN, C., BARRON, A.B., HE, X.J., WANG, Z.L., YAN, W.Y., ZENG, Z.J., 2013. A comparison of digital gene expression profiling and methyl DNA immunoprecipitation as methods for gene discovery in honeybee (*Apis mellifera*) behavioural genomic analyses. *PLoS One* 8(9), e73628.

DRAPEAU, M.D., ALBERT, S., KUCHARSKI, R., PRUSKO, C., MALESZKA, R., 2006. Evolution of the Yellow/Major Royal Jelly Protein family and the emergence of social behavior in honeybees. *Genome Res.* 16 (11), 1385-1394.

ELANGO, N., HUNT, B.G., GOODISMAN, M.A., YI, S.V., 2009. DNA methylation is widespread and associated with differential gene expression in castes of the honeybee, *Apis mellifera. Proc. Nat. Acad. Sci. USA.* 106 (27), 11206-11211.

EVANS, J.D., WHEELER, D.E., 2000. Expression profiles during honeybee caste determination. Genome Biol. 2 (1), research0001.1-0006.

FORET, S., KUCHARSKI, R., PITTELKOW, Y., LOCKETT, G.A., MALESZK, R., 2009. Epigenetic regulation of the honey bee transcriptome: unravelling the nature of methylated genes. *BMC Genomics* 10, 472.

FORET, S., KUCHARSKI, R., PELLEGRINI, M., FENG, S., JACOBSEN, S.E., ROBINSON, G.E., MALESZKA, R., 2012. DNA methylation dynamics, metabolic fluxes, gene splicing, and alternative phenotypes in honeybees. *Proc. Nat. Acad. Sci. USA.* 109 (13), 4968-4973.

HARTFELDER, K., TOZETTO, S.O., RACHINSKY, A., 1993. Sex-specific developmental profiles of juvenile hormone synthesis in honeybee larvae. *Roux's Arch Dev. Biol.* 202 (3), 176-180.

HEPPERLE, C., HARTFELDER, K., 2001. Differentially expressed regulatory genes in honeybee caste development. *Naturwissenschaften* 88 (3), 113-116.

KAMAKURA, M., 2011. Royalactin induces queen differentiation in honeybees. *Nature* 473 (7348), 478-483.

KUCHARSKI, J.R., MALESZKA, J., FORE, S., MALESZKA, R., 2008. Nutritional control of reproductive status in honeybees via DNA methylation. *Science* 319 (5871), 1827-1829.

KUCHARSKI, R., MALESZKA, J., MALESZKA, R., 2016. A possible role of DNA methylation in functional divergence of a fast evolving duplicate gene encoding odorant binding protein 11 in the honeybee. *Proc. Biol. Sci.* 283 (1833), pii:20160558.

LEIMAR, O., HARTFELDER, K., LAUBICHLER, M.D., PAGE, R.E.JR., 2012. Development and evolution of caste dim orphism in honeybees-a modeling approach. *Ecol. Evol.* 2 (12), 3098-3109.

LI, M., WU, H., LUO, Z., XIA, Y., GUAN, J., WANG, T. LI, R., 2012. An atlas of DNA methylomes in porcine adipose and muscle tissues. *Nat. Commun.* 3, 850.

LI, Z., NIE, F., WANG, S., LI, L., 2011. Histone H4 Lys 20 monomethylation by histone methylase SET8 mediates Wnt target gene activation. *Proc. Natl. Acad. Sci. USA.* 108 (8), 3116-3123.

LYKO, F., FORET, S., KUCHARSKI, R., WOLF, S., FALCKENHAYN, C., MALESZKA, R., 2010. The honeybee epigenomes: differential methylation of brain DNA in queens and workers. *PLoS Biology* 9 (11), e1000506.

PATEL, A., FONDRK, M.K., KAFTANOGLU, O., EMORE, C., HUNT, G., FREDERICK, K., AMDAM, G.V., 2007. The making of a queen: TOR pathway is a key player in diphenic caste development. *PLoS One* 2 (6), e509.

SCHAEFER, M., LYKO, F., 2007. DNA methylation with a sting: an active DNA methylation system in the honeybee. *Bioessays* 29 (3), 208-221.

SEELEY, T.D., 1989. The honey bee colony as a superorganism. *American Scientist* 77, 546-553.

SHI, Y.Y., HUANG, Z.Y., ZENG, Z.J., WANG, Z.L., WU, X.B., YAN, W.Y., 2011. Diet and cell size both affect queen-worker differentiation through DNA methylation in honeybees (*Apis mellifera*, Apidae). *PLoS One* 6 (4), e18808.

SHI, Y.Y., ZHENG, H.J., PAN, Q.Z., WANG, Z.L., ZENG, Z.J., 2015. Differentially expressed microRNAs between queen and worker larvae of honey bee (*Apis mellifera*). *Apidologie* 46 (1), 35-45.

WANG, Y., JORDA, M., JONES, P.L., MALESZK, R., LING, X., ROBERTSON, H.M., MIZZEN, C.A., PEINADO, M.A., ROBINSON, G.E., 2006. Functional CpG methylation system in a social insect. *Science* 314 (5799), 645-647.

WEAVER, N., 1957. Effects of larval honey bee age on dimorphic differentiation of the female honeybee. *Ann. Entomol. Soc. Am.* 50, 283-294.

WEAVER, N., 1966. Physiology of caste determination. *Annu. Rev. Entomol.* 11, 79-102.

WEBER, M., HELLMANN, I., STADLER, M.B., RAMOS, L., PÄÄBO, S., REBHAN, M., SCHÜBELER, D., 2007. Distribution, silencing potential and evolutionary impact of promoter DNA methylation in the human genome. *Nat. Genet.* 39 (4), 457-466.

WEINER, S.A., TOTH, A.L., 2012. Epigenetics in social insects: a new direction for understanding the evolution of castes. *Genet. Res. Int.* 2012, 609810.

WINSTON, M.L., 1987. Biology of the honey bee. Harvard University Press, Cambridge.

WOLSCHIN, F., MUTTI, N.S., AMDAM, G.V., 2011. Insulin receptor substrate influences female caste development in honeybees. *Biol. Lett.* 7 (1), 112-115.

WU, J., LI, J.K., 2010. Proteomic analysis of the honeybee (*Apis mellifera L.*) caste differentiation between worker and queen larvae. *Scientia Agricultura Sinica* 43 (5), 176-181.

ZENG, Z.J., 2009. Apiculture. China Agriculture Press, Beijing.

ZHANG, X., YAZAKI, J., SUNDARESAN, A., 2006. Genome-wide high-resolution rapping and functional analysis of DNA methylation in *Arabidopsis*. *Cell* 126 (6), 1189-1201.

# 6. H3K4me1 Modification Functions in Caste Differentiation in Honey Bees

Yong Zhang [1,2,†], Zhen Li [1,2,†], Xujiang He [1,2], Zilong Wang [1,2] and Zhijiang Zeng [1,2,*]

1. Honeybee Research Institute, Jiangxi Agricultural University, Nanchang, Jiangxi 330045, P. R. of China
2. Jiangxi Province Honeybee Biology and Beekeeping Nanchang, Jiangxi 330045, P. R. of China

**Abstract**

Honey bees are important species for the study of epigenetics. Female honey bee larvae with the same genotype can develop into phenotypically distinct organisms (sterile workers and fertile queens) depending on conditions such as diet. Previous studies have shown that DNA methylation and histone modification can establish distinct gene expression patterns, leading to caste differentiation. It is unclear whether the histone methylation modification H3K4me1 can also impact caste differentiation. In this study, we analyzed genome-wide H3K4me1 modifications in both queen and worker larvae and found that H3K4me1 marks are more abundant in worker larvae than in queen larvae at both the second and fourth instars, and many genes associated with caste differentiation are differentially methylated. Notably, caste-specific H3K4me1 in promoter regions can direct worker development. Thus, our results suggest that H3K4me1 modification may act as an important regulatory factor in the establishment and maintenance of caste-specific transcriptional programs in honey bees; however, the potential influence of other epigenetic modifications cannot be excluded.

**Keywords:** histone modification, gene expression, methylation, caste-specifific

## Introduction

Honey bees are eusocial insects and an important model organism for studies of caste development and caste differentiation in social insects. Their division of labor is mainly based on the differentiation of castes (queen and worker) [1]. After divergence, queens and workers have different morphological, physiological, behavioral, and longevity-related traits, despite sharing the same genome [2]. The mechanism underlying caste differentiation is not fully understood. However, there is evidence that differences in nutritional status between queens and workers modulate caste differentiation by altering DNA methylation patterns [3-6]. In addition, various signaling pathways, such as the Wnt signaling pathway [7], the target of rapamycin (TOR) nutrient sensing pathway [8,9], and the mitogen-activated protein kinase (MAPK) signaling pathway [10], are related to

† These two authors contributed equally to this paper.
*Corresponding author:bees@jxau.edu.cn (ZJZ).
注：此文发表在 *International Journal of Molecular Sciences* 2023,24,6217。

honey bee caste differentiation.

DNA and histone modifications are thought to primarily affect transcriptional events. The establishment, maintenance, and regulation of transcriptional programs during development depend on chromatin plasticity [11]. Recent evidence suggests that chromatin-based epigenetic mechanisms can influence nutrient-mediated caste differentiation in honey bees. RNAi knockdown of the DNA methyltransferase DNMT3 has a jelly-like effect on developmental trajectories, resulting in a significantly higher proportion of queens with fully developed ovaries [12]. Differences in DNA methylation also influence the alternative development of queens and workers [13]. H3K27ac has been shown to be a key chromatin modification, and the caste-specific region of intronic H3K27ac directs the worker caste [11]. However, for H3K4me3 and H3K36me3, there is no evidence that modification in specific regions can direct caste development [11]. Histone methylation is methylation that occurs on the N-terminal lysine (K) [14] or arginine [15] residues of H3 and H4 histones. Like histone acetylation, histone methylation contributes to almost all biological processes, from DNA repair, the cell cycle, stress responses, and transcription to development, differentiation, and aging [16-20]. It can also regulate the lifespan of model organisms such as rats [21], Caenorhabditis elegans [22], and Drosophila melanogaster [23], and even have a transgenerational effect on lifespan [24,25]. However, there is no evidence that histone methylation contributes to honey bee development.

Hundreds of genes involved in caste differentiation have been identified in honey bees [13,26]. Differences in chromatin levels may lead to differences in gene expression. Histone acetylation influences caste differentiation in honey bees [27]. However, the role of histone methylation in honey bee caste differentiation has not been determined. We performed the first genome-wide analysis of the distribution of H3K4me1 before (2nd instar) and after (4th instar) critical time points (3rd instar) [1,28] in honey bee caste differentiation. Compared with the 2nd instar, the numbers of differentially expressed genes (DEGs) and H3K4me1 markers were significantly increased at the 4th instar, and gene expression was negatively correlated with H3K4me1. Further analysis revealed that the chromatin patterns of queens and workers differed significantly at the 4th instar. H3K4me1 modification promotes larval development towards worker bees. These findings illustrate the important role of H3K4me1 in honey bee larval development and caste differentiation.

## Results

### *H3K4me1 modifications in the honey bee are enriched in transcribed regions*

To investigate the chromatin structure of honey bees, we determined, for the first time, the genome-wide distribution of H3K4me1. We detected H3K4me1 enrichment around the transcriptional start sites (TSS) of genes in both queen larvae and worker larvae (Fig. S1). These results are consistent with those for other species, including mammals, invertebrates, and plants [29-31]. Over 8 G of data were generated for all samples, with a mapping value and Q30 value of greater than 90% (Table S1). These results indicate that the sequencing results are reliable.

We further evaluated caste-specific patterns in the distribution of H3K4me1. In particular, we divided H3K4me1 into unique peaks (i.e., peaks detected in all three replicates in one sample, but not in any of the three replicates in the other sample) and differential peaks ($p < 0.05$ and fold change $>2$). In 2Q (2nd instar queen) vs. 2W (2nd instar worker), we identified 36 unique H3K4me1 peaks in 2Q and 380 in 2W (Fig. I-6-1A, see page 509). In 4Q vs. 4W, we identified 347 unique H3K4me1 peaks in 4Q and 1185 in 4W (Fig. I-6-1A). There were significantly more unique peaks for workers than queens at both the 2nd and 4th instars (Fig. S2). There was no significant difference in the relative proportions of the positions of unique peaks between 2Q and 2W (Fig. I-6-1B; $\chi^2 = 2.66$, $p = 0.27$, chi-squared test). However, there was a difference in the distribution of unique peaks between 4Q and 4W (Fig. I-6-1C; $\chi^2 = 80.18$, $p = 3.88 \times 10^{-18}$, chi-squared test); 60% of the unique peaks in 4W were enriched in the promoter region, while 61% of the unique peaks in 4Q were enriched in intronic regions. The data for differential peaks between queens and workers were consistent with the results for unique peaks. In 2Q vs. 2W, we identified 18 differential H3K4me1 peaks in 2Q and 326 in 2W (Fig. I-6-1D). In 4Q (4th instar queen) vs. 4W (4th instar worker), we identified 432 differential H3K4me1 peaks in 4Q and 2233 in 4W (Fig. I-6-1D). There was no significant difference in the distribution of differential peaks between 2Q and 2W (Fig. I-6-1E; $\chi^2 = 4.57$, $p = 0.10$, chi-squared test). However, there was a difference in the distribution of differential peaks between 4Q and 4W (Fig. I-6-1F; $\chi^2 = 45.32$, $p = 1.44 \times 10^{-10}$, chi-squared test); 55% of unique peaks in workers were found in promoter regions, while 54% of unique peaks in queens were found in intronic regions. In order to study the functional significance of the caste-specific promoter H3K4me1, we performed a motif enrichment analysis using the MEME website (https://meme-suite.org/meme/tools/meme; accessed on 13 December 2022). The sequence of *Drosophila melanogaster* was used to annotate the honey bee motifs. We identified 2Q and 2W transcription factor binding sites that were significantly enriched, and the top three transcription factor binding sites were identical (Fig. S3A). However, the top five enriched transcription factor binding sites of 4Q and 4W were not the same (Fig. S3B).

### *Caste-specific H3K4me1 modification patterns correlate with differential gene expression*

After identifying caste-specific differences in the distribution of H3K4me1 modifications, we next evaluated whether the caste-specific distributions were associated with gene expression patterns according to caste. Principal component analysis of our RNA-seq data revealed strong separation between the two castes (Fig. S4). In 2Q vs. 2W, we identified 666 DEGs and 296 differential peak-associated genes (DPGs), of which 28 genes were commonly differentially expressed in RNA-seq and ChIP-seq (Fig. I-6-2A, see page 510). In 4Q vs. 4W, 2306 DEGs and 2004 DPGs were identified, of which 384 genes were commonly differentially expressed in RNA-seq and ChIP-seq (Fig. I-6-2A). In 2Q vs. 2W, there were 18 up-regulated peaks and 326 down-regulated peaks (Fig. I-6-2B), and 224 up-regulated genes and 452 down-regulated genes (Fig. S4). In 2Q vs. 2W, there were 432 up-regulated peaks and 2233 down-regulated peaks (Fig. I-6-2B), and 1355 up-regulated genes and 951 down-regulated genes (Fig. S4). We found that the genes enriched for the H3K4me1 modification showed significant expression differences based on transcriptional data (Figure 2C; 2Q vs. 2W, $\rho = 0.07$, $p = 1.35 \times 10^{-27}$, Spearman test; 4Q vs. 4W, $\rho = 0.31$, $p = 3.93 \times 10^{-13}$, Spearman test). In addition, there were significant correlations between differential H3K4me1 peak signals and transcript levels (Fig. S5A) and between DEGs and H3K4me1 peak signals (Fig. S5B). Based on the enrichment with enhancer-associated histone H3K4me1 modifications, transcription factor binding sites, promoter sites, and changes in gene expression, worker-specific H3K4me1-enriched regions are markers of active enhancers and play an important role in caste differentiation.

To investigate the effect of H3K4me1-modified genes on caste differentiation, a KEGG analysis was performed. We detected enrichment for distinct developmental processes in the two castes at both the 2nd and 4th instars. Among the unique peak-associated genes of 2W and 4W, eight KEGG pathways were related to honey bee caste differentiation, while among the unique peak-associated genes of 2Q and 4Q, there were only three and five KEGG pathways related to honey bee caste differentiation, respectively (Fig. I-6-3A-D, see page 511). This was consistent with our previous analysis [32], again showing that H3K4me1 modification favors the development of honey bee larvae into worker bees. More DPGs in 4th instar larvae than in 2nd instar larvae were involved in caste differentiation (Fig. I-6-3E-F).

***Caste features of worker bees may be induced by H3K4me1***

A representative gene involved in highly significant caste-specific changes in both H3K4me1 enrichment and gene expression analysis is shown in Fig. I-6-4A(see page 512). *Juvenile hormone esterase* (*JHe*; LOC406066) is shown as an example of a physio-metabolic worker-specific gene [33], in which H3K4me1 enrichment differences are associated with the TSS. Similarly, the other two caste differentiation-related genes, *P450-6a17* and *IGF*, showed significant differences in gene expression and H3K4me1 enrichment in both TSS and gene ontology regions (Fig. S6).

In addition, we selected five genes with established roles in the caste differentiation of honey bees for verification, namely *JHe* [33], *Vg* (vitellogenin, LOC406088) [34], *JHAMT* (juvenile hormone acid O-methyltransferase, LOC724216) [35], *Hex70a* (hexamerin 70a, LOC726848) [36], and *Hsp90* (*heat shock protein 90*, LOC408928) [37]. Our transcriptome results were consistent with those of previous studies [33,36-38], and trends in H3K4me1 enrichment were consistent with trends in transcriptome data, thus supporting the reliability of our results. Juvenile hormone (JH) is a master regulator of caste differentiation in honey bees [39]. Vitellogenin is an antagonist of JH. H3K4me1 enrichment in the *Vg* genes of workers was significantly higher than that in queens at the 2nd and 4th instars. These results further support the role of H3K4me1 modification in regulating transcription, and reveal the effect of H3K4me1 and transcript co-regulation on caste differentiation in honey bees.

## Discussion

We used ChIP-seq to characterize genome-wide caste-specific chromatin patterns in honey bees and revealed the chromatin patterns related to reproductive division of labor in social insects. Combined with an RNA-seq analysis, the specific modification of H3K4me1 appeared after the critical time point for caste differentiation. There was no significant difference in H3K4me1 modification patterns between worker larvae and queen larvae until the irreversible stage of caste differentiation. A significant number of queen-worker chromatin differences are associated with caste-specific transcription. Importantly, a number of enhancers identified in caste-specific regions may be involved in honey bee caste differentiation.

Previous studies of *Drosophila melanogaster* [40,41], mice [42,43], and embryonic stem cells from humans and mice have demonstrated that changes in histone methylation affect development and regulate cell fate outcomes [44-46]. In honey bees, DNA methylation is predominantly present in gene body regions and the 5′ ends of genes [3,47,48]. Many previous studies have demonstrated the role of DNA methylation in caste differentiation. Different patterns of CpG methylation were detected between the queens and workers [3]. The results of the present study suggest that the developmental asymmetry between queen and worker larvae is associated with an asymmetry in H3K4me1 modification patterns. This is consistent with the distribution of DNA methylation in queen and worker bees [3]. Studies have shown that the degree of enrichment of H3K4me1 differs [49]. In addition, the differential histone methylation of transcription factors that regulate development is thought to alter cell fate decisions [45]. At the same time, the present study also showed that H3K4me1 modification can significantly regulate the transcript level of regulatory factors related to honey bee caste differentiation (such as *JHe*, *Vg*, *Hex70a*, and *JHAMT*). Signaling pathways such as FoxO and TOR are known to influence caste differentiation in honey bees [39,50], and the present study clearly shows that differential H3K4me1 levels also affect the transcription of genes in these signaling pathways. Notably, caste-specific H3K4me1-modified genes (such as *Hex70a* and *Vg*) are related to reproductive division of labor and nutrient metabolism [38,51,52]. These results suggest that H3K4me1 marks contribute to honey bee caste differentiation.

We found that H3K4me1 modification is closely involved in the modification of worker larvae, and is mainly involved in pathways related to honey bee caste differentiation. Caste-specific differences in H3K4me1 were mainly detected in the promoter and intronic regions. Queen-specific H3K4me1 was mainly located in intronic regions. In contrast, worker-specific H3K4me1 was mainly localized in promoter regions, close to transcription initiation sites. In addition, abundant caste-specific H3K4me1 in promoter regions was associated with high levels of caste-specific gene expression, suggesting that these regions play an important *cis*-regulatory role. Monomethylation of histone H3 at lysine 4 (H3K4me1) is a hallmark of activated enhancers in both vertebrates and invertebrates [29,53]. Enhancers are *cis*-regulatory DNA sequences and can increase the transcription of target genes. The opening of repressive enhancer-promoter loops leads to transcriptionally active enhancer-promoter regulation as a fundamental mechanism underlying differential transcriptional regulation [54,55]. The spatial organization of chromatin, including long-range enhancers adjacent to target promoters in *cis*, also modulates gene expression [56]. Therefore, we analyzed the caste-specific promoter region H3K4me1 for conserved transcription factor binding motifs. GATA2 and SPIB accounted for the most transcription factors in 4th instar queens and workers, respectively. Both GATA2 and SPIB can regulate development and cell differentiation [57-60]. Therefore, H3K4me1 modification may mediate caste-specific enhancer activation, thus directing larval development.

Taken together, we speculate that the worker-specific promoter H3K4me1 region has the hallmarks of an active enhancer. In addition, most worker genes enriched in the promoter H3K4me1 region are also transcription factors, suggesting that enhancers are associated with upstream and downstream genes during the development of worker castes. Queen-specific regions may also be caste-specific enhancers; however, further identification is needed.

## Materials and Methods

### *Insects*

Honey bees (*Apis mellifera*) were obtained from Jiangxi Agricultural University in 2020. Queens were restricted for 6 h (8 am to 2 pm) to a plastic frame designed by Pan *et al.* [61] to lay eggs in worker cells. The queen laid eggs on a removable plastic base, which was transferred to a plastic queen cell without touching the egg itself. Half of the eggs were transferred to queen cells at 2 pm on the second day after laying and before hatching, while the other half remained in the worker cells. All eggs (in both queen and worker cells) were cared for by workers. Eggs hatched on the third day after laying. To collect the queen and worker larvae at instars 2 and 4, larvae were sampled from both queen and worker cells in each of three colonies at 4 pm on days 5 and 7 after laying. Larvae were picked with sterilized tweezers and rinsed in ddH$_2$O three times. Filter paper was used to drain the water from the larvae, and larvae were placed immediately in liquid nitrogen. Whole larvae were used for sequencing, as in previous studies, since they were very small (2nd instar worker and queen larvae: 2 mg/larva; 4th instar worker larvae: 12 mg/larva; 4th instar queen larvae: 25 mg/larva).

### *ChIP-seq assay and analysis*

Chromatin immunoprecipitation was performed as described by Wojciechowski *et al.* [11], with slight modifications. Approximately 800 mg of larvae (320 larvae per sample in 2Q and 2W; 32 larvae per sample in 4Q and 4W) were cross-linked for 10 min in 1% ChIP-seq-grade formaldehyde. The H3K4me1 antibody (ab195391; Abcam, Cambridge, UK) was used for immunoprecipitation. The H3K4me1 library was sequenced (50 bp single-end reads or 150 bp paired-end reads) on an Illumina HiSeq3000 sequencer.

The genome assembly Amel_HAv3.1 (GCF_003254395.2) was downloaded from NCBI and indexed using Bowtie 2 (v2.3.3). ChIP-seq samples were mapped to this indexed genome using Bowtie 2 with the default parameters. Detailed mapping statistics for each sample are available in Tables S2 and S3.

### *RNA-seq analysis*

Hierarchical Indexing for Spliced Alignment of Transcripts (HISAT; version 2.1.5) was used to align the RNA-seq reads to the reference genome (Amel_HAv3.1). Expression levels were reported as FPKM values to normalize the length of the annotated transcripts and the total number of reads aligned to the transcriptome. Differential expression was analyzed using the DESeq2 R package (1.30.0). *p*-values were corrected for multiple comparisons with a false discovery rate (value < 0.05). Genes with *p*-adj ≤ 0.05 were defined as DEGs. Detailed statistics are available in Tables S4 and S5.

### *Verification of gene expression differences by qRT-PCR*

The 2nd and 4th instar larvae of queens and workers were sampled. Three samples were collected for each type of larvae, with each sample coming from a different colony. Therefore, a total of 12 larval samples across three different colonies were evaluated.

qRT-PCR was performed according to previously described methods [62]. Briefly, total RNAs were extracted using the TransZol Up Plus RNA Kit (TransGen Biotech, Beijing, China), and then transcribed into cDNA using a PrimeScript RT Reagent Kit (Takara, Kusatsu, Japan). *GAPDH* was used as a reference gene.

The primer sequences were designed using Prime Primer 6.0 (Table I-6-1). A 10 μL reaction system (5 μL of SYBR®Premix Ex Taq™ II, 3 μL of H$_2$O, 1 μL of cDNA, 0.4 μL each of forward and reverse primers, and 0.2 μL of ROX) was established. The PCR conditions were as follows: 95 °C, 5 min; 94 °C, 2 min; 40 cycles (95 °C, 10 s, Tm, 15 s, 72 °C, 15 s); 72 °C, 10 min. To establish the melting curve of the qRT-PCR product, the primers were heated slowly with a gradual increase of 1 °C every 5 s from 72 °C to 99 °C. The data were analyzed using the 2$^{-\Delta\Delta CT}$ method.

**Table I-6-1 Primer sequences for quantitative qRT-PCR**

| Genes | Forward Primer | Reverse Primer |
| --- | --- | --- |
| GAPDH | GCTGGTTTCATCGATGGTTT | ACGATTTCGACCACCGTAAC |
| JHe | CTTTTCTCGCTTCCACAACC | TCCTGGTCCAGCAATGTGTA |
| VG | AAGACCAATCCACCGTTGAG | TGGTTCACGCTCCTAGCTTT |
| JHAMT | GGATTTGCCCAAAGACACAT | CGAGGATTCGCGTACAATTT |
| Hex70a | GAGGGTCAAGCATGGAACAT | GTTGTTCTTCGCCCAGAGAG |
| Hsp90 | CTGAGAGTGACGCGAAGCTA | CTCCGGCATCTTTTCACAAT |

## Conclusions

Significant differences in H3K4me1 modification between queen and worker larvae were observed during caste differentiation in honey bees. Chromatin modification can regulate the transcription of the genes that determine the caste. Furthermore, H3K4me1 modification was closely involved in the regulation of the development of worker larvae, and may be an important modification in the worker development pathway. These findings clearly establish the contribution of histone methylation to honey bee development, and may contribute to further research on caste differentiation and developmental plasticity.

## Supplementary Materials

The following supporting information can be downloaded at: www.mdpi.com/xxx/s1.

## Author Contributions

Conceptualization, Y.Z., Z.L., Z.W., X.H., and Z.Z.; methodology, Y.Z.; software, Y.Z.; validation, Z.W. and Z.Z.; formal analysis, Y.Z. and Z.L.; investigation, Z.Z.; resources, Z.Z.; data curation, Y.Z. and Z.L.; writing—original draft preparation, Y.Z.; writing—review and editing, Y.Z. and Z.L.; visualization, Y.Z. and Z.L.; supervision, Z.Z.; project administration, Z.Z.; funding acquisition, Z.Z. All authors have read and agreed to the published version of the manuscript.

## Funding

This work was supported by the National Natural Science Foundation of China (32172790, 31872432) and the Earmarked Fund for China Agriculture Research System (CARS-44-KXJ15).

## Institutional Review Board Statement

Not applicable.

## Informed Consent Statement

Not applicable.

## Data Availability Statement

Raw sequencing reads for RNA-seq are available at SRA; accession: PRJNA770835. Raw sequencing reads for ChIP-seq are available at SRA1 accession: PRJNA891375.

## Acknowledgments

Thanks to Nogogene for providing the raw data analysis.

## Conflicts of Interest

The authors declare that there are no conflicts of interest.

## References

1. WEAVER, N. Physiology of caste determination. *Annual Reviews* 1966, *11*, 79-102.
2. MICHENER, C.D.; MICHENER, C.D. *The social behavior of the bees: a comparative study*; Harvard University Press: 1974.
3. ELANGO, N.; HUNT, B.G.; GOODISMAN, M.A.; YI, S.V. DNA methylation is widespread and associated with differential gene expression in castes of the honeybee, *Apis mellifera*. *PNAS* 2009, *106*, 11206-11211.
4. FORD, D. Honeybees and cell lines as models of DNA methylation and aging in response to diet. *Experimental Gerontology* 2013, *48*, 614-619.
5. YAGOUND, B.; REMNANT, E.J.; BUCHMANN, G.; OLDROYD, B.P. Intergenerational transfer of DNA methylation marks in the honey bee. *PNAS* 2020, *117*, 32519-32527.
6. YAGOUND, B.; SMITH, N.M.; BUCHMANN, G.; OLDROYD, B.P.; REMNANT, E.J. Unique DNA methylation profiles are associated with cis-variation in honey bees. *Genome Biol. Evol.* 2019, *11*, 2517-2530.
7. SHI, Y.Y.; YAN, W.Y.; HUANG, Z.Y.; WANG, Z.L.; WU, X.B.; ZENG, Z.J. Genomewide analysis indicates that queen larvae have lower methylation levels in the honey bee (Apis mellifera). *Naturwissenschaften* 2013, *100*, 193-197.
8. WHEELER, D.; BUCK, N.; EVANS, J. Expression of insulin/insulin-like signalling and TOR pathway genes in honey bee caste determination. *Insect Mol. Biol.* 2014, *23*, 113-121.
9. PATEL, A.; FONDRK, M.K.; KAFTANOGLU, O.; EMORE, C.; HUNT, G.; FREDERICK, K.; AMDAM, G.V. The making of a queen: TOR pathway is a key player in diphenic caste development. *PLoS One* 2007, *2*, e509.
10. HE, X.J.; WEI, H.; JIANG, W.J.; LIU, Y.B.; WU, X.B.; ZENG, Z. Honeybee (Apis mellifera) maternal effect causes

alternation of DNA methylation regulating queen development. *Sociobiology* 2021, *68*, e5935-e5935.

11. WOJCIECHOWSKI, M.; LOWE, R.; MALESZKA, J.; CONN, D.; MALESZKA, R.; HURD, P.J.J.G.r. Phenotypically distinct female castes in honey bees are defined by alternative chromatin states during larval development. 2018, *28*, 1532-1542.

12. KUCHARSKI, R.; MALESZKA, J.; FORET, S.; MALESZKA, R. Nutritional control of reproductive status in honeybees via DNA methylation. *Science* 2008, *319*, 1827-1830.

13. FORET, S.; KUCHARSKI, R.; PELLEGRINI, M.; FENG, S.; JACOBSEN, S.E.; ROBINSON, G.E.; MALESZKA, R. DNA methylation dynamics, metabolic fluxes, gene splicing, and alternative phenotypes in honey bees. *PNAS* 2012, *109*, 4968-4973.

14. MURRAY, K. The occurrence of iε-N-methyl lysine in histones. *Biochemistry* 1964, *3*, 10-15.

15. BYVOET, P.; SHEPHERD, G.; HARDIN, J.; NOLAND, B. The distribution and turnover of labeled methyl groups in histone fractions of cultured mammalian cells. *Arch. Biochem. Biophys.* 1972, *148*, 558-567.

16. KOUZARIDES, T. Chromatin modifications and their function. *Cell* 2007, *128*, 693-705.

17. EISSENBERG, J.C.; SHILATIFARD, A. Histone H3 lysine 4 (H3K4) methylation in development and differentiation. *Dev. Biol.* 2010, *339*, 240-249.

18. GREENBERG, R.A. Histone tails: Directing the chromatin response to DNA damage. *FEBS Lett.* 2011, *585*, 2883-2890.

19. NOTTKE, A.; COLAIÁCOVO, M.P.; SHI, Y. Developmental roles of the histone lysine demethylases. 2009.

20. PEDERSEN, M.T.; HELIN, K. Histone demethylases in development and disease. *Trends Cell Biol.* 2010, *20*, 662-671.

21. SARG, B.; KOUTZAMANI, E.; HELLIGER, W.; RUNDQUIST, I.; LINDNER, H.H. Postsynthetic trimethylation of histone H4 at lysine 20 in mammalian tissues is associated with aging. *J. Biol. Chem.* 2002, *277*, 39195-39201.

22. MAURES, T.J.; GREER, E.L.; HAUSWIRTH, A.G.; BRUNET, A. The H3K27 demethylase UTX-1 regulates *C. elegans* lifespan in a germline-independent, insulin-dependent manner. *Aging Cell* 2011, *10*, 980-990.

23. SIEBOLD, A.P.; BANERJEE, R.; TIE, F.; KISS, D.L.; MOSKOWITZ, J.; HARTE, P.J. Polycomb Repressive Complex 2 and Trithorax modulate Drosophila longevity and stress resistance. *PNAS* 2010, *107*, 169-174.

24. GUTTMAN, M.; AMIT, I.; GARBER, M.; FRENCH, C.; LIN, M.F.; FELDSER, D.; HUARTE, M.; ZUK, O.; CAREY, B.W.; CASSADY, J.P. Chromatin signature reveals over a thousand highly conserved large non-coding RNAs in mammals. *Nature* 2009, *458*, 223-227.

25. GREER, E.L.; MAURES, T.J.; UCAR, D.; HAUSWIRTH, A.G.; MANCINI, E.; LIM, J.P.; BENAYOUN, B.A.; SHI, Y.; BRUNET, A. Transgenerational epigenetic inheritance of longevity in Caenorhabditis elegans. *Nature* 2011, *479*, 365-371.

26. BARCHUK, A.R.; CRISTINO, A.S.; KUCHARSKI, R.; COSTA, L.F.; SIMÕES, Z.L.; MALESZKA, R. Molecular determinants of caste differentiation in the highly eusocial honeybee *Apis mellifera*. *BMC developmental biology* 2007, *7*, 1-19.

27. WOJCIECHOWSKI, M.; LOWE, R.; MALESZKA, J.; CONN, D.; MALESZKA, R.; HURD, P.J. Phenotypically distinct female castes in honey bees are defined by alternative chromatin states during larval development. *Genome Res.*

2018, *28*, 1532-1542.

28. WADDINGTON, C.; KACSER, H. The strategy of the genes. A discussion of some aspects of theoretical biology. With an appendix by H. Kacser. Strateg. genes. A Discuss. some Asp. Theor. Biol. With an Append. by H. Kacser 1957.

29. LOCAL, A.; HUANG, H.; ALBUQUERQUE, C.P.; SINGH, N.; LEE, A.Y.; WANG, W.; WANG, C.; HSIA, J.E.; SHIAU, A.K.; GE, K. Identification of H3K4me1-associated proteins at mammalian enhancers. *Nat. Genet.* 2018, *50*, 73-82.

30. SIMOLA, D.F.; YE, C.; MUTTI, N.S.; DOLEZAL, K.; BONASIO, R.; LIEBIG, J.; REINBERG, D.; BERGER, S.L. A chromatin link to caste identity in the carpenter ant Camponotus floridanus. *Genome Res.* 2013, *23*, 486-496.

31. VAN DIJK, K.; DING, Y.; MALKARAM, S.; RIETHOVEN, J.-J.M.; LIU, R.; YANG, J.; LACZKO, P.; CHEN, H.; XIA, Y.; LADUNGA, I. Dynamic changes in genome-wide histone H3 lysine 4 methylation patterns in response to dehydration stress in Arabidopsis thaliana. *BMC Plant Biol.* 2010, *10*, 1-12.

32. ZHANG, Y.; HE, X.J.; BARRON, A.; LI, Z.; JIN, M.J.; WANG, Z.; HUANG, Q.; ZHANG, L.Z.; WU, X.; YAN, W.Y. The diverging epigenomic landscapes of honeybee queens and workers revealed by multiomic sequencing. 2022.

33. BOMTORIN, A.D.; MACKERT, A.; ROSA, G.C.C.; MODA, L.M.; MARTINS, J.R.; BITONDI, M.M.G.; HARTFELDER, K.; SIMÕES, Z.L.P. Juvenile hormone biosynthesis gene expression in the corpora allata of honey bee (*Apis mellifera L.*) female castes. *PLoS One* 2014, *9*, e86923.

34. ZHANG, W.; WANG, L.; ZHAO, Y.; WANG, Y.; CHEN, C.; HU, Y.; ZHU, Y.; SUN, H.; CHENG, Y.; SUN, Q. Single-cell transcriptomic analysis of honeybee brains identifies vitellogenin as caste differentiation-related factor. *Iscience* 2022, *25*, 104643.

35. LI, W.; HUANG, Z.Y.; LIU, F.; LI, Z.; YAN, L.; ZHANG, S.; CHEN, S.; ZHONG, B.; SU, S. Molecular cloning and characterization of juvenile hormone acid methyltransferase in the honey bee, Apis mellifera, and its differential expression during caste differentiation. *PLoS One* 2013, *8*, e68544.

36. MARTINS, J.R.; NUNES, F.M.; CRISTINO, A.S.; SIMÕES, Z.L.; BITONDI, M.M. The four hexamerin genes in the honey bee: structure, molecular evolution and function deduced from expression patterns in queens, workers and drones. *BMC Mol. Biol.* 2010, *11*, 1-20.

37. LAGO, D.C.; HUMANN, F.C.; BARCHUK, A.R.; ABRAHAM, K.J.; HARTFELDER, K. Differential gene expression underlying ovarian phenotype determination in honey bee, Apis mellifera L., caste development. *Insect Biochem. Mol. Biol.* 2016, *79*, 1-12.

38. WANG, M.; XIAO, Y.; LI, Y.; WANG, X.; QI, S.; WANG, Y.; ZHAO, L.; WANG, K.; PENG, W.; LUO, G.-Z. RNA m6A modification functions in larval development and caste differentiation in honeybee (Apis mellifera). *Cell Reports* 2021, *34*, 108580.

39. MUTTI, N.S.; DOLEZAL, A.G.; WOLSCHIN, F.; MUTTI, J.S.; GILL, K.S.; AMDAM, G.V. IRS and TOR nutrient-signaling pathways act via juvenile hormone to influence honey bee caste fate. *J. Exp. Biol.* 2011, *214*, 3977-3984.

40. INGHAM, P.; WHITTLE, R. Trithorax: a new homoeotic mutation of Drosophila melanogaster causing transformations of abdominal and thoracic imaginal segments. *Molecular and General Genetics MGG* 1980, *179*, 607-614.

41. BYRD, K.N.; SHEARN, A. ASH1, a Drosophila trithorax group protein, is required for methylation of lysine 4 residues on histone H3. *PANS* 2003, *100*, 11535-11540.

42. ANCELIN, K.; SYX, L.; BORENSZTEIN, M.; RANISAVLJEVIC, N.; VASSILEV, I.; BRISENO-ROA, L.; LIU, T.; METZGER, E.; SERVANT, N.; BARILLOT, E. Maternal LSD1/KDM1A is an essential regulator of chromatin and transcription landscapes during zygotic genome activation. *Elife* 2016, *5*, e08851.

43. ANDREU-VIEYRA, C.V.; CHEN, R.; AGNO, J.E.; GLASER, S.; ANASTASSIADIS, K.; STEWART, A.F.; MATZUK, M.M. MLL2 is required in oocytes for bulk histone 3 lysine 4 trimethylation and transcriptional silencing. *PLoS Biol.* 2010, *8*, e1000453.

44. DODGE, J.E.; KANG, Y.-K.; BEPPU, H.; LEI, H.; LI, E. Histone H3-K9 methyltransferase ESET is essential for early development. *Mol. Cell. Biol.* 2004, *24*, 2478-2486.

45. JAMBHEKAR, A.; DHALL, A.; SHI, Y. Roles and regulation of histone methylation in animal development. *Nat. Rev. Mol. Cell Biol.* 2019, *20*, 625-641.

46. ARNEY, K.L.; BAO, S.; BANNISTER, A.J.; KOUZARIDES, T.; SURANI, M.A. Histone methylation defines epigenetic asymmetry in the mouse zygote. *Int. J. Dev. Biol.* 2002, *46*, 317-320.

47. SUZUKI, M.M.; BIRD, A. DNA methylation landscapes: provocative insights from epigenomics. *Nat. Rev. Genet.* 2008, *9*, 465-476.

48. GLASTAD, K.M.; HUNT, B.G.; GOODISMAN, M.A. Epigenetics in insects: genome regulation and the generation of phenotypic diversity. *Annu. Rev. Entomol* 2019, *64*, e203.

49. SHARIFI-ZARCHI, A.; GEROVSKA, D.; ADACHI, K.; TOTONCHI, M.; PEZESHK, H.; TAFT, R.J.; SCHÖLER, H.R.; CHITSAZ, H.; SADEGHI, M.; BAHARVAND, H. DNA methylation regulates discrimination of enhancers from promoters through a H3K4me1-H3K4me3 seesaw mechanism. *BMC Genomics* 2017, *18*, 1-21.

50. HE, X.J.; BARRON, A.B.; YANG, L.; CHEN, H.; HE, Y.Z.; ZHANG, L.Z.; HUANG, Q.; WANG, Z.L.; WU, X.B.; YAN, W.Y. Extent and complexity of RNA processing in honey bee queen and worker caste development. *Iscience* 2022, *25*, 104301.

51. MARTINS, J.R.; NUNES, F.M.F.; SIMÕES, Z.L.P.; BITONDI, M.M.G. A honeybee storage protein gene, hex 70a, expressed in developing gonads and nutritionally regulated in adult fat body. *J. Insect Physiol.* 2008, *54*, 867-877.

52. MARTINS, J.R.; ANHEZINI, L.; DALLACQUA, R.P.; SIMOES, Z.L.; BITONDI, M.M. A honey bee hexamerin, HEX 70a, is likely to play an intranuclear role in developing and mature ovarioles and testioles. *PLoS ONE* 2011, *6*, e29006.

53. HERZ, H.-M.; MOHAN, M.; GARRUSS, A.S.; LIANG, K.; TAKAHASHI, Y.-H.; MICKEY, K.; VOETS, O.; VERRIJZER, C.P.; SHILATIFARD, A. Enhancer-associated H3K4 monomethylation by Trithorax-related, the Drosophila homolog of mammalian Mll3/Mll4. *Genes Dev.* 2012, *26*, 2604-2620.

54. DALVAI, M.; BELLUCCI, L.; FLEURY, L.; LAVIGNE, A.; MOUTAHIR, F.; BYSTRICKY, K. H2A. Z-dependent crosstalk between enhancer and promoter regulates Cyclin D1 expression. *Oncogene* 2013, *32*, 4243-4251.

55. DALVAI, M.; FLEURY, L.; BELLUCCI, L.; KOCANOVA, S.; BYSTRICKY, K. TIP48/Reptin and H2A. Z requirement for initiating chromatin remodeling in estrogen-activated transcription. *PLoS Genet.* 2013, *9*, e1003387.

56. MORA, A.; SANDVE, G.K.; GABRIELSEN, O.S.; ESKELAND, R. In the loop: promoter-enhancer interactions and bioinformatics. *Brief. Bioinform.* 2016, *17*, 980-995.

57. COSTA, R.M.; SOTO, X.; CHEN, Y.; ZORN, A.M.; AMAYA, E. spib is required for primitive myeloid development in Xenopus. *Blood, The Journal of the American Society of Hematology* 2008, *112*, 2287-2296.

58. WILLIS, S.N.; TELLIER, J.; LIAO, Y.; TREZISE, S.; LIGHT, A.; O'DONNELL, K.; GARRETT-SINHA, L.A.; SHI, W.; TARLINTON, D.M.; NUTT, S.L. Environmental sensing by mature B cells is controlled by the transcription factors PU. 1 and SpiB. *Nat. Commun.* 2017, *8*, 1-14.

59. LUGUS, J.J.; CHUNG, Y.S.; MILLS, J.C.; KIM, S.-I.; GRASS, J.A.; KYBA, M.; DOHERTY, J.M.; BRESNICK, E.H.; CHOI, K. GATA2 functions at multiple steps in hemangioblast development and differentiation. *Development* 2007.

60. KAZENWADEL, J.; BETTERMAN, K.L.; CHONG, C.-E.; STOKES, P.H.; LEE, Y.K.; SECKER, G.A.; AGALAROV, Y.; DEMIR, C.S.; LAWRENCE, D.M.; SUTTON, D.L. GATA2 is required for lymphatic vessel valve development and maintenance. *The Journal of Clinical Investigation* 2015, *125*, 2979-2994.

61. PAN, Q.-Z.; WU, X.-B.; GUAN, C.; ZENG, Z.-J. A new method of queen rearing without grafting larvae. *Am. Bee J.* 2013, *153*, 1279-1280.

62. ZHANG, Y.; LI, Z.; WANG, Z.-L.; ZHANG, L.-Z.; ZENG, Z.-J. A Comparison of RNA Interference via Injection and Feeding in Honey Bees. *Insects* 2022, *13*, 928.

# 优质蜂王培育及其分子机制

# High Quality Queen Bee Breeding and its Molecular Mechanism

蜂王是蜂群中的核心。蜂王的质量优劣直接关系到蜂群群势的强弱以及产量的高低，而且可能与"蜂群衰竭失调病"（colony collapse disorder，CCD）相关。显然育王技术也是养蜂生产中的一个关键技术。进一步解析蜂王发育机理，可为培育优质蜂王提供理论指导和技术支撑，有重要的理论和实际意义，同时也为CCD防控提供新思路。我们最新研究发现：蜂王在王台中产下比工蜂巢房中更大的受精卵，并且由王台中受精卵发育的蜂王，个体大，卵巢管数多，质量好，大量与寿命、繁殖力和免疫相关的基因出现显著变化。这是首次发现母体效应对优质蜂王培育有重要意义，同时突破了蜜蜂级型分化仅受食物和发育空间影响的学术观点。

利用工蜂巢房中的受精卵、1、2和3日龄幼虫分别培育蜂王，比较分析各组蜂王的形态、基因表达与DNA甲基化差异。结果表明：随着移虫日龄的增加，其蜂王和3日龄幼虫的体重逐步下降，其与卵发育组蜂王幼虫的差异基因数量逐步增加，也逐步提高了其蜂王幼虫的全基因组甲基化水平。

移卵、1日龄幼虫、2日龄幼虫和3日龄幼虫进行人工培育蜂王，测定与蜂王发育相关的4个指标（封盖王台长度、王台中王浆剩余量、蜂王初生重、蜂王卵巢管数），同时分别利用全基因组

Bisulfite 技术和 Illumina 高通量测序技术对培育的蜂王进行甲基化测序和转录组测序。结果表明：蜂王质量随着移虫日龄的增加而降低；随着移虫日龄的增加，移虫培育的蜂王与移卵培育的蜂王之间的差异甲基化（differentially methylated genes，DMGs）基因数量逐步增多，移虫培育的蜂王与移卵培育的蜂王之间的差异表达（Differentially expressed genes，DEGs）基因数量逐步增多。

移卵、1 日龄幼虫和 2 日龄幼虫进行连续 4 代的人工培育蜂王，利用全基因组 Bisulfite 甲基化测序技术，对 4 代人工培育的蜂王进行全基因组 DNA 甲基化测序。结果表明：以卵培育的连续 4 代蜂王卵巢管数最多，质量最高，而移虫育王的蜂王质量逐代下降。差异的 DNA 甲基化基因也在移虫育王组中逐代增加，而在卵培育的蜂王组无明显变化，这是首次发现因营养水平匮乏而引起的蜜蜂 DNA 甲基化具有累代效应。

以单雄受精蜂王（G0 代）产的受精卵、1 日龄工蜂幼虫和 2 日龄幼虫分别培育蜂王（G1 代），并且分别产卵培育雄蜂（G2 代）。利用全基因组 Bisulfite 甲基化测序技术，对 G0 代蜂王、G1 代蜂王、G2 代雄蜂进行全基因组 DNA 甲基化测序。结果表明：原有高甲基化位点和突变高甲基化位点均存在表观遗传印迹现象；同时发现随着移虫日龄增加，原有高甲基化位点和突变高甲基化位点遗传印迹数量都是逐步增加。这是在无脊椎动物中，首次发现 DNA 甲基化突变遗传印迹。

利用全基因组 DNA 甲基化测序技术检测了 QE、WE 与 2L 培育的刚羽化蜂王甲基化，结果表明：QE、WE 与 2L 刚羽化蜂王整体甲基化水平显著不差异。三组间均存在大量的差异甲基化基因（DMGs），QE 蜂王与 2L 蜂王对比组差异甲基化基因数为 614 个数量最多；其次是 WE 蜂王与 2L 蜂王对比组差异甲基化基因 473 个，QE 蜂王与 WE 蜂王对比组差异甲基化基因 371 个。42 个 DMGs 存在于 mTOR、MAPK、Wnt、Notch、Hedgehog、FoxO 和 Hippo 信号通路中，这些信号通路参与调控级型分化、繁殖和寿命。证明蜜蜂母体效应会引起表观遗传差异，进而影响蜂王的级型分化。

通过转录组测序发现：刚羽化出房蜂王和性成熟蜂王有 1340 个基因表达显著差异，其中有 667 个基因上调，673 个基因下调。利用不同直径王台（9.4mm、9.6mm、9.8mm 与 10.0mm）培育蜂王，结果表明：随着王台直径增加，蜂王初生重和卵巢管数等指标均显著上升，蜂王腹部卵黄蛋白原基因（$Vg$）表达量也逐步上升。通过控制囚王时间（0 d、2 d 与 4 d）来培育蜂王，结果表明：囚禁蜂王时间越长，其产下卵的卵重、卵长与卵宽均显著上升，培育的蜂王初生重和卵巢管数等指标均显著上升，蜂王腹部的 $Vg$ 表达量也显著提高。

通过人工培养方式，培养中意蜂营养杂交蜂王，并利用全转录组测序技术分析了非编码 RNA 在该过程中的作用。发现了 8 个关键的 lncRNA 和 3 个 miRNA 协同参与调控 7 个参与黑色素合成的关键基因，从而改变营养杂交蜂王体色，揭示了蜜蜂跨物种表观遗传调控的分子机制。

# 1. A Maternal Effect on Queen Production in Honey Bees

Hao Wei[1†], Xujiang He[1†], Chunhua Liao[1†], Xiaobo Wu[1], Wujun Jiang[1], Bo Zhang[1], Linbin Zhou[1], Lizhen Zhang[1], Andrew B. Barron[2], Zhijiang Zeng[1,3*]

1. Honeybee Research Institute, Jiangxi Agricultural University, Nanchang, Jiangxi 330045, P. R. of China
2. Department of Biological Sciences, Macquarie University, North Ryde, NSW 2109, Australia

**Abstract**

Influences from the mother on offspring phenotype, known as maternal effects, are an important cause of adaptive phenotypic plasticity [1,2]. Eusocial insects show dramatic phenotypic plasticity with morphologically distinct reproductive (queen) and worker castes [3,4]. The dominant paradigm for honey bees (*Apis mellifera*) is that castes are environmentally, rather than genetically determined, with the environment and diet of young larvae causing caste differentiation [5-9]. A role for maternal effects has not been considered, but here we show that egg size influences queen development also. Queens laid significantly bigger eggs in the larger queen cells than in the worker cells. Eggs laid in queen cells (QE), laid in worker cells (WE), and 2-day old larvae from worker cells (2L) were transferred to artificial queen cells to be reared as queens in a standardized environment. Newly emerged adult queens from QE were heavier than the other two groups and had more ovarioles, indicating a consequence of egg size for adult queen morphology. Gene expression analyses identified several significantly differentially expressed genes between newly emerged queens from QE and the other groups. These included a disproportionate number of genes involved in hormonal signalling, body development and immune pathways, which are key traits differing between queens and workers. That egg size influences emerging queen morphology and physiology and that queens lay larger eggs in queen cells demonstrates both a maternal effect on the expression of the queen phenotype, and a more active role for the queen in gyne production than has been realized previously.

**Results and Discussion**

Honey bee queens and workers are radically different phenotypes. While both are female and develop from fertilised eggs, queens are typically the sole female reproductive in the colony [10,11]. No genetic difference separates queens and workers: instead the differentiation is controlled epigenetically [5-7]. Thus far attention has focused on the role of the larval developmental environment in the differentiation of workers and queens.

---

† These three authors contributed equally to this paper.
*Corresponding author: bees1965@sina.com (ZJZ).
注：此文发表在 *Current Biology* 2019 年 13 期。

Workers and queens develop in wax cells of different sizes and are fed different diets. Both diet and the amount of space available to developing larvae cause changes in methylation of the larval genome [5,6,12]. The resulting differences in gene regulation (particularly involving signal transduction, gland development, carbohydrate metabolism and immune function pathways [13-15]) establish the divergent queen and worker developmental paths [16]. Here we examined whether the queen herself might influence caste development via maternal effects.

Maternal effects are a causal influence of the maternal genotype or phenotype on the offspring phenotype [1,2], and are an important mechanism of adaptive phenotypic plasticity [1]. Vertebrate examples have shown females can adaptively vary investment in eggs according to the perceived quality of their mate in order to invest more in young of higher quality males [17-19]. Insects can also adjust their investment in their eggs [20,21], or even egg colouration [22], to better adapt offspring to their environment. Flanders in 1945 [23] proposed that maternal effects could influence caste development in social insects via differential investment in eggs, but surprisingly there are very few reports of maternal effects from the hymenoptera.

Passera [24] reported that queens the ant *Pheidole pallidula* tended to lay larger eggs at the time of year colonies raised a generation that included sexuals, and Schwander et al [25] reported a maternal effect on female caste determination in Pogonomyrmex ants. A suggestion of a possible maternal effect on queen production in honey bees came from Borodacheva [26] in 1973 with the observation that some of the variation in the size of adult queens could be attributed to variation in egg size. Honey bee queens lay between 1500-2000 eggs a day [27] in small worker cells, that develop as the next generation of workers. When a colony is ready to reproduce by swarming a few (10-20) larger queen cells are constructed [28]. Eggs laid in these are fed more and richer food and develop as queens [10]. Here we tested whether queens lay larger eggs in queen cells.

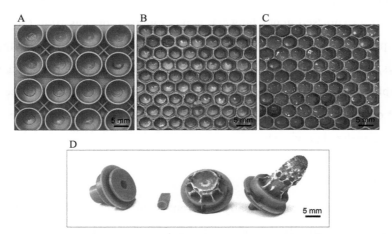

**Figure II-1-1** To sample QE and WE, arrays of standardised plastic cells the same size and shape as queen cells (A) or worker cells (B) were placed in colonies. Queens and attendant workers were restricted to these arrays for 6h to lay. After that time the arrays were removed. To sample 2L, arrays remained in the colony for five days by which time eggs hatched and 2-day-old larvae occupied each cell (C). The base of each plastic cell was removable, which allowed easy transfer of either eggs or larvae to new artificial queen cells (D). Queen cells containing QE, WE or 2L were arranged randomly on a common rack and inserted into a queenless colony where the workers fed and raised each as a queen (D).

*Honey bee queens lay larger eggs in queen cells*

To test for an effect of the queen-laid egg on caste development we provided queens with artificial standardised plastic cells that were the size and shape of either worker cells or queen cells (Fig. II-1-1). After six hours eggs laid in the two cell types were collected and weighed. This study was repeated across 3 colonies, in total 152 eggs were measured. Eggs laid in queen cells (QE) were 13.26 % heavier (157.51 ± 12.37 vs 138.93 ± 10.90, Mean ± SD, μg) and 2.43 % longer (1.56 ± 0.04 vs 1.52 ± 0.05, Mean ± SD, mm) and 4.18 % thicker (0.374 ± 0.010 vs 0.359 ± 0.013, Mean ± SD, mm) than eggs laid in worker cells (WE) (Fig. II-1-2).

*Adult queens from QE are heavier than queens from WE and 2L*

To determine whether this difference in egg size had any consequence for adult queen morphology, six hours after laying QE and WE were transferred by moving the base of each plastic cell (so the egg was not touched, Fig. II-1-1) into artificial queen cells. Some WE remained in worker cells for 5 days until the larvae were 2 days old (2L). The larvae were then similarly transferred to artificial queen cells. All queen cells were placed in a common queenless colony to be reared as queens by workers. Sixteen days later adult queens were collected on emergence from the sealed cells (Fig. II-1-1D) and weighed. This study was replicated using five colonies across two years (Fig. II-1-3). Adult queens from QE were heaviest in all five colonies, and queens from QE were significantly heavier than queens from WE (258.65 ± 22.82 vs 234.50 ± 36.00, Mean ± SD, mg) in three colonies out of five (Fig. II-1-3).

*Adult queens from QE had the greatest number of ovarioles*

The number of ovarioles is an important index of queen fecundity [29]. Our haematoxylin-eosin staining (HE) results (Fig. II-1-3C) showed that five day old adult queens from QE had the greatest number of ovarioles in the right ovary, significantly more than queens from 2L (165.50 ± 10.65 vs 145.90 ± 14.89, Mean ± SD) in all three colonies, but there was no significant difference between queens from QE and WE (165.50 ± 10.65 vs 160.00 ± 9.48, Mean ± SD).

*Differences in gene expression among queens from QE, WE and 2L*

To further examine the consequences of egg source on adult queen phenotype we compared the gene expression profiles of newly emerged adult queens from QE, WE and 2L using RNA-Seq. The heads and thoraces of two newly emerged queens from each group were collected, and pooled for RNA-Seq. This experiment was repeated twice using two colonies and both repeats were considered together in our analyses of gene expression differences in 2016. This RNA-Seq experiment was repeated again in 2018 with two extra colonies using same methods. Methods for sample preparation, mRNA isolation, and sequencing followed those of He *et al* [14].

A small number of differentially expressed genes (DEGs) were detected in comparisons between groups from both 2016 and 2018 RNA-Seq results (Fig. II-1-4, see page 513, Table S1 and S2). Of the 121 DEGs identified across all comparisons in 2016 RNA-Seq experiment, 6 genes with a high expression level were selected and gene expression differences assessed with qRT-PCR (following methods in [14]) to affirm the results of our RNA-Seq analyses (Fig. S2).

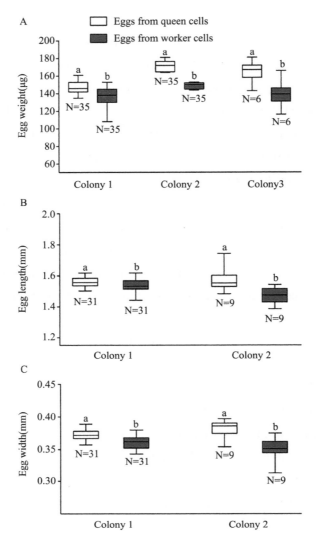

**Figure Ⅱ-1-2** QE and WE differed in size and weight. For experiment 1, egg weights were recorded from 3 different colony replicates (A). For experiment 2 we recorded egg size (B, C) from two of the colony replicates, but the third colony failed and could no longer be used. Eggs from these colonies were used for queen rearing (Fig. Ⅱ-1-3) also. For all panels boxplots show median, quartiles and range. Sample size for each group marked below boxes. Data from three colonies were combined for analysis. The egg weight, length and width of QE and WE were compared using ANOVA tests followed with Fisher's PLSD test. The critical $p$ value was adjusted to 0.0167 according to the Bonferroni correction. Sample sizes shown. Groups that did not differ ($p > 0.0167$) are marked with the same superscript.

Two years' RNA-Seq results both showed that the greatest differences were detected in comparisons between queens reared from QE and 2L, followed by WE/2L and QE/WE comparisons (Fig. Ⅱ-1-4, Table S1 and S2). This is of interest because raising queens from 2L rather than WE has already been shown to have a significant impact on queen reproductive development and morphology [14]. Of the DEGs identified in the QE/WE comparison, 31 (2016, Fig. Ⅱ-1-4A) and 19 (2018, Fig. Ⅱ-1-4B) have been documented previously in comparisons of queen and worker honey bees, or queen honey bees varying in caste development or reproductive condition (Fig. Ⅱ-1-4, Table S1 and S2). Besides, 59 of 191 DEGs from three comparisons of the 2016 RNA-Seq results were also identified in the 2018 RNA-Seq results (Table S1).

This suggests the gene expression differences between adult queen from QE and WE are reflective of variation in the caste development process. Our DEGs contained a disproportionately large number of genes such as *juvenile hormone methyltransferase, abaecin* and *hexamerin* genes involved in hormone synthesis, ovary development, cuticle development and immune functions (Fig. II-1-4, Table S1 and S2) [14, 30-33].

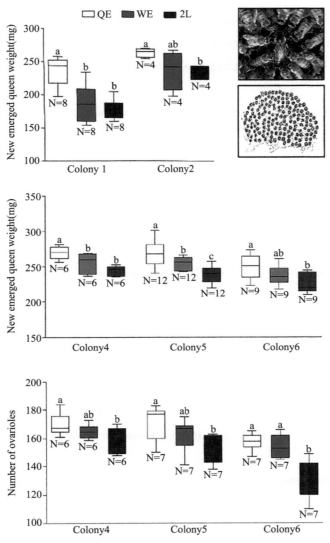

**Figure II-1-3** Queens collected on emergence from queen cells from QE, WE and 2L differed in weight. A: data from two colonies measured in 2016; B: data from three colonies repeated in 2018. Newly emerged queen shown in inset. C: measurements of queen ovaries from QE, WE and 2L. The right ovary of each queen was used (cross section shown in inset). For all panels boxplots show median, quartiles and range. Sample size for each group marked below boxes. Data from each group were compared with ANOVA test followed by Fishers PLSD test. Groups that did not differ ($p > 0.05$) are marked with the same superscript.

## Conclusions

In summary our data demonstrate a maternal effect on honey bee queen size and physiology, which is caused by queens laying larger eggs in queen cells compared to worker cells. This could have significant consequences for colony function since various authors [14, 26, 29, 34-36] have reported a relationship between queen weight and queen ovariole number and fecundity. Bilash [34] has even reported an influence of queen weight on colony honey production.

We do not here propose there is a special class of queen-destined eggs. The distribution of egg masses sampled from queen and worker cells was continuous, normal and unimodal. Rather we propose that fecund queens at any one time have more than one egg ready for laying [10], and that queens may lay the largest available egg in queen cells. Alternatively, queens may pause oviposition prior to laying in queen cells, since delaying oviposition causes bigger eggs with more yolk protein [37], but this possibility needs to be investigated.

An important inference of our data, however, is that queens can actively select larger fertilized eggs for oviposition in queen cells. It has been demonstrated previously that queens can control and withhold fertilization of eggs prior to laying in male (haploid) drone cells, and that queens measure the larger drone cell with their foreleg prior to laying [38]. This is the first evidence that queens can select among fertilized eggs and that they differentiate between queen and worker cells.

We feel the differences observed between QE and WE are attributable to the queen, and not to interactions with workers for the following reasons. In honeybee colony, worker-laid unfertilized eggs in worker cells will be removed by worker policing [39], however the queen-laid eggs we studied here have queen egg-marking pheromones and usually avoid worker policing [40,41]. Workers are not expected to police these eggs. Even so, workers will sometimes consume queen-laid eggs if a colony is stressed [42]. However, there has been no report of selective queen egg removal based on size. There was no evidence of selective egg destruction by workers in our study. It is believed that the QE with more nutrition possibly results in better queens compared to WE, which is determined by the maternal effect.

Our data indicate the *in ovo* environment influences queen development (Fig. II-1-3), but the *in ovo* environment is not necessary for queen formation. The queen developmental pathway proves to be quite robust [14] and queens can be reared from eggs or even larvae transplanted from worker cells [14]. Indeed, this capacity underlies commercial queen-rearing practices. Not all queens are the same quality, however, and queens reared from transplanted worker larvae are smaller and have less well developed reproductive systems [14, 29, 43]. Rangel *et al.* also reported that rearing queens from older worker larvae results in significantly lower production of worker comb, drone comb and stored food compared to those by eggs [43]. Here we provide the first evidence that the *in ovo* environment also influences adult queen morphology and physiology. It feels remarkable for a social insect as intensively studied as the honey bee that the possibility of maternal effects on caste has been overlooked until now. It has perhaps been assumed that the enormous difference in food provision to developing worker and queen larvae must swamp any differential provisioning during the egg. We now recognise, however, that the epigenetically differentiated worker and queen developmental pathways are sensitive to the early larval environment [14], and our data indicate a sensitivity to the *in ovo* environment also. This adds a new perspective on colony function and indicates the queen has a more active role in the production of the next generation of queens than has been previously recognised. It will be important to assess whether similar maternal effects are at play in other social eusocial insects also.

STAR Methods

KEY RESOURCES TABLE

CONTACT FOR REAGENT AND RESOURCE SHARING

EXPERIMENTAL MODEL AND SUBJECT DETAILS

METHOD DETAILS

    Egg collecting and queen rearing methods

    Histolological analyses of queen ovarioles

    RNA-Seq analysis

    Real-time PCR validation

QUANTIFICATION AND STATISTICAL ANALYSIS

    Data access

## Supplemental Information

Supplemental Information can be found online at t https://doi.org/10.1016/j.cub.2019.05.059.

## Acknowledgements

We thank Professor Gene E. Robinson for helpful suggestions for the manuscript. This work was supported by the National Natural Science Foundation of China (31572469, 31872432) and the Earmarked Fund for China Agriculture Research System (CARS-44-KXJ15).

## Author Contributions

H. W., X. J. H., C. H. L., and W. J. J. conducted all experiments. Z. J. Z. designed experiments. X. J. H., A. B. B. and Z. J. Z. wrote the paper. X. B. W., B. Z., L. B. Z. and L. Z. Z. participated in experiments.

## Declaration of Interests

All authors declare no competing interests.

## References

1. MOUSSEAU, T.A., and FOX, C.W. (1998). The adaptive significance of maternal effects. *Trends Ecol. Evol. 13*, 403-407.

2. WOLF, J.B., and WADE, M.J. (2009). What are maternal effects (and what are they not)? *Philosophical Transactions of the Royal Society 364*, 1107-1115.

3. OSTER, G.F., and WILSON, E.O. (1978). Caste and ecology in the social insects. Princeton: Princeton University Press.

4. SEELEY, T.D. (1985). Honeybee Ecology. Princeton: Princeton University Press.

5. KUCHARSKI, R., MALESZKA, J., FORET, S., AND MALESZKA, R. (2008). Nutritional control of reproductive status in honeybees via DNA methylation. *Science 319*, 1827-1830.

6. LYKO, F., FORET, S., KUCHARSKI, R., WOLF, S., FALCKENHAYN, C., and MALESZKA, R. (2010). The honey

bee epigenomes: differential methylation of brain DNA in queens and workers. *PLoS Biol 8*.

7. MALESZKA, R. (2014). Epigenetic integration of environmental and genomic signals in honey bees: the critical interplay of nutritional, brain and reproductive networks. *Epigenetics 3*, 188-192.

8. LINKSVAYER, T.A., KAFTANOGLU, O., AKYOL, E., BLATCH, S., AMDAM, G.V., and PAGE JR, R.E. (2011). Larval and nurse worker control of developmental plasticity and the evolution of honey bee queen-worker dimorphism. *J. Evol. Biol. 9*, 1939-1948.

9. LEIMAR, O., HARTFELDER, K., LAUBICHLER, M.D., and PAGE, R.J. (2012). Development and evolution of caste dimorphism in honeybees - a modeling approach. *Ecol. Evol. 2*, 3098-3109.

10. WINSTON, M.L. (1987). The Biology of the Honey Bee. Cambridge: Harvard University Press.

11. SEELEY, T.D. (1995). The Wisdom of the Hive. Cambridge: Harvard University Press.

12. SHI, Y.Y., HUANG, Z.Y., ZENG, Z.J., WANG, Z.L., WU, X.B., and YAN, W.Y. (2011). Diet and cell size both affect queen-worker differentiation through DNA methylation in honey bees (*Apis mellifera*, Apidae). *PLoS One 6*, e18808.

13. WOODARD, S.H., FISCHMAN, B.J., VENKAT, A., HUDSON, M.E., VARALA, K., CAMERON, S.A., CLARK, A.G., and ROBINSON, G.E. (2011). Genes involved in convergent evolution of eusociality in bees. *Proceedings of the National Academy of Science USA 108*, 7472-7477.

14. HE, X.J., ZHOU, L.B., PAN, Q.Z., BARRON, A.B., YAN, W.Y., and ZENG, Z.J. (2017). Making a queen: an epigenetic analysis of the robustness of the honey bee (*Apis mellifera*) queen developmental pathway. *Mol. Ecol. 26*, 1598-1607.

15. BLOCH, G., and GROZINGER, C.M. (2011). Social molecular pathways and the evolution of bee societies. *Phil Trans Roy. Soc. B 366*, 2155-2170.

16. SIMOLA, D.F., WISSLER, L., DONAHUE, G., WATERHOUSE, R.M., HELMKAMPF, M., ROUX, J., NYGAARD, S., GLASTAD, K.M., HAGEN, D.E., VILJAKAINEN, L., et al. (2013). Social insect genomes exhibit dramatic evolution in gene composition and regulation while preserving regulatory features linked to sociality. *Genome Res. 23*, 1235-1247.

17. CUNNINGHAM, E.J.A., and RUSSELL, A.F. (2000). Egg investment is influenced by male attractiveness in the mallard. *Nature 404*, 74-77.

18. KOLM, N. (2001). Females produce larger eggs for large males in a paternal mouthbrooding fish. *Proc. Roy. Soc. Lond. B 268*, 2229-2234.

19. TYLER, C.R., POTTINGER, T.G., SANTOS, E., SUMPTER, J.P., PRICE, S.A., BROOKS, S., and NAGLER, J.J. (1996). Mechanisms controlling egg size and number in the rainbow trout, *Oncorhynchus mykiss*. *Biol. Reprod. 54*, 8-15.

20. KAWECKI, T.J. (2013). Adaptive plasticity of egg size in response to competition in the cowpea weevil, *Callosobruchus maculatus* (Coleoptera: Bruchidae). *Oecologia 102*, 81-85.

21. FISCHER, K., BRAKEFIELD, P.M., and ZWAAN, B.J. (2003). Plasticity in butterful egg size: why larger offspring at lower temperature? *Ecology 84*, 3188-3147.

22. ABRAM, P.K., GUERRA-GRENIER, E., DESPRÉS-EINSPENNER, M.L., ITO, S., WAKAMATSU, K., BOIVIN, G., and BRODEUR, J. (2015). An insect with selective control of egg coloration. *Curr. Biol. 25*, 2007-2011.

23. FLANDERS, S.E. (1945). Is caste differentiation in ants a function of the rate of egg deposition? *Science 101*, 2618-2619.
24. PASSERA, L. (1980). The laying of biased eggs by the ant *Pheidole pallidula* (Nyl,) (Hymenoptera, Formicidae). *Ins. Soc. 27*, 79-95.
25. SCHWANDER, T., HUMBERT, J.Y., BRENT, C.S., CAHAN, S.H., CHAPUIS, L., RENAI, E., and KELLER, L. (2008). Maternal effect on female caste determination in a social insect. *Curr. Biol. 18*, 265-269.
26. BORODACHEVA, V.T. (1973). Weight of eggs and quality of queens and bees (in Russian). Pchelovodstvo 93, 12-13.
27. WINSTON, M.L. (1992). The honey bee colony: life history. In The Hive and the Honey Bee, J.M. Graham, ed. (Hamilton, Illinois: Dadant & Sons), pp. 73-93.
28. GARY, N.E. (1992). Activites and behavior of honey bees: activities of the queen. In The Hive and the Honey Bee, J.M. Graham, ed. (Hamilton Illinois: Dadant & Sons), pp. 269-361.
29. WOYKE, J. (1971). Correlations between the age at which honeybee brood was grafted, characteristics of the resultant queens, and results of insemination. *J Apic Res 10*, 45-55.
30. CORONA, M., VELARDE, R.A., REMOLINA, S., MORAN-LAUTER, A., WANG, Y., HUGHES, K.A., and ROBINSON, G.E. (2007). Vitellogenin, juvenile hormone, insulin signaling, and queen honey bee longevity. *Proc Nat Acad Sci 104*, 7128-7133.
31. ENGELS, W., and IMPERATRIZ-FONSECA, V.L. (1990). Caste development, reproductive strategies, and control of fertility in honey bees and stingless bees. In Social Insects An Evolutionary Approach to Castes and Reproduction, W. Engels, ed. (Berlin: Springer Verlag), pp. 166-230.
32. WIRTZ, P., and BEETSMA, J. (1972). Induction of caste differentiation in the honeybee (Apis mellifera) by juvenile hormone. *Entomol. Exp. Appl. 15*, 517-520.
33. MARTINS, J.R., NUNES, F.M.F., CRISTINO, A.S., SIMÕES, Z.L.P., and BITONDI, M.M.G. (2010). The four hexamerin genes in the honey bee: structure, molecular evolution and function deduced from expression patterns in queens, workers and drones. *BMC Molecular Biology 11*, 23.
34 BILASH, G.D., BORODACHEVA, V.T., and TIMOSINOVA, A.E. (1983). Quality of artificially reared queen bees. In Proceedings of the XXIXth International Congress of Apiculture. (Bucharest: Apimondia Publishing House), pp. 114-118.
35. HUANG, W.C., and ZHI, C.Y. (1985). The relationship between the weight of the queen honeybee at various stages and the number of ovarioles eggs laid and sealed brood produced (in Japanese). *Honey Bee Science 6*, 113-116.
36. SHAO, R.Y. (1988). Honey Bee Breeding Science., Beijing, China: Chinese Agricultural Press.
37. TORRES, J.J. (1980). A stereological analysis of developing egg chambers in the honeybee queen *Apis mellifera*. *Cell Tissue Res. 208*, 29-33.
38. KOENIGER, N. (1970). Über die fähigkeit der bienenkonigin (*Apis mellifica* L.) zwischen arbeiterinnen-und drohnenzellen zu unterscheiden. *Apidologie 1*, 115-142.
39. RATNIEKS, F.L.W.,and VISSCHER, P.K. (1989). Worker policing in the honeybee. *Nature, 362*, 796-797.
40. RATNIEKS, F.L.W. (1995). Evidence for a queen-produced egg-marking pheromone and its use in worker policing in the honey bee. *Journal of Apicultural Research, 34*, 31-37.

41. KATZAV-GOZANSKY, T., SOROKER, V., IBARRA, F., FRANCKE, W., and HEFETZ, A. (2001). Dufour's gland secretion of the queen honeybee (*Apis mellifera*): an egg discriminator pheromone or a queen signal? *Behavioral Ecology & Sociobiology, 51*, 76-86.

42. SCHMICKL, T., and CRAILSHEIM K. (2001). "Cannibalism and early capping: strategy of honeybee colonies in times of experimental pollen shortages." Journal of comparative physiology. *A, Sensory, Neural, and Behavioral Physiology 187*, 541-547.

43. RANGEL, J. KELLER, J.J. and TARPY, D.R. (2012). The effects of honey bee (*Apis mellifera* L.) queen reproductive potential on colony growth. *Insectes Sociaux 60*, 65-73.

44. PAN, Q.Z., WU, X.B., GUAN, C., and ZENG, Z.J. (2013). A new method of queen rearing without grafting larvae. *Amer. Bee J. 153*, 1279-1280.

45. WOYKE, J. (1998). Size change of *Apis mellifera* eggs during the incubation period. *Journal of Apicultural Research, 37*, 239-246.

46. TABER, S., and ROBERTS, W.C. (1963). Egg weight variability and its inheritance in the honey bee. *Annals of the Entomological Society of America, 56*, 473-476.

47. ZOU C.B., ZHOU L.B., HU J.H., XI F.G., YUAN F., and YAN W.Y. (2016). Effects of queen-rearing without larvae-grafting and two esters of brood pheromone on the queen quality of *Apis cerana cerana*. *Scientia Agricultura Sinica, 49,* 18, 3662-3670. (in Chinese)

48. EWING, B., HILLIER, L., WENDL, M.C., GREEN, P. (1998). Base-calling of automated sequencer traces using phred. I. Accuracy assessment. *Genome Res. 8*, 175-185.

49. KIM, D., PERTEA, G., TRAPNELL, C., PIMENTEL, H., KELLEY, R., and SALZBERG, S.L. (2013). TopHat2: accurate alignment of transcriptomes in the presence of insertions, deletions and gene fusions. *Genome Biol. 14*, R36.

50. TRAPNELL, C., WILLIAMS, B.A., PERTEA, G., MORTAZAVI, A., KWAN, G., VAN BAREN, M.J., SALZBERG, S., WOLD, B.J., and PACHTER, L. (2010). Transcript assembly and quantification by RNA-Seq reveals unannotated transcripts and isoform switching during cell differentiation. *Nat. Biotechnol. 28*, 511-515.

51. LENG, N., DAWSON, J.A., THOMSON, J.A., RUOTTI, V., RISSMAN, A.I., SMITS, B.M.G., HAAG, J.D., GOULD, M.N., STEWART, R.M., and KENDZIORSKI, C. (2013). EBSeq: an empirical Bayes hierarchical model for inference in RNA-seq experiments. *Bioinformatics, 29*, 8, 1035-1043.

52. XIE, C., MAO, X., HUANG, J., DING, Y., WU, J., DONG, S., KONG, L., GAO, G., LI, C.H., WEI, L. (2011). KOBAS 2.0: a web server for annotation and identification of enriched pathways and diseases. *Nucleic Acids Res. 39*, W316-W322.

53. LOURENÇO, A.P., MACKERT, A., DOS SANTOS CRISTINO, A., and SIMÕES, Z.L.P. (2008). Validation of reference genes for gene expression studies in the honey bee, Apis mellifera, by quantitative real-time RT-PCR. *Apidologie 39*, 372-385.

54. LIU. W., and SAINT, D.A. (2002). A new quantitative method of real time reverse transcription polymerase chain reaction assay based on simulation of polymerase chain reaction kinetics. *Anal. Biochem. 1*, 52-59.

## STAR+METHODS
## KEY RESOURCES TABLE

| REAGENT or RESOURCE | SOURCE | IDENTIFIER |
|---|---|---|
| Biological Samples | | |
| Honeybee (*Apis mellifera*) eggs and adult queens | This paper | N/A |
| Chemicals, Peptides, and Recombinant Proteins | | |
| 4% Paraformaldehyde Fix | BBI Life Sciences (China) | Lot#: E205fa0004 |
| Hematoxylin–Eosin | Boster biological Technology (USA) | Lot#: 12K01A80 |
| Neutral Balsam Mounting Medium | BBI Life Sciences (China) | Lot#: E122FD0271 |
| Xylene | Tianjin Damao Reagent Factory (China) | Analytical Reagent |
| Ethanol | Xilong Scientific (China) | Analytical Reagent |
| Paraffin with Ceresin | Sinopharm Chemical Reagent Co., Ltd (China) | Lot#: 69019461 |
| Critical Commercial Assays | | |
| TRlzolTM Reagent kit | Life technologies (USA) | Cat#:15596–026 |
| NEBNext® Poly(A) mRNA Magnetic Isolation Module | NEB (USA) | Cat#:E7490 |
| Oligo (dT) | NEB (USA) | Cat#:E7500 |
| Fragmentation Buffer | NEB (USA) | Cat#:E7530B/E7530L |
| Agencourt AMPure XP | Beckman Coulter (USA) | Cat#:A63880 |
| PrimeScriptTM Reagent Kit | Takara (Japan) | Cat#: RR047A |
| Deposited Data | | |
| Egg weight, length and width; queen weight and ovarioles | Mendeley datasets | http://dx.doi.org/10.17632/ 3xmkwh79gj.3 |
| RNA–Seq data collected in 2016 | NCBI database | BioProject: PRJNA310321 |
| RNA–Seq data collected in 2018 | NCBI database | BioProject: PRJNA530116 |
| Experimental Models: Organisms/Strains | | |
| High royal jelly producing honeybee strain (*Apis mellifera*) | Hangzhou Dexing bee industry Co., Ltd (China) | http://hzdexingtang.cn.gongxuku.com/ |
| Software and Algorithms | | |
| Statview | SAS Institute, Cary, NC, USA. | Version: 5.01 |
| EBSeq | R package. | DOI: 10.18129/B9.bioc.EBSeq |
| KOBAS software | Center for Bioinformatics, Peking University. | Version: 2.0 |
| Cufflinks software | Cole–trapnell–lab (GitHub) | Version: 2.2.1 |
| Other | | |
| Queen cell frame and excluder | This paper | N/A |
| Worker cell frame and excluder | This paper | N/A |

## Contact for Reagent and Resource Sharing

Further information and requests for resources and reagents should be directed to and will be fulfilled by the Lead Contact, Zhi Jiang Zeng (bees1965@sina.com).

## Experimental Model and Subject Details

Six western honey bee (*Apis mellifera*) colonies, which are a high royal jelly producing honeybee strain, were used in this experiment. Honey bee colonies were maintained at the Honeybee Research Institute, Jiangxi Agricultural University, Nanchang, China (28.46 oN, 115.49 oE), according to standard beekeeping techniques. Each colony had 8 frames with approximately 12,000 bees and a mated queen. Three colonies were use for egg weight, length and width measurement, queen weight measurement and RNA-Seq in 2016 (RNA-Seq was performed by Beijing Biomarker Technologies Co., Ltd.); Other three were use for queen weight and ovariole measurement and repeated RNA-Seq in 2018 (RNA-Seq was performed by Guangzhou Gene Denovo Co., Ltd.).

## Method Details

### *Egg collecting and queen rearing methods*

Mated queens were caged for six hours to lay in either a plastic frame of worker cells or a plastic frame of queen cells (Fig. II-1-1A). One side of the box was a queen excluder that allowed workers to pass through and attend the queen as normal. The plastic frame of worker cells was developed by Pan *et al* [44] and designed such that the base of each cell could be removed allowing the egg or larvae within to be transferred to other plastic queen cells or worker cells without touching them (Fig. II-1-1B). Generally, queens were caged in the morning to lay queen cells eggs for 6 hrs and were removed immediately to worker cell frames to lay worker cell eggs for 6 hrs in the afternoon. Egg size changes during the incubation period and varies across inbred lines [45,46], therefore queens were restricted to plastic frames of either queen-sized (internal diameter 9.7 mm) cells or worker-sized (internal diameter 4.9 mm) cells for only 6 h to lay (Fig. II-1-1) and measured immediately on collection. In total 152 eggs from three colonies were sampled and weighed.

For weighing the eggs, a plastic pen with a very thin and soft needle was developed to individually transplant eggs from cells to an analytical balance (Ax26 Comparator (Max=22g, d=1μg), Switzerland Mettler Toledo Co., Ltd.), and data was shown in Fig. II-1-2. Their width and length were measured with a zoom-stereo microscope system (Panasonic Co., Ltd.) according to the manufacturer's instructions and were shown in Figure 2. Since queens laid only dozens of eggs into queen cells, all eggs were measured. In worker cells, about 30 eggs from among 250-300 eggs laid were measured from each colony. We excluded the possibility that workers differentially cannibalize or remove eggs after they are laid by queens, since fertilized eggs laid by queens had queen egg-marking pheromones to avoid worker policing[40,41].

Eggs sampled in this way were also used to rear queens. Eggs sampled from queen cells and worker cells were transplanted into standard plastic queen cells (Fig. II-1-1) Queen cell bars were placed into a strong queenless hive with 8 frames for queen rearing. For the 2L group, eggs laid in worker cells were allowed to develop for 30-36 hours after hatching. The larvae were subsequently transferred to queen cells and added to the same queenless hive to be reared. Cells from the three groups (QE, WE, 2L) were mixed randomly. After 11 days, queens were harvested immediately on emergence. Queen weights were measured using the methods above and were shown in Fig. Ⅱ-1-3. Six of newly emerged queens were collected immediately on emergence for RNA-Seq in 2016, and other six queens were collected from other three colonies for RNA-Seq in 2018.

*Histolological analyses of queen ovarioles*

For measuring the queen ovarioles, 60 newly emerged queens from those three groups were caged and kept into a colony for 5 days until their ovaries were fully developed. The methods of histopathologic observation were according to Zou et al [47]. Ovaries from the right side of queen were collected and fixed in 4% paraformaldehyde for 18h at room temperature. These tissues were then embedded in paraffin after dehydration and permeabilization. Paraffin-embedded ovaries were sectioned serially at 4 μm on a microtome and dried. Subsequently, the slices were stained with HE after deparaffinization and rehydration. Histomorphology was assessed using a microscope (Qlympus-DP80, Olympus Corporation, Tokyo, Japan), data were shown in Fig. Ⅱ-1-3.

*RNA-Seq analysis*

For RNA-Seq, we sampled in total 12 newly emerged queens. Two queens of each of the QE, WE and 2L groups were taken from two different colonies. Each RNA-Seq sample combined 2 queens from the same group from the same colony. Each experimental group had two biological replicates. Only the heads and thoraces of two queens were used and mixed for RNA extraction and sequencing, since microorganisms and food in queen midgut could interfere RNA-Seq analysis. All samples were immediately flash-frozen in liquid nitrogen. Methods for sample preparation, mRNA isolation, RNA sequencing and data analysis followed those of He et al [14]. Firstly, total RNA was extracted using a TRlzol Reagent kit (Life technologies, California, USA) from each sample individually. Total RNA of each sample (around 6 μg) were used for RNA sequencing. The RNA quality was further checked using Agilent 2100 Bioanalyzer (Agilent Technologies, Inc., Santa Clara, CA, USA). mRNA was isolated from total RNA using a NEBNext Poly(A) mRNA Magnetic Isolation Module (NEB, E7490) with Oligo（dT）(NEB, E7500). Then the enriched mRNA was randomly fragmented leading to approximately 200 nt RNA inserts by a fragmentation buffer (NEB, E7530B/E7530L). Fragmented RNA inserts were used to synthesize cDNA, which were purified with AMPure XP beads (Beckman Coulter, Inc.) for End-repair/dA-tail and adaptor ligation. Finally the constructed cDNA libraries were sequenced by an Illumina HiSeq™ 2500 sequencing platform.

The reads with over 50 % of its base pairs had a Q-score of less than 10 ($Q = -10 * \lg Pe$) were filtered [48]. All clean reads were mapped to honeybee (*Apis mellifera*) reference genome (Amel 4.5) using Tophat2 package [49]. Gene expression levels were calculated and analyzed using read counts by the Cufflinks software [50] and normalized using FPKM values (fragments per kilobase of exon per million fragments mapped). Gene expression among three experimental treatments were evaluated and compared by using EBSeq [51]. Only those genes with an absolute value of $\log_2$ ratio ≥ 1 and *P* value < 0.05 were defined as significantly differentially expressed genes (DEGs), which were shown in Figure 4, Table S1 and S2.

The identified DEGs peptide sequences were aligned to NCBI non-redundant database (NCBI Nr), gene ontology database (GO), cluster of orthologous groups of proteins database (COG), kyoto encyclopedia of genes and genomes database (KEGG), Swiss-Prot database, using BLASTX and BLASTn with a cut-off E-value of $10^{-5}$. The Enrichment analysis of DEGs in KEGG pathways was performed using KOBAS 2.0 software [52]. The similarity of DEG results between each comparison (2L/QE, 2L/WE and WE/QE) were shown in Figure S1 and the number of DEGs in each section were marked with star key.

Twelve cDNA libraries were generated from our experimental groups. The Q30 of each sample was higher than 87 % indicating the high quality in the saturation of RNA sequencing (2016: Table S3; 2018: Table S4). The Pearson correlation coefficient among two biological replicates of each experimental group were all ≥ 0.80 (2016: Table S5; 2018: Table S6), which is a conventionally accepted threshold for valid replicates indicating that there was acceptable sequencing quality and repeatability among the biological replicates of each group.

***Real-time PCR validation***

RNA for qRT-PCR was taken from the RNA samples used for the RNA-Seq, and was used as templates to synthesis cDNA by MLV reverse transcriptase (Takara Japan) according to the manufacturer's instructions. Six genes identified as highly and significantly differentially expressed among 2L, WE and QE were chosen for confirmation of expression differences with real-time PCR (Bio-Rad IQ2, USA). The gene Apr-1 was selected as an appropriate internal control [53]. Real-time PCR Primers of these six target genes were designed using Primer 5.0 software (Table S7). The internal standard and each target gene were run in the same plate to eliminate interplate variations. The qRT-PCR cycling conditions were as follows: preliminary 94 °C for 2 min, 40 cycles including 94 °C for 15 sec, xx°C (varied according to the best annealing temperatures of each target gene, Table S7) for 30 sec, and 72 °C for 30 sec. For each gene, two biological replicates with five technical replicates were performed. The Ct value for each biological replicate was obtained by calculating the mean of five technical replicates. The relative gene expression was calculated by $2^{-\Delta\Delta Ct}$ formula reported by Liu and Saint [54]. The results are presented in Fig. S2.

## Quantification and Statistical Analysis

As the weigh, thorax width and length of eggs in this study were highly corrected, we therefore used a Bonferroni correction for the data analysis according to the format: α'≤ α/K where α is the critical value ($p_{critical}$ = 0.05) and K is the number of hypotheses. The adjusted significance value ($p_{adjusted}$=0.05/3=0.0167) was employed as the critical p value. For the egg weight analysis, the weight, length and width of each egg was the response variable, two treatments were the explanatory factors and three colonies were the covariants. Data from three honey bee colonies was integrated together and analyzed by using ANOVA test (StatView 5.01) followed by fisher's PLSD, since there was no significant difference among three colonies in weight, length and width ($p$=0.1206, $p$=0.2563 and $p$=0.1918 respectively). For the analysis of queen weight and number of ovarioles, the data were analyzed by ANOVA test using StatView 5.01 followed by a Fisher's PLSD test, and $p$-value < 0.05 was considered as significance. The data from qRT-PCR of each group were analyzed by ANOVA test using StatView 5.01 followed by a Fisher's PLSD test.

## Data and Software Availability

The raw data of egg weight, egg length and width, queen weight and ovarioles are accessible through Mendeley database: http://dx.doi.org/10.17632/3xmkwh79gj.3.

2016 RNA-Seq raw data are accessible through NCBI' database:

BioProject: PRJNA310321; BioSamples: QE (SAMN04450256), WE (SAMN04450254), 2L (SAMN04450253)

2018 RNA-Seq raw data are accessible through NCBI' database:

BioProject: PRJNA530116 (SRP190001); BioSamples: QE-replicate 1 (SRR8823608), QE-replicate 2 (SRR8823607), WE-replicate 1 (SRR8823606), WE-replicate 2 (SRR8823605), 2L-replicate 1 (SRR8823604), 2L-replicate 2 (SRR8823603).

## 2. Making a Queen: an Epigenetic Analysis of the Robustness of the Honey Bee (*Apis mellifera*) Queen Developmental Pathway

Xujiang He[1], Linbin Zhou[1], Qizhong Pan[1], Andrew B. Barron[2], Weiyu Yan[1], Zhijiang Zeng[1*]

1. Honeybee Research Institute, Jiangxi Agricultural University, Nanchang, Jiangxi 330045, P. R. of China
2. Department of Biological Sciences, Macquarie University, North Ryde, NSW 2109, Australia

**Abstract**

Specialised castes are considered a key reason for the evolutionary and ecological success of the social insect lifestyle. The most essential caste distinction is between the fertile queen and the sterile workers. Honey bee (*Apis mellifera*) workers and queens are not genetically distinct, rather these different phenotypes are the result of epigenetically regulated divergent developmental pathways. This is an important phenomenon in understanding the evolution of social insect societies. Here we studied the genomic regulation of the worker and queen developmental pathways, and the robustness of the pathways by transplanting eggs or young larvae to queen cells. Queens could be successfully reared from worker larvae transplanted up to 3 days age, but queens reared from older worker larvae had decreased queen body size and weight compared to queens from transplanted eggs. Gene expression analysis showed that queens raised from worker larvae differed from queens raised from eggs in the expression of genes involved in the immune system, caste differentiation, body development and longevity. DNA methylation levels were also higher in 3-day queen larvae raised from worker larvae compared to that raised from transplanted eggs identifying a possible mechanism stabilizing the two developmental paths. We propose that environmental (nutrition and space) changes induced by the commercial rearing practice result in a suboptimal queen phenotype via epigenetic processes, which may potentially contribute to the evolution of queen-worker dimorphism. This also has potentially contributed to the global increase in honeybee colony failure rates.

**Keywords:** honey bee, queen, epigenetic analysis, DNA methylation, gene expression, immunity

## Introduction

The evolution of cooperation, cooperative living and animal societies has been an enduring subject of fascination for evolutionary biologists (Wilson 1975). Key insights into the processes of social evolution have come from studies of the advanced social insects (Eilson 1971; Andersson 1984; Robinson 1999).

---

\* Corresponding author: bees1965@sina.com (ZJZ).
注：此文发表在 *Molecular Ecology* 2017 年第 6 期。

These have shaped our understanding of the genetic and ecological factors that can promote the evolution of sociality (Queller & Strassmann 1998; Linksvayer and Wade 2005; Foster *et al.* 2006). Oster and Wilson have particularly emphasized the importance of caste in the evolution of social insect societies (Oster & Wilson 1978). Different castes within the society specialize on different functions. This specialization promotes efficiencies, which provides a key selective advantage to social living. Oster and Wilson (1978) argue castes are one key reason for the ecological success of the social insect lifestyle.

Queens and workers are the defining caste distinction for the social insects. Queens have multiple morphological and behavioural specialisations for extreme fecundity, whereas workers show a similar degree of specialization for social roles supporting the queens' reproduction, and in many social insects workers are sterile. This is the case for honey bees (*Apis mellifera*). A typical colony contains a single reproductive queen supported by up to 50,000 sterile workers (Winston 1991). Studies of the bee have shown how the distinction between queens and workers is not genetic: rather these two phenotypes are the outcome of different developmental pathways (Nijhout 2003; Linksvayer *et al.* 2011).

Both queens and workers develop from fertilized eggs, but differences in nutrition and the amount of food given to young larvae trigger different epigenetically regulated developmental pathways (Kucharski *et al.* 2008; Maleszka 2014; Maleszka *et al.* 2014). Kucharski *et al.* (2008) reported that nutritional differences between queen and worker at their larval stage control their development via DNA methylation. Shi *et al.* (2011) showed that the amount of space in which a larva can develop alters the DNA methylation level of the larval genome and contributes to the process of caste differentiation. Changes in gene regulation caused by these epigenetic mechanisms then establish divergent developmental paths (Simola *et al.* 2013), involving particularly genes involved in signal transduction, gland development, and carbohydrate metabolism (Woodard *et al.* 2011).

Since the 19th century in commercial beekeeping, it has been a standard practice to raise queens by transplanting eggs or young larvae into artificial queen cells, which triggers workers to raise a queen (Doolittle 1888; Büchler *et al.* 2013). Within the commercial queen rearing practice there is variation in the age at which eggs or worker larvae are transplanted to queen cells to be raised as queens. It is not clear how well the honey bees' developmental processes are able to tolerate this kind of intervention. Woyke (1971) reported that rearing queens from young worker larvae resulted in decreased body size, a smaller spermatheca and fewer ovarioles. Rangel *et al.* (2012) reported that colonies from queens reared from older worker larvae had significantly lower production of worker comb, drone comb and stored food compared to colonies from queens reared from young worker larvae. In fact, concern over the long-term consequences of commercial queen rearing for bee stocks is not new. In 1923 Rudolf Steiner predicted that honey bees would become extinct within 100 years as a consequence of commercial queen rearing progressively weakening bee stocks (Thomas 1998). In the current environment of increased honey bee colony failure rates, mass deaths of colonies and declining honey bee stocks there is a great deal of concern as to whether a decline in queen bee quality might be a factor in these problems (vanEngelsdorp *et al.* 2010; Delaney *et al.* 2011). Therefore, we hypothesize that environmental (nutrition and space) changes induced by the commercial rearing practice may potentially affect queen development via epigenetic processes. Here we explored the consequence of age of transplant from worker cells to queen cells on DNA methylation, gene expression and queen morphology. We found that the domestic rearing practice altered queen morphology and induced epigenetic changes in developing queens, which supports our hypothesis.

## Materials and Methods

Three European honey bee colonies (*Apis mellifera*) each with a single drone inseminated queen (SDI) were used throughout this study. These colonies were maintained at the Honeybee Research Institute, Jiangxi Agricultural University, Nanchang, China (28.46 °N, 115.49 °E), according to the standard beekeeping techniques.

## Queen Rearing Methods

Queens were restricted for six hours to a plastic honey bee frame developed by Pan *et al*. (2013) for laying. The frame is designed such that the plastic base of this frame with eggs or larvae can be transferred to plastic queen cells directly (Pan *et al*. 2013). Eggs that queen laid in worker cells were transplanted into queen cells for rearing new queens when eggs were less than six hours old (QWE). For the other experimental groups day-1, day-2 and day-3 worker larvae were transplanted into queen cells for rearing QWL1, QWL2 and QWL3 respectively. Queen cells with worker eggs or larvae were returned into their natal colonies (the SDI colonies) for queen rearing.

For the morphological measurements, new emerging queens were collected and their weight measured using an analytical balance (FA3204B, Shanghai Precision Scientific Instrument Co., Ltd.). Their thorax width and length were measured with a zoom—stereo microscope system (Panasonic Co., Ltd.) according to the manufacturer instructions.

For epigenetic analysis, we sampled three 3-day-old larvae from QWE, QWL1 and QWL2 respectively from their queen cell. The fourth group QWL3 sampled 3-day-old worker larvae directly from worker cells. Each sample group collected 3 larvae and there were three biological replicates, each from different colonies, for each group. We weighed each larva from these four treatment groups with an analytical balance. All samples were immediately flash-frozen in liquid nitrogen. The DNA and RNA from each sample were both extracted for further DNA methylation and RNA sequencing analysis. DNA and RNA were extracted from the same samples.

## RNA-Seq Analysis

Total RNA was extracted from larvae according to the standard protocol for the TRIzol Reagent (Life technologies, California, USA). RNA integrity and concentration were checked using an Agilent 2100 Bioanalyzer (Agilent Technologies, Inc., Santa Clara, CA, USA).

mRNA was isolated from total RNA using a NEBNext Poly(A) mRNA Magnetic Isolation Module (NEB, E7490). A cDNA library was constructed following the manufacturer's instructions for the NEBNext Ultra RNA Library Prep Kit (NEB, E7530) and the NEBNext Multiplex Oligos (NEB, E7500) from Illumina. In brief: enriched mRNA was fragmented into approximately 200 nt RNA inserts, which were used as templates to synthesize the cDNA. End-repair/dA-tail and adaptor ligation were then performed on the double-stranded cDNA. Suitable fragments were isolated by Agencourt AMPure XP beads (Beckman Coulter, Inc.), and enriched by PCR amplification. Finally, the constructed cDNA libraries were sequenced on a flow cell using an Illumina HiSeq™ 2500 sequencing platform.

Low quality reads, such as adaptor-only reads or reads with > 5% unknown nucleotides were filtered from subsequent analyses. Reads with a sequencing error rate less than 1% (Q20 > 98%) were retained. These remaining clean reads were mapped to the honey bee (*Apis mellifera*) official genes (OGSv3.2) using Tophat2 (Kim *et al.*, 2013) software. The aligned records from the aligners in BAM/SAM format were further examined to remove potential duplicate molecules. Gene expression levels were estimated using FPKM values (fragments per kilobase of exon per million fragments mapped) by the Cufflinks software (Trapnell *et al.* 2010).

DESeq2 and $Q$-value statistical methods were used to evaluate differential gene expression among the four experimental treatments (Love *et al.* 2014). The false discovery rate (FDR) control method was used to determine the appropriate threshold of $P$-values in multiple tests comparing gene expression differences by read counts. Only genes with an absolute value of $\log_2$ ratio $\geq 1$ and FDR significance score < 0.05 were used for subsequent analysis. However, gene expression levels of each gene in all samples were present by using their ratio of FPKM values.

Sequences differentially expressed between sample groups were identified by comparison against various protein databases by BLASTX, including the National Center for Biotechnology Information (NCBI) non-redundant protein (Nr) database, Swiss-Prot database with a cut-off E-value of $10^{-5}$. Furthermore, genes were searched against the NCBI non-redundant nucleotide sequence (Nt) database using BLASTn by a cut-off $E$-value of $10^{-5}$. Differentially expressed genes (DEGs) were mapped to KEGG protein database by BLAST ($E$-value < 1e-5), and used KOBAS 2.0 software to test the statistical enrichment of differential expression genes in KEGG pathways (Xie *et al.* 2011).

## DNA-Methylation Analysis by Bisuphite Sequencing

The DNA of each larval sample was extracted using the Universal Genomic DNA Extraction Kit (TaKaRa, DV811A). DNA concentration was measured and adjusted to the same level. Genomic DNA was sheared with Covaris ultrasonicator (Life Technology). The fragmented DNA was purified using AMPure XP beads and end repaired. A single 'A' nucleotide was added to the 3' ends of the blunt fragments followed by ligation to methylated adapter with T overhang. 200-300 bp insert size targets were purified by 2% agarose gel electrophoresis. Bisulfite conversion was conducted using a ZYMO EZ DNA Methylation-Gold™ Kit (ZYMO, Irvine, CA, USA). The final libraries were generated by PCR amplification. Bisulfite libraries were analyzed by an Agilent2100 Bioanalyzer (Agilent Technology) and quantified by QPCR (Agilent QPCR NGS Library Quantification Kit). The construction of bisulfite libraries and paired-end sequencing using Illumina HiSeq™ 2500 (Illumina, San Diego, CA, USA) were performed at Beijing Biomarker Technology Co., Ltd (Beijing, China).

After filtering adaptor sequences and PCR duplicated reads, genomic fragments from bisulfite libraries were mapped against the honey bee genome (*Apis mellifera. Amel* 4.5) using Bowtie 2 software (Langmead & Salzberg 2012). The bismark methylation extractor (Krueger & Andrews 2011) was used to predict all methylation sites. Only uniquely mapped reads were retained. The ratio of C to CT was used to indicate methylation level. Three methods for DNA methylation level analysis were used: fraction of methylated cytosines, mean methylation level and weighted methylation level (Schultz *et al.* 2012). The results are presented in Fig. S8.

## Data Analysis

### *Morphological analysis of queens and 3-day-old queen larvae*

All data from morphological experiments of each group were analyzed by ANOVA using StatView 5.01 followed by a Fisher's PLSD test (SAS Institute, Cary, NC, USA).

### *Correlation analysis between expression and methylation and map construction of genes and chromosome*

Methylated regions were deemed significantly differentially methylated across QWE, QWL1, QWL2 and QWL3 with a false discovery rate (FDR) < 0.05 and log2 fold change ≥ 1.5 in sequence counts using the BSmooth method in R package 3.1.1 (Hansen *et al.* 2012). Significantly different methylated regions (DMRs) of each gene were mapped to the 16 honeybee chromosomes regions (*Apis mellifera. Amel 4.5*) using integrative genomics viewer (IGV, http://www.broadinstitute.org/igv/).

### *Analysis of RNA-Seq quality and DNA methylation sequencing*

In RNA-Seq, four libraries were generated from our experimental groups, and summaries of RNA sequencing analyses are shown in Table S1. In each library, more than 98 % clean reads were unique reads of which more than 89 % reads were paired reads. Very few clean reads (<1.4 %) were multiple mapped reads. Each library had a sufficient coverage of the expected number of distinct genes (stabilized at 3M reads, Fig. S1). The Pearson correlation coefficient among three biological replicates of each experimental group were all ≥ 0.80 (Table S2), a conventionally accepted threshold for valid replicates (Tarazona *et al.* 2011) indicating that there was acceptable sequencing quality and repeatability among the biological replicates of each group. The majority of methylation sites of all samples (77.55 %) were the CG type which was considerably more than other two types (CHH: 20.5% and CHG: 1.95% respectively; Fig. S2).

## Results

As the age of transplant of the worker larvae increased, the size and mass of the emergent adult queen decreased (Fig. II-2-1, see page 514). QWE had the highest thorax length (4.90±0.24 mm, mean±SE), thorax width (4.78±0.21 mm, mean±SE) and weight (267.21±2.49 mg, mean±SE), whereas the QWL3 had lowest, with 4.60±0.13 mm, 4.45±0.26 mm and 226.00±2.82 mg respectively. All morphological indices were differed significantly across the four treatments ($p < 0.05$, ANOVA test followed with Fisher's PLSD test).

RNA-Seq analyses comparing gene expression between QWE and the three larvae transplanted groups (QWL1, QWL2 and QWL3) showed that the number of differentially expressed genes increased as the age of the transplanted worker larva increased (Fig. II-2-2, see page 514 and Fig. S3). In all comparisons the differentially expressed genes contained a high proportion of genes involved in immunity, body development, metabolism, reproductive ability and longevity (Fig. II-2-2 and Table S3). In particular, one of cytochrome P450 family gene (CYP450 6a14-like) was significantly up-regulated in QWL1 compared to QWE; Six of ten immunity related DEGs such as CYP450 6a14-like and CYP450 305a1 were up-regulated in QWL2 compared to QWE, whereas seven of ten body development related genes were down-regulated in QWL2; In QWL3 and QWE comparison, 13 of 23 immunity related DEGs were up-regulated in QWL3 whist 29 of 41 body development related genes were down-regulated in QWL3 respectively. Interestingly, the hormone biosysnthesis genes [*vitellogenin precursor* (Vg), *juvenile hormone esterase precursor* (JH), *juvenile hormone esterase-like* (JH-like) and *ecdysteroid-regulated 16 kDa protein-like*] were up-regulated in QWL3 compared with QWE, whereas the *major royal jelly protein 1* (MRJP1) was down-regulated in QWL3 (Table S3). Similarly, the results of GO enrichment analysis showed that the number of categories of DEGs enriched between the QWE group and the other experimental groups increased with the increasing age of the grafted larva: from 27 categories (QWL1 vs QWE) to 33 (QWL2 vs QWE) and 44 (QWL3 vs QWE) respectively (Fig. S4-6). Categories of DEG included growth, development process, reproductive process and immune system process. The results of COG enrichment analysis showed a similar pattern in that the number of categories between QWE and QWLs increased with their grafting age (Fig. S7).

Furthermore, queens from older grafted worker larvae had a higher global DNA methylation level than QWE (Fig. S8 and Table S4). The QWE vs QWL3 (totally 146 DMRs) comparison had the greatest number of differentially methylated regions (DMRs) than QWE vs QWL2 (108) and QWE vs QWL1 (99) comparisons. Mapping these DMRs to gene regions identified 2, 6 and 23 differentially methylated genes in the QWE vs QWL1, QWE vs QWL2 and QWE vs QWL3 comparisons respectively, and no gene was overlapped among these three comparisons. These genes were different from the DEGs identified by RNA-Seq. Most of them were involved in substance metabolism (Table S5). A correlation analysis of gene expression and DNA methylation developed by (Lou *et al.* 2014) (Fig. S9) suggested a very weak correlation between DNA methylation and gene expression in honeybees for all comparisons (pearson correlation values were -0.0018, 0.0016, -0.004 and -0.0034 in QWE, QWL1, QWL2 and QWL3 respectively, and all $p$-values were > 0.05). We also compared the expression of DGEs involved in immune system and hormone biosynthesis and their DNA methylation among the four treatment groups. The difference in expression of these genes increased as the difference in age of the transplanted larvae increased (Fig. II-2-4, see page 516) (greater when comparing QWE vs QWL3 than when comparing QWE vs QWL1), but the DNA methylation of very few of these genes showed a clear negative correlation with expression (Fig. II-2-4 and Table S6).

## Discussion

Honey bee workers and queens are two very different phenotypes that come about as the result of divergent developmental pathways. Environmental differences in growing space and nutrition cause the divergence, and the two pathways are organized by epigenetic processes (Haydak 1970; Shi *et al.* 2011; Foret *et al.* 2012). Here we explored the mechanisms and stability of the queen developmental path by transplanting worker larvae of different ages into queen cells causing a redirection of the worker developmental path onto the queen developmental path.

Our results speak to both the plasticity and limitations of honey bee phenotypic plasticity. While it is known that worker larvae that are more than 3.5 days old when transplanted fail to develop into queens (Weaver 1966), here we found that transplanting even 3-day-old worker larvae to queen cells resulted in functional adult queens, but with reduced size and weight. Moreover our results showed that the number of differentially expressed genes in QWLs compared to QWE (Fig. II-2-2) increased with the age of the transplanted larvae. Many of these genes were involved in immunity, body development, metabolism, reproductive ability and longevity (Fig. II-2-2 and Table S3). These are all functions that differentiate queens from workers, and are critical to the quality and longevity of the queen. We also showed that queens raised from transplanted older worker larvae had a higher global genomic DNA methylation level when compared to QWE (Table S4) identifying a possible epigenetic mechanism for these differences. In summary, we could conclude that the queen developmental pathway is quite robust. The queen phenotype could be attained even by 3-day old worker larvae (the larval developmental period is just 5 days long) and therefore developmental trajectory is certainly not fixed until relatively late in larval development. However, interfering with the normal developmental process by switching larvae from a worker to a queen path clearly had consequences for the resulting adults that were detectable at morphological and genetic levels.

Our study identified many genetic and epigenetic changes related to the age at which worker larvae were transplanted to queen cells to be raised as queens (Fig. II-2-2, Fig. II-2-3 and Table S3, S4). Of note, the differentially expressed genes included *Insulin-like peptide A chain* (ILP-A), a gene involved in the mTOR pathway. The mTOR pathway is involved in queen ovary development, and caste differentiation (Patel *et al.* 2007; de Azevedo & Hartfelder 2008; Mutti *et al.* 2011). The ILP-A was significantly down-regulated in QWL3 compared to QWE. Hence this identifies a candidate pathway for why transplant age alters adult queen ovariole number and spermatheca size (Woyke 1971). The gene MRJP-1 (down-regulated in QWL3 compared to QWE), JH (up-regulated in QWL3) and Vg (up-regulated in QWL3) were also differentially expressed between QWE and QWL3. These genes are also involved in the regulation of honeybee caste differentiation and longevity (Amdam & Omholt 2002; Kamakura 2011), perhaps indicating why queens from late-stage larval grafts are undersized. The GO enrichment results also confirmed this result that DEGs between QWE and QWLs were enriched in growth, reproductive and development processes (Fig. S4-6).

Previous studies have demonstrated that DNA methylation is widespread in social Hymenoptera (Kronforst *et al.* 2008; Kucharski *et al.* 2008; Foret *et al.* 2012; Lyko & Maleszka 2011). While the level of DNA methylation in honey bees is quite low compared to mammals, differential methylation of the genome has a key role in establishing the divergent worker and queen developmental pathways (Wang *et al.* 2006; Gabor Miklos & Maleszka 2011; Shi *et al.* 2013; Maleszka 2014). We found that increasing the age at which worker larvae were transplanted to queen cells resulted in an increasing number of differentially methylated regions of the genome. We propose this reflects the epigenetic processes underlying the reorientation of a worker-destined developmental pathway to a queen developmental pathway.

There was not a strong negative correlation between the gene expression and DNA methylation in this study (Fig. S9), however, in honey bees most methylation sites are located in gene body regions rather than the upstream and downstream regulatory regions (Fig. S8). Foret *et al.* (2012) showed that honey bee DNA methylation is correlated with gene alternative splicing. Table S7 showed that the alternative splicing also exists in differentially methylated genes. It is still unclear what role these DNA methylation have for gene expression, though DNA methylation plays an important role in the regulation of honeybee caste differentiation (Wang *et al.* 2006; Gabor Miklos & Maleszka 2011; Shi *et al.* 2013; Maleszka 2014). Our results lend further credence to the view that the functions of DNA methylation of sites in gene body regions in hymenoptera have more complex functions than simply inhibiting expression.

In commercial apiculture rearing queens from transplanted worker larvae is a standard commercial practice, and the age of the worker larvae used can be anything up to and including 3 days old. In practice there is a preference to use older worker larvae for transplant since these are more hardy, easier to handle and give a higher success rate. But these larvae will have been fed worker jelly (brood food) rather than queen jelly in their early life. Worker jelly is a very different diet to queen jelly. It differs in sugar content (Asencot & Lensky 1977), amino acid (Brouwers 1984), vitamin (Brouwers *et al.* 1987), juvenile hormone (Asencot & Lensky 1984), and major royal jelly protein content (Kamakura 2011). Therefore the commercial queen rearing practice alters the nutritional environment of the queen larvae, likely resulting in development and epigenetic changes. This is consistent with the previous studies that nutritional differences control the caste differentiation of queen and workers by DNA methylation (Kucharski *et al.* 2008; Shi *et al.* 2011).

There have been long-standing concerns about the consequences of this queen rearing method for queen quality and colony productivity (Thomas 1998; Woyke 1971; Rangel *et al.* 2012). Our results have clearly shown that queens raised from older worker larvae are smaller, and it is also telling that we found a difference in expression between several genes involved in immune function between QWE and QWLs (Fig. II-2-2, table S3, and Fig. S4-6 for GO enrichment results). These included the *Cytochrome P450* family (CYP450s, 11 of 15 were up-regulated in QWL3 compared to QWE). CYP450s may contribute to both disease and insecticide resistance in queen honey bees (Claudianos 2006; Boncristiani *et al.* 2012). We propose that queens reared from older larvae may suffer reduced immune function and insecticide resistance.

Evolutionary theories on the development of eusociality demonstrated that maternal care (nutrition and developmental space) dramatically contribute to the evolution of queen-worker dimorphism in honey bees (Linksvayer *et al.* 2011; Leimar *et al.* 2012). Our study clearly showed that the domestic rearing practice artificially transformed the nutrition and developmental space of queen larvae and resulted in a partial intercaste between queen and workers. In Leimar's model (2012), honey bee caste dimorphism is produced by maternal care rather than a switch-controlled polyphenism, and our results are consistent with this model. We also demonstrated that the domestic rearing practice altered natural honey bee maternal behavior, inducing various epigenetic changes. This is particularly interesting since epigenetic changes such as DNA methylation can introduce environmental effects into the following generations (Bird 2002; Klironomos 2013) and potentially influence evolutionary processes (Dickins & Rahman 2012). If it is possible for epigenetic markers of the genome to persist through gamete formation and operate transgenerationally in honey bees, as occurs in some mammals (Klironomos 2013), then the evolution of queen-worker dimorphism might be influenced by these mechanisms. Moreover, we propose that the often used commercial queen rearing practice results in queens of lower quality. As a proximal remedy, rearing queens from eggs or very young larvae may yield a better outcome for queen performance and colony function.

## Acknowledgments

We thank Dr. Ying Wang for suggesting on experimental design, Mr. Xue Chuan Zhang and Li Sha Huang for helping on epigenetic data analysis. This work was supported by the National Natural Science Foundation of China (No. 31572469 and No. 31460641) and the Earmarked Fund for China Agriculture Research System (CARS-45-KXJ12).

## Author Contributions

ZJZ designed the experiments, XJH, LBZ and QZP performed the experiments, XJH, ABB and WYY analyzed the data, XJH, ABB and ZJZ written the paper. We have declared that no conflict of interests exist.

## Data Accessibility

The raw data for queen morphology are available from the Dryad Digital Repository: http://dx.doi.org/10.5061/dryad.bg4t9. The raw Illumina sequencing data are accessible through NCBI's database:

RNA-Seq and DNA methylation data of QWE: NCBI Bioproject: PRJNA308280/SAMN04390202

RNA-Seq and DNA methylation data of QWL1: NCBI Bioproject: PRJNA308280/SAMN04390203

RNA-Seq and DNA methylation data of QWL2: NCBI Bioproject: PRJNA308280/SAMN04390236

RNA-Seq and DNA methylation data of QWL3: NCBI Bioproject: PRJNA308280/SAMN04390201

## References

AMDAM GV, OMHOLT SW (2002) The regulatory anatomy of honeybee lifespan. *Journal of Theoretical Biology*, 216, 209-228.

ANDERSSON M (1984) The evolution of eusociality. *Annual Review of Ecology and Systematics*, 15, 165-189.

ASENCOT M, LENSKY Y (1977) The effect of sugar crystals in stored royal jelly and juvenile hormone on the differentiation of female honey bee larvae (*Apis mellifera* L.) to queens.*In Proc. 8th Int. Congr. IUSSI, Wageningen*.pp: 3-5.

ASENCOT M, LENSKY Y (1984) Juvenile hormone induction of 'queenliness' on female honey bee (*Apis mellifera* L.) larvae reared on worker jelly and on stored royal jelly. *Comparative Biochemistry and Physiology B-Biochemistry & Molecular Biology*, 1, 109-117.

BIRD A (2002). DNA methylation patterns and epigenetic memory. *Genes & Development*, 16, 6-21.

BONCRISTIANI H, UNDERWOOD R, SCHWARZ R, EVANS JD, PETTIS J, VANENGELSDORP D (2012) Direct effect of acaricides on pathogen loads and gene expression levels in honey bees *Apis mellifera*. *Journal of Insect*

*Physiology,* 58, 613-620.

BROUWERS E (1984) Glucose/fructose ratio in the food of honeybee larvae during caste differentiation. *Journal of Apicultural Research,* 2, 94-101.

BROUWERS EVM, EBERT R, BEETSMA J (1987) Behavioural and physiological aspects of nurse bees in relation to the composition of larval food during caste differentiation in the honeybee. *Journal of Apicultural Research,* 1, 11-23.

BÜCHLER R, ANDONOV S, BIENEFELD K, *et al.* (2013) Standard methods for rearing and selection of *Apis mellifera* queens. *Journal of Apicultural Research,* 52, 1-30.

CLAUDIANOS C, RANSON H, JOHNSON RM, *et al.* (2006) A deficit of detoxification enzymes: pesticide sensitivity and environmental response in the honeybee. *Insect Molecular Biology,* 15, 615-636.

DE AZEVEDO SV, HARTFELDER K (2008) The insulin signaling pathway in honey bee (*Apis mellifera*) caste development - differential expression of insulin-like peptides and insulin receptors in queen and worker larvae. *Journal of Insect Physiology,* 54, 1064-1071.

DELANEY DA, KELLER JJ, CAREN JR, TARPY DR (2011) The physical, insemination, and reproductive quality of honey bee queens (*Apis mellifera* L.). *Apidologie,* 42, 1-13.

DICKINS TE, RAHMAN Q (2012) The extended evolutionary synthesis and the role of soft inheritance in evolution. *Proceedings of the Royal Society B-Biological Sciences,* 279, 2913-2921.

DOOLITTLE GM (1888) Scientific queen-rearing.*American Bee Journal,* USA.

EILSON E (1971) *The Insect Societies.* Harvard University Press, Cambridge MA. USA.

FORET S, KUCHARSKI R, PELLEGRINI M, *et al.* (2012) DNA methylation dynamics, metabolic fluxes, gene splicing, and alternative phenotypes in honey bees. *Proceeding of the National Academy of Science USA,* 109, 4968-4973.

FOSTER KR, WENSELEERS T, RATNIEKS FL (2006) Kin selection is the key to altruism. *Trends in Ecology & Evolution,* 21, 57-60.

GABOR MIKLOS GL, MALESZKA R (2011) Epigenomic communication systems in humans and honey bees: from molecules to behavior. *Hormones and Behavior,* 59, 399-406.

HANSEN KD, LANGMEAD B, IRIZARRY RA (2012) BSmooth: from whole genome bisulfite sequencing reads to differentially methylated regions. *Genome Biology,* 13, R83.

HAYDAK MH (1970) Honey bee nutrition. *Annual Reviewof Entomology,* 15, 143-156.

KAMAKURA M (2011) Royalactin induces queen differentiation in honeybees. *Nature,* 473, 478-483.

KIM D, PERTEA G, TRAPNELL C, PIMENTEL H, KELLEY R, SALZBERG SL (2013) TopHat2: accurate alignment of transcriptomes in the presence of insertions, deletions and gene fusions. *Genome Biology,* 14, R36.

KLIRONOMOS FD, BERG, J., COLLINS, S (2013) How epigenetic mutations can affect genetic evolution: Model and mechanism. *Bioessays,* 35, 571-578.

KRONFORST MR, GILLEY DC, STRASSMANN JE, QUELLER DC (2008) DNA methylation is widespread across social Hymenoptera. *Current Biology,* 18, R287-288.

KRUEGER F, ANDREWS SR (2011) Bismark: a flexible aligner and methylation caller for Bisulfite-Seq applications. *Bioinformatics,* 27, 1571-1572.

KUCHARSKI R, MALESZKA, J., FORET, S., MALESZKA, R (2008) Nutritional Control of Reproductive Status in

Honeybees via DNA Methylation. *Science,* 319, 1827-1830.

LANGMEAD B, SALZBERG SL (2012) Fast gapped-read alignment with Bowtie 2. *Nature Methods,* 9, 357-359.

LEIMAR O, HARTFELDER K, LAUBICHLER MD, PAGE RJ (2012) Development and evolution of caste dimorphism in honeybees - a modeling approach. *Ecology and Evolution,* 2, 3098-3109.

LINKSVAYER TA, WADE MJ (2005) The evolutionary origin and elaboration of sociality in the aculeate Hymenoptera: maternal effects, sib-social effects, and heterochrony. *Quarterly Review of Biology,* 80, 317-336.

LINKSVAYER TA, KAFTANOGLU O, AKYOL E, BLATCH S, AMDAM GV, PAGE JR RE (2011) Larval and nurse worker control of developmental plasticity and the evolution of honey bee queen-worker dimorphism. *Journal of Evolutionary biology,* 9, 1939-1948.

LOU S, LEE HM, QIN H, et al. (2014) Whole-genome bisulfite sequencing of multiple individuals reveals complementary roles of promoter and gene body methylation in transcriptional regulation. *Genome Biology,* 15, 408.

LOVE M, ANDERS S, HUBER W (2014) Differential analysis of count data-the DESeq2 package. *Genome Biology,* 15, 550.

LYKO F, MALESZKA R (2011) Insects as innovative models for functional studies of DNA methylation. *Trends in Genetics,* 27, 127-131.

MALESZKA R (2014) Epigenetic integration of environmental and genomic signals in honey bees: the critical interplay of nutritional, brain and reproductive networks. *Epigenetics,* 3, 188-192.

MALESZKA R, MASON PH, BARRON AB (2014) Epigenomics and the concept of degeneracy in biological systems. *Briefings in Functional Genomics,* 13, 191-202.

MUTTI NS, DOLEZAL AG, WOLSCHIN F, MUTTI JS, GILL KS, AMDAM GV (2011) IRS and TOR nutrient-signaling pathways act via juvenile hormone to influence honey bee caste fate. *Journal of Experimental Biology,* 214, 3977-3984.

NIJHOUT HF (2003) Development and evolution of adaptive polyphenisms. *Evolution Anddevelopment,* 5, 9-18.

PAN QZ, WU XB, GUAN C, ZENG ZJ (2013) A new method of queen rearing without grafting larvae. *American Bee Journal,* 153, 1279-1280.

PATEL A, FONDRK MK, KAFTANOGLU O, et al. (2007) The making of a queen: TOR pathway is a key player in diphenic caste development. *PLoS One,* 2, e509.

QUELLER D, STRASSMANN J (1998) Kin Selection and Social Insects Social insects provide the most surprising predictions and satisfying tests of kin selection. *Bioscience,* 3, 165-175.

RANGEL J, KELLER JJ, TARPY DR (2012) The effects of honey bee (*Apis mellifera* L.) queen reproductive potential on colony growth. *Insectes Sociaux,* 60, 65-73.

ROBINSON GE (1999) Integrative animal behaviour and sociogenomics. *Trends in Ecology & Evolution,* 14, 202-205.

SCHULTZ MD, SCHMITZ RJ, ECKER JR (2012) 'Leveling' the playing field for analyses of single-base resolution DNA methylomes. *Trends in Genetics,* 28, 583-585.

SHI YY, HUANG ZY, ZENG ZJ, WANG ZL, WU XB, YAN WY (2011) Diet and cell size both affect queen-worker differentiation through DNA methylation in honey bees (*Apis mellifera*, Apidae). *PLoS One,* 6, e18808.

SHI YY, YAN WY, HUANG ZY, WANG ZL, WU XB, ZENG ZJ (2013) Genomewide analysis indicates that queen larvae have lower methylation levels in the honey bee (*Apis mellifera*). *Naturwissenschaften,* 100, 193-197.

THOMAS, B (1998) Bees-lecturers by Rudolf Steiner, *Anthroposophic Press*, pp 222.

SIMOLA DF, WISSLER L, DONAHUE G, *et al.* (2013) Social insect genomes exhibit dramatic evolution in gene composition and regulation while preserving regulatory features linked to sociality. *Genome Research,* 23, 1235-1247.

TARAZONA S, GARCÍA-ALCALDE F, DOPAZO J, FERRER A, CONESA A (2011) Differential expression in RNA-seq: a matter of depth. *Genome Research,* 21, 2213-2223.

TRAPNELL C, WILLIAMS BA, PERTEA G, *et al.* (2010) Transcript assembly and quantification by RNA-Seq reveals unannotated transcripts and isoform switching during cell differentiation. *Nature Biotechnology,* 28, 511-515.

OSTER G, WILSON E (1978) *Caste and ecology in the social insects.* Princeton University Press, Princeton, USA.

VANENGELSDORP D, HAYES JR J, UNDERWOOD RM, PETTIS JS (2010) A survey of honey bee colony losses in the United States, fall 2008 to spring 2009. *Journal of Apicultural Research,* 49, 7-14.

WANG Y, JORDA M, JONES PL, MALESZKA R, LING X, ROBERTSON HM *et al.* (2006) Functional CpG methylation system in a social insect. *Science,* 314, 645-647.

WEAVER N (1966) Physiology of caste differentiation. *Annual Review of Entomology,* 11, 79-102.

WILSON E (1975) *Sociobiology: The New Synthesis.* Harvard University Press, Cambridge MA. USA.

WINSTON M (1991) The biology of the honey bee. Harvard University Press.

WOODARD SH, FISCHMAN BJ, VENKAT A, *et al.* (2011) Genes involved in convergent evolution of eusociality in bees. *Proceeding of the National Academy of Science USA,* 108, 7472-7477.

WOYKE J (1971) Correlations between the age at which honeybee brood was grafted, characteristics of the resultant queens, and results of insemination. *Journal of Apicultural Research,* 10, 45-55.

XIE C, MAO X, HUANG J, *et al.* (2011) KOBAS 2.0: a web server for annotation and identification of enriched pathways and diseases. *Nucleic Acids Research,* 39, W316-W322.

# 3. Transgenerational Accumulation of Methylome Changes Discovered in Commercially Reared Honey Bee (*Apis mellifera*) Queens

Yao Yi[a,c†], Xujiang He[a†], Andrew B. Barron[b], Yi Bo Liu[a], Zilong Wang[a], Weiyu Yan[a], Zhijiang Zeng[a*]

a. Honeybee Research Institute, Jiangxi Agricultural University, Nanchang, Jiangxi 330045, P. R. of China
b. Department of Biological Sciences, Macquarie University, North Ryde, NSW 2109, Australia
c. Jiangxi University of Traditional Chinese Medicine, Nanchang, Jiangxi 330004, P. R. of China

**Abstract**

Whether a female honey bee (*Apis mellifera*) develops into a worker or a queen depends on her nutrition during development, which changes the epigenome to alter the developmental trajectory. Beekeepers typically exploit this developmental plasticity to produce queen bee by transplanting worker larvae into queen cells to be reared as queens, thus redirecting a worker developmental pathway to a queen developmental pathway. We studied the consequences of this manipulation for the queen phenotype and methylome over four generations. Queens reared from worker larvae consistently had fewer ovarioles than queens reared from eggs. Over four generations the methylomes of lines of queens reared from eggs and worker larvae diverged, accumulating increasing differences in exons of genes related to caste differentiation, growth and immunity. We discuss the consequences of these cryptic changes to the honey bee epigenome for the health and viability of honey bee stocks.

**Keywords:** DNA methylation, epigenome, caste differentiation, sociogenomics, differential methylation

**Introduction**

Epigenomics is revealing how genomic developmental systems are themselves sensitive to the developmental environment (Cavalli and Heard, 2019). A consequence of this is the possibility of developmental stressors to rewrite the epigenome with profound, and potentially enduring consequences for animal development (Burggren, 2015; Cavalli and Heard, 2019). The western honey bee (*Apis mellifera*) presents a dramatic natural example of developmental plasticity that is epigenomically regulated. The nutritional environment during development selectively changes methylation of the bee genome which establishes the very different worker and queen phenotypes (He *et al.*, 2017; Kucharski *et al.*, 2008). This provides a natural system for study of how the epigenome can be affected by developmental stress. Here we

---

† Authors contributed equally to this work.
*Corresponding Author: bees1965@sina.com(ZJZ).
注：此文发表在 *Insect Biochemistry and Molecular Biology* 2020, 127: 103476。

studied how a current developmental stress routinely applied in contemporary agriculture influenced the honey bee queen epigenome over both long and short timescales.

An interaction of developmental systems with the environment has long been assumed, but it was also a common conception that genomics mechanisms shaping development were themselves isolated from environmental influences, with stressors subverting an ideal genomic developmental pattern. Epigenomics has overturned this view and highlighted how numerous epigenomic systems are directly sensitive to the environment (Burggren, 2015). Indeed, this can be a vital aspect of their functionality (He *et al.*, 2017; Jung-Hoffmann, 1966; Maleszka, 2008), but it can also result in dysfunction (Cavalli and Heard, 2019).

Classic studies with the honey bee have shown how the sensitivity of epigenomic systems to the environment can be an essential mechanism of developmental plasticity (Kucharski *et al.*, 2008; Lyko *et al.*, 2010; Maleszka, 2008). There are two very distinct developmental outcomes for female honey bee: large reproductive queen bee and small sterile worker bee (Evans and Wheeler, 2001; Hartfelder K, 1998; Jung-Hoffmann, 1966). These different castes are key to the success of the honey bee eusocial and colonies lifestyle, but there are no genetic differences between worker and queen bee despite the major morphological differences between them (Evans and Wheeler, 2001; Hartfelder K, 1998; Jung-Hoffmann, 1966). The two castes develop in different nutritional environments. Queen-destined larvae are fed far more richer food (royal jelly) than worker-destined larvae, and the developmental pathways for workers and queens diverge during early larval development (Jung-Hoffmann, 1966; Maleszka, 2008).

The honey bee methylome is sensitive to the nutrition of the development larvae so that the early nutritional environment establishes the larva on either a worker or queen developmental pathway (Maleszka, 2008). Experimental manipulations of DNA methylation early in larval development can switch worker-destined larvae to a queen developmental pathway, revealing the key role of changes in the DNA methylome in the natural phenotypic plasticity of the honey bee (Kucharski *et al.*, 2008; Shi *et al.*, 2013).

Environmental stressors can also disrupt the epigenome leading to developmental dysfunction, However. This is increasingly being recognized as an important component of many diseases (Cavalli and Heard, 2019; Pembrey *et al.*, 2014). An emerging concern is the possibility for stress-induced changes in the epigenome to be passed on to offspring (Cavalli and Heard, 2019; Skvortsova *et al.*, 2018). Until recently this was considered highly unlikely, but more and more cases are emerging. These include numerous examples from humans of transgenerational inheritance of epigenomic changes induced by smoking, nutritional stresses and toxins (Pembrey *et al.*, 2014). Inherited epigenomic changes resulting from environmental stress on the parent have now been linked to pathologies and phenotypic changes in plants, worms, flies, fish, birds, rodents, pigs, and humans (Nilsson *et al.*, 2018)(Anway *et al.*, 2005; Dias and Ressler, 2014; Nilsson *et al.*, 2018). For example, if male rats were exposed to the endocrine disruptor vinclozolin during embryonic gonadal sex determination their fertility and behavior was affected, as was the methylation state of their sperm such that the changes persisted over four generations (Anway *et al.*, 2005). Male mice maintained on a high fat diet for three generations accumulate changes in epigenetic systems regulating lipogenesis altering susceptibility to obesity (Li *et al.*, 2012).

The honey bee provides a fortuitous natural system to explore how the epigenome might respond to sustained developmental stress. In the natural process of queen development the queen lays an egg in an especially large queen cell made by the workers (Wei *et al.*, 2019). The workers fill the cell with royal jelly proving the hatchling with abundant rich food. By contrast, in contemporary commercial beekeeping, most queens are raised by artificially transplanting young worker larvae from worker cells into artificial queen cells, which the workers then provision with royal jelly to produce a queen (Büchler *et al.*, 2013; Doolittle, 1888). A consequence of this manipulative queen rearing method is that larvae begin development on a worker-destined trajectory and later switch to a queen-destined developmental trajectory. Queens reared from older worker larvae are smaller, lighter, have smaller ovaries and show changes in the methylation of many genes important for caste differentiation when compared to queens reared from honey bee eggs, (which more closely matches the natural process of queen production) (He *et al.*, 2017; Woyke, 1971). These differences influence colony growth and performance (Rangel *et al.*, 2012), and deteriorating queen quality is increasingly being recognized as an important factor in the recent declines in honey bee health (Brodschneider *et al.*, 2019).

Here we examined the consequences of rearing queens from worker larvae for repeated generations. We found not only that this influenced the phenotype and epigenome of the adult queens, but that repeated manipulations across successive generations caused an accumulation of changes to the honey bee methylome, affecting particularly genes involved the differentiation of worker and queen phenotypes. We argue such epigenomic attrition of developmental systems might be contributing to a decline in quality of honey bee stocks.

## Materials and methods

### *Animals*

The Western honey bee, *Apis mellifera*, was used throughout this study. honey bee colonies were maintained at the Honeybee Research Institute, Jiangxi Agricultural University, Nanchang, China (28.46°N, 115.49°E), according to standard beekeeping techniques. All experiments were performed in accordance with the guidelines from the Animal Care and Use Committee of Jiangxi Agricultural University, China.

### *Queen rearing and sampling*

Our queen rearing strategy is summarized in Fig. II-3-1(see page 517). Our initial founding queen (Fig. II-3-1) was a standard commercially available queen instrumentally inseminated with semen from a single unrelated drone. She was restricted for 6h (10 am - 4 pm) to a plastic honey bee frame to lay eggs in worker cells. This frame was designed such that the plastic base of each cell holding the egg or larva could be transferred to plastic honey bee queen cells (Pan *et al.*, 2013). 20-30 eggs or larvae were transferred to queen cells at 4pm on the $2^{nd}$, $4^{th}$, and $5^{th}$ day after laying. Thus, three types of daughter queen groups were established.

G1E were generation 1 queens reared from eggs transferred to queen cells on the $2^{nd}$ day after laying. G1L1 were G1 queens reared from one-day old larvae transferred to queen cells on the $4^{th}$ day after laying. G1L2 were G1 queens reared from two-day old larvae transferred to queen cells on the $5^{th}$ day after laying. The queen cells were placed in racks in two queenless honey bee colony to be tended by workers, fed royal jelly and reared as queens. In each generation, half of each queen rearing group was assigned to each queenless colony.

Of the G1 queens, three queen cells of each group were selected randomly on the 14$^{th}$ day after laying, and were each placed in a small queenless hive to emerge and mate naturally. The remaining G1 queen cells were numbered, the length of each queen cell was measured and then they were placed in a dark incubator (35 ℃, 80%) to emerge. From the 15$^{th}$ day post laying queen cells were checked every 2 h for queen emergence, and hourly after the first G1 queen emerged. The four queens in each group to emerge were taken for methylation analysis. These were immediately flash frozen in liquid nitrogen when collected after emergence and stored in a -80 ℃ refrigerator.

Remaining queens were sampled to measure ovariole number. These queens were transferred to queen cages, which were placed in queenless colonies for 4-5 days where they could be fed and tended by workers through the cage, since the ovaries of 4-5 day-old queens are easier to stain and count than newly emerged queens (Berger *et al.*, 2016; Patricio and Cruz-Landim, 2002). When 4-5 days old these queens were flash frozen in liquid nitrogen and stored in a -80 ℃ freezer. To score ovariole number we created paraffin sections of the stained and dissected ovary (Gan *et al.*, 2012). We counted the number of ovarioles in the left ovary by identifying slides in which the ovarioles were very clear and counting slides until we found at least two giving exactly the same number of ovarioles（fiji-win64 software, Fig. Ⅱ-3-2B, see page 517）.

We used two-way ANOVA to investigate the effects of queen type and generation on ovariole number, and Fisher's PLSD test to analyze differences between queen types within each generation. The number of queens in each sample group varied and was affected by queen larvae and queenless colony survival. We sampled 3-11 queens for ovariole analysis in each group (Fig. Ⅱ-3-2).

To rear the second generation queens (G2) we selected one of the mated and laying G1 queens from each group (G1E, G1L1 and G1L2). Each G1 queen was restricted for 6 h (10 am-4 pm) to a plastic worker honey bee frame for laying. We then created three different types of G2 queens as for G1.

Eggs from the G1E queen were transferred as eggs to queen cells on the 2$^{nd}$ day post laying to create the G2E group. Eggs from the G1L1 queens were transferred to queen cells on the 4$^{th}$ day after laying to create the G2L1 group. Eggs from the G1L2 queens were transferred to queen cells on the 5$^{th}$ day after laying to create the G2L2 group (Fig. Ⅱ-3-1). Queen cells of the G2 groups were treated the same way as the G1 queen cells. The emerging G2 queens were reared and sampled as for the G1 groups.

We repeated this process to create the 3$^{rd}$ and 4$^{th}$ generation queen groups: G3E, G3L1, G3L2 and G4E, G4L1 and G4L2. Three types of queens were sampled in each generation for methylation and ovariole anlaysis. 12 queen groups were sampled in total.

### *Paraffin section of the queen ovary*

Briefly, queens from each group were thawed to room temperature and both ovaries dissected from the abdomen. Ovaries were fixed in 4% paraformaldehyde fix solution (BBI Life Sciences) for 12 hours. For dehydration and fixing we used an Automatic dehydrator (Leica, TP1020). Ovaries were dehydrated in a graded ethanol series (70% - 100%). Ovaries were then cleared using xylene and samples then placed in a 1:1 absolute ethanol / xylene mixture for 30 min, then changed to xylene for 10 min, followed by fresh xylene for 5 min. They were transferred to a 1:1 xylene paraffin mixture for 30 min, and then paraffin wax for more than 2 hours.

Ovaries were embedded and blocked in paraffin wax using a Heated Paraffin Embedding Station (Leica). 5-7 μm sections were cut using a Leica RM 2245 microtome. Sections were placed on histological slides (Autostainer XL), stained with HE Staining Kit (BOSTER AR1180), mounted with neutral balsam mounting medium (BBI Life Sciences) and covered with a coverslip. The slides were then imaged and photographed using a 100x transmission light microscope (OLYMPUS, DP80)(Fig. II-3-2B).

*Genome wide methylation analysis*

Queens sampled for methylation analysis were flash frozen in liquid nitrogen when collected after emergence and stored in a -80 °C refrigerator. Four queen bee were sampled in each group. The brain, thorax and ovary of each queen was dissected over ice as one sample. Tissues from each queen were pooled for genomic DNA extraction. In total, 12 queen groups were sampled for methylation analysis. Of these, samples that did not meet our quality control requirements for genomic sequencing were deleted. Nine out of 48 samples were unsuitable. One sample from the queen groups G1E, G1L1, G2E, G2L1, G3E, G3L1, and G3L2 were unsuitable, and two samples from the queen group G2L2 were unsuitable. (Details shown in Supplementary Table S1).

*Genomic DNA extraction and quantification*

Genomic DNA was extracted using the StarSpin Animal DNA Kit (GenStar). Genomic DNA degradation and contamination was assessed by running the DNA on agarose gels. DNA purity was assessed using a NanoPhotometer® (IMPLEN, CA, USA). DNA concentration was measured using a Qubit® DNA Assay Kit (Life Technologies, CA, USA) and a Qubit® 2.0 Flurometer (Life Technologies, CA, USA). DNA samples were then sent for whole-genome bisulfite sequencing analysis by the Novogene Bioinformatics Technology Co., Ltd/www.novogene.cn using the method summarized below.

*Library preparation and quantification*

Then these DNA fragments were treated twice with bisulfite using EZ DNA Methylation-GoldTM Kit (Zymo Research), before the resulting single-strand DNA fragments were PCR amplified using KAPA HiFi HotStart Uracil + ReadyMix (2X).

100 ng genomic DNA was spiked with 0.5 ng lambda DNA and fragmented by sonication to 200-300 bp using a Covaris S220 DNA Sequencing/gene analyzer. These DNA fragments were treated with bisulfite using EZ DNA Methylation -GoldTM Kit (Zymo Research, CA, USA). Bisulfite converted DNA were processed by the Accel-NGS Methyl -Seq DNA Library Kit to create dual-indexed Methyl-Seq libraries. All libraries were amplified in a 9-cycle indexing PCR reaction. Library DNA concentration was quantified by a Qubit® 2.0 Flurometer (Life Technologies, CA, USA) and quantitative PCR. The insert size was assayed on an Agilent Bioanalyzer 2100 system.

*Data Analysis*

The library preparations were sequenced on an Illumina Hiseq XTen and 125 bp to 150 bp paired end-reads were generated. Image analysis and base identification were performed with Illumina CASAVA pipeline.

*Quality control*

FastQC (fastqc_v0.11.5) was used to perform basic statistics on the quality of the raw reads. Read sequences produced by the Illumina pipeline in FASTQ format were pre-processed through Trimmomatic (Trimmomatic-0.36) software use the parameter (SLIDINGWINDOW:4:15, LEADING:3, TRAILING:3, ILLUMINACLIP:adapter.fa:2, 30:10, MINLEN:36). Reads that passed all of these filtering steps were counted as clean reads and all subsequent analyses were performed on these. Finally, we used FastQC to perform basic statistics on the quality of the clean reads data.

*Reference data preparation before analysis*

Before the analysis, we prepared the reference data for *Apis mellifera*, including the reference sequence (as a fasta file), the annotation file in gtf format, the GO annotation file, a description of genes in the *Apis mellifera* genome (downloaded from NCBI) and the gene region file (also from NCBI, in BED format).

*Mapping reads to the reference genome*

Bismark software (version 0.16.3) was used to perform alignments of bisulfite-treated reads to a reference genome (Amel_HAv3.1 (GCF_003254395.2)) (Krueger and Andrews, 2011). For alignment of the library reads to the reference genome, the reference genome and library reads were firstly transformed into bisulfite-converted versions of the sequences (C-to-T and G-to-A) and then assigned to a digital index using bowtie2, so that the index information included data on the sequences, their origin, and the experiment (Langmead and Salzberg, 2012). Sequence reads from the bisulphite-sequenced samples were aligned to fully bisulfite-converted versions (C-to-T and G-to-A converted) versions of the genome in a directional manner. Sequence reads that produced a unique best alignment from the two alignment processes (original top and bottom strand) were then compared to the normal genomic sequence and the methylation state of all cytosine positions in the read was thus inferred. Reads that aligned to the same regions of the genome were regarded as duplicates. The sequencing depth and coverage were calculated assessing number of overlapping reads relative to number of duplicate reads.

*Estimating methylation level*

To identify the level of methylation at each site, we modeled the count of methylated cytosines (mC) at a site as a binomial (Bin) random variable with methylation rate r: mC ~ Bin (mC + umC,r) (http://www.stat.yale.edu/Courses/1997-98/101/binom.htm).

In order to calculate the methylation level of a sequence, we divided the sequence into multiple bins, of 10kb. The sum of methylated and unmethylated read counts in each bin were calculated. Methylation level (ML) for each bin or C site shows the fraction of methylated Cs, and is defined as: ML (C) = (reads mC) / (reads (mC) + reads (C)).

Calculated ML was further corrected with the bisulfite non-conversion rate according to previous studies (Lister *et al*. 2013). Given the bisulfite non-conversion rate r, the corrected ML was estimated as: ML (corrected) = (ML - r) / (1-r).

*Differential methylation analysis*

Differentially methylated regions (DMRs) were identified using the DSS software (v 2.28.0) (Feng *et al.*, 2014). DSS is an R library performing differential analysis for count-based sequencing data. It detects differentially methylated loci or regions (DML/DMRs) from bisulfite sequencing (BS-seq). The core of DSS is a new dispersion shrinkage method for estimating the dispersion parameter from Gamma-Poisson or Beta-Binomial distributions. DMRs were identified using the parameters: smoothing = TRUE, smoothing.span = 200, delta = 0, p.threshold = 1e-05, minlen = 50, minCG = 3, dis.merge = 100, pct.sig = 0.5 (Feng *et al.*, 2014; Park and Wu, 2016; Wu *et al.*, 2015). The final list of DMRs were assigned genomic positions, or overlapped between comparisons. We defined the genes related to DMRs as genes whose gene body region (from TSS to TES) or promoter region (upstream 2 kb from the TSS) overlapped with the DMRs.

### *GO and KEGG enrichment analysis of DMR-related genes (DMGs)*

A DMR related gene is defined as a gene within which the DMR is located. If there is differential methylation in any region of the gene body, the gene will be considered as the differentially methylated gene (DMG). Gene Ontology (GO) enrichment analysis of genes related to DMRs was implemented by the GOseq R package (Young *et al.*, 2010), in which gene length bias was corrected. GO terms with corrected P-value less than 0.05 were considered significantly enriched among DMR-related genes. The KEGG (Kanehisa *et al.*, 2008) database related genes to high-level functions and utilities of a biological system, (http://www.genome.jp/kegg/). We used KOBAS software (Mao *et al.*, 2005) to test the statistical enrichment of DMR related genes to different KEGG pathways.

## Results

### *DNA methylation sequence quality*

From our four generations of queens of each rearing type (Fig. II-3-1) we also assessed the methylation status of the genome with bisulphite sequencing. From each sample the average number of clean reads was 38,751,257, with 10.2 G of clean base sequences (Table S1). The average phred scores Q30% (Ewing and Green, 1998) was 92.4% (Table S1). The average bisulfite conversion rate was 99.7% (Table S1). The average site coverage rate was 25.06 (Table S2), indicating that there was acceptable sequencing quality (NIH roadmap epigenomics project, http://www.roadmapepigenomics.org/protocols). The pearson correlation coefficient among biological replicates of each experimental group were all ≥0.97 (Table S3), indicating good repeatability among the biological replicates of each group.

### *Effect of queen rearing method on queen morphology and the methylome*

Ovariole number differed significantly between queens of different rearing types (Fig. II-3-2, Two-way ANOVA, $F = 18.869$, $DF = 2$, $P < 0.001$), but no effect of generation (G1 - G4, Fig. II-3-2) on ovariole number (Two-way ANOVA, $F = 0.321$, $DF = 3$, $P = 0.809$), and no interaction (Two-way ANOVA, $F = 0.326$, $DF = 6$, $P = 0.921$). Consistently ovariole number was reduced in L2 queens compared to E queens, with L1 queens intermediate between these groups (Fig. II-3-2).

When comparing queens from different rearing types we noted an increase in the number of DMGs with each generation of rearing. In each generation we compared DMGs between L1 with E queens. The number of DMGs increased with each generation of rearing (Fig. II-3-3A, see page 518, Table S4). We observed a similar phenomenon when we compared L2 with E queens in each generation (Fig. II-3-3A, Table S4).

To account for any possible effect of season or time on number of DMGs in our study we analyzed the number of DMGs in successive generations of each queen rearing type. For E queens, numbers of DMGs were extremely stable when we compared G2 with G1, G3 with G2 and G4 with G3 (Fig. II-3-3F, Table S5). By contrast when we examined L1 and L2 queens we observed the numbers of DMGs increased in each comparison of successive generations (Fig. II-3-3F).

With each successive generation the relatedness of queens between our rearing groups decreased. This could also cause the number of DMGs to increase each generation. To explore this possibility, we analyzed the number of DMGs shared by comparisons of different rearing groups within each generation and unique to comparisons of different rearing groups within each generation (Fig. II-3-3B-E), and unique to and shared between comparisons of successive generations (Fig. II-3-3G-I). Our data show that DMGs increased with successive generations of comparison (Fig. II-3-3A-E, Fig. S1). But DMGs were consistently greater for L2-E comparisons than L1-E comparisons (Fig. II-3-3A-E). Further when comparing DMGs between successive generations of E queens (Fig. II-3-3G) the number was very stable, suggesting that a decline in relatedness did not increase DMG number much. But when comparing L1 and E queens or L2 and E queens the number of DMGs increased with each generation (Fig. II-3-3 F, H and I). This suggests that for L1 and L2 queens' methylation differences accumulated with each generation of repeating L1 or L2 rearing.

When considering the lists of DMGs in each generation in more detail, we focused on genes previously identified as related to reproduction or longevity (Corona and Robinson, 2006; He *et al*., 2017; Yin *et al*., 2018), immunity (Barribeau *et al*., 2015; Boutin *et al*., 2015; He *et al*., 2017), metabolism (He *et al*., 2017), and DMGs identified in our own data as related to these groups from KEGG pathway analyses. Some genes may be related to multiple functions, but we only counted them once. Our classification criteria was: reproduction or longevity > immunity > metabolism. We used this functional annotation of target genes to numbers of DMGs in different functional groups in our different comparisons (Fig. II-3-4, see page 519). We saw an increasing number of genes involved in key caste differentiation processes (body development, immunity and reproduction / longevity) differentially methylated between different rearing types with each generation of rearing (Fig. II-3-4 and Table S6).

Consistently in the DMG gene lists comparing our queen groups (Table S7) we noted 106 genes with functions that have been related to caste differentiation (Beltran *et al*., 2007; Buttstedt *et al*., 2016; Guan *et al*., 2013; Marshall *et al*., 2019; Tian *et al*., 2018), body development and metabolism (Bull *et al*., 2012; Davis *et al*., 2002; Evans *et al*., 2006; Mao *et al*., 2017; Miller *et al*., 2012; Parker *et al*., 2012; Shi *et al*., 2011; Zufelato *et al*., 2004), and gene regulatory pathways related to caste differentiation (Amdam, 2011; Barchuk *et al*., 2007; Foret *et al*., 2012; He *et al*., 2017; Yin *et al*., 2018). Of these, we analyzed 40 genes that appeared most consistently in our DMG lists across generations (Fig. II-3-3 and Table S8). From G1 - G4 there was an increase in both the number of DMGs (Fig. II-3-5), and the ratio of methylation differences between the compared sequences (Fig. II-3-5, see page 519).

From the 40 genes in Fig. II-3-5, we selected two genes from gene regulatory pathways already implicated in the epigenomic mechanism of queen / worker differentiation (Foret *et al.*, 2012). The *Cat* gene is involved in FoxO pathway and longevity (Klichko *et al.*, 2004). *S6k1* is involved in mTOR signaling pathway and TGF-beta pathway (Chen *et al.*, 2012; Foret *et al.*, 2012). For these two genes we examined where in the gene sequence changes in methylation occurred (Fig. II-3-6, see page 520 and Table S9). We focused on exons (Lyko *et al.*, 2010), and limited Fig. II-3-6 to displaying only exons in which significant DMRs occurred. We observed accumulating changes in the amount of methylation at specific sites in our two genes from G1 - G4.

## Discussion

The development of both the worker and queen honey bee castes is dependent on differential methylation of the bee genome (Barchuk *et al.*, 2007; Kucharski *et al.*, 2008; Lyko *et al.*, 2010; Maleszka, 2008; Maleszka *et al.*, 2014). This epigenomic "developmental switch" allows workers to control which eggs develop as future queens for their colonies by controlling the nutrition of larvae (Jung-Hoffmann, 1966). But here we show this epigenomic developmental system is itself compromised by the developmental stress inherent in contemporary apicultural methods of queen rearing.

In commercial queen rearing it is common to transplant larvae up to 3-days old from worker cells into queen cells where they will be subsequently provisioned as queens. This practice has been in very widespread use in apiculture since 1888 (Doolittle, 1888). 2-day old worker larvae transplanted to queen cells could be successfully raised as queens, but there were consequences from this developmental manipulation for the queen phenotype. We consistently found that queen ovariole number was lower in L1 and L2 queens compared with E queens (Fig. II-3-2A). The number of ovarioles determines how many eggs can be produced and matured by the queen. This difference would be expected to have consequences for colony growth and function since the queen is the sole reproductive in a honey bee colony, and the mother of the entire worker population. Indeed, Rangel *et al.* (Rangel *et al.*, 2012) reported slower growth of bee colonies headed by queens reared from older worker larvae. Our findings confirmed earlier studies reporting an effect of queen rearing type on queen reproductive organs (Woyke, 1971).

We found many differences in the methylome of queens of different rearing types, but for the first time we tracked how these differences changed if rearing types repeated for successive generations (Fig. II-3-3 to Fig. II-3-6). We found that methylation differences between the different rearing types increased with each successive generation (Fig. II-3-3 to Fig. II-3-6). In effect, we observed a progressive divergence in the methylome of our queen rearing types as we sustained the different methods of queen rearing.

Our analyses focused on pathways and genes that have previously been related to the process of caste differentiation in honey bee (He *et al.*, 2017) such that after four generations of rearing the methylation differences between L1 or L2 queens and E queens were far greater than after one generation (Fig. II-3-4 to Fig. II-3-6).

Confounded with our inter-generational sampling is the passage of time, change in season at the time of sampling and reduced genetic relatedness of between our rearing groups, and so we must consider how each of these might have contributed to the increase in DMGs across generations we report. When we compared L1 with E or L2 with E within our four generations, we saw an increasing number of DMGs with each generation of rearing (Fig. II-3-3A), but when we measured DMGs between mother and daughter E queens across our four generations the number of DMGs was very stable and did not increase (Fig. II-3-3F, G). This result shows that for E queens (the condition that most closely matches the evolved and natural method of queen rearing) the number of DMGs did not increase significantly with each generation of sampling. For E queens we did not see the confounds of reduced genetic relatedness, time or season increasing the number of DMGs significantly in successive intergenerational comparisons (Fig. II-3-3F,G).

For both L1 of L2 queen groups, the number of DMGs in successive mother / daughter comparisons increased with each generation of rearing (Fig. II-3-3F, H, and I) indicating an increase of DMGs with each generation. The number of DMGs between E queens and L2 queens increased from 448 in generation 1 to 2065 in generation 4 (Fig. II-3-3A). Many of these changes were associated with genes with functional characterizations linked to caste determination (Fig. II-3-4B). It seems unlikely that season, time or reduced relatedness would have increased DMG number in our L2 and L1 groups but not our E group. Hence we feel our data includes a signature of repeated rearing of queens from larvae increasing gene methylation differences with each generation.

Our G1 queens were all daughters of one queen mated with the sperm of single drone (Fig. II-3-1), consequently in G1 our queens were all full sisters. In each subsequent generation, however, queens mated naturally with the local population of drones to create the next generation of queens. While we could not prevent relatedness between our rearing groups decreasing with each generation, this strategy should have prevented our developmental lines from diverging into distinct genetic lines by repeated introgression with the same genetic background.

There are about 70,000 methylated cytosine sites in the *Apis mellifera* genome. Most of these are CpG dinucleotides in exons (Lyko *et al.*, 2010). In L1 and L2 queen groups methylation changes accumulated in exons. In insects methylation of exons has been related to functional changes in gene expression, and / or may mediate splice variation (Cingolani *et al.*, 2013; Foret *et al.*, 2012; Foret *et al.*, 2009; Li-Byarlay *et al.*, 2013; Wojciechowski *et al.*, 2014).

If we consider the L1 and L2 queen rearing types as experiencing a form of developmental stress, then we report an accumulation of methylation changes with sustained stress across generations. Similar findings have been reported for nematodes and rodents where rearing an organism under stress for repeated generations induced more methylation changes than rearing under stress for a single generation (Li *et al.*, 2012; Remy, 2010).

Burggren (2015) in a review of this general phenomenon highlights how epigenetic changes should be recognized as graded time related changes that can both "wash in" and "wash out" of the genome over time. As examples, rearing mice on a fatty diet across three generations has been shown to "wash in" epigenetic changes (DNA methylation, modification in histones) contributing to obesity susceptibility (Li *et al.*, 2012). In nematodes, repeated exposure to an odour across 4 generations resulted in what had originally been an induced behavioral change to this odour to become a stable inherited behavioral change (genomic imprinting) (Remy,

2010). The distinction between acquired and inherited characteristics is not absolute (Burggren, 2015; Furrow and Feldman, 2014; Robinson and Barron, 2017).

The distinct worker and queen developmental trajectories are dependent on epigenomic regulation, hence the cryptic accumulation of changes to the queen methylome seen here is troubling.

In this study we did not see any progressive change in ovariole number resulting from rearing queens from L2 larvae for four generations, but rearing queens from larvae has been a standard practice in apiculture for decades, and commercially this practice could sustain a developmental stress on a queen stock for many generations. We now recognize that queens reared from worker-destined larvae are of lower quality than queens reared from eggs (Rangel et al., 2012). Our work highlights the more concerning possibility of this practice causing a progressive and cryptic erosion of the epigenetically regulated queen developmental pathway. This could reduce queen quality without any detectable changes in bee genetics or conventional inbreeding.

The effect of methylation modification on insect caste differentiation is a controversial debate. Libbrecht et al. (2016) suggest that differential methylation in insects is weak or absent. Indeed DNA methylation is unstable and can be altered by various environmental factors such as nutrition or environmental stress (Kucharski et al., 2008; Cavalli and Heard, 2019). However, knocking down the Dnmt3 gene significantly increased the probability of newly hatched L1 larvae developing into queen bee (Kucharski et al., 2008). Shi et al (2013) reported a significantly higher global DNA methylation level in queen larvae compared to worker larvae. The present study showed a great number of DMGs detected in the comparisons of G1L1 with G1E (430) and G1L2 with G1E (448) (Fig. II-3-3A). The DMG number was similar to that reported by Lyko et al. (2011), who showed 560 DMGs in the brains of queens and workers. Interestingly, of those 560 DMGs 25 and 29 genes also appeared in the comparisons of G1L1 with G1E and G1L2 with G21E respectively (Table II-3-1). Moreover, the number of overlapping genes increased with generations, which also indicates the cumulative epigenetic effect of each rearing type in successive generations (Table II-3-1). These results together with the previous data support the hypothesis that DNA methylation plays an important role in honey bee queen / worker differentiation.

Guan et al (2013) showed that the relationship between DNA methylation and gene expression is quite weak in honey bee. We compared our DMGs from two comparisons (G1L1 with G1E, G1L2 with G1E) to the related DEGs from He et al. (2017) and Yi et al. (2020). Similarly, there were 0 and only 3 overlapping genes between our DMGs and DEGs from the above two studies respectively (Table II-3-2). Our results with previous findings indicate that honey bee gene expression may not be regulated directly by the modification of methylation. DNA methylation may interact with other regulatory means such as alternative splicing (Foret et al., 2012; Li-Byarlay et al., 2013) and chromatin state to influence the gene expression. Clearly, the subject requires further investigation.

In summary, we provide the first evidence of accumulating methylation changes arising from domestic rearing of the honey bee. We draw attention to an important potential mechanism for cryptic genomic change (changes in genomic function that could not be detected as changes in DNA sequence caused by inbreeding, for example) in this important species.

Table II-3-1  Comparison between significantly DMGs in queens/workers

| Generations | Comparisons | DMGs in this study | DMGs (Brains in queens/workers) in Lyko et al. (2011) | overlapping genes |
| --- | --- | --- | --- | --- |
| G1 | L1 vs E | 430 | 560 | 25 |
|  | L2 vs E | 448 |  | 29 |
| G2 | L1 vs E | 1021 |  | 83 |
|  | L2 vs E | 1151 |  | 98 |
| G3 | L1 vs E | 1598 |  | 146 |
|  | L2 vs E | 2073 |  | 160 |
| G4 | L1 vs E | 2235 |  | 159 |
|  | L2 vs E | 2065 |  | 153 |

Table II-3-2  Comparison of DEGs with those in other related articles

| Comparisons | DMGs in G1 in this study | He et al. (2017) | | Yi et al. (2020) | |
| --- | --- | --- | --- | --- | --- |
| | | DEGs (3d queen larvae) | overlapping genes | DEGs (Queens) | overlapping genes |
| L1 vs E | 430 | 11 | 0 | 176 | 1 |
| L2 vs E | 448 | 75 | 0 | 218 | 2 |

## Author contributions

ZJZ, XJH and WYY designed research. YY and YBL performed research. XJH, ABB and ZLW provided guidance for data. YY analyzed data. ABB and YY wrote the paper.

## Funding

This work was supported by the National Natural Science Foundation of China (31872432), the Earmarked Fund for China Agriculture Research System (CARS-44-KXJ15) and China Scholarship Council (201808360213).

## Data availability

The raw Illumina sequencing data are accessible through NCBI's database: DNA methylation data of E, L1, L2 in generations (G1 - G4): NCBI Bioproject: PRJNA598779 (SUB6726989).

*Ethics approval and consent to participate*

Ethics approval was not applicable.

*Consent for publication*

Not applicable.

*Declaration of competing interest*

The authors declare that they have no competing interests.

## Acknowledgements

We thank to prof. Yuan Mei Guo of Jiangxi Agricultural University for his guidance in data statistics, Theotime Colin, a doctoral student at Macquarie University, for teaching me to use the fiji-win64 software, and Xiao Feng He and Peng Fei Yu from Novogene Bioinformatics Technology Co., Ltd.

## Appendix A. Supplementary Data

Supplementary data to this article can be found online at Mendeley Data, V2, doi: http://dx.doi.org/10.17632/y46zb27tcg.3

## References

AMDAM, G.V., 2011. Social context, stress, and plasticity of aging. *Aging Cell* 10, 18-27.

ANWAY, M.D., CUPP, A.S., UZUMCU, M., SKINNER, M.K., 2005. Epigenetic transgenerational actions of endocrine disruptors and male fertility. *Science* 308, 1466-1469.

BARRIBEAU, S.M., SADD, B.M., DU PLESSIS, L., BROWN, M.J.F., BUECHEL, S.D., CAPPELLE, K., CAROLAN, J.C., CHRISTIAENS, O., COLGAN, T.J., ERLER, S., EVANS, J., HELBING, S., KARAUS, E., LATTORFF, HM., MARXER, M., MEEUS, I., NAPFLIN, K., NIU, J., SCHMID-HEMPEL, R., SMAGGHE, G., WATERHOUSE, R.M., YU, N., ZDOBNOV, E.M., SCHMID-HEMPEL, P., 2015. A depauperate immune repertoire precedes evolution of sociality in bees. *Genome Biol.* 16, 83.

BARCHUK, A.R., CRISTINO, A.S., KUCHARSKI, R., COSTA, L.F., SIMOES, Z.L., MALESZKA, R., 2007. Molecular determinants of caste differentiation in the highly eusocial honeybee *Apis mellifera*. *BMC Dev. Biol.* 7, 70.

BELTRAN, S., ANGULO, M., PIGNATELLI, M., SERRAS, F., COROMINAS, M., 2007. Functional dissection of the ash2 and ash1 transcriptomes provides insights into the transcriptional basis of wing phenotypes and reveals conserved protein interactions. *Genome Biol.* 8, R67.

BERGER, B., POIANI, S.B., CRUZ-LANDIM, C.D., 2016. Beekeeping practice: effects of *Apis mellifera* virgin queen management on ovary development. *Apidologie* 47, 589-595.

BOUTIN, S., ALBURAKI, M., MERCIER, P.L., GIOVENAZZO, P., DEROME, N., 2015. Differential gene expression between hygienic and non-hygienic honeybee (*Apis mellifera* L.) hives. *BMC genomics* 16, 500.

BRODSCHNEIDER, R., GRAY, A., ADJLANE, N., BALLIS, A., BRUSBARDIS, V., 2019. Loss rates of honey bee colonies during winter 2017/18 in 36 countries participating in the COLOSS survey, including effects of forage sources., Vol. 58,ed., pp. 479-485.

BÜCHLER, R., ANDONOV, S., BIENEFELD, K., COSTA, C., HATJINA, F., KEZIC, N., KRYGER, P., SPIVAK, M., UZUNOV, A., WILDE, J., 2013. Standard methods for rearing and selection of *Apis mellifera* queens. *J. Apicul. Res.* 52, 1-30.

BULL, J.C., RYABOV, E.V., PRINCE, G., MEAD, A., ZHANG, C.J., BAXTER, L.A., PELL, J.K., OSBORNE, J.L.,

CHANDLER, D., 2012. A strong immune response in young adult honeybees masks their increased susceptibility to infection compared to older bees. *PLoS Pathog.* 8, e1003083.

BURGGREN, W.W., 2015. Dynamics of epigenetic phenomena: intergenerational and intragenerational phenotype 'washout'. *J. Exp. Biol.* 218, 80-87.

BUTTSTEDT, A., IHLING, C.H., PIETZSCH, M., MORITZ, R.F.A., 2016. Royalactin is not a royal making of a queen. *Nature* 537, E10-2.

CAVALLI, G., HEARD, E., 2019. Advances in epigenetics link genetics to the environment and disease. *Nature* 571, 489-499.

CHEN, X., HU, Y., ZHENG, H., CAO, L., NIU, D., YU, D., SUN, Y., HU, S., HU, F., 2012. Transcriptome comparison between honey bee queen- and worker-destined larvae. *Insect Biochem. Mol. Biol.* 42, 665-673.

CINGOLANI, P., CAO, X., KHETANI, R.S., CHEN, C.C., COON, M., SAMMAK, A., BOLLIG-FISCHER, A., LAND, S., HUANG, Y., HUDSON, M.E., GARFINKEL, M.D., ZHONG, S., ROBINSON, G.E., RUDEN, D.M., 2013. Intronic non-CG DNA hydroxymethylation and alternative mRNA splicing in honey bees. *BMC Genomics* 14, 666.

CORONA, M., ROBINSON, G.E., 2006. Genes of the antioxidant system of the honey bee: annotation and phylogeny. *Insect Mol. Biol.* 15, 687-701.

DAVIS, M.B., SUN, W., STANDIFORD, D.M., 2002. Lineage-specific expression of polypyrimidine tract binding protein (PTB) in Drosophila embryos. *Mech. Dev.* 111, 143-147.

DIAS, B.G., RESSLER, K.J., 2014. Parental olfactory experience influences behavior and neural structure in subsequent generations. *Nat. Neurosci.* 17, 89-96.

DOOLITTLE, G.M., 1888. Scientific Queen-rearing. *Am. Bee J.*, 132.

EVANS, J.D., WHEELER, D.E., 2001. Gene expression and the evolution of insect polyphenisms. *Bioessays* 23, 62-68.

EVANS, J.D., ARONSTEIN, K., CHEN, Y.P., HETRU, C., IMLER, J.L., JIANG, H., KANOST, M., THOMPSON, G.J., ZOU, Z., HULTMARK, D., 2006. Immune pathways and defence mechanisms in honey bees *Apis mellifera*. *Insect Mol. Biol.* 15, 645-656.

EWING, B., GREEN, P., 1998. Base-calling of automated sequencer traces using phred. II. Error probabilities. *Genome Res.* 8, 186-194.

FENG, H., CONNEELY, K.N., WU, H., 2014. A Bayesian hierarchical model to detect differentially methylated loci from single nucleotide resolution sequencing data. *Nucleic Acids Res.* 42, e69.

FORET, S., KUCHARSKI, R., PITTELKOW, Y., LOCKETT, G.A., MALESZKA, R., 2009. Epigenetic regulation of the honey bee transcriptome: unravelling the nature of methylated genes. *BMC Genomics* 10, 472.

FORET, S., KUCHARSKI, R., PELLEGRINI, M., FENG, S., JACOBSEN, S.E., ROBINSON, G.E., MALESZKA, R., 2012. DNA methylation dynamics, metabolic fluxes, gene splicing, and alternative phenotypes in honey bees. *PNAS.* 109, 4968-4973.

FURROW, R.E., FELDMAN, M.W., 2014. Genetic variation and the evolution of epigenetic regulation. *Evolution.* 68, 673-683.

GAN, H.Y., TIAN, L.Q., YAN, W.Y., 2012. Paraffin section method of queen ovary. *J. bee.* 32, 9.

GUAN, C., BARRON, A.B., HE, X.J., WANG, Z.L., YAN, W.Y., ZENG, Z.J. 2013. A Comparison of Digital Gene

Expression Profiling and Methyl DNA Immunoprecipitation as Methods for Gene Discovery in Honeybee (*Apis mellifera*) Behavioural Genomic Analyses. *PLoS One 9*, e73628.

GUAN, C., ZENG, Z.J., WANG, Z.L., 2013. Expression of Sir2, Hdac1 and Ash2 in Honey Bee (*Apis mellifera* L.) Queens and Workers. *J. Apic. Sci.* 57, 67-73.

HARTFELDER K, E.W., 1998. Social insect polymorphism: Hormonal regulation of plasticity in development and reproduction in the honeybee. *Curr. Top Dev. Biol.* 40, 45-77.

HE, X.J., ZHOU, L.B., PAN, Q.Z., BARRON, A.B., YAN, W.Y., ZENG, Z.J., 2017. Making a queen: an epigenetic analysis of the robustness of the honeybee (*Apis mellifera*) queen developmental pathway. *Mol. Ecol.* 26, 1598-1607.

JUNG-HOFFMANN, I., 1966. Die Determination von Königin und Arbeiterin der Honigbiene. *Z. Bienenforsch* 8, 296-322.

KANEHISA, M., ARAKI, M., GOTO, S., HATTORI, M., HIRAKAWA, M., ITOH, M., KATAYAMA, T., KAWASHIMA, S., OKUDA, S., TOKIMATSU, T., YAMANISHI, Y., 2008. KEGG for linking genomes to life and the environment. *Nucleic Acids Res.* 36, D480-4.

KLICHKO, V.I., RADYUK, S.N., ORR, W.C., 2004. Profiling catalase gene expression in Drosophila melanogaster during development and aging. *Arch. Insect Biochem. Physiol.* 56, 34-50.

KRUEGER, F., ANDREWS, S.R., 2011. Bismark: a flexible aligner and methylation caller for Bisulfite-Seq applications. *Bioinformatics* 27, 1571-1572.

KUCHARSKI, R., MALESZKA, J., FORET, S., MALESZKA, R., 2008. Nutritional control of reproductive status in honeybees via DNA methylation. *Science* 319, 1827-1830.

LANGMEAD, B., SALZBERG, S.L., 2012. Fast gapped-read alignment with Bowtie 2. *Nat. Methods* 9, 357-359.

LI, J., HUANG, J., LI, J.S., CHEN, H., HUANG, K., ZHENG, L., 2012. Accumulation of endoplasmic reticulum stress and lipogenesis in the liver through generational effects of high fat diets. *J. Hepatol.* 56, 900-907.

LIBBRECHT, R., OXLEY, P.R., KELLER, L., & KRONAUER, D.J.C., 2016. Robust DNA Methylation in the Clonal Raider Ant Brain. *Curr. Biol.*, 26, 391-395.

LI-BYARLAY, H., LI, Y., STROUD, H., FENG, S., NEWMAN, T.C., KANEDA, M., HOU, K.K., WORLEY, K.C., ELSIK, C.G., WICKLINE, S.A., JACOBSEN, S.E., MA, J., ROBINSON, G.E., 2013. RNA interference knockdown of DNA methyl-transferase 3 affects gene alternative splicing in the honey bee. *PNAS.* 110, 12750-12755.

LISTER, R., MUKAMEL, E.A., NERY, J.R., URICH, M., PUDDIFOOT, C.A., JOHNSON, N.D., LUCERO, J., HUANG, Y., DWORK, A.J., SCHULTZ, M.D., YU, M., TONTI-FILIPPINI, J., HEYN, H., HU, S., WU, J.C., RAO, A., ESTELLER, M., HE, C., HAGHIGHI, F.G., SEJNOWSKI, T.J., BEHRENS, M.M., ECKER, J.R., 2013. Global epigenomic reconfiguration during mammalian brain development. *Science* 341, 1237905.

LYKO, F., FORET, S., KUCHARSKI, R., WOLF, S., FALCKENHAYN, C., MALESZKA, R., 2010. The honey bee epigenomes: differential methylation of brain DNA in queens and workers. *PLoS Biol.* 8, e1000506.

MALESZKA, R., 2008. Epigenetic integration of environmental and genomic signals in honey bees: the critical interplay of nutritional, brain and reproductive networks. *Epigenetics* 3, 188-192.

MALESZKA, R., MASON, P.H., BARRON, A.B., 2014. Epigenomics and the concept of degeneracy in biological systems. *Brief. Funct. Genomics* 13, 191-202.

MAO, W., SCHULER, M.A., BERENBAUM, M.R., 2017. Disruption of quercetin metabolism by fungicide affects energy production in honey bees (*Apis mellifera*). *Proc. Natl. Acad. Sci. U. S. A.* 114, 2538-2543.

MAO, X., CAI, T., OLYARCHUK, J.G., WEI, L., 2005. Automated genome annotation and pathway identification using the KEGG Orthology (KO) as a controlled vocabulary. *Bioinformatics* 21, 3787-3793.

MARSHALL, H., LONSDALE, Z.N., MALLON, E.B., 2019. Methylation and gene expression differences between reproductive and sterile bumblebee workers. *Evol. Lett.* 3, 485-499.

MILLER, D., HANNON, C., GANETZKY, B., 2012. A mutation in Drosophila Aldolase causes temperature-sensitive paralysis, shortened lifespan, and neurodegeneration. *J Neurogenet* 26, 317-327.

NILSSON, E.E., SADLER-RIGGLEMAN, I., SKINNER, M.K., 2018. Environmentally Induced Epigenetic Transgenerational Inheritance of Disease. *Environmental Epigenetics* 4, dvy016.

PAN, Q.Z., WU, X.B., GUAN, C., ZENG, Z., 2013. A new method of queen rearing without grafting larvae. *Am. Bee J.* 153, 1279-1280.

PARK, Y., WU, H., 2016. Differential methylation analysis for BS-seq data under general experimental design. *Bioinformatics* 32, 1446-1453.

PARKER, R., GUARNA, M.M., MELATHOPOULOS, A.P., MOON, K.M., WHITE, R., HUXTER, E., PERNAL, S.F., FOSTER, L.J., 2012. Correlation of proteome-wide changes with social immunity behaviors provides insight into resistance to the parasitic mite, Varroa destructor, in the honey bee (*Apis mellifera*). *Genome Biol.* 13, R81.

PATRICIO, K., CRUZ-LANDIM, C., 2002. Mating influence in the ovary differentiation in adult queens of *Apis mellifera* L. (Hymenoptera, Apidae). *Braz. J. Biol.* 62, 641-649.

PEMBREY, M., SAFFERY, R., BYGREN, L.O., 2014. Human transgenerational responses to early-life experience: potential impact on development, health and biomedical research. *J. Med. Genet.* 51, 563-572.

RANGEL, J., KELLER, J.J., TARPY, D.R., 2012. The effects of honey bee (*Apis mellifera* L.) queen reproductive potential on colony growth. *Insectes Sociaux* 60, 65-73.

REMY, J.J., 2010. Stable inheritance of an acquired behavior in Caenorhabditis elegans. *Curr. Biol.* 20, R877-8.

ROBINSON, G.E., BARRON, A.B., 2017. Epigenetics and the evolution of instincts. *Science.* 356, 26-27.

SHI, Y.Y., HUANG, Z.Y., ZENG, Z.J., WANG, Z.L., WU, X.B., YAN, W.Y., 2011. Diet and cell size both affect queen-worker differentiation through DNA methylation in honey bees (*Apis mellifera*, Apidae). *PLoS One* 6, e18808.

SHI, Y.Y., YAN, W.Y., HUANG, Z.Y., WANG, Z.L., WU, X.B., ZENG, Z.J., 2013. Genome wide analysis indicates that queen larvae have lower methylation levels in the honey bee (*Apis mellifera*). *Naturwissenschaften* 100, 193-197.

SKVORTSOVA, K., IOVINO, N., BOGDANOVIĆ, O., 2018. Functions and mechanisms of epigenetic inheritance in animals. *Nat. Rev. Mol. Cell Bio.* 19, 774-790.

TIAN, W.L., LI, M., GUO, H.Y., PENG, W.J., XUE, X.F., HU, Y.F., LIU, Y., ZHAO, Y.Z., FANG, X.M., WANG, K., LI, X.T., TONG, Y.F., CONLON, M.A., WU, W., REN, F.Z., CHEN, Z.Z., 2018. Architecture of the native major royal jelly protein 1 oligomer. *Nat. Commun.* 9, 3373.

WEI, H., HE, X.J., LIAO, C.H., WU, X.B., JIANG, W.J., ZHANG, B., ZHOU, L.B., ZHANG, L.Z., BARRON, A.B., ZENG, Z.J., 2019. A maternal effect on queen production in honeybees. *Curr. Biol.* 29, 2208-2213.

WOJCIECHOWSKI, M., RAFALSKI, D., KUCHARSKI, R., MISZTAL, K., MALESZKA, J., BOCHTLER, M.,

MALESZKA, R., 2014. Insights into DNA hydroxymethylation in the honeybee from in-depth analyses of TET dioxygenase. *Open Biol.* 4.

WOYKE, J., 1971. Correlations between the age at which honeybee brood was grafted, characteristics of the resultant queens, and results of insemination. *J. Apicult. Res.* 10, 45-55.

WU, H., XU, T., FENG, H., CHEN, L., LI, B., YAO, B., QIN, Z., JIN, P., CONNEELY, K.N., 2015. Detection of differentially methylated regions from whole-genome bisulfite sequencing data without replicates. *Nucleic Acids Res.* 43, e141.

YI, Y., LIU, Y.B., BARRON, A.B., ZENG Z.J., 2020. Transcriptomic, morphological and developmental comparison of adult honey bee queens reared from eggs or worker larvae of differing ages. *J. Econ. Entomol.* doi: 10.1093/jee/toaa188.

YIN, L., WANG, K., NIU, L., ZHANG, H., CHEN, Y., JI, T., CHEN, G., 2018. Uncovering the Changing Gene Expression Profile of Honeybee (*Apis mellifera*) Worker Larvae Transplanted to Queen Cells. *Front. Genet.* 9, 416.

YOUNG, M.D., WAKEFIELD, M.J., SMYTH, G.K., OSHLACK, A., 2010. Gene ontology analysis for RNA-seq: accounting for selection bias. *Genome Biol.* 11, R14.

ZUFELATO, M.S., LOURENCO, A.P., SIMOES, Z.L., JORGE, J.A., BITONDI, M.M., 2004. Phenoloxidase activity in *Apis mellifera* honey bee pupae, and ecdysteroid-dependent expression of the prophenoloxidase mRNA. *Insect Biochem. Mol. Biol.* 34, 1257-1268.

# 4. Transcriptomic, Morphological and Developmental Comparison of Adult Honey Bee Queens Reared from Eggs or Worker Larvae of Differing Ages

Yao Yi[1,2,†], Yibo Liu[1,†], Andrew B. Barron[3], Zhijiang Zeng[1*]

1. Honeybee Research Institute, Jiangxi Agricultural University, Nanchang, Jiangxi 330045, P. R. of China
2. Jiangxi University of Traditional Chinese Medicine, Nanchang, Jiangxi 330045, P. R. of China
3. Department of Biological Sciences, Macquarie University, North Ryde, NSW 2109, Australia

**Abstract**

Queens and workers are very distinct phenotypes that develop from the same genome. Larvae from worker cells up to 3.5 days old can be transferred to larger queen cells and will subsequently be reared as queens, and develop into functional queens. This has become a very popular queen rearing practice in contemporary apiculture. Here, we used RNA-seq to study the consequences of rearing queens from transplanted worker larvae on the transcriptome of the adult queens. We found that queens reared from transferred older larvae developed slower, weighted less, and had fewer ovarioles than queens reared from transferred eggs, indicating queens were cryptically intercaste. RNA-Seq analysis revealed differentially expressed genes between queens reared from transferred larvae compared to queens reared from transferred eggs: the older the larvae transferred, the greater the number of differentially expressed genes (DEGs). Many of the differentially expressed genes had functions related to reproduction, longevity, immunity, or metabolism, suggesting that the health and long-term viability of queens was compromised. Our finds verify the previous studies that adult queens reared from older transferred larvae were of lower quality than queens reared from transferred eggs or younger larvae.

**Keywords:** *Apis mellifera*, RNA-Seq, Caste differentiation, Reproduction, Immunity

## Introduction

Honey bee workers and queens are a remarkable example of developmental plasticity (Weaver 1957, Shell and Rehan 2018). These two very different phenotypes are the result of divergent developmental pathways, which are themselves triggered by the different rearing environments and diets provided to queen-destined or worker-destined larvae in the colony (Weaver 1957, Shell and Rehan 2018). This natural plasticity has

---

† Authors contributed equally to this work.
*Corresponding Author: bees1965@sina.com(ZJZ).
注：此文发表在 *Journal of Economic Entomology* 2020 年第 6 期。

been exploited in apiculture, where the most common practice of rearing queen bees involves transplanting larvae from a worker rearing environment to a queen rearing environment (Weaver 1966). This changes the larvae onto a queen developmental path and allows queens to be reared in large numbers. There is, however, increasing evidence that queens reared in this way are partially intercaste and not as robust as queens that developed from the egg stage in the queen rearing environment (Woyke 1971, HatchTarpy and Fletcher 1999, He *et al.* 2017). Here we explored the phenotypes of queens that were reared from transplanted worker larvae of different ages, and for the first time we quantified gene expression differences in adult queens from different rearing types.

Both queen and worker honey bees develop from fertilized eggs, but these two phenotypes differ in morphology, physiology and lifespan: queens are larger and fertile and can live for 3-5 years, while the workers are smaller, infertile and usually have a lifespan of just 45 days (Jung-Hoffmann 1966, Hartfelder K 1998, Evans and Wheeler 2001).

Queen and worker developmental pathways diverge during early larval development as a result of different nutritional environments (Asencot and Lensky 1977, Asencot and Lensky 1984, Brouwers 1984, Kamakura 2011, Wang *et al.* 2016) and space (Asencot and Lensky 1977, Asencot and Lensky 1984, Brouwers 1984, Kamakura 2011, Wang *et al.* 2016) provided to workers and queens. The queen pathway is quite robust such that larvae from worker cells up to 3.5 days old can be transferred to queen cells, and will then develop into functional queens (Weaver 1966) (Fig. 1). Such plasticity is used in commercial queen rearing practices and it is now most common to raise queens from transplanted larvae (Weaver 1966). This method of commercial queen rearing has undoubtedly brought many practical benefits for beekeepers, such as providing replacement queens for colonies that have suffered a queen death or have failing queens, and bringing new income sources for beekeepers through selling queens or royal jelly. Shifting larvae from the worker's developmental pathway to the queen's developmental pathway part way through development has some adverse consequences for the resulting queens, however.

Several researchers have documented that queens reared from transplanted worker larvae are lighter, have smaller spermathecae, smaller thorax length and thorax width, and have fewer ovarioles than queens reared from fertile eggs (Woyke 1971, Wei *et al.* 2019, He *et al.* 2017). The older the transferred worker larvae, the bigger the difference between queens reared from larvae and queens reared from eggs. Comparative functional genomics recently revealed differences in both gene methylation and gene expression between 3 or 4 day old larvae that had been transplanted to queen cells from worker cells at different ages (Woyke 1971, Wei *et al.* 2019, He *et al.* 2017). He *et al.* 2017 and Yin *et al.* 2018 reported differences in the transcription of functional genes related to reproduction, longevity and immunity (He *et al.* 2017, Yin *et al.* 2018).

The larval stage is only the first stage of queen development. After about six days of development larvae pupate, and queens spend a further eight days developing as pupae before emerging as adult queens. No study has yet determined what differences in gene expression caused by different larval rearing environments might persist to the adult stage. The physiology of the adult queen, has more direct relevance to the quality of the queen in apiculture. Hence, here we studied the biology and transcriptome of newly emerged adult queens from different experimental queen rearing groups. Newly emerged queens reared from transferred eggs, one-day old larvae, 2-day old larvae or 3-day old larvae were sampled in this experiment (Fig. II-4-1). We analysed differences in gene expression between queens reared from transferred worker larvae and queens reared from transferred eggs by RNA-Seq. We focused on genes related to traits that differ between queens and workers to assess the impacts of different rearing treatments on the phenotype of emergent queens.

**Materials and Methods**

Three colonies of the western honeybee, *Apis mellifera*, from the Honeybee Research Institute, Jiangxi Agricultural University, Nanchang, China (28.46° N, 115.49° E) were used in this experiment. One colony, containing a single-drone inseminated queen, was the mother of all queen samples used in this study. Another two colonies with similar populations of bees were used to rear the queen samples. The resident queens in these colonies were caged for the duration of the study to prevent them destroying the queen cells of our sample queens. Experiments were performed under the supervision of the Animal Care and Use Committee of Jiangxi Agricultural University, China.

Following methods in the single-drone inseminated queen was caged on a special plastic frame of worker cells (Pan *et al.* 2013) (Fig. S2A and S2B) to lay eggs for 6h (9 am - 3 pm). At noon on the $2^{nd}$ day after laying, we transplanted 40-60 eggs to plastic queen cells (Pan *et al.* 2013) (worker eggs, E group) (Fig. S2C, S2D, and S2E). On the $4^{th}$ day after laying we transferred 40-60 1-day-old larvae to queen cells (worker larvae, L1 group). Similarly, on the $5^{th}$ and $6^{th}$ day after laying we transferred 2-day-old larvae (worker larvae, L2 group) and 3-day-old larvae (worker larvae, L3 group) from worker cells to queen cells (Fig. II-4-1). These queen cells from each group were all intermixed and placed in the two rearing colonies to be reared as queen samples. The details of the artificial frame has been described in Pan *et al* (Pan *et al.* 2013).

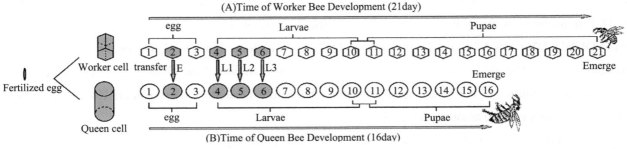

**Figure II-4-1** Timeline of worker bee (A) and queen bee (B) development, showing the points at which worker eggs and larvae were sampled and transplanted to queen cells.

On the $14^{th}$ day after laying, these queen cells (Fig. S2F) were transferred to an incubator (35 °C, 80% relative humidity) to emerge. Sealed queen cells were numbered and their length was measured (Fig. S1). Cells were checked every 2 h before the first queen emerged, and then checked hourly after the first emergence. The specific time of each queen emergence, the weight of each queen, and the weight of the remaining royal jelly in each queen cell were measured after queen emergence. The duration of queen development (h) was measured as the time from laying to the time of emergence.

The first six queens in each group were killed in liquid nitrogen immediately on emergence and stored in a -80 ℃ refrigerator until processing for RNA-Seq. The remaining queens were caged and placed in the rearing colonies for 4 - 5 days. Their ovaries were then dissected to score ovariole number, because the ovarioles of queens at this age are easier to score than those of newly emerged queens (Patricio and Cruz-Landim 2002).

Queens were killed in liquid nitrogen, stored in a -80 ℃ refrigerator, and thawed to room temperature before processing. Paraffin embedded sections of the left ovary were made according to Gan *et al.* (GanTian and Yan 2012). The number of ovarioles on two different sections taken from different positions in the ovary were counted. If the ovariole numbers on the two sections were the same, this value is valid. One way ANOVA followed with Tukey test was used to analyse differences in ovariole number between queen groups.

Of the first six queens sampled from each group for RNA-seq, four of them were selected randomly. The brain, thorax and ovary of each was dissected over wet ice and pooled as one sample for total RNA extraction. Total RNA was extracted using TRIzol reagent (Life technologies, California, USA). The integrity and concentration of all samples were checked using an Agilent 2100 Bioanalyzer (Agilent Technologies, Inc., Santa Clara, CA, USA) in CapitalBio.

The cDNA library was constructed using the NEBNext Ultra RNA Library Prep Kit for Illumina (NEB, E7530S), NEBNext Poly(A) mRNA Magnetic Isolation Module (NEB, E7490S), NEBNext Multiplex Oligos for Illumina (Index Primers Set1) (NEB, E7335S) and the NEBNext Multiplex Oligos for Illumina (NEB, E7500S). The main steps for building cDNA library were: 1) Use a QUBIT RNA BR ASSAY KIT (Invitrogen, Q10211) to accurately quantify the starting amount of total RNA; 2) Purify mRNA from total RNA using Oligod(T)-containing beads; 3) Fragment the mRNA to the target length of 200 nucleotides. These fragments were then used as templates to synthesize the cDNA; 4) Synthesize $1^{st}$ Strand of cDNA; 5) Synthesize complementary $2^{nd}$ Strand of cDNA; 6) Perform End Repair/dA-tail; 7) Adaptor Ligation; 8) PCR amplification; 9) PCR product quality control. This involved quantifying PCR products accurately using QUBIT DNA HS ASSAY KIT (Invitrogen, Q32854). 10) Identifying PCR fragment size using a 2100 Bioanalyzer chip. The absolute quantification of the molar concentration of the library was calculated by using the molar concentration of the q-PCR standard sample in a KAPA Quantitative Kit (Cat no. KK4602), so as to ensure the accuracy of the amount used in the library.

A mixture of 10μL 0.1mol/L NaOH and 10μL 2mol/L library was centrifuged and placed at room temperature for 5 min to denature the DNA to single-strands, then they were placed on ice. 20μL denatured DNA library was added to 980μL precooled HT1 Hybridization buffer, and the final concentration of the library was adjusted to 20 pmol/L and placed on ice before processing. The molecules in the library were combined with primers fixed on the Flowcell to perform bridge PCR amplification, followed by RNA sequencing on the platform (ILLUMINA, HiSeq X Ten).

The raw reads obtained by Illumina was in fastq format with Phred33 quality score. The quality of the sequence data was evaluated by using FastQC software (Andrews 2010). From this analysis we filtered out of the sequence data unqualified reads. We removed from the raw data any reads that were: 1) adapter reads, 2) reads where more than 5% of the bases had not been read, 3) Reads with more than 30% low quality bases (Q value < 20), 4) the entire pair of reads with length less than 50 bp, 5) unpaired reads.

HISAT2 software were used for sequence alignment (KimLangmead and Salzberg 2015). The reference genome file for alignment was GCF_000002195.4_Amel_4.5_genomic.fa. The reference gene annotation file was GCF_000002195.4_Amel_4.5_genomic.gtf. Then, we homogenized the reads (Dohm *et al.* 2008). A random number of reads was selected, and the number of genes covered by those reads was checked with reference to genome annotation to determine whether the data were saturated or not. Saturation analysis was mainly to estimate whether the measured dataset could meet the needs of analysis.

StringTiesoftware was used to assemble each sample (Dohm *et al.* 2008), and combine the results of all sample assembly to obtain a total transcript annotation file using StringTie Merge. The results were presented in gtf format (https://ccb.jhu.edu/software/stringtie/gff.shtml), which mainly reflected the structural location information of the transcripts.

We calculated the read count for each expressed gene using StringTie software (Pertea *et al.* 2015). Then, we counted the Fragments Per Kilobases per Million fragments (FPKM) value for each gene (Mortazavi *et al.* 2008) as:

FPKM = Total exon fragments / (Mapped reads (Millions) × exon length (kb)). We used the DESeq2 package for comparative analysis of differences between sample groups. The screening criteria for differentially expressed genes (DEGs) were:

$|\log_2 FC| \geq 1$ and q-Value $\leq 0.05$.

$\log_2 FC \geq 1$ refers to the up-regulated differentially expressed genes, and $\log_2 FC \leq -1$ refers to down-regulated differentially expressed genes.

All genes in the sample were mapped to the GO database (Http://www.geneontology.org/), and the number of DEGs and non differentially expressed genes in each term were counted. For both GO functions and KEGG Pathways, p-values of significant enrichment analysis were calculated using hypergeometric tests, and p-values were adjusted (refer to q-values) using the Benjamini-Hochberg method. We estimated all KEGG pathways that differed between queen groups from pathways that were enriched for DEGs.

## Results

As the age of the larva transferred to the queen cell increased, queen cell length (Fig. II-4-2A, ANOVA: $F = 62.926$, $DF = 3$, $P < 0.0001$), remaining royal jelly (Fig. II-4-2B, ANOVA: $F = 136.428$, $DF = 3$, $P < 0.0001$), queen weight (Fig. II-4-2D, ANOVA: $F = 11.797$, $DF = 3$, $P < 0.0001$) and number of ovarioles in the left ovary (Fig. II-4-2E, ANOVA: $F = 10.563$, $DF = 3$, $P = 0.001$) all decreased significantly. Duration of queen development increased significantly (Fig. II-4-2C, ANOVA: $F = 17.337$, $DF = 3$, $P < 0.0001$). The average emergence time of the E group was about 18 hours earlier than L3 group.

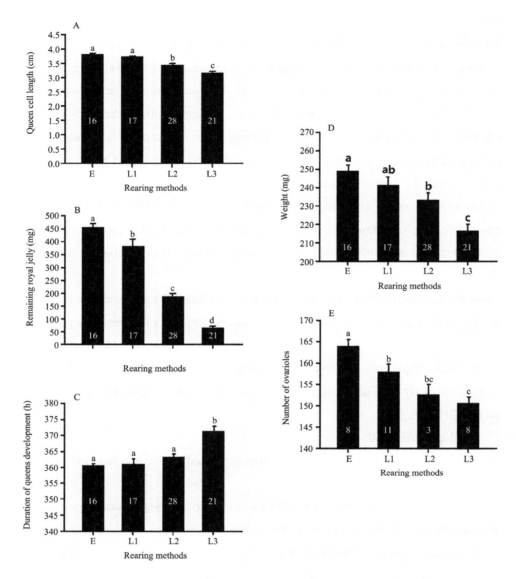

**Figure II-4-2** (A) Queen cell length, (B) Remaining royal jelly, (C) Duration of queen bee development, (D) Weight, and (E) Number of ovarioles of the four queen groups (E, L1, L2, and L3). Bars show mean ± SEM. Sample size of each queen group shown in the bar. One-way ANOVA test followed with Tukey test ($P < 0.05$) were performed and same letters above bars identify groups that did not differ significantly with each other.

Sixteen libraries for RNA-Seq were constructed in this experiment, and all of them met the sequencing requirements. For all libraries, clean bases were ≥ 6,976,904,016, Q30 were ≥ 95.50% (Ewing and Green 1998) (Table S1), uniquely mapped reads were ≥ 97.89%, multiple mapped reads were ≤ 2.11% (Table S2), indicating a high quality of sequencing. All the Pearson correlation coefficients of biological replicates within each queen group were ≥ 0.97 (Table S3), indicating a high repeatability among the biological replicates of each group.

The number of DEGs of the L1, L2 and L3 groups compared to the E group increased as the age of the larvae transferred to the queen cell increased (Fig. II-4-3 and Table S4). 64.86% of the DEGs of all larvae groups (L1, L2 and L3) were down-regulated when compared with E queens. Of these DEGs, many had functions related to reproduction, longevity (Barry and Camargo 2013, Corona and Robinson 2006, He *et al.* 2017, Yin *et al.* 2018, Walton *et al.* 2020), immunity (Fig. 3 and Table S4) (Barry and Camargo 2013, Corona and Robinson 2006, He *et al.* 2017, Yin *et al.* 2018, Walton *et al.* 2020), and metabolism (Fig. II-4-4, see page 521, and Table S5).

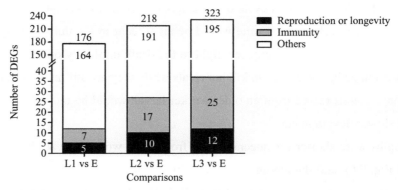

**Figure II-4-3** Number of differentially expressed genes (DEGs) in each comparison of queens reared from larvae (L1, L2 and L3) with E group (Table S4). Genes with functional classes related to reproduction, longevity, or immunity are shown in different shades and the number of these genes are shown in each bar section.

KEGG analysis indicated that DEGs of larvae groups compared with E group were mainly involved in metabolic pathways, and the number of total DEGs increased (N = 7, 11 and 15) as the age of the transferred larvae increased (L1, L2 and L3) (Fig. II-4-4 and Table S5).

All DEGs (N = 46) with functional annotation in reproduction, longevity, immunity, or metabolism between L3 group and E group were selected for further analysis (Table S4 and Table S5). Of them, 28 genes showed consistent up-regulated or down-regulated trends when we compared larvae groups (L1, L2 and L3) with E group (Fig. II-4-5, see page 522, and Table S6). Among them, 80% of the genes related to function in reproduction or longevity were down-regulated in expression. 53% of genes related to functions in immunity were down-regulated in expression.

## Discussion

Fertile queens can be produced by the commercial queen breeding practice of rearing queens from transplanted young worker larvae, but our research shows that adult queens of different larval rearing types (E, L1, L2 and L3) differed in their morphology, development time and transcriptome.

Queens reared from older worker brood are lighter (Fig. II-4-2D) and had fewer ovarioles (Fig. II-4-2E) than queens reared from eggs. This result is consistent with earlier studies (Woyke 1971, He *et al.* 2017, HatchTarpy and Fletcher 1999), and successfully replicates an important result. The reduced ovariole number of queens reared from later age worker larvae is a concern because ovariole number is a key factor for queens reproductive capacity (Bouletreau-Merle 1978, Woyke 1971), which has a direct impact on colony growth and productivity (RangelKeller and Tarpy 2012).

Our data showed that there were differences in development between our rearing groups post transfer to the same uniform queen cells. The duration of development until emergence differed between our groups: developmental time increased significantly as the age of the transferred larva increased (Fig. II-4-2C), therefore queens reared from older larvae developed more slowly than the queens reared from younger brood. Typically workers have a significantly longer developmental period than queens (despite being smaller), and so this result suggests that the developmental cycle of queens reared from worker larvae is partially intercaste. The consequences of a longer developmental time would be significant in a natural colony. A colony would naturally produce several queens who, on emergence, fight to the death until only one remains. The first queen to emerge has a major advantage because she can kill her rivals while they are still in their queen cells (Winston 1991). In a natural colony, queens reared from an older worker larvae would be unlikely to survive because of their lighter weight and slower development.

The queen cell lengths were shorter for queens reared from older worker larvae than queens reared from eggs (Fig. II-4-2A and Fig. S1), and the amount of royal jelly remaining in the queen cell post emergence of the adult queen also decreased with age of transplant (Fig. II-4-2B). We propose the decline in residual royal jelly with increasing age of the transplanted larva could be due to workers having less time to provision the queen larva prior to pupation (Guo *et al.* 2015). This factor, along with the larvae obtaining a smaller size by pupation, likely contributes to the decrease in size of the sealed queen cell. The important point is that there are consequences of the age of larval transfer for the entire developmental cycle of the queen. The developmental environment did not become uniform once eggs or larvae were transferred to the artificial queen cell. This likely contributed to the differences in adult queen phenotype we observed, and the differences in transcriptomes.

Number of DEGs of larvae (L1, L2, and L3) groups compared with E group, increased with the transferred age of worker larvae. Further analyses showed that many of these genes had functional classifications related to reproduction, longevity (Barry and Camargo 2013, Corona and Robinson 2006, He *et al.* 2017, Yin *et al.* 2018, Walton *et al.* 2020), immunity (Boutin *et al.* 2015, Barribeau *et al.* 2015), or metabolism (Fig. II-4-3, Fig. II-4-5, Table S4 and Table S6). This is concerning because these are key traits that differentiate queens and workers, and directly contribute to queen health, longevity and vigour. If queens are compromised in these traits it threatens the survival and growth of whole colonies.

KEGG analysis indicated that DEGs of all comparisons between larvae groups and E group were mainly involved in metabolic pathways (Fig. II-4-4 and Table S5), while the greatest and most abundant differences of DEGs in the comparison of queens and workers were also in genes related to metabolic processes (Corona Estrada and Zurita 1999, SeversonWilliamson and Aiken 1989, Evans and Wheeler 1999). This suggests that queens reared from worker larvae differ in metabolic function compared to queens reared from worker eggs.

In summary our data reinforce the interpretation that queens reared from older worker larvae are partially intercaste (CoronaEstrada and Zurita 1999, SeversonWilliamson and Aiken 1989, Evans and Wheeler 1999). Queens reared from older worker larvae, while viable and appearing normal, had reduced reproductive capacity. Transcriptomic analyses suggested compromised immunity and longevity in queens reared from older worker larvae. In view of this, we recommend transferring eggs or larvae as young as possible from worker cells to queen cells for the rearing of high quality queens.

## Acknowledgements

The authors would like to thank Dr. Xu Jiang HE for giving us help in the data upload. This work was supported by the National Natural Science Foundation of China (31872432), the 2018 Graduate Innovation Fund Project of JiangXi Province (YC2018-BO33) and China Scholarship Council (201808360213).

## Data Availability

The Illumina sequencing data are accessible through NCBI's database: RNA-Seq data of queens reared from eggs (E): PRJNA627872 (SUB7324217); RNA-Seq data of queens reared from 1-d-old worker larvae (L1): PRJNA627910 (SUB7324536); RNA-Seq data of queens reared from 2-d-old worker larvae (L2): PRJNA627983 (SUB7324753); RNA-Seq data of queens reared from 3-d-old worker larvae (L3): PRJNA628012 (SUB7326988).

## Authors' Contribution

ZJZ and ABB conceived research. YY and YBL conducted experiments. ZJZ and YY contributed material. YY and YBL analysed data and conducted statistical analyses. YY and, ABB wrote the manuscript. ZJZ and YY secured funding. All authors read and approved the manuscript.

## Declaration of Competing Interest

The authors declare that they have no known competing financial interests or personal relationships that could have appeared to influence the work reported in this paper.

## References

ANDREWS, S. 2010. FastQC: A quality control tool for high throughput sequence data. http://www.bioinformatics.babraham.ac.uk/projects/fastqc/

ASENCOT, M., AND Y. LENSKY. 1977. The effect of sugar crystals in stored royal jelly and juvenile hormone on the differentiation of female honey bee larvae, pp. 3-5 In: Proc. 8th Int. Congr., IUSSI, Wageningen.

ASENCOT, M., AND Y. LENSKY. 1984. Juvenile hormone induction of 'queenliness' on female honey bee (*Apis mellifera* L.) larvae reared on worker jelly and on stored royal jelly. *Comp. Biochem. Physiol*. 78B:109-117.

BARRIBEAU, S.M., B.M. SADD, L. DU PLESSIS, M.J. BROWN, S.D. BUECHEL, K. CAPPELLE, J.C. CAROLAN, O. CHRISTIAENS, T.J. COLGAN, S. ERLER, et al. 2015. A depauperate immune repertoire precedes evolution of sociality in bees. *Genome Biol*. 16:83.

BARRY, E.R., and F.D. CAMARGO. 2013. The Hippo superhighway: signaling crossroads converging on the Hippo/Yap pathway in stem cells and development. *Curr. Opin. Cell. Biol*. 25:247-253.

BOULETREAU-MERLE, J. 1978. Ovarian activity and reproductive potential in a natural population of Drosophila melanogaster. *Oecologia*. 53:323-329.

BOUTIN, S., M. ALBURAKI, P.L. MERCIER, P. GIOVENAZZO, and N. DEROME. 2015. Differential gene expression between hygienic and non-hygienic honeybee (*Apis mellifera* L.) hives. *BMC Genomics*. 16:500.

BROUWERS, E. 1984. Glucose/fructose ratio in the food of honeybee larvae during caste differentiation. *J. Apicult. Res.* 23:94-101.

CORONA, M., and G.E. ROBINSON. 2006. Genes of the antioxidant system of the honey bee: annotation and phylogeny. *Insect Mol. Biol.* 15:687-701.

CORONA, M., E. ESTRADA, and M. ZURITA. 1999. Differential expression of mitochondrial genes between queens and workers during caste determination in the honeybee *Apis mellifera*. *J. Exp. Biol.* 202:929-938.

DOHM, J.C., C. LOTTAZ, T. BORODINA, and H. HIMMELBAUER. 2008. Substantial biases in ultra-short read data sets from high-throughput DNA sequencing. *Nucleic Acids. Res.* 36:e105.

DOOLITTLE, G.M. 1888. Scientific queen-rearing. *Am. Bee J.*:132.

EVANS, J.D., and D.E. WHEELER. 1999. Differential gene expression between developing queens and workers in the honey bee, *Apis mellifera*. *PNAS*. 96:5575-5580.

EVANS, J.D., and D.E. WHEELER. 2001. Gene expression and the evolution of insect polyphenisms. *Bioessays*. 23:62-68.

EWING, B., and P. GREEN. 1998. Base-calling of automated sequencer traces using phred. II. Error probabilities. *Genome Res.* 8:186-94.

GAN, H.Y., L.Q. TIAN, and W.Y. YAN. 2012. Paraffin section method of queen ovary. *J. bee* 32:9.

GUO, Y.H., L. ZHOU, Q.Z. PAN, L.Z. ZHANG, Y. YI, and Z.J. ZENG. 2015. Effect of different harvesting times on the yield and composition of royal jelly. *Acta Agriculturae Universitatis Jiangxiensis*. 37:120-125.

HARTFELDER K, E.W. 1998. Social insect polymorphism: Hormonal regulation of plasticity in development and reproduction in the honeybee. *Curr. Top. Dev. Biol.* 40:45-77.

HATCH, S., D. R. TARPY, D. J. C. FLETCHER. 1999. Worker regulation of emergency queen rearing in honey bee colonies and the resultant variation in queen quality. *Insectes Soc*. 46: 372-377.

HE, X.J., L.B. ZHOU, Q.Z. PAN, A.B. BARRON, W.Y. YAN, and Z.J. ZENG. 2017. Making a queen: an epigenetic analysis of the robustness of the honeybee (*Apis mellifera*) queen developmental pathway. *Mol. Ecol.* 26:1598-1607.

JUNG-HOFFMANN, I. 1966. Die Determination von Königin und Arbeiterin der Honigbiene. *Z. Bienenforsch.* 8:296-322.

KAMAKURA, M. 2011. Royalactin induces queen differentiation in honeybees. *Nature*. 473:478-483.

KIM, D., B. LANGMEAD, and S.L. SALZBERG. 2015. HISAT: a fast spliced aligner with low memory requirements. *Nat. Methods*. 12:357-360.

KUCHARSKI, R., J. MALESZKA, S. FORET, and R. MALESZKA. 2008. Nutritional control of reproductive status in honeybees via DNA methylation. *Science*. 319:1827-1830.

MORTAZAVI, A., B.A. WILLIAMS, K. MCCUE, L. SCHAEFFER, and B. WOLD. 2008. Mapping and quantifying mammalian transcriptomes by RNA-Seq. *Nat. Methods*. 5:621-628.

PAN, Q.Z., X.B. WU, C. GUAN, and Z. ZENG. 2013. A new method of queen rearing without grafting larvae. *Am. Bee J.* 153:1279-1280.

PATRICIO, K., and C. CRUZ-LANDIM. 2002. Mating influence in the ovary differentiation in adult queens of *Apis mellifera* L. (Hymenoptera, Apidae). *Braz. J. Biol.* 62:641-649.

PERTEA, M., G.M. PERTEA, C.M. ANTONESCU, T.C. CHANG, J.T. MENDELL, and S.L. SALZBERg. 2015. StringTie enables improved reconstruction of a transcriptome from RNA-seq reads. *Nat. Biotechnol.* 33:290-295.

RANGEL, J., J.J. KELLER, and D.R. TARPY. 2012. The effects of honey bee (*Apis mellifera* L.) queen reproductive potential on colony growth. *Insect Soc.* 60:65-73.

SEVERSON, D., J. WILLIAMSON, and J. AIKEN. 1989. Caste-specific transcription in the female honey bee. *Insect Biochem.* 19:215-220.

SHELL, W.A., and S.M. REHAN. 2018. Behavioral and genetic mechanisms of social evolution: insights from incipiently and facultatively social bees. *Apidologie.* 49:13-30.

SHI, Y.Y., Z.Y. HUANG, Z.J. ZENG, Z.L. WANG, X.B. WU, and W.Y. YAN. 2011. Diet and cell size both affect queen-worker differentiation through DNA methylation in honey bees (*Apis mellifera*, Apidae). *PLoS One.* 6:e18808.

WALTON, A., J.P. TUMULTY, A.L. TOTH, and M.J. SHEEHAN. 2020. Hormonal modulation of reproduction in Polistes fuscatus social wasps: Dual functions in both ovary development and sexual receptivity. *J. Insect Physiol.* 120:103972.

WANG, Y., L. MA, W. ZHANG, X. CUI, H. WANG, and B. XU. 2016. Comparison of the nutrient composition of royal jelly and worker jelly of honey bees (*Apis mellifera*). *Apidologie.* 47:48-56.

WEAVER, N. 1957. Effects of larval honey bee age on dimorphic differentiation of the female honeybee. *Ann. Entomol. Soc. Am.* 50:283-294.

WEAVER, N. 1966. Physiology of caste determination. *Annu. Rev. Entomol.* 11:79-102.

WEI, H., X.J. HE, C.H. LIAO, X.B. WU, W.J. JIANG, B. ZHANG, L.B. ZHOU, L.Z. ZHANG, A.B. BARRON, and Z.J. ZENG. 2019. A maternal effect on queen production in honeybees. *Curr. Biol.* 29:2208-2213.e3.

WINSTON, M.L. 1991. The Biology of the Honey Bee Harvard University Press, Cambridge, MA.

WOYKE, J. 1971. Correlations between the age at which honeybee brood was grafted, characteristics of the resultant queens, and results of insemination. *J. Apicult. Res.* 10:45-55.

YIN, L., K. WANG, L. NIU, H. ZHANG, Y. CHEN, T. JI, and G. CHEN. 2018. Uncovering the Changing Gene Expression Profile of Honeybee (*Apis mellifera*) Worker Larvae Transplanted to Queen Cells. *Front Genet.* 9:416.

# 5. Effects of Commercial Queen Rearing Methods on Queen Fecundity and Genome Methylation

Yao Yi[1,2†], Yibo Liu[1†], Andrew B. Barron[3], Zhijiang Zeng[1*]

1. Honeybee Research Institute, Jiangxi Agricultural University, Nanchang, Jiangxi 330045, P. R. of China
2. Jiangxi University of Traditional Chinese Medicine, Nanchang, Jiangxi 330004, P. R. of China
3. Department of Biological Sciences, Macquarie University, North Ryde, NSW 2109, Australia

**Abstract**

The queen and worker castes of the honey bee are very distinct phenotypes that result from different epigenomically regulated developmental programs. In commercial queen rearing it is common to produce queens by transplanting worker larvae to queen cells to be raised as queens. Here we examined the consequences of this practice for queen ovary development and genome wide methylation. Queens reared from transplanted older worker larvae weighed less and had fewer ovarioles than queens reared from transplanted eggs. Methylome analyses revealed a large number of genomic regions in comparisons of egg reared and larvae reared queens. The methylation differences became more pronounced as the age of the transplanted larva increased. Differentially methylated genes had functions in reproduction, longevity, immunity and metabolic functions suggesting that the methylome of larval reared queens was compromised and more worker-like than the methylome of queens reared from eggs. These findings caution that queens reared from worker larvae are likely less fecund and less healthy than queens reared from transplanted eggs.

**Keywords:** *Apis mellifera*, queen rearing, fecundity, genome methylation, caste differentiation

## Introduction

The honey bee presents one of the most impressive documented cases of developmental plasticity (Shell and Rehan 2018; Wheeler 1986). There are no genetic differences between the queen and worker castes. Both castes develop from divergent developmental pathways (Winston 1991). Recent and rapid progress in comparative functional genomics has revealed how the developmental and nutritional environment in the early larval stages establishes a cascade of epigenomic changes that alter patterns of both gene methylation and gene expression that ultimately result in different developmental systems and two radically different phenotypes (He *et al.* 2017; Kucharski *et al.* 2008; Yin *et al.* 2018).

---

† Authors contributed equally to this work.
*Corresponding Author: bees1965@sina.com(ZJZ).
注：此文发表在 *Apidologie* 2021 年第 1 期。

In nature the developmental environment is established by both the queen and the workers. Workers create just a few large queens cells, and many small hexagonal worker cells. The queen lays eggs in both cell types, but she lays slightly larger eggs in queen cells ensuring that even prior to hatching a queen-destined bee is provisioned with more food (Wei *et al.* 2019). Workers supply queen cells with an abundance of royal jelly, and provide worker cells with far less worker jelly. The two food types differ in nutrient content and provide very different nutrition for the two types of developing larvae (Wang *et al.* 2016).

Commercial queen rearing practices in contemporary apiculture exploit this plastic developmental system by creating artificial queen cells into which eggs or young larvae from worker cells are transplanted (Doolittle 1888). If these artificial queen cells are placed in a recently queenless colony they will be provisioned and reared by the workers as queens. Honey bee eggs are very fragile and transplanting eggs is technically difficult. For this reason it is common for queen breeders to transfer larvae from worker cells. Transplanting larvae from worker to queen cells up to the 3rd instar stage will result in a functional queen phenotype (Doolittle 1888). While the queen developmental pathway is robust enough to tolerate this kind of manipulation, diverting larvae from a worker developmental pathway to a queen developmental pathway in the mid larval stage is not without consequence.

Several authors have now documented how queens reared from transplanted worker larvae are smaller, weighed less, have smaller spermathecae and fewer ovarioles than queens reared naturally or from transplanted eggs (He *et al.* 2017; Wei *et al.* 2019; Woyke 1971). The differences get more pronounced as the age of the transplanted larva increases. More recent analyses have documented differences in gene expression in the developing queens depending on whether they were produced from transplanted larvae or eggs (He *et al.* 2017; Yin *et al.* 2018), and He at al reported even changes to the methylation of some genes (He *et al.* 2017).

The gene methylation changes are of particular interest since differential methylation is considered a key part of the mechanism that both enables the genome to be sensitive to the nutritional and developmental environment, and that establishes the different developmental pathways (Kucharski *et al.* 2008). Here we examined the changes to the honey bee methylome resulting from rearing queens from eggs and larvae of different ages. To assess the impacts of different methods of queen rearing on the queen epigenome our analysis considered changes across the whole genome, and also focused on genes that may be functionally related to caste differentiation processes that differ between workers and queens.

## Materials and Methods

### *Queen rearing and sampling*

Colonies of the Western honeybee, *Apis mellifera*, were maintained at the Honeybee Research Institute, Jiangxi Agricultural University, Nanchang, China (28.46 °N, 115.49 °E). Experiments were performed under the supervision of the Animal Care and Use Committee of Jiangxi Agricultural University, China.

In this study we compared four groups of queens: queens raised from worker eggs transplanted to queen cells on the $2^{nd}$ day after laying, and queens raised from worker larvae transplanted to queen cells on the $4^{th}$, $5^{th}$ and $6^{th}$ day after laying.

To produce these queens a naturally mated laying queen was caged for 6 h on a frame of plastic worker cells to lay (Pan et al. 2013). This frame was moved to an incubator and 30-50 eggs or larvae were transferred to plastic queen cells on the 2$^{nd}$ day (egg queens raised from eggs: E), 4$^{th}$ day (one-day-old larvae queens: L1), 5$^{th}$ day (two-day-old larvae queens: L2), and 6$^{th}$ day after laying (three-day-old larvae queens: L3).

The queen cells were placed in racks in two queenless colonies to be reared. Half of each queen rearing group was assigned to each colony. Once sealed, queen cells were numbered, the length of each queen cell was measured and they were then relocated to a dark incubator (35°C, 80%) to emerge. From the 15$^{th}$ day post laying the cells were checked every 2 h to see if any queens had emerged, and hourly after the first queen emerged. We measured the weight of royal jelly remaining in the queen cell after queen emerged.

The first four queens in each rearing group to emerge were flash frozen in liquid nitrogen and stored in a -80 °C freezer for later genomic methylation analysis. The remaining queens were transferred to queen cages, which were placed in queenless colonies for 4-5 days and then flash frozen in liquid nitrogen and stored in a -80 °C freezer for later ovariole scoring.

### *Scoring ovariole number*

Sampled queens were thawed to room temperature and the ovaries dissected from the abdomen. For dehydration and fixing we used an Automatic dehydrator (Leica, TP1020). First, ovaries were fixed in 4% paraformaldehyde fix solution (BBI Life Sciences) for 12 hours. They were then dehydrated in a graded ethanol series (70% - 100%). Following dehydration ovaries were placed in a 1:1 absolute ethanol / xylene mixture for 30 min, then 100% xylene for 10 min, followed by immersion in fresh xylene for 5 min. Finally, they were transferred to a 1:1 xylene-paraffin mixture for 30 min, and embedded in paraffin wax using a heated paraffin embedding station (Leica).

Once blocked in paraffin wax, 5-7 μm sections of the ovary were cut with a Leica RM 2245 microtome. Sections were placed on histological slides (Autostainer XL), stained with HE Staining Kit (BOSTER AR1180), mounted with neutral balsam mounting medium (BBI Life Sciences) and covered with a coverslip. Slides were imaged and photographed using a 100× transmission light microscope (OLYMPUS, DP80) (Fig. S1).

We counted the number of ovarioles in the left ovary by identifying sections in which the ovarioles were very clear and counting sections (fiji-win64 software, Fig. S1) until we found at least two giving exactly the same number of ovarioles (Gan et al. 2012). We used a one way ANOVA followed with Fisher's PLSD test to analyse differences between queen types.

### *Genome wide methylation analysis*

Four queens in each group were used for genome-wide methylation testing. Brains, ovaries and thorax of the queens were dissected over ice and pooled as one sample for genomic DNA extraction.

Genomic DNA was extracted (StarSpin Animal DNA Kit, GenStar). DNA samples were then sent for whole-genome bisulfite sequencing analysis (Novogene Bioinformatics Technology Co., Ltd/www.novogene.cn) using the method summarised below.

### *Library preparation and quantification*

100 ng genomic DNA was spiked with 0.5 ng lambda DNA and sonicated using a Covaris S220 DNA Sequencing/gene analyzer to break the genomic DNA into short 200-300 base pair fragments. Fragmented DNA was treated with bisulfite (EZ DNA Methylation -GoldTM Kit, Zymo Research), and the bisulfite converted sequences were then processed using an Accel -NGS Methyl -Seq DNA Library Kit to add to each fragment a truncated adapter sequence, including an index sequence.

Library DNA concentration was quantified by a Qubit® 2.0 Flurometer (Life Technologies, CA, USA) and quantitative PCR. The insert size was assayed on an Agilent Bioanalyzer 2100 system.

The prepared genomic DNA library samples were sequenced (Illumina Hiseq XTen) and 125 - 150 bp paired end-reads were generated. Image analysis and base identification were performed with the Illumina CASAVA pipeline.

The quality of the raw reads was assessed with FastQC (fastqc_v0.11.5). Read sequences produced by the Illumina pipeline in FASTQ format were pre-processed through Trimmomatic (Trimmomatic-0.36) software use the parameter (SLIDINGWINDOW:4:15, LEADING:3, TRAILING:3, ILLUMINACLIP:adapter. fa:2, 30:10, MINLEN:36). Reads that passed these filtering steps were counted as clean reads and all subsequent analyses were performed on clean reads only. For all samples, two samples with low mapping values did not meet our requirements. Consequently we lost one E sample and one L3 sample. The average number of clean reads was 41,485,779, with 10.89 G clean base sequences (the two eliminated samples are not included in these statistics). The average Phred scores Q30% (Ewing and Green 1998) was 92.5% (Table SI).

### *Mapping reads to the Apis mellifera reference genome*

Bismark software (version 0.16.3) was used to align bisulfite-treated library reads to the reference honey bee genome (Amel_HAv3.1 (GCF_003254395.2)) (Krueger and Andrews 2011). For alignment both the reference genome and library reads were transformed into bisulfite-converted versions (C-to-T and G-to-A) and then given a digital index with bowtie2, that included data on the sequences, and sample (Langmead and Salzberg 2012).

Aligned sequence reads were then compared with the normal (non bisulfite converted) genomic sequence so that the methylation state of cytosines could be inferred. Reads that aligned to the same regions of the genome were regarded as duplicates. Sequencing depth and coverage were calculated assessing number of overlapping reads relative to number of duplicate reads.

The conversion rate was obtained by aligning to the lambda spike. The average bisulfite conversion rate was 99.7% (Table SI). The average site coverage rate was 26.74 (Table SII).

### *Methylation analysis*

To identify the level of methylation at each site, we modeled the count of methylated cytosines (mC) at a site as a binomial (Bin) random variable with methylation rate r: mC $\sim$ Bin (mC + umC*r) (Wang *et al.* 2019).

To calculate the methylation level of a sequence, we divided the sequence into multiple bins, of 10kb and measured the sum of methylated and unmethylated read counts in each bin. Methylation level (ML) for each bin or C site shows the fraction of methylated Cs, and is defined as:

ML (C) = (reads mC) / (reads (mC) + reads (C)).

Calculated ML was further corrected with the bisulfite non-conversion rate according to Listers' method (Lister *et al.* 2013). Given the bisulfite non-conversion rate r, the corrected ML was estimated as: ML (corrected) = (ML - r) / (1 - r).

When comparing samples, differentially methylated regions (DMRs) were identified using DSS software (Feng *et al.* 2014; Park and Wu 2016; Wu *et al.* 2015) with the parameters: smoothing = TRUE, smoothing.span = 200, delta = 0, p.threshold = 1e-05, minlen = 50, minCG = 3, dis.merge = 100, pct.sig = 0.5. These parameters mean regions with the minimal length of 50 bp, more than 3 CG sites, and more than 50% of differentially methylated loci in the regions with $P < 1 \times 10^{-05}$ were considered as DMRs. DSS is an common used R library performing differential analysis for count-based sequencing data (MacKay *et al.* 2019; Mendizabal *et al.* 2019; Wang *et al.* 2019). The core of DSS is a new dispersion shrinkage method for estimating the dispersion parameter from Gamma-Poisson or Beta-Binomial distributions. Adjacent DMRs were combined when the distance between two DMRs was less than 100 bp. We analyzed the DMRs in all contrasts (L1 vs E, L2 vs E, L3 vs E, L2 vs L1, L3 vs L2, L3 vs L1).

We related DMRs to genes if the DMR overlapped at all with the gene body or promoter region. The KEGG database classifies genes according to high-level functions (http://www.genome.jp/kegg/). We used KOBAS software (Mao *et al.* 2005) to test the statistical enrichment of DMGs to different KEGG pathways. We analyzed the 20 most enriched KEGG pathways to explore the most differ pathway between the comparisons of L1-3 with E.

## Results

### *Effect of queen rearing method on queen morphology*

As the age of the larva transferred to the queen cell increased, queen cell length (Fig. II-5-1a, ANOVA: $F = 84.240$, $DF = 3$, $P < 0.0001$), remaining royal jelly (Fig. II-5-1b, ANOVA: $F = 95.408$, $DF = 3$, $P < 0.0001$), and queen weight (Fig. II-5-1c, ANOVA: $F = 14.772$, $DF = 3$, $P < 0.001$) all decreased significantly. There was no significant difference in the number of ovarioles between experimental groups (Fig. II-5-1d, ANOVA: $F = 2.918$, $DF = 3$, $P = 0.0607$).

### *Effect of queen rearing method on genome methylation*

DMGs in all comparisons (L1 vs E, L2 vs E, L3 vs E, L2 vs L1, L3 vs L2, L3 vs L1) were shown in Table SIII. When considering the DMGs in different functional classes, we focused on genes identified as related to the function of reproduction or longevity (Corona and Robinson 2006; He *et al.* 2017; Yin *et al.* 2018), immunity (Barribeau *et al.* 2015; Boutin *et al.* 2015; He *et al.* 2017), metabolism (He *et al.* 2017), and DMGs identified in our own data related to these groups from KEGG pathway analyses. If genes were related to multiple functions they were only counted once. Our classification criteria was: reproduction or longevity > immunity > metabolism (Table SIV). Then, we mainly focused on the DMGs of L1-3 vs E for further study.

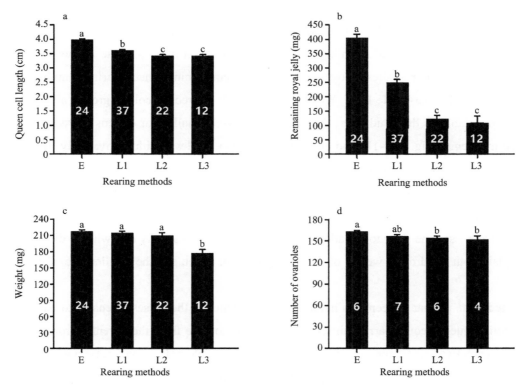

**Figure II-5-1** Queen cell length (a), Remaining royal jelly (b), Weight (c) and number of ovarioles (d) of the four queen groups (E, L1, L2 and L3). Bars show mean ± SEM. Sample size of each queen group shown in each bar. One-way ANOVA was performed and letters above bars identify groups that did and did not differ significantly with each other ($P < 0.05$, ANOVA test followed with Fisher's PLSD test).

The 20 most enriched KEGG pathways indicated that for all comparisons of queens reared from larvae with E queens most DMGs were involved in metabolic pathway, and as the age of the transferred larvae increased, the number of DMGs and degree of difference between treatment groups increased (Fig. II-5-2, see page 523). In these three comparisons, there were 79 DMGs (22, 34, and 52 DMGs in the comparisons of L1 with E, L2 with E, L3 with E respectively) related to the function of metabolism (Table SV). Of these, some genes were involved in multiple important pathways like caste differentiation, and immunity (Lyko et al. 2010).

We analyzed the methylation differences at CG sites in the exons of all the DMGs (L1-L3 vs E) functions related to reproduction or longevity, immunity, body development or metabolism, and found 55 genes that showed a uniform increasing or decreasing trend with the age of larval transfer. This analysis revealed that then comparing queens reared from larvae (L1-L3) with E queens, the number of DMGs (Fig. II-5-3, see page 524), and the ratio of methylation differences between the compared sequences (Fig. II-5-3) increased with the age of larval transfer to queen cells.

## Discussion

While it appears that functional queens can be produced by transplanting larvae from worker cells up to three days old, our study revealed significant consequences of this for queen development and the epigenome of the adult queen.

Queens raised from older worker larvae weighed less (Fig. II-5-1c), had a tendency but not significantly fewer ovarioles than queens reared from eggs (Fig. II-5-1d) which might be due to low sample size. This confirms reports from earlier studies (Woyke 1971), and is concerning since ovariole number will limit the queens fecundity (Bouletreau-Merle 1978; Woyke 1971), which can reduce colony growth and productivity.

It was also notable in this study that the queen cells were smaller for queens reared from larvae than queens reared from eggs (Fig. II-5-1b) and on emergence contained less residual royal jelly. This demonstrates that the developmental and nutritional environment of queens reared from transplanted larvae is not the same as queens reared from transplanted eggs. We propose that since queens raised from larvae have less time in the queen cell before pupation than E queens there is less time for workers to provision them with an abundance of food, and they have less time prior to pupation to consume the rich royal jelly diet with its distinctive nutrition. This may explain both the drop in weight of the emerging adult queen and the smaller size of the sealed queen cell.

In this study, the parental queen was naturally mated, which is the closest scenario to natural conditions. Thus the offspring queens could carry a different set of chromosome from the parental drones. Therefore, the variance within each treatment group may be increased, but this could reduced any false positive loci. However, we have to acknowledge that allowing the introduction of multiple genotypes might have affected the results because the exactly intragenic DNA methylation remain unclear in social insects genomes.

In our data, the degree of difference of DMRs in some genes increased with the age at which the worker larva was transplanted, when compared to queens reared from transplanted eggs (Fig. II-5-2 to Fig. II-5-3).

Our analyses focused on the functions of the identified DMGs. Many of the DMGs have known functions in immunity, reproduction and longevity. Some of these 55 genes may play an important role in caste differentiation or the development of queens and deserve further study. These are processes that might be compromised by immune stress during development (Butler and McGraw 2011; Searcy *et al.* 2004). Moreover, we found some DMGs identified in our comparisons of queens from the E group and the L groups were also identified in comparison of queens and workers (Lyko *et al.* 2010).

Lyko *et al.* (2010) found 561 DMGs between brains of queens and workers, out of which 36 genes were also found in the DMGs when comparing the L1 and E groups. 39 genes occurred the DMG comparison of L2 with E. 69 DMGs from Lyko *et al* (2010) were also identified in the DMG comparison of L3 with E (Table SVI). Furthermore, some genes with functions related to reproduction, longevity, immunity, body development, metabolism, or others identified as differentially methylated in the comparisons in the brains of queens and workers by Lyko *et al* 2010 were also found in our comparison of queens from different rearing treatments (Table SVI). In this case it is especially concerning because these traits might compromise the health, fecundity and longevity of the queen and thereby the vigour of the whole colony. However, since DNA methylation might not always be associated to change in gene expression, the role of these differentially methylated genes remains to be further studied.

These characteristics also distinguish queens and workers, and our epigenomic analyses revealed that for these key traits queens reared from worker larvae might be more worker-like than queens reared from eggs. In effect our epigenomic analyses suggest that the adult phenotype of queens reared from larvae is partially intercaste and more worker-like than naturally reared queens.

Further supporting this interpretation, KEGG analyses revealed that the most common functional class for the DMGs in our comparisons of queen of different rearing types was metabolic processes (Fig. II-5-2). Of the 79 DMGs that were related to the function of metabolism, some genes were involved in other important functions like caste differentiation, and immunity (Lyko *et al.* 2010). For example, the gene *Cat* (LOC443552) was involved in Peroxisome, Glyoxylate, Dicarboxylate metabolism, and the FoxO signaling pathway. The gene *Plc* (LOC408791) was involved in the Metabolic pathways, Phototransduction, Inositol phosphate metabolism, and the Wnt signaling pathway. The FoxO signaling pathway and Wnt signaling pathway have been related to caste differentiation (Yin *et al.* 2018). Moreover, some DMGs we identified in the Metabolic pathway functional group were also identified by Lyko (2010) in their comparison of the metyhylation of brains of queens and workers (Table SVI). *Ndufs1* (LOC726035) and *Ndufs2* (LOC413891) were all involved in Metabolic pathways, and these two genes have also been related to immune functions (Boutin *et al.* 2015).

Metabolic processes normally differ strongly between workers and queens (Corona *et al.* 1999; Evans and Wheeler 1999; Severson *et al.* 1989). In comparisons of gene expression differences between workers and queens, the greatest and most abundant differences in gene expression are also in genes related to metabolic processes (Corona *et al.* 1999; Evans and Wheeler 1999; Severson *et al.* 1989). The above suggests that queens reared from larvae are shifted away from the metabolic program of queens reared from transplanted eggs.

To summarize, our analyses revealed the epigenomic mechanisms that enable the differentiation of the worker and queen phenotypes are compromised if queens are reared from transplanted worker larvae, with the issue becoming more severe as the transplanted larvae become older. This is significant because the most common commercial practice in apiculture is to rear queens from larvae rather than eggs because larvae are far easier than eggs to transplant. Our data argue, however, that rearing queens from eggs gives a better developmental outcome and a more fecund and robust developmental outcome queen than rearing queens from larvae.

## Acknowledgements

We thank Xu-Jiang He and Peng-Fei Yu for their guidance in DNA methylation data analysis, Yong Zhang for the help with the experiment, and Qiang Huang and Xiao-Feng He for helping upload the data.

## Authors' Contribution

ZJZ conceived this research and designed it. YY and YBL performed research. ABB provided guidance for data analysis. YY analyzed data. ABB, YY and ZJZ wrote the paper.

## Funding

This work was supported by the National Natural Science Foundation of China (31872432), the 2018

Graduate Innovation Fund Project of JiangXi Province (YC2018-BO33), and China Scholarship Council (201808360213).

## References

BARRIBEAU, S.M., SADD, B.M., DU PLESSIS, L., BROWN, M.J., BUECHEL, S.D., *et al.* (2015) A depauperate immune repertoire precedes evolution of sociality in bees. *Genome Biol.* 16, 83.

BOULETREAU-MERLE, J. (1978) Ovarian activity and reproductive potential in a natural population of Drosophila melanogaster. *Oecologia* 53, 323-329.

BOUTIN, S., ALBURAKI, M., MERCIER, P.L., GIOVENAZZO, P. AND DEROME, N. (2015) Differential gene expression between hygienic and non-hygienic honeybee (*Apis mellifera* L.) hives. *BMC Genomics* 16, 500.

BUTLER, M.W. and MCGRAW, K.J. (2011) Past or present? Relative contributions of developmental and adult conditions to adult immune function and coloration in mallard ducks (Anas platyrhynchos). *J. Com. Physiol. B*, 181, 551-563.

CAKOUROS, D., DAISH, T.J., MILLS, K. and KUMAR, S. (2004) An arginine-histone methyltransferase, CARMER, coordinates ecdysone-mediated apoptosis in Drosophila cells. *J. Biol. Chem.* 279, 18467-1871.

CORONA, M. and ROBINSON, G.E. (2006) Genes of the antioxidant system of the honey bee: annotation and phylogeny. *Insect Mol. Biol.* 15, 687-701.

CORONA, M., ESTRADA, E. and ZURITA, M. (1999) Differential expression of mitochondrial genes between queens and workers during caste determination in the honeybee *Apis mellifera*. *J. Exp. Biol.* 202, 929-938.

DOOLITTLE, G.M. (1888) Scientific Queen-rearing. *Am. Bee J.* 132.

EVANS, J.D. and WHEELER, D.E. (1999) Differential gene expression between developing queens and workers in the honey bee, *Apis mellifera*. *Proc. Natl. Acad. Sci. U.S.A.* 96, 5575-5580.

EWING, B. and GREEN, P. (1998) Base-calling of automated sequencer traces using phred. II. Error probabilities. *Genome Res.* 8, 186-194.

FENG, H., CONNEELY, K.N. and WU, H. (2014) A Bayesian hierarchical model to detect differentially methylated loci from single nucleotide resolution sequencing data. *Nucleic Acids Res.* 42, e69.

GAN, H.Y., TIAN, L.Q. and YAN, W.Y. (2012) Paraffin section method of queen ovary. *J. bee*, 32, 9.

HE, X.J., ZHOU, L.B., PAN, Q.Z., BARRON, A.B., YAN, W.Y. and ZENG, Z.J. (2017) Making a queen: an epigenetic analysis of the robustness of the honeybee (*Apis mellifera*) queen developmental pathway. *Mol. Ecol.* 26, 1598-1607.

KRUEGER, F. and ANDREWS, S.R. (2011) Bismark: a flexible aligner and methylation caller for Bisulfite-Seq applications. *Bioinformatics* 27, 1571-1572.

KUCHARSKI, R., MALESZKA, J., FORET, S. and MALESZKA, R. (2008) Nutritional control of reproductive status in honeybees via DNA methylation. *Science* 319, 1827-1830.

LANGMEAD, B. and SALZBERG, S.L. (2012) Fast gapped-read alignment with Bowtie 2. *Nat. Methods* 9, 357-359.

LISTER, R., *et al.* (2013) Global epigenomic reconfiguration during mammalian brain development. *Science* 341, 1237905.

LYKO, F., FORET, S., KUCHARSKI, R., WOLF, S., FALCKENHAYN, C. and MALESZKA, R. (2010) The honey bee epigenomes: differential methylation of brain DNA in queens and workers. *PLoS Biol.* 8, 1-12.

MACKAY, H., *et al.* (2019) DNA methylation in AgRP neurons regulates voluntary exercise behavior in mice. *Nat. Commun.* 10, 5364.

MAO, X., CAI, T., OLYARCHUK, J.G. and WEI, L. (2005) Automated genome annotation and pathway identification using the KEGG Orthology (KO) as a controlled vocabulary. *Bioinformatics* 21, 3787-3793.

MENDIZABAL, I., *et al.* (2019) Cell type-specific epigenetic links to schizophrenia risk in the brain. *Genome Biol.* 20, 135.

PAN, Q.Z., WU, X.B., GUAN, C. and ZENG, Z. (2013) A new method of queen rearing without grafting larvae. *Am. Bee J.* 153, 1279-1280.

PARK, Y. and WU, H. (2016) Differential methylation analysis for BS-seq data under general experimental design. *Bioinformatics* 32, 1446-1453.

SEARCY, W.A., PETERS, S. and NOWICKI, S. (2004) Effects of early nutrition on growth rate and adult size in song sparrows Melospiza melodia. *J. Avian Biol.* 35, 269-279.

SEVERSON, D., WILLIAMSON, J. and AIKEN, J. (1989) Caste-specific transcription in the female honey bee. *Insect Biochem.* 19, 215-220.

SHELL, W.A. and REHAN, S.M. (2018) Behavioral and genetic mechanisms of social evolution: insights from incipiently and facultatively social bees. *Apidologie* 49, 13-30.

WANG, H., ZONG, Q., WANG, S., ZHAO, C., WU, S. and BAO, W. (2019) Genome-Wide DNA methylome and transcriptome analysis of porcine intestinal epithelial cells upon deoxynivalenol exposure. *J Agric. Food Chem.* 67, 6423-6431.

WANG, Y., MA, L., ZHANG, W., CUI, X., WANG, H. and XU, B. (2016) Comparison of the nutrient composition of royal jelly and worker jelly of honey bees (*Apis mellifera*). *Apidologie* 47, 48-56.

WEI, H., HE, X.J., LIAO, C.H., WU, X.B., JIANG, W.J., *et al.* (2019) A maternal effect on queen production in honeybees. *Curr. Biol.* 29, 2208-2213.

WHEELER, D.E. (1986) Developmental and physiological determinants of caste in social Hymenoptera: evolutionary implications. *Am. Nat.* 128, 13-34.

WINSTON, M.L. (1991) The Biology of the Honey Bee. Harvard University Press, Cambridge, MA.

WOYKE, J. (1971) Correlations between the age at which honeybee brood was grafted, characteristics of the resultant queens, and results of insemination. *J. Apic. Res.* 10, 45-55.

WU, H., XU, T., FENG, H., CHEN, L., LI, B., YAO, B., QIN, Z., JIN, P. and CONNEELY, K.N. (2015) Detection of differentially methylated regions from whole-genome bisulfite sequencing data without replicates. *Nucleic Acids Res.* 43, e141.

YIN, L., WANG, K., NIU, L., ZHANG, H., CHEN, Y., JI, T. and CHEN, G. (2018) Uncovering the Changing Gene Expression Profile of Honeybee (*Apis mellifera*) Worker Larvae Transplanted to Queen Cells. *Front. Genet.* 9, 416.

# 6. Honeybee (*Apis mellifera*) Maternal Effect Causes Alternation of DNA Methylation Regulating Queen Development

Xu Jiang He[1,3], Hao Wei[1], Wu Jun Jiang[2], Yi Bo Liu[1], Xiao Bo Wu[1], Zhi Jiang Zeng[1,3*]

1. Honeybee Research Institute, Jiangxi Agricultural University, Nanchang, Jiangxi 330045, P. R. of China
2. Honeybee Research Institute of Jiangxi Province, Nanchang, Jiangxi 330045, P. R. of China
3. Jiangxi Key Laboratory of Honeybee Biology and Bee Keeping, Nanchang, Jiangxi 330045, P. R. of China

**Abstract**

Queen-worker caste dimorphism is a typical trait for honeybees (*Apis mellifera*). We previously showed a maternal effect on caste differentiation and queen development, where queens emerged from queen-cell eggs (QE) had higher quality than queens developed from worker cell eggs (WE). In this study, newly-emerged queens were reared from QE, WE, and 2-day worker larvae (2L). The thorax size and DNA methylation levels of queens were measured. We found that queens emerging from QE had significantly larger thorax length and width than WE and 2L. Epigenetic analysis showed that QE/2L comparison had the most different methylated genes (DMGs, 612) followed by WE/2L (473), and QE/WE (371). Interestingly, a great number of DMGs (42) were in genes belonging to mTOR, MAPK, Wnt, Notch, Hedgehog, FoxO, and Hippo signaling pathways that are involved in regulating caste differentiation, reproduction and longevity. This study proved that honeybee maternal effect causes epigenetic alteration regulating caste differentiation and queen development.

**Keywords:** honeybees, maternal effect, development, caste differentiation, DNA methylation

## Introduction

Maternal effects have been recognized in animals and plants for a long time (Bernardo 1996; Roach & Wulff, 1987). Maternal effects are that the phenotype of offspring is influenced by maternal phenotype rather than its own genotype (Van Dooren *et al.*, 2016; Schwabl & Groothuis, 2019). Vertebrate females can adjust their investment in eggs to get higher quality offspring (Cunningham & Russell, 2000). Insects can also adaptively vary investment in their eggs (Passera, 1980). Honeybees are a typical eusocial insect species, and the queen is the only fertile female individual among sterile workers (Winston, 1991). Previous studies showed that female larvae fed with richer nutritional diet develop into queens, whereas lower nutritional diet results in the development of workers (Haydak, 1970). The space in which larvae develop within the cell also contributes to the queen-worker differentiation (Shi *et al.*, 2011). However, our previous study showed a

---

* Corresponding author: bees1965@sina.com (ZJZ).
注：此文发表在 *Sociobiology* 2021 年第 1 期。

maternal effect on honeybee caste differentiation, resulting in high quality queens. Queens lay larger eggs into queen cells compared to worker cells, which results in queens with heavier body, more ovarioles and different gene expression patterns (Wei *et al*., 2019). This maternal effect likely depends on the nutritional content of the fertilized queen cell egg, and not on genomic differences between queen-cell eggs and worker cell eggs. However, the underlying epigenetic mechanism for this maternal effect remains unclear.

More importantly, the quality of the queen has strong effects on the fitness of a colony (Winston, 1991; Gilley *et al*., 2003; Amiri *et al*., 2017). In recent years, honeybee colony loss has been frequently reported in public media, and the farming industry has been significantly impacted by the death of honeybees (Michael *et al*., 2009; Steinhauer *et al*., 2013; Antúnez *et al*., 2016). For example, scientists showed a high rate of honeybee winter colony loss over 13 years from 2006-2019 (Milius, 2019). The poor quality of queens was considered as one of main factors for honeybee winter loss (van Engelsdorp *et al*., 2010; Delaney *et al*., 2011). Rudolf Steiner argued in 1923 that honeybees would become extinct within 100 years due to the weakening of queens in the commercial queen rearing industry (Thomas, 1998). Commercial beekeepers rear new queens by transplanting young worker larvae into queen cells, altering the larva's diet and depriving the maternal effect from the queen (Doolittle, 1888; Büchler *et al*., 2013). There are many concerns about the adverse consequences of this queen rearing technology for queen quality and colony health. Woyke (1971) showed that queens reared from young worker larvae were smaller and had a smaller spermatheca and fewer ovarioles compared to queens from worker eggs. Rangel *et al*., (2013) reported that queens reared from old worker larvae produced fewer workers and drone combs and less honey than colonies with queens reared from young worker larvae. He *et al*., (2017) and Wei *et al*., (2019) found that queens developed from queen cells are better than queens reared from worker cell eggs and young larvae. A recent study clearly showed that honeybee queens have an ability to alter egg size in response to both genetic and environmental factors (Amiri *et al*., 2020). Therefore, honeybee maternal effects may potentially influence the quality of the queen.

DNA methylation plays an important role in regulating honeybee queen development (Kucharski *et al*., 2008; Shi *et al*., 2011; Maleszka, 2008). Honeybee queens have lower genome-wide methylation levels than workers (Shi *et al*., 2013). The DNA methyltransferase Dnmt3 profound shifts honeybee reproductive status (Kucharski *et al*., 2008). Li-Byarlay *et al*., (2013) showed that knocking down DNA methyltransferase 3 (Dnmt3) changes gene alternative splicing in honeybees. We previously found that transplanting younger larvae from worker cells result in better queens with lower DNA methylation levels (He *et al*. 2017). Cardoso-Júnior *et al*., (2018) reported that DNA methylation altered queen lifespan by regulating the expression of *vitellogenin* gene. Therefore, we used honeybee eggs laid in queen cells to rear queens and compare DNA methylation among queens reared from queen-cell eggs (QE), worker-cell eggs (WE) and 2-day worker larvae (2L). Our results showed that maternal effect caused DNA methylation alternation in honeybee queens.

## Materials and methods

### *Insects*

Three honeybee colonies (*Apis mellifera*) were used throughout this study. Virgin queens were sisters reared from the same mother queen in order to minimize differences in genetic background. Each colony had nine frames with approximately 30,000 workers and a single drone inseminated queen (SDI). These colonies were kept at the Honeybee Research Institute, Jiangxi Agricultural University, Nanchang, China (28.46 °N, 115.49 °E). Queens were collected from the same colonies as in a previous study (Wei et al., 2019).

### *Queen rearing*

Queens were caged in a plastic queen-cell frame for 6 hrs, and then transferred into a plastic worker-cell frame to lay worker-cell eggs following methods in Wei et al., (2019). Plastic queen cells frames were arranged horizontally. Generally, 20-50 eggs were harvested from queen cells after 6 hrs laying. Some worker-cell eggs were placed into plastic queen cells by transferring the base of each cell (Pan et al., 2013). The rest of the worker-cell eggs were kept in their native colonies until they reached 2d worker larvae stage. These 2d worker larvae were removed and placed into plastic queen cells. Cells from QE, WE and 2L were mixed together and placed into the same colony to rear queens. Newly emerged queens were harvested immediately for morphological measurement and DNA methylation sequencing.

### *Morphological measurements*

Thoraxes of queens were collected and placed under a zoom-stereo microscope system (Panasonic Co., Ltd., accuracy: 0.01 mm). The width and length were measured according to the manufacturer's instructions. Each queen group had seven biological replicates, therefore, totally 21 queens were used for morphological measurements.

### *DNA methylation sequencing*

Newly emerged queens were placed into liquid nitrogen, and heads and thoraxes were collected for total DNA extraction. Each sample contained tissue from two queens, and three biological replicates were conducted for each rearing condition. The QE group had two biological replicates since the sequencing of the third sample failed. Total DNA was extracted using Universal Genomic DNA Extraction Kit (TaKaRa, DV811A) following the manufacturer's protocol. Concentrations of all samples were measured and adjusted to the same level, and DNA samples (300 ng DNA for each sample) were used for DNA methylation sequencing by Illumina HiSeq™ 2500 (Illumina Inc., CA, USA). The detailed methods are listed in He et al., (2017). Briefly, DNA was sheared into 200-300 bp insert size targets using Covaris ultrasonicator (Life Technology) and then purified using AMPure XP beads and end repaired. A single 'A' nucleotide was added to the 3'ends of the blunt fragments followed by ligation to methylated adapter with a T overhang. Insert size targets (200-300 bp) were purified by 2% agarose gel electrophoresis. A ZYMO EZ DNA Methylation-Gold™ Kit (ZYMO, Irvine, CA, USA) was used to conduct bisulfite conversion. Bisulfite libraries were generated by PCR amplification and quantified by qPCR (Agilent QPCR NGS Library Quantification Kit). In total, eight libraries were prepared and sequenced. These processes were performed in Beijing Biomarker Technology Co., Ltd (Beijing, China).

Low quality reads (reads containing adapter sequences, reads containing unknown nucleotide "N" over 10 % and reads with a quality value lower than 10 occupying more than 50% of the whole read) from eight bisulfite libraries were filtered. These filtered genomic fragments were then mapped to the honeybee genome (*Apis mellifera. Amel* 4.5), transformed into bisulfite-converted versions of the sequences (C-to-T and G-to-A) and then assigned to a digital index using bowtie2 according to Langmead and Salzberg (2012). Methylation sites were predicted using a bismark methylation extractor (Krueger & Andrews, 2011). Only uniquely mapped reads (clean reads) were retained. The ratio of C to CT was used to measure methylation levels.

The Bismark package was used to test 5mC (same to Krueger & Andrews, 2011) and different methylation regions (DMRs) were identified using Bisulfighter (Yutaka *et al.*, 2014), according to our previous study (He *et al.*, 2017). DMR related genes (DMGs) were annotated by mapping the target regions [gene body region (from transcription start sites to transcription end sites) or promoter region (upstream 2kb from the transcription start sites)] to the honeybee genome (*Amel 4.5*), since DNA methylation data was used to compare with our previous RNA-Seq data (Wei *et al.*, 2019).

## Data Analysis

For queen thorax length and width, data were analyzed using an analysis of variance (ANOVA) followed by Fisher's PLSD test in Statview 5.01 package (SAS, USA), where $p < 0.05$ was considered as significantly different.

## Results

### *QE were larger than WE and 2L*

The thorax length and width of QE were significantly higher than that of WE and 2L. In addition, the thorax size of WE was significantly higher than 2L (Fig. II-6-1, $p < 0.05$).

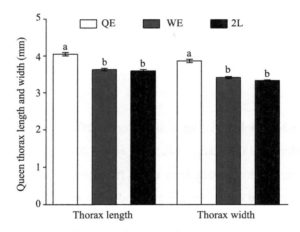

**Figure II-6-1** Thorax length and width of newly emerged queens from QE, WE and 2L. Each bar shows mean ± SE of thorax length or width. Each group had seven biological replicates. Data were analyzed by ANOVA test followed by Fisher's PLSD test. Different characters on the top of bars represent significant difference ($p < 0.05$), same character indicates no difference ($p > 0.05$).

### *Quality of methylation sequencing*

In total, 43.44 G clean bases were detected from eight libraries. Q30 values of samples were over 85 % and the average ratio of bisulfite conversion reached to 99 % (Table S1). These values indicate a considerably high sequencing quality.

***DMRs and DMGs contribute to queen development***

There were hundreds of DMRs among three queen types (Fig. II-6-2, see page 525, and S1). DMRs (CG type) were mapped to all 16 chromosomes. Most chromosomes, such as Chr4, Chr5, Chr10 and Chr14, had more DMRs in QE/2L when comparing to WE/2L and QE/WE (Fig. II-6-2). The DMRs in CHH type were less than the CG type but showed a similar pattern (Fig. S1). There were no DMRs of CHG type could map to the chromosomes.

QE/2L had more DMGs (612) than WE/2L (473) and QE/WE (371) comparisons (Fig. II-6-3, see page 525, Table S2-4). These DMGs were enriched in 72 KEGG pathways which belong to four KEGG categories including metabolism, gene information processing, environmental information processing, cellular processes and organismal systems (Fig. II-6-4, see page 526). Moreover, 42 DMGs enriched in environmental information processing were genes belonging to important signaling pathways, such as mTOR, MAPK, Notch, and Wnt signaling (Fig. II-6-5, see page 527), which are involved in regulating queen development and reproduction (Chen *et al*., 2012; Chen *et al*., 2017). Similarly, QE/2L comparison had more DMGs (21) of these 42 genes than WE/2L (20) and QE/WE (17) comparisons (Fig. II-6-5).

***DNA methylation had a low relationship with previous RNA-Seq data***

By comparing the DNA methylation data to our previous RNA-Seq data from Wei *et al*., (2019), only 2 (GB41428 and GB46222), 1 (GB41428) and 0 DMGs were mapped to DEGs in QE/2L, WE/2L and QE/WE comparisons respectively (Fig. S2). We also compared 35 immunity and development DEGs with our DNA methylation data, and the result (Fig. II-6-6, see page 528) showed that the differences of gene expression and DNA methylation of these genes both increased as the age of transplant of the worker larvae increased. However, there wasn't a clear negative correlation between gene expression and DNA methylation in these genes (Fig. II-6-6).

## Discussion

Environmental factors, such as larval diet and maternal effect, contribute towards honeybee caste differentiation and queen development (Haydak, 1970; Wei *et al*., 2019). In this study, queens developed from queen-cells had significantly larger thorax size than those from worker eggs and young worker larvae (Fig. II-6-1). Many studies have shown that queen weight and body size are strongly correlated with queen ovariole number which influences queen fecundity and quality (Borodacheva, 1973; Bilash *et al*., 1983; Huang & Zhi, 1985; He *et al*., 2017; Wei *et al*., 2019). Consequently, these results from the present study are consistent with our previous study (Wei *et al*., 2019) where we showed that QE had significantly higher weight and ovariole number than both WE and 2L. We conclude there is a strong maternal effect on honeybee queen development and potentially contributes to the queen quality.

Previous studies showed that queens have lower global DNA methylation level than workers, and DNA methylated genes such as genes involved in mTOR pathways can control queen-worker dimorphism (Kucharski *et al*., 2008; Chen *et al*., 2012; Shi *et al*., 2013). Here, we found that QE/2L had the most DMRs and DMGs, followed by WE/2L and QE/WE (Fig. II-6-2 and 3). In addition, 42 DMGs were involved in 11 signaling pathways such as mTOR, MAPK, Wnt and Notch pathways (Fig. II-6-4 and Fig. II-6-5) that are known to regulate honeybee caste differentiation and queen development (Chen et al., 2012; Chen et al., 2017). These DMGs induced by maternal effect are strongly related to queen-worker differentiation, reflecting that honeybee maternal effect could cause various epigenetic alteration in caste differentiation and queen development.

How is DNA methylation regulated by maternal effects? Different diet during larval stage alters DNA methylation (Wang et al., 2006; Kucharski et al., 2008; Shi et al., 2011; Foret et al., 2012; Maleszka, 2008). We previously showed that queens deposited larger eggs into queen cells (Wei et al., 2019). During development, eggs absorb nutrients such as vitellin from the ovary epidermal cells for 21 days (Torres, 1980; Li & Zhang, 2017). Larger eggs found in queen cells may contain more nutrients than eggs in worker cells, therefore, the nutrients available to the developing embryo may alter DNA methylation and promote queen development. However, the kind of nutrients in the eggs and how queens control egg size require further investigations.

Moreover, DNA methylation plays an important role in soft inheritance and animal evolution (Bird, 2002; Dickins & Rahman, 2012; Klironomos, 2013). Maternal effect also dramatically contributes to the process of animal evolution (Galloway et al., 2009; Marshall & Uller, 2007). The commercial queen rearing technology started in the 19 century (Doolittle, 1888) artificially transplants young worker larvae into queen cells to rear queens rather uses queen cell eggs. The widely use of commercial queen rearing technology for over 100 years continuously deprives the investment from mother queens (namely maternal effect) and may consistently weakens the queen and honeybee colony via altering DNA methylation. As a proximal remedy, rearing queens from queen-cell eggs may be a good strategy to yield high quality queens and healthy colonies.

Similar to our previous study (He et al., 2017), our DMGs had a low relationship with previous RNA-Seq results (Wei et al., 2019). Only 3 of 1456 DMGs were mapped to our previous DEGs data (Fig. S2). Evidence showed that honeybee DNA methylation has an association with alternative splicing and gene duplication (Elango et al., 2009; Dyson & Goodisman, 2020), but there isn't a completely direct link between DNA methylation and gene expression. The present study and our previous study showed a same pattern that QE/2L comparison had more DMGs and DEGs than QE/WE and WE/2L comparisons (see Fig. II-6-3 from this study and Fig. II-6-3 from Wei et al., 2019). Many DMGs and DEGs were involved in queen-worker differentiation, body development and immunity etc., however they didn't show a clear negative or positive relationship with our DNA methylation data (Fig. II-6-6). How DNA methylation regulating gene expression remains essentially unknown in honeybees. DNA methylation has been shown to associate with chromosome structure and histone modifications (Hunt et al., 2013; Dmitrijeva et al., 2018). Perhaps the alternative splicing, chromosome structure, histone modification and other undiscovered factors jointly contribute to the regulation on honeybee gene expression by DNA methylation, which needs further investigations.

In conclusion, this study firstly indicates that honeybee maternal effect causes DNA methylation changes in queen rearing and potentially contributes to the queen development and colony health, due to the function of DNA methylation in soft inheritance and animal evolution. Since the poor quality has been frequently reported as a main factor for colony losses (van Engelsdorp et al., 2010; Delaney et al., 2011), therefore, the maternal effect should be deeply considered and used in commercial queen rearing.

## Acknowledgments

This work was supported by the National Natural Science Foundation of China (31702193 and 31960685), Natural Science Foundation of Jiangxi Province (20171BAB214018), Key Research and Development Project of Jiangxi province (20181BBF60019) and the Earmarked Fund for China Agriculture Research System (CARS-44-KXJ15).

## Author Contributions and Declarations

ZJZ and XJH designed the experiments; XJH, WJJ and HW performed the experiments; YBL and XBW participated in experiments; XJH analyzed the data; XJH and ZJZ written and revised the paper. All authors read and approved the final manuscript. We have declared that no conflict of interests exists.

## Data Accessibility

The DNA methylation sequencing data refer to NCBI database (BioProject: PRJNA310321): QE: SAMN04450251; WE: SAMN04450250; 2L:SAMN04450248.

## References

AMIRI, E., STRAND, M., RUEPPELL, O., TARPY, D. (2017) Queen quality and the impact of honey bee diseases on queen health: potential for interactions between two major threats to colony health. *Insects*, 8, 2, 48. https://doi.org/10.3390/insects8020048.

AMIRI, E., LE, K., MELENDEZ, C.V., STRAND, M.K., TARPY, D.R., RUEPPELL, O. (2020) Egg-size plasticity in *Apis mellifera*: honey bee queens alter egg size in response to both genetic and environmental factors. *J. Evolution. Biol.* 33, 4, 534-543. https://doi.org/10.1111/jeb.13589.

ANTÚNEZ, K., INVERNIZZI, C., MENDOZA, Y., VANENGELSDORP, D., ZUNINO, P. (2016) Honeybee colony losses in Uruguay during 2013-2014. *Apidologie*, 48, 3, 1-7. https://doi.org/10.1007/s13592-016-0482-2.

BERNARDO, J. (1996) Maternal Effects in Animal Ecology. *Integr. Comp. Biol.* 36, 2, 83-105. https://doi.org/10.1093/icb/36.2.83.

BIRD, A. (2002) DNA methylation patterns and epigenetic memory. *Gene. Dev.* 16, 6-21. https://doi.org/10.1101/gad.947102.

BILASH, G.D., BORODACHEVA, V.T., TIMOSINOVA, A.E. (1983) Quality of artificially reared queen bees. In Proceedings of the XXIXth International Congress of Apiculture. (Bucharest: Apimondia Publishing House), pp. 114-118.

BORODACHEVA, V.T. (1973) Weight of eggs and quality of queens and bees (in Russian). *Pchelovodstvo*, 93, 12-13.

BÜCHLER, R., ANDONOV, S., BIENEFELD, K., COSTA, C., HATJINA, F., KEZIC, N., KRYGER, P., SPIVAK, M., UZUNOV, A., WILDE, J. (2013) Standard methods for rearing and selection of *Apis mellifera* queens. *J. Apicult. Res.* 52, 1, 1-30. http://dx.doi.org/10.3896/IBRA.1.52.1.07.

CARDOSO-JÚNIOR, C.A.M., GUIDUGLI-LAZZARINI, K.R., HARTFELDER, K. (2018) DNA methylation affects the lifespan of honey bee (*Apis mellifera* L.) workers-evidence for a regulatory module that involves vitellogenin expression but is independent of juvenile hormone function. *Insect Biochem. Molec.* 92, 21-29. https://doi.org/10.1016/j.ibmb.2017.11.005.

CHEN, X., HU, Y., ZHENG, H.Q., CAO, L.F., NIU, D.F., YU, D.L., SUN, Y.Q., HU, S.N., HU, F.L. (2012) Transcriptome comparison between honey bee queen- and worker-destined larvae. *Insect Biochem. Molec.* 42, 9, 665-673. https://doi.org/10.1016/j.ibmb.2012.05.004.

CHEN, X., MA, C., CHEN, C., LU, Q., SHI, W., LIU, Z.G., WANG, H.H., GUO, H.K. (2017) Integration of lncRNA-miRNA-mRNA reveals novel insights into oviposition regulation in honey bees. *Peer J*, 5, e3881. https://doi.org/10.7717/peerj.3881.

CUNNINGHAM, E.J.A., RUSSELL, A.F. (2000) Egg investment is influenced by male attractiveness in the mallard. *Nature*, 404, 74-77. https://doi.org/10.1038/35003565.

DELANEY, D.A., KELLER, J.J., CAREN, J.R., TARPY, D.R. (2011) The physical, insemination, and reproductive quality of honey bee queens (*Apis mellifera* L.). *Apidologie*, 42, 1-13. https://doi.org/10.1051/apido/2010027.

DMITRIJEVA, M., OSSOWSKI, S., SERRANO, L., SCHAEFER, M.H. (2018) Tissue-specific DNA methylation loss during ageing and carcinogenesis is linked to chromosome structure, replication timing and cell division rates. *Nucleic Acids Res.* 46, 14, 7022-7039. https://doi.org/10.1093/nar/gky498.

DOOLITTLE, G.M. (1888) Scientific queen-rearing.*Am. Bee* J. USA.

DICKINS TE, RAHMAN Q (2012) The extended evolutionary synthesis and the role of soft inheritance in evolution. *P. Roy. Soc. B-Biol. Sci.* 279, 2913-2921. https://doi.org/10.1098/rspb.2012.0273.

DYSON, C.J., GOODISMAN, M.A.D. (2020) Gene duplication in the honeybee: patterns of DNA methylation, gene expression, and genomic environment. *Mol. Biol. Evol.* msaa088. https://doi.org/10.1093/molbev/msaa088.

ELANGO, N., HUNT, B.G., GOODISMAN, M.A.D., YI, S.V. (2009) DNA methylation is widespread and associated with differential gene expression in castes of the honeybee, *Apis mellifera*. *P. Natl. Acad. Sci.* USA. 106, 27, 11206-11211. https://doi.org/10.1073/pnas.0900301106.

FORET, S., KUCHARSKI, R., PELLEGRINI, M., FENG, S., JACOBSEN, S.E., ROBINSON, G.E., MALESZKA, R. (2012) DNA methylation dynamics, metabolic fluxes, gene splicing, and alternative phenotypes in honey bees. *P. Natl. Acad. Sci. USA*, 109, 4968-4973. https://doi.org/10.1073/pnas.1202392109.

GALLOWAY, L.F., ETTERSON, J.R., MCGLOTHLIN, J.W. (2009) Contribution of direct and maternal genetic effects to life-history evolution. *New Phytol.* 183, 3, 826-838. https://doi.org/10.1111/j.1469-8137.2009.02939.x.

GILLEY, D.C., TARPY, D.R., LAND, B.B. (2003) Effect of queen quality on interactions between workers and dueling queens in honeybee (*Apis mellifera* L.) colonies. *Behav. Ecol. Sociobiol.* 55, 2, 190-196. https://doi.org/10.1007/s00265-003-0708-y.

HAYDAK, M.H. (1970) Honey bee nutrition. *Ann. Rev. Entomol.* 15, 143-156. https://doi.org/10.1146/annurev.en.15.010170.001043.

HE, X.J., ZHOU, L.B., PAN, Q.Z., BARRON, A.B., YAN, W.Y., ZENG, Z.J. (2017) Making a queen: an epigenetic analysis of the robustness of the honey bee (*Apis mellifera*) queen developmental pathway. *Mol. Ecol.* 26, 6, 1598-1607.

https://doi.org/10.1111/mec.13990.

HUANG, W.C., ZHI, C.Y. (1985) The relationship between the weight of the queen honeybee at various stages and the number of ovarioles eggs laid and sealed brood produced (in Japanese). *Honey Bee Science*, 6, 113-116.

HUNT, B.G., GLASTAD, K.M., YI, S.V., GOODISMAN, M.A.D. (2013) Patterning and regulatory associations of DNA methylation are mirrored by histone modifications in insects. *Genome Biol. Evol.* 5, 3, 591-598. https://doi.org/10.1093/gbe/evt030.

KLIRONOMOS, F.D., BERG, J., COLLINS, S. (2013) How epigenetic mutations can affect genetic evolution: Model and mechanism. *Bioessays* 35, 571-578. https://doi.org/10.1002/bies.201200169.

KRUEGER, F., ANDREWS, S.R. (2011) Bismark: a flexible aligner and methylation caller for Bisulfite-Seq applications. *Bioinformatics* 27, 1571-1572. https://doi.org/10.1093/bioinformatics/btr167.

KUCHARSKI, R., MALESZKA, J., FORET, S., MALESZKA, R. 2008. Nutritional control of reproductive status in honeybees via DNA methylation. *Science* 319, 1827-1830. https://doi.org/10.1126/science.1153069.

LI-BYARLAY, H., LI, Y., STROUD, H., FENG, S., ROBINSON, G.E. (2013) RNA interference knockdown of DNA methyl-transferase 3 affects gene alternative splicing in the honey bee. *P. Natl. Acad. Sci. USA.* 110, 31, 12750-12755. https://doi.org/10.1073/pnas.1310735110.

LI, H., ZHANG, S. (2017) Functions of vitellogenin in eggs. In Oocytes; Springer: Cham, Switzerland, pp. 389-401. https://doi.org/10.1007/978-3-319-60855-6_17.

LANGMEAD, B., SALZBERG, S.L. (2012) Fast gapped-read alignment with Bowtie 2. *Nat. Methods* 9, 357-359. https://doi.org/10.1038/nmeth.1923.

MARSHALL, D., ULLER, T. (2007) When is a maternal effect adaptive?. *Oikos* 116, 12, 1957-1963. https://doi.org/10.1111/j.2007.0030-1299.16203.x.

MALESZKA, R. (2008) Epigenetic integration of environmental and genomic signals in honey bees: the critical interplay of nutritional, brain and reproductive networks. *Epigenetics* 3, 188-192. https://doi.org/10.4161/epi.3.4.6697.

MICHAEL, B., RANDAL, R., WALTER, T. (2009) Honey bee colony mortality in the pacific northwest (USA) winter 2007/2008, *Am. bee J.* 149, 6, 573-575. https://doi.org/10.1111/j.1365-3113.2009.00474.x.

MILIUS, S. (2019) U.S. honeybees had the worst winter die-off in more than a decade.[online].https://www.sciencenews.org/article/us-honeybees-had-worst-winter-die-more-decade (accessed on 20 June 19).

PAN, Q.Z., WU, X.B., GUAN, C., ZENG, Z.J. (2013) A new method of queen rearing without grafting larvae. *Am. Bee J.* 153, 1279-1280.

RANGEL, J., KELLER, J.J., TARPY, D.R. (2013) The effects of honey bee (*Apis mellifera* L.) queen reproductive potential on colony growth. *Insect. Soc.* 60, 65-73. https://doi.org/10.1007/s00040-012-0267-1.

PASSERA, L. (1980) The laying of biased eggs by the ant *Pheidole pallidula* (Nyl,) (Hymenoptera, Formicidae). *Insect. Soc.* 27, 79-95. https://doi.org/10.1007/BF02224522.

ROACH, D.A., WULFF, R.D. (1987) Maternal effects in plants. *Annu. Rev. Ecol. Evol. S.* 18, 1, 209-235. https://doi.org/10.1146/annurev.es.18.110187.001233.

SCHWABL, H., GROOTHUIS, T.G.G. (2019) Maternal Effects on Behavior. In Choe, J. C. (ed) Encyclopedia of Animal Behavior (Second Edition). *Academic Press*, pp 483-494.

SHI, Y.Y., HUANG, Z.Y., ZENG, Z.J., WANG, Z.L., WU, X.B., YAN, W.Y. (2011) Diet and cell size both affect queen-worker differentiation through DNA methylation in honey bees (*Apis mellifera*, Apidae). *PLoS One* 6, e18808. https://doi.org/10.1371/journal.pone.0018808.

SHI, Y.Y., YAN, W.Y., HUANG, Z.Y., WANG, Z.L., WU, X.B., ZENG, Z.J. (2013) Genomewide analysis indicates that queen larvae have lower methylation levels in the honey bee (*Apis mellifera*). *Naturwissenschaften* 100, 193-197. https://doi.org/10.1007/s00114-012-1004-3.

STEINHAUER, N., RENNICH, K., WILSON, M.E., CARON, D., LENGERICH, E.J., PETTIS, J.S., ROSE, R., SKINNER, J.A., TARPY, D.R., WILKES, J.T., VANENGELSDORP, D. (2014) A national survey of managed honey bee 2012-2013 annual colony losses in the USA: results from the bee informed partnership. *J. Apicult. Res.* 53, 1, 1-18. https://doi.org/10.3896/ibra.1.53.1.01.

THOMAS, B. (1998) Bees-lecturers by Rudolf Steiner, *Anthroposophic* Press, pp 222.

TORRES, J. (1980) A stereological analysis of developing egg chambers in the honeybee queen, *Apis mellifera*. *Cell Tissue Res*. 208, 1, 29-33. https://doi.org/10.1007/BF00234170.

VANENGELSDORP, D., HAYES JR, J., UNDERWOOD, R.M., PETTIS, J.S. (2010) A survey of honey bee colony losses in the United States, fall 2008 to spring 2009. *J. Apicult. Res.* 49, 7-14. https://doi.org/10.3896/IBRA.1.49.1.03.

VAN DOOREN, T.J.M., HOYLE, R.B., PLAISTOW, S.J. (2016) Maternal Effects. In Kliman, R. M. (ed) Encyclopedia of Evolutionary Biology. *Academic Press*, pp 446-452.

WEI, H., HE, X.J., LIAO, C.H., WU, X.B., JIANG, W.J., ZHANG, B., ZHOU, L.B., ZHANG, L.Z., BARRON, A.B., ZENG, Z.J. (2019) A maternal effect on queen production in the honey bee. *Curr. Biol.* 29, 13, 2208-2213. https://doi.org/10.1016/j.cub.2019.05.059.

WINSTON, M. (1991) The Biology of the Honey Bee. Harvard University Press, Cambridge, MA, USA.

WOYKE, J. (1971) Correlations between the age at which honeybee brood was grafted, characteristics of the resultant queens, and results of insemination. *J.Apicult. Res.* 10, 45-55. https://doi.org/10.1080/00218839.1971.11099669.

WANG, Y., JORDA, M., JONES, P.L., MALESZKA, R., LING, X., ROBERTSON, H.M., MIZZEN, C.A., PEINADO, M.A., ROBINSON, G.E. (2006) Functional CpG methylation system in a social insect. *Science* 314, 645-647. https://doi.org/10.1126/science.1135213.

YUTAKA, S., JUNKO, T., TOUTAI, M. (2014) Bisulfighter: accurate detection of methylated cytosines and differentially methylated regions. *Nucleic Acids Res.* 42, 6, e45. https://doi.org/10.1093/nar/gkt1373.

# 7. Transcriptome Comparison between Newly Emerged and Sexually Matured Bees of *Apis mellifera*

Xiaobo Wu, Zilong Wang, Haiyan Gan, Shuyun Li, Zhijiang Zeng*

Honeybee Research Institute, Jiangxi Agricultural University, Nanchang 330045, China

**Abstract**

In order to understand the transcriptome characteristics of queens and drones of honeybee *Apis mellifera*, the transcriptome differences between newly emerged stage and sexually matured stage of queens and drones of *A. mellifera* L. were compared using high-throughput RNA-Seq. In drones, a total of 1618 DEGs were detected between the two stages. Out of these, 782 genes were up-regulated and 836 genes were down-regulated in sexually matured drones compared to newly emerged drones. In queens, the DEGs between the two stages were 1340, with 667 up-regulated and 673 down-regulated genes in matured queens compared to newly emerged queens. 411 genes showed the same expression trend in drones and queens during sexual maturing, with 233 (56.69%) up-regulated genes and 178 (43.31%) down-regulated at newly emerged stage. We found that genes encoding cuticular proteins (CP), cytochrome P450 (CYP), odorant binding proteins (OBP) and odorant receptor (OR), which are related to developments of bones, reproductive system and olfaction system, were differentially expressed between the sexually matured bees and the newly emerged bees. The results indicated that the expression levels of a large number of genes changed during sexual maturing of *A. mellifera* L. bees, which give us an insight into the characteristics of the gene expression during sexual maturing of adult queens and drones.

**Keywords:** *Apis mellifera*, Queen, Drone, Development, Transcriptome

Honeybee is a typical social insect. Usually, a honeybee colony comprises a queen, hundreds of drones and tens of thousands of workers. The drones are to mate with virgin queens and the mated queens are mainly responsible for reproduction. In order to improve the bee colony continuity and adaptability, honeybee has developed its special competitive mating mechanism, named mating flight. Queens and drones will choose the right opportunity for mating when they both reach sexual maturity which often accompanied with obvious physiological changes in the body of queens and drones. It has been confirmed at the physiological, biochemical and molecular levels that the physiological status of honeybee at different development stages of

---

* Corresponding author: bees1965@sina.com (ZJZ).
注：此文发表在 *Journal of Asia-Pacific Entomology* 2016 年第 3 期。

the sexual maturity are different (Kocher *et al.*, 2008, 2010; Colonello-Frattini and Hartfelder, 2009; Behura and Whitfield, 2010; Chen *et al.*, 2012; Fang *et al.*, 2012; Zhang and Yuan, 2013 ).

Transcriptome changes during mating process have been conducted by several studies, a large number of genes showed expression changes during the mating flight (Wu *et al.*, 2013a, 2013b, 2014). Through transcriptome comparison, 1615 DEGs were detected between ovaries of virgin and mated queens, moreover, a similar set of genes were found to be participated in the ovary activation of both queens and workers (Niu *et al.*, 2014).

As two kinds of sex, the queens and drones have different physiological status during maturing. At present, the related regulation mechanism during sexual maturing of queens and drones has not yet been in-depth studied, especially the gene expression information of adult bees during sexual maturing are lacking. In this study, we used the high-throughput RNA sequencing technology to find out the difference of gene expression between newly emerged and sexually matured bees of drones and queens. All the DEGs during maturing were then subjected to Gene Ontology category (GO) and Kyoto Encyclopedia of Genes and Genomes (KEGG) pathway enrichment analysis. Our results showed that a large number of genes changed during sexual maturing, and a lot of transcript sequences with important function were acquired for future gene expression or regulation research about growth, development and reproduction of *A. mellifera*.

## Materials and Method

### Bee rearing and sample collection

*Apis mellifera* were sampled from the Honey bee Research Institute, Jiangxi Agricultural University, China. Queens were artificially bred in normal conditions according to standard rearing practices (Zeng, 2009). After emergence, twelve young queens were sampled as newly emerged queens immediately. Others were caged and introduced into prepared individual nucleus colonies without other queens and prevented from taking mating flights by a strip of 'queen excluder' material. 12 days later, these queens reached sexual maturity and were sampled. Drones of *A. mellifera* L were obtained from hives. The laying queens were only allowed to get access to empty drone-combs and laid haploid eggs in drone cells. Then, the drone-combs were taken out of the hives and placed in an incubator before emerging. After emergence, twelve newly emerged drones were sampled, the remained drones were paint-marked and put in natural colonies, 16 days later, sexually matured drones were sampled.

The intestines of all the samples were removed for preventing contamination, and all these samples were stored at -80 °C until use.

### RNA extraction, library preparation and sequencing

Total RNAs were isolated from each sample respectively and the quality and quantity of the RNA were determined by a Qubit fluorometer and an Agilent 2100 Bioanalyzer. After DNase I treatment, mRNAs were isolated from total RNAs using Oligo (dT) magnetic beads and fragmented into short sequences in the fragmentation buffer. Then cDNA was synthesized using the mRNA fragments as templates. Short cDNA fragments were purified and resolved with EB buffer for end reparation and single nucleotide A (adenine) addition. After that, the short cDNA fragments were connected with adapters. The suitable fragments were selected for PCR amplification as templates. During the QC steps, Agilent 2100 Bioanaylzer and ABI StepOnePlus Real-Time PCR System were used in quantification and qualification of the sample libraries. At last, the libraries were sequenced using Illumina HiSeq™ 2000.

*Raw data processing and statistical analysis*

The original image data produced by the sequencer were transformed into sequence data by base calling and the clean reads were obtained by discarding the dirty raw reads. The clean reads were mapped to reference gene sequences of *Apis mellifera* (ftp://ftp.ncbi.nih.gov/genomes/Apis_mellifera) using the SOAP2 alignment algorithm, with a tolerance of no more than two mismatches. The sequencing saturation, distribution and coverage of reads were used to assess sequencing quality. The clean reads were submitted to the Sequence Read Archive (SRA) database in NCBI under accession numbers SRR3569809, SRR3569811, SRR3569812, and SRR3569813.

*Identification and functional analysis of DGEs*

The level of gene expression was calculated as RPKM, the formula is shown below:

$$RPKM = \frac{10^6 C}{NL/10^3}$$

Set RPKM to be the expression of unigene A, and C to be number of reads that uniquely aligned to unigene A, N to be total number of reads that uniquely aligned to all unigenes, and L to be the base number in the CDS of unigene A. The RPKM method is able to eliminate the influence of different gene length and sequencing level on the calculation of gene expression. Therefore the calculated gene expression can be directly used for comparing the difference of gene expression between samples (Mortazavi *et al.*, 2008).

We can identify DEGs between two samples referring to "The significance of digital gene expression profiles" which has been cited hundreds of times (Audic and Claverie, 1997). Gene Ontology analysis, and pathway enrichment analysis were performed to investigate functional enrichment among up- or down-regulated genes using the DFCI Honey Bee Gene Index database(http://compbio.dfci.harvard.edu/cgi-bin/tgi/gimain.pl?gudb=honeybee), Cluster (http://bonsai.hgc.jp/~mdehoon/software/cluster/), AmiGO (http://amigo.geneontology.org/amigo), and the KEGG database (http://www.genome.jp/kegg/).

*Validation of RNA-seq data by qRT-PCR*

To verify the data obtained by RNA-seq, qRT-PCR was performed in triplicate and GAPDH was selected as the reference gene to correct for sample variation in qRT-PCR efficiency and errors in sample quantification. The qRT-PCR data were expressed relative to the expression of GAPDH using the $2^{-\Delta\Delta Ct}$ method, an independent-sample t-test available in SPSS software. The primers were designed with Prime Premer 5.0 software and synthesized by Generay Biotech (Generay, PRC) based on the mRNA sequences obtained from the GenBank database (Table II-7-1).

Table II-7-1  Primer Sequence

| Official Symbol | Forward primer | Reverse primer |
|---|---|---|
| Obp13 | CCTCGTTGGTATTCTGGC | CTTCTTAACGTCGTCTGCT |
| Obp14 | GCAAATGACGTTATTGAAGGC | TCCATGACTGCTTTGATTCC |
| Obp15 | AATTAATCACCGAATGTTCAGC | GTCGTCGTATTCTGAATAGTCT |
| Obp17 | ATGTTGTTGATGAGAATGCAAA | CAGAAATAGGTGAACATTCGG |
| Obp21 | TTACAAATTGGGCTACGTGC | TTTCTACATCAATGATGCCGTT |
| GAPDH | TTCATGCTGTTACTGCTACAC | GAAGGCCATACCAGTCAAT |

## Results and Discussion

### Statistic of sequencing results

Illumina HiSeq™ 2000 platform was used to identify DEGs amongst the four groups. More than 31,567,186 raw reads per library were obtained and over 87.80% of these reads were identified as clean reads before they were mapped to the reference database. More than 93.10% of the clean reads successfully matched to either unique or multiple locations of the honey bee genome. Of them, >98.01% of the mapped reads are matched to unique gene (Table II-7-2).

Table II-7-2  Summary of read numbers based on the RNA-sequencing data

| Sample | Raw read | Clean read | Mapped read | 1-hit |
|---|---|---|---|---|
| 1-A | 34782266 | 30537378 (87.80%) | 28627695(93.75%) | 98.54% |
| 1-B | 34926978 | 31359796 (89.79%) | 29458935(93.94%) | 98.62% |
| 2-A | 37840852 | 33795620 (89.31%) | 31571133(93.42%) | 98.20% |
| 2-B | 38897574 | 34786524 (89.43%) | 32659377(93.89%) | 98.01% |
| 3-A | 39260028 | 35216552 (89.70%) | 32840526(93.25%) | 98.62% |
| 3-B | 36371084 | 32440540 (89.19%) | 30200883(93.10%) | 98.57% |
| 4-A | 34171446 | 30139946(88.20%) | 28068106(93.13%) | 98.26% |
| 4-B | 31567186 | 28031394(88.80%) | 26184906(93.41%) | 98.36% |

1: Newly emerged drone. 2: Newly emerged queen 3: Sexually matured drone 4: Sexually matured queen. A and B are the two biological replicates

### DEGs between the newly emerged drones and sexually matured drones

Very few studies have investigated the molecular mechanism of sexual maturation in male insects. To identify changes in gene expression associated with sexual maturity, we used strict statistical criteria [false discovery rate (FDR) ≤ 0.001 and the absolute value of log2Ratio ≥ 1] to screen for genes that showed significant expression difference between the newly emerged drones and sexually matured drones. A total of 1618 DEGs were detected between the two stages of drones, with 782 up-regulated genes and 836 down-regulated genes in sexually matured drones compared to newly emerged drones (Document S1). GO analysis of these DEGs indicated that 139 GO items were significantly enriched ($p<0.05$), Document S5A lists the top 47 significantly enriched terms.

KEGG pathway analysis indicated that 21 pathways were significantly enriched ($p < 0.05$), including "Parkinson's disease", "Oxidative phosphorylation", "Alzheimer's disease", "Circadian rhythm - fly" and "Linoleic acid metabolism" and "NOD-like receptor signaling pathway", and so on. Document S5B lists the top 10 significantly enriched pathways. The enrichment of the "Circadian rhythm" related genes might suggest that the mating behavior of drones is under control of circadian rhythm.

Of the 1618 DEGs, many olfactory related genes showed expression difference between newly emerged drones and sexually matured drones, including 12 OBP genes, 26 OR genes and the chemosensory protein 5 (CSP5) gene. Of the 12 OBPs, 10 of them were up-regulated in the sexually matured drones and 2 down-regulated. For the 26 ORs, 16 of them were up-regulated in the sexually matured drones and 10 down-regulated. The chemosensory protein 5 gene was up-regulated in the sexually matured drones. Odorant binding proteins are a kind of water soluble proteins distributed in olfactory neurons and they can transmit the small odor molecules to the sensory receptors of the branch diaphragm of the olfactory nerve. The up-regulation of OBP, OR and CSP genes in the sexually matured drones may be associated with the mating behavior of drones which are often attracted by pheromones released by virgin queens.

Of the DEGs, there are 19 cuticular protein coding genes, most of them were expressed higher at the newly emerged stage. Insect cuticle proteins are a kind of structure proteins which can resist the external environment together with chitin as a barrier which forms the stratum corneum (Liu *et al.*, 2010). The higher expression of cuticle protein genes in the newly emerged drones might be because the newly emerged bees are weak and their physique is soft, they need to synthesize more cuticle proteins to be transported to the epithelial cells of integument to build the exoskeleton.

We found 20 cytochrome P450 genes in the DEGs, most of them were up-regulated in the matured drones. Cytochrome P450s are widely distributed in all aerobic organisms, such as insects, plants, bacteria, etc. They not only participate in the biological metabolism, but also in synthesis or decomposition the hormone (Li *et al.*, 2004). The higher expression of most cytochrome P450s at sexually matured stage may be associated with pheromone synthesis, such as 9-ODA. In addition, as a class of detoxifying enzyme, higher expression of cytochrome P450s will lead to a better resistance of adult bees to some toxic and harmful substances compared to newly emerged bees.

We found the gene JH (GB15327) and a yolk protein Vitellogenin (GB13999) were up-regulated in matured drones compared to newly emerged drones (Document S1). A previous study indicated that juvenile hormone (JH) is involved in the regulation of male reproductive maturation by coordinating pheromonal gland development in Caribbean fruit flies (Teal *et al.* 2000). It suggests that these two genes might also coordinate the sexual maturation of honeybee drones.

We found three circadian rhythm related genes (GB10192, GB19264, GB16884) in the DEGs, two of them were up-regulated in the sexually matured drones, including period, which is a key gene in circadian rhythm pathway, indicating that the behavior of sexually matured drones is strictly regulated by circadian rhythm.

We also analyzed DEGs associated with insect behavior and found two such genes, octopamine receptor (GB11266) and serotonin receptor 7 (GB14021), which are important factors involved in insect behavior. Of them, octopamine receptor was up-regulated in the sexually matured drones, which may be able to improve the locomotion ability of drones, thus further improve the probability of a successful mating.

### DEGs between the newly emerged queens and sexually matured queens

A total of 1340 DEGs were detected between newly emerged queens and sexually matured queens, with 667 up-regulated genes and 673 down-regulated genes in emerged queens (Document S2).

GO analysis of these DEGs indicated significantly enrichment of 114 GO terms ($p<0.05$), Document S5C lists the top 46 significantly enriched terms. The "metabolic process" and "cellular process" in biological process, "cell "and "cell part" in cellular component, "binding" and "catalytic activity" in molecular function were dominant items in each categories respectively.

KEGG pathway analysis indicated that 19 pathways were significantly enriched ($p < 0.05$), including "Parkinson's disease", "Oxidative phosphorylation", "Alzheimer's disease", "ECM-receptor interaction", "Citrate cycle (TCA cycle)", and so on. Document S5D lists the top 9 significantly enriched pathways.

Of the 1340 DEGs, 5 OBP genes and 40 OR genes showed expression difference between the two stages of queens. Of the 5 OBPs, 3 were up-regulated and 2 were down-regulated in the sexually matured queens. Most of the ORs were up-regulated at the sexually matured stage. The up-regulation of OBP and OR in sexually matured queens might be associate with the fact that queens need to feel the stimulation from the environment.

Three chemosensory proteins (CSP) differentially expressed between these two stages. Of them, CSP1 (GB17875) was up-regulated in the sexually matured queens, while CSP2 (GB18819) and CSP3 (GB19453) were opposite. CSPs are a kind of water-soluble proteins with small molecular weight widely distributed in lymph in a variety of insect chemoreceptors. Their main function is feeling environment chemical stimulation, carrying and protecting non-volatile odor molecules reaches the corresponding receptors through lymph in chemoreceptor. The up-regulation of CSP1 indicates that sexual mature queens need to synthesize chemosensory protein to feel the changes of the environment.

We found 21 cuticular protein genes, most of them were expressed higher at the newly emerged queens which might also need to synthesize more cuticle proteins to build the exoskeleton.

Besides, two JH related genes, juvenile hormone esterase (GB15327) and juvenile hormone acid methyltransferase (GB10517), were significantly up-regulated in matured honeybee queens compared with newly emerged queens. It was reported that reproductive maturation in female insects normally correlates with their ovarian maturity and egg production, and many hormonal genes such as JH and Vg dominate the reproductive maturation in female insects (Teal *et al.* 2000; Ringo *et al.* 1991; Soller *et al.* 1999; Yano *et al.* 1994). The up-regulation of JH related genes in matured queens suggests that JH might regulate the reproductive maturation of honeybee queens.

### Comparison of sexual maturity related DEGs between queens and drones

597 genes were expressed differently between the newly emerged stage and sexually matured stage both in

drones and queens, of them, 411 showed the same expression trend between drones and queens (Document S3), while 186 showed opposite expression trend (Document S4). Of the 411 co-expressed genes, 178 (43.31%) were up-regulated and 233 (56.69%) were down-regulated at the sexually matured stage (Fig. II-7-1A and B, see page 528). Gene expression clustering of all the up-regulated and down-regulated co-expression DEGs showed that samples from the same sex were clustered together (Document S5E and F).

We found that both matured drones and queens had significantly higher expression of JH gene compared to the newly-emerged bees. Moreover, 7 serine protease RpSs which involve in the development, immunity and digestion in insects (Krem and Di Cera 2002; Rawlings and Barrett 1993; Zou et al. 2006; Wang et al., 2007), as well as other 34 DEGs (6 cytochrome P450s, 4 OBPs, 10 ORs and 14 cuticular proteins), were expressed differentially between the two developmental stages of both drones and queens. It revealed that honeybee sexual maturation, as in other insect species (Soller et al. 1999; Teal et al. 2000), may undergo a hormonally regulated period and accelerate the development of honeybee body, olfactory, immunity and secondary sexual characters.

It is interesting that 37 ribosomal proteins were highly expressed at the newly emerged stage of drones, while just 3 ribosomal proteins showed expression difference between the two stages in queens. The reason might be that the newly emerged drones need to synthesize large amounts of proteins for individual development, while the matured drones no longer need. Thus in drones, these ribosomal proteins showed a higher expression at the newly emerged stage. However, in queens, not only the newly emerged queens need to synthesize a large amount of protein, the sexually matured queens also need a lot of protein for the development of eggs, therefore, in queens the expression level of most ribosomal protein genes showed no difference between the two stages.

### qRT-PCR validation

To confirm the results of the deep sequencing data, five differentially expressed OBP genes were chosen for qRT-PCR verification. The reason for choosing OBPs is that they are important proteins in the chemical information exchange between insects and the environment, moreover, the five OBPs genes were significantly differentially expressed between newly emerged and sexually matured bees in both queen and drones. The results showed as in Table II-7-3 and Table II-7-4. The variation trend of genes expressed between newly emerged- and sexually matured bees were similar with the result of RNA-seq. These data demonstrate the reliability of the RNA-Seq results.

Table II-7-3 Quantitative analysis of 5 odorant binding protein genes in drones

| Gene | Newly emerged drone | Sexually matured drone | $P$-value |
|---|---|---|---|
| OBP13 | 1 | 0.0066 | $P>0.05$ |
| OBP14 | 1 | 2.2227 | $P<0.05$ |
| OBP15 | 1 | 5.1696 | $P<0.05$ |
| OBP17 | 1 | 1.7318 | $P>0.05$ |
| OBP21 | 1 | 7.2986 | $P<0.05$ |

Table II-7-4  Quantitative analysis of 5 odorant binding protein genes in queens

| Gene | Newly emerged queen | Sexually matured queen | $P$-value |
| --- | --- | --- | --- |
| OBP13 | 1 | 0.0124 | $P>0.05$ |
| OBP14 | 1 | 9.1843 | $P<0.05$ |
| OBP15 | 1 | 0.7193 | $P>0.05$ |
| OBP17 | 1 | 4.1043 | $P<0.05$ |
| OBP21 | 1 | 5.2963 | $P<0.05$ |

## Conclusions

This study described the gene expression pattern during honeybee sexual maturation. Thousands of genes, including hormonal genes, serine proteases, cytochrome P450s, OBPs, ORs and cuticular proteins, were significantly differentially expressed during this process, indicating that honeybee sexual maturation are accompanied with huge physiology and gene transcription changes. These results provide valuable information for identifying key genes involved in regulating honeybee sexual maturation.

Supplementary data to this article can be found online at http://dx.doi.org/10.1016/j.aspen.2016.08.002.

## Acknowledgments

We thank Doctor Qiang Huang for improving the English of this manuscript. This work was supported by the Earmarked Fund for China Agriculture Research System (CARS-45-KXJ12), the National Natural Science Foundation of China (No. 31060327, No.31360587, No. 31260584), and the Science and Technology Support Program of Jiangxi Province (No. 20141BBF60033).

## References

AUDIC,S., CLAVERIE,J.M., 1997. The significance of digital gene expression profiles. *Genome Res.* 7(10), 986-995.

BEHURA,S.K., WHITFIELD,C.W., 2010. Correlated expression patterns of microRNA genes with age-dependent behavioural changes in honeybee. *Insect Mol. Biol.* 19(4), 431-439.

CHEN,X., HU,Y., ZHENG,H., CAO,L., NIU,D., YU,D., SUN,Y., HU,S., HU,F., 2012. Transcriptome comparison between honeybee queen-and worker-destined larvae. *Insect Biochem and Mol. Biol.* 42,665-673.

COLONELLO-FRATTINI,N.A., HARTFELDER,K., 2009. Differential gene expression profiling in mucus glands of honeybee (*Apis mellifera*) drones during sexual maturation. *Apidologie* 40,481-495.

FANG,Y., SONG,F., ZHANG,L., ALEKU,D.W., HAN,B., FENG,M., LI,J., 2012. Differential antennal proteome comparison of adult honeybee drone, worker and queen (*Apis mellifera* L.). *J.Proteome* 75, 756-773.

KOCHER,S.D., RICHARD,F.J., TARPY,D.R., GROZINGER,C.M., 2008. Genomic analysis of post-mating changes in the honey bee queen (*Apis mellifera*). *BMC Genomics* 9, 232.

KOCHER,S.D., TARPY,D.R., GROZINGER,C.M., 2010. The effects of mating and instrumental insemination on queen honey bee flight behaviour and gene expression. *Insect Mol. Biol.* 19(2),153-162.

KREM,M.M., DI CERA,E., 2002. Evolution of enzyme cascades from embryonic development to blood coagulation. *Trends Biochem.Sci.* 27(2),67-74.

LI,B., XIA,Q.Y., LU,C., ZHOU,Z.Y., XIANG,Z.H., 2004. Analysis of cytochrome P450 in silkworm genome. *Sci.China Ser.C Life Sci.* 34(6),517-521.

LIU,Q.M., YUAN,Y.Y., LIN,J.R., ZHONG,Y.S., 2010. Advance of researches on insect cuticular proteins and the regulation mechanism of their gene expression. *Chin. Bull.Entomol* 47(2),247-255.

MORTAZAVI,A.,WILLIAMS,B.A.,MCCUE,K., SCHAEFFER,L.,WOLd,B.,2008. Mapping and quantifying mammalian transcriptomes by RNA-Seq. *Nat.Methods* 5, 621-628.

NIU,D., ZHENG,H., CORONA,M., LU,Y., CHEN,X., CAO,L., SOHR,A., HU,F.,2014. Transcriptome comparison between inactivated and activated ovaries of the honey bee *Apis mellifera* L. *Insect Mol.Biol.* 23(5),668-681.

RAWLINGS,N.D., BARRETT,A.J., 1993. Evolutionary families of peptidases. *Biochem.J.* 290 (Pt 1),205-218.

RINGO,J., WERCZBERGER,R., ALTARATZ,M., SEGAL,D., 1991. Female sexual receptivity is defective in juvenile hormone-deficient mutants of the apterous gene of *Drosophila melanogaster*. *Behav. Genet.* 21(5),453-469.

SOLLER,M., BOWNES,M., KUBLI,E., 1999. Control of oocyte maturation in sexually mature Drosophila females. *Dev. Biol.* 208(2),337-351.

TEAL,P.E., GOMEZ-SIMUTA,Y., PROVEAUX,A.T., 2000. Mating experience and juvenile hormone enhance sexual signaling and mating in male Caribbean fruit flies. *Proc.Natl.Acad.Sci.U.S.A.* 97(7),3708-3712.

WANG,S.H., WANG,W.Y., HUANG,Y.Z., LIN,L., SHA,L., 2007. On serine protease. *Fujian J.Agric.Sci.* 22(4),453-456

WU,X.B., WANG,Z.L., SHI,Y.Y., ZHANG,F., ZENG,Z.J., 2013a. Effects of mating flight on sRNAs expression in sexual matured virgin queens (*Apis cerana cerana*). *Sci.Agric.Sin.* 46(17), 3721-3728.

WU,X.B., WANG,Z.L., ZHANG,F., SHI,Y.Y., ZENG,Z.J.,2013b. Mating flight behaviour affects gene expression in matured virgin queens of *Apis cerana cerana* (Hymenoptera: Apidae). *Acta Entomol.Sin.* 56(5),486-493.

WU,X.B., WANG,Z.L., ZHANG,F., SHI,Y.Y., ZENG,Z.J.,2014.Mating flight causes genome-wide transcriptional changes in sexually mature honeybee queens. *J.Asia Pac.Entomol* 17(1),37-43.

YANO,K., SAKURAI,M.T., IZUMI,S., TOMINO,S., 1994. Vitellogenin gene of the silkworm, *Bombyx mori*: structure and sex-dependent expression. *FEBS Lett.* 356(2-3),207-211.

ZENG,Z.J., 2009. Apiculture. China Agriculture Press, Beijing.

ZHANG,Q.L.,YUAN,M.L.,2013. Progress in insect transcriptomics based on the next-generation sequencing technique. *Acta Entomol.Sin.* 56(12),1489-1508.

ZOU,Z.,LOPEZ,D.L.,KANOST,M.R.,EVANS,J.D.,JIANG,H.,2006.Comparative analysis of serine protease-related genes in the honey bee genome: possible involvement in embryonic development and innate immunity. *Insect Mol. Biol.* 15(5),603-614.

# 8. Effects of Queen Cell Size and Caging Days of Mother Queen on Rearing Young Honey Bee Queens *Apis mellifera* L.

Xiaobo Wu, Linbin Zhou, Chuibin Zou, Zhijiang Zeng*

Honeybee Research Institute, Jiangxi Agricultural University, Nanchang, Jiangxi 330045, China

**Abstract**

This study aims to investigate the effect of queen cell size (9.4mm, 9.6mm, 9.8mm and 10.0mm) and mother queen caged time (0 day, 2 days and 4 days) on rearing young queens without grafting larvae. The birth weight, ovarian tubes, thorax length and width were significantly increased with the increasing diameter of queen cell size. The expression level of Vitellogenin ($Vg$) in young queen ovaries was also up-regulated with the increased queen cell size diameter. These results indicate that the queen cell size can strongly affect the rearing queen quality and reproductive ability. Moreover, the weight, length and width of laying eggs rose with the mother queen caging time, and young queens reared with the hatched larvae from these eggs were also increased in terms of birth weight, ovarian tubes, thorax length and width. Furthermore, the expression level of $Vg$ in reared queen ovaries was also up-regulated with the caged time. These results reveal that the caged time of queens could significantly influence egg size and their relative queen quality.

**Keywords:** bee queens, queen cell size, imprisoned days of mother queen, quality of reared queen

## Introduction

Queens are vital for the survival of honey bee colonies, not only because of their ability to lay a large numbers of eggs but also because of the social coherence of their pheromones (Amiri *et al.*, 2017). Queens with high reproductive potential produce colonies that exhibit high growth and survival (Rangel, Keller, & Tarpy, 2013). Beekeepers have developed techniques to rear large numbers of young queens and typically replace their old queens, because of the critical importance of a vigorous queen to colony survival and productivity (Büchler *et al.*, 2013; Amiri *et al.*, 2017). When rearing young honey bee queens, 1st star larvae are usually grafted into queen cells and are inserted into a queenless colony. However, the quality of reared queens are affected by many such factors as genetic background, rearing season, queen breeding methods and the age of larvae used for rearing queen (Gilley, Tarpy, & Land, 2003; Koç & Karacaoglu, 2004; Liu *et al.*, 2011; Zhang *et al.*, 2013; Hatjina *et al.*, 2014; He *et al.*, 2017). Beyond that, external environmental factors also affect the quality of reared queens, such as the intensity of the colony, food-storage and weather (Nabors, 2000; Liao *et*

---

\* Corresponding author: bees1965@sina.com (ZJZ).
注：此文发表在 *Journal of Apicultural Science* 2018 年第 2 期。

*al.*, 2016).

Queen cells are used to rear new queens in a honeybee colony and the natural size of the swarming queen cells of *Apis mellifera* ranges between 8-10 mm. Büchler *et al.* (2013) claimed that queen cells should be 8-9 mm in diameter at the rim and the most common diameter of a plastic queen cell for queen rearing is about 9.4mm. However, a cell could be 9.6, 9.8 or 10.0 mm, the size has been reported to affect queen-worker differentiation (Shi *et al.*, 2011). Moreover, when the mother queen was prohibited to lay eggs for a few days in a queen cage and released for laying, increased weight and size of laid eggs were observed, which then developed into better quality queens with more ovarian tubes (Liu *et al.*, 2012; 2014).

We reared young queens with different queen cell sizes in the same colonies at the same time to determine the effect of different size of queen cell on the quality of reared queens. Meanwhile, we reared queens with hatched larvae from the first eggs which had been laid by the mother queen who was forbidden laying in queen cage with different caged time and released for laying, to determine the effect of caged time of mother queen on the quality of reared queens.

## Material and Methods

### Experimental honey bee colonies

With the use of standard beekeeping techniques, the *Apis mellifera ligustica* colonies were kept at the Honey-bee Research Institute, Jiangxi Agricultural University, Nanchang, China (28.46 °N, 115.49 °E).

### Rearing queens

According to the new method for queen rearing without larvae grafting as described by Pan QZ *et al.* (2013), the mother queen was restricted from laying eggs for six hours on a built-up comb and then was removed to a normal comb. After three days, the eggs hatched into larvae and were inserted into queen cells with an inner diameter of 9.4 mm, 9.6 mm, 9.8 mm and 10.0 mm, respectively. The larvae-containing queen cells were then inserted into queen rearing frames according to the layout shown in Fig. II-8-1. Next two frames were placed in a breeding colony to be further looked after by nurse bees for rearing queens, and so the effect of queen cell distribution was avoided. When the newly reared queens emerged, their birth weight, ovarian tube number, thorax length and width, and Vitellogenin (*Vg*) expression level in ovaries were tested. The effect of queen cells with different diameters on the quality of rearing queens was analysed to determine the best inner diameter. All the experiments were replicated three times on three bee colonies.

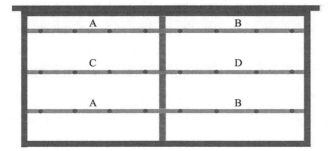

**Figure II-8-1** Layout of different queen cell diameters in the queen rearing frames. A, B, C and D indicate a queen cell with a diameter of 9.4, 9.6 9.8 and 10.0mm respectively and the points represent the position of queen cells.

Queens were caged and prohibited for laying eggs for four days, and released to lay eggs on the built-up comb for six hours, and then removed to a normal comb. At the same time, thirty newly laid eggs on the built-up comb were immediately measured for their weight, length and width with an analytical balance and microscope system, while the built-up combs with the remaining newly laid eggs were transferred to the hatching area without the queen. Three days later the supporting devices with newly hatched larvae were inserted into queen cells with an inner diameter of 10.0 mm and then were added to the breeding colonies for queen rearing. The birth weight, thorax length and width, ovarian tubes numbers and the *Vg* expression level of reared queens were measured. Similarly, the mother queens were kept in a queen cage and forbitten to lay eggs for 0-2 days and released to lay eggs for six hours. The weight, length and width of newly laid eggs were recorded and new queens were reared with the hatched larvae from the remaining newly laid eggs with the same method. When the new queens emerged, their birth weight, ovarian tube number, thorax length and width and the *Vg* gene expression level in reared queen ovaries were measured. The effect of caged time of mother queen on the quality of newly reared queens *Apis mellifera* L with the best queen cell diameter was analysed.

### *The birth weight*

On the first day of emerging, queen cells were transferred to a box of at a constant temperature of 35°C and humidity of 80%. The birth weight of twelve newly emerged queens in each group was weighed with an electronic scale.

### *Ovarian tubes numbers*

Nine newly emerged queens from each group were starved for five hours and their abdomens were removed with scissors. The abdomens were fixed in 10% paraformaldehyde for four hours. The ovarian tissues were then taken out and fixed in 10% paraformaldehyde for twenty hours with embedding cassettes, washed with phosphate buffered saline (PBS), dehydrated in ethanol and embedded in paraffin. Five-micrometer sections were cut on a microtome and after the removal of paraffin stained with hematoxylin & eosin (H&E) for general morphology. After staining, the slides were analyzed with a light microscopy and the ovarian tubes were counted (Gan, Tian, & Yan, 2012; Zhang *et al.*, 2015).

### *Thorax length and width*

After the ovarian tissues were taken out, the queen's head and abdomen were removed with scissors and the thorax length and width of the queens were measured.

### *The expression level of Vitellogenin gene*

The ovarian tissues of nine newly emerged queens in each group were taken out with scissors and rinsed with DEPC water. The ovarian tissue was put into a 1.5 mL EP tube after being rinsed, and kept in liquid nitrogen until used for RNA extraction. Total RNA was extracted using Trizol reagent (Invitrogen, Carlsbad, CA, USA), according to the manufacturer's instructions. RNA was reverse-transcribed to cDNA with Prime Script™ RT Master Mix kit (TaKaRa). Real-time quantitative PCR was performed on real-time PCR system with the Real Time SYBR master mix kit. Gene-specific primers were listed in Table Ⅱ-8-1 and the β-actin gene was used as an internal control. All samples were analyzed in triplicate.

## Table II-8-1 Gene-specific primers used in real time quantitative PCR

| Target gene | Type Name | Primer sequence (5'–3') |
|---|---|---|
| Vg | Forward Primer | CGCATCACGAATACGACTAAGA / |
|  | Reverse Primer | ACGCTCCTCAGGCTCAACTC |
| β–actin | Forward Primer | GCTGGTTTCATCGATGGTTT/ |
|  | Reverse Primer | ACGATTTCGACCACCGTAAC |

### Data analyses

Data was analyzed by analysis of variance (ANOVA) through StatView (v 5.01, SAS Institute, Gary, NC, USA). Multiple comparisons of the means were carried out using Fisher's protected least significant difference only after ANOVA showed a significant effect ($P < 0.05$). Data are Means ± SE.

## Results

### Effect of queen cell diameter on birth weight, ovaries and thorax length and width of reared queens

The results showed that the method of queen rearing without grafting larvae is feasible and more than 85% of larvae were accepted. As shown in Table II-8-2, birth weight, ovarian tube number and thorax length and width were all significantly increased with larger queen cell diameters ($P<0.05$).

### Table II-8-2 Effect of queen cell diameter on birth weight, ovaries and thorax length and width of reared queen

| Queen cell diameters (mm) | Birth weight (mg) $N=12$ | Ovaries $N=9$ | Thorax length (mm) $N=9$ | Thorax width (mm) $N=9$ |
|---|---|---|---|---|
| 9.4 | 194.83 ± 12.25$^a$ | 201.00 ± 6.87$^a$ | 4.52 ± 0.05$^a$ | 4.38 ± 0.07$^a$ |
| 9.6 | 215.33 ± 21.45$^b$ | 213.83 ± 6.74$^b$ | 4.61 ± 0.04$^b$ | 4.48 ± 0.03$^b$ |
| 9.8 | 231.50 ± 6.63$^c$ | 237.17 ± 7.03$^c$ | 4.69 ± 0.04$^c$ | 4.59 ± 0.03$^c$ |
| 10.0 | 247.67 ± 6.71$^d$ | 251.33 ± 6.65$^d$ | 4.79 ± 0.07$^d$ | 4.69 ± 0.03$^d$ |

Note: Data are reported as mean±SE. Values in the same column with different letter superscripts indicate significant differences ($P<0.05$). The same notation is used in Table II-8-3 and Table II-8-4.

### Effect of queen cell diameter on expression level of Vg gene in reared queen's ovarian tissues

As shown in Fig. II-8-2, the Vg expression level in reared queen ovarian tissues was higher as the queen cell size increased.

### Effect of mother queen caged days on weight, length and width of laid eggs

As shown in Table II-8-3, an increase in the number of days that the mother queen was caged was associated with a significant increase in the weight and dimensions of the laid eggs ($P<0.05$).

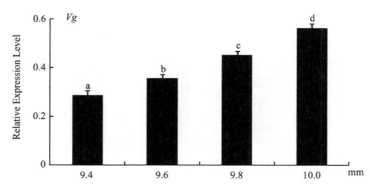

**Figure II-8-2** Effect of queen cell diameter on gene expression of $Vg$ in queen's ovarian tissues. Histogram shows the relative $Vg$ expression level. Each bar corresponds to a single group and shows the mean±SE of three biological replicates. $N$=9. Different letters above bars mean a significant difference between groups ($P < 0.05$). The same as below.

**Table II-8-3  Effect of mother queen caging days on weight and dimensions of laid eggs**

| Caged days (d) | Weight of laid eggs (μg) $N$=30 | length of laid eggs (mm) $N$=30 | Width of laid eggs (mm) $N$=30 |
| --- | --- | --- | --- |
| 0 | 160.17 ± 2.48$^a$ | 1.58 ± 0.01$^a$ | 0.30 ± 0.02$^a$ |
| 2 | 177.33 ± 2.34$^b$ | 1.62 ± 0.02$^b$ | 0.32 ± 0.01$^b$ |
| 4 | 183.33 ± 4.63$^c$ | 1.71 ± 0.01$^c$ | 0.33 ± 0.02$^c$ |

*Effect of mother queen caging days on birth weight, ovaries and thorax length and width of reared queens*

As shown in Tab.4, birth weight, ovarian tubes number and thorax length and width all significantly increased for the greater duration of mother queen caging ($P<0.05$).

**Table II-8-4  Effect of mother queen caging days on birth weight, ovaries and thorax length and width of reared queen**

| Caged days (d) | Birth weight (mg) $N$=12 | Ovaries $N$=9 | Thorax length (mm) $N$=9 | Thorax width (mm) $N$=9 |
| --- | --- | --- | --- | --- |
| 0 | 238.60 ± 6.80$^a$ | 225.57 ± 6.63$^a$ | 4.67 ± 0.04$^a$ | 4.56 ± 0.07$^a$ |
| 2 | 248.00 ± 4.53$^b$ | 249.00 ± 7.28$^b$ | 4.74 ± 0.06$^b$ | 4.68 ± 0.06$^b$ |
| 4 | 262.20 ± 4.87$^c$ | 264.29 ± 4.68$^c$ | 4.84 ± 0.03$^c$ | 4.79 ± 0.04$^c$ |

*Effect of mother queen caging days on Vg gene expression level in reared queen's ovarian tissues*

As shown in Fig. II-8-3, increasing the number of mother queen-caging days led to an increase in the $Vg$ expression level in reared queen ovarian tissues.

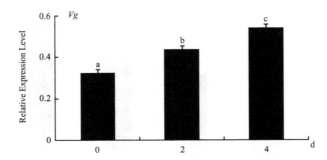

**Figure II-8-3** Effect of mother queen caging days on Vg gene expression in queen's ovarian tissues. $N=9$.

## Discussion

Queen weight at emergence, thorax length and width, ovarioles and Vg expression level in queen ovarian tissues are critical physical characteristics for evaluating the quality of honey bee queens (Tarpy *et al.*, 2012; Nunes *et al.*, 2013; Amiri *et al.*, 2017). The experimental results showed that birth weight, ovarian tubes number, thorax length and width and Vg expression level in ovarian tissues of reared queens all significantly increased when larger-diameter queen cells were used. The newly reared queens with 10.0 mm queen cells were clearly better than those reared with the other three cell sizes. 10.0 mm cells found to be better and the natural swarming cells were between 8.0-10.0 mm. The best swarming queens probably came from the 10.0 mm diameter cells. This might be because the larger queen cells had more space, which promoted rearing queen development and also the reared queen's birth weight, ovarian tubes number and other measured dimensions increased significantly. Shi *et al.* (2011) found different size of nest can significantly affect the DNA methylation level of larvae, affecting the Dnmt3 enzyme activity which plays an important regulating role in the development of larvae.

At the same time, we observed that the height of the royal jelly in different size queen cells was the same. The height of royal jerry is probably information for nurse bees to feed the larvae. So the 10.0 mm queen cell could contain more royal jelly than the other three groups because of the large size cell. The nutrition and the amount of food given to young larvae cause the caste determination of worker and queen regulated developmental pathways. The more the larvae are given, the better the reared queen would be (Kucharski *et al.*, 2008; Maleszka, 2008; He *et al.*, 2017). However, when the larvae had enough royal jelly for food, the quality of food was very important for the development of larvae. With the increase of larvae, the food intake also increased. The bigger the queen cells were, the more fresh royal jelly was given. So the older larvae in bigger queen cells had enough fresh royal jelly for food everyday, while the older larvae in smaller queen cells did have enough who went on to ingest the remaining jelly left before. Guo *et al.* (2015) found that the composition of royal jelly from different harvesting times differed. This study found that queen cell diameter affects reared-queen growths which may be due to increased development space and more abundant food, but the specific molecular mechanism still requires further research.

Our results also showed that the weight, length and width of laid eggs significantly increased with an increase in mother queen-caging time, while birth weight, ovarian tube number and other dimensions of reared queens with larvae hatched from eggs also increased significantly. The quality of eggs or larvae affects the quality of reared queens (Woyke, 1971; Chuda-Mickiewicz & Prabucki, 1998). The results suggest that caging lets the queen lay larger eggs and thus the hatched larvae from the eggs are reared as better new queens. The longer the queen was caged, the more it accumulated such nutrients in the body as vitellogenin and other antioxidant enzymes (Alaux *et al.*, 2011). Queen-delaying oviposition causes eggs to be bigger with more yolk protein (Torres, 1980). When the caged queen is released, her first eggs are large, and the quality of a queen reared from such eggs are better. We have shown that caging the queen can be used to increase the quality of fertilized eggs and improve the quality of queens reared from eggs laid by a caged queen.

In conclusion, our results indicate that in the conditions of our rearing colonies caging a mother queen for two to four days and then releasing increases the quality of eggs, which is conducive for queen rearing. When rearing new queens, we should use the bigger queen cell whose diameter is about 10.0 mm, which can increase the quality of reared queens.

## Acknowledgements

We thank Dr. Frederick Partridge, Dr Qiang Huang and Dr Cui Guan for improving the English of this manuscript. This work was supported by the technology support program of Jiangxi Province (20141BBF60033), the outstanding young talent program of Jiangxi Province (20162BCB23029) and the Earmarked Fund for China Agriculture Research System (CARS-44-KXJ15).

## References

ALAUX, C., FOLSCHWEILLER, M., MCDONNELL, C., BESLAY, D., COUSIN, M., DUSSAUBAT, C., BRUNET, J.L., LE CONTE, Y. (2011). Pathological effects of the microsporidium *Nosema ceranae* on honey bee queen physiology (*Apis mellifera*). *Journal of Invertebrate Pathology*, 106, 380-385. DOI: 10.1016/j.jip.2010.12.005.

AMIRI, E., STRAND, M.K., RUEPPELL, O., & TARPY, D.R. (2017). Queen quality and the impact of honey bee diseases on queen health: potential for interactions between two major threats to colony health. *Insects*, 8, 48. DOI: 10.3390/insects8020048.

BÜCHLER, R., ANDONOV, S., BIENEFELD, K., COSTA, C., HATJINA, F., KEZIC, N., ... WILDE, J. (2013). Standard methods for rearing and selection of *Apis mellifera* queens. *Journal of Apicultural Research*, 52, 1-29. DOI: 10.3896/IBRA.1.52.1.07.

CHUDA-MICKIEWICZ, B., & PRABUCKI, J. (1998). The effect of rearing queens from eggs and larvae. *Pszczelnicze Zeszyty Naukowe*, 42, 27-28.

GAN, H.Y., TIAN, L.Q., & YAN, W.Y. (2012). Queen ovary slice and dye technology. *Bee Journal*, 32, 9.

GILLEY, D.C., TARPY, D,R., & LAND, B.B. (2003). Effect of queen quality on interactions between workers and dueling queens in honeybee (*Apis mellifera* L.) colonies. *Behavioral Ecology Sociobiology*, 55, 190-196. DOI: 10.1007/s00265-003-0708-y.

GUO, Y.H., ZHOU, L.B., PAN, Q.Z., ZHANG, L.Z., YI, Y., ZENG, Z.J. (2015). Effect of different harvesting times on the yield and composition of royal jerry. *Acta Agriculturae Universitatis Jiangxiensis*, *37*, 120-125.

HATJINA, F., BIENKOWSKA, M., CHARISTOS, L., CHLEBO, R., COSTA, C., DRAZIC, M.M., ... WILDE, J. (2014). A review of method used in some European countries for assessing the quality of honeybee queens through their physical characters and the performance of their colonies. *Journal of Apicultural Research*, 53, 337-363. DOI: 10.3896/IBRA.1.53.3.02.

HE, X.J., ZHOU, L.B, PAN, Q.Z., BARRON, A.B., YAN, W.Y., ZENG, Z.J. (2017). Making a queen: an epigenetic analysis of the robustness of the honeybee (*Apis mellifera*) queen developmental pathway. *Molecular Ecology*, *26*, 1598-1607. DOI:10.1111/mec.13990.

KOÇ, A.U., & KARACAOGLU, M. (2004). Effects of rearing season on the quality of queen honeybees (*Apis mellifera* L.) raised under the conditions of Aegean region. *Mellifera*, *4*, 34-37.

KUCHARSKI, R., MALESZKA, J., FORET, S., & MALESZKA, R. (2008). Nutritional control of reproductive status in honeybees via DNA methylation. *Science, 319*, 1827-1830. DOI: 10.1126/science.1153069.

LIAO, C.H., ZOU, C.B., XIE, G.X., WU, X.B., HUANG, J.S. (2016). Effect of dietary crude protein level on quality of rearing queens for *Apis cerana cerana*. *Chinese Journal of Animal Nutrition*, *28*, 2998-3004.

LIU, G.N., WU, X.B., ZENG, Z.J., & YAN, W.Y. (2011). Effects of different methods for queen rearing on queen-cell accepted rate and queen quality in *Apis mellifera ligustica*. *Shandong Agricultural Science*, *3,* 106-108.

LIU, Y.Q., DONG, K., ZHANG, L.Y., & HE, S.Y. (2012). Comparisons on the weights and sizes of eggs before and after the queen of *Apis cerana cerana* caged. *Apiculture of China*, *63*, 18-20.

LIU, Y.Q., DONG, K., ZHOU, D.Y., HE, S.Y., LIN, Z.C. (2014). Prisoners queen spawning impact on virgin queens small body size and the number of ovarian tubes. *Apiculture of China, 65*, 30-32.

MALESZKA, R. (2008). Epigenetic integration of environmental and genomic signals in honey bees: the critical interplay of nutritional, brain and reproductive networks. *Epigenetics, 3*, 188-192. DOI: 10.4161/epi.3.4.6697.

NABORS, R. (2000). The effect of spring feeding pollen substitute to colonies of *Apis mellifera*. *American Bee Journal*, *140*(4), 322-323.

NUNES, F.M.F., IHLE, K.E., MUTTI, N.S., SIMÕES, Z.L.P., AMDAM, G.V. (2013). The gene vitellogenin affects microRNA regulation in honey bee (*Apis mellifera*) fat body and brain. *The Journal of Experimental Biology, 216*, 3724-3732. DOI: 10.1242/jeb.089243.

PAN, Q.Z., WU, X.B., GUAN, C., & ZENG, Z.J. (2013). A new method of queen rearing without grafting larvae. *American Bee Journal*, *12*, 1279-1280.

RANGEL, J., KELLER, J.J., & TARPY, D.R. (2013). The effects of honey bee (*Apis mellifera* L.) queen reproductive potential on colony growth. *Insectes. Sociaux*, *60*, 65-73. DOI: 10.1007/s00040-012-0267-1.

SHI, Y.Y., HUANG, Z.Y., ZENG, Z.J., WANG, Z.L., WU, X.B., YAN,W.Y. (2011). Diet and cell size both affect queen-worker differentiation through DNA methylation in honey bees (*Apis mellifera*, Apidae). *PLoS One*, *6*,e18808. DOI: 10.1371/journal.pone.0018808.

TARPY, D.R., KELLER, J.J., CAREN, J.R., & DELANEY, D.A. (2012). Assessing the mating health of commercial honey bee queens. *Journal of Economic Entomology*, *105*, 20-25. DOI: http://dx.doi.org/10.1603/EC11276.

TORRES, J.J. (1980). A stereological analysis of developing egg chambers in the honeybee queen *Apis mellifera*. *Cell*

*Tissue Research, 208*, 29-33.

WOYKE, J. (1971). Correlations between the age at which honeybee brood was grafted, characteristics of the resultant queens, and results of insemination. *Journal of Apicultural Research, 10*, 45-55.

ZHANG, F., GAN, H.Y., LI, S.Y., & ZENG, Z.J. (2013). Preliminary study on the bionic technology of rearing queen bees without transferring larvae. *Heilongjiang Animal Science and Veterinary Medicine, 10*, 144-146.

ZHANG, H., GUO, X., ZHONG, S., GE, T., PENG, S., YU, P., ZHOU, Z. (2015). Heterogeneous vesicles in mucous epithelial cells of posterior esophagus of Chinese giant salamander (Andrias davidianus). *European Journal of Histochemistry, 59*, 2521. DOI: 10.4081/ejh.2015.2521.

# 9. Feeding Asian Honeybee Queens with European Honeybee Royal Jelly Alters Body Color and Expression of Related Coding and Non-coding RNAs

Amal Abdelmawla[1,2†], Chen Yang[1†], Xin Li[1], Mang Li[1], Changlong Li[1], Yibo Liu[1], Xujiang He[1,3*], Zhijiang Zeng[1,3*]

1. Honeybee Research Institute, Jiangxi Agricultural University, Nanchang, 330045, China
2. Faculty of Agriculture, Fayoum University, Fayoum 63514, Egypt
3. Jiangxi Key Laboratory of Honeybee Biology and Bee Keeping, Nanchang, Jiangxi 330045, China

## Abstract

The Asian honeybee (*Apis cerana*) and the European honeybee (*Apis mellifera*) are reproductively isolated. Previous studies reported that exchanging the larval food between the two species, known as nutritional crossbreeding, resulted in obvious changes in morphology, physiology and behavior. This study explored the molecular mechanisms underlying the honeybee nutritional crossbreeding. This study used full nutritional crossbreeding technology to rear *A. cerana* queens by feeding them with an *A. mellifera* royal jelly-based diet in an incubator. The body color and the expression of certain genes, microRNA, lncRNA, and circRNA among nutritional crossbred *A. cerana* queens (NQ), and control *A. cerana* queens (CQ) were compared. The biological functions of two target genes, *TPH1* and *KMO*, were verified using RNA interference. Our results showed that the NQ's body color turned yellow compared to the black control queens. Whole transcriptome sequencing results showed that a total of 1484, 311, 92, and 169 DEGs, DElncRNAs, DEmiRNAs, and DEcircRNAs, respectively, were identified in NQ and CQ, in which seven DEGs were enriched for three key pathways (tryptophan, tyrosine, and dopamine) involved in melanin synthesis. Interestingly, eight DElncRNAs and three DEmiRNAs were enriched into the key pathways regulating the above key DEGs. No circRNAs were enriched into these key pathways. Knocking down two key genes (*KMO* and *TPH1*) resulted in altered body color, suggesting that feeding NQ's an RNAi-based diet significantly down-regulated the expression of *TPH1* and *KMO* in 4-day-old larvae, which confirmed the function of key DEGs in the regulation of honeybee body color. These findings reveal that the larval diets from *A. mellifera* could change the body color of *A. cerana*, perhaps by altering the expression of non-coding RNAs and related key genes. This study serves as a model of

† These two authors contributed equally to this paper.
*Corresponding author: hexujiang3@163.com (XJH); bees1965@sina.com (ZJZ).
注：此文发表在 *Frontiers in Physiology* 2023, 14: 1073625。

epigenetic regulation in insect body color induced by environmental factors.

**Keywords**: Honeybees, nutritional crossbreeding, body color alteration, gene expression, non-coding RNA expression

## Introduction

The honey bee is one of the most important and beneficial insects, pollinating over 85% of global crops (Klein *et al*., 2007). There are nine honeybee species around the world, of which the European honeybee (*Apis mellifera*) and the Asian honeybee (*Apis cerana*) are economically the most important. Some European honeybee subspecies, for example, the Italian honeybee (*A. mellifera* ligustica), have a "yellow" body color and exhibit advantages in brood rearing, nectar collecting, and royal jelly production (Graham, 1993). On the other hand, Asian honeybees (*A. cerana*) are smaller and are primarily black body color. They are more sensitive to smell, better at exploiting sporadic nectar sources, and more suited to locating a diversity of honey plants (He and Liu, 2011). Asian honeybees have a significantly stronger *Varroa destructor* mite resistance, greater low-temperature tolerance, and decreased food consumption, compared to European bees (Chen, 2001; Zeng *et al*., 2006; Diao *et al*., 2018; Wang *et al*., 2020). Even though these two species are closely related, they have clear reproductive isolation and can't produce filial generations. However, previous studies showed that feeding European honeybee larvae with Asian honeybee royal jelly or vice versa, could alter their body color, and other morphological and behavioral characteristics. This phenomenon is known as honeybee nutritional crossbreeding, and was first seen in European honeybee subspecies (Smaragdova, 1963; Rinderer *et al*., 1985) and is frequently reported in Asian and European honeybees. For instance, studies showed body color changes in *A. cerana* and *A. mellifera* nutritional crossbreeding (Li, 1982; Zhu, 1985; Xie *et al*., 2008). Xie *et al*. showed that nutritional crossbreeding could decrease proboscis length, wing area, and the length of the 3$^{rd}$ and 4$^{th}$ dorsal plate of the abdomen and alter the mite-defending ability of *A. mellifera* workers (Xie *et al*., 2008). It also increased the body size, birth weight, and ovariole numbers of *A. cerana* workers (Chen *et al*., 2015; Zeng *et al*., 2005). A few studies revealed that *A. cerana-A. mellifera* nutritional crossbreeding can also induce genetic changes (Xie *et al*., 2005; He *et al*., 2010). Shi *et al*. showed that feeding *A. cerana* larvae with *A. mellifera* royal jelly could alter the genome-wide alternative splicing of *A. cerana* (Shi *et al*., 2014).

One commonly observed phenomenon of *A. mellifera* and *A. cerana* nutritional crossbreeding is a change in body color. In honeybees, body color is considered a heritable morphological trait used in taxonomy. However, body color changes in nutritional crossbreeding are epigenetic alterations induced by larval diets. Diet-induced body color alternation is common in many animals. For instance, feeding turtles (*Pelodiscus sinensis*) with a xanthophyll-b diet enhances the "yellowness" of their body color (Wang *et al*., 2020). Adding plant carotenoids to the diet of the Ornamental dwarf cichlid (*Microgeophagus ramirezi*) affects their color enhancement (Padowicz and Harpaz, 2007). Diet can also affect body color in the Baikal endemic amphipod (*Eulimnogammarus cyaneus*) (Saranchina *et al*., 2021). A recent study also confirmed that environmental factors could alter the body color of fruit flies, and this change can be passed to the next generation (Golic *et al*., 1998).

Honeybee royal jelly is secreted by the hypopharyngeal and mandibular glands of nurse bees and contains major-royal-jelly-family proteins (MRJPs), amino acids, sugars, vitamins, organic acids as well as DNA and RNA (Yeung and Argüelles, 2019). Previous studies have shown differences between *A. cerana* and *A. mellifera* royal jelly with regard to 10-hydroxy-α-decenoic acid, total carbohydrate, proteins, amino acids, DNA, and microRNA components (Zeng *et al.*, 2006; Zou *et al.*, 2007; Oldroyd and Wongsiri, 2009; Yu *et al.*, 2010; Liu *et al.*, 2014). The royal jelly is powerful and important for honeybee caste differentiation so that larvae with the same genetic background fed with royal jelly develop into queens compared to larvae fed with worker jelly who develop into workers (Winston, 1991). Notably, Shi *et al.* (2012) reported that there are 23 microRNAs (miRNAs) specific to *A. mellifera* royal jelly, two miRNAs specific to *A. cerana* royal jelly, and 46 miRNAs that are significantly differentially expressed in both types of royal jelly. However, it is still unclear how the differences between *A. cerana* and *A. mellifera* royal jelly affects nutritional crossbreeding and the resultant altered body color.

Melanins are commonly associated with black and brown pigmentation, and play a key role in the determination of insect body color (Sugumaran and Barek, 2016). Dopamine melanin is the most common melanin pigment found in insects, and is synthesized from tyrosine in the epidermal cells (Wang *et al.*, 2022; Chen *et al.*, 2019). Dopamine is converted to N-β-alanyldopamine (NBAD) and N-acetyldopamine (NADA) by NBAD synthase and arylalkylamine N-acetyltransferase (aaNAT), respectively. The synthetic pathway of NBAD requires β-alanine, which is derived from L-aspartic acid by aspartate 1-decarboxylase (ADC). In addition, NBAD can be converted to dopamine and β-alanine by NBAD hydrolase (NBADH), which catalyzes the reverse reaction of NBADS (Futahashi and Osanai-Futahashi, 2021). Sasaki *et al.* (2021) quantified the functional monoamines (dopamine, tyramine, octopamine, and serotonin) and their precursors found in bumble bees (*Bombus ignitus*). These monoamines are synthesized from amino acids (Sasaki *et al.*, 2021; Arakane *et al.*, 2016; Noh *et al.*, 2016), where tyrosine or tryptophan are important for the determination of body color (Wang *et al.*, 2022).

Furthermore, key pathways and related key genes involved in the synthesis of melanins are an insect's main molecular mechanisms in body color alteration. Previous studies suggest that the expression patterns and levels of *yellow* and *ebony* genes together, determine the patterns and intensity of melanization (Wittkopp *et al.*, 2002). Borycz *et al.* also explain that *ebony* and *tan*, two cuticle melanizing mutants, regulate the conjugation (*ebony*) of β-alanine to dopamine or hydrolysis (*tan*) of the β-alanyl conjugate to liberate dopamine in fruit flies (*D. melanogaster*) (Borycz *et al.*, 2002). In these fruit flies, two key genes namely *tyrosine hydroxylase* (*TH*) and *dopa decarboxylase* (*DDC*) participate in the conversion of tyrosine to dopamine (Tang, 2009), contributing to the formation of body color. In butterflies (*Bicyclus anynana*), deletion of the *yellow* and *DDC* melanin genes alters both the body color and the wing morphology (Matsuoka and Monteiro, 2018). In the pea aphid (*Acyrthosiphon pisum*), the biosynthetic pathways of amino acids (phenylalanine, tyrosine, and dopamine pathways) played a key role in cuticle formation during parthenogenetic development (Rabatel *et al.*, 2013).

In honeybees, knocking out the *yellow-y* gene decreased the amount of black pigment in the cuticle of mosaic workers of *A. mellifera*. The expression of *Amyellow-y* and *aaNATA* in mutant drones, which have a dramatic body pigmentation defect, was lower than in wild-type drones, whereas the expression of *laccase2* was significantly up-regulated (Nie et al., 2021). Additionally, seven genes involved in the biosynthesis of melanin and sclerotizing compounds are up-regulated in the pharate-adults and newly-emerged bees. The gene *dopamine N-acetyltransferase* (*Dat*) and *ebony* might also contribute to body color changes in the eusocial *Frieseomelitta varia* and the solitary *Centris analis*, where the *TH* gene also showed a significantly higher expression level in the early developmental phase (Ne) than in the later phase of the three bee species (Falcon et al., 2019). By using RNA-Seq, 17 genes in *A. mellifera*, and 18 genes in *Frieseomelitta varia* and *Centris analis*, including the *cardinal, scarlet, brown, vermillion, light, sepia,* and *henna* genes, were predicted to contribute to the formation of the adult cuticle (Falcon et al., 2019). Furthermore, Elias-Neto et al., (2010) identified an *Amlac2* gene that encodes for a laccase2 in *Apis mellifera*. The *Amlac2* is highly expressed in the adult integument of pharate adults and is present before cuticle coloring and sclerotization intensify. The exoskeleton's structural defects driven by post-transcriptional *Amlac2* gene knockdown had a significant impact on adult eclosion. Together, these findings show that ecdysteroids regulate *Amlac2* expression and are essential for the development of the adult honey bee exoskeleton. However, the molecular mechanisms of body color alteration in honeybees remain unclear, especially the color alteration induced by *A. cerana* and *A. mellifera* nutritional crossbreeding which achieves body color alternation epigenetically.

In this study, we artificially reared *A. cerana* queens in an incubator using an *A. mellifera* royal jelly based diet to get fully nutritionally crossbred queens (NQs). We also compared gene expression between NQs and control *A. cerana* queens (CQs), to identify the body color controlling pathways and related genes. Since the smallRNAs in *A. cerana* and *A. mellifera* royal jelly are different, and research has shown that the body color of fruit flies can be regulated by microRNA (Kennell et al., 2012), we hypothesized that small RNAs play a key role in *A. cerana-A. mellifera* nutritional crossbreeding related color changes. Here we compared both coding and non-coding RNA expression between NQ and CQ, to identify the body color controlling pathways, related key genes, and non-coding RNAs. This study allowed us to explore the epigenetic mechanism of nutritional crossbreeding in two honeybee species and serves as a model of epigenetic modification and phenotypic plasticity induced by nutritional diets.

## Materials and Methods

### Insects

Six healthy Asian honeybee colonies (*A. cerana*) were used as larvae suppliers. Each colony had a mated queen and 12,000 worker bees. Three strong European honeybee (*A. mellifera*) colonies, each with a mated queen and more than 30,000 worker bees, were used to produce fresh royal jelly. All colonies were kept in the Honeybee Research Institute, Jiangxi Agricultural University, China.

### Diet format

Fresh 2nd-day *A. mellifera* royal jelly (RJ) was produced according to (Wu et al., 2015). The diet formula was as follows: fresh *A. mellifera* RJ 53%, 6% D-glucose (purity: analytical reagent, Xilong Scientific, China), 6% D-fructose (purity: ≥ 99%, Solarbio life sciences, China), 1% yeast extract (Lot: 2194133, Oxoid Ltd, UK), and 34% distilled water.

***Queen rearing***

Six healthy *A. cerana* and three *A. mellifera* colonies were used for egg-laying, and the queens were caged in an empty comb for 6h according to (Zhu *et al.*, 2017). The queens were released and combs with eggs were placed into a queenless area of the hive. Similarly, half of the newly hatched *A. cerana* larvae (6h) were transplanted into 24-cell tissue culture plates and then incubated at 35°C and 95% ± 3% relative humidity (RH). Each cell with one *A. cerana* larva was primed with 200-400 μL food formula (increased daily according to the larval age). For pupation, 6-day-old larvae were transferred to 6-cell tissue culture plates lined with a piece of Kimwipe and kept in an incubator at 35°C and 80% RH. Fully developed queens with a body size exceeding 220 mg, notches in their mandibles and at least 16 days of development were sampled. The rest of the hatched *A. cerana* larvae were transplanted into wax queen cells to rear as natural queens in their naive colonies for the control group. The *A. cerana* queens artificially reared by feeding with an *A. cerana* royal jelly (AcRJ) diets was considered as the control group, however, the limitation of AcRJ production could not provide enough AcRJ for artificially rearing. Additionally, it is also extremely difficult to artificially rear *Apis cerana* queens based on the AcRJ diet in an incubator. Consequently, we used the natural *A.cerana* queens as the control group. After emergence, 12 fully developed NQ queens and 12 CQ queens were sampled for body color measuring and whole transcriptome sequencing. The body color was measured using Ruttner's color scale (Ruttner, 1988). This scale classifies the amount of light pigmentation present based on an empirical series of pigmentation patterns; It is comparable with tergites, and ranges from class 0 (completely dark) to class 9 (completely yellow). The data of color scales from NQs ($n=4$) and CQs ($n=4$) were compared by independent-sample *t*-test (2-tailed) using the SPSS package (v25), and $p < 0.05$ was considered as significant difference.

***RNA extraction, library preparation, and whole transcriptome sequencing***

All collected samples (NQ and CQ queens) were immediately stored in liquid nitrogen for future RNA extraction. Three queens from the same treatment were fixed together as one sample, and each treatment had three biological replicates. The total RNA of each sample was extracted using TRIzol reagent (Invitrogen, USA). The quality and quantity of the RNA were assessed using the RNA Nano 6000 Assay Kit of the Bioanalyser 2100 system (Agilent Technologies, CA, USA) (Liu *et al.*, 2019). Three μg of RNA per sample was used as input material for RNA sample preparations. After removing the ribosomal RNA and rRNA-free residue, we used rRNA-depleted RNA to construct sequencing libraries using the NEBNext® Ultra™ Directional RNA Library Prep Kit from Illumina® (NEB, USA). First-strand cDNA was synthesized using a random hexamer primer and Reverse Transcriptase. Second-strand cDNA synthesis was subsequently completed using DNA polymerase I and RNase H. Any remaining overhangs were converted into blunt ends via exonuclease/polymerase activity. After adenylation of the 3' ends of the DNA fragments, the NEBNext Adaptor with hairpin loop structure was ligated to prepare for hybridization.

The clustering of the index-coded samples was performed on the act Cluster Generation System using TruSeq PE Cluster Kitv3-cBot-HS (Illumia) according to the manufacturer's instructions. After cluster generation, the library preparations were sequenced on an Illumina Hiseq platform and paired-end reads were generated.

*Analysis of raw data*

Raw data (raw reads) in fastq format were firstly processed in-house per scripts. In this step, clean reads were obtained by removing reads containing adapters, or Ploy-N segments and reads of low quality from the raw data. At the same time, Q20, Q30, GC-content, and sequence duplication levels of the clean data were calculated. All the downstream analyses were based on clean data with high quality. The clean reads from each sample were sequence-aligned with the designated reference *A. cerana* genome (ASM1110058v1), where the efficiency of the alignment varied from 99.69% to 99.89%. The correlation values of the three biological replicates for each sample calculated for lncRNA, miRNA, sRNA, and circRNA are presented in Table S3.

Hereafter, sequences were aligned with the specified reference genome to obtain mapped data. Based on the mapped data, the quality of sequencing libraries was evaluated with the insert length test and the randomness test. The transcriptome was assembled using StringTie (Kovaka *et al.*, 2019), based on the reads mapped to the reference genome. RNA-Seq was used for quality control of the clean reads from lncRNA, miRNA, sRNA, and circRNA. The assembled transcripts were annotated using the Cuffcompare program (Pertea and Pertea, 2020). Unknown transcripts were used to screen for putative lncRNAs. The different types of lncRNAs including lncRNA, intronic lncRNA, anti-sense lncRNA, and sense lncRNA were selected using Cuffcompare (Trapnell *et al.*, 2010a). Bowtie (version v1.0.0) is a short sequence comparison software (Langmead *et al.*, 2009), especially suitable for high-throughput sequencing. This software uses the Silver database, GtRNAdb database, Rfam database, and Repbase database for sequence comparison and filtering of ribosomal RNA (rRNA), transport RNA (tRNA), Intranuclear small RNA (snRNA), nucleolar small RNA (snoRNA) and other ncRNAs and repeat sequences to get unannotated reads containing miRNAs. The percentage of Q30 base in each sample (Table II-9-1) is evidence that the high-quality RNA-Seq data acquired could be used for further analysis.

*Correlation analysis*

The correlation values of the three biological replicates for each sample calculated for lncRNA, mRNA, sRNA, and circRNA are presented in Table S1. We excluded the NQ1 and CQ3 samples from further analysis due to their low Pearson's correlation coefficient.

*Gene and non-coding RNA expression analysis*

StringTie (v1.3.1) was used to calculate FPKMs (fragments per kilo-base of exon per million fragments mapped calculated based on the length of the fragments and reads count mapped to this fragment) of both lncRNAs and coding genes in each sample. Gene FPKMs were computed by summing the FPKMs of transcripts in each gene group. Differential expression analysis of two treatments based on read counts was performed using the DESeq2 R package (v1.10.1) (Sun *et al.*, 2013). DESeq2 provides statistical routines for determining differential expression in digital gene expression data using a model based on the negative binomial distribution. The resulting *p*-values were adjusted using Benjamini and Hochberg's approach for controlling the false discovery rate (FDR). Genes with an adjusted *p*-value < 0.05 and the absolute value of log2 (fold change) >1 found by DESeq, were labelled as "differentially expressed". Before differential gene expression analysis, the read counts were adjusted through one scaling normalized factor for each sequenced library using the edgeR program package (Robinson *et al.*, 2010) through one scaling normalized factor. Differential expression analysis of two samples was performed using the EBseq (2010) R package. The resulting FDR (false discovery rate) was adjusted using the PPDE (posterior probability of being DE). The FDR < 0.05 and |log$_2$(Fold Change)| ≥1 were set as the threshold for significant differential expression. Similarly, the significantly differentially expressed miRNAs (DEmiRNAs) and circRNAs (DEcircRNAs) were identified using DESeq2 for differential expression analysis, and log$_2$(FC)| ≥ 1.00 was used as a threshold. The FDR values < 0.05 was used as the key indicator for differential expression miRNA and circRNA screening according to previous studies (Love *et al.*, 2014; Robinson *et al.*, 2010).

*GO and Kyoto Encyclopedia of genes and genomes (KEGG) enrichment analysis*

Genes were annotated to various protein and nucleotide sequence databases using BLASTX (version 2.2.28), including the Nr (NCBI non-redundant protein sequences), Pfam (Protein family), KOG/COG (Clusters of Orthologous Groups of proteins), Swiss-Prot (A manually annotated and reviewed protein sequence database) and non redundant nucleotide sequence (Nt) databases with a cutoff *E*-value of $10^{-5}$. GO enrichment analysis of DEGs, DElncRNAs, and DEmiRNAs was implemented using the topGO R packages ($p < 0.01$ indicates significance). The top 20 GO terms were selected. The KOBAS 2.0 (Xie *et al.*, 2011) software was used to test the statistical enrichment (*Q*-value < 0.05) of differentially expressed genes and non-coding RNAs in KEGG pathways.

*RNAi experiment*

We investigated the function of *TPH1* and *KMO* genes in cuticle pigmentation in *A. cerana* queens, according to the methodology of Mao *et al.*, (2015). One-day-old worker larvae were fed with a semi-artificial diet in a petri dish, and were incubated at 34℃ and 95%±3% humidity. Artificially manufactured siRNA for *TPH1* (F: GCGACAACUGGGCCAUUAATT; R: UUAAUGGCCCAGUUGUCGCTT) and *KMO* (F: GGUUGUGGUCGAUCACCAUTT; R: AUGGUGAUCGACCACAACCTT) were added to the semi-artificial diet, with a final concentration of 20 μg/ml. Similarly, the negative siRNA (F: UUCUUCGAACGUGUCACGUTT; R: ACGUGACACGUUCGGAGAATT) was added to a semi-artificial diet and fed to the control group larvae. Each group had 40 biological replicates. In total, 18 larvae (each treatment had 6 larvae) fed with the above *TPH1* (TPH1-RNAi treatment), *KMO* (KMO-RNAi treatment), and control siRNA (negative control) diets were collected on day 4 for qRT-PCR validation to verify the effect of RNAi on target gene expression. Each groups had three biological replicates and each biological replicate contained two mixed larvae. Each cDNA library had four technical replicates. The rest of the larvae were reared until they were fully-developed queens and their body color was measured according to the methodology of Ruttner *et al.* (2000) and Ruttner (1988). Data were analyzed using One Way ANOVA in SPSS package (v25), and $p < 0.05$ was considered as significant difference also.

Total RNA of each sample was extracted using TRIzol reagent (Tiangen, Beijing). The cDNA of each sample was synthesized from the total RNA using the Primer-Script RT reagent Kit (TaKaRa) according to the manufacturer's instructions. Each cDNA library had 4 technical replicates. The *β*-actin gene of *A. cerana* was selected as the reference gene. The primers were designed using Primerpremier 5 (version 5.0) and produced by Shanghai GenePharma Co., Ltd (Shanghai, China) (see Table S6). An ABI 7500 real-time PCR machine (Applied Biosystems, USA) was used for amplification. Each 10 mL reaction tube contained 5 μL TB Green Premix Ex Taq II+Probe Master Mix V1 μL cDNA, 0.8 μL Primer Set, 0.2 μL ROX Reference Dye II and 3 μL RNase Free ddH$_2$O. The qPCR experiment was performed with an initial denaturation step of 10 min at 95 °C, followed by 40 cycles of 94 °C for 15 s, 60 °C for 40 s, and 72 °C for 35 s. A cycle threshold (Ct) was calculated by determining the point at which the fluorescence exceeded a threshold limit. For the data analysis, the relative expression of these two genes (*TPH1* and *KMO*) was calculated using the $2^{-\Delta\Delta Ct}$ comparative Ct method and was transformed by taking their root to be normally distributed. Data were analyzed by Independent-sample *t*-test (2-tailed) using the SPSS package (v25), and $p < 0.05$ was considered as a significant difference.

## Results

### Body-color alteration in NQ queens

Fig. II-9-1(see page 529) clearly shows that *A. cerana* queens reared on an *A. mellifera* royal jelly based diet had a significantly lighter body color compared to their black mother queens and control queens ($p < 0.0001$, Fig. II-9-1B).

### Data quality of whole transcriptome sequencing

Six whole transcriptome sequencing libraries of NQ queens and CQ queens were established. In total, 108.0 2 GB of clean reads from mRNAs and lncRNAs were obtained, resulting in 16.09 GB of clear data per sample after quality control. The percentage of Q30 base of each sample was more than 93.27% (Table II-9-1). The same six RNA-seq library samples were used to construct non-coding RNA libraries for miRNAs and circRNAs. A total of 162.29 M and 366.01 M of clean reads were obtained in miRNAs and circRNAs, respectively, and the percentage of Q30 base for each sample was over 95.90% (Table II-9-1). All results indicate high-quality RNA-Seq data. Pearson's correlation coefficient of mRNA in all biological replicates of each group was above 0.8, except for NQ1 and CQ3 (Table S1). Pearson's correlation coefficient of miRNA in all biological replicates was above 0.8 and the correlation of lncRNA and circRNA in all replicates (Table S1).

Table II-9-1  The statistics of coding and non-coding RNAs

| Samples | LncRNA and mRNA | | | circRNA | | | miRNA | | |
|---|---|---|---|---|---|---|---|---|---|
| | Total Reads | Mapped Reads | Q30(%) | Total Reads | Mapped Reads | Q30(%) | Total Reads | Mapped Reads | Q30(%) |
| NQ-1 | 118,764,536 | 90,975,067 (76.60%) | 95.09 | 118,764,536 | 118,715,250 (99.96%) | 98.47% | 30,413,276 | 22,484,016 (73.93%) | 96.93 |
| NQ-2 | 109,272,510 | 69,090,736 (63.23%) | 95.13 | 109,272,510 | 109,145,690 (99.88%) | 98.46% | 23,026,242 | 15,309,482 (66.49%) | 96.96 |
| NQ-3 | 113,155,188 | 73,727,115 (65.16%) | 95.33 | 113,155,188 | 112,900,024 (99.77%) | 98.49% | 18,722,227 | 12,006,471 (64.13%) | 96.62 |
| CQ-1 | 113,048,850 | 86,497,271 (76.51%) | 93.27 | 113,048,850 | 112,997,938 (99.95%) | 97.64% | 26,516,925 | 18,886,608 (71.22%) | 95.9 |

(continued)

| Samples | LncRNA and mRNA | | | circRNA | | | miRNA | | |
|---|---|---|---|---|---|---|---|---|---|
| | Total Reads | Mapped Reads | Q30(%) | Total Reads | Mapped Reads | Q30(%) | Total Reads | Mapped Reads | Q30(%) |
| CQ-2 | 131,028,148 | 94,865,818 (72.40%) | 94.11 | 131,028,148 | 130,761,452 (99.80%) | 97.99% | 16,800,308 | 10,104,645 (60.15%) | 96.16 |
| CQ-3 | 146,759,188 | 131,831,300 (89.83%) | 93.55 | 146,759,188 | 146,566,742 (99.87%) | 97.70% | 17,412,842 | 12,242,110 (70.31%) | 97.59 |

### *DEGs and differentially expressed non-coding RNAs*

In total, 1484, 311, 92, and 169 DEGs, DElncRNAs, DEmicroRNAs, and DEcircRNAs were identified between NQ and CQ respectively (Table II-9-2, Table S2), with 782, 209, 45, and 99 of these genes up-regulated in NQ and 702, 102, 47, and 70 up-regulated in CQ (Fig. II-9-2A-D, see page 529). These results were constant in the sample clustering (heat map) using the full gene set (Fig. S1).

**Table II-9-2** Number of differentially expressed coding and non-coding RNAs identified from *Apis cerana* artificially nutritional cross queens vs. *Apis cerana* natural queens

| NQ vs. CQ | DELncRNA | | DEGs | | DEmiRNA | | DEcricRNA | |
|---|---|---|---|---|---|---|---|---|
| Regulation | Up | Down | Up | Down | Up | Down | Up | Down |
| Number of DEGs or non-coding RNAs | 209 | 102 | 782 | 702 | 45 | 47 | 99 | 70 |

### *GO and KEGG enrichment*

The DEGs, DElncRNAs, DEmicroRNAs, and DEcircRNAs were enriched into 53, 53, 749, and 47 GO terms respectively (Table S3). The enriched and classification of GO terms are summarized in Fig. S2. Here the top 10 GO terms of DEGs and differentially expressed non-coding RNAs were transmembrane transport, an integral component of membrane, and NADH dehydrogenase (ubiquinone activity). KEGG enrichment analysis showed a total of DEGs, DElncRNAs, DEmiRNAs, and DEcircRNAs enriched into 123, 151, 38, and 8 KEGG pathways, respectively (Fig. II-9-2E-H, Table S4), and one key pathway, (tryptophan metabolism), formed part of the top ten KEGG pathways in DEGs (Fig. II-9-2E) and DEmiRNA (Fig. II-9-2G).

### *Key KEGG pathways for body color regulation and related DEGs, DElncRNAs and DEmiRNAs*

The phenylalanine, dopamine, tryptophan, and tyrosine pathways are the most important KEGG pathways involved in insect pigmentation (Lambrus *et al.*, 2015; Ahmad *et al.*, 2020; Sasaki *et al.*, 2021; Rabatel *et al.*, 2013; Zhang *et al.*, 2020). More interestingly, our results showed that seven DEGs were enriched in the above three key pathways, including *ALDH, ALDH7, KMO, GCDH, HADHA, FAH* and *TDC* (Fig. II-9-3, see page 531, details see Table S5). Three key genes (*DDC, TH* and *TPH1*) were also selected and presented in Fig. II-9-3 due to their important functions in insect body color regulation (Tang, 2009; Lambrus *et al.*, 2015; Futahashi and Osanai-Futahashi, 2021), even though their FDR or log2(FC) values did not reach the threshold of significant difference. Moreover, a total of eight DElncRNAs and three DEmiRNAs were also enriched into these key pathways (Fig. II-9-3, Table S5), including MSTRG.24103.16, MSTRG.37819.1, MSTRG.40443.5, MSTRG.23925.4, MSTRG.23925.2, MSTRG.11791.1, MSTRG.18658.4, MSTRG.19099.2 as well as three DEmiRNAs (novel miR195, novel miR11 and novel miR123). More interestingly, the key DEGs, DElncRNAs, and DEmiRNAs showed a regulating network (Fig. II-9-3B). One of the DEGs was regulated by multiple DElncRNAs or by both DElncRNAs and DEmiRNAs together (Fig. II-9-3B).

*RNAi effect on body color key genes*

The RNAi results revealed that feeding RNAi-based food significantly down-regulated the expression of *TPH1-2* and *KMO* in 4-day-old larvae ($p < 0.01$, Fig. II-9-4A, see page 532), resulting in a clear and significant color change in newly emerged queens (Fig. II-9-4B and II-9-4C). The RNAi-treated queens exhibited a notable increase in yellow pigment and lacked black color, whereas the control queens developed normally and exhibited the normal black body pigmentation similar to their mother queens (Fig. II-9-4C; df for all traits =2, while $P$ value <0.0001 for Thorax, Scutum, Scutellum, and Sternum. For Thorax $F$ value =109, Scutum $F$ value =112, Scutellum $F$ value =289, Tergum: $F$ value =35.45, $p$ value = 0.0005, and Sternum $F$ value =103.20, Fig. II-9-4 and Fig. S3). This confirms that the *TPH1* and *KMO* genes play an important role in the formation of honeybee body color and pigmentation.

## Discussion

Nutritional crossbreeding between Asian and European honey bees induces morphological, physiological, and behavioral changes in the adults (Shi *et al.*, 2014; Smaragdova, 1963; Ruttner, 1988; Rinderer *et al.*, 1985), which renders it an optimal model for studies on the epigenetic mechanisms of animal phenotypic plasticity. The change in body color is one of the most observed changes in honeybee nutritional crossbreeding. In this study, we artificially reared fully nutritionally crossbred *A. cerana* queens on *A. mellifera* RJ diets. Our results showed a clear change in body color from black to yellow (Fig. II-9-1). We also identified a regulation network of eight DElncRNAs and three DEmiRNAs that regulate key DEGs involved in the melanin synthesis pathways. These results revealed that non-coding RNAs presumably participate in honeybee body color alteration induced by *A. cerana-A. mellifera* nutritional crossbreeding by regulating the expression of related key genes. We note that the CQ is not an optimal control, since different rearing methods for producing NQs and CQs might, to same extent, influence the results of whole transcriptome sequencing. However, recently it is too difficult to artificially rear *A. cerana* queens using a diet format based on the *A. cerana* royal jelly. Therefore, the CQs produced by the same mother queen as NQs were selected as the control group.

*The effect of nutritional crossbreeding on NQ body color alteration*

Many previous studies have reported a body color alteration in Asian and European honeybee nutritional crossbreeds by using partial nutritional crossbreeding methods. This involves rearing nutritionally crossbred queens in their native colonies by adding other species' royal jelly in queen cells or rearing queens in a different honeybee species colony for 1-2 days and then returning them to their native colonies until emergence (Li, 1982; Zhu, 1985; Xie *et al.*, 2008). The body color of the queen or worker bees in these studies was only partly altered (He *et al.*, 2010 and 2011; Zeng *et al.*, 2005; Xie *et al.*, 2005a and 2005b). By using a complete nutritional crossbreeding method that rears the *A. cerana* queens on an *A. mellifera* royal jelly-based diet, a clear color change of the whole body was observed. Fig. II-9-1 illustrates how the head, thorax, abdomen, and all six legs of the NQ queens were yellow, whereas the mother and control queens had black bodies and legs. This reflects the large-scale body color changes induced by nutritional crossbreeding. Consequently, the complete nutritional crossbreeding method firstly developed in this study demonstrates the powerful effects of nutritional crossbreeding and allows us to explore its underlying epigenetic mechanisms.

*Key pathways and genes involved in body color alteration*

The formation of insect body color is linked to the synthesis of melanin, which is mainly regulated by a few key KEGG pathways and related key genes. The most important KEGG pathways involved in insect pigmentation are phenylalanine, tryptophan, dopamine, tyrosine, and tryptophan (Ahmad *et al.*, 2020; Rabatel *et al.*, 2013). We subsequently compared gene expression in NQs compared to CQs through whole transcriptome sequencing. Interestingly, seven DEGs were enriched into the above four key pathways (Fig. II-9-3), which have been shown to participate in the regulation of insect body colors (Liu *et al.*, 2020; Cole *et al.*, 2005; Eisner *et al.*, 1997; Moraes *et al.*, 2022). The *KMO* gene in silkworms acts as a transgenic marker, which turns the integument of the first instar larvae brown (Kobayashi *et al.*, 2007). Therefore, these genes possibly play a key role in color alteration in NQs. Moreover, three genes (*TPH1*, *DDC*, and *TH*) are also key genes enriched into the tryptophan and tyrosine pathways and have been previously reported as key genes for the determination of insect body color, even though their FDR and log2 FC values did not reach the threshold of significant difference (Fig. II-9-3). Previous studies indicate that these three genes are essential for the formation of insect body color: The *TPH1* gene determines the eye pigmentation in the planarian *Schmidtea mediterranea* (Lambrus *et al.*, 2015), while the *TH* and *DDC* genes participate in dopamine biosynthesis of bees and fruit flies such as *D. melanogaster* (Futahashi and Osanai-Futahashi, 2021; Tang, 2009). Moreover, two DElncRNAs (MSTRG.24103.16 and MSTRG.19099.2) and two DEmiRNAs (miR195 and miR123) were related to these three genes (Fig. II-9-3). Eventually, our RNAi results confirmed that knocking down two key genes (*KMO* and *TPH1*) resulted in the clear alteration of honeybee body color (Fig. II-9-4). Consequently, the effects of honeybee nutritional crossbreeding possibly influence bee body color by altering the expression of the above key genes. Here we note that the whole body of queens rather special tissues was used for whole transcriptome sequencing, which might conceal some other key genes. Further studies could verify our results using special tissues (for example: color-changed cuticle).

*Epigenetic modification in body color alteration of NQ*

Body color is the most distinguishing and conservative morphological trait of honeybees, and have been used as a selection parameter and diagnostic character in breeding and taxonomy. The pattern of light (yellow, orange) and dark (black, brown) color differs between species as well as between the casts of the same colony (Woyke, 1997; Tilahun *et al.*, 2016). However, in the present study, the body-color alteration in honeybees induced by nutritional crossbreeding is an epigenetic phenomenon, since this alteration was based on exchanged larval diets. Nutrients can reverse or change epigenetic phenomena such as DNA methylation and histone modification in insects, altering the expression of critical genes related to development (Choi and Friso, 2010), phenotype, and body color regulation (Golic *et al.*, 1998; Saranchina *et al.*, 2021). The royal jelly of honeybees is a powerful food source that determines queen-worker caste differentiation (Winston, 1991; Wojciechowski *et al.*, 2018; Slater *et al.*, 2020) through epigenetic modifications (He *et al.*, 2017; Maleszka, 2008). Indeed, our results showed that thousands of non-coding RNAs were significantly differentially expressed between NQ and CQ queens (Fig. II-9-2 and Table II-9-2). Here, eight DElncRNAs and three DEmiRNAs were highly related to the gene expression regulation of key genes that are vital to body color determination (Fig. II-9-3), although no circRNAs were involved. Our results are supported by many fruit fly studies showing that miRNAs regulate insect body color (Kennell *et al.*, 2012; Bejarano and Lai, 2021). As one of the most powerful epigenetic modifications, non-coding RNAs such as lncRNAs, circRNAs, and miRNAs play a vital role in animal phenotypic plasticity by regulating gene expression (Legeai and Derrien, 2015; Wen *et al.*, 2020; Choudhary *et al.*, 2021; Zhu *et al.*, 2017; Shi *et al.*, 2014). Such epigenetic mechanism possibly also applies in *A. cerana-A. mellifera* nutritional crossbreeding. Therefore, it is believed that non-coding RNAs act as a vital epigenetic modification underlying *A. cerana-A. mellifera* nutritional crossbreeding, resulting in whole-body color alteration by regulating key pathways and related key genes. Shi *et al.* (2012) showed that dozens of miRNAs differ between *A. cerana* and *A. mellifera* royal jellies but differences were not detected in this study, perhaps due to the short actuation duration of miRNAs that do not persist in adult queens. Exactly which key components of honeybee royal jelly affect the expression of non-coding RNAs still requires further investigation.

In summary, the study explored the epigenetic mechanisms underlying nutritional crossbreeding of two honeybee species and revealed that lncRNAs, and, miRNAs contribute to body color alteration in NQs by regulating the expression of key genes and pathways that are related to melanin synthesis. This study not only demonstrated an epigenetic mechanism underlying honeybee nutritional crossbreeding but also serves as a model for studies on the epigenetic mechanisms of animal phenotypic plasticity induced by environmental factors.

## Data Availability Statement

The datasets presented in this study can be found in online repositories. The names of the repository/repositories and accession number(s) can be found in the article/Supplementary Material.

## Authors' Contributions

ZJZ and XJH conceived and designed the experiments. AM and CY performed the experiments. ML, XL, YBL, and CLL helped experience. AM and XJH wrote the paper. All authors read and approved the final manuscript. We would like to thank MogoEdit (https://www.mogoedit.com) for its English editing during the preparation of this manuscript.

## Funding

This work was supported by the National Natural Science Foundation of China (32172790 and 32160815), the Major Discipline Academic and Technical Leaders Training Program of Jiangxi Province (20204BCJL23041), and the Earmarked Fund for China Agriculture Research System (CARS-44-KXJ15).

## Conflict of Interest

The authors declare that the research was conducted in the absence of any commercial or financial relationships that could be construed as a potential conflict of interest.

## Supplementary Material

The Supplementary Material for this article can be found online at: https://www.frontiersin.org/articles/10.3389/fphys.2023.1073625/full#supplementary-material.

## References

AHMAD, S., MOHAMMED, M., PRASUNA, M., CH, S., and CH, R. (2020). Tryptophan, a non-canonical melanin precursor: New L-tryptophan based melanin production by Rubrivivax benzoatilyticus JA2. *Sci. Rep.* 10(1):8925. doi: 10.1038/s41598-020-65803-6.

ARAKANE, Y., NOH, M. Y., ASANO, T., and KRAMER, J., K. (2016). "Tyrosine metabolism for insect cuticle pigmentation and sclerotization," in Extracellular composite matrices in arthropods. Editors E. Cohen and B. Moussian (Switzerland, 156-220. doi:10.1007/978-3- 319-40740-1_6.

BEJARANO, F., and LAI, E. C. (2021). A comprehensive dataset of microRNA misexpression phenotypes in the Drosophila eye. *Data Br.* 36, 107037.

BORYCZ, J., BORYCZ, J. A., LOUBANI, M., and MEINERTZHAGEN, I. A. (2002). Tan and Ebony Genes Regulate a Novel Pathway for Transmitter Metabolism At Fly Photoreceptor Terminals. *J. Neurosci.* 22, 10549-10557. doi: 10.1523/jneurosci.22-24-10549.2002.

CHEN. H., LI, F. M., ZHENG M. J., CAO, D. D., HE, Z. M., and YANG, M. X. (2015). The effect of nutritional hybridization on the reproductive potential of the queen bee in China. *Sichuan Anim. Husb. Vet. Med.* 42(09), 24-26. (in Chinese).

CHEN, S. L. (2001). The apicultural science in China. *China Agric. Beijing*, 1-16.

CHEN, X., XIAO, D., DU, X., GUO, X., ZHANG, F., DESNEUX, N., et al. (2019). The role of the dopamine melanin pathway in the ontogeny of elytral melanization in harmonia axyridis. *Front. Physiol.* 10, 1-8. doi: 10.3389/fphys.2019.01066.

CHOI, S.-W., and FRISO, S. (2010). Epigenetics: a new bridge between nutrition and health. *Adv. Nutr.* 1, 8-16. doi:10.3945/an.110.1004.

CHOUDHARY, C., SHARMA, S., MEGHWANSHI, K. K., PATEL, S., MEHTA, P., SHUKLA, N., et al. (2021). Long non-coding rnas in insects. *Animals* 11, 1-20. doi: 10.3390/ani11041118.

COLE, S. H., CARNEY, G. E., MCCLUNG, C. A., WILLARD, S. S., TAYLOR, B. J., and HIRSH, J. (2005). Two Functional but Noncomplementing *Drosophila Tyrosine Decarboxylase* Genes: distinct roles for neural tyramine and octopamine in female fertility*. *J. Biol. Chem.* 280, 14948-14955. doi: https://doi.org/10.1074/jbc.M414197200.

DIAO, Q., SUN, L., ZHENG, H., ZENG, Z., WANG, S., XU, S., *et al.* (2018). Genomic and transcriptomic analysis of the Asian honeybee *Apis cerana* provides novel insights into honeybee biology. *Sci. Rep.* 8, 1-14.

EISNER, T., MEINWALD, MORGAN, R. C., ATTYGALLE, A. B., SMEDLEY, S. R., HERATH, K. B. (1997). Defensive production of quinoline by a phasmid insect (*Oreophoetes peruana*). *J. Exp. Biol. 200,* 200, 2493-2500. doi:10.1242/jeb.200.19.2493.

ELIAS-NETO, M., SOARES, M. P. M., SIMÕES, Z. L. P., HARTFELDER, K., and BITONDI, M. M. G. (2010). Developmental characterization, function and regulation of a Laccase2 encoding gene in the honey bee, *Apis mellifera* (Hymenoptera, Apinae). *Insect Biochem. Mol. Biol.* 40, 241-251. doi: 10.1016/j.ibmb.2010.02.004.

FALCON, T., PINHEIRO, D. G., FERREIRA-CALIMAN, M. J., TURATTI, I. C. C., PINTO DE ABREU, F. C., GALASCHI-TEIXEIRA, J. S., *et al.* (2019). Exploring integument transcriptomes, cuticle ultrastructure, and cuticular hydrocarbons profiles in eusocial and solitary bee species displaying heterochronic adult cuticle maturation. *PLoS ONE* 14(3): e0213796. doi: 10.1371/journal.pone.0213796.

FUTAHASHI, R., and OSANAI-FUTAHASHI, M. (2021). Pigments in insects. *Pigment. Pigment Cells Pigment Patterns*, 3-43.

GOLIC, K. G., GOLIC, M. M., and PIMPINELLI, S. (1998). Imprinted control of gene activity in Drosophila. *Curr. Biol.* 8, 1273-1276. doi:10.1016/s0960-9822(07)00537-4.

GRAHAM, J. M. (1993). *The Hive and the Honey Bee.* Dadant & Sons press. Chelsea, USA.

HE, X. J., WANG, Z. P., CHEN, L. H., DAI, S. Z., and YAN, W. Y. (2010). Effects of nutritional hybridization of Chinese bee and Italian bee on the mite resistance and hygienic behavior of Italian bee. *J. Jiangxi Agric. Univ.* 32, 1245-1247. (in Chinese).

HE, X. J., WANG, Z. P., QIN, Q. H., WU, X. B., and CHEN, L.H. (2011). Effects of microsatellite genetic polymorphisms of *Apis mellifera ligustica* on nutritional crossbreed between *Apis cerana cerana* and *Apis mellifera ligustica*. *China Anim. Husb. Vet. Med.* 38, 107-110. (in Chinese).

HE, X. J., ZHOU, L. BIN, PAN, Q. Z., BARRON, A. B., YAN, W. Y., and ZENG, Z. J. (2017). Making a queen: an epigenetic analysis of the robustness of the honeybee (*Apis mellifera*) queen developmental pathway. *Mol. Ecol.* 26, 1598-1607. doi:10.1111/mec.13990.

KENNELL, J. A., CADIGAN, K. M., SHAKHMANTSIR, I., and WALDRON, E. J. (2012). The microRNA miR-8 is a positive regulator of pigmentation and eclosion in Drosophila. *Dev. Dyn.* 241, 161-168. doi: 10.1002/dvdy.23705.

KLEIN, A.-M., VAISSIERE, B. E., CANE, J. H., STEFFAN-DEWENTER, I., CUNNINGHAM, S. A., KREMEN, C., *et al.* (2007). Importance of pollinators in changing landscapes for world crops. *Proc. R. Soc. B Biol. Sci.* 274, 303-313. doi:10.1098/rspb.2006.3721.

KOBAYASHI, I., UCHINO, K., SEZUTSU, H., IIZUKA, T., and TAMURA, T. (2007). Development of a new piggyBac vector for generating transgenic silkworms using the kynurenine 3-mono oxygenase gene. *J. Insect Biotechnol. Sericology* 76, 145-148. doi:10.11416/jibs.76. 3_145.

KOVAKA, S., ZIMIN, A. V, PERTEA, G. M., RAZAGHI, R., SALZBERG, S. L., and PERTEA, M. (2019). Transcriptome assembly from long-read RNA-seq alignments with StringTie2. *Genome Biol.* 20(1),1-13. doi:10.1186/s13059-019-1910-1.

LAMBRUS, B. G., COCHET-ESCARTIN, O., GAO, J., NEWMARK, P. A., COLLINS, E.-M. S., and COLLINS, J. J. 3rd (2015). Tryptophan hydroxylase Is Required for Eye Melanogenesis in the Planarian Schmidtea mediterranea. *PLoS One* 10, e0127074. doi: 10.1371/journal.pone.0127074.

LANGMEAD, B., TRAPNELL, C., POP, M., and SALZBERG, S. L. (2009). Ultrafast and memory-efficient alignment of short DNA sequences to the human genome. *Genome Biol. 2009;* 10, R25. doi:10.1186/gb-2009-10-3-r25.

LEGEAI, F., and DERRIEN, T. (2015). Identification of long non-coding RNAs in insects genomes. *Curr. Opin. Insect Sci.* 7, 37-44. doi: 10.1016/j.cois.2015.01.003.

LI, M. S. (1982). Discuss on breeding of cross-feeding between *Apis cerana* cerana and *Apis mellifera* ligustica. *Apic. Sci. Technol.* 1, 42-44.

LIU, H., WANG, Z.L., TIAN, L.Q., QIN, Q.H., WU, X.B., YAN, W.Y., et al. (2014). Transcriptome differences in the hypopharyngeal gland between Western Honeybees (*Apis mellifera*) and Eastern Honeybees (*Apis cerana*). *BMC Genomics* 15, 744. doi:10.1186/1471-2164-15-744.

LIU, F., SHI, T., QI, L., SU, X., WANG, D., DONG, J., et al. (2019). LncRNA profile of *Apis mellifera* and its possible role in behavioural transition from nurses to foragers. BMC Genomics 20(1),1-13. doi:10.1186/s12864-019-5664-7.

LIU, X. L., HAN, W. K., ZE, L. J., PENG, Y. C., YANG, Y. L., ZHANG, J., et al. (2020). Clustered Regularly Interspaced Short Palindromic Repeats/CRISPR-Associated Protein 9 Mediated Knockout Reveals Functions of the yellow-y Gene in Spodoptera litura. *Front. Physiol.* 11. doi: 10.3389/fphys.2020.615391.

LOVE, M. I., HUBER, W., and ANDERS, S. (2014). Moderated estimation of fold change and dispersion for RNA-seq data with DESeq2. *Genome Biol.* 15, 550. doi: 10.1186/s13059-014-0550-8.

MALESZKA, R. (2008). Epigenetic integration of environmental and genomic signals in honey bees: the critical interplay of nutritional, brain and reproductive networks. *Epigenetics* 3, 188-192. doi:10.4161/epi.3.4.6697.

MAO, W., SCHULER, M. A., and BERENBAUM, M. R. (2015). A dietary phytochemical alters caste-associated gene expression in honey bees. *Sci. Adv.* 1, e1500795. doi:10.1126/sciadv.1500795.

MATSUOKA, Y., and MONTEIRO, A. (2018). Melanin Pathway Genes Regulate Color and Morphology of Butterfly Wing Scales. *Cell Rep.* 24, 56-65. doi: 10.1016/j.celrep.2018.05.092.

MORAES, B., BRAZ, V., SANTOS-ARAUJO, S., OLIVEIRA, I. A., BOM, L., RAMOS, I., et al. (2022). Deficiency of Acetyl-CoA Carboxylase Impairs Digestion, Lipid Synthesis, and Reproduction in the Kissing Bug Rhodnius prolixus. 13(4), 1-13. doi: 10.3389/fphys.2022.934667.

NIE, H.Y., LIANG, L.Q., LI, Q.F., LI, Z.H.Q., ZHU, Y.N., GUO, Y.K., et al. (2021). CRISPR/Cas9 mediated knockout of Amyellow-y gene results in melanization defect of the cuticle in adult *Apis mellifera*. *J. Insect Physiol.* 132, 104264. doi: https://doi.org/10.1016/j.jinsphys.2021.104264.

NOH, M. Y., KOO, B., KRAMER, K. J., MUTHUKRISHNAN, S., and ARAKANE, Y. (2016). Arylalkylamine N-acetyltransferase 1 gene (TcAANAT1) is required for cuticle morphology and pigmentation of the adult red flour beetle, Tribolium castaneum. *Insect Biochem. Mol. Biol.* 79, 119-129. doi: 10.1016/j.ibmb.2016.10.013.

OLDROYD, B. P., and WONGSIRI, S. (2009). *Asian honey bees: biology, conservation, and human interactions*. Harvard University Press.

PADOWICZ, D., and HARPAZ, S. (2007). Color enhancement in the ornamental dwarf cichlid Microgeophagus ramirezi by addition of plant carotenoids to the fish diet. *Isr. J. Aquac.* 59, 20536. doi:10.46989/001c.20536.

PERTEA, G., and PERTEA, M. (2020). GFF Utilities: GffRead and GffCompare [version 2; peer review: 3 approved]. *F1000Research* 9, 1-20. Available at: https://f1000research.com/articles/9-304/v2. doi:10.12688/f1000research.23297.1.

RABATEL, A., FEBVAY, G., GAGET, K., DUPORT, G., BAA-PUYOULET, P., SAPOUNTZIS, P., et al. (2013). Tyrosine pathway regulation is host-mediated in the pea aphid symbiosis during late embryonic and early larval development. *BMC Genomics* 14, 235. doi: 10.1186/1471-2164-14-235.

RINDERER, T. E., HELLMICH, R. L., DANKA, R. G., and COLLINS, A. M. (1985). Male reproductive parasitism: a factor in the Africanization of European honey-bee populations. *Science (80-. ).* 228, 1119-1121. doi:10.1126/science.228.4703.1119.

ROBINSON, M. D., MCCARTHY, D. J., and SMYTH, G. K. (2010). edgeR: a Bioconductor package for differential expression analysis of digital gene expression data. *Bioinformatics* 26, 139-140. doi: 10.1093/bioinformatics/btp616.

RUTTNER, F. (1988). "Morphometric Analysis and classification," in *Biogeography and Taxonomy of Honeybees* (Springer), 66-78. doi: DOI 10.1007/978-3-642-72649-1.

RUTTNER, F., ELMI, M. P., and FUCHS, S. (2000). Ecoclines in the Near East along 36°N latitude in *Apis mellifera* L. *Apidologie* 31, 157-165. doi: 10.1051/apido:2000113.

SARANCHINA, A., DROZDOVA, P., MUTIN, A., and TIMOFEYEV, M. (2021). Diet affects body color and energy metabolism in the Baikal endemic amphipod Eulimnogammarus cyaneus maintained in laboratory conditions. *Bio. Comm.* 66 (3), 245-255. doi:10.21638/spub03.2021.306.

SASAKI, K., YOKOI, K., and TOGA, K. (2021). Bumble bee queens activate dopamine production and gene expression in nutritional signaling pathways in the brain. *Sci. Rep.* 11(2), 1-14. doi: 10.1038/s41598-021-84992-2.

SHI, Y. Y., WU, X. B., HUANG, Z. Y., WANG, Z. L., YAN, W. Y., and ZENG, Z. J. (2012). Epigenetic modification of gene expression in honey bees by heterospecific gland secretions. *PLoS One* 7, e43727. doi:10.1371/journal.pone.0043727.

SHI, Y. Y., HUANG, Z. Y., WU, X. B., WANG, Z. L., YAN, W. Y., and ZENG, Z. J. (2014). Changes in alternative splicing in *Apis mellifera* bees fed *Apis cerana* royal jelly. *J. Apic. Sci.* 58, 25-31. doi: 10.2478/jas-2014-0019.

SLATER, G. P., YOCUM, G. D., and BOWSHER, J. H. (2020). Diet quantity influences caste determination in honeybees (*Apis mellifera*). *Proceedings. Biol. Sci.* 287, 20200614. doi: 10.1098/rspb.2020.0614.

SMARAGDOVA, N. (1963). Study on the brood food of worker of the bees *Apis mellifera* L., *Apis mellifera* caucasica Gorb., and of their crossbreds. in *Proceeding of International Beekeeping Congress*, 109-110.

SUGUMARAN, M., and BAREK, H. (2016). Critical analysis of the melanogenic pathway in insects and higher animals. *Int. J. Mol. Sci.* 17, 1-24. doi: 10.3390/ijms17101753.

SUN, J., NISHIYAMA, T., SHIMIZU, K., and KADOTA, K. (2013). TCC: an R package for comparing tag count data with robust normalization strategies. *BMC Bioinformatics* 14, 219. doi: 10.1186/1471-2105-14-219.

TANG, H. (2009). Regulation and function of the melanization reaction in Drosophila. *Fly (Austin).* 3, 105-111. doi: 10.4161/fly.3.1.7747.

TILAHUN, M., ABRAHA, Z., GEBRE, A., and DRUMOND, P. (2016). Beekeepers' honeybee colony selection practice in Tigray, Northern Ethiopia. *Livest. Res. Rural Dev. 28 2016* 28. Available at: http://www.lrrd.org/lrrd28/5/cont2805.htm.

TRAPNELL, C., WILLIAMS, B. A., PERTEA, G., MORTAZAVI, A., KWAN, G., SALZBERG, M. J. VAN B. S. L., *et al.* (2010). Transcript assembly and abundance estimation from RNA-Seq reveals thousands of new transcripts and switching among isoforms. *Nat Biotechnol.* 28, 511-515. doi:10.1038/nbt.1621.

WANG, P., LI, X., WANG, X., PENG, N., and LUO, Z. (2020a). Effects of Dietary Xanthophyll Supplementation on Growth Performance, Body Color, Carotenoids, and Blood Chemistry Indices of Chinese Soft-Shelled Turtle Pelodiscus sinensis. *N. Am. J. Aquac.* 82, 394-404. doi:10.1002/naaq.10161.

WANG, Z.L., ZHU, Y.Q., YAN, Q., YAN, W.Y., ZHENG, H.J., and ZENG, Z.J. (2020b). A Chromosome-Scale Assembly of the Asian Honeybee *Apis cerana* Genome. *Front. Genet.* 11, 279. doi:10.3389/fgene.2020.00279.

WANG, Q., ZHONG, L., WANG, Y., ZHENG, S., BIAN, Y., DU, J., *et al.* (2022). Tyrosine Hydroxylase and DOPA Decarboxylase Are Associated With Pupal Melanization During Larval-Pupal Transformation in Antheraea pernyi. *Front. Physiol.* 13, 832730-832813. doi:10.3389/fphys.2022.832730.

WEN, J., LUO, Q., WU, Y., ZHU, S., and MIAO, Z. (2020). Integrated analysis of long non-coding RNA and mRNA expression profile in myelodysplastic syndromes. *Clin. Lab.* 66, 825-834. doi: 10.7754/CLIN.LAB.2019.190939.

WINSTON, M. L. (1991). *The biology of the honey bee.* harvard university press.

WITTKOPP, P. J., TRUE, J. R., and CARROLL, S. B. (2002). Reciprocal functions of the Drosophila Yellow and Ebony proteins in the development and evolution of pigment patterns. *Development* 129, 1849-1858. doi: 10.1242/dev.129.8.1849.

WOJCIECHOWSKI, M., LOWE, R., MALESZKA, J., CONN, D., MALESZKA, R., and HURD, P. J. (2018). Phenotypically distinct female castes in honey bees are defined by alternative chromatin states during larval development. *Genome Res.* 28, 1532-1542. doi:10.1101/gr.236497.118.

WOYKE, J. (1997). Expression of body colour patterns In three castes of four asian honeybees. *Proc. Int. Conf. Trop. Bees Eviron.*, 85-95.

WU, X.B., ZHANG, F., GUAN, C., PAN, Q.Z., ZHOU, L.B., YAN, W.Y., *et al.* (2015). A new method of royal jelly harvesting without grafting larvae. *Entomol. News* 124, 277-281. doi:10.3157/021.124.0405.

XIE, X.B., SU, S. K., YAN, W. Y., GUO, D. S., and ZENG, Z. J. (2005a). Study on Random Amplified Polymorphic DNA of Nutrient Hybrid Progeny of Chinese Bee and Italian Bee. *J. Zhejiang Univ. (Agriculture Life Sci. 2005* 31, 741-744. Available at: https://www.zjujournals.com/agr/CN/abstract/abstract23453.shtml.

XIE, X.B., SU, S. K., YAN, W. Y., GUO, D. S., and ZENG, Z. J. (2005b). Random amplified polymorphic DNA study on the vegetative hybrid offspring of Chinese bee and Italian bee. *J. Zhejiang Univ. Agric. Life Sci.* 31, 741-741. (in Chinese).

XIE, X.B., PENG, W. J., and ZENG, Z.J. (2008). Breeding the mite-resistant honeybee by nutritional crossbreed technology. *Sci. Agric. Sin.* 7, 762-767. (in Chinese) doi:10.1016/ s1671-2927(08)60112-1.

XIE, C., MAO, X., HUANG, J., DING, Y., WU, J., DONG, S., *et al.* (2011). KOBAS 2.0: a web server for annotation and identification of enriched pathways and diseases. *Nucleic Acids Res.* 39, W316-W322. doi:10.1093/nar/gkr483.

YEUNG, Y. T., and ARGÜELLES, S. (2019). *Nonvitamin and nonmineral nutritional supplements. Bee products: royal jelly and propolis.* 475-484. doi: 10.1016/B978-0-12-812491-8.00063-1.

YU, F., MAO, F., and LI, J. K. (2010). Royal jelly proteome comparison between *A. mellifera ligustica* and *A. cerana cerana*. *J. Proteome Res.* 9, 2207-2215. doi:10.1021/ pr900979h.

ZENG, Z.J., XIE, X.B., and YAN, W.Y. (2005). Effects of nutritional hybridization on the birth weight of worker bees. *J. Econ. Zool.* 3, 149-151. (in Chinese).

ZENG, Z.J., ZOU, Y., GUO, D.S., and YAN, W.Y. (2006). Comparative studies of DNA and RNA from the royal jelly of *Apis mellifera* and *Apis cerana*. *Indian Bee J.* 68, 18-21.

ZHANG, Y., WANG, X. X., FENG, Z. J., CONG, H. S., CHEN, Z. S., LI, Y. D., *et al.* (2020). Superficially Similar Adaptation Within One Species Exhibits Similar Morphological Specialization but Different Physiological Regulations and Origins. *Front. Cell Dev. Biol.* 8, 1-12. doi: 10.3389/fcell.2020.00300.

ZHU, K., LIU, M., FU, Z., ZHOU, Z., KONG, Y., LIANG, H., *et al.* (2017). Plant microRNAs in larval food regulate honeybee caste development. *PLoS Genet.* 13. doi: 10.1371/journal.pgen.1006946.

ZHU, M. (1985). Study on cross-feeding between *Apis cerana cerana* and *Apis mellifera* ligustica. *J. Bee* 3, 14. (in Chinese).

ZOU, Y., HUANG, K., YU, YAN, W. Y., AND ZENG, Z. J. (2007). A Study on extraction of DNA from fresh royal jelly of *Apis cerana*. *Acta Agric. Univ. Jiangxiensis* 29, 279-281 (in Chinese).

## Additional Information

Whole transcriptome sequencing data have been deposited in the NCBI database under the accession ID: PRJNA889707.

Fig. S1 The heat map of expression profiles of mRNAs, lncRNAs, circRNAs and sRNAs in NQs and CQs. (A) The heat map of expression profiles of lncRNAs.; (B) the heat map of expression profiles of mRNAs; (C) the heat map of expression profiles of sRNAs. (D) the heat map of expression profiles of circRNAs. NQ and CQ are nutritional crossbreed queens and control queens, and each had two biological replicates. The horizontal coordinates represent the sample name and the clustering results of the sample, and the ordinates represent the difference genes and the clustering results of the genes. The different columns in the figure represent different samples, and the different rows represent different genes. Colors represent the gene expression levels [$\log_{10}$ (FPKM plus 0.000001)] of each sample.

Fig. S2 The top enriched and classification of Go terms identified from (A) DEcircRNAs, (B) DElncRNAs, (C) DEGs, and (D) DEmiRNAs. The results are summarized in three main categories: biological process, cellular component and molecular function. The lighter colored bars in Y-axis incident the percentages of coding or non-coding RNAs in a category, and the deeper colored ones indicate the present ages of differentially expressed coding or non-coding RNAs.

Fig. S3 The body color of full developed queens from TPH1, KMO and control groups.

Table II-9-1 The statistics of coding and non-coding RNAs. Note: Total clean reads: The number of clean reads, as single-ended; Mapped reads: the number of reads on the reference genome and the percentage of them in clean reads. Q30(%): Percentage of bases with a clean data mass value greater than Q30. In sRNA, Total reads: the number of uncommented reads used to compare with the reference genome; Mapped reads: clean reads to the reference genome.

Table S1 (A)Pearson correlation coefficient of mRNA. (B) correlation coefficient of lncRNA. (C) Pearson correlation coefficient of circRNA. (D) Pearson correlation coefficient of sRNA.

Table S2 The list of DEGs, DELncRNA, DEmiRNA, DEcircRNAs.

Table S3 GO enrichment of DEGs, DELncRNA, DEmiRNA, DEcircRNAs.

Table S4 Enriched KEGG pathways of DEGs, DELncRNA, DEmiRNA, DEcircRNAs.

Table S5 The lists of DEGs, DElncRNAs and DEmiRNAs and key KEGG pathways involved in body color.

Table S6 The primers of the β-actin gene, TPH1 and KMO in RNAi.

# 10. 全基因组 DNA 甲基化揭示西方蜜蜂表观遗传印迹

李震[1,2]，易瑶[3]，何旭江[1,2]，黄强[1,2]，刘一博[1,2]，曾志将[1,2*]

（1. 江西农业大学蜜蜂研究所，南昌 330045；
2. 江西省蜜蜂生物学与饲养重点实验室，南昌 330045；3. 江西中医药大学，南昌 330004）

**摘要**：旨在研究全基因组甲基化在蜜蜂体内的表观遗传印迹。以西方蜜蜂（*Apis mellifera*）作为试验材料，对 G0 代蜂王进行单雄授精，待 G0 代蜂王产卵后，以卵（E）、1 日龄工蜂幼虫（L1）和 2 日龄幼虫（L2）分别培育蜂王（G1 代，以卵培育的蜂王简称 G1E，以 1 日龄幼虫培育的蜂王简称 G1L1，以 2 日龄幼虫培育的蜂王简称 G1L2），选取 G1 代的 3 种蜂王分别产卵培育雄蜂（G2 代，G1E 产的雄蜂简称 G2DE，G1L1 产的雄蜂简称 G2DL1，G1L2 产的雄蜂简称 G2DL2）。利用全基因组 Bisulfite 甲基化测序技术，对 G0 代蜂王、G1 代蜂王、G2 代雄蜂进行全基因组 DNA 甲基化测序。使用 ANOVA 方差分析和高甲基化位点比对揭示西方蜜蜂 DNA 甲基化印迹特性以及移虫日龄对西方蜜蜂 DNA 甲基化遗传的影响。西方蜜蜂原有高甲基化位点和突变高甲基化位点均存在表观遗传印迹现象；同时发现随着移虫日龄增加，原有高甲基化位点和突变高甲基化位点遗传印迹数量都是逐步增加；3 个调控蜜蜂级型分化的高甲基化遗传印迹模式进一步表明移虫日龄会调控蜂王的高甲基化水平，且蜂王高甲基化位点数目越多，所育雄蜂后代也将获得更多的高甲基化位点。研究表明 DNA 甲基化遗传印迹能够在西方蜜蜂这一模式生物体内稳定遗传，为解析蜜蜂 DNA 甲基化遗传调控机制提供科学依据。

**关键词**：蜜蜂；DNA 甲基化；表观遗传印迹

# Genome-wide DNA Methylation Reveals Epigenetic Inheritance Marks in Honeybee (*Apis mellifera*)

LI Zhen[1,2], YI Yao[3], HE Xujiang[1,2], HUANG Qiang[1,2], LIU Yibo[1,2], ZENG Zhijiang[1,2*]

(1. Honeybee Research Institute, Jiangxi Agricultural University, Nanchang 330045, China;
2. Jiangxi Province Key Laboratory of Honeybee Biology and Beekeeping, Nanchang 330045, China;
3. Jiangxi University of Traditional Chinese Medicine, Nanchang 330004, China)

---

\* 通信作者：bees1965@sina.com.
注：此文发表在《江西农业大学学报》2022 年第 4 期。

## 蜜蜂发育及行为生物学理论研究
Study on Developmental and Behavioral Theories of Honeybee Biology

**Abstract**: This experiment aims to explore the epigenetic imprinting of genome-wide methylation in honey bees. *Apis mellifera* was used as the model organism in this study. A single male inseminated queen (G0 queen) was used to rear offspring queens (G1 queen). The G1 queen was reared from eggs (E), 1 day old larvae (L1) and 2 day old larvae (L2) respectively (in G1 generation, G1E represents the queen from eggs. G1L1 represents the queen reared from 1-day-old larvae and G1L2 represents the queen reared from 2-day-old larvae). The three types of G1 queens were further used to breed drones (G2DE) represents drones produced by G1E. G2DL1 represents drones produced by G1L2 and G2DL2 represents drones produced by G1L2). The methylated loci in G0 queens, G1 queens and G2 drones were quantified using the whole genome Bisulfite methylation sequencing. Analysis of variance (ANOVA) and comparison of high-methylation loci was used to reveal the DNA methylation inheritance marks of honeybee (*Apis mellifera*) and the effect of the aging of larvae for the transplantation on DNA methylation inheritance in honeybee (*Apis mellifera*). The results showed that both the highly methylated loci and the mutant methylated loci could be epigenetically inherited. The number of high-methylation and mutant highly methylated loci gradually increased with the aging of larvae for transplantation. Three genetic imprinting patterns of high-methylation that regulated the caste differentiation of honey bees further indicated that the aging of larvae for transplantation the high-methylation level of the queen. The higher the number of high-methylation loci in the queen, the more hypermethylation loci the drones offspring will acquire. The marks of DNA methylation can be stably inherited in the honeybee model organism.

**Keywords:** Honeybee, DNA methylation, epigenetic inheritance

蜜蜂蜂王是蜂群的核心，蜂王质量优劣不仅直接关系到蜂群群势强弱以及蜂产品产量高低[1-2]。蜜蜂的授粉行为对全球农业生产都具有极其重要的促进作用[3-4]。但2006年美国暴发的蜂群衰竭失调病（Colony Collapse Disorder，CCD）引发了北美授粉危机，具体表现为蜂群内大量工蜂突然失踪，只留有蜂王、大量幼虫和少量工蜂[5]。研究表明，CCD的暴发与蜂王质量密切相关[6]。幼虫日龄、食物来源、生活空间、DNA甲基化和母体效应等因素均参与了蜂王的发育调节[7-10]。尽管关于DNA甲基化调节蜂王发育已有诸多报道[9,11-12]，但DNA甲基化在蜜蜂群体内的遗传规律尚不清晰。而深入解析西方蜜蜂DNA甲基化遗传调控机制，能够为优育蜂王提供理论指导，同时也为解决CCD问题提供新思路。Kucharski等[10]发现，蜂王食物中的营养物质通过介导DNA甲基化来调节蜜蜂级型分化，这首次在分子水平上揭示了DNA甲基化对蜜蜂表观遗传现象。此外，Shi等[9]报道王台大小也会影响蜜蜂的DNA甲基化，进而调节蜜蜂的级型分化。本实验室发现：随着移植蜂王幼虫日龄的增加，所培育出蜂王的出生重、胸部长度和宽度均逐渐减小；此外，与低日龄蜂王幼虫培育出的蜂王相比，由3日龄的幼虫培育出的蜂王的整体DNA甲基化水平最高[13]。本研究组还发现蜂王的差异性DNA甲基化区域（Differential DNA Methylated Regions，DMRS）在连续几代中有累积效应[14]。蜜蜂DNA甲基化可通过父系遗传印迹[15]。DNA甲基化是表观遗传信息稳定跨带传递的媒介，也是最早开始研究的表观遗传遗传信息载体[16-17]。目前还不清楚蜜蜂DNA甲基化是否与斑马鱼（*Brachydanio rerio*）、果蝇（*drosophilid*）、长牡蛎（*Crassostrea gigas*）、拟南芥（*Arabidopsis thaliana*）等模式生物一样具有母系表观遗传印迹特性[18-21]，也不清楚环境（移虫日龄）对蜜蜂DNA甲基化表观遗传印迹的影响[22]。本研究利用蜂王单雄授精技术成功培育出1只单雄授精产卵蜂王G0，并人工建立了1个具有3种蜂王后代的品系（分别利用3日龄卵、1日龄幼虫和2日龄幼虫所培育出的G1蜂王）。待G1代蜂王交尾产卵后，采集G0代蜂王、G1代蜂王和G1代蜂王产的雄蜂进行全基因组DNA甲基化测序来系统探究西方蜜蜂DNA甲基化表观遗传印迹特性，以及移虫日龄对西方蜜蜂DNA甲基化表观遗传印迹影响，以期探明DNA甲基化调控西方蜜蜂蜂王发育的分子机理，进一步揭示移虫日龄调控蜂王及后代雄蜂发育的内在机制。

## 1 材料与方法

### 1.1 试验材料

试验蜂群：江西农业大学蜜蜂研究所饲养的西方蜜蜂（*Apis mellifera*），试验时间：2018年3～10月，按标准方法进行饲养。试验试剂：OMEGA组织DNA提取试剂盒（Omega Bio-Tek，美国）。

### 1.2 试验方法

1.2.1 蜜蜂取样方法　将一只单雄人工授精并成功产卵的蜂王（G0），限制在免移虫塑料巢脾上产卵6 h。对G0代蜂王进行单雄授精，G0代蜂王产卵后，以卵（E）、1日龄工蜂幼虫（L1）和2日龄幼虫（L2）分别培育蜂王（G1代，以卵培育的蜂王简称G1E，以1日龄幼虫培育的蜂王简称G1L1，以2日龄幼虫培育的蜂王简称G1L2），然后以G1代蜂王产卵培育雄蜂（G2代，G1E产的雄蜂简称G2DE，G1L1产的雄蜂简称G2DL1，G1L2产的雄蜂简称G2DL2）。采样方法见图Ⅱ-10-1。

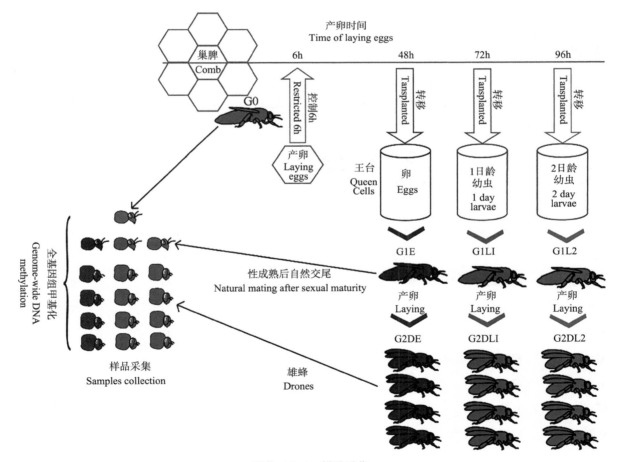

图Ⅱ-10-1　样品采集

**1.2.2 全基因 DNA 甲基化测定方法** 取 G0、G1E、G1L1、G1L2 蜂王以及 G2DE、G2DL1、G2DL2 雄蜂，将其头、胸组织提取 DNA 后送往北京诺禾致源科技股份有限公司进行全基因组甲基化测序。上机测序后得到的原始数据均已上传 NCBI，提交序列号为 SUB9847854，数据序列号为 PRJNA737625。将原始数据删除低质量数据和测序接头来得到高质量的 clean data。以西方蜜蜂基因组 (Amel_HAv3.1 版本) 为参考序列，采用 Bismark 软件 (0.16.3 版) 将甲基化数据与参考基因组进行比对分析[23]。首先去除甲基化数据中的重复部分；然后以比对上的参考基因组的 C 位点的碱基类型：若比对上的碱基为 C，则判定该位点发生了甲基化；若为 T，则未发生甲基化。

甲基化水平的计算公式为：$ML=mC/(mC+umC)$

ML 为甲基化水平；

mC 为甲基化 C 的数量；

umC 未甲基化 C 的数量。

**1.2.3 原有高甲基化位点和突变高甲基化位点统计** 参照 Liu 的方法[24]，选取 $ML \geq 0.8$ 的位点为高甲基化位点。原有高甲基化位点指的是高甲基化位点从 G0 传递到 G1、G2 时共有的个数；突变高甲基化位点的研究对象是 G0 代蜂王的低甲基化位点遗传到 G1 代蜂王和 G2 代雄蜂时变为高甲基化位点。

**1.3 数据分析与统计**

利用 SPSS25.0 进行分析雄蜂甲基化的数据。数值用平均值 ± 标准误（Mean±SE）表示，并通过单因素方差分析（one-way ANOVA）和 Fisher's LSD 检验进行分析。$P<0.05$ 表示数据差异显著。

## 2 结果与分析

### 2.1 基因组 DNA 整体甲基化测序质量

18 个样品的平均 clean reads 数为 38 204 92，平均 clean base 为 9.90 G，Q20% 的平均为 96.70%，Q30 的平均值为 91.69%，亚硫酸氢盐的平均转化率为 99.70%（表Ⅱ-10-1）。

表 Ⅱ-10-1 甲基化测序结果统计表

| 样品 | Clean Reads | Clean Bases/G | Q20/% | Q30/% | GC 含量/% | BS 平均转化率/% |
|---|---|---|---|---|---|---|
| G0 | 36 334 972 | 9.13 | 95.93 | 90.02 | 15.9 | 99.75 |
| G1E | 41 712 296 | 10.72 | 96.77 | 91.51 | 16.17 | 99.728 |
| G1L1 | 38 136 689 | 9.68 | 96.31 | 90.75 | 16.3 | 99.667 |
| G1L2 | 40 357 273 | 10.56 | 97.36 | 92.95 | 19.64 | 99.473 |
| G2DE_1 | 35 315 072 | 9.03 | 96.47 | 91.05 | 16.03 | 99.752 |
| G2DE_2 | 33 863 197 | 8.68 | 96.52 | 91.22 | 15.11 | 99.76 |
| G2DE_3 | 37 530 239 | 9.6 | 96.47 | 91.05 | 15.82 | 99.758 |
| G2DE_3 | 37 475 558 | 9.58 | 96.46 | 90.79 | 19.57 | 99.682 |
| G2DL1_1 | 41 691 908 | 11.09 | 97.43 | 93.39 | 15.56 | 99.737 |
| G2DL1_2 | 42 029 140 | 11.08 | 97.28 | 93.05 | 18 | 99.674 |
| G2DL1_3 | 43 650 685 | 11.62 | 97.45 | 93.44 | 15.4 | 99.729 |

（续表）

| 样品 | Clean Reads | Clean Bases/G | Q20/% | Q30/% | GC 含量/% | BS 平均转化率/% |
|---|---|---|---|---|---|---|
| G2DL1_4 | 38 081 700 | 9.97 | 96.59 | 91.95 | 15.52 | 99.729 |
| G2DL2_1 | 34 218 724 | 8.81 | 96.67 | 91.53 | 15.41 | 99.696 |
| G2DL2_2 | 37 338 380 | 9.72 | 96.3 | 91.29 | 14.97 | 99.724 |
| G2DL2_3 | 35 455 948 | 9.08 | 96.53 | 91.14 | 16.85 | 99.684 |
| G2DL2_4 | 38 081 700 | 9.97 | 96.59 | 91.95 | 15.52 | 99.729 |

## 2.2　DNA 甲基化遗传表观印迹特性

**2.2.1 原有高甲基化位点遗传印迹**　从图Ⅱ-10-2 可见：以卵和 1 日龄幼虫培育的 G1 代蜂王（G1E、G1L1），分别从 G0 代蜂王中遗传了 57.6% 和 57.4% 的高甲基化位点，远低于以 2 日龄幼虫培育的 G1 代蜂王（G1L2）遗传的 79.5% 高甲基化位点。同样，G1E 和 G1L1 所产的雄蜂（G2DE，G2DL1）分别遗传了来自 G0 的 36.5% 和 39.2% 高甲基化位点，也远低于 G2DL2 从 G0 遗传的 51.3% 高甲基化位点。

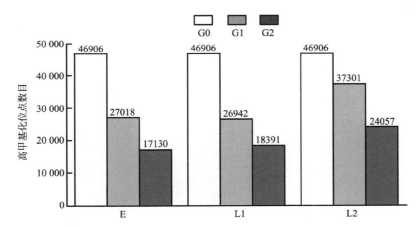

G0: G0 代蜂王，G1: G1 代蜂王，G2: G2 代蜂王，E: G0-G1E-G2DE 总原有高甲基化位点遗传，L1: G0-G1L1-G2DL1 总原有高甲基化位点遗传，L2: G0-G1L2-G2DL2 总原有高甲基化位点遗传。

**图Ⅱ-10-2　原有 DNA 高甲基化位点遗传印迹**

**2.2.2 突变高甲基化位点遗传印迹**　从图Ⅱ-10-3 可见：G1 代蜂王（G1E，G1L1，G1L2）从 G0 代蜂王遗传的突变高甲基化位点随着移虫日龄的增加而增加；G2DE、G2DL1 和 G2DL2 从各自母本蜂王（G1E、G1L1 和 G1L2）中遗传了 40.1%、38.7% 和 35.4% 的高甲基位点。虽然遗传的高甲基化位点比例随着移虫日龄增加而减少，但 G2 代雄蜂（G2DE，G2DL1，G2DL2）从 G1 代蜂王遗传的高甲基化位点随着移虫日龄的增加而增加。

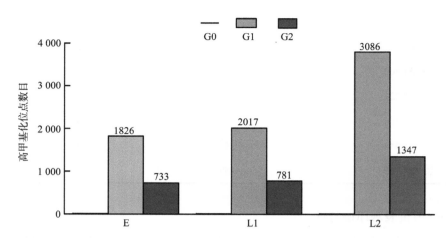

G0: G0 代蜂王，G1: G1 代蜂王，G2：G2 代蜂王，E: G0-G1E-G2DE 总突变高甲基化位点遗传，L1: G0-G1L1-G2DL1 总突变高甲基化位点遗传，L2: G0-G1L2-G2DL2 总突变高甲基化位点遗传。

图 Ⅱ-10-3　突变高甲基化位点遗传印迹

### 2.3　移虫日龄对 DNA 甲基化遗传表观印迹影响

2.3.1 移虫日龄对原有高甲基化位点遗传的影响　从图 Ⅱ-10-4 可见：G2DL2 从 G0 代蜂王遗传到的原有高甲基化位点显著高于 G2DE 和 G2DL1 从 G0 代蜂王遗传到的原有高甲基化位点（$P<0.05$）。

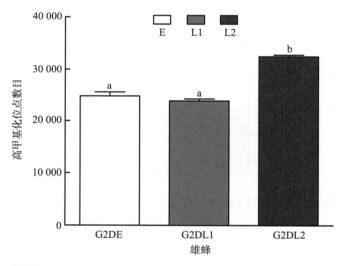

E: 卵，L1: 1 日龄幼虫，L2: 2 日龄幼虫，G2DE: G0 代蜂王遗传给 G2 代雄蜂（雄蜂由卵培育的蜂王所产）的原有高甲基化位点，G2DL1: G0 蜂王遗传给 G2 雄蜂（雄蜂由 1 日龄幼虫培育的蜂王所产）的原有高甲基化位点，G2DL2: G0 蜂王遗传给 G2 代雄蜂（雄蜂由 2 日龄幼虫培育的蜂王所产）的原有甲基化位点。

图 Ⅱ-10-4　移虫日龄对原有高甲基化位点遗传的影响

2.3.2 移虫日龄对突变高甲基化位点遗传的影响　从图Ⅱ-10-5 可见：随着移虫日龄的增加，从 G0 代蜂王遗传印迹到 G2 代雄蜂的突变高甲基化位点数目随之增加。且 G2DL2 遗传到的突变高甲基化位点显著高于 G2DE 和 G2DL1（$P<0.05$）。

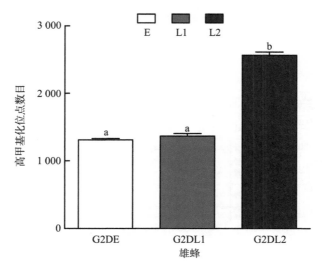

E: 卵，L1: 1 日龄幼虫，L2: 2 日龄幼虫，G2DE: G0 代蜂王遗传给 G2 代雄蜂（雄蜂由卵培育的蜂王所产）的突变高甲基化位点，G2DL1: G0 蜂王遗传给 G2 雄蜂（雄蜂由 1 日龄幼虫培育的蜂王所产）的突变高甲基化位点，G2DL2: G0 蜂王遗传给 G2 代雄蜂（雄蜂由 2 日龄幼虫培育的蜂王所产）的突变甲基化位点。

图Ⅱ-10-5　移虫日龄对突变高甲基化位点遗传的影响

### 2.4　高甲基化位点遗传印迹模式图

研究结果深入揭示了 3 个调控蜜蜂级型分化的关键基因高甲基化位点遗传印迹模式（图Ⅱ-10-6）。LOC409393 基因高甲基化位点遗传印迹模式图揭示了一些原有高甲基化位点能够从 G0 代蜂王稳定遗传印迹到 G2 代雄蜂（图Ⅱ-10-6A）。LOC409393 和 LOC408438 高甲基化位点遗传印迹模式图则进一步证明了随着移虫日龄增加，所培育 G1 代蜂王从 G0 代蜂王获得的高甲基化位点数目增多，且 G2 代雄蜂从 G1 代蜂王处获得的高甲基化位点数目同样增多（图Ⅱ-10-6A，图Ⅱ-10-6B，图Ⅱ-10-6C）。

## 3　讨论与结论

DNA 甲基化普遍存在于动植物和真菌等基因组中，是目前研究最广泛的表观修饰机制。大量研究表明：DNA 甲基化调控蜜蜂级型分化过程，进而影响蜂王发育质量[11-15]，但仍然有许多研究问题还不清楚，比如蜜蜂 DNA 甲基化与蜂王质量关系；蜜蜂 DNA 甲基化能否母系表观遗传印迹。

Shi 等[11]发现工蜂 DNA 甲基化水平是随着幼虫日龄增高，而蜂王 DNA 甲基化水平是先升高然后下降。Yi 等[14]首次发现在逐代培育蜂王过程中，DNA 甲基化存在累代现象。本实验首次发现原有高甲基化位点和突变甲基化位点都存在表观遗传印迹现象，有助于解释蜂王与工蜂 DNA 甲基化水平差异及累代现象。而 Yi 等[13]发现移虫日龄的增加，所培育的蜂王初生重、胸长和胸宽逐步下降，而 3 日龄蜂王整体 DNA 甲基化水平却逐步提高；Yi 等[25]进一步研究发现：蜂王质量随着移虫日龄的增加而降低，同时随着移虫日龄的增加，移虫培育的蜂王与移卵培育的蜂王之间的差异甲基化基因数量逐步增多，差异甲基化基因涉及蜜蜂级型分化、寿命、免疫、发育等重要途径。本实验首次发现：现随着移虫日龄增加，原有高甲基化位点和突变高甲基化位点遗传印迹数量都是逐步增加。

DNA 甲基化位于启动子区时抑制基因的表达，但当甲基化位于基因区时，它不仅不抑制基因的表达，相反还会起到促进作用[26]。有研究表明：基因区甲基化程度小于 70% 时正向调控基因表达，大于

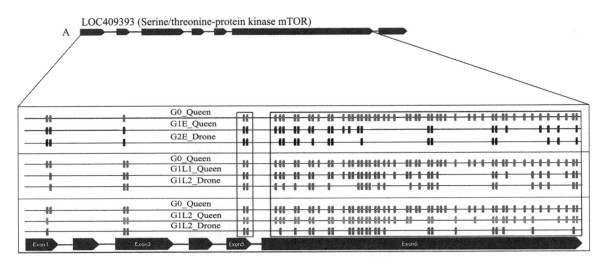

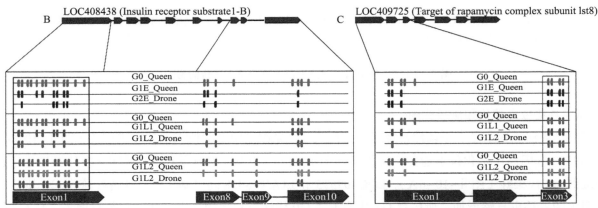

A：丝氨酸/苏氨酸蛋白激酶 mTOR，B：胰岛素受体底物 1-B，C：雷帕霉素复合物亚单位 LST8 的靶点；左框标注说明高甲基化位点能从 G0 代蜂王稳定遗传 G1 代蜂王和 G2 代雄蜂；右框标注说明高甲基化位点从 G0 代蜂王遗传到 G1 代蜂王和 G2 代雄蜂的数目随着移虫日龄的增加而增加。

**图Ⅱ-10-6　3 个基因的高甲基化位点遗传印迹模式**

70% 时会起负调控作用[27]。本次实验选取外显子上甲基化程度大于和等于 80% 的位点作为高甲基化位点，因此基因区高甲基化位点数目越多，其自身表达将会进一步受到抑制。LOC408438 和 LOC409393 基因被干扰或剪切时，蜂王幼虫会在实验室培育条件下转向完整工蜂形态发育，并出现卵巢管数减少这一生殖退化现象[28-29]。而 LOC408438 和 LOC409393 基因的高甲基化位点遗传印迹模式图揭示了在使用 2 日龄以上（包括 2 日龄）的幼虫人工培育蜂王时，会造成蜂王和后代雄蜂获得的更多的高甲基化位点，这或许会导致培育的蜂王出现质量下降现象[14]，进而降低整个蜂群繁衍。

因此，在当前西方蜜蜂大规模商业饲养的大背景下，建议养蜂者应优先利用卵和 1 日龄小幼虫来培育优质蜂王，提高蜂群的繁殖和存续能力，达到预防发生 CCD 的目的本研究结果进一步解释了蜜蜂 DNA 甲基化与蜂王质量关系，为在养蜂生产中人工培育优质蜂王和 CCD 防控提供了全新思路。

## 参考文献

[1] VANENGELSDORP D, HAYES J, UNDERWOOD R M, et al. A survey of honey bee colony losses in the U.S., fall 2007 to spring 2008[J]. PLoS One, 2008, 3(12): e4071.

[2] PETTIS J S, DELAPLANE K S. Coordinated responses to honey bee decline in the USA[J]. Apidologie, 2010, 41: 256-263.

[3] 李震, 刘志勇, 张永, 等. 天然蜂粮对高脂血症大鼠肠道微生物多样性的影响 [J]. 江西农业大学学报, 2019, 41(6): 1159-1166.

[4] 张祖芸, 李震, 吴志豪, 等. 2种杀虫剂对舞蹈蜜蜂大脑转录组中可变剪切的影响. 江西农业大学学报, 2021, 43(1): 175-181.

[5] SPIVAK M, Le Conte Y. Editorial, special issue on bee health[J]. Apidologie, 2010, 41: 225-226.

[6] 张波, 吴小波, 廖春华, 等. 蜜蜂免移虫技术研究与应用 [J]. 中国农业科学, 2018, 51(22): 4387-4394.

[7] 李震, 刘志勇, 江武军, 等. 天然蜂粮对高脂血症大鼠血脂、抗氧化及免疫功能的影响 [J]. 中国农业科学, 2019, 52(16): 2912-2920.

[8] SHI Y Y, HUANG Z Y, ZENG Z J, et al. Diet and cell size both affect queen-worker differentiation through DNA methylation in honey bees (*Apis mellifera, Apidae*) [J]. PLoS One, 2011, 6: e18808.

[9] WEI H, HE X J, LIAO C H, et al. A maternal effect on queen production in honeybees[J]. Current Biology, 2019, 29(13): 2208-2213.

[10] KUCHARSKI R, MALESZKA J, FORET S, et al. Nutritional control of reproductive status in honeybees via DNA methylation[J]. Science, 2008, 319(5871): 1827-1830.

[11] SHI Y Y, YAN W Y, HUANG Z Y, et al. Genomewide analysis indicates that queen larvae have lower methylation levels in the honey bee (*Apis mellifera*) [J]. Naturwissenschaften, 2013, 100(2), 193-197.

[12] HE X J, ZHOU L B, PAN Q Z, et al. Making a queen: an epigenetic analysis of the robustness of the honeybee (*Apis mellifera*) queen developmental pathway[J]. Molecular Ecology, 2017, 26(6): 1598-1607.

[13] YI Y, LIU Y B, BARRON A B, et al. Transcriptomic, morphological, and developmental comparison of adult honey bee queens (*Apis mellifera*) reared from eggs or worker larvae of differing ages[J]. Journal of Economic Entomology, 2020, 113(6): 2581-2587.

[14] YI Y, HE X J, BARRON A B, et al. Transgenerational accumulation of methylome changes discovered in commercially reared honey bee (*Apis mellifera*) queens[J]. Insect Biochemistry and Molecular biology, 127: 103476.

[15] YAGOUND B, REMNANT E J, BUCHMANN G, et al. Intergenerational transfer of DNA methylation marks in the honey bee[J]. Proceedings of the National Academy of Sciences, 2020, 117(51): 32519-32527.

[16] 徐鹏, 于文强. 表观遗传信息的跨代传递 [J]. 科学通报, 2016, 61(32): 3405-3412.

[17] 荐思婧, 宁超, 高磊, 等. 表观遗传信息在动物中的跨代遗传和重编程 [J]. 中国科学: 生命科学, 2021, 51(05): 556-566.

[18] JIANG L, ZHANG J, WANG J J, et al. Sperm, but not oocyte, DNA methylome is inherited by zebrafish early embryos[J]. Cell, 2013, 153(4): 773-784.

[19] YAN H, BONASIO R, SIMOLA D F, et al. DNA methylation in social insects: how epigenetics can control behavior and longevity[J]. Annual Review of Entomology, 2015, 60: 435-452.

[20] JIANG Q, Li Q, Yu H, et al. Inheritance and variation of genomic DNA methylation in diploid and triploid pacific oyster (*Crassostrea gigas*)[J]. Marine Biotechnology, 2016, 18(1): 124-132.

[21] HOFMEISTER B T, LEE K, ROHR N A, et al. Stable inheritance of DNA methylation allows creation of epigenotype

maps and the study of epiallele inheritance patterns in the absence of genetic variation[J]. Genome biology, 2017, 18(1): 1-16.

[22] SHARMA U, RANDO O J. Father-son chats: inheriting stress through sperm RNA[J]. Cell Metabolism, 2014,19(6):894-895.

[23] 易瑶. 移虫日龄对西方蜜蜂（Apis mellifera）蜂王发育的影响 [D]. 南昌：江西农业大学, 2020.

[24] LIU G, WANG W, HU S, et al. Inherited DNA methylation primes the establishment of accessible chromatin during genome activation[J]. Genome Research, 2018, 28(7): 998-1007.

[25] YI Y, LIU Y B, BARRON A B, et al. Effects of commercial queen rearing methods on queen fecundity and genome methylation[J]. Apidologie, 2021, 52: 282-291.

[26] SUZUKI M M, BIRD A. DNA methylation landscapes: provocative insights from epigenomics[J]. Nature Reviews Genetics, 2008, 9: 465.

[27] ZILBERMAN D, GEHRING M, TRAN R K, et al. Genome-wide analysis of Arabidopsis thaliana DNA methylation uncovers an interdependence between methylation and transcription[J]. Nature Genetics, 2007, 39(1): 61-69.

[28] PATEL A, FONDRK M K, KAFTANOGLU O, et al. The making of a queen: TOR pathway is a key player in diphenic caste development[J]. PLoS One, 2007, 2(6): e509.

[29] WOLSCHIN F, MUTTI N S, AMDAM G V. Insulin receptor substrate influences female caste development in honeybees. [J]. Biology Letters, 2011, 7(1): 112-115.

# 蜜蜂行为生物学及其分子机制

## Behavioral Biology of Honeybee and its Molecular Mechanism

  行为生物学是生物学中的一个热门领域。传统动物行为生物学主要局限于对动物行为作表象分析，而现代动物行为生物学则侧重对动物行为作全面解析。蜜蜂是一种真社会性昆虫，具有很多复杂有趣的个体和社会行为，因此，蜜蜂行为生物学研究一直是国内学者研究关注主题，原因是蜜蜂行为生物学研究结果对整个社会生物学有重要的借鉴意义。

  利用江西农业大学蜜蜂研究所与广州市远望谷信息技术股份有限公司合作研发的蜜蜂无线射频识别（RFID）系统研究了工蜂采集及回巢行为，首次从理论上证实蜜蜂具有预测未来天气的能力。

  利用 Needle trap 技术与气质联用技术分析鉴定出 9 种工蜂幼虫信息素，并推测 E-β-罗勒烯为工蜂幼虫的饥饿信息素，之后通过行为学实验进一步证实了 E-β-罗勒烯为工蜂幼虫饥饿信息素。利用 RNA-Seq 技术分析鉴定出 E-β-罗勒烯在幼虫体内的生物合成通路。

  利用 Needle trap 热脱附技术和气质联用分析技术调查了蜜蜂采集和不采集的 15 种植物花朵的花香物质，发现壬醛是唯一一种在所有蜜蜂授粉植物的花香物质中均存在，且未在蜜蜂不授粉的植物花香物质发现的花香物质。同时，发现花朵中壬醛的含量与花朵的开花状态，花蜜分泌量呈正相关。利用空心尼龙网罩住花朵 2 h，发现这些花朵比未罩网的花朵分泌更多的花蜜和释放更多的壬醛。昆虫触角电位检测发现蜜蜂触角对壬醛的反应也显著高于其他主要花香物质。行为学实验进一步证实壬醛是对采集蜂最具吸引力的一种花香物质。该研究揭示了壬醛可能作为一种广谱的食物指引信号参与蜜蜂与植物花朵的信息交流，这一研究结果不仅丰富了人们对传粉昆虫和植物之间"化学通讯"的认知，也可为蜜蜂人工诱导授粉技术的发展提供新思路。

利用气质联用色谱仪分析了这4种封盖信息素在4日龄、正在封盖和已封盖的工蜂幼虫和雄蜂幼虫中的含量，并利用转录组测序分析了工蜂幼虫和雄蜂幼虫在这3个封盖阶段的基因表达变化。结果表明：雄蜂幼虫的封盖信息素含量比工蜂幼虫高，推测这可能与雄蜂幼虫更吸引蜂螨有关；此外还鉴定出这4种封盖信息素在蜜蜂中可能的生物学合成通路。

通过同位素相对标记与绝对定量技术（iTRAQ）分析了东方蜜蜂嗅觉学习前后大脑蛋白质表达量变化，结果表明：从训练组和非训练组中共鉴定得到2406个蛋白，2组之间共有147个蛋白的表达量显著差异，其中87个蛋白在训练组中上调表达，60个蛋白下调表达。

用亚致死剂量溴氰菊酯或吡虫啉在距蜂箱300m、500m和1000m的地点饲喂蜜蜂，并用q-PCR检测跳舞蜜蜂调控学习记忆相关基因表达特性。结果表明：与对照组相比，饲喂农药的蜜蜂表达蜜源方向、距离、质量的舞蹈行为特征发生改变，说明蜜蜂通过调整舞蹈交流行为将蜜源潜在"危险信号"传递给巢内同伴。与学习和记忆相关基因表达下调，表明蜜蜂定位和导航能力受到损伤。

开展了蜂王上颚腺信息素对雄蜂选择行为影响及气味受体基因表达特性研究工作，结果表明：中蜂雄蜂对不同剂量的QMP的选择率均显著高于意蜂雄蜂，低剂量9-ODA对两种雄蜂均具有吸引作用；雄蜂对蜂王上颚腺信息素3种主要成分9-ODA、9-HDA、HOB的选择行为不明显；发现信息素受体基因Ors在雄蜂选择行为中有重要作用。

开展了蜂王侍从工蜂行为特性研究工作，结果表明：蜂王侍从工蜂的日龄范围一般是2~23日龄，但是主要集中在6~18日龄；蜂王侍从工蜂组与"对照蜜蜂"组在各亚家庭的分布有显著差异，即工蜂的蜂王侍从行为会受到其遗传背景的影响。

# 1. RFID Monitoring Indicates Honeybees Work Harder before a Rainy Day

Xujiang He[†], Liuqing Tian[†], Xiaobo Wu, Zhijiang Zeng[*]

Honeybee Research Institute, Jiangxi Agricultural University, Nanchang, Jiangxi 330045, P. R. of China

**Dear Editor,**

Storms are usually accompanied by a drop in temperature, and an increase in wind and barometric pressure and rainfall, which have negative impacts on most activities, survival and reproduction in insects (Gillot, 2005). The majority of studies mainly focused on how the flight activity of various flying insects such as honeybees, bumble bees, horse flies and leafminer were directly influenced by intraday weather changes (Burnett & Hays, 1974; Lundberg, 1980; Casas, 1989; Vicens & Bosch, 2000). However, accumulating evidences showed that animals can make behavioral changes before storms, which is enormously important for their survival in severe weather condition. Before upcoming storms birds unusually chirp and bathe with sand; native frogs croak and hide their eggs masses; spiders spin shorter and produce thicker webs and wasps hide their comb before rains (Galacgac & Balisacan, 2009; Acharya, 2011). In early 1893, honeybees were reported more active before storms (Inwards, 1893). In this study, we compared the working habits of foragers on days that were followed by a sunny day and those that followed by a rainy day using the Radio Frequency Identification (RFID) which was developed and manufactured by the Honeybee Research Institute of Jiangxi Agricultural University in collaboration with the Guangzhou Invengo Information Technology Co., Ltd., and we firstly showed that honeybees worked harder before a rainy day.

Three honeybee (*Apis mellifera*) colonies were maintained at the Honeybee Research Institute, Jiangxi Agricultural University, Nanchang, China. Each colony had four full frames, with approximately 6,000 workers and a laying queen. Newly emerged workers ($N$=300) of each experimental colony were each glued with an RFID tag on the dorsal surface of thorax with shellac. Tagged workers were introduced to their natal colonies, and monitored 24 hours per day during a period of 34 days by an RFID reader (Fig. III-1-1). When tagged workers entered and exited a colony, the RFID system recorded their unique ID, time stamp and direction. Duration for each trip was calculated based on the exit and entrance time stamps after sorting the data based on each RFID. Flights with duration longer than 5 minutes were considered as foraging trips (Calderone & Page,

---

† These two authors contributed equally to this paper.
*Corresponding author: bees1965@sina.com (ZJZ).
注：此文发表在 *Insect Science* 2016 年第 1 期。

1988). To be conservative, we used only data from bees that were older than 21 days. We used foraging data only from good foraging days: those that had temperatures from 15 to 34 °C, sunny, no precipitation and wind < 12 km/h. Rainy days were defined as those with daily precipitation > 5 mm. We summed the duration of all foraging trips per day of each forager as "foraging duration", and defined "quitting time" as the last returning trip. For further statistical analysis, we collected a delta time of each bee by using daily sunset to subtract their quitting time. Sunset time for each day was obtained from the website (http://sunrise.supfree.net/). We classified the foraging data in two categories, days that were followed by a sunny day, and days that followed by a rainy day. Foraging duration and the delta time between sunset and quitting time were transformed by taking their square root, which made the data to be normally distributed (Bartlett 1947). The transformed data were then analyzed with ANOVA in Statview 5.0 (SAS Institute, Gary, NC, USA). In addition, data of meteorological factors including humidity, temperature, barometric pressure and precipitation during the experimental period were recorded by Jiangxi Agricultural University weather station which is approximately 500 meters away from experimental hives. Foraging data and its relative weather data of next day were analyzed using generalized linear model analysis in R statistical package for testing the correlation between them.

Our results showed that there was a significant difference in foraging duration depending on next day's weather ($F_{1,832} = 47.38$, $P < 0.001$; Fig. III-1-2A), with bees spending longer total times outside the hive on days that were followed by a rainy day compared to those followed by sunny days. Bees also quitted later on days that were followed by a rainy day (delta time followed by a rainy day was significantly shorter than that was followed by a sunny day, $F_{1,819} = 26.10$, $P < 0.001$; Fig. III-1-2B). These suggest that honeybees may attempt to collect more food before rainy days as a food shortage prevention. These behavioral changes possibly can lead to increasing colony productivity prior to inclement weather. Moreover, there was a strongly correlation linked the honeybees' foraging duration and the delta barometric pressure, delta humidity delta temperature and precipitation of next day ($P < 0.01$), while the delta time between sunset and quitting time was in strong relation only with the delta barometric pressure ($P < 0.01$) and precipitation ($P < 0.05$) (Table III-1-1).

**Figure III-1-1** One of the experiment colony, honeybee RFID system and tag-marked workers. a: tag; b: recorder; c: reader. The tag has a diameter of 3 mm and is 0.08 mm thick and weighs 1 mg.

Pervious studies revealed that cats may use their hair to detect slight changes of static electricity in the air during the weather change for weather forecasting (Acharya, 2011). Sharks are extremely sensitive to the barometric pressure even it drops just a few millibars before rain (Toothman, 2008). Honeybees have been shown able to detect changes in carbon dioxide levels, relative humidity, temperature and air pressure (Southwick & Moritz, 1987), and even electric changes before thunderstorms (Warnke, 1976). These findings, and our results in Table III-1-1, are consistent with the interpretation that honeybees may have an ability of measuring changes of meteorological factors such as barometric pressure, humidity and temperature to forecast the coming weather.

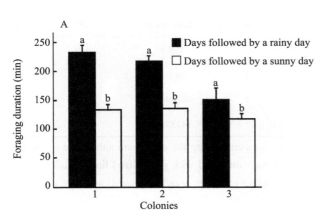
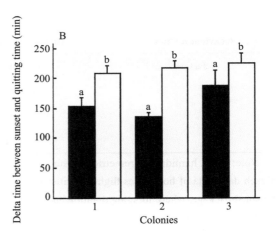

**Figure III-1-2** A. Mean (+SE) foraging duration (summed total time spent on foraging) by foragers on days that were followed by a rainy day (solid) or those by a sunny day (open). There were significant differences between the two types of days and among colonies ($F= 47.38, P < 0.001$; $F= 9.61, P < 0.001$) but interactions between the between foraging duration and colonies were not significant ($F= 2.63, P > 0.05$). Data were analyzed after square root transformation but presented here without transformation. B. Mean (+SE) delta time between sunset and quitting time on days that were followed by a rainy day (solid) or a sunny day (open). Quitting time was defined as the return time of the last foraging trip. Delta time was collected by using daily sunset to subtract the quitting time of each forager. There were significant differences between bees working on the types of days ($F= 26.10, P < 0.001$), but differences among colonies and interactions between the colony and type of days were not significant ($F= 0.41, P > 0.05$; $F= 1.97, P > 0.05$). Data were analyzed after square root transformation but presented here without transformation. Different letters "a" and "b" on the top of each bar indicate significant differences ($P<0.05$, ANOVA test).

The honeybees are widely acknowledged as a model species for the studies of social behavior, social organization and even neurology (Robinson et al., 2005; Qin et al., 2014). Our results showed that honeybees change their habits by increasing foraging duration and working later in the afternoons. Because these changes are at the social level, and not just at individual level, the question remains how honeybee workers can communicate the need to work "harder" after perceiving the predictors for the impending weather changes. Future studies should focus on what factors in weather system are perceived and how bees communicate the need of behavioral changes to the whole colony.

**Table III-1-1  Correlation analysis of weather factors of next day with honeybees foraging duration and delta time between sunset and quitting time**

| Behavioural types | Weather factors | df | Mean square | F | P-Value |
| --- | --- | --- | --- | --- | --- |
| Foraging duration | Δ Humidity | 1 | 283166.06 | 17.85 | <0.01 |
|  | Δ Temperature | 1 | 461890.63 | 29.11 | <0.01 |
|  | Δ Pressure | 1 | 201768.67 | 12.72 | <0.01 |
|  | Precipitation | 1 | 759915.05 | 47.89 | <0.01 |

|  |  |  |  |  | (continued) |
| --- | --- | --- | --- | --- | --- |
| Behavioural types | Weather factors | df | Mean square | F | P–Value |
| Delta time | Δ Humidity | 1 | 279.31 | 0.01 | 0.92 |
|  | Δ Temperature | 1 | 401.37 | 0.02 | 0.90 |
|  | Δ Pressure | 1 | 488494.09 | 18.87 | <0.01 |
|  | precipitation | 1 | 142385.63 | 5.50 | <0.05 |

Note: Delta humidity, barometric pressure and temperature were calculated by the maximum subtracted to the minimum of each day. Data of honeybees flight activities and weather factors were analyzed with generalized linear model analysis in R statistical package, where foraging duration and delta time were dependent variables, weather types of next day were factors and weather factors were covariates.

This work was supported by the Earmarked Fund for China Agriculture Research System (No. CARS-45-KXJ12) and the National Natural Science Foundation of China (No.31360587). We thank Prof. Dr. Zachary Y. Huang and Dr. Qiang Huang for revising the paper, and thank Hao Liu and Haiyan Gan for helping the experiment.

## References

ACHARYA, S. (2011) Presage biology: Lessons from nature in weather forecasting. *Indian Journal of Traditional Knowledge,* 10, 114-124.

BARTLETT, M. S. (1947) The use of transformations. *Biometrics*, 3, 39-52.

BURNETT, A. M., HAYS, K. L. (1974) Some influences of meteorological factors on flight activity of female horse flies (Diptera: Tabanidae). *Ecological Entomology,* 3, 515-521.

CALDERONE, N. W., PAGE, R. E. (1988) Genotypic variability in age polyethism and task specialization in the honeybee, *Apis mellifera* (Hymenoptera: Apidae). *Behavioral Ecology and Sociobiology,* 22, 17-25.

CASAS, J., 1989. Foraging behaviour of a leafminer parasitoid in the field. *Ecological Entomology,* 14, 257-265.

GALACGAC, E. S., BALISACAN, C. M. (2009) Traditional weather forecasting for sustainable agroforestry practices in Ilocos Norte Province, Philippines. *Forest Ecology and Management,* 257, 2044-2053.

GILLOT, C. (2005) Entomology. Springer, Dordrecht, The Netherlands. pp. 678-688.

INWARDS, R. (1893) Weather lore: a collection of proverbs, saying and rules concerning the weather. Elliot stock, London (reprinted 1994, Studio Editions Ltd).

LUNDBERG, H. (1980) Effects of weather on foraging-flights of bumblebees (Hymenoptera, Apidae) in a subalpine/alpine area. *Ecography,* 3, 104-110.

QIN, Q. H., WANG, Z. L., TIAN, L. Q., GAN, H. Y., ZHANG, S. W., ZENG, Z. J. (2014) The integrative analysis of microRNA and mRNA expression in *Apis mellifera* following maze-based visual pattern learning. *Insect Science,* 21(5), 619-636.

ROBINSON, G.E., GROZINGER C.M. AND WHITFIELD C.W. (2005) Sociogenomics: social life in molecular terms. *Nature Reviews Genetics,* 6 (4), 257-270.

SOUTHWICK, E. E., MORITZ, R. F. A. (1987) Effects of meteorological factors on defensive behaviour of honeybees. *International Journal of Biometeorology,* 31, 259-265.

TOOTHMAN, J. (2008) Can animals predict the weather? http://science.howstuffworks.com/nature/climate-weather/storms/animals-predict-weather1.htm.

VICENS, N., BOSCH, J. (2000) Weather-dependent pollinator activity in an apple orchard, with special reference to *Osmia cornuta* and *Apis mellifera* (Hymenoptera: Megachilidae and Apidae). *Environmental Entomology,* 29, 413-420.

WARNKE, U. (1976) Effects of electric changes on honeybees. *Bee World,* 57, 50-56.

## 2. Starving Honey Bee (*Apis mellifera*) Larvae Signal Pheromonally to Worker Bees

Xujiang He[1,3,†], Xuechuan Zhang[2,†], Wujun Jiang[1,†], Andrew B. Barron[3], Jianhui Zhang[2], Zhijiang Zeng[1,*]

1. Honeybee Research Institute, Jiangxi Agricultural University, Nanchang, 330045, China
2. Biomarker Technologies Co., Ltd. Beijing, 101300, China
3. Department of Biological Sciences, Macquarie University, North Ryde, NSW 2109, Australia

Cooperative brood care is diagnostic of animal societies. This is particularly true for the advanced social insects, and the honey bee is the best understood of the insect societies. A brood pheromone signaling the presence of larvae in a bee colony has been characterised and well studied, but here we explored whether honey bee larvae actively signal their food needs pheromonally to workers. We show that starving honey bee larvae signal to workers via increased production of the volatile pheromone E-β-ocimene. Analysis of volatile pheromones produced by food-deprived and fed larvae with gas chromatography-mass spectrometry showed that starving larvae produced more E-β-ocimene. Behavioural analyses showed that adding E-β-ocimene to empty cells increased the number of worker visits to those cells, and similarly adding E-β-ocimene to larvae increased worker visitation rate to the larvae. RNA-seq and qRT-PCR analysis identified 3 genes in the E-β-ocimene biosynthetic pathway that were upregulated in larvae following 30 minutes of starvation, and these genes also upregulated in 2-day old larvae compared to 4-day old larvae (2-day old larvae produce the most E-β-ocimene). This identifies a pheromonal mechanism by which brood can beg for food from workers to influence the allocation of resources within the colony.

Communication between parents and young offspring for food provisioning presents an area of potential conflict between the amount of food requested by young and the optimal resource allocation by the parents. This has become an area of increasing interest to evolutionary biologists[1]. Recent studies mainly focused on parent-offspring conflicts in mammals and avian species[2-4]. In these species since parents are equally related to their offspring they are expected to favour an even division of food to their young[1], however, varying condition of young offspring will also influence allocation decisions. In many species parental provision is in relation to begging intensity as a proxy of offspring need and condition rather than the real food needs of hungry sibs[1]. For example, pigeons (*Columba livia*) and sows (*Sus scrofa*) allocated more food to loudspeakers playing begging calls rather than the hungry young[5,6]. On the other hand, young offspring often try to acquire

---

† These three authors contributed equally to this paper.
*Corresponding author: bees1965@sina.com (ZJZ).
注：此文发表在 *Scientific Reports* 2016, 6:22359。

a disproportionately larger investment from their parents than their sibs by making louder begging calls and attempt to manipulate their parent to obtain more food than the parents' optimal resource allocation[3,4]. Nevertheless the costs of begging calls curtail call exaggeration and help to ensure their honesty[7]. Furthermore, environmental factors also influence the degree of parent-offspring conflict: for instance, a shortage of food could reduce the honesty of begging signals and increase sibling scramble competition[4]; Adding unrelated broodmates increases barn swallow chicks' begging intensity[8]. These influential studies of parent-offspring conflict in mammals and birds illustrate the potential complexity of communication over resource allocation.

As an advanced eusocial insect, honey bees have a radically different social organisation for brood rearing. The single queen in the colony lays all the eggs but young larvae are reared by their adult sisters rather than their parents[9]. During the larval stage, each larva is fed $1.5 \pm 0.2$ times/h by nurse bees not randomly[10,11], and thousands of larvae need to be fed daily[9]. Brood rearing is the primary function of a honey bee colony, but the potential for conflict between brood and nurses over resource allocation in this specialized society has thus far not been explored because it has not been clear whether the largely immobile larvae are capable of begging for food.

Generally, begging signals in vertebrates are body movements and acoustic signals[12-15]. Pervious studies showed that some insect larvae such as wasp larvae (*Vespa orientalis* F) and beetle larvae (*Nicrophorus vesoilloides*) produce acoustic or tactile signals to beg for food[16,17]. It is well documented that larval honey bees produce pheromones capable of altering adult worker behaviour[18-20]. Therefore, it is possible that honey bees could use volatile pheromones as their begging signal to attract workers to them when hungry. This has been demonstrated for bumble bees, although the exact composition of the pheromone remains unknown[21], and the possibility has been suggested for honey bees[22]. Here we explored whether honey bee larvae signal their hunger pheromonally.

Our findings suggest that a volatile component of brood pheromone, E-β-ocimene, is a candidate begging signal for young honey bee larvae. It is actively synthesized by hungry larvae and attracts workers to inspect cells.

## Results

Food deprivation affected 2-day old larvae far more severely than 4-day old larvae (Fig. III-2-1). Because of the high mortality observed in 2-day old larvae after more than 2 h starvation, we limited starvation duration to less than this time for all other experiments.

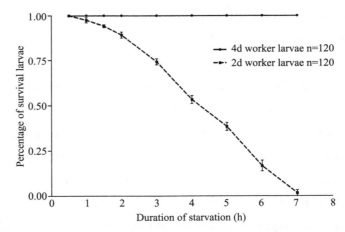

**Figure III-2-1** Survival of worker larvae with different durations of starvation from 0 to 7 hours. 4-day old larvae (solid line and diamonds) survived longer than 2-day old larvae (dash line and squares). Forty 2- and 4-day old larvae were tested and the study was replicated three times. The mean + SE of each time point were presented.

## Experiment 2: NT and GC-MS

Nine chemicals were detected in 2- and 4-day worker larva groups: E-β-ocimene, Myristic acid, Palmitic acid, Methyl palmitic ester, Stearic acid, Palmitoleic acid, Pentadecanoic acid, Acetic acid and Ethyl acetate. These chemicals mapped to the standard chemicals in the NIST32I database with a relative coefficients > 80% for all except Stearic acid which was 78% (Table S1). E-β-ocimene was the only chemical which was significantly higher in all starved larvae groups (SL) compared to the respective fed larvae groups (FL) (Fig. III-2-2, Fig. S1), suggesting that it might be the candidate for the larval hunger signal.

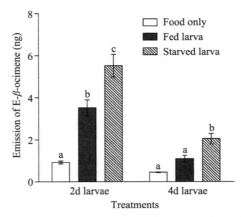

**Figure III-2-2** Emission of E-β-ocimene from SL (striped bar), FL (gray bar) and food only (open bar). Mean + SE presented in each bar. 2- and 4-day old SL released significantly more E-β-ocimene compared to FL (2-day old comparison: $n$=6, $p$=0.013; 4-day old comparison: $n$=6, $p$=0.009). Different letters "a" and "b" on top of bars indicate significant difference (Fisher's PLSD test, $P$<0.05, after ANOVA showed an significant effect), whereas same letter indicates non significant difference (ANOVA $P$>0.05).

## Experiment 3: Effect of E-β-ocimene on worker behaviour

E-β-ocimene increased worker attendance of cells in a dose-dependent manner (Fig. III-2-3A). In addition, workers attended living larvae supplemented with E-β-ocimene more than the larvae, E-β-ocimene alone or wax only control groups (Fig. III-2-3B).

## Experiment 4: RNA sequence analysis

Four libraries were generated from our experimental groups, and summaries of RNA sequencing analyses are shown in Table S2. In each library, more than 97% clean reads were unique reads of which more than 86% reads were paired reads. Very few clean reads (< 2.1%) were multiple mapped reads. Each library had a sufficient coverage of the expected number of distinct genes (stabilized at 3M reads, Table S2). The Pearson correlation coefficient among three biological replicates of each experimental group were all > 0.80 (Table S3), which is a conventionally accepted threshold for replicates indicating that there was acceptable sequencing quality and repeatability among the biological replicates of each group[23].

There were only 1 and 12 genes were significantly differentially expressed between SL and FL in 2-day old and 4-day old comparisons respectively (log2 ratio ≥2 and FDR <0.01, Table S4). We noticed that 3 and 5 genes putatively involved in E-β-ocimene synthesis were slightly upregulated in SL compared to FL in 2-day old and 4-day old comparisons respectively (Fig. III-2-4).

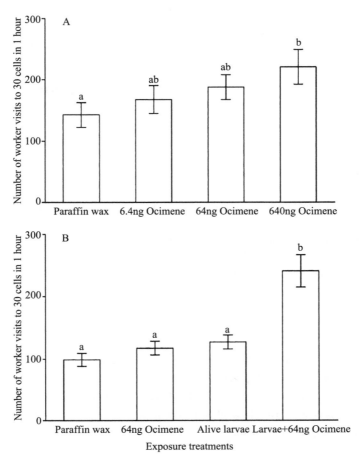

**Figure III-2-3** Total number of worker visits to the cells added with (A) different amounts of E-β-ocimene (Expt 3a) and (B) E-β-ocimene with or without living larvae (Expt 3b). Mean +SE is presented in each bar. Only the 640ng E-β-ocimene treatment has significantly more visits than the control in Expt3a ($n=6$, $p=0.048$), whereas the living larvae supplemented with E-β-ocimene has significantly more visits than the larvae, E-β-ocimene alone or wax only control groups in Expt 3b ($n=8$, $p<0.0001$). Different letters "a" and "b" above bars indicate significant differences ($P<0.05$) between groups (ANOVA followed by Fisher's PLSD test).

Many biosynthetic pathways of E-β-ocimene have been found in animals and plants (Fig. S2). By mapping to the KEGG pathway database, we found that honey bee larvae had a possible biosynthetic pathway for E-β-ocimene involving 9 genes (Fig. III-2-4). It appears that in bees E-β-ocimene might be *de novo* synthesised from a Mevalonate pathway (Fig. III-2-4). Furthermore, a transcription factor (Longitudinals lacking protein-like, *llp-like*) involved in the regulation of genes Farnesyl pyrophosphate synthase *(fps)* and Mevalonate kinase-like *(mk)* could also be involved in the regulation of E-β-ocimene production. None of these 9 genes were differentially expressed between SL and FL when starved for 45 min, nor between 2-day old larvae and 4-day old larvae (Fig. III-2-4). Although differences were not significant, three of the nine genes were expressed far more highly in 2-day old larvae than 4-day old larvae (2-day old larvae produce more E-β-ocimene). Four of these nine genes [Phosphomevalonate kinase-like *(pk-like)*, Geranylgeranyl pyrophosphate synthase-like *(ggps-like)*, *fps* and *aatc-like*] were expressed slightly higher in FL compared to SL (Fig. III-2-4).

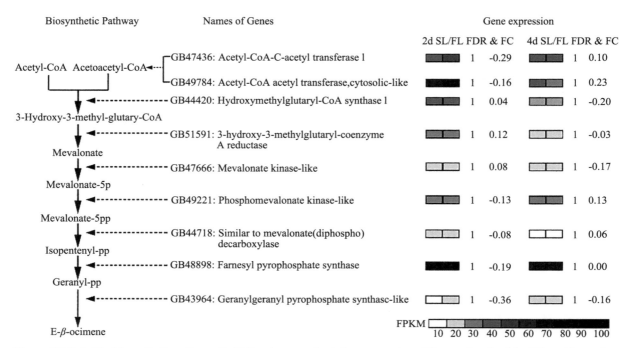

**Figure III-2-4** The biosynthetic pathway for E-β-ocimene and relative expression of 9 genes involved in different honey bee samples. Honey bee larva de novo synthesise E-β-ocimene from Acetyl-CoA and Acetoacetyl CoA (left). The name of each gene (middle) and its relative expression (right) in four worker larva groups (day 2 SL, day 2 FL, day 4 SL and day 4 FL) involved in this pathway were presented. The shade of each gene in each larva group indicates its FPKM value (fragments per kilobase of exon per million fragments mapped).

### Experiment 5: Results of qRT-PCR

Our RT-PCR analyses focused on the putative genes of the E-β-ocimene pathway suggested from Expt 4. None of these genes were significantly differentially expressed between the same age SL and FL that had been starved for 45 minutes (Expt 5a, Fig. III-2-5), which was consistent with the results from Expt 4. However, three of these genes *(llp-like, fps and aatc-like)* were more highly expressed in 2-day old worker larvae compared to 4-day old worker larvae (Fig. III-2-5), which also supported the conclusions of the RNA sequence analyses (Fig III-2-4). Two-day old larvae produce more E-β-ocimene than four-day old larvae (Fig. III-2-2).

We also compared the expression of these nine genes in 2-day old worker larvae that were starved for 0, 15, 30, 45 and 60 min. Three genes were upregulated after 30 min of starvation, and then expression decreased (Fig. III-2-6). The control and other genes showed a different expression pattern (Fig. III-2-6).

## Discussion

In summary, our results have shown that food deprivation increased the amount of the volatile chemical E-β-ocimene produced by worker larvae (Fig. III-2-2), and that 2-day old larvae, which are highly susceptible to starvation, produced more E-β-ocimene than 4-day old larvae. E-β-ocimene increased the number of visits to brood cells by workers (Fig. III-2-3), and hence we propose that E-β-ocimene signals the state of larval hunger to workers.

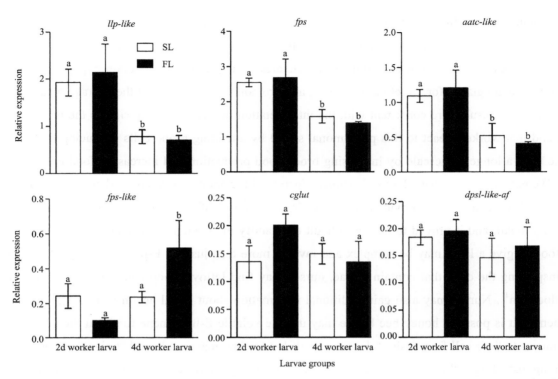

**Figure III-2-5** Comparison of relative gene expression level among 2- and 4-day old SL and FL. Mean + SE presented in each bar. Different letters "a" and "b" on top of bars indicate significant differences ($P<0.05$, ANOVA followed by Fisher's PLSD test).

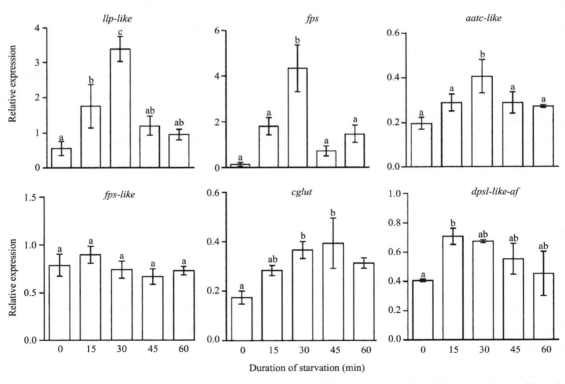

**Figure III-2-6** Expression of 6 genes following different durations of starvation from 0 to 60 min in day-2-old worker larvae. Mean + SE was presented in each bar. Different letters "a" and "b" on top of bars indicate significant differences ($P<0.05$, ANOVA followed by Fisher's PLSD test).

E-β-ocimene was first identified from mated honey bee queens[24] and could promote the acceptance of introduced queens to honey bee colony [25], and then it was detected from young brood and even brood frames[26,27]. A previous study showed that it also accelerates the development of worker hypopharyngeal glands, which are the main glands for secreting royal jelly, and promotes workers to start their foraging earlier[22]. These findings, and our results, are consistent with the interpretation that E-β-ocimene signals the need for food by larvae, and that workers react to this pheromonal signal by attending brood cells producing the pheromone, and also over a longer timescale by increasing brood food production and increasing food collection by the colony. We note, however, that while E-β-ocimene increased the number of visits workers made to brood cells, we did not observe more food provided to these cells. The amount of food given to a worker larva on a single trip is very small, and it proved to be too difficult to quantify the amount in this study. The decision of how much food to give a larva may be complex and involve more than just the E-β-ocimene signal. For example, when inspecting the cells that contain larvae, nurses may find leftover food and this in turn may influence food allocation[11]. Nurses may also gain additional information about larval health and hunger through larval movement[28]. It is possible honey bee larvae may use the volatile E-β-ocimene to signal their food need and attract nurses to their cell from a short distance away, but feeding may involve other signals as well. We noted that adding just 64 ng E-β-ocimene into a cell containing a living 2-day old worker larva dramatically increased worker visits compared to control treatments (Expt 3b, Fig. III-2-3b). However, adding E-β-ocimene alone to a cell was less effective in soliciting a response from workers. Adding 640 ng E-β-ocimene diluted in solid paraffin wax to a cell slightly increased workers' attendance compared to the control (Expt 3a, Fig. III-2-3a). This perhaps suggests the E-β-ocimene and other signals from a living larva may have a synergistic effect on the attendance behaviour of worker bees.

RNA-seq results indicated a KEGG pathway in honey bee larvae for *de novo* synthesis of E-β-ocimene through the mevalonate pathway. This is similar to the monoterpene synthesis pathway of the bark beetle (*Ips pini*)[29]. Three genes were upregulated in 2-day old larvae compared to four day old larvae (Fig. III-2-5), since two-day old larvae produce significantly more E-β-ocimene compared to the four-day olds. Three of the six genes tested from this pathway (Fig. III-2-6) showed elevated expression after 30 minutes of starvation suggesting dynamic genomic regulation of synthesis of E-β-ocimene capable of rapid upregulation in response to a starvation stress. However, there was no significant difference between the gene expression of the FL and SL. This is perhaps because the changes in E-β-ocimene released from a hungry larva and a fed larva (approximately 2 ng and 1 ng in 2-day old and 4-day old larvae respectively) are infinitesimal and not easily reflected in gene expression. Furthermore, both FL and SL were detected producing E-β-ocimene which may have reduced any difference at the gene expression level of analysis. A previous study showed that the highest amount of brood esterifiable pheromones was detected on the anterior part of the larvae, especially high in the salivary gland[30]. As a hydrocarbon and a hunger pheromone, it may be the case that E-β-ocimene is produced from (the anterior part around) the mouth area, however, this needs further investigation.

Begging signals have received a great deal of attention because of their possible roles in parent-offspring conflicts over the resourcing of young, but most of this research has been with vertebrate systems. Examples of begging signals from insects are rare. This is no doubt partly because parental care is also generally rare in insects, however sibling care is common in social insects, and obligatory for advanced social insects. Some examples of begging signals have been reported for social insects. Wasp larvae (*Vespa orientalis* F) use their mandibles to rub against the cell wall to make a scraping sound as a hungry signal[16]. Some ant species larvae such as *Myrmica rubra* and *Gnamptogenys stratula* beg for food by swaying behaviour in which they raise their head and neck, and reach and wave towards workers[31,32]. The cuticular chemicals from hungry bumblebee (*Bombus terrestris*) and earwig (*Forficula auricularia*) larvae attract greater adult attendance suggesting that these cuticular compounds might serve as a pheromonal signal of hunger[21,33]. Our study is the first to chemically identify a pheromonal begging signal in eusocial insects, and to propose an active mechanism whereby larval honey bees signal to workers. While there is no indication of conflict in this signalling system in the present study, it raises the possibility of potential discordance between how much brood signal for food from workers, and how much food is delivered to them, particularly in periods of food shortage for the colony. A previous study showed that hungry larvae were able to attract more inspections and feeding visits than the control group, however, nurses may deposit food only when food quantity within the cell is below a minimum threshold[11]. On the other hand, it is not clear whether brood could use E-β-ocimene to manipulate their food allocation. Therefore, studying the possibility for conflict over larval provisioning in honey bees could prove a fascinating test of general theory of parent-offspring conflict which are very different from those of parent-offspring care in most avian or mammalian systems, and provide important information on how food flow is regulated in honey bee colonies.

## Methods and Materials

### Insects

Experiments 1, 2, 4 and 5 used the standard Chinese commercial strain of western honey bee (*Apis mellifera*). Experiments 1, 4 and 5 used bees from three colonies located at the Honeybee Research Institute of Jiangxi Agricultural University (28.46° N, 115.49 ° E). Experiment 2 used three colonies that were transferred to Nanjing University of Chinese Medicine (32.6° N, 118.56 ° E). Experiment 3 used bees from five colonies of the standard Australian commercial strain of the western honey bee (mostly *Apis mellifera ligustca*), and was conducted at Macquarie University, New South Wales, Australia (33.76 ºS, 151.11 ºE).

### Experiment 1: sensitivity of larvae of different ages to starvation

We examined the impact of food deprivation on 2- and 4-day old worker larvae to determine the severity of starvation treatments and establish the optimal time points for sampling for pheromone and genomic analyses. Forty 2- and 4-day old larvae were placed in a 20 mL plastic tube without food and kept in an incubator at 34 °C. Larvae were inspected under a microscope to check whether they were breathing and moving at 0.5 h, 1 h, 1.5 h, 2 h, 3 h, 4 h, 5 h, 6 h and 7 h. Larvae that were completely still were assumed to be dead. This study was replicated three times.

### Experiment 2: Measurement of volatile chemicals produced by larvae with needle trap sample collection

A needle trap device system from PAS technology, Magdala, Germany was used to sample honey bee larvae pheromones[34]. Twenty 2-day and ten 4-day old worker larvae and their food were collected from their wax cells. Larvae were placed in cells with abundant food (FL), which had been freshly collected from their own cells, or placed in cells with no food (SL). Each group was immediately put in a 20 mL airtight glass tube and kept in an incubator under 35 °C for 45 min after collection from honey bee hives. The whole sample collection procedure took less than five minutes min per group in a 30 °C air conditioning room.

Sampling needles were used to extract 10 mL gas from the sample tube at 5 mL/min, 20 MPa and 35 °C. The first type of needle preferentially absorbed fatty acid and other oxophile pheromones (fatty acid model needles), and the second preferentially adsorbed non-oxophile pheromones [divinylbenzene (DVB), PDMS and carboxen 1000 (CAR 1000) model needles[34]]. In addition to sampling from FL and SL, we also sampled volatiles from cells that contained larval food only as a control group to allow us to distinguish odours from food in our SL and FL samples. Each sample group had six replicates.

For GC-MS analysis two types of chromatographic column (DB-5 and VOC, both were 30 m, 250 μm, 25 μm film thickness, Agilent technolgies) were used for the different needle samples: for the fatty acid needles we used the DB-5 column; for the DVB, PDMS and Carboxen 1000 mixed needles we used the VOC column. The column temperatures were 35 °C for 2 min, then 35 °C to 250 °C at 8 °C/min. then 250 °C for 5 min. The temperature of injection was 250 °C and the pressure of gas helium was 6.7776 psi. Samples were desorbed from extraction needles using 1 mL of helium passed through the needle and into the GC-MS in less than 30 seconds by using a 1 mL sterile syringe (PAS technology). Chemicals identified in GC-MS were then mapped to the NIST32I database.

A preliminary experiment, and previous studies showed that E-β-ocimene was a major volatile chemical from honey bee young larvae[22,26,27]. For calculating the amounts of this pheromone from honey bee samples, we used 1-Nonene (assay ≥ 99.5%, Sigma-Fluka) and E-β-ocimene (assay ≥ 90%, Sigma-Aldrich) as internal and external standard substances respectively. Two μL 1-Nonene injections and six levels of E-β-ocimene (0μL, 1μL, 2μL, 4μL, 8μL and 16μL) were added into 20mL ethyl alcohol to establish a standard curve. Afterward, to determine the efficiency of volatile collection with the needles, 2 μL of ethyl alcohol with 1-Nonene (v/v:10000/1) and six levels of E-β-ocimene (v/v: 10000/0; 10000/0.5; 10000/1; 10000/2; 10000/4 and 10000/8) were injected into 20 mL airtight glass tubes under 35 °C for 45 min and subsequently the volatiles extracted by needles according to the method above. The correlation coefficients of the two standard curves (DB-5 and VOC columns) were 99.16% and 99.27% respectively suggesting that the NT and GC-MS systems (7890A/5975C, Agilent technolgies Inc., Santa Clara, CA, USA) were stable enough for the later quantitative analysis of these pheromone compounds from samples from honey bee larvae.

For the honey bee samples, 2μL ethyl alcohol only with 1-Nonene (v/v:10000/1) was injected into the bottom of the airtight glass tube after adding larval samples and the tube was capped immediately afterward for the duration of the starvation treatment.

### Experiment 3: effect of E-β-ocimene on worker behaviour

Expt 2 suggested E-β-ocimene was produced by hungry larvae. We then examined how worker bees reacted to E-β-ocimene in Expt 3a. E-β-ocimene was diluted in small paraffin wax (Sigma-Aldrich) pellets

20±1.5 mg that resembled the size of a worker larva. These were added individually to cells of empty wax comb. There were four treatments: wax with 6.4 ng E-β-ocimene (which is approximately equal to the E-β-ocimene amount detected from a 2-day old worker larva), wax with 64 ng E-β-ocimene, wax with 640 ng E-β-ocimene and a wax-only control group. For each trial 30 pellets of each type were used: 120 pellets in total. Wax pellets were stuck to the cell bottom so that they could not be easily removed by workers, however a few wax pellets were still cleaned from cells by workers during the assay (Table S5). The frame of comb with these 120 wax pellets was firstly introduced into a bee colony for 10 min to attract workers who climbed onto the frame. The comb frame was then placed in a box along with a frame with 1500 bees, young larvae and a mated queen. The box containing the two frames, approximately 1700-1800 workers and a queen was kept in a dark incubator under 34 °C and the experimental cells containing the wax pellets were observed with an infrared camera. We recorded every incidence when workers placed their heads into cells containing wax pellets for one hour.

In experiment 3b we compared responses of workers to larvae and E-β-ocimene. There were four treatments: comb cells contained either: *i.* 64ng E-β-ocimene diluted in a wax pill as above, *ii.* a wax pill with no E-β-ocimene, *iii.* a living 2-day old worker larva grafted into the cell, *iv.* living 2-day old worker larvae with a 64ng E-β-ocimene pill. For group *iv*, the wax pill was first stuck to the bottom of the cells, and living larvae were then grafted to lie on the pellets. There were 30 cells per group. The experimental cells were then exposed to bees and observed as for Expt 3a above to record numbers of visits by bees to cells. Very few wax pellets and larvae were removed by workers during the observations (Table S5).

### Experiment 4: RNA-Seq analysis of 2 and 4 day old starved and fed larvae

To examine the effects of food deprivation on larval gene expression 2-day and 4-day old worker larvae were placed in a disinfected plastic tissue culture dish in a biochemical incubator (under 35 °C, 75% humidity) for 45 min, and were then immediately flash-frozen in liquid nitrogen. FL lay on their food collected from their cells for this period. SL had no food. For 2-day old larvae samples, approximately 30 larvae were collected for each RNA sequencing sample, while for 4-day larvae samples 6 were collected. 4-day old larvae are much larger than 2-day old larvae and hence fewer larvae were needed for adequate RNA yield. Three RNA samples from pooled larvae were collected for each experimental group (2-day and 4-day old SL and FL). Within a group each sample was collected from a different colony.

Total RNA of each sample was extracted from the honeybee larvae according to the standard protocol of the TRIzol Reagent (Life technologies, California, USA). RNA integrity and concentration were checked using an Agilent 2100 Bioanalyzer (Agilent Technologies, Inc., Santa Clara, CA, USA).

mRNA was isolated from total RNA using a NEBNext Poly(A) mRNA Magnetic Isolation Module (NEB, E7490). A cDNA library was constructed following the manufacturer's instructions for the NEBNext Ultra RNA Library Prep Kit（NEB, E7530）and the NEBNext Multiplex Oligos（NEB, E7500）from Illumina. In brief: enriched mRNA was fragmented into approximately 200nt RNA inserts, which were used as templates to synthesize the cDNA. End-repair/dA-tail and adaptor ligation were then performed on the double-stranded cDNA. Suitable fragments were isolated by Agencourt AMPure XP beads (Beckman Coulter, Inc.), and enriched by PCR amplification. Finally, the constructed cDNA libraries of the honey bee were sequenced on a flow cell using an Illumina HiSeq™ 2500 sequencing platform.

Low quality reads, such as adaptor-only reads or reads with > 5% unknown nucleotides were filtered from subsequent analyses. Reads with a sequencing error rate less than 1% (Q20 > 98%) were retained. These remaining clean reads were mapped to the honeybee (*Apis mellifera*) genome (OGSv3.2) using Tophat2 software[35]. The aligned records from the aligners in BAM/SAM format were further examined to remove potential duplicate molecules. Gene expression levels were estimated using FPKM values (fragments per kilobase of exon per million fragments mapped) by the Cufflinks software[36].

DESeq and *Q*-value were employed and used to evaluate differential gene expression between starved and fed worker larvae[37]. Gene abundance differences between sample groups were calculated based on the ratio of the FPKM values. The false discovery rate (FDR) control method was used to identify the threshold of the *P*-value in multiple tests in order to compute the significance of the differences. Here, only genes with an absolute value of log2 ratio ≥2 and FDR significance score <0.01 were used for subsequent analysis.

Sequences differentially expressed between sample groups were identified by comparison against various protein database by BLASTX, including the National Center for Biotechnology Information (NCBI) non-redundant protein (Nr) database, Swiss-Prot database with a cut-off *E*-value of $10^{-5}$. Furthermore, genes were searched against the NCBI non-redundant nucleotide sequence (Nt) database using BLASTn by a cut-off *E*-value of $10^{-5}$. Genes were retrieved based on the best BLAST hit (highest score) along with their protein functional annotation. KEGG pathways were assigned to the assembled sequences by Perl script.

### Experiment 5a: Quantitative RT-PCR anlaysis of gene expression differences between starved and fed larvae of different ages

Total RNA from the RNA-Seq samples (Expt 4) were used in qRT-PCR validation of putative differences in gene expression between samples. RNA integrity was determined by agarose gel (1%), electrophoresis, and ethidium bromide staining. The purity (260 nm/280 nm ratio between 1.8 and 2.0 for RNA) and the concentration of each RNA sample were measured in triplicate following methods in the limma function of HTqPCR package of R[38]. RNA samples were standardized to a concentration of 1 μg/μL for reverse transcription. cDNA was synthesized using MLV reverse transcriptase (Takara, Japan) according to the manufacturer's instructions. GAPDH-1 was used as an internal 'housekeeping gene' who's expression level was similar within all samples and against which levels of expression of our genes of interest were compared[39,40]. We focused on six target genes for qRT-PCR analysis (Table S6). Four of these genes were from the E-β-ocimene biosynthetic pathway, two control genes unrelated to the E-β-ocimene pathway were selected: Ceramide glucosyltransferase (*cglut*) and Decaprenyl-diphosphate synthase subunit 1-like-*Apis florae* (*dps1-like-af*). Primers for these genes were designed using Primer 5.0 software (Table S6).

qRT-PCR cycling conditions were as follows: preliminary 94 °C for 2 min, 40 cycles including 94 °C for 15 sec, 58.9 °C for 30 sec, and 72 °C for 30 sec. The specificity of the PCR products was verified by melt curve analysis for each sample. For each gene, three biological replicates (with five technical replicate for each biological replicate) were performed.

Control and target genes for each sample were run in the same plate to control for interplate variation. The Ct value for each biological replicate was obtained by calculating the mean of three technical replicates. The relative expression level between FL and SL was calculated using the $2^{-\Delta\Delta Ct}$ formula reported by Liu and Saint[41].

***Experiment 5b: Examination of influence of duration of starvation on gene expression in 2-day old larvae***

Three hundred 2-day old worker larvae were sampled from the one honey bee colony. After sampling, 300 larvae were equally divided into five groups: larvae in the first group were immediately flash-frozen in liquid nitrogen without starving as control. The remaining four groups were placed in a plastic tissue culture dish in an incubator (under 35 °C, 75% relative humidity) and deprived of food for different intervals: 15 min, 30 min, 45 min and 60 min. Afterward, samples were immediately flash-frozen in liquid nitrogen. Each sample had three biological replicates taken from the same three honeybee colonies used for Expt 5a. The RNA extraction, qRT-PCR procedures and 6 target genes were the same as for Expt 5a.

## Statistics

All data from GC-MS, behavioural observation and qRT-PCR of each group were analyzed by ANOVA followed by a Fisher's PLSD test in StatView 5.01 (SAS Institute, Cary, NC, USA).

## Acknowledgement

This work was supported by the Science and Technology Project of Colleges and Universities of Jiangxi Province (KJLD13028), the Earmarked Fund for China Agriculture Research System (CARS-45-KXJ12) and China Scholarship Council (No. 201408360073).

## Author Contributions

ZJZ and XJH designed the experiments, XJH and WJJ performed the experiments, XCZ, XJH and JHZ analyzed RNA-Seq data, XJH, ZJZ and ABB written the paper.

## Additional Information

**Competing financial interests:** the authors declare no competing financial interests.

**Data accessibility:**

RNA-Seq data of 2-day old starved honey bee worker larva: NCBI SRA: SRS1025814

(ftp://ftp-trace.ncbi.nih.gov/sra/sra-instant/reads/BySample/sra/SRS/SRS102/SRS1025814/).

RNA-Seq data of 2-day old fed honey bee worker larva: NCBI SRA: SRS1025817

(ftp://ftp-trace.ncbi.nih.gov/sra/sra-instant/reads/BySample/sra/SRS/SRS102/SRS1025817/).

RNA-Seq data of 4-day old starved honey bee worker larva: NCBI SRA: SRS1025772

(ftp://ftp-trace.ncbi.nih.gov/sra/sra-instant/reads/BySample/sra/SRS/SRS102/SRS1025772/).

RNA-Seq data of 4-day old fed honey bee worker larva: NCBI SRA: SRS1025798

(ftp://ftp-trace.ncbi.nih.gov/sra/sra-instant/reads/BySample/sra/SRS/SRS102/SRS1025798/).

# References

1. KILNER, R., & JOHNSTONE, R. A Begging the question: are offspring solicitation behaviours signals of need?. *Trends Ecol. Evol.* 12, 11-15 (1997).

2. HARPER, A. B. The evolution of begging: sibling competition and parent-offspring conflict. *Am. Nat.* 99-114 (1986).

3. GODFRAY, H. C. J. Signaling of need between parents and young: parent-offspring conflict and sibling rivalry. *Am. Nat.* 1-24 (1995).

4. ROYLE, N. J., HARTLEY, I. R., & PARKER, G. A. Begging for control: when are offspring solicitation behaviours honest?. *Trends Ecol. Evol.* 17, 434-440 (2002).

5. REDONDO, T., & CASTRO, F. Signalling of nutritional need by magpie nestlings. *Ethology*, 92, 193-204 (1992).

6. Mondloch, C. J. Chick hunger and begging affect parental allocation of feedings in pigeons. *Anim. Behavi.* 49, 601-613 (1995).

7. LEECH, S. M., & LEONARD, M. L. Begging and the risk of predation in nestling birds. *Behavi. Ecol.* 8, 644-646 (1997).

8. BONCORAGLIO, G., & SAINO, N. Barn swallow chicks beg more loudly when broodmates are unrelated. *J. Evolution. Biol.* 21, 256-262 (2008).

9. WINSTON, M. L. The biology of the honey bee. Harvard University Press.181-198 (1991).

10. HUANG, Z. Y., & OTIS, G. W. Nonrandom visitation of brood cells by worker honey bees (Hymenoptera: Apidae). *J. Insect. Behav.* 4, 177-184 (1991).

11. HUANG, Z. Y., & OTIS, G. W. Inspection and feeding of larvae by worker honey bees (Hymenoptera: Apidae): effect of starvation and food quantity. *J. Insect. Behav.* 4, 305-317 (1991).

12. KILNER, R. When do canary parents respond to nestling signals of need?. *P. Roy. Soc. Lond. B. Bio.* 260, 343-348 (1995).

13. PRICE, K., HARVEY, H., & YDENBERG, R. O. N. Begging tactics of nestling yellow-headed blackbirds, *Xanthocephalus xanthocephalus*, in relation to need. *Anim. Behav.* 51, 421-435 (1996).

14. LEONARD, M. L., & HORN, A. G. Need and nestmates affect begging in tree swallows. *Behav. Ecol. Sociobiol.* 42, 431-436 (1998).

15. MANSER, M. B., MADDEN, J. R., KUNC, H. P., ENGLISH, S., & Clutton-Brock, T. Signals of need in a cooperatively breeding mammal with mobile offspring. *Anim. Behavi.* 76, 1805-1813 (2008).

16. ISHAY, J., & LANDAU, E. M. Vespa larvae send out rthythmic hunger signals. *Nature* 237, 286-287 (1972).

17. RAUTER, C. M., & MOOREF, A. J. Do honest signalling models of offspring solicitation apply to insects? *P. Roy. Soc. B.* 266, 1691-1696 (1999).

18. LE CONTE, Y. *et al.* Chemical recognition of queen cells by honey bee workers *Apis mellifera* (Hymenoptera: Apidae). *Chemoecology* 5, 6-12 (1994).

19. LE CONTE, Y., SRENG, L., & POITOUT, S. H. Brood Pheromone Can Modulate the Feeding Behavior of *Apis mellifera* Workers (Hytnenoptera: Apidae). *J. Econ. Entomol.* 88, 798-804 (1995).

20. SLESSOR, K. N., WINSTON, M. L., & LE CONTE, Y. Pheromone communication in the honeybee (*Apis mellifera*

L.). *J. Chem. Ecol.* 31, 2731-2745 (2005).

21. DEN BOER, S. P. A. & DUCHATEAU, M. J. H. M. A larval hunger signal in the bumblebee *Bombus terrestris. Insect. Soc.* 53, 369-373 (2006).

22. MAISONNASSE, A., LENOIR, J. C., BESLAY, D., CRAUSER, D., & LE CONTE, Y. *E*-β-ocimene, a volatile brood pheromone involved in social regulation in the honey bee colony (*Apis mellifera*). *PLoS One* 5, e13531 (2010).

23. TARAZONA, S., GARCÍA-ALCALDE, F., DOPAZO, J., FERRER, A., & CONESA, A. Differential expression in RNA-seq: a matter of depth. *Genome res.* 21, 2213-2223 (2011).

24. GILLEY, D. C., DEGRANDI-HOFFMAN, G., & HOOPER, J. E. Volatile compounds emitted by live European honey bee (*Apis mellifera L.*) queens. *J. Insect. Physiol.* 52, 520-527 (2006).

25. DEGRANDI-HOFFMAN, G., GILLEY, D., & HOOPER, J. The influence of season and volatile compounds on the acceptance of introduced European honey bee (*Apis mellifera*) Queens into European and Africanized colonies. *Apidologie* 38, 230-237 (2007).

26. MAISONNASSE, A. *et al*. A scientific note on E-β-ocimene, a new volatile primer pheromone that inhibits worker ovary development in honey bees. *Apidologie* 40, 562-564 (2009).

27. CARROLL, M. J., & DUEHL, A. J. Collection of volatiles from honeybee larvae and adults enclosed on brood frames. *Apidologie* 43, 715-730 (2012).

28. HEIMKEN, C., AUMEIER, P., & KIRCHNER, W. H. Mechanisms of food provisioning of honeybee larvae by worker bees. *J. Exp. Biol.* 212, 1032-1035 (2009).

29. SEYBOLD, S. J., BOHLMANN, J., & RAFFA, K. F. Biosynthesis of coniferophagous bark beetle pheromones and conifer isoprenoids: evolutionary perspective and synthesis. *Can. Entomol.* 132, 697-753 (2000).

30. LE CONTE, Y. *et al*. Larval salivary glands are a source of primer and releaser pheromone in honey bee (*Apis mellifera L.*). Naturwissenschaften 93, 237-241 (2006).

31. CREEMERS, B., BILLEN, J., & GOBIN, B. Larval begging behaviour in the ant *Myrmica rubra. Ethol. Ecol. Evol.* 15, 261-272 (2003).

32. KAPTEIN, N., BILLEN, J., & GOBIN B. Larval begging for food enhances reproductive options in the ponerine ant *Gnamptogenys striatula. Anim. Behav.* 69, 293-299 (2005).

33. MAS, F., HAYNES, K. F., & KOLLIKER, M. A chemical signal of offspring quality affects maternal care in a social insect. *P. Roy. Soc. B.* 276, 2847-2853 (2009).

34. TREFZ, P. *et al*. Needle trap micro-extraction for VOC analysis: effects of packing materials and desorption parameters. *J. Chromatogr. A.* 1219, 29-38 (2012).

35. KIM, D. *et al*. TopHat2: accurate alignment of transcriptomes in the presence of insertions, deletions and gene fusions. *Genome Biol.* 14, R36 (2013).

36. TRAPNELL, C. *et al*. Transcript assembly and quantification by RNA-Seq reveals unannotated transcripts and isoform switching during cell differentiation. *Nat. Biotechnol.* 28, 511-515 (2010).

37. ANDERS, S. & Huber, W. Differential expression analysis for sequence count data. *Genome biol.* 11, R106 (2010).

38. DVINGE, H., & BERTONE, P. HTqPCR: high-throughput analysis and visualization of quantitative real-time PCR data in R. *Bioinformatics* 24, 3325-3326 (2009).

39. BAIER, G. *et al.* Improved specificity of qRT-PCR amplifications using nested cDNA primers. *Nucleic. Acids. Res.* 5, 1329 (1993).

40. MCDONALD, L. J., & MOSS, J. Stimulation by nitric oxide of an NAD linkage to glyceraldehyde-3-phosphate dehydrogenase. *P. Natl. Acad. Sci. USA* 13, 6238-6241 (1993).

41. LIU. W., & SAINT, D. A. A new quantitative method of real time reverse transcription polymerase chain reaction assay based on simulation of polymerase chain reaction kinetics. *Anal. Biochem.* 1, 52-59 (2002).

# 3. The Involvement of a Floral Scent in Plant-honeybee Interaction

Yibo Liu[1,2†], Zhijiang Zeng[1,2†], Andrew B. Barron[3], Ye Ma[1], Yuzhu He[1], Junfeng Liu[1], Zhen Li[1], Weiyu Yan[1,2], Xujiang He[1,2*]

1. Honeybee Research Institute, Jiangxi Agricultural University, Nanchang, Jiangxi 330045, P. R. of China
2. Department of Biological Sciences, Macquarie University, North Ryde, NSW 2109, Australia
3. Jiangxi Key Laboratory of Honeybee Biology and Bee Keeping, Nanchang, Jiangxi 330045, P. R. of China

**Abstract**

Volatile odours from flowers play an important role in plant-pollinator interaction. The honeybee is an important generalist pollinator of many plants. Here, we explored whether any components of the odours of a range of honeybee pollinated plants are commonly involved in the interaction between plants and honeybees. We used a needle trap system to collect floral odours, and GC-MS analysis revealed nonanal was the only component scent detected in 12 different honeybee pollinated flowers and not present in anemophilous plant species. For *Ligustrum compactum* blooming flowers released significantly more nonanal than buds and faded flowers. For *Sapium sebiferum* nonanal release through the day correlated with nectar secretion. Experimentally increasing nectar load in flowers of *Sapium sebiferum*, *Ligustrum compactum* and *Castanea henryi* increased nonanal levels also. Nonanal was also detected in flower nectar and honeys from experimental colonies. Elecroantennogram recordings and behavioral observations showed that untrained honeybees could detect and were strongly attracted to nonanal. We argue that nonanal persists in both honey and nectar odours facilitating a learned association between nonanal and food reward in honeybees.

**Keywords:** honeybees, floral scents, electroantennogram response, nectar, pollination

## Introduction

Most flowering plant species rely on animals for pollination, and have developed a range of mechanisms to attract pollinators to them, commonly offering a form of food reward to the animals (Kiester *et al.* 1984). Many plant species use floral scents or flower colors and as a cue to attract their pollinators, improving successful pollen transfer between flowers and pollination (Chittka and Raine 2006; Belsare *et al.* 2009; Wright and Schiestl 2009; Leonard *et al.* 2010; Zhang *et al.* 2012), particularly in habitats with high diversities of pollinators and plants (Pauw *et al.* 2009). A variety of floral odours have been shown to attract pollinators or

---

† These two authors contributed equally to this paper.
*Corresponding author: hexujiang3@163.com (XJH).
注：此文发表在 *Science of Nature* 2022 年第 3 期。

repel enemies, facilitating the identification of host flowers for pollinators and improvement of pollination efficiency for plants (Schiestl 2010; Chapurlat et al. 2019). For instance, volatile odour phenylacetaldehyde from flowers attracts bumble bees (*Bombus terrestris*) (Knauer and Schiestl 2015), and benzenoids produced by *Dianthus inoxianus* attract their pollinating hawkmoths (*Hyles livornica*) (Balao et al. 2011).

The honeybee (*Apis mellifera*) is an important insect pollinator with excellent intelligence, and also an extreme floral generalist (Menzel and Muller 1995; Klein et al. 2007). Many agriculturally important flower species are visited and pollinated by honeybees, including over 85% of crop plants (Klein et al. 2007). Here we explored whether there was any evidence for a general reliable signal of floral nectar reward status produced by honeybee pollinated flowers.

The most plants' nectar is colorless and concealed within the flower (Rodriguez-Girones and Santamaria 2005, 2006). Floral scents, however, play an important role in attracting bee pollinators to flowers from a distance (Daumer 1956, 1958; Glaettli and Barrett 2008; Wright and Schiestl 2009; Schiestl 2010; Krishna and Keasar 2018). Honeybees have excellent olfactory acuity and outstanding olfactory learning ability and sensitivity (Menzel and Muller 1995; Joerges et al. 1997; Laska et al. 1999; Galizia et al. 1999; Komischke et al. 2002; Scheiner et al. 2005; Galán et al. 2014). Honeybees could not detect flower nectar status through visual at a long distance, therefore, floral odours could be an important cue for honeybees to detect flower nectar status. Honeybees are also excellent learners and possessed of robust long-term memory (Müller 2002; Hourcade et al. 2009; Eisenhardt 2014). They rapidly learn to identify the most rewarding flowers (Bhagavan and Smith 1997; Laska et al. 1999; Chittka and Raine 2006; Wright et al. 2002; Wright et al. 2009). As a highly eusocial species bees develop inside a colony feeding on honey which has been generated from nectars gathered by previous generations of foragers (Seeley 1985). If a nectar odour was able to persist in honey stored within a colony, bees may learn to associate the components of honey odours with feeding even before they begin to forage. This could predispose forager bees toward flowers possessed of that odour in their nectar. Hence the social structure of the hive could facilitate transgenerational fidelity of a plant pollinator relationship via long-lasting odour cues in nectars.

To explore this hypothesis, we examined whether there were any specific odours that were common to honeybee pollinated flowers and that were also present in stored honey, and how honeybees reacted to those odours. We collected floral scents from 15 honeybee pollinated plant species and 3 anemophilous species using a needle trap and a gas chromatography-mass spectrometry (GC-MS) system to investigate whether there were any common components to the odours of in honeybee visited plants. We investigated the odour profiles of buds, blooming flowers and faded flowers, and related flower odour profiles to time of day and nectar amount. We also sampled honey odours. Finally, we tested how honeybees responded behaviourally and electrophysiology to candidate odour cues. Honeybees' behavioral and electrophysiological response to candidate odour cues were tested.

## Materials and Methods

### *Insects and plants*

Honeybees (*Apis mellifera*) were sourced from five colonies at Jiangxi Agricultural University, Nanchang,

China (28.46 °N, 115.49 °E). Each colony contained a mated queen and 9 frames with approximately 30,000 workers.

Plant species were selected for this study from the list of known honeybee pollinated species described in "China nectar and pollen plants" (Xu 1992). *Litchi chinensis* and *Mangifera indica* were sourced from Fuzhou city, Fujian province, China. All other plant species were sourced from Nanchang city, Jiangxi province, China.

### *Experiment 1: qualitative analysis on floral scents*

All experiments were performed on sunny days (temperature: 20-30 °C, wind speed < 5 m/s). Prior to sampling, needles from the needle trap system (fatty acid type, PAS technology, Germany) were inserted into the GC-MS at 250 °C for 2 hrs to desorb any chemical residues. Female or hermaphroditic flowers from all plant species were wrapped in a plastic bag (25cm * 35cm) for 15 min. Five to hundreds of flowers were bagged depending on the flower size and structure for each species. A needle was used to adsorb floral scents. Needles were inserted into the plastic bag surrounding the flower bract to extract 50 mL of air with an airflow speed of 15 mL/min to sample floral odours. Three independent samples were taken for each plant species, each sample from a different bag of flowers.

The needle with adsorbed floral scent was injected into a GC-MS system (5977B-7890B, Agilent Technologies) with a DB-5MS chromatographic column (30 m, 0.25 mm, 0.25 μm film thickness, 112-5532, Agilent technologies). GC-MS analysis methods are fully described in He *et al.* (2016). Briefly, needles were connected to a sterile syringe (PAS technology). The syringe drove 1 mL pure helium through the needle to desorb floral scents into the injection port of the GC-MS system at 250 °C. The column temperature profile was 35 °C for 2 min, then increasing from 35 °C to 240 °C at 5 °C /min. Then the column temperature held at 240 °C for 5 min. The helium pressure was 6.7776 psi and the electron impact ion source (EI) was 70 eV. GC-MS data were mapped to the NIST 17.0 database. Gas from plastic bags without flowers was also sampled (replicated three times) as controls.

### *Experiment 2: quantitative analysis of nonanal from flowers*

Data from experiment 1 indicated nonanal as the only odour common to all bee pollinated flowers sampled. To measure the daily variation in amount of nonanal from *Sapium sebiferum*, flowers were sampled at 6:00 a.m. (temperature: 25°C), 9:00 a.m. (27°C), 12:00 a.m. (29°C), 3:00 p.m. (32°C) and 6:00 p.m. (29°C) on 19th June 2019. In Experiment 2 each sample contained 12 flowers, which were placed into a 250 mL glass bottle containing 2 μL 1-octanol (2 mL/L, purity > 99.8%, Xilong Science, China) as an internal standard. Flowers in bottles were kept under room temperature (25 °C) for 15 min. A needle was used to extract 50 mL gas from the bottle at an airflow speed of 15 mL/min. And floral scents were injected into GC-MS system as in Experiment 1 for quantitative analysis, but here we used a different temperature profile for the column to shorten detection time. The column temperature was programmed as follows: 40 °C for 2 min, then rising to 150 °C at a rate of 8 °C/min, and then rising to 250 °C at a rate of 16 °C/min, finally held constant at 250 °C for 5 min. Other GC-MS procedures were same as above, and are detailed in He *et al.* (2016).

Using this method, we also compared the amount of nonanal released from blooming flowers (250 flowers per sample), buds (250) and faded flowers (250) of *Ligustrum compactum* (Fig. 1B). This study was repeated three times.

A standard curve was established for quantitative analysis of nonanal. Two standard chemicals nonanal (purity: 95%) and 1-octanol were purchased from Sigma-Aldrich (the United States of America) and Xilong Science (China). One μL 1-octanol injections and six levels of nonanal (0 μL, 0.01 μL, 0.1 μL, 1 μL, 10 μL and 100 μL) were added into 5 ml ethanol (purity > 99.8 %, Xilong Science). Selected ion monitoring chromatograms (SIM) were reconstructed for nonanal and the internal standard 1-octanol. These generated major peaks at the mass-to-charge ratio (m/z) 56 and 57 respectively. The correlation coefficient of the standard curve was 0.9998 (Fig. S1). By reference to the standard curve, we were able to quantify the amount of nonanal in floral odour samples.

### Experiment 3: measurement of nonanal and nectar from flowers

To explore the relationship between amount of nectar and amount of nonanal in flowers, for two plant species (*Sapium sebiferum* and *Ligustrum compactum*) we manipulated nectar amount and measured nonanal amount. Branches of flowers were wrapped with plastic mesh net (mesh size 5 mm$^2$) for 2 hrs to exclude insect visits. This prevented pollinators to visit the wrapped flowers but allowed air flow to across the mesh net. Control branches were not wrapped. Twelve *Sapium sebiferum* and 250 *Ligustrum compactum* flowers were cut and placed into 250 mL glass bottles with 2 μL 1-octanol internal standard as in Experiment 2. Air from these samples was collected and analyzed by needle trap and GC-MS systems as in Experiment 2. Three measures were made for each treatment of each species.

To confirm whether the nectar volume in *Ligustrum compactum* flowers would increase after branches enveloped by plastic mesh net for 2 hrs, 30 flowers were picked at random in the mesh net and control branches and the nectaries were imaged with a dissecting microscope (Nanjing Jiangnan novel optics Co., Ltd, SE2200). Each flower has one nectary. We scored how many of the 30 nectaries had nectar droplets. Each group had 6 replicates. *Sapium sebiferum* flowers have many nectaries. For this species we randomly sampled 30 nectaries from a flower and scored how many nectaries had visible nectar droplets. Each group had 30 replicates.

To exam whether nonanal is directly from nectar droplets, we selected *Castanea henryi* flowers (this plant had the highest proportion of nonanal in floral scents, see Table III-3-1) as experimental materials. Blooming flowers of the experimental group were wrapped by plastic mesh net as in Experiment 2 for 2 hrs from 7:00 a.m. to 9:00 a.m., and nectar droplets were collected using precalibrated pipettes (5 μL; inner diameter: 0.4 mm; A. Hartenstein GmbH, Germany). Nectar from 5 clusters of flowers was collected as one sample and 7 biological replicates were sampled. Nectar from unwrapped flowers from same trees was sampled as control. Nectar droplets were blown into $H_2O$ (5 mL) with $N_2$ gas, added with 2 μL 1-octanol as internal standard. The floral scents in nectar were extracted with $CH_2Cl_2$ (5 mL) and shaken for 10 min using an oscillator (200 r/min). The $CH_2Cl_2$ layer was transferred into a new vial and concentrated under nitrogen flow to about 100 μL. A 1 μL sample was injected into the GC/MS. The GC-MS method (SIM method) was same as Experiment 2. A standard curve for this analysis was established and the correlation coefficient was 0.9989. Using the same methods, we also examined nonanal in *Brassica campestri*s nectar droplets.

*Experiment 4: measurement of nonanal from honeys*

Honey samples were collected from 5 research hives. For each sample we extracted 5 mL honey into a 20 mL headspace vial. These vials were placed in a water bath (Tianjin Laiyuenage Co.Ltd.) under 80 °C for 15 min. A needle was used to extract 10 mL of air with an airflow speed of 2 mL/min from a honey sample and were injected into GC-MS system for qualitative analysis same as Experiment 1. The column temperature profile used here was 40 °C for 2 min, then 40 °C to 250 °C at 8 °C /min, then 250 °C for 5 min. From each hive we took three honey samples. In total 15 samples were collected.

*Experiment 5: honeybee antennal response to nonanal*

To determine whether honeybees could detect nonanal, pollen forager honeybees were captured at the hive entrance, and the left antenna was cut for electrophysiological recording by electoantennogram (using a Syntech EAG platform, Germany). The five odours used in Experiment 4 were presented to each antenna: nonanal, benzaldehyde diethyl acetal (purity > 98 %, Tokyo chemical industry), benzaldehyde (purity > 99 %, Tokyo chemical industry), linalool (purity ≥ 95 %, Sigma-Aldrich), and phenylacetaldehyde (purity ≥ 95 %, Sigma-Aldrich). Chemicals were dissolved in ethanol (Brockmann *et al.* 1998), each at five different concentrations (0.08 mL/L, 0.4 mL/L, 2 mL/L, 10 mL/L and 50 mL/L). Ethanol alone served as a control. A filter paper strip was added 10 μl of an odour solution and placed in a 1 cm Pasteur pipette (15 cm). The honeybee antenna was connected to detection electrodes and placed 0.5 cm away from the tip of the pipette in a high humidity continuous air flow at 10 ml/s into which a stimulus pulse of the odour was added for 2 s at 5 mL/s. Two antennal responses to each odour solution were recorded, and five different antennae were used for each chemical. In total 25 antennae were used.

Antennae had the strongest responses to 2 mL/L concentration of the odourants (Fig. Ⅲ-3-4A-E), in consequence, we repeated the study focusing on this dose only using further 10 antennae. Each honeybee antenna was tested with five compounds, as above, and each compound repeated twice.

*Experiment 6: behavioral responses of honeybees to floral odours*

To assess honeybee preferences for different floral odours, color marked honeybee foragers were trained to feed on sucrose solution (30% - 50% depending to the weather and temperature) from an artificial flower placed in the center of a 50 cm diameter circular platform. Artificial flowers were similar to those used by Andrew *et al.* (2014): an inverted orange bottle cap (diameter: 34.4 mm) with four holes (diameter: 4.4 mm). Sucrose could be drunk from a shallow well in the middle of each cap, and odours dispersed from 10 μl droplets of odour solution placed beneath each cap.

Forager bees were caught randomly at the entrance of one of the same five research hives from which we sampled honey (Experiment 4) and relocated to the artificial flower on the circular platform. Once the focal bee had returned by themselves to the platform, they were marked with coloured paint on the thorax. For an odour test five artificial flowers were placed equidistant around the edge of the platform. All artificial flowers offered sucrose, but each artificial flower offered a different odour: nonanal, benzaldehyde diethyl acetal, benzaldehyde, linalool, and phenylacetaldehyde. These were the main floral compounds from honeybee-visiting flowers (Table S1). Odours were diluted in paraffin oil (Spectrum pure, Sigma-Aldrich) at a concentration of 10%, following

Andrew et al. (2014).

Bees were tested individually. The first artificial flower each marked forager landed on was recorded. Artificial flowers were replaced after each visit to avoid any marking pheromones from previous foragers. The platform was randomly rotated to avoid any possible direction preference of foragers. Foragers were captured after their first landing so each visit scored was from a different bee, and bees could not influence each other's choices.

*Data anaylsis*

Nonanal amounts and numbers of flowers or nectaries with nectar drops (Fig. III-3-2 and Fig. III-3-3) were analyzed with one-way ANOVA (Statview 5.01 package, SAS Institute Inc., the United States of America). EAG response data (Fig. III-3-4 and S2) were analyzed by ANOVA test followed by Fisher's PLSD test (Statview 5.01). Artificial flower preferences of bees (Fig. III-3-5) were analyzed with chi-square tests (SPSS 17.0 package, IBM, the United States of America). Nonanal amounts (Fig. III-3-1) were normalized by square-root transformation and were analyzed by ANOVA test followed by Fisher's PLSD test (Statview 5.01).

# Results

*Nonanal was the only one floral scent these detected in all 12 honeybee-visited flower species share*

Qualitative analysis of flower scents of 12 honeybee visited plant species indicated that nonanal was the only common component detected in 12 plants' floral odours and not identified in the three anemophilous flowers, which are not insect pollinated (Table III-3-1 and S1).

*Nonanal levels peaked at 9:00 am and 15:00 pm and in blooming flowers*

For *Sapium sebiferum*, nonanal release peaked at 9:00 a.m. and 3:00 p.m.. Nonanal at 15:00 pm was significantly higher than that at 6:00 p.m. (Fig. III-3-1A). Open flowers of *Ligustrum compactum* released significantly more nonanal than either buds or faded flowers (Fig. III-3-1B).

Table III-3-1 The relative percentage composition of nonanal detected from floral odours and honey odours by GC-MS

| Plant species | Nectar secretion | Relative percentage composition (%, mean ± SE) |
|---|---|---|
| *Brassica campestris* | +++ | 11.96±5.33 |
| *Astragalus sinicus* | +++ | 28.21±5.21 |
| *Litchi chinensis* | +++ | 16.10±7.70 |
| *Castanea henryi* | ++ | 34.03±18.44 |
| *Lamium album* | ++ | 8.56±2.46 |
| *Ligustrum compactum* | ++ | 4.96±3.06 |
| *Citrus sinensis* | ++ | 2.13±0.27 |
| *Ligustrum quihoui* | + | 1.57±0.63 |
| *Mangifera indica* | + | 1.73±1.15 |

(continued)

| Plant species | Nectar secretion | Relative percentage composition (%, mean ± SE) |
|---|---|---|
| *Rosa multiflora* | + | 1.31 ± 0.62 |
| *Cinnamomum bodinieri* | + | 2.91 ± 0.69 |
| *Lycopersicon esculentum* | + | 0.20 ± 0.087 |
| *Pinus massoniana* | – | NA |
| *Zea mays* | – | NA |
| *Juglans cathayensis* | – | NA |
| Honey | NA | 6.24 ± 0.93 |
| *Brassica campestris* nectar | +++ | 6.95 ± 0.58 ng/uL |

Note: Each plant species had three biological replicates. "+++" indicates high nectar secreting plants that can harvest > 10 kg honey per hive, "++" indicates medial nectar secreting plants that can harvest several kilogram honey per hive, "+" indicates low nectar secreting plants that can harvest very little honey and "-" indicates non-nectar plants, according to a book "China nectar and pollen plants"(Xu 1992). "NA" indicates not available. Nonanal amounts are expressed as relative peak area, and represent as Mean ± SE. Data is from Table S1.

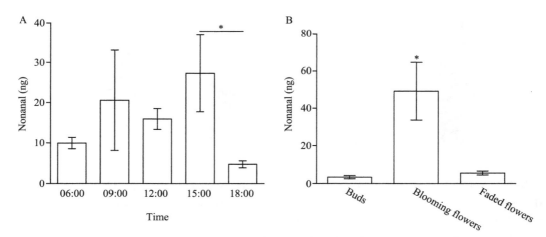

**Figure III-3-1** A: Nonanal amount from *Sapium sebiferum* flowers at five time points on a sunny day. Each bar means mean ± SE of nonanal from samples of 12 cluster of flowers. Each group had three biological replicates. Data were normalized by square-root transformation and were analyzed by ANOVA test followed by Fisher's PLSD test. "*" represents significant difference ($p < 0.05$). B: Nonanal amounts from buds, blooming and faded flowers of *Ligustrum compactum*. Each bar shows mean ± SE of nonanal amount from samples of 250 flowers, and each group had three biological replicates. Data were normalized by square-root transformation and were analyzed by ANOVA test followed by Fisher's PLSD test. "*" represents significant difference ($p < 0.05$), and no "*" represents no significant difference.

### *Netted flowers released significantly more nonanal*

For *Sapium sebiferum* and *Ligustrum compactum*, flowers wrapped with a mesh net for 2 hrs had significantly more nectar droplets than unwrapped controls (Fig. III-3-2C). Nonanal amounts were higher in wrapped flowers than unwrapped flowers for both species (Fig. III-3-2D). Furthermore, nonanal was directly detected from *Castanea henryi* nectar droplets, and experimentally manipulated flowers had significantly higher nectar and nonanal amounts compared to the control (Fig. III-3-3). These results suggest that nonanal was correlated with nectar status of flowers.

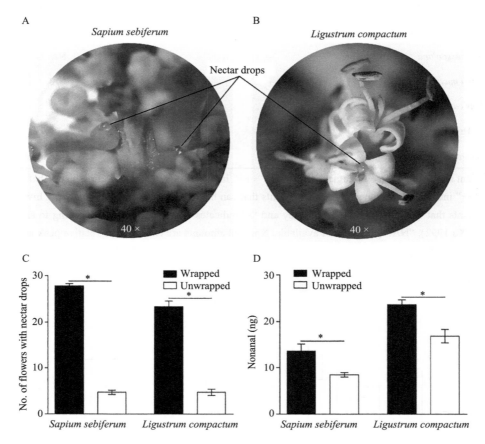

**Figure III-3-2** A: Nectar drop on *Sapium sebiferum* nectaries viewed under microscope. Flowers were magnified 40 times. B: A nectar drop in a *Ligustrum compactum* flower under microscope. Flowers were magnified 40 times also. C: Number of flowers or nectaries of *Sapium sebiferum* and *Ligustrum compactum* with nectar drops from samples of 30 nectaries studied. Each group in *Sapium sebiferum* and *Ligustrum compactum* flowers had 30 and 6 replicates respectively. D: Nonanal amounts from netted and control flowers. Twelve cluster of *Sapium sebiferum* flowers and 250 *Ligustrum compactum* flowers of the netted and control groups were sampled and analyzed by needle trap and GC-MS. Each group had three biological replicates. Each bar shows the mean ± SE. Data were analyzed using one-way ANOVA test. "*" represents significant difference ($p < 0.05$) between two groups.

### Nonanal detected in honey odour and flower nectar

In total 13 volatile components were identified from honey samples (Table S1). The largest component was phenylacetaldehyde (27.32±4.79, mean±SE, percentage composition). Nonanal was detected from all samples and ranked as the seventh most abundant component of the odours (6.24±0.93, mean±SE, %). In additon, nonanal also existed in *Brassica campestris* nectar droplets and was 6.95±0.58 μg/mL (Table III-3-1).

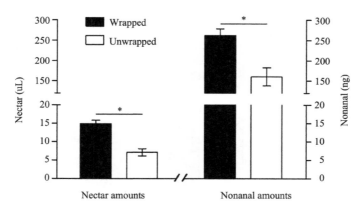

**Figure III-3-3** Nectar and nonanal amounts from *Castanea henryi* flowers. The left bars were nectar amounts from 5 clusters of flowers and the right bars were nonanal amounts from the relative nectar droplets. Each bar shows the mean±SE. Data were analyzed using one-way ANOVA test. "*" represents significant difference ($p < 0.05$) between two groups.

### *Honeybee antennae are sensitive to nonanal*

EAG responses to nonanal and benzaldehyde were greatest at the 2 mL/L odour concentration (Fig. S2). EAG responses to nonanal were greater than to four other common floral odours and the control (Fig. III-3-4).

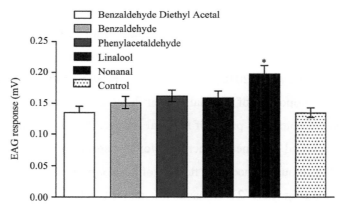

**Figure III-3-4** EAG response of honeybees to 5 floral components. Antenna response to the 2 mL/L concentration of each of the five compounds. Each bar shows the mean ± SE of EAG response. Data were analyzed using ANOVA tests followed with Fisher's PLSD test. Different letters represent significant difference ($p < 0.05$), and same letters represent no significant difference.

### *Nonanal was the most attractive odour to honeybees*

Five main floral components were selected for a preference test with honeybee foragers. Artificial flowers scented with nonanal were more attractive to foragers than four other common floral scents (Fig. III-3-5).

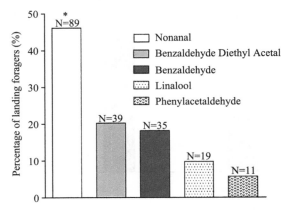

**Figure III-3-5** The attractiveness of 5 floral odours to honeybees using mimic flowers. Each bar shows the number of foragers that landed on each of the five odours in the preference test expressed as a percentage of the 193 foragers tested in this study. Choice data were compared to the nonanal group with a chi-square test. "*" represents significant difference ($p < 0.05$) between nonanal and other four groups.

## Discussion

Here we explored whether honeybee pollinated plants have a common odour that may play a role in plant and honeybee interaction. We analyzed all floral scents from 12 honeybee pollinated plant flowers. We found nonanal was the only component common to the floral bouquets of all 12 honeybee pollinated plants measured (Table III-3-1 and S1). Nonanal is common in floral scents of honeybee visited plants (Jørgensen *et al.* 2000; Alissandrakis *et al.* 2011; Twidle *et al.* 2018). We noted a correlation between amount of nectar offered by different plant species and the proportion of nonanal in their floral scents (Table III-3-1 and S1).

For *Sapium sebiferum* we noted two peaks of nonanal release: 09:00 a.m. and 3:00 a.m. (Fig. III-3-1A), which coincided with the recorded peak nectar secretion for this species (Xu 1992). Moreover, we found that experimentally increasing the amount of nectar in flowers of both *Sapium sebiferum* and *Ligustrum compactum* increased amounts of nonanal (Fig. III-3-2). Nonanal was directly detected from *Castanea henryi* nectar, and experimentally manipulating flowers significantly increased both nectar and nonanal amounts (Fig. III-3-3). Therefore, both within and between honeybee pollinated flower species we documented a relationship between the amount of nonanal in floral odours and nectar availability. Nonanal elicited a robust EAG response (Fig. III-3-4). A similar EAG result was previously reported for *Apis cerana* that is more sensitive to nonanal than other floral odours such as linalool and decanal (Luo *et al.* 2013). Nonanal was also found at significant levels in honey odours (Table III-3-1 and S1) and it was found to be highly attractive to untrained honeybees sampled from these colonies (Fig. III-3-5). Therefore, we are confident that honeybees can perceive nonanal and respond behaviorally to it.

Previously, nonanal has been found in both flowers of *Castanea sativa* and *Eucalyptus globus* and the honeys produced from these flowers (Alissandrakis *et al.* 2011). Naef *et al.* (2004) showed that nonanal exists in both nectar and the content of the bee honey stomach. Our results also detected nonanal in nectar droplets

(Fig. III-3-3 and Table III-3-1) and honey samples (Table III-3-1), which is consistent with many previous studies (Alissandrakis *et al.* 2007; Jerkovi *et al.* 2009; Bavraktar and Ur 2011). Therefore, these indicate that nonanal is present within nectar, and persists in honey, but the role of nonanal in the plant-pollinator interaction wasn't noticed because of the previous studies might not have sampled that many species simultaneously.

Interestingly, nonanal was not the highest component of honey odour by any means: phenylacetaldehyde was (Table S1). Even so, nonanal was more attractive to untrained forager honeybees than phenylacetaldehyde (Fig. III-3-4 and Fig. III-3-5). Phenylacetaldehyde was detected in the floral bouquet of just 2 of the 12 flower species we sampled, whereas nonanal was common to all (Table S1). It appears that a range of diverse honeybee pollinated plants advertise their current nectar status by the same volatile chemical that is highly attractive to honeybees.

Nonanal is a very common volatile substance existing in the world, and it can be produced by plants and microorganisms etc (Abanda-Nkpwatt *et al.* 2006; Baldwin 2010; Wu *et al.* 2014). Our results detected nonanal in nectar droplets (Fig. III-3-3 and Table III-3-1) and honey samples but not in blank control and flower samples from anemophilous plant species (Table III-3-1). Nonanal has been detected in dozens of flowers and honey samples (Jørgensen *et al.* 2000; Schade *et al.* 2001; Alissandrakis *et al.* 2007, 2011; Jerkovi *et al.* 2009; Bianchi *et al.* 2011; Bavraktar and Ur 2011; Twidle *et al.* 2018). Nonanal has been found in both flowers of Castanea sativa and Eucalyptus globus and the honeys produced from these flowers (Alissandrakis *et al.* 2011). Naef *et al* (2004) showed that nonanal exists in both nectar and the content of the bee honey stomach, suggesting that nonanal can be transferred from nectar into honey in beehives. In addition, the concentration of nonanal in nectar ($6.95\pm0.58$ μg/mL) is dramatically higher than that in an in-door air pollution ($<100$ μg/$m^3$) (Daisey and Hopke 2007; Iwashita and Hibino 2011; Iwashita and Tokunaga 2012), suggesting nonanal of honey couldn't come from air. Therefore, we are confident that nonanal is present within nectar, and persists in honey rather an environmental pollutant.

Honeybees are excellent olfactory learners and can learn both complex odour blends and components contained therein (Bhagavan and Smith 1997; Laska *et al.* 1999; Wright *et al.* 2002, 2009). With this capacity bees could certainly learn features of specific flower types to maintain floral constancy (Wright and Schiestl 2009). A generalizable cue associated with food information that operated across flower types could increase the speed of gathering food at which a bee could maximize foraging efficiency on a new flower species. This would benefit the forager bee, and also improve pollination by enhancing fidelity to the new flower type (Arenas *et al.* 2019; Wright and Schiestl 2009). Honeybees also share nectar after foraging and recruit to rewarding flowers via their dance language (Farina *et al.* 2007). Moreover, nurse bees deliver characteristic information of food to young larvae and young bees in the whole colony. New bees thereby obtain this food characteristic information by excellent learning and memory abilities before foraging (Bhagavan and Smith 1997; Laska *et al.* 1999; Wright *et al.* 2002, 2009). Invididually distinctive flower odours would allow new recruits to identify the advertised flower type, but a generalizable cue of nectar availability would facilitate the new recruits to obtain reward from the flowers. In this case the response of foragers to nonanal could be facilitated by the shared social environment of successive generations of forager bees, and their excellent learning abilities. It is possible forager bees begin foraging having already learned to associate nonanal with food by being fed with honey which contains nonanal. The ubiquitous association of nonanal and honeybee pollinated flower species would consistently reinforce that association.

In conclusion, this study demonstrated that the floral scent nonanal is correlated with nectar status across a range of honeybee pollinated plants. The attraction of bees to nonanal appears to be facilitated by nonanal persisting in the odours of honeybees feed on in the hive. Nonanal seems to play an important role in the plant-honeybee interaction, but this interaction is far more complex and poorly understood. Not only floral scents but also flower shapes, colors and ultraviolet signal (Moyroud *et al.* 2017) may also involve in this complex plant-pollinator interaction, which needs deeper investigations.

## Acknowledgements

We thank Dr. Guangli Wang for suggesting on EAG experimental design, Dr. Shengliang Liao for help in GC-MS experiments, and Dr. Xinjian Xu for help in floral scent collection.

## Funding

This work was supported by the National Natural Science Foundation of China (No. 31702193), Natural Science Foundation of Jiangxi province (No. 20171BAB214018), Key Research and Development Project of Jiangxi province (No. 20181BBF60019) and the Earmarked Fund for China Agriculture Research System (No. CARS-44-KXJ15).

## Author Contributions

XJH and ZJZ designed the experiments, YBL and XJH conducted the experiments, YM, YZH, JFL, ZL and WYY for help in performing the experiments, XJH, YBL, ABB and ZJZ wrote and revised the paper.

## Ethics Declarations

The study was performed in accordance with the relevant legal requirements in China.

## Conflict of Interest

We have declared that no conflict of interests exists.

## References

ABANDA-NKPWATT D, KRIMM U, COINER HA, SCHREIBER L, SCHWAB W (2006) Plant volatiles can minimize the growth suppression of epiphytic bacteria by the phytopathogenic fungus *Botrytis cinerea* in co-culture experiments. *Environ. Exp. Bot.* 56, 108-119. https://doi.org/10.1016/j.envexpbot.2005.01.010.

ALISSANDRAKIS E, TARANTILIS PA, HARIZANIS PC, POLISSIOU M (2007) Comparison of the volatile composition in thyme honeys from several origins in Greece. *J. Agric. Food Chem.* 55, 8152-8157. https://doi.org/10.1021/jf071442y.

ALISSANDRAKIS E, TARANTILIS PA, PAPPAS C, HARIZANIS PC, POLISSIOU M (2011) Investigation of organic extractives from unifloral chestnut (*Castanea sativa* L.) and eucalyptus (*Eucalyptus globulus* Labill.) honeys and flowers to identification of botanical marker compounds. *LWT--Food Science and Technology* 44, 1051. https://doi.org/10.1016/j.lwt.2010.10.002.

ANDREW SC, PERRY CJ, BARRON AB, BERTHON K, PERALTA V, CHENG K (2014) Peak shift in honey bee olfactory learning. *Anim. Cogn.* 17, 1177-86. https://doi.org/10.1007/s10071-014-0750-3.

ARENAS A, KOHLMAIER MG (2019) Nectar source profitability influences individual foraging preferences for pollen and pollen-foraging activity of honeybee colonies. *Behav. Ecol. Sociobiol.* 73, 34. https://doi.org/10.1007/s00114-006-0176-0.

BALAO F, HERRERA J, TALAVERA S, DOTTERL S (2011) Spatial and temporal patterns of floral scent emission in *Dianthus inoxianus* and electroantennographic responses of its hawkmoth pollinator. *Phytochem.* 72, 601-609. https://doi.org/10.1016/j.phytochem.2011.02.001.

BALDWIN IT (2010) Plant volatiles. *Curr. Biol.* 20, 392-397. https://doi.org/10.1016/j.cub.2010.02.052.

BAYRAKTAR D, UR TAO (2011) Investigation of the aroma impact volatiles in Turkish pine honey samples produced in Marmaris, Data and Fethiye regions by SPME/GC/MS technique. *Int. J. Food Sci. Tech.* 46, 1060-1065. https://doi.org/10.1111/j.1365-2621.2011.02588.x.

BELSARE PV, SRIRAM B, WATVE MG (2009) The co-optimization of floral display and nectar reward. *J. Biosci.* 34, 963-967. https://doi.org/10.1007/s12038-009-0110-7.

BHAGAVAN S, SMITH BH (1997) Olfactory conditioning in the honey bee, *Apis mellifera*: effects of odor intensity. *Physiol. Behav.* 61, 1-117. https://doi.org/10.1016/s0031-9384(96)00357-5. https://doi.org/10.1016/s0031-9384(96)00357-5.

BIANCHI F, MANGIA A, MATTAROZZI M, MUSCI M (2011) Characterization of the volatile profile of thistle honey using headspace solid-phase microextraction and gas chromatography-mass spectrometry. *Food Chem.* 129, 1030-1036. https://doi.org/10.1016/j.foodchem.2011.05.070.

BROCKMANN A, BRÜCKNER D, CREWE RM (1998) The EAG response spectra of workers and drones to queen honeybee mandibular gland components: the evolution of a social signal. *Naturwissenschaften* 85, 283-285. https://doi.org/10.1007/s001140050500.

CHAPURLAT E, ÅGREN J, ANDERSON J, FRIBERG M, SLETVOLD N (2019) Conflicting selection on floral scent emission in the orchid *Gymnadenia conopsea*. *New Phytologist* 222:2009-2022. https://doi.org/10.1111/nph.15747.

CHITTKA L, RAINE NE (2006) Recognition of flowers by pollinators. *Curr. Opin. Plant Biol.* 9, 428-435. https://doi.org/10.1016/j.pbi.2006.05.002.

DAISEY JM, HOPKE PK (1991) Potential for ion-induced nucleation of volatile organic compounds by radon decay in indoor environments. *Aerosol Sci. Tech.* 19,1, 80-93. https://doi.org/10.1080/02786829308959623.

DAUMER K (1956) Reizmetrische Untersuchung des Farbensehens der Bienen. *Z. Vgl. Physiol.* 38, 413-478.

DAUMER K (1958) Blumenfarben, wie sie die Bienen sehen. *Z. Vgl. Physiol.* 49-110. https://doi.org/10.1007/BF00340242.

EISENHARDT D (2014) Molecular mechanisms underlying formation of long-term reward memories and extinction memories in the honeybee (*Apis mellifera*). *Learn. Memory* 21, 534-542. https://doi.org/10.1101/lm.033118.113.

FARINA WM, GRUTER C, ACOSTA L, MC CS (2007) Honeybees learn floral odors while receiving nectar from foragers within the hive. *Sci. Nat-Heidelberg* 94, 55-60. https://doi.org/10.1007/s00114-006-0157-3.

GALÁN RF, WEIDERT M, MENZEL R, HERZ AV, GALIZIA CG (2014) Sensory memory for odors is encoded in spontaneous correlated activity between olfactory glomeruli. *Neural. Comput.* 18, 10-25. https://doi.org/10.1162/089976606774841558.

GALIZIA CG, SACHSE S, RAPPERT A, MENZEL R (1999) The glomerular code for odor representation is species specific in the honeybee *Apis mellifera*. *Nat Neurosci.* 2, 473-478. https://doi.org/10.1038/8144.

GLAETTLI M, BARRETT SC (2008) Pollinator responses to variation in floral display and flower size in dioecious *Sagittaria latifolia* (Alismataceae). *New Phytol.* 179, 1193-1201. https://doi.org/10.1111/j.1469-8137.2008.02532.x.

HE XJ, ZHANG XC, JIANG WJ, BARRON AB, ZHANG JH, ZENG ZJ (2016) Starving honey bee (*Apis mellifera*) larvae signal pheromonally to worker bees. *Sci. Rep.* 6, 22359. https://doi.org/10.1038/srep22359.

HOURCADE B, PERISSE E, DEVAUD JM, SANDOZ JC (2009) Long-term memory shapes the primary olfactory center of an insect brain. *Learn. Memory* 16,10, 607-615. https://doi.org/10.1101/lm.1445609.

IWASHITA G, HIBINO T (2011) Assessment on odor intensity of bioeffluents by VOC concentration perceived air pollution caused by human bioeffluents (part 2). *J. Environ. Eng. AIJ* 76,664,539-545. https://doi.org/10.3130/aije.76.539.

IWASHITA G, TOKUNAGA N (2012) Discussion on the index of body odor pollution based on VOC concentration in classrooms perceived air pollution caused by human Bioeffluents (part 3). *J. Environ. Eng. AIJ* 77, 672, 65-70. https://doi.org/10.3130/aije.77.65.

JERKOVI I, TUBEROSO CIG, MARIJANOVI Z, JELI M, KASUM A (2009) Headspace, volatile and semi-volatile patterns of *Paliurus spina-christi* unifloral honey as markers of botanical origin. *Food Chem.* 112, 239-245. https://doi.org/10.1016/j.foodchem.2008.05.080.

JOERGES J, KÜTTNER A, GALIZIA CG, MENZEL R (1997) Representations of odours and odour mixtures visualized in the honeybee brain. *Nature* 387, 285-287. https://doi.org/10.1038/387285a0.

JØRGENSEN U, HANSEN M, CHRISTENSEN LP, JENSEN K, KAACK K (2000) Olfactory and quantitative analysis of aroma compounds in elder flower (*Sambucus nigra* L.) drink processed from five cultivars. *J. Agric. Food Chem.* 48, 2376-2383. https://doi.org/10.1021/jf000005f.

KIESTER AR, LANDE R, SCHEMSKE D (1984) Models of coevolution and speciation in plants and their pollinators. *The American Naturalist* 2, 220-243. https://doi.org/10.1086/284265.

KLEIN AM, VAISSIERE BE, CANE JH, STEFFAN-DEWENTER I, CUNNINGHAM SA, KREMEN C, TSCHARNTKE T (2007) Importance of pollinators in changing landscapes for world crops. *Proc. Biol. Sci.* 274, 303-313. https://doi.org/10.1098/rspb.2006.3721.

KNAUER AC, SCHIESTL FP (2015) Bees use honest floral signals as indicators of reward when visiting flowers. *Ecol. Lett.* 18, 135-143. https://doi.org/10.1111/ele.12386.

KOMISCHKE B, GIURFA M, LACHNIT H, MALUN D (2002) Successive olfactory reversal learning in honeybees. *Learn. Memory* 9, 122. https://doi.org/10.1101/lm.44602.

KRISHNA S, KEASAR T (2018) Morphological Complexity as a Floral Signal: From Perception by Insect Pollinators to Co-Evolutionary Implications. *Int. J. Mol. Sci.* 19. https://doi.org/10.3390/ijms19061681.

LASKA M, GALIZIA CG, GIURFA M, MENZEL R (1999) Olfactory discrimination ability and odor structure-activity relationships in honeybees. *Chem. Senses* 24, 429-438. https://doi.org/10.1093/chemse/24.4.429.

LEONARD AS, DORNHAUS A, PAPAJ DR (2010) Flowers help bees cope with uncertainty: signal detection and the function of floral complexity. *J. Exp. Biol.* 214, 113-121. https://doi.org/10.1242/jeb.047407.

LUO C, HUANG ZY, LI K, CHEN X, CHEN Y, SUN Y (2013) EAG responses of Apis cerana to floral compounds of a biodiesel plant, *Jatropha curcas* (Euphorbiaceae). *J. Econ. Entomol.* 106, 1653-1658. https://doi.org/10.1603/EC12458.

MENZEL R, MULLER U (1995) Learning and memory in honeybees: from behavior to neural substrates. *Annual Annu. Rev. Neurosci.* 19, 379-404. https://doi.org/annurev.ne.19.030196.002115.

MOYROUD E, WENZEL T, MIDDLETON R, RUDALL PJ, BANKS H, REED A, MELLERS G, KILLORAN P, WESTWOOD MM, STEINER U, VIGNOLINI S, GLOVER BJ (2017) Disorder in convergent floral nanostructures enhances signalling to bees. *Nature* 550, 469-474. https://doi.org/10.1038/nature24285.

MÜLLER U (2002) Learning in honeybees: from molecules to behaviour. *Zoology* 105, 313-320. https://doi.org/10.1078/0944-2006-00075.

NAEF R, JAQUIER A, VELLUZ A, BACHOFEN B (2004) From the linden flower to linden honey-volatile constituents of linden nectar, the extract of bee-stomach and ripe honey. *Chem. Biochem.* 1, 1870-1879. https://doi.org/10.1002/cbdv.200490143.

PAUW A, STOFBERG J, WATERMAN RJ (2009) Flies and flowers in Darwin's race. *Evolution* 63, 268-279. https://doi.org/10.1111/j.1558-5646.2008.00547.x.

RODRIGUEZ-GIRONES MA, SANTAMARIA L (2005) Resource partitioning among flower visitors and evolution of nectar concealment in multi-species communities. *Proc. Biol. Sci.* 272, 187-192. https://doi.org/10.1098/rspb.2005.2936.

RODRIGUEZ-GIRONES MA, SANTAMARIA L (2006) Models of optimum foraging and resource partitioning: deep corollas for long tongues. *Behav. Ecol.* 17, 905-910. https://doi.org/10.1093/beheco/arl024.

SEELEY TD (1985) Honeybee Ecology: A Study of Adaptation in Social Life. New Jersey: Princeton University Press.

SCHADE F, LEGGE RL, THOMPSON JE (2001) Fragrance volatiles of developing and senescing carnation flowers. *Phytochemistry* 56, 703-710. https://doi.org/10.1016/S0031-9422(00)00483-0.

SCHEINER R, SCHNITT S, ERBER J (2005) The functions of antennal mechanoreceptors and antennal joints in tactile discrimination of the honeybee (*Apis mellifera* L.). *J. Comp. Physiol. A Neuroethol.* 191, 857-864. https://doi.org/10.1007/s00359-005-0009-1.

SCHIESTL FP (2010) The evolution of floral scent and insect chemical communication. *Ecol. Lett.* 13, 643-656. https://doi.org/10.1111/j.1461-0248.2010.01451.x.

TWIDLE AM, BARKER D, SEAL AG, FEDRIZZI B, SUCKLING DM (2018) Identification of Floral Volatiles and Pollinator Responses in Kiwifruit Cultivars, *Actinidia chinensis var*. chinensis. *J. Chem. Ecol.* 44, 406-415. https://doi.org/10.1007/s10886-018-0936-2.

WRIGHT GA, SKINNER BD, SMITH BH (2002) Ability of honeybee, *Apis mellifera*, to detect and discriminate odors of varieties of canola (*Brassica rapa* and *Brassica napus*) and snapdragon flowers (*Antirrhinum majus*). *J. Chem. Ecol.* 28, 721-740. https://doi.org/10.1023/a:1015232608858.

WRIGHT GA, CARLTON M, SMITH BH (2009) A honeybee's ability to learn, recognize, and discriminate odors depends upon odor sampling time and concentration. *Behav. Neurosci.* 123, 1, 36-43. https://doi.org/10.1037/a0014040.

WRIGHT GA, SCHIESTL FP (2009) The evolution of floral scent: the influence of olfactory learning by insect pollinators on the honest signalling of floral rewards. *Funct. Ecol.* 23, 841-851. https://doi.org/10.1111/j.1365-2435.2009.01627.x.

WU Q, LING J, XU Y (2014) Starter culture selection for making Chinese sesame-flavored liquor based on microbial metabolic activity in mixed-culture fermentation. *Appl. Environ. Microb.* 2014, 80, 4450-4459. https://doi.org/10.1128/AEM.00905-14.

XU WL (1992) Nectar and pollen plants of China. Harbin, Heilongjiang, China: Heilongjiang science and technology press.

ZHANG FP, LARSON-RABIN Z, LI DZ, WANG H (2012) Colored nectar as an honest signal in plant-animal interactions. *Plant Signal. Behav.* 7, 811-812. https://doi.org/10.4161/psb.20645.

# 4. The Capping Pheromones and Putative Biosynthetic Pathways in Worker and Drone Larvae of Honey Bees *Apis mellifera*

Qiuhong Qin[1,2,†], Xujiang He[1,†], Andrew B. Barron[3], Lei Guo[4], Wujun Jiang[1], Zhijiang Zeng[1,*]

1. Honeybee Research Institute, Jiangxi Agricultural University, Nanchang, Jiangxi 330045, P. R. of China
2. Guangxi Liuzhou Animal Husbandry and Veterinary School, Liuzhou, Guangxi 545003, P. R. of China
3. Department of Biological Sciences, Macquarie University, North Ryde, NSW 2109, Australia
4. Research and Development Centre, China Tobacco Jiangxi Industrial Co., LTD., Nanchang, Jiangxi 330096, P.R. of China

**Abstract**

In honey bees (*Apis mellifera*), methyl palmitate (MP), methyl oleate (MO), methyl linoleate (ML) and methyl linolenate (MLN) are important pheromone components of the capping pheromones triggering the capping behaviour of worker bees. In this study, we compared the amounts of these four pheromone components in the larvae of workers and drones, prior to be capped, in the process of being capped and had been capped. The amounts of MP, MO, and MLN peaked at the capping larval stage and ML was highest at capped larvae in worker larvae, whereas in drone larvae the amounts of the four pheromone components were higher overall and increased with aging. Furthermore, we proposed de novo biosynthetic pathways for MP, MO, ML, and MLN, from acetyl-CoA. Besides, stable isotope tracer $^{13}$C and deuterium were used to confirm that these capping pheromone components were de novo synthesized by larvae themselves rather than from their diets.

**Keywords:** Honey bee, larvae, capping, pheromones, pathways

## Introduction

Honey bees have been instrumental in revealing both the significance and the complexity of pheromonal communication systems. Pheromones help to coordinate a colony's collective behaviour, such as foraging, defensive behaviour and brood-rearing activity (Free and Winder 1983; Free 1987; Vallet *et al.* 1991; Breed *et al.* 2004; Hunt 2007; Stout and Goulson 2001; Slessor *et al.* 2005; Maisonnasse *et al.* 2010). The subtleties and complexities of bee pheromonal communication are perhaps best illustrated by the pheromonal communication between workers and developing larvae.

Honey bee larvae are entirely dependent on their attendant adult workers. Larvae release capping

---

† These two authors contributed equally to this paper.
*Corresponding author: bees1965@sina.com (ZJZ).
注：此文发表在 *Apidologie* 2019 年第 6 期。

pheromone components, form their salivary glands containing ten methyl or ethyl fatty acid esters (Le Conte et al. 1989, 1990, 2006; Trouiller et al. 1991). The composition of this pheromone varies with larval age and sex (worker or drone), and workers adjust their behavioral responses to larvae accordingly (Free and Winder 1983; Le Conte et al. 1994; Slessor et al. 2005). When larvae reach the stage of pupation, their cells are closed off by workers constructing a thin wax cap over the cell so that the larvae pupate in a clean and stable environment. Four of the components of BEP are particularly important in triggering the capping behaviour of workers. These are methyl palmitate (MP), methyl oleate (MO), methyl linoleate (ML), and methyl linolenate (MLN) (Le Conte et al. 1990). Production of these components by larvae increases quite dramatically when the larva reaches the developmental stage prior to pupation (Trouiller et al. 1991).

Currently, our knowledge of how pheromones are synthesized and released by worker and drone larvae at different developmental stages are limited. To fulfill the gaps, here we examined changes in the BEP components related to cell capping in both worker and drone larvae pre, during and post cell capping. In parallel, we analyzed gene expression changes in larvae across the cell capping stage as a first step to the identification of biosynthetic pathways in bees for pheromone components.

## Materials and Methods

### Experimental honeybee colonies

Honey bee colonies (*Apis mellifera*) used throughout this study were maintained at the Honeybee Research Institute, Jiangxi Agricultural University, Nanchang, China (28.46°N, 115.49°E), according to standard beekeeping techniques.

### Gas chromatograph-mass spectroscopy (GC/MS) analysis

Amounts of capping pheromone components were compared in the following sample groups: i. non-capped larvae of 4-day old (4th) instar of workers and drones, ii. capping-stage worker and drone larvae sampled as wax caps were being constructed over their cells, iii. capped worker and drone larvae sampled from cells that had been capped completely and had a thin silk cocoon under the wax cover, which were approximately 8-day old instar for worker and 9-day old instar for drones. There were four biological replicates of each group, with each replicate sourced from a different colony. Each sample contained groups of larvae to provide enough biological material for GC/MS analysis. For each sample we collected 40 4th instar worker and drone larvae, 20 capping-stage and capped worker larvae and 5 capping-stage and capped drone larvae from brood frames.

The whole larvae of each sample group freshly collected from their wax cells were placed in (without body washes) glass vials with 3 ml dichloromethane respectively. 20 μL Methyl nonadecanoate (≥98%, AccuStandard, 10 μg/mL) was added to each vial as the internal standard. Afterwards, the glass vials with larvae were shaken lightly for 30 min using an oscillator (120 r/min), and then the solution was transferred into a clean vial and concentrated under nitrogen flow to about 20 μL. A 1-μL sample was injected into GC/MS.

A GC/MS system (7890A/5975C, Agilent Technolgies Inc., Santa Clara, CA, USA) with an HP-Innowax chromatographic column (60-m, 0.25-mm, 0.25-μm film thickness, 19091N-136, Agilent Technolgies) was used. Helium was used as carrier gas at a constant flow rate of 1.0 mL/min. Samples were analyzed with time split injection mode and split ratio of 2∶1. The column temperature was programmed as follows: 120 °C for 1 min, then rising to 180 °C at a rate of 5 °C/min, and then rising to 230 °C at a rate of 2 °C/min, finally cooling to 250 °C at a rate of 5 °C/min and held constant for 10 min. The temperature of the sample injection port was 250 °C, and the pressure of helium was 13.5552 psi. Sample ionization was performed in electron impact ionization mode at 70 eV. The temperature of ion source, electrode stem and the transmission line were 250 °C, 150 °C, and 250 °C, respectively. Single ion monitoring chromatograms were reconstructed at the base peak for MP, MO, ML, and MLN and the internal standard: mass-to-charge ratio (m/z) 270, 264, 294, 292, and 312.

For each compound of interest, peak areas were converted to amounts by reference to calibration curves for each compound. MP, MO, ML, and MLN (≥99.5%, Sigma) were used as external standards. Seven concentrations of each external standard (0.25 μg/mL, 0.5 μg/mL, 1 μg/mL, 2 μg/mL, 5 μg/mL, 10 μg/mL, and 25 μg/mL) were used to construct calibration curves. All of the correlation coefficients of the four calibration curves were > 99.96% (Fig. S1), which suggested that the GC/MS system was stable enough for the later quantitative analysis of MP, MO, ML, and MLN from honey bee larvae.

### *RNA-Seq data analysis*

Samples from the groups described above were also used for RNA-sequence analysis to analyze differences in gene expression between groups. For each sample, we collected six workers or drone larvae and created three biological replicates with each replicate sourced from a different colony. The RNA was extracted with Trizol. The samples were flash frozen in liquid nitrogen and stored at -80 °C until use.

RNA extraction, RNA sequencing, and data analysis were performed according to the methods described in He *et al.* (2016). Briefly, the raw counts for each gene were normalized with FPKM to calculate the *P* value for significantly regulated genes with DESeq packages. The *P* value was further corrected for multiple comparisons with FDR. Significantly differentially expressed genes (DEGs) between different samples were defined as genes with a FDR≤0.05 and the absolute value of log2 Ratio≥1.

### *Identification of pheromone biosynthesis pathway*

The longest transcripts of significantly differentially expressed genes were converted to protein sequences, and then compared to the KEGG protein database by using BLAST with $E$-value $<10^{-5}$ set as the cut-off criterion. To identify putative enriched metabolic pathways in the RNA-seq data, the outcome of the KEGG mapping was analyzed with the KEGG Orthology Based Annotation System 2.0 (KOBAS 2.0). KOBAS 2.0 (Xie *et al.* 2011) references the provided gene database against multiple existing databases of metabolic and signaling pathways to identify pathways that are enriched in the RNA-seq samples.

### *Verification of DEGs with qRT-PCR analysis*

Six genes, which participate in the inferred biosynthetic pathways of MP, MO, ML, and MLN (Table SI) and a "housekeeping gene" (β-actin), were selected for qPCR analysis. For each group, 3 larvae from each of three colonies were sampled, yielding three biological replicates per group. Samples were flash frozen in liquid nitrogen and stored at -80 °C until used. Detailed methods for qRT-PCR are described in our previous study (Qin *et al.* 2014). Primers for these genes were designed by Primer 5.0 (Table SI). For each gene, three technical replicates were performed for each sample. The Ct value for each sample was obtained by calculating the mean of three technical replicates. Relative expression levels between samples were calculated using the formula $2^{-\Delta\Delta Ct}$ reported by Liu and Saint (2002), and then square-root transformed to normalize the distribution of the data before ANOVA analysis.

### *Stable isotopic tracing*

While insects are able to synthesize fatty acids de novo, but they can also get them from their diets. In order to provide further evidence to demonstrate that these capping pheromone components were de novo synthesized by larvae themselves rather than from their diets, stable isotope tracer method was used.

In this experiment, the basic larval food (BLD) consists of 50% royal jelly, 6% fructose, 6% glucose, 1% yeast extract, and 37% ddH$_2$O (Vandenberg and Shimanuki 1987). For the glucose in the BLD was consist of 20% $^{13}$C-glucose (Sigma) and 80% $^{12}$C-glucose. Pheromone components were collected from the following sample groups: i. larvae reared with BLD as control group (C); ii. larvae reared with BLD and $^{13}$C-glucose (Sigma) as treatment group 1 (T1); iii. larvae reared with BLD with 0.25% $^2$H-methanol (Sigma) as treatment group 2 (T2). Larval food of all groups was stored at -80 °C to keep fresh and was pre-warmed at 35 °C for 15 min before usage.

One-day worker larvae were removed from the comb and transferred to 24-well cell culture plates with 200 μl of larval food per well, and then were placed in the incubator at 35 °C and 90% relative humidity. Every 12 hours, larvae were fed with 100 μl fresh food per well. When began to defecate, larvae were removed from the cell culture plates, dried on Kimwipe™ tissues, and transferred to the pupation plates with Kimwipe™ tissues on the bottom (Kucharski *et al.*, 2008). Six hours later, every 10 larvae of the same group were freshly collected from the pupation plates to a glass vial with 3 ml dichloromethane respectively. The abundances of $^{13}$C and $^2$H of the capping pheromone were analyzed using GC/MS system (7890A/5975C, Agilent Technolgies Inc., Santa Clara, CA, USA) by Shanghai Research Institute of Chemical Industry CO., LTD. Sample treatment method and experiment condition were the same as GC/MS analysis as described previously. Three biological replicates were made for each group. The abundances of $^{13}$C and $^2$H in control and treated groups were calculated by amounts of $^{13}$C and $^2$H marked capping pheromone against the total amount of related pheromone, respectively. Three capping pheromones were detected with $^{13}$C and $^2$H and one pheromone, ML, was failed.

### *Statistics*

StatView 5.01 (SAS Institute, Cary, NC, USA) was used to analyze all data from GC/MS and qRT-PCR of each group, by using ANOVA followed by a Fisher's PLSD test. Differences between groups were considered to be significant at the probability level of 0.05.

## Results

### *Levels of capping pheromone components in larvae of different developmental stages*

The amounts of MP, MO, ML, and MLN in samples of worker larvae are shown in Fig. III-4-1. The levels of MP, MO, and MLN were significantly higher in capping-stage worker larvae compared to the 4th instar and capped larvae. Additionally, the amount of ML was significantly higher in capped larvae than 4th instar and capping-stage larvae.

Results for drone larvae are shown in Fig. III-4-2. In drones, the amounts of all four pheromone components increased with age from 4th instar to capped larvae and were significantly higher in capped larvae than in 4th instar and capping-stage larvae.

*RNA sequence analysis in larvae of different capping stage*

Summaries of RNA sequence sample quality are shown in Table SII. In the libraries, the alignment rate between the reads and reference genome were all above 90.00%. Due to the Pearson correlation coefficient of capped WL-R2 and capped DL-R1 were too low with other two replicate samples, thus they were eliminated in following analysis (Table SIII). But beyond that, the Pearson correlation coefficients among the other three biological replicates of all experimental groups were all ≥ 0.75 (Table SIII), which is a conventionally accepted threshold for valid replicates (Tarazona *et al.* 2011) indicating that there was acceptable sequencing quality and repeatability among the biological replicates of each group.

The number of DEGs varied dramatically among developmental stages (using a threshold of log2 ratio ≥ 1, FDR < 0.01, Table SIV). The top 5 significantly enriched KEGG pathways of DEGs were Glycolysis / Gluconeogenesis (ko00010); Citrate cycle (TCA cycle, ko00020); Pentose phosphate pathway (ko00030); Valine, leucine and isoleucine biosynthesis (ko00290); and Glutathione metabolism (ko00480). These five pathways were not involved in pheromone biosynthesis pathways.

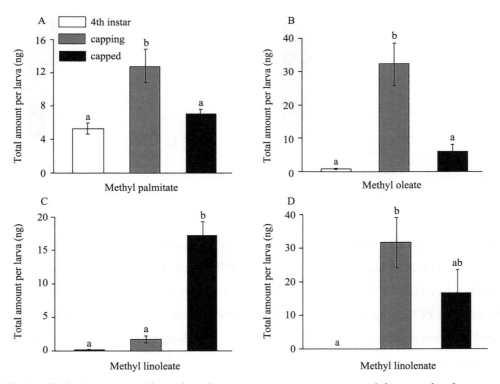

**Figure III-4-1** Amounts of capping pheromone components extracted from worker larvae at different capping stages. Each bar corresponds to a single group represented as the mean ± S.E. of its biological. Each larval group has 4 replicates from 4 different honey bee colonies.

However, two KEGG pathways were identified possibly relating to the long carbon chain biosynthesis: one for fatty acid elongation (ko00062) and one for biosynthesis of unsaturated fatty acid (ko01040), which we predicted might be involved in the biosynthesis of four capping pheromone components. The fatty acid elongation pathway involves in the de novo synthesis of hexadecanoate (C16) from Acetyl-Coenzyme A (acetyl-CoA), and the biosynthesis of unsaturated fatty acid pathway is involved in the formation of one or more double-bond unsaturated fatty acids such as oleic acid and linoleic acid by using C16 as a precursor. This pathway has been reported to be involved in the pheromone biosynthesis of some insect species (Roelofs and Bjostad 1984; Ando *et al.* 1988; Tang *et al.* 1989).

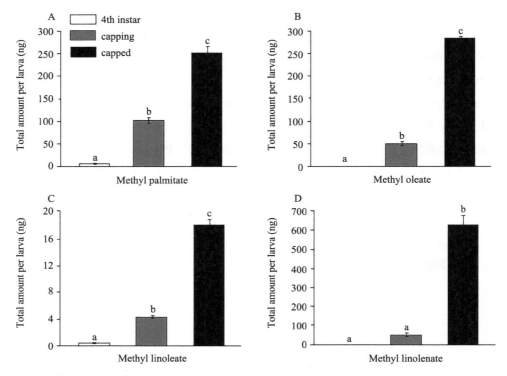

**Figure III-4-2** Amounts of capping pheromone components extracted from drone larvae at different capping stages. Each bar corresponds to a single group represented as the mean ± S.E. of its biological replicates. Each larval group has 4 replicates from 4 different honey bee colonies.

In this study, 6 genes *Kat* (GB50970: 3-ketoacyl-CoA thiolase, mitochondrial), *Hadha* (GB45128: trifunctional enzyme subunit alpha, mitochondrial), *Echs1* (GB40460: enoyl Coenzyme A hydratase short chain 1, mitochondrial), *Mecr* (GB43706: probable trans-2-enoyl-CoA reductase, mitochondrial), *Ppt2* (GB40083: lysosomal thioesterase PPT2 homolog), and *Ppt1* (GB49609: palmitoyl-protein thioesterase 1) were enriched in the fatty acid elongation pathway, and *Vlcecr* (GB46772: very-long-chain enoyl-CoA reductase), *Vlchacd2* (GB54258: very-long-chain (3R)-3-hydroxyacyl-CoA dehydratase 2), and four *acyl-CoA* Δ11 desaturase like genes (Δ11-desaturases, GB48193, GB48194, GB48195, and GB51238) were enriched in the biosynthesis of unsaturated fatty acid pathway. All these 12 genes were significantly differentially expressed in at least one comparison among 4th instar, capping and capped larvae (Table SV). In particular, GO annotation showed that two Δ11-desaturase genes (GB48193 and GB48195) had Δ9-desaturase activity (Table SV). We therefore proposed these as candidate genes for the process of biosynthesis of MP, MO, ML, and MLN in honey bee larvae (Fig. III-4-3).

We compared our four candidate Δ11-desaturase genes GB48193, GB48194, GB48195, and GB51238 with domestic silkwarm (*Bombyx mori*) acyl-CoA delta (11) desaturase by protein BLAST using the NCBI website. Protein blast results showed that these four genes were high homology with *Bombyx mori* acyl-CoA delta (11) desaturase with positives of 62%, 46%, 63%, and 67%, respectively.

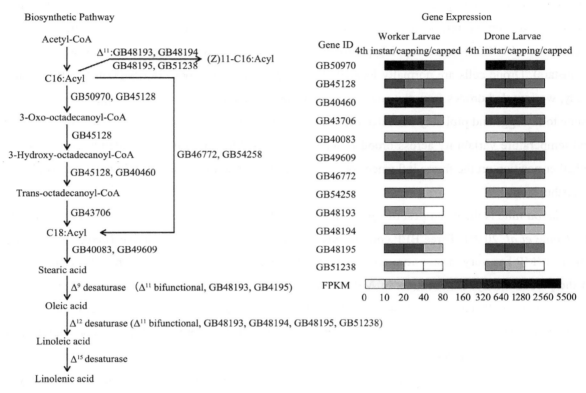

**Figure III-4-3** Proposed biosynthetic pathways involved in the synthesis of capping pheromone components, and expression of related genes in worker and drone larvae of different capping stages. The grids illustrate the gene expression levels for 4th instar, capping and capped larvae of workers and drones. Greyscale density of each cell shows the absolute expression magnitude of honey bee larvae, with the FPKM values 0–10, 10–20, 20–40, 40–80, 80–160, 160–320, 320–640, 640–1280, 1280-2560 and 2560-5500 represented by grey scale levels respectively.

### Results of qRT-PCR

Fig. III-4-4 shows the results of qRT-PCR analyses alongside the results of RNA sequence analyses for four genes from the fatty acid elongation pathway and two of the Δ11-desaturase-like genes. The results of qRT-PCR analysis were concordant with the results of the RNA sequence. Furthermore, the general expression tendency showed that these genes had a higher relative expression level both in 4th instar and capping larvae than that in capped larvae.

### Isotope abundance analysis

The abundances of $^{13}C$ and $^{2}H$ from different capping pheromone components are shown in Table III-4-1. On account of the natural isotopic abundance, $^{13}C$ and $^{2}H$ were detected from control groups as well.

Nevertheless, for MP, MO, and MLN (ML was not detected for its low abundance), there was at least one of the abundances of $^{13}$C or $^{2}$H in treatment groups were significantly higher than that in control groups.

## Discussion

MP, MO, ML, and MLN have been reported to be important signals for honey bee capping behaviour (Le Conte *et al.* 1990). Our data is consistent with Trouiller *et al.* (1992), indicating that honey bee larvae increase the amounts of capping pheromone components at a critical larval stage for wax capping. Yan *et al.* (2009) also reported a similar releasing trend of pheromone levels in *Apis cerana* worker and drone larvae. However, Trouiller *et al.* (1992) found capping pheromone decreased in *Apis mellifera* drone larvae post-capping (8-10 instars). Drone cells are normally located at the edge of natural brood frames (Winston 1991), and in this study, we sampled drones from the edges of natural frames. Trouiller *et al.* (1992) caged a queen onto a drone frame to lay eggs and probably collected drone samples from central areas. There are differences in temperature and temperature variability across brood frames, as well as differences in larval attendance rates by workers. Whether position on the frame influences pheromone accumulation is presently unknown, but would be worthy of further study.

*Varroa* mite is the worst pest of *Apis mellifera*, resulting in high death rate of honey bees around the world (Le Conte *et al.* 2010). Three BEP components have been reported to be attractive *Varroa* mites in honey bee colonies (MP has very strong attractive response, EP and ML have weaker response, Le Conte *et al.* 1989). In the present study, the amounts of MP and ML in drone larvae were dramatically higher compared with worker larvae at all stages (Fig. III-4-1 and Fig. III-4-2). Our study supports earlier suggestions that these BEP components may be used by *Varroa* to differentiate and select drone brood at the capping stage from workers brood (Le Conte *et al.* 1989).

In many insect species, acetyl-CoA is the precursor of several divergent metabolic pathways leading to the biosynthesis of fatty acids and esters (McGee and Spector 1972; Cripps *et al.* 1986; Moshitzky *et al.* 2003). For example, in silkworm and some species, methyl esters (including methyl hexadecanoate) and the sex pheromone bombykol were de novo synthesized from acetyl-CoA to hexadecanoate or octadecanoate respectively (Ando *et al.* 1988; Tang *et al.* 1989). Furthermore, similar pathways were found in the methyl esters of leptospira and soybean (Stern *et al.* 1969; Bachlava *et al.* 2009). These studies support the reliability of our results on the biosynthesis pathway of four honey bee capping pheromone components.

In many insects and animals, stearic acid is generally synthesized via two pathways (Wang *et al.* 2002): in mitochondria, stearic acid can be biosynthesized from palmitic acid by mainly five steps (condensation, reduction, dehydration, reduction, and release); in endoplasmic reticulum, stearic acid can be directly biosynthesized by using palmitic acid as substrate. For the unsaturated methyl esters, it is well known that $\Delta 9$-desaturase, $\Delta 12$-desaturase, and $\Delta 15$-desaturase are the relative desaturases for the synthesis of MO, ML, and MN (Wang *et al.* 2002). In this study, 4 $\Delta 11$-desaturase genes were enriched in the proposed biosynthetic pathway of MO, ML, and MN in honey bees. We suspect that these four $\Delta 11$-desaturase genes may play the same function as Δ9-desaturase, Δ12-desaturase, and Δ15-desaturase in the biosynthetic process of these three unsaturated methyl esters. Firstly, the Δ11-desaturase is a biofunctional enzyme having for example Δ11 and Δ12 desaturation activities (Serra *et al.* 2006), playing an important role in insect pheromone biosynthesis (Lofstedt *et al.* 1986; Rodriguez *et al.* 1992; Foster 1998). Secondly, GO annotation showed that two Δ11-desaturase gene (GB48193 and GB48195) had stearoyl-CoA 9-desaturase activities (Table SV). Moreover, there are no Δ9-desaturase, Δ12-desaturase, and Δ15-desaturase genes in the GenBank database, but 12 Δ11-desaturase genes have been annotated. We mapped all 12 honey bee Δ11-desaturase genes to house cricket's (*acheta domesticus*) Δ9-desaturase which biological function has been confirmed (Riddervold *et al.* 2002). The protein of GB48195 had the highest homology of 66% with the Δ9-desaturase of house cricket and could not map to its Δ11-desaturase. Therefore, the gene GB48195 may be the Δ9-desaturase ortholog of honey bees. Eventually, Roelofs and Bjostad (1984) showed that the oleate, linoleate, and linolenate in the red banded leaf roller moth, cabbage looper moth, and the domestic silkworm, are de novo synthesized, and desaturated by a delta (11) desaturase enzyme. Consequently, it is believed that these four Δ11-desaturase genes are involved in the biosynthetic pathways of honey bee capping pheromone components. For the last step of fatty acids becoming methyl esters, Castillo *et al.* (2012) showed that honey bee ethyl oleate was synthesized by dehydration of oleic acid and free ethanol under regulating of genes GB11403 and GB13365. In our study, the gene GB13365 (updated BeeBase: GB43508, *lipase member H-A-like*) was significantly differentially expressed among three larval groups. However, whether this gene participates in regulating the biosynthetic expressed process of honey bee methyl esters from fatty acid and free methanol still requires further investigations. While insects are able to synthesize fatty acids de novo, but they can also get them from their diets. Results of stable isotope tracing of $^{13}C$ and $^{2}H$ provided further evidence to demonstrate that these capping pheromone components were de novo synthesized by larvae themselves rather than from their diets and make a definite connection between the gene expression seen and the biosynthesis of the methyl esters. However, it is a complex process from gene regulating to pheromone producing. The regulation of gene expression can be carried out on multiple levels including gene, transcription, post-transcription, translation, and post-translation. According to our results, we proposed these 12 genes as candidate genes for the process of biosynthesis of MP, MO, ML, and MLN in honey bee larvae, but the exact regulatory mechanism still requires further investigations.

In summary, this study showed that honey bee *Apis mellifera* worker and drone larvae increased their capping pheromone components during the wax capping process. Drone larvae released significantly higher pheromones than worker larvae, which attracts more *Varroa* mites during wax capping. We also identified putative biosynthetic pathways for these pheromones and predicted that these pheromones are de novo from acetyl-CoA. Our findings contribute information on how brood pheromones are biosynthesized and released in honey bees and how these pheromones influence the host selection of honey bee pests.

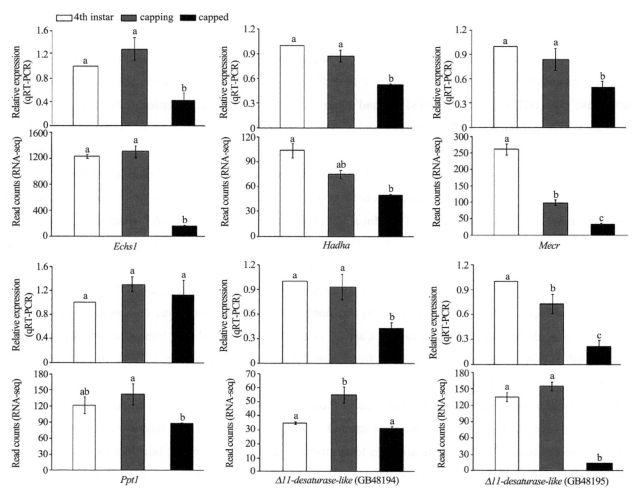

**Figure III-4-4** Expression levels of six genes possibly implicated in the biosynthesis of capping pheromone components in larvae of different capping stages (4th instar, capping and capped). Estimates of relative gene expression level assessed with RNA-seq and qRT-PCR are shown for each gene. Different superscripts (a,b and c) to bars indicate significant differences in expression between capping stages (qRT-PCR: $P<0.05$, RNA-seq: FDR$\leq 0.05$). Relative expression data of qRT-PCR were transformed by square root transformation, and were presented here after transformation. Each bar corresponds to a single group represented as the mean ± S.E. of its biological replicates.

Table III-4-1 The abundances of $^{13}$C and $^{2}$H from different capping pheromones

| Pheromone | $^{13}$C | | | $^{2}$H | | |
|---|---|---|---|---|---|---|
| | C (%) | T1 (%) | P | C (%) | T2 (%) | P |
| MP | 1.80±0.10 | 1.90±0.28 | >0.05 | 0.93±0.06 | 4.20±0.20 | <0.01 |
| MO | 1.17±0.06 | 2.80±0.00 | <0.01 | 0.60±0.00 | 0.90±0.10 | <0.01 |
| MLN | 1.87±0.21 | 4.50±0.85 | <0.01 | 1.10±0.10 | 1.23±0.23 | >0.05 |

# Acknowledgements

We thank Prof. Junwu Ma for help on biosynthetic pathway analysis and Dr. Qiang Huang for revising the manuscript.

## Contributions

QHQ, XJH, and WJJ conducted all experiments; ZJZ designed the experiments; QHQ, XJH, and ABB wrote the paper; and LG participated in experiments and data analysis.

## Funding Information

This work was supported by the National Natural Science Foundation of China (31572469, 31872432) and the Earmarked Fund for China Agriculture Research System (CARS-44-KXJ15).

All sequencing data has been submitted to GenBank under accession numbers SRR4051855, SRR4051820, bioproject PRJNA339509.

## References

ANDO, T., HASE, T., ARIMA, R., UCHIYAMA, M. (1988). Biosynthetic pathway of bombykol, the sex pheromone of the female silkworm moth. *Agric. Bioi. Chern.* 52(2), 473-478.

BACHLAVA, E., DEWEY, R.E., BURTON, J.W., CARDINAL, A.J. (2009). Mapping candidate genes for oleate biosynthesis and their association with unsaturated fatty acid seed content in soybean. *Mol. Breeding* 23(2), 337-347.

BREED, M.D., GUZMAN-NOVOA, E., HUNT, G.J. (2004) Defensive behavior of honey bees: organization, genetics, and comparisons with other bees. *Annu. Rev. Entomol.* 49, 271-298.

CASTILLO, C., CHEN, H., GRAVES, C., MAISONNASSE, A., LE, C.Y., PLETTNER, E. (2012). Biosynthesis of ethyl oleate, a primer pheromone, in the honey bee (*Apis mellifera L.*). *Insect Biochem. Molec.* 42(6), 404-416.

CRIPPS, C., BLOMQUIST, G.J., RENOBALES, M.D. (1986). De novo biosynthesis of linoleic acid in insects. *BBA-Lipid Lipid Met.* 876(3), 572-580.

FOSTER, S.P. (1998). Sex pheromone biosynthesis in the tortricid moth planotortrix excessana (walker) involves chain-shortening of palmitoleate and oleate. *Arch. Insect Biochem.* 37(2), 158-167.

FREE, J.B., WINDER, M.E. (1983) Brood recognition by honey bee *Apis mellifera* workers. *Anim. Behav.* 31, 539-545.

FREE, J.B. (1987) Pheromones of Social Bees. Chapman & Hall, London.

HE, X.J., ZHANG, X.C., JIANG, W.J., BARRON, A.B., ZHANG, J.H., ZENG, Z.J. (2016) Starving honey bee *(Apis mellifera)* larvae signal pheromonally to worker bees. *Sci. Rep.* 6, 22359.

HUNT, G.J. (2007) Flight and fight: a comparative view of the neurophysiology and genetics of honey bee defensive behavior. *J. Insect Physiol.* 53, 399-410.

KUCHARSKI, R. MALESZKA, J., FORET, S., MALESZKA, R. (2008) Nutritional control of reproductive status in honey bees via DNA methylation. *Science* 319(5871): 1827-1830.

LE CONTE, Y., ARNOLD, G., TROILER, J., MASSON, C., CHAPPE, B., OURISSON, G. (1989) Attraction of the parasitic mite Varroa to the drone larvae of honey bees by simple aliphatic esters. *Science*, 245, 638-639.

LE CONTE, Y., ARNOLD, G., TROUILER, J., MASSON, C. (1990) Identification of a brood pheromone in honeybees. *Naturwissenschaften.* 77, 334-336.

LE CONTE, Y. (1994) The Recognition of Larvae by Worker Honeybees. *Naturwissenschaften*. 81, 462-465.

LE CONTE Y., BÉCARD, J.M., COSTAGLIOLA G., DE VAUBLANC, G., EL MAÂTAOUI, M., CRAUSER, D., PLETTNER, E., SLESSOR, K.M. (2006) Larva salivary glands are a source of primer and releaser pheromone in honey bee *(Apis mellifera L.)*. *Naturwissenschaften*. 93, 237-241.

LE CONTE, Y., ELLIS, M., RITTER, W. (2010). *Varroa* mites and honey bee health: can *Varroa*, explain part of the colony losses? *Apidologie*, 41(3), 353-363.

LIU, W., SAINT, D.A. (2002) A new quantitative method of real time reverse transcription polymerase chain reaction assay based on simulation of polymerase chain reaction kinetics. *Anal. Biochem*. 1, 52-59.

LOFSTEDT, C., ELMFORS, A., SJÖGREN, M., WIJK, E. (1986). Confirmation of sex pheromone biosynthesis from (16-d3) palmitic acid in the turnip moth using capillary gas chromatography. *Cell Mol. Life Sci*. 42(9), 1059-1061.

MAISONNASSE, A., LENIOR, J.C., BESLAY, D., CRAUSER, D., LE CONTE, Y. (2010) E-β-ocimene, a volatile brood pheromone involved in social regulation in the honey bee colony *(Apis mellifera)*. *PLoS One*. 5, 1-7.

MCGEE, R., SPECTOR, A. A. (1975). Fatty acid biosynthesis in erlich cells. the mechanism of short term control by exogenous free fatty acids. *J. Biol. Chem*. 250(14), 5419.

MOSHITZKY, P., MILOSLAVSKI, I., AIZENSHTAT, Z., APPLEBAUM, S. W. (2003). Methyl palmitate: a novel product of the medfly *(ceratitis capitata)* corpus allatum. Insect Biochem. *Mol. Biol*. 33(12), 1299-1306.

QIN, Q.H., HAN, X., LIU, H., ZHANG, S.W., ZENG, Z.J. (2014) Expression levels of glutamate and serotonin receptor genes in the brain of different behavioural phenotypes of worker honeybee *(Apis mellifera)*.Türk. *Entomol. Derg*. 38 (4), 431-441.

RIDDERVOLD, M. H., TITTIGER, C., BLOMQUIST, G. J., BORGESON, C. E. (2002). Biochemical and molecular characterizaton of house cricket *(acheta domesticus*, orthoptera: gryllidae) Δ9 desaturase. *Insect Biochem. Mol. Biol*. 32(12), 1731-1740.

RODRIGUEZ, F., HALLAHAN, D. L., PICKETT, J. A., CAMPS, F. (1992). Characterization of the Δ11-palmitoyl-coa-desaturase from *spodoptera littoralis* (lepidoptera: noctuidae). *Insect Biochem. Mol. Biol*. 22(2), 143-148.

ROELOFS, W., BJOSTAD, L. (1984). Biosynthesis of lepidopteran pheromones. *Bioorg. Chem*. 12(4), 279-298.

SLESSOR, K.N., WINSTON, M.L., LE CONTE, Y. (2005) Pheromone communication in the honey bee *(Apis mellifera L.)*. *J. Chem. Ecol*. 31, 2731-2745.

SERRA, M., PIÑA, B., BUJONS, J., CAMPS, F., FABRIÀS, G. (2006). Biosynthesis of 10, 12-dienoic fatty acids by a bifunctional Δ11desaturase in Spodoptera littoralis. *Insect Biochem. Mol. Biol*. 36(8), 634-641.

STERN, N., SHENBERG, E., TIETZ, A. (1969). Studies on the metabolism of fatty acids in leptospira: the biosynthesis of Δ9 - and Δ11 -monounsaturated acids. *European J. Biochem*. 8(1), 101-108.

STOUT, J.C., GOULSON, D. (2001) The use of conspecific and interspecific scent marks by foraging bumble- bees and honeybees. *Anim. Behav*. 62, 183-189.

TANG, J.D., CHARLTON, R.E., JURENKA, R.A., WOLF, W.A., PHELAN, P.L., SRENG,L., ROELOFS, W.L. (1989). Regulation of pheromone biosynthesis by a brain hormone in two moth species. *Proc. Natl. Acad. Sci. USA*. 86(6), 1806-10.

TARAZONA, S., GARCÍA-ALCALDE, F., DOPAZO, J., FERRER, A., CONESA, A. (2011). Differential expression in

RNA-seq: a matter of depth. *Genome Res.* 21, 2213-2223.

TROUILLER, J., ARNOLD, G., LE CONTE, Y., MASSON, C., CHAPPE, B. (1991) Temporal pheromonal and kairomonal secretion in the brood of honeybees. *Naturwissenschaften.* 78(8), 368-370.

TROUILLER, J., ARNOLD, G., CHAPPE, B., LE CONTE, Y., MASSON, C. (1992). Semiochemical basis of infestation of honey bee brood by *Varroa jacobsoni. J. Chem. Ecol.* 18(11), 2041-2053.

VALLET, A., CASSIER, P., LENSKY, Y. (1991) Ontogeny of the fine structure of the mandibular glands of the honeybee (*Apis mellifera L.*) workers and the pheromonal activity of 2-heptanone. *J. Insect Physiol.* 37, 789-804.

VANDENBERG, J.D., SHIMANUKI, H. (1987) Technique for rearing worker honeybees in the laboratory. *J. Apicult. Res.* 26(2), 90-97.

WANG, J. Y., ZHU, S. G., XU, C. F. Biological chemistry (Third edition). China higher education press, Beijing, 2002.

WINSTON, M. (1991) The Biology of the Honey Bee. Harvard University Press, Cambridge, MA, USA.

XIE, C., MAO, X., HUANG, J., DING, Y., WU, J.M., DONG, S., KONG, L., GAO, G., LI, C.Y., WEI, L.P. (2011). KOBAS 2.0: a web server for annotation and identification of enriched pathways and diseases. *Nucleic. Acids. Res.* 39, W316-W322.

YAN, W.Y., LE CONTE, Y., BESLAY, D., ZENG, Z.J. (2009). Identification of brood pheromone in Chinese honeybee [*Apis cerana cerana* (Hymenoptera: apidae)]. *Scientia Agricultura Sinica* 42(6), 2250-2254. (In Chinese).

# 5. Differential Protein Expression Analysis following Olfactory Learning in *Apis cerana*

Lizhen Zhang[1], Weiyu Yan[1], Zilong Wang[1], Yahui Guo[1], Yao Yi[1], Shaowu Zhang[2], Zhijiang Zeng[1*]

1. Honeybee Research Institute, Jiangxi Agricultural University, Nanchang 330045, P.R. China
2. Research School of Biology, Australian National University, Canberra, ACT 0200, Australia

**Abstract**

Studies of olfactory learning in honeybees have helped to elucidate the neurobiological basis of learning and memory. In this study, protein expression changes following olfactory learning in *Apis cerana* were investigated using isobaric tags for relative and absolute quantification (iTRAQ) technology. A total of 2406 proteins were identified from the trained and untrained groups. Among these proteins, 147 were differentially expressed, with 87 up-regulated and 60 down-regulated in the trained group compared with the untrained group. These results suggest that the differentially expressed proteins may be involved in the regulation of olfactory learning and memory in *A. cerana*. The iTRAQ data can provide information on the global protein expression patterns associated with olfactory learning, which will facilitate our understanding of the molecular mechanisms of learning and memory of honeybees.

**Keywords:** *Apis cerana*, Proboscis extension response, Learning and memory, Isobaric tags for relative and absolute quantification (iTRAQ)

## Abbreviations

| | |
|---|---|
| *A. cerana* | *Apis cerana* |
| *A. mellifera* | *Apis mellifera* |
| AKT | RAC serine/threonine-protein kinase |
| CAMK | Calcium/calmodulin-dependent protein kinase |
| iTRAQ | Isobaric tags for relative and absolute quantification |
| MAP2 | Microtubule-associated protein 2 |
| NCDN | Neurochondrin |
| NMDAR1 | Glutamate [NMDA] receptor-associated protein 1 |
| PER | The proboscis extension reflex |

\* Corresponding author: bees1965@sina.com (ZJZ).

注：此文发表在 *Journal of Comparative Physiology A* 2015 年第 11 期。

| | |
|---|---|
| RGN | Regucalcin |
| SGMS1 | Phosphatidylcholine: ceramide cholinephosphotransferase 1 |
| SLC6A15 | Orphan sodium- and chloride-dependent neurotransmitter transporter NTT73 |
| SNAP25 | Synaptosomal-associated protein 25 |
| STX1 | Syntaxin-1A |
| VAChT | Vesicular acetylcholine transporter |

## Introduction

Honeybees, an important model organism for neuroethological studies, exhibit high behavioral plasticity and a remarkable ability to learn. Previous studies revealed that honeybees not only accurately remember the odor (Menzel *et al.* 1996), color (Frisch 1914) and shape (Srinivasan 1994) of a target but also learn the characteristics and sequences of landmarks to ensure a safe return to the hive (Collett *et al.* 2003). Moreover, they can generate associative memory to facilitate the search for a food source (Srinivasan *et al.* 1998). More recent study has also shown that honeybees exhibit cross-modal interaction between visual and olfactory learning (Zhang *et al.* 2014).

Honeybees exhibit strong olfactory abilities to ensure intra-specific communication and search for food (Sandoz *et al.* 2007; Wright *et al.* 2002). At present, the olfactory learning ability of honeybees is assessed in the laboratory using the proboscis extension reflex (PER) (Bitterman *et al.* 1983; Smith *et al.* 1992; Giurfa and Sandoz 2012). This behavioral response was initially used in gustative physiology studies of honeybees (Frings 1944). Specifically, it was first developed by Takeda (1961) as a research method of olfactory learning based on associating a sucrose reward with odorant.

In the last two decades, the molecular mechanisms of olfactory learning and memory in honeybees have been investigated substantially. The reduced expression of Protein Kinase A (PKA) and N-methyl-D-aspartate (NMDA) receptors were found to impair long-term memory during the olfactory learning of honeybees (Fiala *et al.* 1999; Si *et al.* 2004). In addition, an acetylcholine receptor (AChRs) (Dacher and Gauthier 2008), a metabotropic glutamate receptor (AmGluRA) (Kucharski *et al.* 2007), calcium/calmodulin-dependent kinase II (CaMKII) (Matsumoto *et al.* 2014), octopamine receptors (Farooqui *et al.* 2003) and an AmDOP1 receptor (Blenau *et al.* 1998) have also been demonstrated to be involved in the learning and memory processes of honeybees.

High-throughput sequencing and microarray technology are important methods to comprehensively unravel the underpinnings of olfactory learning in honeybees. Using a tag-based digital gene expression (DGE) and microarray transcriptome analysis, Wang *et al.* (2013a) and Cristino *et al.* (2014), respectively, demonstrated a general down-regulation of protein-coding genes after associative olfactory learning in *Apis mellifera*. Qin *et al.* (2014) found that 88.40 % of differentially expressed mRNAs are down-regulated after maze learning, as evidenced by DGE. The above-mentioned studies focused on RNAs and found many more down-regulated than up-regulated coding RNAs. However, very few studies have examined which proteins are involved in olfactory learning and memory in honeybees.

Isobaric tags for relative and absolute quantification (iTRAQ) is the latest, highly sensitive and accurate technique for the quantitative examination of proteomics. The technology, in combination with multidimensional liquid chromatography and tandem mass spectrometry, can simultaneously relatively or absolutely quantify up to eight protein samples. Moreover, iTRAQ can separate and identify a variety of proteins, including membrane proteins, proteins of high molecular weight, insoluble proteins, acidic proteins and alkaline proteins. This technology has been widely applied in various life science fields such as analyses of orange leaf proteomics (Song *et al.* 2012), molecular mechanisms underlying the regulation of the plant flowering phase (Ai *et al.* 2012), the formation mechanism of mollusk shells (Zhang *et al.* 2012), the protein expression spectrum of cancer cells (Wang *et al.* 2013b) and mammalian organelle assessment (Hakimov *et al.* 2009).

*Apis cerana*, an Asian honeybee, is found in China, Japan and Pakistan. Compared with *A. mellifera*, *A. cerana* has actually been shown to learn better in a controlled laboratory setting (Chen 2001; Qin *et al.* 2012). Wang and Tan (2014) showed that *A. cerana* is as amenable as *A. mellifera* to the study of olfactory learning using the PER assay. However, little is known about the molecular mechanisms of learning and memory in *A. cerana*. In this study, the iTRAQ approach was used to identify the protein expression associated with the olfactory learning of *A. cerana*.

## Materials and Methods

### Insect

A honeybee (*Apis cerana*) colony was maintained at the Honeybee Research Institute of Jiangxi Agricultural University in Nanchang, China (28.46 °N, 115.49 °E). The colony consisted of 4 frames and approximately 6000 bees. Frames with hundreds of 11-day-old pupae were packaged in a nylon net in the evening.

The next morning, newly emerged bees were removed from the nylon net and placed into a rectangular box containing 1 mol/L sucrose and bee-bread (pollen and sugar solution). After 1 week of incubation in the box, honeybees were collected from the box for the experiment.

### PER experiment

The honeybees were collected from the rectangular box in the morning of the eighth day for the PER experiment. The PER experimental procedure was based on the reports of Letzkus *et al.* (2006) and Wang *et al.* (2013a). The honeybees were briefly immobilized on ice for 3-5 min. Subsequently, each honeybee was mounted in a thin-walled copper tube (6 mm inner diameter) using a thin strip of GAFFA tape to immobilize the whole body, leaving the head and prolegs exposed. The fixed honeybees were randomly divided into two groups, the trained group and untrained (control) group. They were then allowed to recover in an incubator at a constant temperature of 34 °C and a constant humidity of 90 %.

In the afternoon, the honeybees were conditioned to both a rewarded odorant (CS+) and a punished odorant (CS-). The rewarded odorant contained a lemon essence plus 1 M sucrose solution and the aversive odorant contained a vanilla essence plus saturated NaCl solution, constituting a punishment. The odorant used in the present study are natural flavoring essences used for food (Queen Fine Foods Pty Ltd., Australia). The stimuli were presented as drops emerging from a 2.5 mL needle. A suction fan (20 cm × 23 cm) attached to a pipe was placed behind the honeybees to ensure both a continual stream of scented air during stimulus presentation and the quick removal of the residual odorant traces before the next bee was trained. On the first trial, the rewarded stimulus was presented approximately 1-2 cm away from the antennae of each honeybee for 5 s until the honeybee extended its proboscis and consumed a small amount of the stimulus. If the honeybee did not extend its proboscis within 5 s, the antennae were briefly touched with the stimulus to ensure that the scent was associated with the sucrose reward. Subsequently, the same procedure was performed with the unrewarded stimulus. During training, each honeybee was allowed to consume some of stimulus drop so they could learn to distinguish between lemon and vanilla odorants. This training trial for each bee was repeated three times with intervals of 5 min. The rewarded stimulus was always presented prior to the unrewarded stimulus. After training, the honeybees were fed using a 1 mol/L sucrose solution and then returned to the incubator (34 °C, humidity of 90 %) overnight. To reduce death, the honeybees were fed twice; at 23:00 h on the eighth day and at 6:00 h on the ninth day.

Retention tests were performed in the afternoon of the ninth day (24 h after training). The order of the presentation stimuli was reversed with respect to the training sessions—the unrewarded stimulus was offered first, followed by the rewarded stimulus. The drops of unrewarded stimulus and rewarded stimulus were presented at a distance of 1-2 cm in front of the antennae and for 5 s without touching the antennae. The test trial was repeated three times at an interval of 5 min for each bee. When all testing was done, only trained bees that showed a correct proboscis extension response (extended their proboscis when the rewarded stimulus was presented) in all three retention tests and control bees that were active and extended their proboscis in response to sucrose water were flash frozen in liquid nitrogen and stored at -80 °C.

### Protein extraction and iTRAQ labeling with iTRAQ reagents

Approximately 150 samples were collected from each group after the PER experiment. Pools of 50 brain tissues served as a biological replicate for protein extraction, and three biological replicates were employed for each group.

Whole brain tissue was manually dissected to obtain protein lysates. The protein was extracted according to the methods reported by Chen *et al.* (2012). The protein concentration of each sample was determined using Bovine serum albumin (BSA) as a protein standard based on the Bradford method, and the quality of proteins was analyzed by SDS-PAGE (120 V, 120 min). The extracted proteins were then digested using Trypsin Gold (Promega, Madison, WI, USA) at 37 °C for 16 h at a protein:trypsin ratio of 30:1.

After trypsin digestion, the peptides were dried and resuspended in 0.5 M TEAB (Applied Biosystems, Milan, Italy). The peptides were then labeled using 8-plex iTRAQ reagent according to the manufacturer's protocol (Applied Biosystems, Foster City, CA, USA). The labeling reaction was incubated at room temperature for 2 h. The samples were then labeled as follows: 114 trained group; 116 untrained group; 117 trained group; 118 untrained group; 119 trained group; 121 untrained group. Subsequently, the labeled peptides were pooled, dried by vacuum centrifugation and stored at -80 °C for mass spectrometry (MS) analyses.

### SCX Fractionation of Peptides

The iTRAQ-labeled peptide mixtures were fractionated with an Ultremex SCX column (Phenomenex) according to the description by Kuss *et al.* (2012). A total of 20 fractions were collected, desalted with an Ultremex SCX column (Phenomenex) and vacuum-dried for LC-MS/MS analysis.

### LC-ESI-MS/MS analysis based on Q EXACTIVE

The iTRAQ-labeled peptides were analyzed using a LTQ Orbitrap velos instrument (Thermo Fisher Scientific, San Jose, CA) coupled with a LC-20AD nanoHPLC (Shimadzu, Kyoto, Japan). Each fraction was resuspended, then separated with a 2 cm C18 trap column and finally packed with a 10 cm analytical C18 column. The peptides were eluted with an acetonitrile gradient from 5 to 80 % for 44 min at a velocity 300 nL/min. The peptides eluted from the column directly entered the ESI-MS/MS at a resolution of 17,500. For MS scans, the m/z scan range was 100-2000, and the electrospray voltage was 1.6 kV. High-energy collision dissociation (HCD) operating mode and automatic gain control (AGC) were used to select peptides and optimize the spectra from the orbitrap, respectively.

### Mass spectrometric data analysis

The raw data files obtained from the orbitrap were converted into MGF files using proteome and the proteins were identified using the Mascot search engine (Matrix Science, London, UK; version 2.3.02) against a database containing Apis_cerana (5594 sequences) (http://www.ncbi.nlm.nih.gov/protein?term=txid7461[Organism]). The following search parameters were employed: peptide mass tolerance at 10 ppm; fragment mass tolerance at 0.05 Da; trypsin as the enzyme with allowance for one missed cleavage; Carbamidomethyl (C), iTRAQ 8plex (N-term), and iTRAQ 8plex (K) as fixed modifications; Gln->pyro-Glu (N-term Q), Oxidation (M), and Deamidated (NQ) as the potential variable modifications; and a peptide charge of 2+ or 3+. To reduce the probability of false peptide identification, the peptides were filtered with significance scores ($\geq 20$) at the 99% confidence interval and involved at least one unique peptide.

For protein quantification, a protein was required to contain at least two unique peptides. The quantitative protein ratios were weighted and normalized by the median ratio in Mascot.

### Identification of differentially expressed proteins

The differentially expressed proteins were selected according to the following cut-off criteria: the protein ratio in at least one biological replicate meets the fold change ($\geq 1.2$ or $\leq 0.833$ at $p < 0.05$), and the tendency of protein expression (the fold change $\geq 1.0$ or $\leq 1.0$) is consistent with the three biological replicates.

A GO annotation analysis can determine the main biological function of the differentially expressed proteins by searching for significantly enriched GO terms in differentially expressed proteins compared with the enrichment in all identified proteins. Specifically, a GO enrichment analysis applies a hypergeometric test to map all differentially expressed proteins to terms (molecular function, cellular component and biological process) in the GO database (http://www.geneontology.org/). The test employs the following formula:

$$P = 1 - \sum_{i=0}^{m-1} \frac{\binom{M}{i}\binom{N-M}{n-i}}{\binom{N}{n}}$$

where N is the number of all proteins with GO annotation; n is the number of differentially expressed proteins in N; M is the number of all proteins that are annotated to the certain GO terms; and m is the number of differentially expressed proteins that are annotated to certain GO terms.

The KEGG pathway enrichment analysis (http://www.genome.jp/kegg/) was utilized to identify significantly enriched biochemical pathways or signal transduction pathways in differentially expressed proteins compared with all identified proteins. The formula used for the pathway analysis is the same as that used for the GO analysis.

## Results

### *iTRAQ analysis of protein identification*

Using a PER assay, *A. cerana* was trained to associate one odorant with a sugar reward and another with a salt water punishment. After the third training trial, the bees showed a good level of learning for the odorant as nearly 75 % of bees responded correctly (Supplemental Figure S1a). Approximately 14 % of the trained bees also showed correct proboscis extension responses in all three retention trials the following day (Supplemental Figure S1b). Consequently, these bees were considered for long-term memory training and further sampled for iTRAQ analysis.

The iTRAQ technique was performed to obtain a global view of the proteome differences between the trained and untrained groups of *A. cerana*. In the mass spectrum experiment, a total of 307,252 MS spectra were obtained. A total of 65,256 spectra were successfully matched to peptide fragments, and 60,662 spectra were matched to unique peptide fragments with a Mascot analysis (Table III-5-1). Among these spectra, 1.5 % showed multiple matches, and 78.76 % did not match the peptide. Moreover, 2406 proteins were identified from 13,995 unique peptide sequences deduced from 60,662 spectra based on the *A. cerana* (5594 sequences) database (Table III-5-1). The identified proteins were used to further analyze the differential expression of proteins.

Table III-5-1 Summary statistics for iTRAQ analyses of brain proteins of trained and untrained *Apis cerara*

| Group name | Total spectra | Spectra | Unique spectra | Peptide | Unique Peptide | Protein |
|---|---|---|---|---|---|---|
| *Apis cerara* | 307252 | 65256 | 60662 | 14500 | 13995 | 2406 |

A statistical analysis showed that 45.14 % of the identified proteins had a coverage greater than 20, and 14 % of proteins had a coverage below 5 % (Supplemental Fig. S2). In addition, approximately 1806 (75.47 %) of all identified proteins were identified with at least two peptides per protein, and 2049 (85.16 %) were identified within ten peptides (Supplemental Fig. S3).

### *Differentially expressed proteins*

All identified proteins were filtered based on a ratio ≥1.2-fold or ≤0.833-fold at $p < 0.05$ in at least one biological replicate and the consistency of protein expression in the three biological replicates. Moreover, the three biological replicates of each sample exhibited a mean CV (coefficient of variation) of 0.10 (Supplemental

Figure S4), suggesting the high reliability of the results. A total of 147 proteins were differentially expressed between the trained and untrained groups, 87 of which (59.18 %) were up-regulated and 60 of which (40.82%) were down-regulated in the trained group compared with the untrained group, based on the above criteria (Fig. III-5-1, Supplemental file 1).

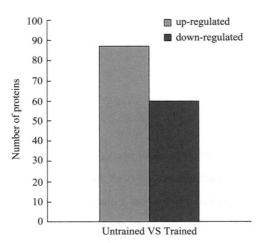

**Figure III-5-1** Differential expression analysis of proteins in the trained and untrained groups. A total of 147 differentially expressed proteins were detected, with 87 up-regulated proteins and 60 down-regulated proteins in the trained group compared with the untrained group.

### *Function of the differentially expressed proteins*

The biological functions of these differentially expressed proteins were investigated based on the Gene Ontology database. Of the 147 differentially expressed proteins, 98 proteins were successfully mapped to one or more GO terms, whereas 49 proteins were not classified. Among the 98 proteins mapped to the GO terms, 77 (78.57 %), 63 (64.28 %), and 83 (84.69 %) are involved in biological processes, cellular components and molecular functions, respectively (Fig. III-5-2). Of the 77 proteins involved in biological processes, 57, 45, and 41 are implicated in the cellular processes, metabolic processes and single-organism processes, respectively (Fig. III-5-2a). Similarly, 39, 39, and 29 of the proteins involved in cellular components are related to cells, cell parts, and membranes of the cellular components, respectively (Fig. III-5-2b). A total of 53 proteins exhibit catalytic activity, and 46 proteins function in molecular binding (Fig. III-5-2c). Moreover, the GO significant enrichment analysis indicated that 37, 19 and 6 terms were significantly enriched ($Q$ value <0.05) from the biological process ontology, the molecular function ontology and the cellular components ontology, respectively, compared with all identified proteins (Supplementary file 2).

The biochemical pathways of the differentially expressed proteins were investigated based on the KEGG database. Of the 147 differentially expressed proteins, 114 proteins were associated with a KO ID and involved in 132 pathways (Supplementary file 3). Compared with the background of all identified proteins, three pathways (peroxisome, vitamin digestion and absorption, riboflavin metabolism) were significantly enriched ($Q$ value <0.05).

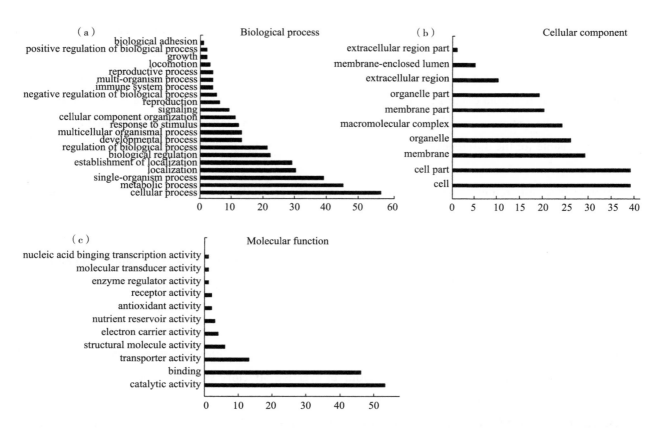

**Figure III-5-2** Functional categorization of the differentially expressed proteins between the trained and untrained groups. The proteins were arranged in terms of GO classification, and the number of proteins in each category is displayed based on a biological process, b cellular component, and c molecular function.

## Discussion

The results of the iTRAQ analysis quality assessment, including the distribution of protein sequence coverage (Supplementary Fig. S2), peptide number distribution (Supplementary Fig. S3) and reproducibility analysis (Supplementary Fig. S4) showed that our iTRAQ data were accurate and reliable. Although the gene and protein databases of *A. cerana* were incomplete, 147 proteins were found to be differentially expressed between the trained and untrained groups, and some of these proteins are reportedly related to learning and memory, such as neurotransmitter transporter proteins, synaptic proteins, neurotransmitter receptor proteins and neurochondrin.

Intriguingly, although the expression levels of two proteins (gi|151368183| and gi|373194840|) were in line with the results reported by Wang *et al*. (2013a), 87 (59.15 %) differentially expressed proteins were up-regulated after learning in *A. cerana*. This observation was not consistent with the reported down-regulation of coding genes after learning in *A. mellifera* by Wang *et al*. (2013a) and Cristino *et al*. (2014). The present experimental procedure was similar to that utilized by Wang *et al*. (2013a), but different results were obtained. This difference may be due to significant changes in the mRNA expression levels after the test that are not necessarily accompanied by corresponding changes in protein expression during the same period because of delays in protein translation related to mRNA transcription. In addition, the protein expression reported herein also differs from the findings reported by Cristino *et al.* (2014), and these differences may be attributed to differences in the experimental procedures, including the odorant presentation, feeding appetitive and aversive stimuli and treatment of the control group. Recent studies have indicated that changes in gene expression are not frequently reflected at the protein level (Guo *et al.* 2008; Vogel and Marcotte 2012). This phenomenon may also be responsible for the differences between the reported mRNA and protein expression levels in response to bee learning.

Among these differentially expressed proteins, two neurotransmitter transporter proteins; orphan sodium- and chloride-dependent neurotransmitter transporter NTT73 (SLC6A15, gi|373212984|) and vesicular acetylcholine transporter (VAChT, gi|373195337|) were found to show significant expression differences after memory formation (Table III-5-2). Acetylcholine (ACh) is major neurotransmitter in the central and peripheral nervous system, whereas VAChT mediates the storage and release of Ach by synaptic vesicles (de Castro *et al.* 2009). Furthermore, the elimination of the VAChT gene from the forebrain impairs synaptic plasticity and causes deficits that interfere with learning and memory (Martyn *et al.* 2012; de Castro *et al.* 2009). SLC6A15 mRNA is widely expressed in neurons of the olfactory bulb, cerebral cortex and hippocampus (Inoue *et al.* 1996). SLC6A15-null mice are less likely to avoid an aversive olfactory stimulus in initial experiments, but these data have not been reproducible (Drgonova *et al.* 2007). In addition, two synaptic proteins; synaptosomal-associated protein 25 (SNAP25, gi|373201736|) and syntaxin-1A (STX1, gi|373212420|) were found to be significantly down-regulated after learning (Table III-5-2). HPC-1/syntaxin-1 is involved in the synaptic plasticity of the hippocampus in vivo. Furthermore, this protein participates in the consolidation of conditioned fear memory as a member of the syntaxin super-family (Fujiwara *et al.* 2006). Previous studies have also suggested that SNAP-25 is involved in cognitive dysfunction, verbal memory and memory consolidation (Spellmann *et al.* 2008; Golimbet *et al.* 2010; Hou *et al.* 2004). Moreover, STX1 interacts with SNAP-25 protein, which plays an essential role in the regulation of neurotransmitter release (Lin and Scheller 2000). Thus, SLC6A15, VAChT, STX1 and SNAP-25 may be required during the olfactory learning process of *A. cerana*.

Table III-5-2 **List of part differentially expressed proteins implicated the learning and memory between trained and untrained group by iTRAQ analysis**

| Accession ID | Protein name | Fold change | | | No. of peptides |
|---|---|---|---|---|---|
| | | 116/114 | 118/117 | 121/119 | |
| up-regulated | | | | | |
| gi|373198053| | Phosphatidylcholine:ceramide cholinephosphotransferase 1 | 1.29 * | 1.34 | 1.197 | 2 |
| gi|373195337| | Vesicular acetylcholine transporter | 1.183 | 1.221 * | 1.173 | 4 |
| gi|373217289| | RAC serine/threonine-protein kinase | 1.216 * | 1.305 * | 1.183 * | 4 |
| gi|373208367| | Regucalcin | 1.38 * | 1.595 * | 1.6 | 5 |
| gi|373204511| | Neurochondrin | 1.285 * | 1.404 * | 1.007 | 1 |

(continued)

| Accession ID | Protein name | Fold change | | | No. of peptides |
|---|---|---|---|---|---|
| | | 116/114 | 118/117 | 121/119 | |
| down-regulated | | | | | 2 |
| gi\|373197014\| | Synaptosomal-associated protein 25 | 0.82 | 0.776 * | 0.94 | 3 |
| gi\|373198326\| | Microtubule-associated protein 2 | 0.926 | 0.811 * | 0.952 | 19 |
| gi\|373194043\| | Calcium/calmodulin-dependent protein kinase | 0.792 * | 0.928 | 0.942 * | 1 |
| gi\|373212420\| | Syntaxin-1A | 0.876 | 0.707 | 0.778 * | 11 |
| gi\|373205317\| | Glutamate [NMDA] receptor-associated protein 1 | 0.899 * | 0.549 * | 0.918 | 1 |
| gi\|373212984\| | Orphan sodium- and chloride-dependent neurotransmitter transporter NTT73 | 0.833 * | 0.777 * | 0.937 | 1 |

ID represents accession numbers

* indicates significance of differential expression; Fold change represents the ratios of trained/untrained group; No. of peptides represents number of peptides identified for each protein.

In our study, glutamate [NMDA] receptor-associated protein 1 (NMDAR1, gi|373205317|), calcium/calmodulin-dependent protein kinase (CAMK, gi|373194043|) and microtubule-associated protein 2 (MAP2, gi|373198326|) were found to be significantly down-regulated after memory formation. Conversely, phosphatidylcholine: ceramide cholinephosphotransferase 1 (SGMS1, gi|373198053|), regucalcin (RGN, gi|373208367|), RAC serine/threonine-protein kinase(AKT, gi|373217289|)and neurochondrin (NCDN, gi|373204511|) were significantly up-regulated after memory formation (Table III-5-2). Previous studies also demonstrated that these proteins are directly or indirectly involved in learning and memory (Si *et al*. 2004; Matsumoto *et al*. 2014; Woolf *et al*. 1999; Lim and Suzuki 2008; Yamaguchi 2000; Yao *et al*. 2011; Wang *et al*. 2009). NMDAR1 is expressed throughout the brain, in neurons and in glial cells (Zannat *et al*. 2006), and plays a critical role in olfactory long-term memory formation in Drosophila (Xia *et al*. 2005) as well as *A. mellifera* (Si *et al*. 2004). CaMKII has also been reported to play an essential role in memory formation and storage (Cammarota *et al*. 1998; Coultrap and Bayer 2012). Recently, a pharmacological experiment showed that CaMKII was also required for the formation of LTM during the olfactory conditioning of honeybees (Matsumoto *et al*. 2014). The changes in the MAP-2 levels may reflect dendritic remodeling related to contextual memory storage (Woolf *et al*. 1999). The reduced expression of MAP-2 and synaptophysin in the hippocampus of rats is thought to contribute to cognitive impairment (Hai *et al*. 2010). SGMS1 is required to convert sphingomyelin and diacylglycerol to phosphatidylcholine and ceramide, and phosphatidylcholine can improve the maze-learning performance of adult mice (Lim and Suzuki 2008). Regucalcin, a calcium-binding protein, has been demonstrated to play an important role in $Ca^{2+}$ signaling and is implicated in the long-term potentiation of neuronal plasticity (Brocher *et al*. 1992; Yamaguchi 2000). Furthermore, regucalcin may be involved in learning and memory. In addition, AKT is involved in the PI3K/Akt signaling pathway, which is implicated in learning and memory (Musumeci *et al*. 2009; Yao *et al*. 2011). Furthermore, a deficiency in NCDN, which is predominantly expressed in the nervous system, has been shown to cause serious spatial learning defects in neurons (Shinozaki et al. 1997; Dateki et al. 2005) and impaired synaptic plasticity (Wang et al. 2009).

The results of the KEGG pathway enrichment analysis indicate that some metabolic or signaling pathways may be involved in olfactory learning in A. cerana, including SNARE interactions in vesicular transport, the neurotrophin signaling pathway, the MAPK signaling pathway and the dopaminergic synapse signaling pathway (Supplementary file 3). The differentially expressed proteins, such as SNAP25, STX1, AKT, AKT, MAP-2, and AKT, participate in these signaling pathways. These proteins have been documented to be related to learning and memory, as described above.

In conclusion, the iTRAQ technology was used to identify a total of 147 proteins that were differentially expressed in A. cerana in response to olfactory learning. Some of these proteins are thought to be implicated in learning and memory from previous studies. The present study provides the first report of olfactory learning- and memory-related proteins in A. cerana, and these proteins will help to elucidate the detailed molecular mechanisms underlying learning and memory in honeybees in future studies.

# References

AI XY, LIN G, SUN LM, HU CG, GUO WW, DENG XX, ZHANG JZ (2012) A global view of gene activity at the flowering transition phase in precocious trifoliate orange and Its Wild-type [Poncirus trifoliate (L.) Raf.] by transcriptome and proteome analysis. *Gene* 510: 47-48.

BITTERMAN ME, MENZEL R, FIETZ A, SCHAFER S (1983) Classical conditioning of proboscis extension in honeybees (*Apis mellifera*). *J Comp Psychol* 97:107-119.

BLENAU W, ERBER J, BAUMANN A (1998) Characterization of a dopamine D1 receptor from *Apis mellifera*: cloning, functional expression, pharmacology, and mRNA localization in the brain. J Neurochem 70:15-23.

BROCHER S, ARTOLA A, SINGER W (1992) Intracellular injection of $Ca^{2+}$ chelators blocks induction of long-term depresion in rat visual cortex. *Proc Natl Acad Sci USA*. 89:123-127.

CAMMAROTA M, BERNABEU R, LEVI DE STEIN M, IZQUIERDO I, MEDINA JH (1998) Learning-specific, time-dependent increases in hippocampal $Ca^{2+}$/calmodulin-dependent protein kinase II activity and AMPA GluR1 subunit immunoreactivity. *Eur J Neurosci* 10:2669-2676.

CHEN S (2001) The apicultural science in China. China Agriculture Press, Beijing.

CHEN Z, WANG Q, LIN L, TANG Q, EDWARDS JL, LI S, LIU S (2012) Comparative Evaluation of Two Isobaric Labeling Tags, DiART and iTRAQ. *Anal Chem* 84:2908-2915.

COLLETT TS, GRAHAM P, DURIER V (2003) Route learning by insects. *Curr Opin Neurobiol* 13:718-725.

COULTRAP SJ, BAYER KU (2012) CaMKII regulation in information processing and storage. *Trends Neurosci* 35:607-618.

CRISTINO AS, BARCHUK AR, FREITAS FC, NARAYANAN RK, BIERGANS SD, ZHAO Z, SIMOES ZL, REINHARD J, CLAUDIANOS C (2014) Neuroligin-associated microRNA-932 targets actin and regulates memory in the honeybee. *Nat Commun* 5:5529.

DACHER M, GAUTHIER M (2008) Involvement of NO-synthase and nicotinic receptors in learning in the honey bee.

*Physiol Behav* 95:200-207.

DATEKI M, HORII T, KASUYA Y, MOCHIZUKI R, NAGAO Y, ISHIDA J, SUGIYAMA F, TANIMOTO K, YAGAMI K, IMAI H, FUKAMIZU A (2005) Neurochondrin negatively regulates CaMKII phosphorylation, and nervous system-specific gene disruption results in epileptic seizure. *J Biol Chem* 280:20503-20508.

DE CASTRO BM, PEREIRA GS, MAGALHÃES V, ROSSATO JI, DE JAEGER X, MARTINS-SILVA C, LELES B, LIMA P, GOMEZ MV, GAINETDINOV RR, CARON MG, IZQUIERDO I, CAMMAROTA M, PRADO VF, PRADO MA (2009) Reduced expression of the vesicular acetylcholine transporter causes learning deficits in mice. *Genes Brain Behav* 8: 23-35.

DRGONOVA J, LIU QR, HALL FS, KRIEGER RM, UHL GR (2007) Deletion of v7-3 (SLC6A15) transporter allows assessment of its roles in synaptosomal proline uptake, leucine uptake and behaviors. *Brain Res* 1183:10-20.

FAROOQUI T, ROBINSON K, VAESSIN H SMITH BH (2003) Modulation of early olfactory processing by an octopaminergic reinforcement pathway in the honeybee. *J Neurosci* 23:5370-5380.

FIALA A, MULLER U, MENZEL R (1999) Reversible downregulation of protein kinase A during olfactory learning using antisense technique impairs long-term memory formation in the honeybee, *Apis mellifera*. *J Neurosci* 19:10125-10134.

FRINGS H (1944) The loci of olfactory end-organsin the honeybee, *Apis mellifera* Linn. *J. Exp Zool* 97: 123-134.

FUJIWARA T, MISHIMA T, KOFUJI T, CHIBA T, TANAKA K, YAMAMOTO A, AKAGAWA K (2006) Analysis of knock-out mice to determine the role of HPC-1/syntaxin 1A in expressing synaptic plasticity. *J Neurosci* 26:5767-5776.

Giurfa M, Sandoz JC (2012) Invertebrate learning and memory: Fifty years of olfactory conditioning of the proboscis extension response in honeybees. *Learn Mem* 19:54-66.

GOLIMBET VE, ALFIMOVA MV, GRITSENKO IK, LEZHEIKO TV, LAVRUSHINA OM, ABRAMOVA LI, KALEDA VG, BARKHATOVA AN, SOKOLOV AV, EBSTEIN RP (2010) Association between a synaptosomal protein (SNAP-25) gene polymorphism and verbal memory and attention in patients with endogenous psychoses and mentally healthy subjects. *Neurosci Behav Physiol* 40:461-465.

GUO Y, XIAO P, LEI S, DENG F, XIAO GG, LIU Y, CHEN X, LI L, WU S, CHEN Y, JIANG H, TAN L, XIE J, ZHU X, LIANG S, DENG H (2008) How is mRNA expression predictive for protein expression? A correlation study on human circulating monocytes. *Acta Biochim Biophys Sin* (Shanghai), 40:426-436.

HAI J, SU SH, LIN Q, ZHANG L, WAN FJ, LI H, CHEN YY, LU Y (2010) Cognitive impairment and changes of neuronal plasticity in rats of chronic cerebral hypoperfusion associated with cerebral arteriovenous malformations. *Acta Neurol Belg* 110:180-185.

HAKIMOV HA, WALTERS S, WRIGHT TC, MEIDINGER RG, VERSCHOOR CP, GADISH M, CHIU DKY, STROMVIK MV, FORSBERG C W, GOLOVAN SP (2009) Application of iTRAQ to catalogue the skeletal muscle proteome in pigs and assessment of effects of gender and diet dephytinization. *Proteomics* 9:4000-4016.

HOU Q, GAO X, ZHANG X, KONG L, WANG X, BIAN W, TU Y, JIN M, ZHAO G, LI B, JING N, YU L (2004) SNAP-25 in hippocampal CA1 region is involved in memory consolidation. *Eur J Neurosci* 20:1593-1603.

INOUE K, SATO K, TOHYAMA M, SHIMADA S, UHL GR (1996) Widespread brain distribution of mRNA encoding the orphan neurotransmitter transporter v7-3. *Brain Res Mol Brain Res* 37: 217-223.

KUCHARSKI R, MITRI C, GRAU Y, MALESZKA R (2007) Characterization of a metabotropic glutamate receptor in the honeybee (*Apis mellifera*): implications for memory formation. *Invert Neurosci* 7:99-108.

KUSS C, GAN CS, GUNALAN K, BOZDECH Z, SZE SK, PREISER PR (2012) Quantitative proteomics reveals new insights into erythrocyte invasion by Plasmodium falciparum. *Mol Cell Proteomics* 11:M111.010645.

LETZKUS P, RIBI WA, WOOD JT, ZHU H, ZHANG SW, SRINIVASAN MV (2006) Lateralization of olfaction in the honeybee *Apis mellifera*. *Curr Biol* 16: 1471-1476.

LIM SY, SUZUKI H (2008) Dietary phosphatidylcholine improves maze-learning performance in adult mice. *J Med Food* 11:86-90.

LIN RC, SCHELLER RH (2000) Mechanisms of synaptic vesicle exocytosis. *Annu Rev Cell Dev Biol* 16:19-49.

MARTYN AC, DE JAEGER X, MAGALHA˜ES AC, KESARWANI R, GONÇALVES DF, RAULIC S, GUZMAN MS, JACKSON MF, IZQUIERDO I, MACDONALD JF, PRADO MA, PRADO VF (2012) Elimination of the vesicular acetylcholine transporter in the forebrain causes hyperactivity and deficits in spatial memory and long-term potentiation. *Proc Natl Acad Sci USA*. 109:17651-17656.

MATSUMOTO Y, SANDOZ JC, DEVAUD JM, LORMANT F, MIZUNAMI M, GIURFA M (2014) Cyclic nucleotide-gated channels, Calmodulin, adenylyl cyclase and calcium/ calmodulin-dependent protein kinase II are required for late but not early long-term memory formation in the honey bee. *Learn Mem* 21: 272-286.

MENZEL R, HAMMER M, MÜLLER U, ROSENBOOM H (1996) Behavioral, neural and cellular components underlying olfactory learning in the honeybee. *J Physiol Paris* 90:395-398.

MUSUMECI G, SCIARRETTA C, RODRÍGUEZ-MORENO A, AL BANCHAABOUCHI M, NEGRETE-DÍAZ V, COSTANZI M, BERNO V, EGOROV AV, VON BOHLEN UND HALBACH O, CESTARI V, DELGADO-GARCÍA JM, MINICHIELLO L (2009) TrkB modulates fear learning and amygdalar synaptic plasticity by specific docking sites. *J Neurosci* 29:10131-10143.

QIN QH, HE XJ, TIAN LQ, ZHANG SW, ZENG ZJ (2012) Comparison of learning and memory of *Apis cerana* and *Apis mellifera*. *J Comp Physiol A* 198:777-786.

QIN QH, WANG ZL, TIAN LQ, GAN HY, ZHANG SW, ZENG ZJ (2014) The integrative analysis of microRNA and mRNA expression in *Apis mellifera* following maze-based visual pattern learning. *Insect Science* 21:619-636.

SANDOZ JC, DEISIG N, DEBRITO SANCHEZ M G, GIURFA M (2007) Understanding the logics of pheromone processing in the honeybee brain : from labeled-lines to across-fiber patterns. *Front Behav Neurosci* 1:5.

SHINOZAKI K, MARUYAMA K, KUME H, KUZUME H, OBATA K (1997) A novel brain gene, norbin, induced by treatment of tetraethylammonium in rat hippocampal slice and accompanied with neurite-outgrowth in neuro 2a cells. *Biochem Biophys Res Commun* 240:766-771.

SI A, HELLIWELL P, MALESZKA R (2004) Effects of NMDA receptor antagonists on olfactory learning and memory in the honeybee (*Apis mellifera*). *Pharmacol Biochem Behav* 77:191-197.

SMITH BH, ABRAMSON CI, NADEL LI (1992) Insect learning: case studies in comparative psychology. In: Squire LR (ed) Encyclopedian of learning and memory. Macmillan, New York, 276-283.

SONG J, SUN R, LI D, TAN F, LI X, JIANG P, HUANG X, LIN L, DENG Z, ZHANG Y (2012) An Improvement of Shotgun Proteomics Analysis by Adding Next-Generation Sequencing Transcriptome Data in Orange. *PLoS One* 7:e39494.

SPELLMANN I, MÜLLER N, MUSIL R, ZILL P, DOUHET A, DEHNING S, CEROVECKI A, BONDY B, MÖLLER H-J, RIEDEL M (2008) Associations of SNAP-25 polymorphisms with cognitive dysfunctions in Caucasian patients with schizophrenia during a brief trail of treatment with atypical antipsychotics. *Eur Arch Psychiatr Clin Neurosci* 258:335-344.

SRINIVASAN MV (1994) Pattern recognition in the honeybee: recent progress. *J Insect Physiol* 40:183-194.

SRINIVASAN MV, ZHANG S, ZHU H (1998) Honeybees link sights to smells. *Nature* 396:637-638.

TAKEDA K (1961) Classical conditioned response in the honey bee. *J Inseet Physiol* 6:168-179.

VOGEL C, MARCOTTE EM (2012) Insights into the regulation of protein abundance from proteomic and transcriptomic analyses. *Nat Rev Genet*, 13:227-232.

VON FRISCH K (1914) Der Farbensinn und Formensinn der Biene. Fischer, Jena.

WANG Z, TAN K (2014) Comparative analysis of olfactory learning of *Apis cerana* and *Apis mellifera*. *Apidologie* 45:45-52.

WANG H, WESTIN L, NONG Y, BIRNBAUM S, BENDOR J, BRISMAR H, NESTLER E, APERIA A, FLAJOLET M, GREENGARD P (2009) Norbin is an endogenous regulator of metabotropic glutamate receptor 5 signaling. *Science* 326:1554-1557.

WANG Q, WEN B, YAN G, WEI J, XIE L, XU S, JIANG D, WANG T, LIN L, ZI J, ZHANG J, ZHOU R, ZHAO H, REN Z, QU N, LOU X, SUN H, DU C, CHEN C, ZHANG S, TAN F, XIAN Y, GAO Z, HE M, CHEN L, ZHAO X, XU P, ZHU Y, YIN X, SHEN H, ZHANG Y, JIANG J, ZHANG C, LI L, CHANG C, MA J, YAN G, YAO J, LU H, YING W, ZHONG F, HE QY, LIU S (2013a) Qualitative and quantitative expression status of the human chromosome 20 genes in cancer tissues and the representative cell lines. *J Proteome Res* 12:151-161.

WANG ZL, WANG H, QIN QH, ZENG ZJ (2013b) Gene expression analysis following olfactory learning in *Apis mellifera*. *Mol Biol Rep* 40:1631-1639.

WOOLF NJ, ZINNERMAN MD, JOHNSON GVW (1999) Hippocampal microtubule-associated protein-2 alterations with contextual memory. *Brain Res* 821:241-249.

WRIGHT GA, SKINNER BD, SMITH BH (2002) Ability of honeybee, *Apis mellifera*, to detect and discriminate odors of varieties of canola (*Brassica rapa* and *Brassica napus*) and snapdragon flowers (*Antirrhinum majus*). *J Chem Ecol* 28:721-740.

XIA S, MIYASHITA T, FU TF, LIN WY, WU CL, PYZOCHA L, LIN IR, SAITOE M, TULLY T, CHIANG AS (2005) NMDA receptors mediate olfactory learning and memory in Drosophila. *Current Biol.* 15:603-615.

YAMAGUCHI M (2000) Role of regucalcin in brain calcium signaling. *Life Sci* 66(19):1769-1780.

YAO D, HE X, WANG JH, ZHAO ZY (2011) Effects of PI3K/Akt signaling pathway on learning and memory abilities in neonatal rats with hypoxic-ischemic brain damage. *Zhongguo Dangdai Er Ke Za zhi* 13:424-427.

ZANNAT MT, LOCATELLI F, RYBAK J, MENZEL R, LEBOULLE G (2006) Identification and localisation of the NR1 sub-unit homologue of the NMDA glutamate receptor in the honeybee brain. *Neurosci Lett* 398:274-279.

ZHANG G, FANG X, GUO X, LI L, LUO R, XU F, YANG P, ZHANG L, WANG X, QI H *et al* (2012) The oyster genome reveals stress adaptation and complexity of shell formation. *Nature* 90:49-54.

ZHANG LZ, ZHANG SW, WANG ZL, YAN WY, ZENG ZJ (2014) Cross-modal interaction between visual and olfactory learning in *Apis cerana*. *J Comp Physiol A* 200:899-909.

# 6. Deltamethrin Impairs Honeybees (*Apis mellifera*) Dancing Communication

Zuyun Zhang[1,2,†], Zhen Li[1,†], Qiang Huang[1], Xuewen Zhang[2], Li Ke[1], Weiyu Yan[1], Lizhen Zhang[1], Zhijiang Zeng[*]

1. Honeybee Research Institute, Jiangxi Agricultural University, Nanchang, 330045, China;
2. Sericultural and Apicultural Institute, Yunnan Academy of Agricultural Sciences, Mengzi, 661101, China

**Abstract**

As a commonly used pyrethroid insecticide, deltamethrin is very toxic to honeybees, which seriously threatens the managed and feral honeybee population. As deltamethrin is a nerve agent, it may interfere with the nervous system of honeybees such as dance behavior and memory-related characteristics. We found the waggle dances were less precise in honeybees consumed syrup containing deltamethrin (pesticide group) than those consumed normal sucrose syrup (control group). Compared with the control group, honeybees of the pesticide group significantly increased number of circuits per 15 seconds, the divergence angle, return phases in waggle dances, as well as the crop content of the dance followers. Furthermore, six learning and memory related genes were significantly interfered with the gene expression levels. Our data suggest that the sub-lethal dose of deltamethrin impaired the honeybees' learning and memory and resulted in cognitive disorder. The novel results assist in establishing guidelines for the risk assessment of pesticide to honeybee safety and prevention of non-target biological agriculture pesticide poisoning.

**Keywords:** *Apis mellifera*, deltamethrin, waggle dance, learning and memory-related genes, healthy

## Introduction

Honeybees are ecologically and economically vital pollinators for both wild and cultivated flowers. Honeybee population is in decline, caused by multiple factors, including pathogens and parasites (Cornman *et al.* 2012; Francis *et al.* 2013), pesticides (Henry *et al.* 2012; Gill *et al.* 2012), and other human induced stressors (Goulson *et al.* 2015). As an important pesticide, pyrethroids are applied to a wide range of crop plants. Exposure to pyrethroids is known to have deleterious effects on honeybees (Liao *et al.* 2018). Deltamethrin is a type II semi-synthetic pyrethrin, which acts as a potent inhibitor of calcineurin (CN) and affects the cellular immune response, signal transduction pathway and biological function (Enan and

---

† These two authors contributed equally.
*Correspondence author: bees1965@sina.com (ZJZ).
注：此文发表在 *Archives of Environmental Contamination and Toxicology* 2020 年第 1 期。

Matsumura 1992). Deltamethrin can induce sub-lethal effects such as impaired olfaction and disturbed learning (Decourtye et al. 2004), disturbed orientation (Thompson 2003; Vandame et al. 1995), altered foraging activity and reduced memory (Ramirez-Romero et al. 2005). Sub-lethal concentrations of 21.6 mg/mL (sucrose solution) deltamethrin reduced honeybees' fecundity and impaired the development of honeybees (Dai et al. 2010).

Waggle dance is a well-studied and surprisingly sophisticated example of animal communication. The waggle dance was first deciphered by von Frisch (1967), who determined that honeybee foragers communicate the location of profitable flower patches to hive-mates using the dance. The direction and distance to a patch are indicated by the angle and duration of the waggle run, respectively. The quantity and quality of nectar and pollen available from various plant species is communicated by the number of dancing honeybees and frequency of their recruiting behavior (von Frisch 1967). Honeybees are also capable of avoiding flowers containing cues of elevated risk, which seems plausible that experienced foragers modify the waggle dance to facilitate the avoidance of predation risk by naive recruits (Jack-McCollough and Nieh 2015). Also, some protection to the colony is achieved when hazards kill or delay the return of affected foragers to the hive, thus interfering with communication (Abbott and Dukas 2009).

In this study, we investigated honeybees' responses towards nectar with pesticide contamination risk in terms of waggle dance and memory related genes expression. We predict that honeybees modify their recruitment behavior to alarm the nest-mates as indicated by irregular dance with greater variance on the location of food resources.

## Materials and Methods

### Sub-lethal dose of deltamethrin preparation

The deltamethrin solution (J&K scientific and chemical company) was prepared by adding distilled water to deltamethrin, then diluting to 50% sucrose water and 235 μg/kg deltamethrin (Decourtye et al. 2005).

### Honeybees training

Honeybee colonies (A. mellifera) were maintained at the Honeybee Research Institute, Jiangxi Agricultural University, Nanchang, China (28.46 °N, 115.49 °E) according to standard beekeeping techniques.

Nine honeybee colonies were sequentially used in test, each with four frames of honeybees and brood in an observation hive. On each experimental day, about 100 foragers at the entrance were captured and placed into individual opaque tubes. Then the honeybees were released at a feeder placed A (1000 m, 59 °), B (500 m, 75 °) and C (300 m, 166 °) from the hive (supplementary material, Fig. S1). If a released honeybee began to imbibe food, it was marked with color until there were 30 individually marked honeybees. For each colony, 50% sucrose water (as control group) or 50% sucrose syrup with 235 μg/kg deltamethrin (as pesticide group) was offered at the feeder on the alternate days. Marked honeybees were video recorded (Sony FDR-AX700) after they return into their observation hive, and recordings were subsequently analyzed at frame by frame (supplementary material, Fig. S2). In each colony at each distance, 30 recorded honeybees were collected and preserved in liquid nitrogen for later RNA extraction. All marked honeybees were removed from the observation hive under test at the end of each day to avoid interference from honeybees of the previously tested colonies.

When marked honeybees are dancing, they return into their observation hive, the dance followers were marked with different colors. When dance followers intend to leave and forage, approximately 30 honeybees were captured at the entrance of hive and anaesthetized on ice. The honeybees' honey stomach were dissected and weighed by a scale of one parts per 100,000 (HZ-104/35S, Huazhi Scientific Instrument Company, Fuzhou).

*Data collection and statistic analysis*

Ten marked honeybees in each colony at each of three distance and two treatment groups (total 180 honeybees) were randomly examined the recordings at frame by frame, and recorded the number of circuits per 15 s, divergence of angle (direction) and duration of the return phases (supplementary material, Fig. S2). The characteristics of dances between control and pesticide group at each distance were compared using independent sample *t*-test (SPSS Statistics 17.0). Specifically, dance precision was assessed by calculating the within-dance variance in the number of circuits per 15 s and duration of the return phase from *t*-test of dance per colony. Since divergence angle and crop content of data that were not normally distributed, the data was logarithm transformed to perform *t*-test.

*RNA extraction and RT-q PCR analysis*

As the head can better represent the learning and memory regulation, each of three collected honeybee heads was pooled for RNA extraction with TRizol. RNA purity was determined via nucleic acid protein analyzer (the OD260/280 ratio range of 1.9-2.1 met the standards). The integrity of the RNA was evaluated through the bands of 28S, 18S, and 5S r-RNA with agarose gel electrophoresis. Reverse transcription of total RNA was conducted using a reverse transcription kit (PrimeScript ™ RT reagent Kit with g-DNA Eraser). Six learning and memory related genes (*GluRA*, *NMDAR*, *Tyr1*, *DopR2*, *Dop3*, *Amdat*) as well as two reference genes (*GAPDH 1* and *GAPDH 2*) were selected from previous reports (Qin *et al.* 2014; Zhang *et al.* 2014) (supplementary material, Table S1). Three technical replicates were conducted for each gene. The Ct value for each sample was obtained by calculating the mean of three technical replicates. The relative genes expression were analyzed according to the formula $2^{-\Delta\Delta Ct}$ reported by Liu and Saint (2002). Genes showing significant differences at expression level were identified by independent sample *t*-test (SPSS Statistics 17.0) (supplementary material, Table S2).

# Results

*Effects of deltamethrin on the waggle dance at various food resource distance*

At the distance of 300 m from the hive, the circuits per 15 s, the duration of the return phase and crop content were not significantly different between control and pesticide groups for three test colonies (*t*-test, $df = 58$, $p = 0.994$; *t*-test, $df = 58$, $p = 0.970$; *t*-test, $df = 96$, $p = 0.629$) (Table III-6-1). However, the divergence angle showed significantly higher variation in pesticide group than control group (*t*-test, $df = 58$, $p = 0.007$) (Table III-6-1). At the distance of 500 m, the circuits per 15 s, the duration of the return phase and crop content had greater variance in pesticide group than control group (*t*-test, *df* = 60, *p* = 0.030; *t*-test, *df* = 60, *p* < 0.001; *t*-test, *df* = 115, *p* = 0.001). However, the significant difference for divergence angle was not during the experiment (Table III-6-1). At the distance of 1000 m, the duration of the return phase and crop content were significantly longer and heavier in pesticide group compared with control group (*t*-test, *df* = 58, *p* < 0.001; *t*-test, *df* = 124, *p* = 0.002). The circuits per 15 s and the divergence angle were not significantly different between the two groups (*t*-test, *df* = 58, *p* = 0.192; *t*-test, *df* = 58, *p* = 0.087) (Table III-6-1).

**Table III-6-1 Effects of sub-lethal dose of deltamethrin on dancing behavior characteristics and crop content in honeybees**

| Distance (m) | characteristic groups | circuits per 15 s | divergence angle (°) | duration of the return phase (s) | crop content (mg) |
|---|---|---|---|---|---|
| 300 | Suc + Delt | 6.992±0.180$^a$ | 4.534±0.114$^a$ | 1.494±0.032$^a$ | 1.444±0.092$^a$ |
|  | Suc | 6.994±0.197$^a$ | 4.038±0.134$^b$ | 1.492±0.030$^a$ | 1.374±0.114$^a$ |
| 500 | Suc + Delt | 5.619±0.095$^a$ | 3.753±0.148$^a$ | 1.841±0.045$^a$ | 1.765±0.099$^a$ |
|  | Suc | 5.942±0.110$^b$ | 3.544±0.102$^a$ | 1.609±0.036$^b$ | 1.305±0.092$^b$ |
| 1000 | Suc + Delt | 4.805±0.064$^a$ | 3.744±0.107$^a$ | 1.862±0.031$^a$ | 1.620±0.076$^a$ |
|  | Suc | 4.975±0.111$^a$ | 3.491±0.098$^a$ | 1.672±0.040$^b$ | 1.244±0.096$^b$ |

Note: *t*-test of dances performed by thirty bees returning from a feeder containing 50% sucrose syrup with 235 μg/kg deltamethrin (pesticide group, Suc + Delt) and a feeder containing 50% sucrose syrup (control group, Suc). Data were from three replicate colonies from nectar resources of 300 m, 500 m and 1000 m from the hive. The same letter indicates no significant difference between two groups (*p* > 0.05), while different letters indicate significant difference (*p* < 0.05), with Mean ± SE.

### *Effects of deltamethrin on honeybees' learning and memory*

At the distance of 300 m, the expression level of *Dop3* was significantly higher in pesticide group than the control group (*t*-test, *df* = 38, *p* = 0.016). Additionally, the relative gene expression level of *Tyr1* was significantly lower than the control group (*t*-test, *df* = 34, *p* < 0.001) (Fig. III-6-1 C, E). The differences of *GluRA*, *NMDAR*, *DopR2* and *Amdat* were not statistically significant between pesticide group and control group respectively (*t*-test, *df* = 34, *p* = 0.575; *t*-test, *df* = 37, *p* = 0.778; *t*-test, *df* = 42, *p* = 0.588; *t*-test, *df* = 40, *p* = 0.554) (Fig. III-6-1 A, B, D, F). At the distance of 500 m, the gene expression level of *NMDAR* was significantly lower in pesticide group than control group (*t*-test, *df* = 42, *p* = 0.029) (Fig. III-6-1 B). The differences were not statistical significant in *GluRA*, *Tyr1*, *DopR2*, *Dop3* and *Amdat* between the two groups (*t*-test, *df* = 41, *p* = 0.276; *t*-test, *df* = 43, *p* = 0.476; *t*-test, *df* = 42, *p* = 0.120; *t*-test, *df* = 33, *p* = 0.354; *t*-test, *df* = 40, *p* = 0.652) (Fig. III-6-1 A, C, D, E, F). At the distance of 1000 m, the expression level of *GluRA*, *NMDAR*, *DopR2* and *Amdat* were significantly lower in pesticide group than the control group (*t*-test, *df* = 42, *p* = 0.002; *t*-test, *df* = 46, *p* = 0.001; *t*-test, *df* = 41, *p* = 0.005; *t*-test, *df* = 43, *p* = 0.028) (Fig. III-6-1 A, B, D, F). The difference in *Tyr1* and *Dop3* were no significantly different between the two groups (*t*-test, *df* = 46, *p* = 0.064; *t*-test, *df* = 37, *p* = 0.655) (*p*>0.05) (Fig. III-6-1 C, E).

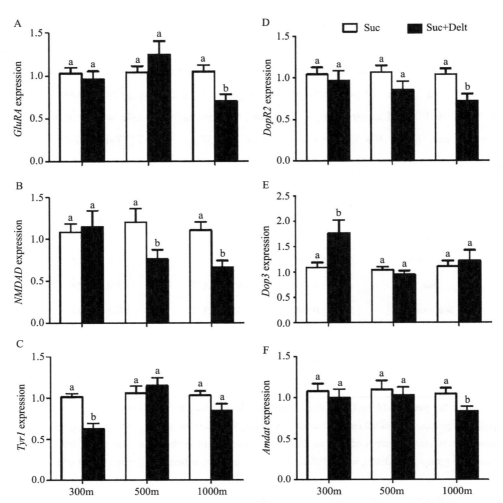

**Figure III-6-1** Effect of 50% sucrose syrup (Suc group, represented in white color) and 50% sucrose syrup with 235 μg/kg deltamethrin (Suc+Delt group, represented in black color) on learn and memory related gene expression. Six target genes of *GluRA*, *NMDAR*, *Tyr1*, *DopR2*, *Dop3* and *Amdat* as well as two reference genes of *GAPDH*1 and *GAPDH*2 were selected. The same letter indicates no significant difference between two groups ($p > 0.05$), while different letters indicate significant difference ($p < 0.05$), with Mean ± SE. The graphics were designed with GraphPad Prism.

## Discussion

### *Effects on distance and direction communication*

The honeybees' waggle dance encodes both the direction and distance to the advertised source. Directional information is contained in the angle of the waggle phase, while distance information is encoded in circuits per 15 s (von Frisch 1967). Honeybees can use multiple information sources to orient (Webb and Wystrach 2016). Distance is estimated from optic flow (Srinivasan 2000; Esch *et al.* 2001), which is the movement of the image of the environment across the eye during flight. Direction is determined using the position of the honeybee relative to the sun (El Jundi *et al.* 2014) or the pattern of polarized light in blue sky (Dovey *et al.* 2013). In our data, honeybees foraging on syrup containing deltamethrin increased number of circuits per 15 s than honeybees foraging on syrup without deltamethrin. The increased circuits per 15 s indicate that the food source is closer to the hive. The honeybees decrease the waggle phase when foraging toxic plant nectar *Tripterygium hypoglaucum*, also indicating that the food source is closer to colony (Tan *et al.* 2007; Tan *et al.* 2012).

The 'waggle' component of the dance (indicating direction) contains an inherent error and this error becomes smaller with increased nectar distance, when a honeybee dances for a resource, the error, or spread in the dance angle decreases with increasing distance to the resource (Preece and Beekman 2014). When resources are nearby, the dance deforms into more of a sickle or 'round dance' shape as the waggle runs are necessarily shorter, and the dance is consequently faster, which may reduce accuracy. The divergence angle was the difference between the average of the waggle run direction encoded in the dance and the actual direction to the goal. In our studies, the divergence angle in dance was smaller in the distance of 1000 m from the hive than 300 m's. Furthermore, the divergence angle was significantly greater in the pesticide group than the control group at the distance of 300 m (Table III-6-1). The sub-lethal dose of deltamethrin might disrupte honeybees' visuospatial memory and orientation at shorter distances from the hive.

*GluRA* is considered a metabotropic glutamate receptor gene that affects long-term learning and memory ability of honeybees (Danbolt 2001; Kucharski et al. 2007). The *NMDAR* is an important ionotropic receptor involved in the processes of learning and memory (Zachepilo *et al.* 2008; Morris *et al.* 1991). Tyramine receptor-like (*Tyr1*) is an important neurotransmitter in insects, which regulates physiological behaviors such as insect flight, learning and memory (Morris et al. 1991). *Tyr1* is also an important gene that affects short-term learning and memory of western honeybees (Blenau *et al.* 2000). In our studies, the relative genes expression levels of *GluRA*, *NMDAR* and *Tyr1* in pesticide group were down-regulate. It suggests that sub-lethal dose of deltamethrin impaired the honeybees' learning and memory and resulted in cognitive disorder.

As an important neurotransmitter in insects, dopamine is involved in regulating a variety of behaviors and physiological processes of insects, such as learning and memory, feeding, mating, development and information transmission of excitement and pleasure, and plays an important role in the regulation of learning and memory (Mustard *et al.* 2010; Zhang *et al.* 2014; Pignatelli *et al.* 2017). In honeybees, the expression level of the transporter gene *Amdat* can reflect the activity of dopaminergic neurons. In our study, the relative gene expression levels of *DopR2* and *Amdat* were down-regulated and the relative gene expression level of *Dop3* was up-regulated in pesticide group. Activation of D1-like receptor (*DopR2*) in vertebrates causes the increase of intracellular *cAMP*, while the activation of D2-like receptor (*Dop3*) reduces the amount of *cAMP* (Blenau *et al.* 1998; Beggs *et al.* 2005; Mustard *et al.* 2010; Razavi *et al.* 2017). In addition, the *cAMP* signaling pathway is necessary for regulating learning and memory. These results suggest that dopamine effect on learning and memory may be mediated by the *cAMP* signaling pathway.

*Honeybees' risk predication*

When flowers are considered dangerous owing to the presence of predators, experienced foragers are less likely to perform waggle dances, thus steering recruits away from potentially dangerous sites (Abbott and Dukas 2009). A colony trades off its need for food with its need to avoid food with toxic components (Tan *et al.* 2012). A good quality of food source is judged by a high concentration of sugar, more quantity of nectar, the shorter time of flight, lower risk and so on (Seeley and Visscher 1988). The speed of the return phase of the waggle dance and the number of waggle phases in a honeybee's dance are positively correlated with the perceived quality of food sources (Seeley et al. 2000). Given that the duration of the return phases of a honeybee's dance circuits are adjusted in relation to the quality of her food source, the question arises whether dance followers can acquire information about the quality of a dancer's food source by measuring the duration of the return phases of her dance circuits, just as dance followers can acquire information about the distance of a dancer's food source by measuring the duration of the waggle phases of her dance circuits (Seeley et al. 2000).

We noted the effect of pesticide on the crop content at three feeding sites. However, the naive foragers in pesticide group carried greater volume of fuel than the control group. We hypothesize there is a certain relationship between the quality of food and the crop content of dance followers. For example, free-flying honeybee foragers mitigate the risk of starvation in the field when foraging on a food source that offers variable rewards by carrying more 'fuel' food on their outward journey. The further away the food source, and the less familiar the forager is with its location, the more fuel their will take with (Beutler 1951; Harano et al. 2013, 2014). Therefore, the amount of 'fuel' taken by a foraging bee on her outward journey may be an objective measure of her perception of the riskiness of the foraging trip she is embarking on (Tan et al. 2015).

## Conclusions

In our studies, the honeybee's waggle dance communication was interfered by the sub-lethal dose deltamethrin, including direction, distance and quality of food sources. As a result, the crop content of the followers was incorrect according to the distance of the food resource. Our data suggest honeybees altered their dance behavior according to their perceptions of the riskiness of resource variability. Furthermore, the expression of genes related to learning and memory suggested that honeybees were impaired in their ability to locate and navigate. Deltamethrin influenced the foraging activity and healthy development of bee colony.

## Declarations

### *Ethics approval and consent to participate*

All animal procedures were performed in accordance with guidelines developed by the China Council on Animal, care and protocols were approved by the Animal Care and Use Committee of Jiangxi Agricultural University, China.

### *Consent for publication*

Not applicable.

### *Competing interests*

The authors declare that they have no competing interests.

## Acknowledgements

This work was supported by the National Natural Science Foundation of China (31572469, 31872432), the Earmarked Fund for China Agriculture Research System (CARS-44-KXJ15) and Science and Technology Project Founded of Education Department of Jiangxi Province (20171BAB204012). Zhijiang Zeng, Zuyun Zhang designed experiments, Zuyun Zhang and Zhen Li conducted all experiments. Li Ke, Xuewen Zhang, Weiyu Yan and Lizhen Zhang participated experiments. Zuyun Zhang, Qiang Huang and Zhijiang Zeng wrote the paper.

## References

ABBOTT KR, DUKAS R (2009) Honeybees consider flower danger in their waggle dance. *Animal Behaviour* 78: 633-635.

BEGGS KT, HAMILTON IS, KURSHAN PT, MUSTARD JA, MERCER AR (2005) Characterization of a D2-like dopamine receptor (*AmDOP3*) in honey bee, *Apis mellifera*. *Insect Biochem Mol Biol* 35 (8): 873-882.

BEUTLER R (1951) Time and distance in the life if the foraging bee. *Bee World* 32: 25-27.

BLENAU W, ERBER J, BAUMANN A (1998) Characterization of a dopamine D1 receptor from *Apis mellifera*: cloning, functional expression, pharmacology, and mRNA localization in the brain. *Journal of Neurochemistry* 70 (1): 15-23.

BLENAU W, BALFANZ S, BAUMANN A (2000) Amtyr1: characterization of a gene from honeybee (*Apis mellifera*) brain encoding a functional tyramine receptor. *Journal of Neurochemistry* 74 (3): 900-908.

CORNMAN RS, TARPY DR, CHEN Y, JEFFREYS L, LOPEZ D, PETTIS JS, VAN ENGELSDORP D, EVANS JD (2012) Pathogen webs in collapsing honey bee colonies. *PLoS One* 7, e43562.

DAI PL, WANG Q, SUN JH, LIU F, WANG X, WU YY, ZHOU T (2010) Effects of sublethal concentrations of bifenthrin and deltamethrin on fecundity, growth, and development of the honeybee *Apis mellifera* ligustica. *Environmental Toxicology and Chemistry* 29 (3): 644-649.

DANBOLT NC (2001) Glutamate uptake. *Progress Neurobiol* 65: 100-105.

DECOURTYE A, DEVILLERS J, CLUZEAU S, CHARRETON M, PHAMDELÈGUE MH (2004) Effects of imidacloprid and deltamethrin on associative learning in honeybees under semi-feld and laboratory conditions. *Ecotoxicol Environ Saf* 57: 410-419.

DECOURTYE A, DEVILLERS J, GENECQUE E, LE MENACH K, BUDZINSKI H, CLUZESU S, PHAM-DELÈGUE MH (2005) Comparative sublethal toxicity of nine pesticides on olfactory learning performances of the honeybee *Apis mellifera*. *Arch Environ Contam Toxicol* 48 (2): 242-250.

DOVEY KM, KEMFORT JR, TOWNE WF (2013) The depth of the honeybee's backup sun-compass systems. *J. Exp. Biol.* 216: 2129-2139.

ENAN E, MATSUMURA F (1992) Specific inhibition of calcineurin by type II synthetic pyrethroid insecticides. *Biochemical Pharmacology* 43 (8): 1777-1784.

ESCH HE, ZHANG SW, SRINIVASAN MV, TAUTZ J (2001) Honeybee dances communicate distances measured by optic flow. *Nature* 411: 581-583.

EL JUNDI B, PFEIFFER K, HEINZE S, HOMBERG U (2014) Integration of polarization and chromatic cues in the insect sky compass. J. *Comp. Physiol. A* 200 (6): 575-589.

FRANCIS RM, NIELSEN SL, PER K, MARTIN SJ (2013) Varroa-virus interaction in collapsing honey bee colonies.

*PLoS One* 8, e57540.

GILL RJ, RAMOS-RODRIGUEZ O, RAINE NE (2012) Combined pesticide exposure severely affects individual and colony-level traits in bees. *Nature* 491: 105-108.

GOULSON D, NICHOLLS E, BOTÍAS C, ROTHERAY EL (2015) Bee declines driven by combined stress from parasites, pesticides, and lack of flowers. *Science* 347 (6229): 1255957.

HARANO K, MITSHUHATA-ASAI A, KONISHI T, SUZUKI T, SASAKI K (2013) Honeybee foragers adjust crop contents before leaving the hive. *Behav. Ecol. Sociobiol* 67: 1169-1178.

HARANO K, MITSUHATA-ASAI A, SASAKI M (2014) Honey loading for pollen collection: regulation of crop content in honeybee pollen foragers on leaving hive. *Naturwissenschaften* 101: 595-598.

HENRY M, BÉGUIN M, REQUIER F, ROLLIN O, ODOUX JF, AUPINEL P, APTEL J, TCHAMITCHIAN S, DECOURTYE A (2012) A common pesticide decreases foraging success and survival in honey bees. *Science* 336 (6079): 348-350.

JACK-MCCOLLOUGH RT, NIEH JC (2015) Honeybees tune excitatory and inhibitory recruitment signaling to resource value and predation risk. *Animal Behaviour* 110: 9-17.

KUCHARSKI R, MITRI C, GRAU Y, MALESZKA R (2007) Characterization of a metabotropic glutamate receptor in the honeybee (*Apis mellifera*): implications for memory formation. *Invertebr Neurosci* 7 (2): 99-108.

LIAO CH, HE XJ, WANG ZL, BARRON AB, ZHANG B, ZENG ZJ, WU XB (2018) Short-term exposure to lambda-cyhalothrin negatively affects the survival and memory-related characteristics of worker bees *Apis mellifera*. *Archives of Environmental Contamination and Toxicology* 75: 59-65.

LIU W, SAINT DA (2002) A new quantitative method of real time reverse transcription polymerase chain reaction assay based on simulation of polymerase chain reaction kinetics. *Analytical Biochemistry* 302: 52-59.

MORRIS RGM, DAVIS S, BUTCHER S (1991) Hippocampal synaptic plasticity and N-methyl-D-aspartate receptors: a role in information storage? Long term potentiation: a debate of current issues, MIT Press, Cambridge, MA, 267-300.

MUSTARD JA, PHAM PM, SMITH BH (2010) Modulation of motor behavior by dopamine and the D1-like dopamine receptor *AmDOP2* in the honey bee. *Journal of Insect Physiology* 56 (4): 422-430.

Pignatelli M, Umanah GKE, Ribeiro SP, Chen R, Karuppagounder SS, Yau HJ, Eacker S, Dawson VL, Dawson TM, Bonci A (2017) Synaptic plasticity onto dopamine neurons shapes fear learning. *Neuron* 93 (2): 425-436.

PREECE K, BEEKMAN M (2014) Honeybee waggle dance error: adaption or constraint? Unravelling the complex dance language of honeybees. *Animal Behaviour* 94: 19-26.

QIN QH, HAN X, LIU H, ZHANG SW, ZENG ZJ (2014) Expression levels of glutamate and serotonin receptor genes in the brain of different behavioral phenotypes of worker honeybee, *Apis mellifera*. *Turkish Journal of Entomology*, 38 (4): 431-441.

RAMIREZ-ROMERO R, CHAUFAUX J, PHAM-DELÈGUE, MINH-HÀ (2005) Effects of Cry1Ab protoxin, deltamethrin and imidacloprid on the foraging activity and the learning performances of the honeybee *Apis mellifera*, a comparative approach. *Apidologie* 36 (4): 601-611.

RAZAVI AM, KHELASHVILI G, WEINSTEIN H, MARKOV A (2017) State-based quantitative kinetic model of sodium release from the dopamine transporter. *Scientific Reports* 7:40-76.

RÉMY VAN DAME, MELED M, COLIN ME, BELZUNCES LP (1995) Alteration of the homing-flight in the honey bee *Apis mellifera* L. exposed to sublethal dose of deltamethrin. *Environmental Toxicology and Chemistry* 14 (5): 855-860.

SEELEY TD, VISSCHER PK (1988) Assessing the benefits of cooperation in honeybee foraging: search costs, forage quality, and competitive ability. *Behavioral Ecology and Sociobiology* 22: 229-237.

SEELEY TD, MIKHEYEV AS, PAGANO GJ (2000) Dancing bees tune both duration and rate of waggle-run production in relation to nectar-source profitability. *Journal of Comparative Physiology A* 186: 813-819.

SRINIVASAN MV (2000) Honeybee navigation: nature and calibration of the 'odometer'. *Science* 287: 851-853.

TAN K, GUO YH, NICOLSON SW, RADLOFF SE, SONG QS, HEPBURN HR (2007) Honeybee (*Apis cerana*) foraging response to the toxic honey of *Tripterygium hypoglaucum* (Celastraceae): changing threshold of nectar acceptability. *Journal of Chemical Ecology* 33: 2209-2217.

TAN K, WANG ZW, YANG MX, FUCHS S, LUO LJ, ZHANG ZY, LI H, ZHUANG D, YANG S, TAUTZ J, BEEKMAN M, OLDROYD BP (2012) Asian hive bees, *Apis cerana*, modulate dance communication in response to nectar toxicity and demand. *Animal Behaviour* 84 (6): 1589-1594.

TAN K, LATTY T, DONG SH, LIU XW, WANG C, OLDROYD BP (2015) Individual honey bee (*Apis cerana*) foragers adjust their fuel load tomatch variability in forage reward. *Scientific Reports* 5 (1): 16418.

TAUTZ J (2008) The Buzz about bees: biology of a super organism. Berlin: Springer-Verlag.

THOMPSON HM (2003) Behavioural effects of pesticides in bees: their potential for use in risk assessment. *Ecotoxicology* 12 (1-4): 317-330.

VON FRISCH K (1967) The dance language and orientation of bees. Cambridge, Massachusetts: Harvard University Press.

WEBB B, WYSTRACH A (2016) Neural mechanisms of insect navigation. *Curr. Opin. Insect Sci.* 15: 27-39.

ZACHEPILO TG, IL'INYKH YF, LOPATINA NG, MOLOTKOV DA, POPOV AV, SAVVATEEVA-POPOVA EV, VAIDO AI, CHESNOKOVA EG (2008) Comparative analysis of the locations of the nr1 and nr2 nmda receptor sub-units in honeybee (*Apis mellifera*) and fruit fly (drosophila melanogaster, canton-s wild-type) cerebral ganglia. *Neurosci Behav Physiol* 38 (4): 369-372.

ZHANG LZ, ZHANG SW, WANG ZL, YAN WY, ZENG ZJ (2014) Cross-modal interaction between visual and olfactory learning in *Apis cerana*. *J. Comp. Physiol. A*. 200: 899-909.

# 7. The Effects of Sublethal Doses of Imidacloprid and Deltamethrin on Honeybee Foraging Time and the Brain Transcriptome

Zuyun Zhang[1,2,3], Zhen Li[1,3], Qiang Huang[1,3], Zhijiang Zeng[1,3]*

1. Honeybee Research Institute, Jiangxi Agricultural University, Nanchang, Jiangxi, 330045, P. R. of China;
2. College of Biological and Agricultural Sciences, Honghe University, Mengzi, Yunnan, 661101, P. R. of China;
3. Jiangxi Province Key Laboratory of Honeybee Biology and Beekeeping, Nanchang 330045, P. R. of China

**Abstract**

Colony collapse disorder (CCD) has occurred in the United States since 2006 and has been reported in many countries with varying levels of severity. Although, the cause of CCD is multi-factorial, pesticide is a major factor that leads to colony collapse. At sublethal doses, pesticide are known to negatively affect honeybee physiological development and behaviour. Previously, we found the insecticides imidacloprid and deltamethrin significantly reduced honeybee dancing and foraging efficiency. In our experiments, the duration of honeybee imbibing food at the feeder declined and the returning period from the feeder to the hive increased in both insecticide groups compared to the control group. As a follow-up, we performed a deep RNA-seq analysis to reveal the gene regulatory mechanisms underlying this altered foraging performance. Genes involved in detoxification were up-regulated in both the imidacloprid and the deltamethrin treatment groups. Gene members in immune pathways, odorant receptors and major royal jelly protein families were significantly down-regulated in the treatment groups compared to the controls. This fluctuating gene expression profile shows that multifaceted aspects of honeybee physiology were affected by the two insecticides, which may lead to inaccurate communication and impaired learning and memory. Our findings reveal candidate molecular mechanisms leading to impaired dance performance in honeybees exposed to insecticides.

**Keywords:** *Apis mellifera* L., insecticides, imidacloprid, deltamethrin, transcriptome

## Introduction

Honeybees (*Apis mellifera* L.) are eusocial insects that play an important role in natural ecosystems by providing global pollination services for crop and wild plants (Klein *et al.* 2007). However, honeybee colonies have declined in recent years (Neumann and Carreck 2010). Multiple factors have been associated with colony

---

* Corresponding author: bees1965@sina.com (ZJZ).
注：此文发表在 *Journal of Applied Entomology* 2022 年第 9 期。

losses, including parasitic mite *Varroa destructor* (Rosenkranz *et al.* 2010; Yang *et al.* 2021), fungal parasite *Nosema ceranae* (Antunez *et al.* 2009), viruses (McMenamin and Genersch 2015), malnutrition, environmental degradation, climate change, queen quality decline (Wei *et al.*, 2019) and widespread insecticide application (Budge *et al.* 2015; Sanchez-Bayo *et al.* 2016).

Although the causes of honeybee loss are multifactorial, the large-scale use of systemic insecticides such as imidacloprid, deltamethrin, and other insecticides has been implicated as a major contributing factor (Henry *et al.* 2012; Farooqui 2013; Woodcock *et al.* 2016). Worldwide, imidacloprid is a commonly used insecticide and has a highly specific affinity for nicotinic acetylcholine receptors (nAChRs) in the honeybee nervous system (Decourtye *et al.* 2004 a; Yang *et al.* 2008; Brandt *et al.* 2016). Deltamethrin is a nerve agent that disturbs the normal physiological function and signal transmission of nerve cells by interfering with the calcium channel of nerve cells. Accumulating evidence indicates that at sublethal doses, insecticides can cause honeybee brain dysfunction and reduce immune competence, leading to impaired navigation, olfactory learning and memory and susceptibility to pathogens (Decourtye *et al.* 2004 a; Desneux *et al.* 2007; Yang *et al.* 2008; Di Prisco *et al.* 2013; Matsumoto 2013; Palmer *et al.* 2013; Williamson and Wright 2013; Brandt *et al.* 2016; Shi *et al.* 2020). Williamson found that imidacloprid, coumaphos and a combination of the two compounds impaired the bees' ability to differentiate the conditioned odor from a novel odor during the memory test (Williamson and Wright 2013). Shi found that gene expression related to immune and detoxification of worker bees exposed to acetamiprid was roughly activated, returned and then inhibited from larval to emerged and to the late adult stage, respectively (Shi *et al.* 2020). Insecticides have potential effects on honeybees.

The waggle dance is a well-studied and sophisticated example of animal communication. The waggle dance was first deciphered by von Frisch (1967), who determined that honeybee foragers communicate the location of proftable fower patches to hive-mates using the dance. The direction and distance to a patch are indicated by the angle and duration of the waggle run, respectively. The quantity and quality of nectar and pollen available from various plant species is communicated by the number of dancing honeybees and frequency of their recruiting behavior (von Frisch 1967). Previously, we found that insecticides disturbed the dance communication system in honeybees. Feeding honeybees insecticide was found to significantly increase the divergence angle, return phases in waggle dances, and the crop content of dance followers. Data suggest that sublethal doses of deltamethrin or imidacloprid impair honeybees' learning and memory and result in the onset of cognitive disorders (Zhang *et al.* 2020 a, b).

However, the global gene expression profile associated with impaired learning and memory process in the bee brain during dance periods remains unclear. In this study, we tested whether sublethal doses of insecticides on honeybee foraging time, and whether affecting gene expression in the brain of honeybee. We predict that sublethal doses of insecticides interfere with honeybee foraging efficiency, and genes involved in immune, detoxification, and chemosensory responses were altered.

## Material and Methods

### *Sublethal dose of insecticides preparation*

Imidacloprid of 99% purity was provided by J&K Scientific and Chemical Company (China). Imidacloprid was dissolved in acetone, and stock solutions of 24 mg/kg in water were then diluted to final concentrations of 24 µg/kg with 50% sucrose solution(Decourtye *et al.* 2004 b). Treatment with imidacloprid at a concentration of 24 µg/kg was adopted because this concentration corresponds to the lowest observed effect concentration (LOEC) affecting the olfactory learning performance of bees after chronic oral treatment and under laboratory conditions (Decourtye *et al.* 2003). Deltamethrin of 98% purity (J&K Scientific and Chemical Company, China) was dissolved, and stock solutions of 235 mg/kg in water were diluted to final concentrations of 235 µg/kg with 50% sucrose solution. Deltamethrin administered at a concentration of 235 µg/kg was applied because it is half the no-observed-effect concentration (NOEC) for the conditioned proboscis extension reflex (PER) (Decourtye *et al.* 2005).

### *Behavioural experiments*

Experiments were conducted in October 2018 at Jiangxi Agricultural University, Nanchang, China (28.46 °N, 115.49 °E) during a period of constant warm temperatures and floral dearth to facilitate bee feeder training.

Three honeybee colonies (*Apis mellifera* L.) were sequentially used for testing, each with four frames of honeybees and broods in an observation hive. The colonies were healthy and free of external pests and disease by visual inspection. On each experimental day, approximately 100 foragers at the entrance were captured and placed into individual opaque tubes. The honeybees were released at a feeder placed approximately 500 metres from the hive. If a released honeybee began to consume food, it was marked with color until there were 75 individually marked honeybees. For each colony, 50% sucrose water (as a control group) or 50% sucrose water with sublethal doses of imidacloprid or deltamethrin was provided at the feeder on alternate days. The honeybees imbibe one pesticide first were marked with one color. Finally, we collected all honeybees with this color mark. Another day, when honeybees imbibed other pesticide first were marked with other color. For approximately 10 marked honeybees in each group, the duration of imbibing food and the return period from the hive to the feeder were recorded with a stopwatch, and each honeybee was recorded 10 times. The experiments were repeated in 2 colonies. Marked honeybees performed waggle dances after they returned to the observation hive. Approximately 75 dancing honeybees from one colony were collected and preserved in liquid nitrogen for subsequent brain dissection grouped by colony.

### *RNA sequencing and analysis*

The honeybee samples were stored in liquid nitrogen until RNA extraction. Individual bees were dissected under microscope on dry ice. Total RNA was extracted from a pool of 25 brains according to the standard protocol of TRIzol Reagent (Life Technologies, Carlsbad, CA, USA). Three biological replicates were used for control, imidacloprid-treated and deltamethrin-treated bees. In total, 9 pooled samples from one colony were collected. RNA libraries were prepared and sequenced using an Illumina HiSeq 4000 device following a standard protocol developed by Gene Denovo Biotechnology Co., Ltd. (Guangzhou, China) (Wei *et al.* 2019).

***Read mapping and bioinformatic analysis***

When total RNA was extracted, the quality level was determined with the Agilent 2100 Bioanalyser system. The quality of the RNA-seq libraries was assessed using FastQC. The raw reads produced by the sequencing instrument were filtered to remove adaptors, low-quality sequences with unknown nucleotides N, and reads with more than 20% low-quality bases (base quality < 10). The high-quality clean reads were assembled into unigenes using short read assembly program Trinity with default parameters (version 2.0.6). Gene functions and classifications were analysed based on searches made against the NCBI nonredundant database and the Kyoto Encyclopedia of Genes and Genomes (KEGG) database. Blast2 GO was used to conduct a GO annotation and enrichment analysis.

***Differentially expressed genes (DEGs)***

The expression level of each gene was calculated and normalized by RPKM (reads per kb per million reads) to calculate significantly differentially expressed genes (Audic and Claverie 1997; Trapnell *et al.* 2014). The false discovery rate (FDR) was used to adjust multiple comparisons. A FDR ≤ 0.05 threshold and an absolute value of $\log_2 \text{Ratio} \geq 1$ were used to determine the significance of gene expression differences in the analysis. All DEGs were mapped to terms in the Kyoto Encyclopedia of Genes and Genomes (KEGG) database. We looked for significantly enriched pathways in DEGs using the hypergeometric test, and *p*-values were adjusted (refer to *q*-values) using the Benjamini-Hochberg method. Pathways with $Q \leq 0.05$ were defined as significantly changed KEGG pathways.

***Statistical analysis***

Two honeybee colonies (*Apis mellifera* L.) were used to test the duration of imbibing food and return period from the hive to the feeder between the control, imidacloprid-treated and deltamethrin-treated groups. Approximately 10 marked honeybees in each group were recorded with a stopwatch, and each honeybee was recorded 10 times. An independent sample t-test (SPSS Statistics 19.0) was used to examine differences in the mean duration of imbibing food and returning periods of the control, imidacloprid-treated and deltamethrin-treated bees.

## Results

***Foraging time analysis***

The duration of honeybee imbibing food for the imidacloprid group was significantly lower than that for the control group (*t*-test, adjust *P*-value < 0.05)(Fig. III-7-1a). The duration of honeybee imbibing food for the deltamethrin group was significantly lower than that for the control group (*t*-test, adjust *P*-value < 0.05)(Fig. III-7-1c). The return period was significantly longer for the imidacloprid group than that for the control group (*t*-test, adjust *P*-value < 0.05)(Fig. III-7-1b). The return period was significantly longer for the deltamethrin group than for the control group (*t*-test, adjust *P*-value < 0.05) (Fig. III-7-1d).

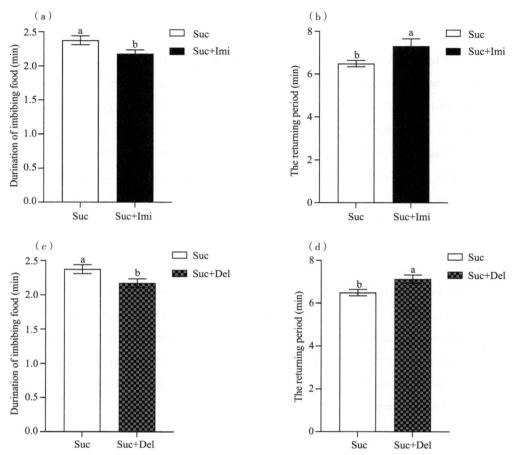

**Figure III-7-1** Comparison of the duration of imbibing food and returning period between the insecticide groups and control group. (a) Duration of imbibing food was compared between imidacloprid-treated group and normal sucrose solution group (as control group). (b) Returning period was compared between imidacloprid-treated group and normal sucrose solution group. (c) Duration of imbibing food was compared between deltamethrin-treated group and normal sucrose solution group. (d) Returning period was compared between deltamethrin-treated group and normal sucrose solution group. The black bars represent the data of imidacloprid-treated group (50% sucrose solution with 24 μg/kg imidacloprid, Suc+Imi). The white bars represent the data of control group (50% sucrose solution, Suc). The dotted bars represent the data of deltamethrin-treated group (50% sucrose solution with 235 μg/kg deltamethrin, Suc+Del). Data were expressed as the mean±SE (standard error), and the different lowercase letters indicate a significant difference (adjust $p$-value <0.05). The graphics were designed by GraphPad prism.

### *Differentially expressed genes detected in all samples*

A total of 85 DEGs were identified in the imidacloprid-treatment group versus the control group (Table S1), and 122 DEGs were identified in the deltamethrin-treatment group versus the control group (Table S2). Additionally, 27 DEGs were shared in both treatment groups, including apidaecin, hymenoptaecin, CYP450 6a14, CYP450 303a1, TpnCI, TpnCIIb, Mrjp1, calmodulin-like protein and other uncharacterized proteins (Fig. III-7-2, see page 533). There were 58 DEGs specific to the imidacloprid-treatment group and 95 DEGs were specific to the deltamethrin-treatment group (Fig. III-7-2). In the imidacloprid-treatment group, 60 genes were upregulated, including 10 detoxification immune genes, 6 metabolism-related genes, 6 sensory and signal transduction genes, and 38 other genes. There were 25 downregulated differentially expressed genes, including

2 genes involved in detoxification and immunity, 1 metabolism-related gene, 2 genes involved in sensory and signaling, and 20 genes involved in other genes (Fig. III-7-3, see page 533, Table S1). In the deltamethrin-treatment group, 76 genes were upregulated, 10 genes facilitated detoxification and immunity, 7 were metabolism-related genes, 6 genes facilitated sensory and signal transduction, and 53 other genes were present. There were 46 downregulated differentially expressed genes, including 5 detoxification genes, 2 metabolically related genes, 11 sensory and signalling genes, and 28 other genes (Fig. III-7-3, Table S2).

*GO enrichment analysis*

In the imidacloprid-treatment group and deltamethrin-treatment group, the DEGs were significantly enriched in 7 categories of biological functions, namely, the response to biotic stimulus (GO: 0009607), response to bacterium (GO: 0009617), response to external biotic stimulus (GO: 0043207), response to other organisms (GO: 0051707), glucose metabolic process (GO: 0006006), defence response (GO: 0006952) and response to external stimuli (GO: 0009605) (Table S1, Table S2). These categories are closely related to biological activities such as metabolism, defence response, chemical sensing and responses to external stimuli.

Furthermore, three categories in the deltamethrin group reached a significant enrichment level (adjusted $P$ value < 0.05) for molecular functions, including nucleotide binding (GO: 0000166), oxidoreductase activity (GO: 0016491) and nucleoside phosphate binding (GO: 1901265) (Table S2). In addition, a biological function analysis also found ten significant enrichment terms. These categories are related to biological activities such as cytoskeleton, signal transduction, receptor activity, ion balance, defence response, metabolism, and biosynthesis processes.

*Pathway analysis of DEGs*

Fifteen DEGs were enriched in 32 pathways in the imidacloprid group relative to the control group. Three pathways were significantly enriched, including lysine biosynthesis (ko00300), lysine degradation (ko00310) and the phosphatidylinositol signalling system (ko04070) (Table III-7-1). These pathways are mainly involved in metabolism, amino acid metabolism, drug metabolism, detoxification and immunity. Eighteen DEGs were enriched in 35 pathways in the deltamethrin group relative to the control group. Three pathways were significantly enriched, namely glycine, serine and threonine metabolism (ko00260), metabolic pathways (ko01100) and Hippo signalling pathway-fly (ko04391) (Table III-7-1). These pathways are mainly related to biological activities such as metabolism, amino acid metabolism, signal transduction, detoxification, drug metabolism and endocytosis.

**Table III-7-1 Pathway enrichment analysis for the significantly expressed genes in imidacloprid-treated bees and deltamethrin-treated bees**

| Control *vs* Imidacloprid Pathway Enrichment | | | | Control *vs* Deltamethrin Pathway Enrichment | | | |
|---|---|---|---|---|---|---|---|
| Pathway ID | Pathway | Differentially expressed genes | $P$-value | Pathway ID | Pathway | Differentially expressed genes | $P$-value |
| ko00300 | Lysine biosynthesis | 724239 | 0.017 | ko00260 | Glycine, serine and threonine metabolism | 551044, 552457 | 0.020 |

(continued)

| Control *vs* Imidacloprid Pathway Enrichment | | | | Control *vs* Deltamethrin Pathway Enrichment | | | |
|---|---|---|---|---|---|---|---|
| Pathway ID | Pathway | Differentially expressed genes | P-value | Pathway ID | Pathway | Differentially expressed genes | P-value |
| ko00310 | Lysine degradation | 102654572, 724239 | 0.020 | ko01100 | tMetabolic pathways | 102653682, 102655219, 412843, 551044, 551726, 552457, 725017, 725215, 727387, 727584 | 0.023 |
| ko04070 | Phosphatidylinositol signaling system | 551726 | 0.031 | ko04391 | Hippo signaling pathway–fly | 552637, 727368 | 0.048 |

Note: For KEGG Pathways, *p*-values of significant enrichment analysis were calculated using hypergeometric tests, and *p*-values were adjusted (refer to *q*-values) using the Benjamini-Hochberg method.

## Discussion

In our study, the duration of honeybee imbibing food at the feeder declined and the return period from the hive to the feeder increased in both insecticide groups relative to the control groups. The insecticide significantly reduced the honeybee foraging duration, possibly because honeybees dislike food containing insecticides. As insecticides disrupt the cognitive system, the honeybees may have deviated from their normal flight routes from the feeder to the hive (Henry *et al*. 2012; Matsumoto 2013). Honeybees may deliver defective recruitment information to hive mates, resulting in longer return times, as we observed. Recruitment declines more rapidly when dances are disoriented, and the next foraging trip can take several hours (Dornhaus 2002; Matsumoto 2013). The average wait time for receiver bees thus provides returning foragers with an indication of nectar level in the colony's environment (Anderson and Ratnieks 1999). When wait times are long, the forager may perform a tremble dance, which discourages foraging (Seeley 1992). Foragers waited for receiver bees returning from a feeder with *T. hypoglaucum* honey longer than when the feeder was supplied with common vetch honey (Tan *et al*. 2012).

In our data, both the up- and down-regulation of genes were found to be involved in detoxification, immunity, sensory processing, signalling pathways, and metabolism. These complex changes in gene expression show that multifaceted aspects of honeybee physiology might have been affected, leading to dysfunctions of honeybee dance communication. For example, 6 genes in the CYP450 family are involved in xenobiotic detoxification (Claudianos *et al*. 2006). In our study, the upregulated expression of CYP450s may have rendered honeybees more sensitive to environmental xenobiotics and may have impacted the metabolic detoxification process of the two insecticides. In addition, esterase, multidrug resistance-associated protein and peroxisomal multifunctional enzyme were upregulated after both insecticide treatments. We also observed that genes encoding cuticular proteins were upregulated in response to insecticide treatments, suggesting that barrier defence was initiated upon insecticide exposure. These results are the same as those of a previous study reporting that cuticular proteins are upregulated when honeybees are exposed to parasite and insecticide treatments (Aufauvre *et al*. 2014).

In the immunity category, 5 genes-encoding antimicrobial peptides, namely hymenoptaecin, abaecin, apidaecin, apisimin, and defensin were found to be altered at the expression level. The expression levels of hymenoptaecin, apidaecin and apisimin were significantly upregulated after imidacloprid treatment (Fig. III-7-4, see page 534, Table S3). The expression of hymenoptaecin, apidaecin and abaecin was significantly upregulated after deltamethrin treatment, while the expression of defensin was significantly downregulated (Fig. III-7-4, Table S3). With respect to immunity, these results contradict those of previous studies showing that insecticides downregulate immunity-related genes, such as lysozyme- and hymenoptaecin-encoding genes (Aufauvre et al. 2014; Brandt et al. 2016). This might be due to the different concentrations of the insecticide exposure. Imidacloprid (1 μg/L or 10 μg/L) reduced hemocyte density, encapsulation response, and antimicrobial activity even at field realistic concentrations. The results suggest that neonicotinoids affect the individual immunocompetence of honeybees, possibly leading to an impaired disease resistance capacity (Brandt et al. 2016). In addition, proteins involved in pathogen recognition and signalling proteins in upstream immunity pathways were not significantly altered after imidacloprid and deltamethrin treatment.

Calmodulin-like protein (CaM) is a major $Ca^{2+}$-binding protein in the central nervous system (Malenka et al. 1989; Margrie et al. 1998). Insecticides are reported to inhibit CaM and affect the cellular immune response, the signal transduction pathway, and biological function (Enan and Matsumura 1992). CaM may directly act on CaMKII, which has been repeatedly related to long-term memory (LTM) formation (Nakazawa et al. 1995; Limback-Stokin et al. 2004). In our study, calmodulin-like protein 4 in the two treatment groups was upregulated (Fig. III-7-4, Table S3). Thus, insecticides may impair honeybees' memory.

Honeybees' brain systems can be easily disrupted, especially because the insecticides found in floral resources directly target key neural pathways (Palmer et al. 2013; Peng and Yang 2016; Klein et al. 2017) and interfere with foraging behaviour and dance communication (Menzel and Greggers 1985; Menzel 1990). The decreased duration of food collection and the increased return period of dancing honeybees after insecticide treatment can affect foraging behaviour. We observed changes in the expression of genes involved in immune, detoxification, learning or memory and chemosensory responses. Our results confirm that sublethal doses of insecticides can affect honeybee foraging time and transcription, implying that pollinator population decline could be the result of a failure of neural function of honeybees exposed to insecticides in agricultural landscapes. Honeybees are frequently exposed to insecticides, which are being discussed as one of the stress factors that may lead to colony failure.

## Author Contributions

ZJZ and ZYZ conceived of and designed the experiments. ZYZ and ZL performed the experiments. ZYZ and QH analysed the data. ZYZ wrote the paper. All authors have read and approved of the final manuscript.

## Acknowledgments

We thank Dr. Frederick Partridge for improving the language of this manuscript. This work was supported by the National Natural Science Foundation of China (31872432, 32172790) and the Earmarked Fund for

China Agriculture Research System (CARS-44-KXJ15).

## Conflict of Interest

The authors have no conflicts of interest to report.

## Data Availability Statement

The raw Illumina sequencing data used are accessible through NCBI Bioproject SUB7653830. Including, RNA-Seq data of the control group (T500-1, T500-2 and T500-3), imidacloprid-treated group (B500-1, B500-2 and B500-3) and deltamethrin-treated group (X500-1, X500-2 and X500-3).

## Supplementary Material

Table S1 DEGs between the imidacloprid-treated group and the sucrose syrup group.
Table S2 DEGs between the deltamethrin-treated group and the sucrose syrup group.
Table S3 Expression levels of 34 honeybee genes in response to insecticides.

## References

ANDERSON C, RATNIEKS FLW, 1999. Worker allocation in insect societies: coordination of nectar foragers and nectar receivers in honey bee (*Apis mellifera*) colonies. *Behav Ecol Sociobiol* 46(2): 73-81.

ANTUNEZ K, MARTIN-HERNANDEZ R, PRIETO L, MEANA A, ZUNINO P, HIGES M, 2009. Immune suppression in the honey bee (*Apis mellifera*) following infection by Nosema ceranae (microsporidia). *Environ Microbiol* 11(9): 2284-2290.

AUDIC S, CLAVERIE JM, 1997. The significance of digital gene expression profiles. *Genome Res* 7(10): 986-995.

AUFAUVRE J, MISME-AUCOUTURIER B, VIGUES B, TEXIER C, DELBAC F, BLOT N, 2014. Transcriptome analyses of the honeybee response to *Nosema ceranae* and insecticides. *PLoS One* 9(3): e91686-91698.

BRANDT A, GORENFLO A, SIEDE R, MEIXNER M, BUCHLER R, 2016. The neonicotinoids thiacloprid, imidacloprid, and clothianidin affect the immunocompetence of honey bees (*Apis mellifera* L.). *J Insect Physiol* 86: 40-47.

BUDGE GE, GARTHWAITE D, CROWE A, BOATMAN ND, DELAPLANE KS, BROWN MA, THYGESEN HH, PIETRAVALLE S, 2015. Evidence for pollinator cost and farming benefits of neonicotinoid seed coatings on oilseed rape. *Sci Rep* 5:12574-12585.

CLAUDIANOS C, RANSON H, JOHNSON RM, BISWAS S, SCHULER MA, BERENBAUM MR, FEYEREISEN R, OAKESHOTT JG, 2006. A deficit of detoxification enzymes: pesticide sensitivity and environmental response in the honeybee. *Insect Mol Biol* 15(5): 615-636.

DECOURTYE A, LACASSIE E, PHAMDELÈGUE MH, 2003. Learning performances of honeybees (*Apis mellifera* L.) are differentially affected by imidacloprid according to the season. *Pest Manage Sci* 59(3): 269-278.

DECOURTYE A, ARMENGAUD C, RENOU M, DEVILLERS J, CLUZEAU S, GAUTHIER M, PHAM-DELÈGUE MH, 2004a. Imidacloprid impairs memory and brain metabolism in the honeybee (*Apis mellifera* L.). *Pestic Biochem Physiol* 78(2): 83-92.

DECOURTYE A, DEVILLERS J, CLUZEAU S, CHARRETON M, PHAM-DELÈGUE MH, 2004b. Effects of imidacloprid and deltamethrin on associative learning in honeybees under semi-field and laboratory conditions. *Ecotoxicol Environ Saf* 57(3): 410-419.

DECOURTYE A, DEVILLERS J, GENECQUE E, LE MENACH K, BUDZINSKI H, CLUZEAU S, PHAM-DELÈGUE MH, 2005. Comparative sublethal toxicity of nine pesticides on olfactory learning performances of the honeybee *Apis mellifera*. *Arch Environ Contam Toxicol* 48(2): 242-250.

DESNEUX N, DECOURTYE A, DELPUECH JM, 2007. The sublethal effects of pesticides on beneficial arthropods. *Annu Rev Entomol* 52(1): 81-106.

DI PRISCO G, CAVALIERE V, ANNOSCIA D, VARRICCHIO P, CAPRIO E, NAZZI F, GARGIULO G, PENNACCHIO F, 2013. Neonicotinoid clothianidin adversely affects insect immunity and promotes replication of a viral pathogen in honey bees. *Proc Natl Acad Sci U S A* 110(46): 18466-18471.

DORNHAUS A, 2002. Significance of honeybee recruitment strategies depending on foraging distance (hymenoptera: apidae: *Apis mellifera*). *Entomol Gen* 26(2): 93-100.

ENAN E, MATSUMURA F, 1992. Specific inhibition of calcineurin by type II synthetic pyrethroid insecticides. *Biochem Pharmacol* 43(8):1777-1784.

FAROOQUI T, 2013. A potential link among biogenic amines-based pesticides, learning and memory, and colony collapse disorder: a unique hypothesis. *Neurochem Int* 62(1): 122-136.

HENRY M, BEGUIN M, REQUIER F, ROLLIN O, ODOUX JF, AUPINEL P, APTEL J, TCHAMITCHIAN S, DECOURTYE A, 2012. A common pesticide decreases foraging success and survival in honey bees. *Science*, 336(6079): 348-350.

KLEIN AM, VAISSIERE BE, CANE JH, STEFFAN-DEWENTER I, CUNNINGHAM SA, KREMEN C, TSCHARNTKE T, 2007. Importance of pollinators in changing landscapes for world crops. *Proc Biol Sci* 274(1608): 303-313.

KLEIN S, CABIROL A, DEVAUD JM, BARRON AB, LIHOREAU M, 2017. Why bees are so vulnerable to environmental stressors. *Trends Ecol Evol* 32 (4): 266-276.

LIMBACK-STOKIN K, KORZUS E, NAGAOKA-YASUDA R, MAYFORD M, 2004. Nuclear calcium/calmodulin regulates memory consolidation. *J Neurosci* 24(48): 10858-10867.

MALENKA RC, KAUER JA, PERKEL DJ, MAUK MD, KELLY PT, NICOLL RA, WAXHAM MN, 1989. An essential role for postsynaptic calmodulin and protein-kinase activity in long-term potentiation. *Nature*, 340:554-557.

MARGRIE TW, ROSTAS JAP, SAH P, 1998. Presynaptic long-term depression at a central glutamatergic synapse: a role for CaMKII. *Nat Neurosci* 1:378-383.

MATSUMOTO T, 2013. Reduction in homing flights in the honey bee *Apis mellifera* after a sublethal dose of neonicotinoid insecticides. *B Insectol* 66(1):1-9.

MCMENAMIN AJ, GENERSCH E, 2015. Honey bee colony losses and associated viruses. *Curr Opin Insect Sci* 8:121-129.

MENZEL R, GREGGERS U, 1985. Natural phototaxis and its relationship to colour vision in honeybees. *J Com Physiol A* 157(3): 311-321.

MENZEL R, 1990. Learning, memory, and 'cognition' in honey bees. *Neurobiol Com Cogn* 273: 292.

NAKAZAWA H, KABA H, HIGUCHI T, INOUE S, 1995. The importance of calmodulin in the accessory olfactory bulb in the formation of an olfactory memory in mice. *Neuroscience* 69(2): 585-589.

NEUMANN P, CARRECK NL, 2010 Honey bee colony losses. *J Apicult Res* 49: 1-6.

PALMER MJ, MOFFAT C, SARANZEWA N, HARVEY J, WRIGHT GA, CONNOLLY CN, 2013. Cholinergic pesticides cause mushroom body neuronal inactivation in honey-bees. *Nat Commun* 4: 1634-1642.

PENG YC, YANG EC, 2016. Sublethal dosage of imidacloprid reduces the micro-glomerular density of honey bee mushroom bodies. *Sci Rep* 6(1): 19298-19311.

ROSENKRANZ P, AUMEIER P, ZIEGELMANN B, 2010. Biology and control of *Varroa destructor*. *J Invertebr Pathol* 103(1): 96-119.

SANCHEZ-BAYO F, GOULSON D, PENNACCHIO F, NAZZI F, GOKA K, DESNEUX N, 2016. Are bee diseases linked to pesticides? - A brief review. *Environ Int* 89: 7-11.

SEELEY TD, 1992. The tremble dance of the honey bee: message and meanings. *Behav Ecol Sociobiol* 31(6): 375-383.

SHI JL, ZHANG RN, PEI YL, LIAO CH, WU XB, 2020. Exposure to acetamiprid influences the development and survival ability of worker bees (*Apis mellifera* L.) from larvae to adults. *Environ Pollut* 266: 115345-115354.

TAN K, WANG ZW, YANG MX, FUCHS S, LUO LJ, ZHANG ZY, LI H, ZHUANG D, YANG S, TAUTZ J, BEEKMAN M, OLDROYD BP, 2012. Asian hive bees, *Apis cerana*, modulate dance communication in response to nectar toxicity and demand. *Anim Behav* 84(6): 1589-1594.

TRAPNELL C, ROBERTS A, GOFF L, PERTEA G, KIM D, KELLEY DR, PIMENTEL H, SALZBERG SL, RINN JL, PACHTER L, 2014. Differential gene and transcript expression analysis of RNA-seq experiments with Tophat and Cufflinks. *Nat Protoc* 9(10): 2513.

VON FRISCH K, 1967. The dance language and orientation of bees. Harvard University Press, Cambridge.

WEI H, HE XJ, LIAO CH, WU XB, JIANG WJ, ZHANG B, ZHOU LB, ZHANG LZ, BARRON AB, ZENG ZJ, 2019. A maternal effect on queen production in the honey bee. *Curr Biol* 29(13): 2208-2213.

WILLIAMSON SM, WRIGHT GA, 2013. Exposure to multiple cholinergic pesticides impairs olfactory learning and memory in honeybees. *J Exp Biol* 216(10): 1799-1807.

WOODCOCK BA, ISAAC NJ, BULLOCK JM, ROY DB, GARTHWAITE DG, CROWE A, PYWELL RF, 2016. Impacts of neonicotinoid use on long-term population changes in wild bees in England. *Nat Commun* 7: 12459-12466.

YANG EC, CHUANG YC, CHEN YL, CHANG LH, 2008. Abnormal foraging behavior induced by sub-lethal dosage of imidacloprid in the honey bee. *J Econ Entomol* 101(6): 1743-1748.

YANG HY, SHI JL, LIAO CH, YAN WY, WU XB, 2021. *Varroa destructor* mite infestations in capped brood cells of honeybee workers affect emergence development and adult foraging ability. *Current Zoology*, 67(5):569-571.

ZHANG ZY, LI Z, HUANG Q, ZHANG XW, KE L, YAN WY, ZHANG LZ, ZENG ZJ, 2020. Deltamethrin impairs honeybees (*Apis mellifera*) dancing communication. *Arch Environ Contam Toxicol* 78(1):117-123.

ZHANG ZY, LI Z, HUANG Q, YAN WY, ZHANG LZ, ZENG ZJ, 2020. Honeybees (*Apis mellifera*) modulate dance communication in response to pollution by imidacloprid. *J Asia-Pac Entomol* 23:477-482.

# 8. Honeybees (*Apis mellifera*) Modulate Dance Communication in Response to Pollution by Imidacloprid

Zuyun Zhang[1], Zhen Li[1], Qiang Huang[1], Weiyu Yan[1], Lizhen Zhang[1], Zhijiang Zeng[1*]

1. Honeybee Research Institute, Jiangxi Agricultural University, Nanchang, 330045, China;
2. Sericultural and Apicultural Institute, Yunnan Academy of Agricultural Sciences, Mengzi, 661101, China

**Abstract**

Imidacloprid, one of the most commonly used insecticides, is highly toxic to honeybees and other beneficial insects. Imidacloprid is a chloronicotinyl insecticide, which has a highly specific affinity to the nicotinic acetylcholine receptors (nAChRs) in the honeybee's nervous system. So it may interfere with dance behavior and memory formation. We found the waggle dances were modulated in honeybees fed sucrose water containing imidacloprid (pesticide group) compared to those fed normal sucrose water (control group). In our data, dancers of the pesticide group significantly increased the variance of divergence angle and the return phases in waggle dances than the control group. And the dance followers in pesticide group significantly increased the variance of crop content than the control group. Furthermore, four learning and memory related genes were significantly regulated at the gene expression levels between pesticide and control group. Our data revealed that the sub-lethal dose of imidacloprid impaired the honeybees' learning and memory and resulted in cognitive disorder. The dancers may adjust their recruitment behavior leading to the observed reduced number of followers. We conclude that modulation of in-hive communication serves to protect the colony from foraging toxic food.

**Keywords:** *Apis mellifera*, imidacloprid, sub-lethal effect, waggle dance, learning and memory

## Introduction

Honeybees are very important insects for human food production. More than one-third of the crops depend on honeybees' pollination, including fruits and vegetables (Glinski *et al.*, 2012). Honeybee population is in decline, numerous suspects have been identified in the hunt for the colony collapse disorder (CCD), from nutritional deficiencies, viral/ pathogen infection, parasites to exposure to genetically modified plants or pesticides (Glinski *et al.*, 2012). Imidacloprid is a pesticide widely used to control plant-sucking agricultural pests (Tomizawa and Casida, 2003), which inhibits receptors for γ-aminobutyric acid (GABA), a major

---

\* Correspondence author: bees1965@sina.com (ZJZ).
注：此文发表在 *Journal of Asia-Pacific Entomology* 2020 年第 2 期。

neurotransmitter in the central nervous system (Thany, 2010). Researchers have also examined the effects of imidacloprid on honeybees' behavior, particularly its effects on foraging, a key aspect of colony fitness (Sherman and Visscher, 2002; Dornhaus and Chittka, 2004; Goulson et al., 2015).

In the honeybees, imidacloprid is a partial agonist of nAChRs in brain areas associated with olfaction, learning and memory: cultured antennal lobe neurons (Barbara et al., 2005; Barbara et al., 2008) and cultured Kenyon cells from mushroom bodies (Déglise et al., 2002). Imidacloprid's metabolites also interfere with habituation and memory formation in honeybees (Guez, 2001; Decourtye et al., 2004a). Imidacloprid can induce sub-lethal effects such as impaired olfaction and disturbed orientation in honeybees (Thompson, 2003; Decourtye et al., 2004b). The sub-lethal dose of imidacloprid decreased the honeybees' foraging activity and homing ability (Henry et al., 2012; Fischer et al., 2014). And the sub-lethal dose of imidacloprid affected the honeybees of learning and memory functions (Fischer et al., 2014; Decourtye et al., 2005; Gauthier, 2010; Yang et al., 2012; Williamson et al., 2013; Van der Sluijs et al., 2013). Sub-lethal concentrations of 20 µg/kg (sucrose solution) imidacloprid cause trembling and affect the dance frequencies of honeybees (Kirchner, 1999; Eiri and Nieh, 2012). Colin (2019) suggested that a trace concentration (5µg/kg) of imidacloprid in sugar water could impact colony function by misbalancing the normal age based division of labor and reducing foraging efficiency. However, it is unclear whether such behaviors are due to the perception of the food being contaminated, or to the ingestion of imidacloprid, which subsequently alters recruitment.

The waggle dance is a well-studied and surprisingly sophisticated example of animal communication. The honeybees' waggle dance encodes both the direction and distance to the advertised source. Directional information is contained in the angle of the waggle phase, while distance information is encoded in circuits per 15 s (von Frisch, 1967). Honeybees can use multiple information sources to orient (Webb and Wystrach, 2016). Distance is estimated from optic flow (Srinivasan, 2000; Esch et al., 2001), which is the movement of the image of the environment across the eye during flight. Direction is determined using the position of the honeybee relative to the sun (el Jundi et al., 2014) or the pattern of polarized light in the sky (Dovey et al., 2013). Honeybees are also capable of avoiding flowers containing cues of elevated risk. The experienced foragers adjust the waggle dance to facilitate the avoidance of predation risk by naive recruits (Jack-McCollough and Nieh, 2015). Also, some protection to the colony is achieved when hazards kill or delay the return of affected foragers to the hive, thus interfering with communication (Abbott and Dukas, 2009).

In this study, we investigated honeybees' responses to sucrose water with pesticide in terms of waggle dance and memory related gene expression. We hypothesize that honeybees modify their recruitment behavior to the nest-mates by irregular dance to information on the quality/risk of the food resources.

## Materials and Methods

### *Sub-lethal dose of imidacloprid preparation*

The imidacloprid was provided by J&K scientific and chemical company. Imidacloprid was dissolved in acetone and stock solutions of 24mg/kg in water, which were then diluted to the concentrations of 24µg/kg in 50% sucrose solution (Decourtye et al., 2004b). Treatment with imidacloprid at concentration of 24µg/kg was chosen because this concentration corresponds to the lowest observed effect concentration (LOEC) affecting the olfactory learning performances of bees after chronic oral treatment and under laboratory conditions (Decourtye et al., 2003).

***Honeybees training***

Experiments were conducted from October 2017 to October 2018 at the Honeybee Research Institute, Jiangxi Agricultural University, Nanchang, China (28.46 °N, 115.49 °E) during a period of constant, warm temperatures and floral dearth, which facilitates feeder training of honeybees. The three feeding sites are in our campus, the distances of food source are 300 m, 500 m and 1000 m from the hive using a global positioning system (GPS).

Nine honeybee colonies were sequentially used in test, each with four frames of honeybees and brood in an observation hive. On each experimental day, about 100 foragers at the entrance were captured and placed into individual opaque tubes. Then the honeybees were released at a feeder placed at 300 m, 500 m and 1000 m from the colony (von Frisch, 1967; Harano et al., 2013). Based on previous studies, 1 km has been a long distance for the controlled feeding study. For our paper, we aim to study the impacts of short (300m), intermediate (500m) and long (1000m) distances on the dancing behavior and gene expression. If a released honeybee began to imbibe food, it was marked with color until there were 30 individually marked honeybees per hive and per distance. For each colony, 50% sucrose water (as control group) or 50% sucrose water with 24μg/kg imidacloprid (as pesticide group) were offered at the feeder on the alternate days. Marked honeybees were video recorded (Sony FDR-AX700) after their return into their observation hive, and recordings were subsequently analyzed at frame by frame. In each colony at each distance, 30 marked honeybees were collected and preserved in liquid nitrogen for later RNA extraction.

When marked honeybees have been dancing after their return into their observation hive, the dance followers were marked with different colors. When dance followers go to forage and leave the hive, approximately 30 dance followers were captured at the entrance of hive and anaesthetized on ice. The honeybees' honey sac were dissected and weighed by an analytical balance (minimum = 0.01 mg) (HZ-104/35S, Huazhi Scientific Instrument Company, Fuzhou). The size of followers' crop content was from three colonies in different feeding sites.

***Data collection and statistic analysis***

Ten marked honeybees in each colony at each of three distance and two treatment groups (total 180 honeybees) were randomly examined the recordings frame by frame, and recorded the number of circuits per 15 s, duration of the waggle phases, divergence of angle (direction), duration of the return phases, the number of dance circuits and the number of followers by referring to the previous reports (von Frisch, 1967; Seeley et al., 2000; Tautz, 2008; Preece and Beekman, 2014). We averaged all the dances per forager with at least three complete circuits. For example, the angle ($\alpha$) that the dancing honeybee is headed during the straight run, relative to gravity (up and down) clockwise from vertical (set to zero line), corresponds to the sun's azimuth direction of the food source outside the hive, which was a modest adjustment during a series of waggle phase (Tautz, 2008). The divergence angle was the difference between the average of the waggle run direction encoded in the dance and the actual direction to the goal (Preece and Beekman, 2014). The dance follower trip along behind a dancer, their antennae always extended toward her. The number of followers was calculated by video frame by frame. The characteristics of dances between control and pesticide groups at each distance were compared using independent sample *t*-test (SPSS Statistics 17.0). Specifically, dance precision was assessed by calculating the within-dance variance in the number of circuits per 15s, duration of the waggle phases, duration of the return phase, the number of dance circuits and the number of followers from *t*-test of dance per colony. Since divergence angle and crop content of data that were not normally distributed, the data were logarithm transformed to perform t-test.

*RNA extraction and RT-q PCR analysis*

As the head can better represent the learning and memory regulation (Giurfa, 2013), each of three collected honeybee heads was pooled for RNA extraction with an RNA extraction kit (TransZol Up Plus RNA Kit). RNA purity was determined via nucleic acid protein analyzer (the OD260/280 ratio range of 1.9-2.1 met the standards). The integrity of the RNA was evaluated through the bands of 28S, 18S, and 5S r-RNA with agarose gel electrophoresis. Reverse transcription of total RNA was conducted using a reverse transcription kit (PrimeScript™ RT reagent Kit with g-DNA Eraser) and the cDNAs were preserved in refrigerator at -80°C. Four learning and memory related genes (*GluRA*, *Tyr1*, *Dop3*, *Amdat*) as well as two reference genes (*GAPDH 1* and *GAPDH 2*) were selected based on previous reports (Qin *et al.*, 2014; Zhang *et al.*, 2014; Shi *et al.*, 2018) (Table III-8-1).

Table III-8-1 Description of primer sequences and genes used for RT-q PCR

| Gene name | GenBank ID | Forward primer (5′–3′) | Reverse primer (5′–3′) | References |
|---|---|---|---|---|
| *GluRA* | NM_001011623.1 | ACTCTGTTCGTCTGTGGGGTG | TTCGTTAGAAGGGCAGCGTA | Qin *et al.*, 2014; Shi *et al.*, 2018 |
| *Tyr 1* | NM_001037318 | CGTCGGGCGAGCGAGATA | GCCAAACCAACCAGCAAAT | Qin *et al.*, 2014; Shi *et al.*, 2018 |
| *Dop3* | NM_001014983.1 | CGGCTTTGTCTGTGACTTTTA | TCTTGTGCTTGGCGTATTTT | Zhang *et al.*, 2014 |
| *Amdat* | NM_001145738.1 | CGAATCAAGGATACAACAGCA | GATAGACCATCAGCAGGCATAAT | Zhang *et al.*, 2014 |
| *GAPDH 1* | NM_001014994.1 | GCTGGTTTCATCGATGGTTT | ACGATTTCGACCACCGTAAC | Qin *et al.*, 2014; Zhang *et al.*, 2014; Shi *et al.*, 2018 |
| *GAPDH 2* | XM_393605.7 | TGCTCAGGTTGTTGCCATT | TTTTTGCCTCCGTTCACTAA | Qin *et al.*, 2014; Zhang *et al.*, 2014; Shi *et al.*, 2018 |

The RT q-PCR reaction mixture (10 μL) was as follows: 1 μL of c-DNA, 5 μL of TB Green Premix Ex Taq II (Tli RNaseH Plus), 0.2 μL of Rox Reference Dye II, 0.4 μL each of forward and reverse primer, and 3 μL of ultrapure sterile water. These were added to the fluorescent quantitative PCR machine (Applied Biosystems ABI 7500) for amplification after being blended. The reaction conditions were as follows: 95°C for 30 s and 60°C for 1 min for 40 cycles, followed by heating from 50 to 60°C (the temperature was increased by

1 °C every 6 s) to construct the dissolution curve (Liu *et al.*, 2019). Three technical replicates were conducted for each gene. The Ct value for each sample was obtained by calculating the mean of three technical replicates.

***Statistics analysis***

All the statistics were analyzed using SPSS package. The impacts of the treatment on the behavior (the circuits per 15 s, the duration of the waggle phases, the duration of the return phase, the number of dance circuits, the number of followers, the divergence angle and the crop content) each of the distance were analyzed using *t*-test and corrected with FDR. In order to analyze the overall impact of the distance and treatment on the phenotypes (the circuits per 15 s, the duration of the waggle phases, the duration of the return phase, the number of dance circuits, the number of followers, the divergence angle and the crop content), the Generalized Linear Models (GLM) was constructed where the treatment and distance as fixed factor and colony as random factor.

The relative gene expression were analyzed according to the formula $2^{-\Delta\Delta Ct}$ reported by Liu and Saint (2002). Genes showing significant differences at expression level were identified by independent sample *t*-test, with FDR correction. In order to analyze the overall impact of the distance and treatment on the genes (*GluRA*, *Tyr1*, *Dop3*, *Amdat*), the GLM was again constructed where the treatment and distance as fixed factor and colony as random factor.

## Results

***Effects of imidacloprid on the waggle dance at various food resource distance***

Overall, the distances showed significant effects on the circuits per 15 s ($P < 0.001$), the duration of the waggle phases ($P < 0.001$), the divergence angle ($P < 0.001$), the duration of the return phase ($P < 0.001$), the number of dance circuits ($P < 0.001$), the number of followers ($P < 0.01$) and the crop content ($P < 0.05$). Treatments showed significant effects on the divergence angle ($P < 0.01$), the duration of the return phase ($P < 0.01$), the number of followers ($P < 0.01$) and the crop content ($P < 0.001$). However, the effect of the treatments on the circuits per 15 s ($P > 0.05$), the duration of the waggle phases ($P > 0.05$) and the number of dance circuits ($P > 0.05$) were not significant. The impacts of the interaction between the distance and the treatment on the circuits per 15 s ($P > 0.05$), the duration of the waggle phases ($P > 0.05$), the divergence angle ($P > 0.05$), the duration of the return phase ($P > 0.05$), the number of dance circuits ($P > 0.05$) and the number of followers ($P > 0.05$) were not significant. However, the interaction between the distance and the treatment showed significant effects on the crop content ($P < 0.01$).

At the distance of 300 m from the hive, the circuits per 15 s (*t*-test, adjust *P*-value > 0.05), the duration of the waggle phases (*t*-test, adjust *P*-value > 0.05), the duration of the return phase (*t*-test, adjust *P*-value > 0.05), the number of dance circuits (*t*-test, adjust *P*-value > 0.05) and the number of followers (*t*-test, adjust *P*-value > 0.05) were not significantly different between control and pesticide groups for three test colonies (Table III-8-2). However, the divergence angle (*t*-test, adjust *P*-value < 0.05) and crop content (*t*-test, adjust *P*-value < 0.05) in pesticide group showed significant difference than control group (Table III-8-2).

At the distance of 500 m, the duration of the return phase had a significant difference in pesticide group than control group ($t$-test, adjust $P$-value $< 0.05$). However, the results showed that there were no significant difference for the circuits per 15 s ($t$-test, adjust $P$-value $> 0.05$), the duration of the waggle phases ($t$-test, adjust $P$-value $> 0.05$), divergence angle ($t$-test, adjust $P$-value $> 0.05$), the number of dance circuits ($t$-test, adjust $P$-value $> 0.05$), the number of followers ($t$-test, adjust $P$-value $> 0.05$) and crop content ($t$-test, adjust $P$-value $> 0.05$) (Table III-8-2).

At the distance of 1000 m, the crop content ($t$-test, adjust $P$-value $< 0.05$) were significantly heavier in pesticide group when compared to control group. The circuits per 15 s ($t$-test, adjust $P$-value $> 0.05$), the duration of the waggle phases ($t$-test, adjust $P$-value $> 0.05$), the divergence angle ($t$-test, adjust $P$-value $> 0.05$), the duration of the return phase ($t$-test, adjust $P$-value $> 0.05$), the number of dance circuits ($t$-test, adjust $P$-value $> 0.05$) and the number of followers ($t$-test, adjust $P$-value $> 0.05$) were not significantly different between the two groups (Table III-8-2).

**Table III-8-2  Effects of sub-lethal dose of imidacloprid on dancing behavior characteristics and crop content in honeybees**

| Distance (m) | groups | circuits per 15 s | duration of the waggle phase (s) | divergence angle (°) | duration of the return phase (s) | the number of dance circuits | the number of followers | crop content (mg) |
|---|---|---|---|---|---|---|---|---|
| 300 | Suc+Imid | 7.240±0.150$^a$ | 0.576±0.019$^a$ | 4.528±0.134$^a$ | 1.492±0.040$^a$ | 20.900±1.859$^a$ | 5.663±0.153$^a$ | 1.732±0.124$^a$ |
|  | Suc | 7.108±0.174$^a$ | 0.580±0.015$^a$ | 4.038±0.134$^b$ | 1.491±0.030$^a$ | 21.750±1.739$^a$ | 6.151±0.175$^a$ | 1.358±0.122$^b$ |
| 500 | Suc+Imid | 6.016±0.093$^a$ | 0.864±0.028$^a$ | 4.072±0.101$^a$ | 1.656±0.038$^a$ | 15.933±0.769$^a$ | 6.232±0.230$^a$ | 1.285±0.111$^a$ |
|  | Suc | 6.223±0.098$^a$ | 0.830±0.023$^a$ | 3.970±0.104$^a$ | 1.498±0.037$^b$ | 15.467±0.863$^a$ | 6.720±0.221$^a$ | 1.305±0.092$^a$ |
| 1000 | Suc+Imid | 4.867±0.104$^a$ | 1.300±0.029$^a$ | 3.808±0.125$^a$ | 1.806±0.047$^a$ | 15.300±0.810$^a$ | 6.216±0.148$^a$ | 1.889±0.098$^a$ |
|  | Suc | 4.975±0.111$^a$ | 1.346±0.034$^a$ | 3.492±0.098$^a$ | 1.671±0.040$^a$ | 15.600±0.875$^a$ | 6.689±0.161$^a$ | 1.244±0.096$^b$ |

Note: $t$-test of dances performed by thirty bees returning from a feeder containing sucrose water with 24μg/kg imidacloprid (pesticide group, Suc+Imid) and a feeder containing 50% sucrose water (control group, Suc). Data were from three replicate colonies from nectar resources of 300 m, 500 m and 1000 m from the hive. The same letter indicates no significant difference between two groups (adjust $P$-value $> 0.05$), while different letters indicate significant difference (adjust $P$-value $< 0.05$), with Mean ± SE.

### *Effects of imidacloprid on gene expression related to learning and memory of honeybees*

The treatments showed significant effects on the expression level of *GluRA* ($P < 0.05$) and *Tyr1* ($P < 0.01$). However, the effect of the treatments of the expression level of *Dop3* ($P > 0.05$) and *Amdat* ($P > 0.05$) was not significant. The distances showed no significant effects on the expression level of *GluRA* ($P > 0.05$), *Tyr1* ($P > 0.05$), *Dop3* ($P > 0.05$) and *Amdat* ($P > 0.05$). Additionally, the interaction between the distance and the treatment showed no significant effects on the expression level of *GluRA* ($P > 0.05$), *Tyr1* ($P > 0.05$), *Dop3* ($P > 0.05$) and *Amdat* ($P > 0.05$).

At the distance of 300 m, the relative gene expression level of *Tyr1* was significantly lower in pesticide group than the control group ($t$-test, adjust $P$-value $< 0.05$) (Fig. III-8-1 B). The expression level of *GluRA* ($t$-test, adjust $P$-value $> 0.05$), *Dop3* ($t$-test, adjust $P$-value $> 0.05$) and *Amdat* ($t$-test, adjust $P$-value $> 0.05$)

were not statistically significant between pesticide group and control group respectively (Fig. III-8-1 A, C, D). At the distance of 500 m, the gene expression level of *Tyr1* (*t*-test, adjust *P*-value < 0.05) was significantly lower in pesticide group than control group (Fig. III-8-1 B). The relative gene expression level was not statistically significant for *GluRA* (*t*-test, adjust *P*-value > 0.05), *Dop3* (*t*-test, adjust *P*-value > 0.05) and *Amdat* (*t*-test, adjust *P*-value > 0.05) between the two groups (Fig. III-8-1 A, C, D). At the distance of 1000 m, the relative gene expression level of *GluRA* (*t*-test, adjust *P*-value > 0.05), *Tyr1* (*t*-test, adjust *P*-value > 0.05), *Dop3* (*t*-test, adjust *P*-value > 0.05) and *Amdat* (*t*-test, adjust *P*-value > 0.05) were not significantly different between the two groups (Fig. III-8-1 A, B, C, D).

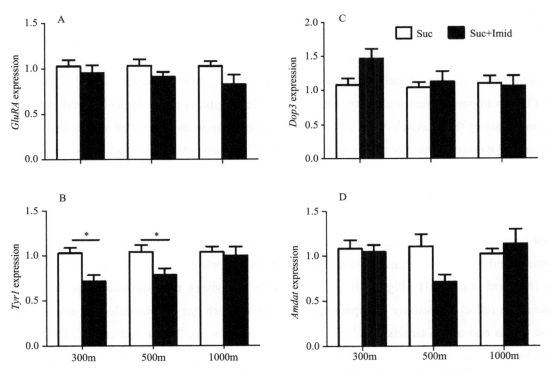

**Figure III-8-1** Effect of 50% sucrose water (Suc group, represented in white color) and 50% sucrose water with 24 μg/kg imidacloprid (Suc+Imid group, represented in black color) on learn and memory related gene expression. Four target genes of *GluRA*, *Tyr1*, *Dop3* and *Amdat* as well as two reference genes of *GAPDH1* and *GAPDH2* were selected. The data was expressed as the Mean ± SE, and the "*" indicates a significant difference (adjust *P*-value < 0.05). The graphics were designed by GraphPad Prism.

## Discussion

### *Effects on waggle dance communication*

Honeybees have evolved excellent memory and navigation skills enabling them to exploit complex and variable foraging environments (Giurfa, 2013; Menzel, 2012). These brain systems can be easily disrupted, and it is especially problematic that many pesticides found in floral resources directly target key neural pathways (Palmer *et al.*, 2013; Peng and Yang, 2016), and interfering the foraging behavior and dance communication (Menzel, 1985, 1990).

The waggle dance encodes both the direction and distance to the advertised source. The variation of sources direction indicated that between each successive waggle phase, the 'error', are diverse and interrelated. The 'waggle' component of the dance (indicating direction) contains an inherent error and this error becomes smaller with increased nectar distance (Preece and Beekman, 2014). When resources are nearby, the dance deforms into more of a sickle or 'round dance' shape as the waggle runs are necessarily shorter, and the dance is consequently faster, which may reduce accuracy (Preece and Beekman, 2014). In our studies, the divergence angle was significantly greater in the pesticide group than the control group at the distance of 300 m. Additionally, treatments and distances showed significant effects on the divergence angle. The sub-lethal dose of imidacloprid might disrupt honeybees' visual and spatial memory and orientation at shorter distances from the hive.

*GluRA* is widely distributed in the central nervous system in honeybees and is considered a metabotropic glutamate receptor gene that affects long-term learning and memory ability of honeybees. It is a key neurotransmitter in learning and memory processes (Danbolt, 2001; Kucharski et al., 2007). Tyramine receptor-like (*Tyr1*) is an important neurotransmitter in insects, which regulates physiological behaviors such as flight, learning and memory (Sinakevitch et al., 2017). *Tyr1* is also an important gene that affects short-term learning and memory of western honeybees (Blenau et al., 2000). In our studies, the treatments showed significant effects on the relative gene expression levels of *GluRA* and *Tyr1*. It suggests that imidacloprid impaired the honeybees' learning and memory and resulted in cognitive disorder.

Dopamine is related to the information transmission of excitement and pleasure, and can regulate a variety of behaviors and physiological processes of insects, such as learning and memory, feeding, mating, development and information transmission, and plays an important role in the regulation of learning and memory (Mustard et al., 2010; Pignatelli et al., 2017). In honeybees, the expression level of the transporter gene *Amdat* can reflect the activity of dopaminergic neurons, which has the physiological effect of reacquiring dopamine and is one of the targets of addictive drugs (Harano et al., 2005). Activation of D2-like receptor (*Dop3*) reduces the amount of *cAMP* (Blenau et al., 1998; Beggs et al., 2005; Mustard et al., 2010; Razavi et al., 2017). In addition, the *cAMP* signaling pathway is necessary for regulating learning and memory. In our study, the effect of the treatments and distances on the relative gene expression level of *Dop3* and *Amdat* were not significant. This may be due to factors such as colony, external nectar conditions and the imidacloprid of LOEC. In the current study, we showed different distances have an effect on dance behavior and intrinsic gene expression profile. However, the mechanism underlying the response towards the food resource difference remains unclear.

### Honeybees' risk predication

When flowers are considered dangerous owing to the presence of predators, experienced foragers are less likely to perform waggle dances, thus steering recruits away from potentially dangerous sites (Abbott and Dukas, 2009). A colony trades off its need for food requirement and the toxic nectar (Tan et al., 2012). In our data, treatments and distances showed significant effects on the number of dance followers, so the dancers may adjust their recruitment behavior to inform the nest-mates, leading to the observed reduced number of followers. Therefore, the sub-lethal dose of imidacloprid acted at both the level of foragers' avoidance behavior and reduction of activity performance.

In addition, we noted the effect of pesticide on the crop content at three feeding sites. The naive foragers in pesticide group carried greater volume of fuel than the control group at 300 m and 1000 m from hive. At the distance of 500 m, the results showed that there was no significant difference for crop content. We need to take into account the factors of the colony, the external conditions of the nectar, imidacloprid of LOEC and so on. However, treatments and distances showed significant effects on the crop content using the GLM of analyses. For example, free-flying foragers mitigate the risk of starvation in the field when foraging on a food source that offers variable rewards by carrying more 'fuel' on their outward journey. The further away the food source, and the less familiar the forager is with its location, the more fuel she will take with (Beutler, 1951; Harano et al., 2013, 2014). Tan (2015) reported the honeybee foragers adjust their fuel load to match variability in forage reward. Therefore, the amount of 'fuel' taken by a foraging honeybee on her outward journey may be an objective measure of her perception of the riskiness of the foraging trip she is embarking on (Tan et al., 2015). Honeybees may adjust the waggle dance form the sucrose water containing imidacloprid to their nest mates to inform the riskiness of foraging trip to the followers.

## Conclusions

In this study, we reported the sub-lethal dose of imidacloprid, which impaired the honeybees waggle dance communication, including interfered accuracy of direction and the indication of food quality. Due to the inaccurate dancing, the crop content of the followers was incorrect according to the distance of the food resource. Furthermore, the expression of genes related to learning and memory suggested that honeybees were impaired in their ability to locate and navigate. Our data suggest honeybees tuned their dance behavior according to their perceptions of the riskiness of the food source. We conclude that imidacloprid influenced the foraging activity and recruitment behavior.

## Declarations

Ethics approval and consent to participate: Not applicable.
Consent for publication: Not applicable.
Competing interests: The authors declare that they have no competing interests.

## Acknowledgements

Zhijiang Zeng, Zuyun Zhang designed experiments, Zuyun Zhang and Zhen Li conducted all experiments. Weiyu Yan and Lizhen Zhang participated experiments. Zuyun Zhang, Qiang Huang and Zhijiang Zeng wrote the paper. This work was supported by the National Natural Science Foundation of China (31572469, 31872432), the Earmarked Fund for China Agriculture Research System (CARS-44-KXJ15) and Science and Technology Project Founded of Education Department of Jiangxi Province (20171BAB204012).

# References

ABBOTT, K.R., DUKAS, R., 2009. Honeybees consider flower danger in their waggle dance. *Anim. Behav.* 78, 633-635. https://doi.org/10.1016/j.anbehav.2009.05.029.

BARBARA, G., ZUBE, C., RYBAK, J., GAUTHIER, M., GRÜNEWALD, B., 2005. Acetylcholine, GABA and glutamate induce ionic currents in cultured antennal lobe neurons of the honeybee, *Apis mellifera. J. Comp. Physiol. A* 191, 823-836. https://doi.org/10.1007/s00359-005-0007-3.

BARBARA, G., GRÜNEWALD, B., PAUTE, S., GAUTHIER, M., RAYMOND-DELPECH, V., 2008. Study of nicotinic acetylcholine receptors on cultured antennal lobe neurons from adult honeybee brains. Invert. *Neurosci.* 8, 19-29. https://doi.org/10.1007/s10158-007-0062-2.

BEGGS, K.T., HAMILTON, I.S., KURSHAN, P.T., MUSTARD, J.A., MERCER, A.R., 2005. Characterization of a D2-like dopamine receptor (*AmDOP3*) in honey bee, *Apis mellifera. Insect Biochem. Mol. Biol.* 35, 873-882. https://doi.org/10.1016/j.ibmb.2005.03.005.

BEUTLER, R. 1951. Time and distance in the life if the foraging bee. *Bee World* 32, 25-27. https://doi.org/10.1080/0005772X.1951.11094669.

BLENAU, W., ERBER, J., BAUMANN, A., 1998. Characterization of a dopamine D1 receptor from *Apis mellifera*: cloning, functional expression, pharmacology, and mRNA localization in the brain. *J. Neurochem.* 70, 15-23. https://doi.org/10.1046/j.1471-4159.1998.70010015.x.

BLENAU, W., BALFANZ, S., BAUMANN, A., 2000. Amtyr1: characterization of a gene from honeybee (*Apis mellifera*) brain encoding a functional tyramine receptor. *J. Neurochem.* 74, 900-908. https://doi.org/10.1046/j.1471-4159.2000.0740900.x.

COLIN, T., MEIKLE, W.G., WU, X.B., BARRON, A.B., 2019. Traces of a neonicotinoid induce precocious foraging and reduce foraging performance in honey bees. *Environ. Sci. Technol.* 53, 8252-8261. https://doi.org/10.1021/acs.est.9b02452.

DANBOLT, N.C., 2001. Glutamate uptake. Prog. Neurobiol. (Oxford, U. K.) 65, 100-105. https://doi.org/10.1016/S0301-0082(00)00067-8.

DECOURTYE, A., LACASSIE, E., PHAMDELÈGUE, M.H., 2003. Learning performances of honeybees (*Apis mellifera* L.) are differentially affected by imidacloprid according to the season. *Pest Manage.* Sci. 59, 269-278. https://doi.org/10.1002/ps.631.

DECOURTYE, A., ARMENGAUD, C., RENOU, M., DEVILLERS, J., CLUZEAU, S., GAUTHIER, M., PHAM-DELÉGUE, M., 2004a. Imidacloprid impairs memory and brain metabolism in the honeybee (*Apis mellifera* L.). *Pestic. Biochem. Phys.* 78, 83-92. https://doi.org/10.1016/j.pestbp.2003.10.001.

DECOURTYE, A., DEVILLERS, J., CLUZEAU, S., CHARRETON, M., PHAMDELÈGUE, M.H., 2004b. Effects of imidacloprid and deltamethrin on associative learning in honeybees under semi-feld and laboratory conditions. *Ecotoxicol. Environ. Saf.* 57, 410-419. https://doi.org/10.1016/j.ecoenv.2003.08.001.

DECOURTYE, A., DEVILLERS, J., GENECQUE, E., LE MENACH, K., BUDZINSKI, H., CLUZESU, S., PHAM-DELÈGUE, M.H., 2005. Comparative sub-lethal toxicity of nine pesticides on olfactory learning performances of the honeybee *Apis mellifera*. *Arch. Environ. Contam. Toxicol.* 48, 242-250. https://doi.org/10.1007/s00244-003-0262-7.

DÉGLISE, P., GRÜNEWALD, B., GAUTHIER, M., 2002. The insecticide imidacloprid is a partial agonist of the nicotinic receptor of honeybee Kenyon cells. *Neurosci. Lett.* 321, 13-16. https://doi.org/10.1016/S0304-3940(02)00283-5.

DORNHAUS, A., CHITTKA, L., 2004. Why do honey bees dance? *Behav. Ecol. Sociobiol.* 55, 395-401. https://doi.org/10.1007/s00265-003-0726-9.

DOVEY, K.M., KEMFORT, J.R., TOWNE, W.F., 2013. The depth of the honeybee's backup sun-compass systems. *J. Exp. Biol.* 216, 2129-2139. https://doi.org/10.1242/jeb.084160.

EIRI, D.M., NIEH, J.C., 2012. A nicotinic acetylcholine receptor agonist affects honey bee sucrose responsiveness and decreases waggle dancing. *J. Exp. Biol.* 215, 2022-2029. https://doi.org/10.1242/jeb.068718.

ESCH, H.E., ZHANG, S.W., SRINIVASAN, M.V., TAUTZ, J, 2001. Honeybee dances communicate distances measured by optic flow. *Nature*, 411, 581-583. https://doi.org/10.1038/35079072.

EL JUNDI, B., PFEIFFER, K., HEINZE, S., HOMBERG, U., 2014. Integration of polarization and chromatic cues in the insect sky compass. *J. Comp. Physiol.* A 200, 575-589. https://doi.org/10.1007/s00359-014-0890-6.

FISCHER, J., MÜLLER, T., SPATZ, A.K., GREGGERS, U., GRÜNEWALD, B., MENZEL, R., 2014. Neonicotinoids interfere with specific components of navigation in honeybees. *PLoS One*, 9, e91364. https://doi.org/10.1371/journal.pone.0091364.

GAUTHIER, M., 2010. State of the art on insect nicotinic acetylcholine receptor function in learning and memory. *Adv. Exp. Med. Biol.*, 683, 97-115. https://doi.org/10.1007/978-1-4419-6445-8_9.

GIURFA, M., 2013. Cognition with few neurons: higher-order learning in insects. *Trends Neurosci.* 36, 285-294. https://doi.org/10.1016/j.tins.2012.12.011.

GLINSKI, Z., MARC, M., CHELMINSKI, A., 2012. Role of varroa destructor as immune suppressor and vector of infections in colony collapse disorder (ccd). *Med. Weter.* 68, 585-588.

GOULSON, D., NICHOLLS, E., BOTÍAS, C., ROTHERAY, E.L., 2015. Bee declines driven by combined stress from parasites, pesticides, and lack of flowers. *Science* 347, 1255957. https://science.sciencemag.org/content/347/6229/1255957.

GUEZ, D., 2001. Contrasting effects of imidacloprid on habituation in 7- and 8-day-old honeybees (*Apis mellifera*). *Neurobiol. Learn. Mem.* 76, 183-191. https://doi.org/10.1006/nlme.2000.3995.

HARANO, K., SASAKI, K., NAGAO, T., 2005. Depression of brain dopamine and its metabolite after mating in European honeybee (*Apis mellifera*) queens. *Naturwissenschaften* 92, 310-313. https://doi.org/10.1007/s00114-005-0631-3.

HARANO, K., MITSHUHATA-ASAI, A., KONISHI, T., SUZUKI, T., SASAKI, K., 2013. Honeybee foragers adjust crop contents before leaving the hive. *Behav. Ecol. Sociobiol.* 67, 1169-1178 http://dx.doi.org/10.1007/s00265-013-1542-5.

HARANO, K., MITSUHATA-ASAI, A., SASAKI, M., 2014. Honey loading for pollen collection: regulation of crop content in honeybee pollen foragers on leaving hive. *Naturwissenschaften* 101, 595-598. https://doi.org/10.1007/s00114-014-1185-z.

HENRY, M., BÉGUIN, M., REQUIER, F., ROLLIN, O., ODOUX, J.F., AUPINEL, P., APTEL, J., TCHAMITCHIAN, S., DECOURTYE, A. 2012. A common pesticide decreases foraging success and survival in honey bees. *Science* 336, 348-350. https://doi.org/10.1126/science.1215039.

JACK-MCCOLLOUGH, R.T., NIEH, J.C., 2015. Honeybees tune excitatory and inhibitory recruitment signaling to resource value and predation risk. *Anim. Behav.* 110, 9-17, https://doi.org/10.1016/j.anbehav.2015.09.003.

KIRCHNER, W.H., 1999. Mad-bee-disease? Sublethal effects of imidacloprid ("Gaucho") on the behavior of honey-bees. *Apidologie* 30, 422.

KUCHARSKI. R., MITRI, C., GRAU, Y., MALESZKA, R., 2007. Characterization of a metabotropic glutamate receptor in the honeybee (*Apis mellifera*): implications for memory formation. *Invertebr. Neurosci.* 7, 99-108. https://doi.org/10.1007/s10158-007-0045-3.

LIU, W., SAINT, D.A., 2002. A new quantitative method of real time reverse transcription polymerase chain reaction assay based on simulation of polymerase chain reaction kinetics. *Anal. Biochem.* 302, 52-59. https://doi.org/10.1006/abio.2001.5530.

LIU, J.F., YANG, L., LI, M., HE, X.J., WANG, Z.L., ZENG, Z.J., 2019. Cloning and expression pattern of odorant receptor 11 in asian honeybee drones, *Apis cerana* (hymenoptera, apidae). *J. Asia-Pac. Entomol.* 22, 110-116. https://doi.org/10.1016/j.aspen.2018.12.014.

MENZEL, R., GREGGERS, U., 1985. Natural phototaxis and its relationship to colour vision in honeybees. *J. Comp. Physiol.* A 157, 311-321. https://doi.org/10.1007/BF00618121.

MENZEL, R., 1990. Learning, memory, and 'cognition' in honey bees. *Neurobiology of Comparative Cognition*, 273-292.

MENZEL, R., 2012. The honeybee as a model for understanding the basis of cognition. *Nat. Rev. Neurosci.* 13, 758-768. https://doi.org/10.1038/nrn3357.

MUSTARD, J.A., PHAM, P.M., SMITH, B.H., 2010. Modulation of motor behavior by dopamine and the D1-like dopamine receptor *AmDOP2* in the honey bee. *J. Insect Physiol.* 56, 422-430. https://doi.org/10.1016/j.jinsphys.2009.11.018.

PALMER, M.J., MOFFAT, C., SARANZEWA, N., HARVEY, J., WRIGHT, G.A., CONNOLLY, C.N., 2013. Cholinergic pesticides cause mushroom body neuronal inactivation in honeybees. *Nat. Commun.* 4, 1634-1642. https://doi.org/10.1038/ncomms2648.

PENG, Y.C., YANG, E.C., 2016. Sublethal dosage of imidacloprid reduces the microglomerular density of honey bee mushroom bodies. *Sci. Rep*, 6, 19298-19311. https://doi.org/10.1038/srep19298.

PIGNATELLI, M., UMANAH, G.K.E., RIBEIRO, S.P., CHEN, R., KARUPPAGOUNDER, S.S., YAU, H.J., EACKER, S., DAWSON, V.L., DAWSON, T.M., BONCI, A., 2017. Synaptic plasticity onto dopamine neurons shapes fear learning. *Neuron* 93, 425-436. https://doi.org/10.1016/j.neuron.2016.12.030.

PREECE, K., BEEKMAN, M., 2014. Honeybee waggle dance error: adaption or constraint? Unravelling the complex dance language of honeybees. *Anim. Behav.* 94, 19-26. https://doi.org/10.1016/j.anbehav.2014.05.016.

QIN, Q.H., HAN, X., LIU, H., ZHANG, S.W., ZENG, Z.J., 2014. Expression levels of glutamate and serotonin receptor genes in the brain of different behavioral phenotypes of worker honeybee, *Apis mellifera*. *Turkish Journal of Entomology* 38, 431-441. https://doi.org/10.16970/ted.54867.

RAZAVI, A.M., KHELASHVILI, G., WEINSTEIN, H., MARKOV, A., 2017. State-based quantitative kinetic model of sodium release from the dopamine transporter. *Sci. Rep.* 7, 40-76. https://doi.org/10.1038/srep40076.

SEELEY, T.D., MIKHEYEV, A.S., PAGANO, G.J., 2000. Dancing bees tune both duration and rate of waggle-run production in relation to nectar-source profitability. *J. Comp. Physiol.* A 186, 813-819. https://doi.org/10.1007/s003590000134.

SHERMAN, G., VISSCHER, P.K., 2002. Honeybee colonies achieve fitness through dancing. *Nature* 419, 920-922. https://doi.org/10.1038/nature01127.

SHI, J.L., LIAO, C.H., WANG, Z.L., WU, X.B., 2018. Effect of royal jelly on longevity and memory-related traits of *Apis mellifera* workers. *J. Asia-Pac. Entomol.* 21, 1430-1433. https://doi.org/10.1016/j.aspen.2018.11.003.

SINAKEVITCH, I.T., DASKALOVA, S.M., SMITH, B.H., 2017. The biogenic amine tyramine and its receptor (Amtyr1) in olfactory neuropils in the honey bee (*Apis mellifera*) brain. *Front. Syst. Neurosci.*, 11, 77. https://doi.org/10.3389/fnsys.2017.00077.

SRINIVASAN, M.V., 2000. Honeybee navigation: nature and calibration of the 'odometer'. *Science* 287, 851-853. https://doi.org/10.1126/science.287.5454.851.

TAN, K., WANG, Z.W., YANG, M.X., FUCHS, S., LUO, L.J., ZHANG, Z.Y., LI, H., ZHUANG, D., YANG, S., TAUTZ, J., BEEKMAN, M., OLDROYD, B.P. 2012. Asian hive bees, *Apis cerana*, modulate dance communication in response to nectar toxicity and demand. *Anim. Beha.* 84, 1589-1594. https://doi.org/10.1016/j.anbehav.2012.09.037.

TAN, K., LATTY, T., DONG, S.H., LIU, X.W., WANG, C., OLDROYD, B.P., 2015. Individual honey bee (*Apis cerana*) foragers adjust their fuel load to match variability in forage reward. *Sci. Rep.* 5, 16418. https://doi.org/10.1038/srep16418.

TAUTZ, J., 2008. The Buzz about Bees: Biology of a Super organism. Berlin: Springer-Verlag https://doi.org/10.1007/978-3-540-78729-7.

THANY, S.H., 2010. Electrophysiological studies and pharmacological properties of insect native nicotinic acetylcholine receptors. In Insect Nicotinic Acetylcholine Receptors, 683, 53-63. New York: Springer. http://dx.doi.org/10.1007/978-1-4419-6445-8_5.

THOMPSON, H.M., 2003. Behavioural effects of pesticides in bees: their potential for use in risk assessment. *Ecotoxicology* 12 (1-4): 317-330. https://doi.org/10.1023/A:1022575315413.

TOMIZAWA, M., CASIDA, J.E., 2003. Selective toxicity of neonicotinoids attributable to specificty of insect and mammalian nicotinic receptors. *Annu. Rev. Entomol.* 48, 339-364. https://doi.org/10.1146/annurev.ento.48.091801.112731.

VAN DER SLUIJS, J.P., SIMON-DELSO, N., GOULSON, D., MAXIM, L., BONMATIN, J.M., BELZUNCES, L.P., 2013. Neonicotinoids, bee disorders and the sustainability of pollinator services. *Curr. Opin. Environ. Sustain.* 5, 293-305. https://doi.org/10.1016/j.cosust.2013.05.007.

VON FRISCH, K., 1967. The Dance language and orientation of bees. Cambridge, Massachusetts: Harvard University Press. https://doi.org/10.1093/ae/40.3.187a.

WEBB, B., WYSTRACH, A., 2016. Neural mechanisms of insect navigation. *Curr. Opin. Insect Sci.* 15, 27-39. https://doi.org/10.1016/j.cois.2016.02.011.

WILLIAMSON, S.M., MOFFAT, C., GOMERSALL, M.A.E., SARANZEWA, N., CONNOLLY, C.N., WRIGHT, G.A., 2013. Exposure to acetyl-cholinesterase inhibitors alters the physiology and motor function of honeybees. *Front. Physiol.* 4, 1-10. http://www.doc88.com/p-9955321258903.html.

YANG, E.C., CHANG, H.C., WU, W.Y., CHEN, Y.W., 2012. Impaired olfactory associative behavior of honeybee workers due to contamination of imidacloprid in the larval stage. *PLoS One* 7, e49472. https://doi.org/10.1371/journal.pone.0049472.

ZHANG, L.Z., ZHANG, S.W., WANG, Z.L., YAN, W.Y., ZENG, Z.J., 2014. Cross-modal interaction between visual and olfactory learning in *Apis cerana*. *J. Comp Physiol. A*. 200, 899-909. https://doi.org/10.1007/s00359-014-0934-y.

# 9. Absence of Nepotism in Waggle Communication of Honeybees (*Apis mellifera*)

Zuyun Zhang[1,2], Zhen Li[1], Wujun Jiang*, Qiang Huang[1], Zhijiang Zeng[1]*

1. Honeybee Research Institute, Jiangxi Agricultural University, Nanchang, Jiangxi, 330045, P. R. of China
2. Sericultural and Apicultural Institute, Yunnan Academy of Agricultural Sciences, Mengzi, Yunnan, 661101, P. R. of China
3. Apicultural Institute of Jiangxi Province, Nanchang, Jiangxi, 330052, P. R. of China

**Abstract**

The polyandrous mating behavior of the honeybee queen increases the genetic variability among her worker offspring and the workers of particular subfamilies tend to have a genetic predisposition for tasks preference. In this study, we intended to understand whether there is nepotism in dance communication of honeybees during natural conditions. Microsatellite DNA analyses revealed a total of fourteen and twelve subfamilies in two colonies. The subfamily composition of the dancer and the followers did not deviate from random. The majority of the subfamilies did not show kin recognition in dance recruit communication in honeybee colonies, but some subfamilies showed significant nepotism for workers to follow their super-sister dancer. Because it seems unlikely that honeybee would change the tendency to follow dancers due to the degree of relatedness, we conclude that honeybees randomly follow a dancer in order to e benefit colony gain and development.

**Keywords:** microsatellite DNA, nepotism, recruits, subfamily, waggle dance

## Introduction

Honeybees are a model organism to study the genetic basis of task specialization in social insects due to polyandrous queen (Robinson *et al.*, 2005). As a result, 6-20 subfamilies (paternity) are within a honeybee colony that are genetically different, and these genetic differences also have translated into different behaviors preference (Page & Robinson, 1991). Genetic variability has been described for several behavioral traits, including nectar and pollen foraging (Page & Robinson, 1991), nest site scouting (Robinson & Page, 1989), plant choice for pollen collection (Oldroyd *et al.*, 1992), foraging distance (Oldroyd *et al.*, 1993), water collecting and scenting (Kryger *et al.*, 2000), fanning (Su *et al.*, 2007), emergency queen-cell building (Xie *et al.*, 2008), mite (*Varroa destructor*) parasitism rate (Liu *et al.*, 2009), swarming (Huang & Zeng, 2009),

---

* Correspondence author: bees1965@sina.com (ZJZ).
注：此文发表在 *Journal of Apicultural Science* 2020 年第 2 期。

survival differences (Wang *et al.*, 2012) and genotypic variability of the queen retinue workers (Yi *et al.*, 2018).

Honeybees have evolved numerous mechanisms for increasing colony-level foraging efficiency, mainly the combined system of scout-recruit division of labor and recruitment communication (von Frisch, 1967; Seeley, 1998). This recruitment process incorporating information about the food reward, the colony food stores and the environmental food availability (George & Brockmann, 2019). At the individual level, the efficiency of a scout depends on its ability to find and informing the nest mates the food source. A successful forager performs waggle dances on the surface of the comb where it interacts with nectar receivers and dance followers (Seeley, 1998). The responses towards the waggle dance signal reflect the contextual information possessed by the various honeybees on the dance floor (Seeley & Towne, 1992). Honeybee workers might be able to behave nepotistic-ally and favor the reproductive success of their subfamily (Getz *et al.*, 1983; Page *et al.*, 1989, 1990).

Oldroyd (1991) showed that a strong tendency for recruits to follow their super-sister dancers. This may increase honeybee colony efficiency in dance communication. Arnold (2002) suggested that sub-familial variance for propensity may vary both for waggle and tremble dances. However, neither in a colony consisting of only two subfamilies nor in a colony consisting of 17 subfamilies, there was any evidence for subfamily discrimination among dancers and their followers under foraging from artificial feeding site providing sucrose solution (Kirchner & Arnold, 2001). However, the result might be biased by different foraging preferences of the subfamilies, such as nectar and pollen foraging (Page & Robinson, 1991). In order to avoid the foraging preference, we used colonies during natural foraging rather than artificial feeder. The aim of the present study is to determine any nepotism in dancing recruitment under natural conditions. We therefore determined the subfamily frequencies of natural mated colonies, as well as the subfamily of the dancers and followers by microsatellite genetic markers.

## Materials and Methods

### Samples collection

The experiments were conducted with two honeybee colonies (colony A and colony B) with naturally mated queens. The colonies were during natural foraging from food resources rather than artificial feeding site providing sucrose solution. Each colony was composed of 4 frames of honeybees, and housed in observation hives. One side of the observation hive was open during the experiments in order to pick honeybees for genetically analysis directly from the combs. The dancing floor area of the observation hive was observed. The foragers which have been observed 10 times in a row to perform a waggle dance after returning from the food source were classified as waggle dancers. The dancing followers were defined as the honeybees attending a waggle dance persistently for several circuits. When the foragers were dancing, the recruits followed her to dance, forming a circle. The outer followers were captured first in a box and the center of dancer was captured later in an EP tube. A dancer and dance followers (about 8 individuals) were frozen immediately on dry ice and kept for a later determination of their subfamily membership. Thirteen dancers and 117 followers were collected from colony A. Twelve dancers and 93 followers were collected from colony B. For each dancer, a minimal of 5 followers were collected. Additionally, 64 (colony A) and 48 (colony B) workers were randomly picked out to represent the overall subfamily composition of the colonies.

### *DNA amplification and genetic analysis*

After the sampling, a genomic DNA extraction kit (StarSpin Animal DNA Kit) was used to extract DNA of each sample individually. Three microsatellite loci (A14, A24 and A113) were selected to determine the subfamilies according to previous reports (Tian *et al.*, 2013; Zhang *et al.*, 2017; Yi *et al.*, 2018) (Table III-9-1). These sequences have a high degree of polymorphism. The PCR reaction mixture (25 μL) was as follows: 1 μL of DNA, 1 μL each of forward and reverse primer, 12.5 μL of SinoBio 2×Master Mix, and 9.5 μL of ultrapure sterile water. The loci were amplified using Bio-RAD T100™ thermal cyclers through 30 cycles consisting of denaturation for 30 s at 94 ℃, annealing for 45 s at 55-60 ℃ (depending on the locus), and elongation for 60 s at 72 ℃ (Table III-9-1). The allele of each microsatellite marker was determined with capillary electrophoresis (QIAxcel Advanced system). Then, the subfamily was analyzed using Mate-soft (Moilanen, *et al.*, 2004).

**Table III-9-1  Description of primer sequences and reactioncondition used for microsatellites**

| Locus | Sequence of primers | Size (bp) | Annealing temperature (℃) | No. cycles |
|---|---|---|---|---|
| A14 | F5'-GTGTCGCAATCGACGTAACC-3'<br>R 5'-GTCGATTACCGATCGTGACG-3' | 200 | 58 | 30 |
| A24 | F5'-CACAAGTTCCAACAATGC-3'<br>R 5'-CACATTGAGGATGAGCG-3' | 100 | 55 | 30 |
| A113 | F5'-CTCGAATCGTGGCGTCC-3'<br>R 5'-CCTGTATTTTGCAACCT CGC-3' | 220 | 60 | 30 |

### *Statistic analysis*

All statistic analysis was conducted using SPSS Statistics 26.0. The distribution of workers in each subfamily between dancing followers and randomly collected workers was analyzed using Fisher's exact test. The distribution of recruits in all super-sister of against all other half-sisters in two colonies respectively was analyzed using Fisher's exact test (Kryger *et al.*, 2000).

## Results

Genotypes of paternity were calculated in each colony respectively (Table III-9-2). As the subfamilies frequencies were unequal in our samples, we extended the statistical analysis by considering the subfamilies separately. Colony A consisted 14 subfamilies, 22 of 117 dance followers were super-sisters of the dancers and 95 were half-sisters (Table III-9-2, 3). Colony B consisted 12 subfamilies, where 23 of 93 analyzed followers were super-sisters of the dancers and 70 were half-sisters (Table III-9-2, 4). One group in colony A, showed significant nepotism for workers to follow super-sister dancer (chi-square test: $\chi^2 = 7.778$, $df = 1$, $P = 0.021$). However, significant preference on nepotism was not observed between the distribution of the waggle dancers and that of the dance followers overall in colony A (chi-square test: $\chi^2 = 3.760$, $df = 6$, $P = 0.768$) (Table III-9-3). For colony B, neither individual group nor whole colony showed significant preference on nepotism between the waggle dancers and that of the dance followers on the whole (chi-square test: $\chi^2 = 5.516$, $df = 7$, $P = 0.498$) (Table III-9-4).

**Table III-9-2  Genotypes of paternity were in each colony**

| Colony A | | | Colony B | | |
|---|---|---|---|---|---|
| Subfamilies | Genotypes of paternity A014/A024/A113 | Random samples | Subfamilies | Genotypes of paternity A014/A024/A113 | Random samples |
| 1 | 225/120/215 | 18 | 1 | 215/120/215 | 6 |
| 2 | 225/120/235 | 12 | 2 | 215/120/210 | 4 |
| 3 | 225/120/225 | 9 | 3 | 235/120/215 | 8 |
| 4 | 215/120/235 | 8 | 4 | 220/120/215 | 7 |
| 5 | 215/120/215 | 6 | 5 | 225/120/215 | 5 |
| 6 | 215/120/225 | 2 | 6 | 215/120/235 | 5 |
| 7 | 245/120/215 | 2 | 7 | 235/110/225 | 5 |
| 8 | 235/120/225 | 2 | 8 | 235/120/235 | 3 |
| 9 | 225/110/215 | 1 | 9 | 245/120/215 | 0 |
| 10 | 235/120/235 | 2 | 10 | 235/120/210 | 1 |
| 11 | 245/110/225 | 0 | 11 | 220/120/235 | 3 |
| 12 | 245/120/235 | 2 | 12 | 220/120/210 | 1 |
| 13 | 235/110/235 | 0 | | | |
| 14 | 235/110/215 | 0 | | | |

Note: Colony A consisted of 14 paternity, colony B of 12.

**Table III-9-3  Distribution of super-sisters and half-sisters in colony A**

| Dancer | | Recruits | | | | |
|---|---|---|---|---|---|---|
| No. | Subfamilies | Super-sister Observed | Half-sisters Observed | Super-sister Expected | Half-sisters Expected | $P$ values |
| A1 | 5 | 1 | 11 | 1.1 | 10.9 | 1.000 |
| A2 | 7 | 0 | 11 | 0.3 | 10.7 | 1.000 |
| A3 | 7 | 1 | 8 | 0.3 | 8.7 | 1.000 |
| A4 | 6 | 0 | 8 | 0.2 | 7.8 | 1.000 |
| A5 | 3 | 3 | 4 | 1.0 | 6.0 | 0.559 |
| A6 | 1 | 1 | 6 | 2.0 | 5.0 | 1.000 |
| A7 | 2 | 2 | 4 | 1.1 | 4.9 | 1.000 |
| A8 | 1 | 3 | 6 | 2.5 | 6.5 | 1.000 |
| A9 | 6 | 5 | 2 | 0.2 | 6.8 | 0.021 |

(continued)

| Dancer | | Recruits | | | | |
|---|---|---|---|---|---|---|
| No. | Subfamilies | Super–sister Observed | Half-sisters Observed | Super–sister Expected | Half-sisters Expected | P values |
| A10 | 10 | 1 | 6 | 0.2 | 6.8 | 1.000 |
| A11 | 3 | 1 | 10 | 1.6 | 9.4 | 1.000 |
| A12 | 6 | 1 | 12 | 0.4 | 12.6 | 1.000 |
| A13 | 5 | 3 | 7 | 0.9 | 9.1 | 0.582 |

Note: The last column gives the $P$-values for Fisher exact tests for the distribution of the number of super-sister against all other half-sisters in each group. Fisher's exact test (two-tailed) for the distribution of the number of super-sister against all other half-sisters on the whole ($\chi^2 = 3.760$, $df = 6$, $P = 0.768$).

**Table III-9-4  Distribution of super-sisters and half-sisters in colony B**

| Dancer | | Recruits | | | | |
|---|---|---|---|---|---|---|
| No. | Subfamilies | Super–sister Observed | Half-sisters Observed | Super–sister Expected | Half-sisters Expected | P values |
| B1 | 1 | 1 | 6 | 0.9 | 6.1 | 1.000 |
| B2 | 4 | 1 | 7 | 1.2 | 6.8 | 1.000 |
| B3 | 1 | 2 | 7 | 1.1 | 7.9 | 1.000 |
| B4 | 6 | 0 | 8 | 0.8 | 7.2 | 1.000 |
| B5 | 1 | 4 | 1 | 0.6 | 4.4 | 0.206 |
| B6 | 2 | 1 | 6 | 0.6 | 6.4 | 1.000 |
| B7 | 9 | 1 | 6 | 0 | 7.0 | 1.000 |
| B8 | 5 | 2 | 5 | 0.7 | 6.3 | 1.000 |
| B9 | 3 | 1 | 8 | 1.5 | 7.5 | 1.000 |
| B10 | 1 | 6 | 4 | 1.3 | 8.7 | 0.057 |
| B11 | 1 | 3 | 4 | 0.9 | 6.1 | 0.559 |
| B12 | 8 | 1 | 8 | 0.6 | 8.4 | 1.000 |

Note: The last column gives the $P$-values for Fisher exact tests for the distribution of the number of super-sister against all other half-sisters in each group. Fisher's exact test (two-tailed) for the distribution of the number of super-sister against all other half-sisters on the whole ($\chi^2 = 5.516$, $df = 7$, $P = 0.498$).

## Discussion

For our study, the evidence for subfamily nepotism among dancers and their followers was not strong. Oldroyd (1991) reported that honeybees dance preferentially with their super-sisters, which suggests a strong tendency to recruit super-sisters, indicating subfamily discrimination. The tendency of a recruit to follow a dancer was affected by the subfamily of the recruit, the subfamily of the dancer, foraging preferences of dancer and what she was carrying (Waddington, 1989; Page & Robinson, 1991). Here we show that recognition between super-sisters and half-sisters is at least not a general feature of the dance communication system.

Honeybees are highly social insects, which are information sharing, division and cooperation. Natural selection therefore should not favor nepotism in the context of foraging and recruitment. Such factors as need of colony, distance of flight, plant species foraged and availability of the food resources should be taken into consideration (Oldroyd et al., 1992, 1993, 1994). Kirchner and Arnold (2001) found no evidence for subfamily discrimination among dancers and their followers in a colony. However, honeybees just foraged from artificial feeding site providing sucrose solution. So, the result might be due to different foraging preferences of the subfamilies. Gilley (2014) also suggest that the waggle-dance hydrocarbons play an important role in honeybee foraging recruitment by stimulating foragers to perform waggle dances. In addition, dance activity for a food sources regulated by an inter-play between individual response thresholds and the social context obtained through interaction with nest mates (George & Brockmann, 2019). Thus, it seems unlikely that honeybee would change the tendency to follow dancers due to the degree of relatedness, and we conclude that in honeybee colonies the majority of subfamilies do not show kin recognition in dance-recruit associations under natural conditions. But the minority of subfamilies showed significant nepotism for workers to follow super-sister dancer. We believe that this result is due to the difficulty of sampling and the low sample size in individual groups. This helps the colony to efficiently utilize its foraging force to effectively exploit food sources in the environment around it.

## Acknowledgements

This work was supported by the National Natural Science Foundation of China (31572469, 31872432), the Earmarked Fund for China Agriculture Research System (CARS-44-KXJ15).

## References

ARNOLD, G., QUENET, B., PAPIN, C., MASSON, C. & KIRCHNER, W.H. (2002). Intra-colonial variability in the dance communication in honeybees (*Apis mellifera*). *Ethology, 108*, 751-761. doi: 10.1046/j.1439-0310.2002.00809.x.

GEORGE, E.A. & BROCKMANN, A. (2019). Social modulation of individual differences in dance communication in honey bees. *Behavioral Ecology and Sociobiology, 73*, 41-55. doi: 10.1007/s00265-019-2649-0.

GETZ, W.M. & SMITH, K.B. (1983). Genetic kin recognition: honey bees discriminate between full and half sisters. *Nature, 302*, 147-148. doi: 10.1038/302147a0.

GILLEY, D.C. (2014). Hydrocarbons emitted by waggle-dancing honey bees increase forager recruitment by stimulating dancing. *PLoS ONE, 9*, e105671. doi:10.1371/journal.pone.0105671.

HUANG, Q. & ZENG, Z.J. (2009). Nepotism in swarming honeybees (*Apis cerana cerana*). *Chinese Bulletin of Entomology, 46*, 107-111.

KIRCHNER, W.H. & ARNOLD, G. (2001). Intracolonial kin discrimination in honeybees: do bees dance with their supersisters? *Animal Behaviour, 61*, 597-600. doi: 10.1006/anbe.2000.1626.

KRYGER, P., KRYGER, U. & MORITZ, R.F.A. (2000). Genotypical variability for the tasks of water collecting and scenting in a honey bee colony. *Ethology, 106,* 769-779. doi: 10.1046/j.1439-0310.2000.00571.x.

LIU, Y.B., HUANG, Q. & ZENG, Z.J. (2009). Study on the selective parasitism of the varroa mite to the different subfamilies by ISSR. *Journal of Bee, 8,* 6-9.

MOILANEN, A., SUNDSTROEM, L. & PEDERSEN, J.S. (2004). Matesoft: a program for deducing parental genotypes and estimating mating system statistics in haplodiploid species. *Molecular Ecology Notes, 4,* 795-797. doi: 10.1111/j.1471-8286.2004.00779.x.

OLDROYD, B.P., RINDERER, T.E. & BUCO, S.M. (1991). Honey bees dance with their super sisters. *Animal Behaviour, 42,* 121-129. doi: 10.1016/s0003-3472(05)80612-8.

OLDROYD, B.P., RINDERER, T.E. & BUCO S.M. (1992). Intra-colonial foraging specialism by honey bees (*Apis mellifera*) (Hymenoptera: apidae). *Behavioral Ecology and Sociobiology, 30,* 291-295. doi: 10.1007/BF00170594.

OLDROYD, B.P., RINDERER, T.E., BUCO S.M. & BEAMAN, L. (1993). Genetic variance in honeybees for preferred foraging distance. *Animal Behaviour, 45,* 323-332. doi: 10.1006/anbe.1993.1037.

OLDROYD, B.P., RINDERER, T.E. SCHWENKE, J.R. & BUCO, S.M. (1994). Subfamily recognition and task specialization in honey bees (*Apis mellifera*) (Hymenoptera: apidae). *Behavioral Ecology and Sociobiology, 34,* 169-173. doi: 10.1007/BF00167741.

PAGE, R.E., ROBINSON, G.E. & FONDRK, M.K. (1989). Genetic specialist, kin recognition and nepotism in honeybee colonies. *Nature, 338,* 576-579. doi: 10.1038/338576a0.

PAGE, R.E., BREED, M.D. & GETZ, W. (1990). Nepotism in the honey bee. *Nature, 346,* 707-707. doi: 10.1038/346707a0.

PAGE, R.E. & ROBINSON, G.E. (1991). The genetics of division of labor in honey bee colonies. *Advances in Insect Physiology, 23,* 117-169. doi: 10.1016/S0065-2806(08)60093-4.

RAYMOND, M. & ROUSSET, F. (1995). An exact test for population differentiation. *Evolution, 49,* 1280-1283. doi: 10.2307/2410454.

ROBINSON, G.E. & PAGE, R.E. (1989). Genetic determination of nectar foraging, pollen foraging, and nest-site scouting in honey bee colonies. *Behavioral Ecology and Sociobiology, 24,* 317-323. doi: 10.1007/BF00290908.

ROBINSON, G.E., GROZINGER C.M. & WHITFIELD, C.W. (2005). Socio-genomics: social life in molecular terms. *Nature Reviews Genetics, 6,* 257-270. doi:10.1038/nrg1575.

SEELEY, T.D. & TOWNE, W.F. (1992). Tactics of dance choice in honey bees: do foragers compare dances? *Behavioral Ecology and Sociobiology, 30,* 59-69. doi: 10.1007/BF00168595.

SEELEY, T.D. (1998). Thoughts on information and integration in honey bee colonies. *Apidologie, 29,* 67-80.

SU, S.K., ALBER, T.S., ZHANG, S., MAIER, S., CHEN, S.L, DU, H. & TAUZT, J. (2007). Non-destructive genotyping and genetic variation of fanning in a honeybee colony. *Journal of Insect Physiology, 53,* 411-417. doi: 10.1016/j.jinsphys.2007.01.002.

TIAN L.Q., HE, X.J., WANG, H. & ZENG, Z.J. (2013). Analysis of genetic background between two types of nurses in honeybee (*Apis mellifera*). *Journal of Bee, 10,* 1-3.

VON FRISCH, K. (1967). The Dance language and orientation of bees. Cambridge, Massachusetts: Harvard University Press.

WANG, H., ZHANG, S.W., ZHANG, F. & ZENG, Z.J. (2012). Analysis of life-spans of workers from different sub-families in a honeybee colony. *Entomological Knowledge, 49*, 1172-1175.

WADDINGTON, K.D. (1989). Implications of variation in worker body size for the honey bee recruitment system. *Journal of Insect Behavior, 2*, 91-103. doi: 10.1007/BF01053620.

XIE, X.B., SUN, L.X., HUANG, K. & ZENG, Z.J. (2008). Worker nepotism during emergency queen rearing in Chinese honeybees *Apis cerana cerana*. *Acta Zoologica Sinica, 54*, 695-700.

YI, Y., YAN, W.Y., LI, Y., GUO, Y.H., ZHANG, L.Z. & ZENG, Z.J., (2018). Genotypic variability of the queen retinue workers in honeybee colonies (*Apis mellifera*). *African Entomology, 26*, 30-35. doi: 10.4001/003.026.0030.

ZHANG, Z.Y., YANG, R.P., CHEN, L., YU, Y.S., SONG, W.F., WANG, Y.H. & ZHANG, X.W. (2017). Genetic diversity analysis of mite-resistance honeybees (*Apis mellifera*) based on microsatellite markers. *Journal of Southwest China Normal University, 42*, 31-37. doi: 10.13718/j.cnki.xsxb.2017.04.006.

# 10. Influence of RNA Interference-mediated Reduction of *Or*11 on the Expression of Transcription Factor *Kr-h*1 in *Apis mellifera* Drones

Junfeng Liu [1,2†], Xiaojuan Wan [3†], Zilong Wang [1], Xujiang He [1], Zhijiang Zeng [1*]

1. Honeybee Research Institute, Jiangxi Agricultural University, Nanchang, Jiangxi, P.R. of China.
2. Environment and Plant Protection Institute, Chinese Academy of Tropical Agricultural Sciences, Haikou, P. R. China.
3. Department of Laboratory Animal Science, Nanchang University, Nanchang, Jiangxi, P.R. of China.

## Abstract

9-oxo-2-decenoic acid (9-ODA, the predominant component of honeybee queen mandibular pheromones) acts as a sex pheromone attracting drones during mating flights in midair. *Odorant receptor* 11 (*Or*11), which is located on the membrane of antennal olfaction receptor neurons in bees, can specifically recognize 9-ODA. At present, it is still unclear about the molecular pathway of honeybee drones responding to 9-ODA. Studies have demonstrated that 9-ODA could down-regulate the expression of *Krüppel-homolog*1 (*Kr-h*1, a transcription factor related to the regulation of reproduction and division of labor mediated by juvenile hormone) gene in the mushroom of honeybee brain. We speculate that *Kr-h1* may be the downstream gene of *Or*11, which is involved in the pathway of drones responding to 9-ODA. Therefore, we analyzed the influence of 9-ODA on the expression of *Or*11 and *Kr-h*1 in the antennae of sexually immature (4 days) and mature (14 days) male honeybees (*Apis mellifera*) by quantitative polymerase chain reaction (qPCR). The results demonstrated that 9-ODA significantly downregulated the expression of *Or*11 and *Kr-h*1 in the antennae of sexually immature and mature drones. Additionally, siRNA-*Or*11 was injected into the antennae and brain tissues of 8-day-old drone pupae, and the expression patterns of *Or*11, *Kr-h*1 and *Broad-Complex* (*Br-c*, downstream gene of *Kr-h*1) were determined by qPCR at 72 h. The RNAi-induced knockdown of *Or*11 significantly decreased the expression of *Or*11, *Kr-h*1 and *Br-c* in the antennae and brains of drones. This study suggests that the transcription factor *Kr-h*1 is downstream of *Or*11, *Kr-h*1, which may play an important role in the signal transduction process of drones responding to 9-ODA.

**Keywords:** *Apis mellifera*, Drone, 9-ODA, *Odorant receptor* 11, *Krüppel-homolog* 1, RNA interference

---

† These two authors contributed equally to this paper.
*Corresponding author: bees1965@sina.com (ZJZ).
注：此文发表在 *Insectes Sociaux* 2020 年第 3 期。

# Introduction

Pheromones, which are one of the major forms of communication for honeybees, can quickly trigger physiological and behavioral changes in honeybees (Grozinger et al. 2007). Queen mandibular pheromone (QMP) acts both as a social pheromone and a sex pheromone (Grozinger and Robinson 2007). As a social pheromone, QMP trigger long-term physiological and behavioral responses in worker honeybees through physiologically related systems (Slessor et al. 2005; Rangel et al. 2016). For example, the pheromone inhibits the development of worker bee ovaries and the brood of a new queen and delays the change of nurse worker to forage worker (Pankiw et al. 1998; Pankiw and Page 2003). QMP also acts as a sex pheromone by inducing rapidly behavioral responses in drones through the nervous system during mating flight (Grozinger et al. 2007). In mating flight, the virgin queen releases QMP to attract sexually mature drones for copulation (Butler et al. 1959, 1962; Gary 1962; Butler 1971; Gary and Marston 1971). QMP is composed of 5 major functional secretions: (E)-9-oxodec-2-enoic acid (9-ODA) and two enantiomers of 9-hydroxydec-2-enoic acid (9-HDA), methyl p-hydroxybenzoate (HOB) and 4-hydroxy-3-methoxyphenyl ethanol (HVA) (Keeling et al. 2003). To date, 9-ODA is the most widely used and studied pheromone (Carlisle and Butler 1956; Butler et al. 1959, 1962; Gary 1962). Previous studies have shown that 9-ODA can produce multiple effects on the behaviors and physiology of drones (Loper et al. 1993; Brockmann et al. 2006). 9-ODA acts as a sex pheromone attracting drones during mating flights in midair (Gary 1962), and delays the development of sexually immature drones, such as the time for initial mating flight and decreases the number of flights (Villar and Grozinger 2017).

Drones have a highly specific and extremely sensitive olfactory system. Specifically, there are rich placoid sensilla on the surface of drone antennae, the number of which in drone is approximately 7 times more than in worker (Sandoz et al. 2007). Brockmann et al. (1998) discovered that drone antennae were more sensitive to 9-ODA than worker antennae, the electroantennography (EAG) response to 9-ODA in drone was higher than others components of QMP. Using a custom chemosensory-specific microarray and qPCR, Wanner et al. (2007) found high expression levels of candidate sex pheromone receptor genes *Or*10, *Or*11, *Or*18 and *Or*170 in the antennae of drones, and proved *Or*11 responds specifically to 9-ODA by injecting the cRNA of the four receptors into *Xenopus* oocytes and examining the sensitivity of each of the QMP components with two-electrode voltage-clamp electrophysiology in vitro. Claudianos et al. (2014) reported that the expression levels of *Or*11 in antennae declined after a proboscis extension reflex (PER) assay in which a worker encountered 9-ODA. The expression of *Ac*Or11 in brain of sexually mature drones was significantly higher than those of immature drones. Additionally, the expression of and *Ac*Or11 in brains of mature flying drones was higher than those of drones in hive, indicating that the expression levels of *Ac*Or11 in drone brains may be associate with sexual maturity and mating flight (Liu et al. 2019). In addition, the structure and functions of the brain tissues of drone are different from those of worker. The antennal lobe (AL) of drone includes both 103 ordinary glomeruli and 4 macroglomeruli (MG), but the AL of worker only consists of conventional glomeruli (Sandoz 2006). Moreover, by using calcium imaging technology, these authors identified that the MG2 of drones specifically responds to 9-ODA. Nonetheless, the mechanism of honeybee drones response to 9-ODA remains unclear.

*Krüppel-homolog* 1 (*Kr-h*1), a nuclear receptor gene that was first identified in a study in *Drosophila melanogaster*, can regulate insect metamorphosis (Schuh *et al.* 1986). *Kr-h*1 is related to the regulation of the behavior by QMP mediated ovary activation and labor division mediated by juvenile hormone (JH) in honeybee (Whitfield *et al.* 2003; Grozinger *et al.* 2007; Shpigler *et al.* 2010; Kilaso *et al.* 2017). Interestingly, 9-ODA, as effective as QMP, can inhibit expression of the transcription factor *Kr-h*1 in the brain of worker bees (Grozinger *et al.* 2007). However, it is still unknown for the function of *Kr-h*1 in drone bees. Therefore, we speculate that *Kr-h*1 (nuclear receptor) may be the downstream gene of *Or*11 (membrane receptor), and involved in modulating the signal transduction process that the *Or*11 responds to 9-ODA in honeybee drones. To further explore the signal transduction pathway behind the responses of drones to 9-ODA, we further tested the expression level of *Broad-Complex* (*Br-c*, transcription factor) after knocking down *Or*11. *Br-c* is the downstream gene of *Kr-h*1 in many insects (Minakuchi *et al.* 2008, 2009; Huang *et al.* 2013; Belles and Santos 2014), and is a key gene in the JH and 20-hydroxyecdysone (20E) signaling pathway of fruit fly and honeybee (Paul *et al.* 2006; Abdou *et al.* 2011).

In this study, the influence of 9-ODA on the expression levels of *Or*11 and *Kr-h*1 in *A. mellifera* drones were detected by qPCR. Then siRNAs targeting the gene *Or*11 was injected into the antennae and head of honeybee drones. We determined the expression characteristics of *Or*11, *Kr-h*1 and *Br-c* by qPCR for exploring the relationship between *Kr-h*1 and *Or*11. The results help interpret the physiological functions of *Or*11 and *Kr-h*1 in drones and exploring the signaling pathway that drones respond to 9-ODA in honeybees.

## Materials and Methods

### Insects

Western honeybee (*A. mellifera*) colonies were maintained at the Honeybee Research Institute, Jiangxi Agricultural University, Nanchang, China (28.46 °N, 115.49 °E) using standard beekeeping techniques.

### General bee rearing, 9-ODA treatment and collection

Four *A. mellifera* colonies with the same genetic background and population were selected. Four-day-old (sexually immature) and 14-day-old (sexually mature) drones were produced by caging the queen on frames with drone-sized honeycomb; the queen was moved to another comb after 24 h. Twenty-three days later, the frames with capped cells were transferred to an incubator (34 °C and 70 % humidity) to obtain emerging drone bees. Upon emergence, the drones were marked on their thorax with marking paint.

Rearing of 4-day-old drones in cages was performed as described (Villar *et al.* 2015). Groups of 10 newly emerged drones and 20 newly emerged workers were established in Plexiglass cages (10 cm × 7 cm × 6 cm); the insects were fed 50 % sucrose, water and crushed pollen. The cages were maintained in a dark incubator at 34 °C and 50 % relative humidity. The treatment group was stimulated by daily administration of a solution of synthetic 9-ODA (Contech International, Victoria, BC) ethanol equivalent to that of 1 virgin queen (approximately 70 μg, according to previously described Plettner *et al.* 1997), which was dissolved in ethanol and placed on filter paper. The blank control group was treated only with the same dose of ethanol on filter paper. There were three replicate cages for both the treatment and control groups. Twenty-four 4-day-old drones were collected from the treatment and control groups, immediately frozen in liquid nitrogen in a centrifugal tube, and stored at -80 °C.

The rearing of 14-day-old drones in cages was performed as previously described (Liu *et al.* 2019). Three hundreds newly emerged drones with marked paint were placed back into their parent colony. When these drones reached 14 days of age, we collected those that returned to the entrance of the hive (hereafter referred to as flying drones); the drones crawling in the wall and comb of the hive (hereafter referred to as hive drones) were caught in hive with a 50-mL centrifugal tube. Moreover, to confirm the biological mating habits of adult drones, we collected drones during the peak flight period, between 13:00-16:00, during good weather that was clear and warm with light, variable breezes. A drone was then placed into the base of a Y-tube olfactometer (35 cm arm length and 5 cm internal diameter, customized by the Glass instrument factory of Nanchang University, China), and tests were performed in a darkroom. For the treatment group, a solution of 9-ODA ethanol that was equivalent to that of 1 virgin queen was placed on filter paper in the left and right bottles of the Y-tube olfactometer. For the blank control group, only 5 μL of ethanol solution was added to the filter paper. After 5 min of testing, the drone was caught and put into a centrifuge tube, in which liquid nitrogen was added for quick freezing; the samples were stored at -80 °C for subsequent isolation of RNA. There were 6 groups for each treatment, with 8 drone repetitions for each group.

### *Design and synthesis of siRNA sequences and qPCR primers*

Three siRNA sequences (siRNA-*AmOr*11) were designed and synthesized by Shanghai GenePharma Co., Ltd (*AmOr*11 GenBank accession number: NM_001242962.1), as was negative control siRNA (siRNA-NC). The siRNA sequences are listed in Table III-10-1. qPCR primers were designed by Primer 5.0 software and synthesized by Shanghai Sangon Biotechnology Co. Ltd.; these primers are shown in Table III-10-2.

**Table III-10-1  siRNA sequence designed for RNA interference**

| siRNA | Sense | Antisense |
| --- | --- | --- |
| siRNA-*Or*11-329 | GCAACGGGCUAAGGAAUUUTT | AAAUUCCUUAGCCCGUUGCTT |
| siRNA-*Or*11-528 | CCGAACAACAUGACAGUAATT | UUACUGUCAUGUUGUUCGGTT |
| siRNA-*Or*11-1143 | GCAGGAAGAAUUAUGGAUUTT | AAUCCAUAAUUCUUCCUGCTT |
| siRNA-NC | UUCUCCGAACGUGUCACGUTT | ACGUGACACGUUCGGAGAATT |

**Table III-10-2  Primers used for qRT-PCR**

| Primer name | Primer sequence | TM (°C) |
| --- | --- | --- |
| *AmOr*11-F | 5'- CTTTTACCGAACAACATGACAG -3' | 54 |
| *AmOr*11-R | 5'- TTATCTCGTAATTAGGTGTGG -3' | |
| *AmKr-h*1-F | 5'- GCACTGGCAGTGACAAGGAA -3' | 60 |
| *AmKr-h*1-R | 5'- CGTGGAGTGTTATCGTAAGTAGCAA-3' | |

(continued)

| Primer name | Primer sequence | TM (°C) |
|---|---|---|
| AmBrc-F | 5'- GACAGGTGGCAACAGCGGTAAC-3' | 60 |
| AmBrc-R | 5'- TGGACGTGTGCTCGGACTCG -3' | |
| Amβ-actin-F | 5'- GGTATTGTATTGGATTCGGGTG -3' | 60 |
| Amβ-actin-R | 5'- TGCCATTTCCTGTTCAAAGTCA -3' | |

### *Injecting siRNA into antennae and head tissues of drone pupae*

Gene knockdown using siRNA is a powerful experimental approach to identify gene function in honeybee. Such as, knockdown of target genes have been successfully achieved by injecting siRNA into eggs, larvae or adults of honeybee (Beye *et al.* 2002; Erezyilmaz *et al.* 2006; Minakuchi *et al.* 2009; Guo *et al.* 2018). Our previous test found that drones were easy to die when they were injected with siRNA or water at the adult stage, the reason is that cutting the ommateum for injecting injured their brains, and these injured adult drones were easily attacked by worker bees. Thus, we used 8-day-old drone pupa to perform the RNAi experiment and found that this method of RNAi is feasible.

Drone pupae from four *A. mellifera* colonies were produced by caging the queen on frames with drone-sized honeycomb for 24 h. Twenty days later, the drone pupae on the frames with capped cells were transferred to 12-well sterile cell culture plates.

Approximately 100 ng/μL siRNA-*Or*11, siRNA-NC and diethyl pyrocarbonate water (DEPC water) water solutions were injected into two antennae (0.25 μL each) and the head (0.5 μL, at the base of antennas) by using a microinjector under 3X bench magnifiers, and the culture plate containing the pupae was placed in a constant-temperature (34 °C) and humidity (50 %) incubator for approximately 72 h until emergence. Statistical analysis of the survival rates of the drone pupae in each group was performed by one-way ANOVA in SPSS 17.0. For each colony, 8 emerged drones were added to a centrifuge tube and quickly frozen by liquid nitrogen; the centrifuge tube was stored at -80 °C.

### *Preparation of antennae and brain tissue*

Preparation of antennae samples occurred as follows. Drones were removed from the liquid nitrogen. One pair of complete antennae was removed from the head using clean tweezers and blades and placed into a 1.5-mL RNase-free EP tube. Eight pairs of antennae from 8 drones of the same colony were pooled as one sample, quickly frozen in liquid nitrogen, and stored at -80 °C for extraction of RNA.

The brain samples were prepared as follows. The remaining heads after the antennae were removed were placed in phosphate-buffered saline (PBS). Head shell and ommateum tissues were eliminated by using clean tweezers and blades, and the tissue was quickly placed in a 1.5-mL RNase-free EP tube. Liquid nitrogen was added to freeze the samples quickly. Eight pairs of brain tissues from 8 drones in the same colony were used as one sample, which was stored at -80 °C for subsequent extraction of RNA.

### *Total RNA isolation, cDNA synthesis and qPCR experiments*

Antennae and brain tissues were collected from each group of samples, and RNA was extracted according to the operating instructions of a TransZol Up kit (Transgen Biotech). The RNA concentration and mass of each sample were measured and tested by spectrophotometry and agarose gel electrophoresis (AGE), respectively. cDNA was synthesized from total antenna RNA using a reverse transcription kit (TaKaRa) according to the manufacturer's instructions.

Relative expression of *AmOr*11 and *AmKr-h*1 in the antennae of *A. mellifera* drones between 4 and 14 days, as well as relative expression of *AmOr*11, *AmKr-h*1 and *AmBr-c* in antennae and brains of drones after injection of siRNA, were tested by qPCR. qPCR was performed with an initial denaturation step of 10 min at 95 °C, followed by 40 cycles of 94 °C for 15 s, 60 °C for 40 s, and 72 °C for 35 s; a melting curve analysis was conducted to verify the specificity of the amplification. Experiments for test samples, an endogenous control, and a negative control were performed in triplicate to ensure reproducibility.

To quantify the expression levels of the target genes, the *actin* gene was often used as a single reliable reference gene in honeybees, *Apis mellifera*, in many studies (Lourenço et al. 2008; Scharlaken et al. 2008; Nunes and Simões 2009; Martins et al. 2010; Wang et al. 2013; Harwood et al. 2019). Especially, in the study performed by Wang et al.(2013), they observed a high consistency between behavioral changes of adult bees and expression changes of target genes after RNAi of *vitellogenin* (*vg*) and *ultraspiracle* (*usp*), in which the *actin* was used as a single reference gene for qRT-PCR. These results suggests that it is reliable using *actin* as a single reference gene for adequate normalization. Therefore, we used *actin* as an endogenous control in this study.

*Statistical analysis*

Statistical analysis of the survival rates of drone pupae was carried out using one-way ANOVA in SPSS 17.0. The CT values of the target genes and β-actin, which was used as an internal control, were collected. The relative expression levels of target genes in honeybees were calculated using the $2^{-\Delta\Delta CT}$ method (Schmittgen and Livak 2008). The statistical analyses of qPCR were performed by *t*-tests and one-way ANOVA. $P < 0.05$ represents a significant difference.

# Results

*Effects of 9-ODA on expression of AmOr11 and AmKr-h1 in antennae of 4 and 14-day-old drones*

According to qPCR results, sustained exposure to 9-ODA significantly inhibited expression of *AmOr*11 and *AmKr-h*1 in the antennae of 4-day-old drones ($P < 0.05$, Fig. III-10-1). Similarly, acute exposure to 9-ODA significantly repressed expression of *AmOr*11 and *AmKr-h*1 in the antennae of flying and hive drones at 14 days ($P < 0.05$, Fig. III-10-2).

*siRNA screen and the effects of injection on the survival of drone pupae*

The effectiveness of RNA interference by 3 siRNAs targeting *Or*11 was compared through a pilot experiment. We found that siRNA-*Or*11-528 downregulated relative expression of *Or*11 in drone antennae at 72 h after injection; in contrast, the expression levels of *Or*11 after the other 2 siRNA-*Or*11s were injected were similar to those of the blank control group (SI Fig. 1). The 3 siRNA-*Or*11s all repressed *Or*11 expression in drone brains at different stages after injection (SI Fig. 2). Therefore, we carried out subsequent RNA interference experiments using siRNA-*Or*11-528.

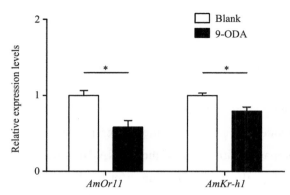

**Figure III-10-1** Effects of 9-ODA on *Or*11 and *Kr-h*1 expression in antennae of 4-day-old *A. mellifera* drones. The open bars represent normalized expression levels. The black bars represent relative expression levels of the 9-ODA tested group. Data are expressed as the mean ± SE, as normalized to a blank sample. Statistical analysis was performed with a *t*-test, and the "*" indicates a significant difference ($P < 0.05$).

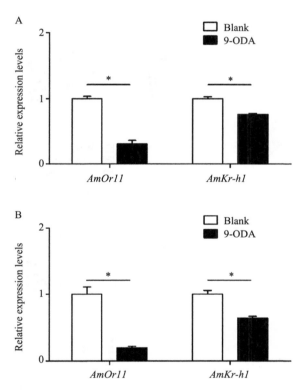

**Figure III-10-2** Effects of 9-ODA on *Or*11 and *Kr-h*1 expression levels in antennae of flying (A) and hive (B) *A. mellifera* drones at 14 days. The open bars represent normalized expression levels. The black bars represent relative expression levels of the 9-ODA tested group. Data are expressed as the mean ± SE, as normalized to the blank sample. Statistical analysis was performed with a *t*-test, and the "*" indicates a significant difference ($P < 0.05$).

siRNA-*Or*11, siRNA-NC and DEPC water solutions were injected into the antennae and brain tissues of *A. mellifera* drone pupae, and the survival rate of the drone pupae at 72 h was observed. The survival rates after the injection of siRNA-*Or*11, siRNA-NC and DEPC water solutions into antennae were 72.92% ± 0.02 %, 77.08% ± 0.02 % and 79.17% ± 0.02 %, respectively, with no significant differences ($P > 0.05$, SI Fig. 3 A). The survival rates after injection of siRNA-*Or*11, siRNA-NC and DEPC water solutions into brain tissues were

75.00% ± 0.05 %, 70.83% ± 0.04 % and 70.83% ± 0.02 %, respectively, also with no significant differences ($P > 0.05$, SI Fig. 3 B).

### *Effects of siRNA-Or11 injection on expression of AmOr11, AmKr-h1 and AmBr-c in the antennae and brains of drones*

We knocked down the *AmOr11* gene by RNAi to further study its function in vivo. Seventy-two hours after siRNA injection in both antennae and brains, the expression levels of *AmOr11*, *AmKr-h1* and *AmBr-c* were significantly reduced compared with those of the siRNA-NC-injected and water-injected controls ($P < 0.05$, Fig. III-10-3; $P < 0.05$, Fig. III-10-4). The transcript levels of *AmOr11*, *AmKr-h1* and *AmBr-c* in the siRNA-NC-injected or water-injected drones remained unchanged.

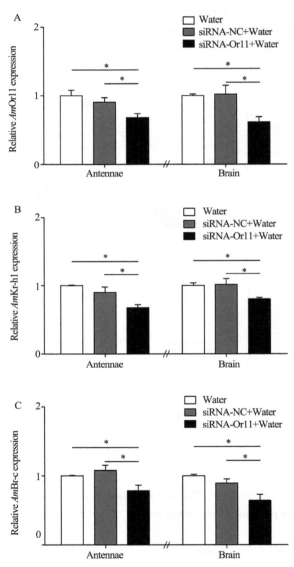

**Figure III-10-3** Effects of injecting antennae with siRNA-*Or11* on expression of *AmOr11* (A), *AmKr-h1* (B) and *AmBr-c* (C) in the antennae and brains of drones. The open bars represent the group injected with water. The gray bars represent the group injected with siRNA-NC. The black bars represent the group injected with siRNA-*Or11*. Data are expressed as the mean ± SE, as normalized to the water control. Statistical analysis was performed by one-way ANOVA, and the "*" indicates a significant difference ($P < 0.05$).

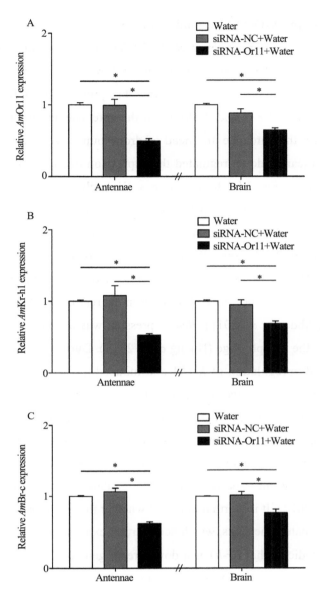

**Figure III-10-4** Effects of injecting the head with siRNA-*Or*11 on expression of *AmOr*11 (A), *AmKr-h*1 (B) and *AmBr-c* (C) in the antenna and brains of drones. The open bars represent the group injected with water. The gray bars represent the group injected with siRNA-NC. The black bars represent the group injected with siRNA-*Or*11. Data are expressed as the mean ± SE, as normalized to the water control. Statistical analysis was performed by one-way ANOVA, and the "*" indicates a significant difference ($P < 0.05$).

## Discussion

According to the results of this study, 9-ODA inhibits gene expression of *Or*11 and *Kr-h*1 in antennae of 4- (Fig. III-10-1) and 14-day-old (Fig. III-10-2) *A. mellifera* drones. This result is consistent with the responses of the honeybee workers (Grozinger *et al*. 2007; Claudianos *et al*. 2014). Wanner *et al*. (2007) found that *Or*11, a sex pheromone receptor, is a specific receptor for 9-ODA. *AmOr*11 expression were significantly down-regulated after honeybee worker were conditioned with 9-ODA in an olfactory learning paradigm, and EAG recordings showed that the neural response of the antenna was similarly reduced after 9-ODA learning

(Claudianos et al. 2014). Moreover, 9-ODA also inhibits the expression of transcription factor *AmKr-h*1 in the brains of *A. mellifera* worker bees (Grozinger et al. 2007). These results indicated that *Or*11 (membrane receptor) recognized 9-ODA and *Kr-h*1 (nuclear receptor) were downregulated by 9-ODA both in immature and mature drone, *Kr-h*1 may involve in the signaling pathway of responding to 9-ODA in drones.

Relative expression levels of *AmOr*11 and *AmKr-h*1 in the antennae and brains of drones at 72 h after injection with siRNA-*Or*11 into the antennae and head of drone pupae were tested by qPCR. Our results showed that siRNA-*Or*11 significantly down-regulated the expression of *AmOr*11 and *AmKr-h*1 in both the antennae (Fig. III-10-3) and brains (Fig. III-10-4) of drones. We preliminarily speculated that *AmOr*11 and *AmKr-h*1 were components of the same signaling pathway responding to 9-ODA. To validate this speculation, we further determined the expression level of *Br-c*, which is the downstream gene of *Kr-h*1 in *D. melanogaster* and *T. castaneum* (Minakuchi et al. 2008, 2009; Belles and Santos 2014). Our results of RNAi experiments substantiated this hypothesis: siRNA-*Or*11 treatment also inhibited expression of *AmBr-c* in the antennae (Fig. III-10-3) and brains (Fig. III-10-4) of drones. *Br-c* plays roles in the regulation of metamorphosis in insects by JH and 20E (Paul et al. 2006; Abdou et al. 2011). *Br-c* expression was downregulated by knocking down of *Kr-h*1 in *Blattella germanica* at the nymph stage (Huang et al. 2013). During the pupal stage of *T. castaneum*, exogenous JH analogs mediated upregulation of *Kr-h*1 and induced transcription of *Br-c* (Minakuchi et al. 2009). Therefore, this result further confirmed our speculation that *Kr-h*1 is the downstream gene of *Or*11. Studies have showed that expression level of *Kr-h*1 in the brains of foraging workers is significantly higher than that in the brains of juvenile workers and brood workers in hives, suggesting that *Kr-h*1 participates in regulating the foraging behavior of workers (Whitfield et al. 2003; Grozinger and Robinson 2007). DNA methylation regulation of *Kr-h*1 plays an important role in the regulatory gene network of ovary activation in honeybee workers (Kilaso et al. 2017). It is worth noting that whether the signal pathway responding to 9-ODA is the same between male and female honeybees, which need further exploration.

In conclusion, our results indicate that *Kr-h*1 is a downstream gene of *Or*11, and is negatively regulated by 9-ODA. It suggests that *Kr-h*1 is a key player in the response of honeybee drones to sex pheromone 9-ODA. This study provides a new insight into the molecular mechanisms of the mating flight of drones.

## Acknowledgements

We thank Frederick Partridge and anonymous reviewers for their comments that improved this manuscript. This work was supported by the National Natural Science Foundation of China(31572469) and the Earmarked Fund for the China Agricultural Research System (CARS-44-KXJ15).

## References

ABDOU MA, HE Q, WEN D, *et al* (2011) *Drosophila* Met and Gce are partially redundant in transducing juvenile hormone action. *Insect Biochemistry and Molecular Biology* 41:938-945. doi: 10.1016/j.ibmb.2011.09.003.

BELLES X, SANTOS CG (2014) The MEKRE93 (Methoprene tolerant-Krüppel homolog 1-E93) pathway in the regulation of insect metamorphosis, and the homology of the pupal stage. *Insect Biochemistry and Molecular Biology*

52:60-68. doi: 10.1016/j.ibmb.2014.06.009.

BEYE M, HARTEL S, HAGEN A, et al (2002) Specific developmental gene silencing in the honey bee using a homeobox motif. *Insect Molecular Biology* 11:527-532. doi: 10.1046/j.1365-2583.2002.00361.x.

BROCKMANN A, BRÜCKNER D, CREWE RM (1998) The EAG Response Spectra of Workers and Drones to Queen Honeybee Mandibular Gland Components: The Evolution of a Social Signal. *Naturwissenschaften* 85:283-285. doi: 10.1007/s001140050500.

BROCKMANN A, DIETZ D, SPAETHE J, TAUTZ J (2006) Beyond 9-ODA: SEX Pheromone Communication in the European Honey Bee *Apis mellifera* L. *Journal of Chemical Ecology* 32:657-667. doi: 10.1007/s10886-005-9027-2.

BUTLER CG (1971) The mating behaviour of the honeybee (*Apis mellifera* L.). *Journal of Entomology Series A, General Entomology* 46:1-11. doi: 10.1111/j.1365-3032.1971.tb00103.x.

BUTLER CG, CALLOW RK, JOHNSTON NC (1962) The Isolation and Synthesis of Queen Substance, 9-oxodec-trans-2-enoic Acid, a Honeybee Pheromone. *Proceedings of the Royal Society B: Biological Sciences* 155:417-432. doi: 10.1098/rspb.1962.0009.

BUTLER CG, CALLOW RK, JOHNSTON NC (1959) Extraction and Purification of "Queen Substance" from Queen Bees. *Nature* 184:1871-1871. doi: 10.1038/1841871a0.

CARLISLE DB, BUTLER CG (1956) The"queen-substance"of honeybees and the ovary-inhibiting hormone of crustaceans. *Nature* 178:323.

CLAUDIANOS C, LIM J, YOUNG M, et al (2014) Odor memories regulate olfactory receptor expression in the sensory periphery. *European Journal of Neuroscience* 39:1642-1654. doi: 10.1111/ejn.12539.

EREZYILMAZ DF, RIDDIFORD LM, TRUMAN JW (2006) The pupal specifier broad directs progressive morphogenesis in a direct-developing insect. *Proceedings of the National Academy of Sciences of the United States of America* 103:6925-6930. doi: 10.1073/pnas.0509983103.

GARY NE (1962) Chemical Mating Attractants in the Queen Honey Bee. *Science* 136:773-774. doi: 10.1126/science.136.3518.773.

GARY NE, MARSTON J (1971) Mating behaviour of drone honey bees with queen models (*Apis mellifera* L.). *Animal Behaviour* 19:299-304. doi: 10.1016/S0003-3472(71)80010-6.

GROZINGER CM, FISCHER P, HAMPTON JE (2007) Uncoupling primer and releaser responses to pheromone in honey bees. *Naturwissenschaften* 94:375-379. doi: 10.1007/s00114-006-0197-8.

GROZINGER CM, ROBINSON GE (2007) Endocrine modulation of a pheromone-responsive gene in the honey bee brain. *Journal of Comparative Physiology A: Neuroethology, Sensory, Neural, and Behavioral Physiology* 193:461-470. doi: 10.1007/s00359-006-0202-x.

GROZINGER CM, SHARABASH NM, WHITFIELD CW, ROBINSON GE (2003) Pheromone-mediated gene expression in the honey bee brain. *Proceedings of the National Academy of Sciences* 100:14519-14525. doi: 10.1073/pnas.2335884100.

GUO X, WANG Y, SINAKEVITCH I, et al (2018) Comparison of RNAi knockdown effect of tyramine receptor 1 induced by dsRNA and siRNA in brains of the honey bee, *Apis mellifera*. *Journal of Insect Physiology* 111:47-52. doi: 10.1016/j.jinsphys.2018.10.005.

HARWOOD G, AMDAM G, FREITAK D (2019) The role of Vitellogenin in the transfer of immune elicitors from gut to hypopharyngeal glands in honey bees (*Apis mellifera*). *Journal of Insect Physiology* 112:90-100. doi: 10.1016/j.jinsphys.2018.12.006.

HUANG JH, LOZANO J, BELLES X (2013) Broad-complex functions in postembryonic development of the cockroach *Blattella germanica* shed new light on the evolution of insect metamorphosis. Biochimica et *Biophysica Acta - General Subjects* 1830:2178-2187. doi: 10.1016/j.bbagen.2012.09.025.

KAATZ HH, HILDEBRANDT H, ENGELS W (1992) Primer effect of queen pheromone on juvenile hormone biosynthesis in adult worker honey bees. *Journal of Comparative Physiology B* 162:588-592. doi: 10.1007/BF00296638.

KEELING CI, SLESSOR KN, HIGO H A, WINSTON ML (2003) New components of the honey bee (*Apis mellifera* L.) queen retinue pheromone. *Proceedings of the National Academy of Sciences of the United States of America* 100:4486-4491. doi: 10.1073/pnas.0836984100.

KILASO M, REMNANT EJ, CHAPMAN NC, et al (2017) DNA methylation of Kr-h1 is involved in regulating ovary activation in worker honeybees (*Apis mellifera*). *Insectes Sociaux* 64:87-94. doi: 10.1007/s00040-016-0518-7.

LIU JF, YANG L, LI M, et al (2019) Cloning and expression pattern of odorant receptor 11 in Asian honeybee drones, *Apis cerana* (Hymenoptera, Apidae). *Journal of Asia-Pacific Entomology* 22:110-116. doi: 10.1016/j.aspen.2018.12.014.

LOPER GM, WOLF WW, TAYLOR OR (1993) Radar detection of drones responding to honeybee queen pheromone. *Journal of Chemical Ecology* 19:1929-1938. doi: 10.1007/BF00983797.

LOURENÇO AP, MACKERT A, CRISTINO ADS, SIMÕES ZLP (2008) Validation of reference genes for gene expression studies in the honey bee, *Apis mellifera*, by quantitative real-time RT-PCR. *Apidologie* 39:372-385. doi: 10.1051/apido:2008015.

MARTINS JR, NUNES FMF, CRISTINO AS, et al (2010) The four hexamerin genes in the honey bee: Structure, molecular evolution and function deduced from expression patterns in queens, workers and drones. *BMC Molecular Biology* 11: 23. doi: 10.1186/1471-2199-11-23.

MINAKUCHI C, NAMIKI T, SHINODA T (2009) Krüppel homolog 1, an early juvenile hormone-response gene downstream of Methoprene-tolerant, mediates its anti-metamorphic action in the red flour beetle *Tribolium castaneum*. *Developmental Biology* 325:341-350. doi: 10.1016/j.ydbio.2008.10.016.

MINAKUCHI C, ZHOU X, RIDDIFORD LM (2008) Krüppel homolog 1 (Kr-h1) mediates juvenile hormone action during metamorphosis of *Drosophila melanogaster*. *Mechanisms of Development* 125:91-105. doi: 10.1016/j.mod.2007.10.002.

NUNES FMF, SIMÕES ZLP (2009) A non-invasive method for silencing gene transcription in honeybees maintained under natural conditions. *Insect Biochemistry and Molecular Biology* 39:157-160. doi: 10.1016/j.ibmb.2008.10.011.

PANKIW T, HUANG Z. ., WINSTON M., ROBINSON G. (1998) Queen mandibular gland pheromone influences worker honey bee (*Apis mellifera* L.) foraging ontogeny and juvenile hormone titers. *Journal of Insect Physiology* 44:685-692. doi: 10.1016/S0022-1910(98)00040-7.

PANKIW T, PAGE RE (2003) Effect of pheromones, hormones, and handling on sucrose response thresholds of honey bees (*Apis mellifera* L.). *Journal of Comparative Physiology A: Sensory, Neural, and Behavioral Physiology* 189:675-684. doi: 10.1007/s00359-003-0442-y.

PAUL RK, TAKEUCHI H, KUBO T (2006) Expression of Two Ecdysteroid-Regulated Genes, Broad-Complex and E75, in the Brain and Ovary of the Honeybee (*Apis mellifera* L.). *Zoological Science* 23:1085-1092. doi: 10.2108/zsj.23.1085.

PLETTNER E, OTIS GW, WIMALARATNE PDC, et al (1997) Species- and Caste-Determined Mandibular Gland Signals in Honeybees (*Apis*). *Journal of Chemical Ecology* 23:363-377. doi: 10.1023/B:JOEC.0000006365.20996.a2.

RANGEL J, BÖRÖCZKY K, SCHAL C, TARPY DR (2016) Honey bee (*Apis mellifera*) queen reproductive potential affects queen mandibular gland pheromone composition and worker retinue response. *PLoS ONE* 11:1-16. doi: 10.1371/journal.pone.0156027.

SANDOZ J-C (2006) Odour-evoked responses to queen pheromone components and to plant odours using optical imaging in the antennal lobe of the honey bee drone *Apis mellifera* L. *Journal of Experimental Biology* 209:3587-3598. doi: 10.1242/jeb.02423.

SANDOZ J-C, DEISIG N, DE BRITO SANCHEZ MG, GIURFA M (2007) Understanding the logics of pheromone processing in the honeybee brain: from labeled-lines to across-fiber patterns. *Frontiers in Behavioral Neuroscience* 1:5. doi: 10.3389/neuro.08.005.2007.

SCHARLAKEN B, DE GRAAF DC, GOOSSENS K, et al (2008) Reference Gene Selection for Insect Expression Studies Using Quantitative Real-Time PCR: The Head of the Honeybee, *Apis mellifera*, After a Bacterial Challenge. *Journal of Insect Science* 8:1-10. doi: 10.1673/031.008.3301.

SCHMITTGEN TD, LIVAK KJ (2008) Analyzing real-time PCR data by the comparative CT method. *Nature Protocols* 3:1101-1108. doi: 10.1038/nprot.2008.73.

SCHUH R, AICHER W, GAUL U, et al (1986) A conserved family of nuclear proteins containing structural elements of the finger protein encoded by Krüppel, a *Drosophila* segmentation gene. *Cell* 47:1025-1032. doi: 10.1016/0092-8674(86)90817-2.

SLESSOR KN, WINSTON ML, LE CONTe Y (2005) Pheromone communication in the honeybee (*Apis mellifera* L.). *Journal of Chemical Ecology* 31:2731-2745. doi: 10.1007/s10886-005-7623-9.

VILLAR G, BAKER TC, PATCH HM, GROZINGER CM (2015) Neurophysiological mechanisms underlying sex- and maturation-related variation in pheromone responses in honey bees (*Apis mellifera*). *Journal of Comparative Physiology A* 201:731-739. doi: 10.1007/s00359-015-1006-7.

VILLAR G, GROZINGER CM (2017) Primer effects of the honeybee, *Apis mellifera*, queen pheromone 9-ODA on drones. *Animal Behaviour* 127:271-279. doi: 10.1016/j.anbehav.2017.03.023.

WANG Y, BAKER N, AMDAM G V. (2013) RNAi-mediated double gene knockdown and gustatory perception measurement in honey bees (*Apis mellifera*). *Journal of Visualized Experiments* 77:1-9. doi: 10.3791/50446.

WANNER KW, NICHOLS AS, WALDEN KKO, et al (2007) A honey bee odorant receptor for the queen substance 9-oxo-2-decenoic acid. *Proceedings of the National Academy of Sciences of the United States of America* 104:14383-14388. doi: 10.1073/pnas.0705459104.

WHITFIELD CW, CZIKO A-M, ROBINSON GE (2003) Gene Expression Profiles in the Brain Predict Behavior in Individual Honey Bees. *Science* 302:296-299. doi: 10.1126/science.1086807.

## Supporting Information

SI Fig. 1 Breeding 4-day-old drones with 9-ODA treatment.

SI Fig. 2 Injecting the antennae (A) and brain (B) of drone pupae.

SI Fig. 3 The developmental state of drone pupae at 0 h (A), 24 h (B), 48 h (C) and 72 h (D) after injection

SI Fig. 4 Effects of injecting 3 siRNAs into antennae on expression of $Or11$ in antennae of $A.\ mellifera$ drones. Data are expressed as the mean ± SE, and the "*" indicates a significant difference ($P < 0.05$).

SI Fig. 5 Effects of injecting 3 siRNAs into the head on expression of $Or11$ in the brains of $A.\ mellifera$ drones. Data are expressed as the mean ± SE, and the "*" indicates a significant difference ($P < 0.05$).

SI Fig. 6 Effects of siRNA injection into the antennae and head on the $A.\ mellifera$ drone survival rate. The open bars represent the group injected with water. The gray bars represent the group injected with siRNA-NC. The black bars represent the group injected with siRNA-$Or11$. Data are expressed as the mean ± SE, and the "*" indicates a significant difference ($P < 0.05$).

# 11. Cloning and Expression Pattern of Odorant Receptor 11 in Asian Honeybee Drones, *Apis cerana* (Hymenoptera, Apidae)

Junfeng Liu [a,b,1], Le Yang [a,c,1], Mang Li [a], Xujiang He [a], Zilong Wang [a], Zhijiang Zeng [a,*]

a. Honeybee Research Institute, Jiangxi Agricultural University, Nanchang, 330045, P.R. China.
b. Environment and Plant Protection Institute, Chinese Academy of Tropical Agricultural Sciences, Haikou, 571101, P. R. China. c. Kaihua County Animal Husbandry and Veterinary Bureau, Quzhou, 324300, P. R. China.

**Abstract**

Odorant receptors play a crucial role in the special recognition of scent molecules in the honeybee olfaction system. The odorant receptor 11 gene (*Am*Or11) in western honeybee drones (*Apis mellifera*) has been demonstrated to specifically bind to 9-oxo-2-decenoic acid (9-ODA) of queens. However, little is known regarding the functions of OR11 Asian honeybee drones (*Apis cerana*) in the context of their mating activities. In this study, the odorant receptor 11 gene (*Ac*Or11) from *A. cerana* was cloned, and its expression profiles were examined during two developmental stages (immature and sexually mature) and different physiological statuses (flying and crawling). The cDNA sequence of *Ac*Or11 was highly similar to that of *Am*Or11, and encoded a membrane-coupled protein of 384 amino acids. The results of qRT-PCR indicated that *Ac*Or11 was expressed at higher levels in drone antennae compared to brains, and the expression was significantly up-regulated in sexually mature drone brains compared to immature brains. Interestingly, *Ac*Or11 expression in brains of mature flying drones was dramatically higher than those of mature crawling drones. To our knowledge, this study demonstrate a link between *Ac*Or11 gene expression in the brain of honeybee drones and behavior associated with sexual maturity and mating flight.

**Keywords:** *Apis cerana*, odorant Receptor 11, drones, mating flight, sexual development

# Introduction

Honeybee olfaction is essential for the perception and discrimination of a variety of odor molecules in external environment (Laska *et al.*, 1999; Hugh M. Robertson and Wanner, 2006). This olfaction ability allows honeybees an efficient method of chemical communication inside and outside of their colonies. Furthermore, this ability had been shown to be especially important in mating flights (Slessor *et al.*, 2005; Sandoz *et al.*, 2007). In mating flights, male bees arrive early at drone congregate area (DCA) that in midair. A virgin queen

---

1 These two authors contributed equally to this paper.
* Corresponding author: bees1965@sina.com (ZJZ).
注：此文发表在 *Journal of Asia-Pacific Entomology* 2019 年第 1 期。

fly through a DCA and release queen mandibular pheromones (QMPs) which mainly including (E)-9-oxodec-2-enoic acid (9-ODA), two enantiomers of (E)-9-hydroxydec-2-enoic acid (9-HDA; 85% (R)-(-), 15% (S-(+)), methyl p-hydroxybenzoate (HOB) and 4-hydroxy-3-methoxyphe-nylethanol (HVA) (Butler *et al.*, 1959; Butler, 1971; Gary and Marston, 1971; Slessor *et al.*, 1988; Keeling *et al.*, 2003). On the other hand, drones use their olfaction capabilities, which is believed to be specific to males, to locate virgin queens in mating flights by scanning for and sensing queen pheromones in the air (Brockmann and Brückner, 2001; Wanner *et al.*, 2007). Evidence has indicated that drones can detect QMPs over an extended distance ( > 800m), indicating very high sensitivity of their olfaction (Loper *et al.*, 1993).

Among these, 9-ODA, is one of the predominantly detected compounds of QMPs, which function as a short-range social pheromone attracting workers within the colony, and as a long-range sex pheromone attracting drones at mating flights (Butler, 1971; Gary and Marston, 1971; Boch *et al.*, 1975; Loper *et al.*, 1993; Brockmann *et al.*, 2006). Electrophysiological recordings have demonstrated that drones have a greater proportion of olfactory neurons in their antennae tuned to QMPs compared to workers. More interestingly, drone's antennae are more sensitive to 9-ODA than any other single component of QMPs (Vetter and Visscher, 1997; Brockmann *et al.*, 1998). This is believed to result from the expression of a special odorant receptor in the antennae of honeybees. *A. mellifera* odorant receptor 11 gene (*Am*Or11) in male bees has been demonstrated to specifically bind 9-ODA (Wanner *et al.*, 2007). Recently, Wu *et al.* (2016) found that 16 ORs were up-regulated in the sexually matured drones of *A. mellifera* by using high-throughput RNA-Seq. Moreover, a subsequent investigation reported that *Am*Or11 expression levels are higher in the antennae of sexually mature drones than immature drones (Villar *et al.*, 2015). This observation suggests a likely association between *Am*Or11 and the process of sexual maturation of honeybee drones. In addition, QMPs not only trigger drone mating behavior, but also exert other primer effects on body development of drone bees. Young male-bees exposed to 9-ODA in hive not only result in delayed initial mating flights, but reduce flight duration as well (Villar and Grozinger, 2017). Other evidence has indicated that *Drosophila melanogaster* specific pheromone receptor participate in their regulation of mating behavior both male and female (Kurtovic *et al.*, 2007). In honeybee brains, the antennal lobes (AL) consist of approximately 160-170 glomeruli which correspond with their ORs (Hansson and Anton, 2000; Hugh M Robertson and Wanner, 2006), although the specific mechanisms of action remain unclear. Therefore, we suspect that odorant receptors expressed in honeybee drones, especially *Am*Or11, may participate in multiple biological functions that aid in the detection of 9-ODA during mating flights, or in the regulation of drone maturation and mating behaviors.

*Apis cerana,* an Asian honeybee species, has been demonstrated to exhibit enhanced searching abilities to locate sparse floral resources (Zeng, 2017), and better color cognition and orientation learning relative to that of *A. mellifera* (Qin *et al.*, 2012; Zhang *et al.*, 2014). Recently, many *A. cerana* odorant receptor genes (*Ac*Ors) have been preliminarily investigated, including annotated, cloning, characterized, mRNA/protein expression patterns, and localization within the organisms (Zhang *et al.*, 2012; Zhao *et al.*,2012; Zhao *et al.*, 2013, 2014; Park *et al.*, 2015; Zhao *et al.*, 2015; Zhang *et al.*, 2016; Du *et al.*, 2017a, 2017b; Yang *et al.*, 2017). These mRNA sequences were found to be highly similar with those of *A. mellifera*. Park *et al.* (2015)

have characterized 119 Ors by whole *A. cerana* genome sequencing. The *Ac*Or2 and *Ac*Or3 in Asian honeybee antennae both reveal male-bias, and are expressed at the highest levels at sexual maturity (Zhao et al., 2014; Zhang et al., 2016). However, information is scarce regarding the *A. cerana* odorant receptor 11 gene (*Ac*Or11) at the time of publication. In this study, we identified the *Ac*Or11, determined the gene's DNA sequence, and characterized expression patterns across sexual developmental stages and different physiological statuses in both antenna and brain of *A. cerana* drones. This allows for thorough investigation of the biological functions of *Ac*Or11 in male mating behavior and progression through sexual maturity, and eventually to provide a physiologic background leading to mating flights of honeybee drones.

## Materials and Methods

### *Insects*

Three Asian honeybee (*A. cerana*) colonies were maintained at the Honeybee Research Institute of Jiangxi Agricultural University (28.46°N, 115.49°E). Each colony had a mature egg-laying queen and 5 frames.

### *Sample collection*

In first experiment, 30 drones were randomly collected at entrance of the hive, upon returning home, using forceps. The drones were immediately stored in liquid nitrogen for subsequent cloning experiment of target genes. In the second experiment, antennae and brains were collected from 4-day-old (sexually immature) and 14-day-old (sexually mature) drones. This included crawling drones inside of colonies as well as those flying back to the hive. These insects were used for gene expression analysis of the *Ac*Or11 and *Ac*Or2 (odorant receptor co-receptor, ortholog Or83b family) genes. This detailed methods used for our study of collected drones was referenced (Villar et al., 2015). For each group, 30 pairs of antennae and brains each, with 3 biological replicates from 3 different colonies were examined. These studies were conducted during the spring of 2018.

### *Cloning of the AcOr11 gene*

Total RNA was extracted from 30 pairs of drone antennae using the TransZol reagent (Transgen Biotech, www.transgen.com.cn) according to the manufacturer's instructions, and stored in a freezer at -80°C until use. Since honeybee odorant receptors are distributed mainly in antennae, and their expression patterns are especially enriched in antennae of honeybee, we therefore used antennae for cloning (Claudianos et al., 2014; Hugh M Robertson and Wanner, 2006). The cDNA was synthesized from the total RNA isolated from antenna using the Primer-Script RT reagent Kit (TaKaRa, www.takara-bio.com) according to the manufacturer's instructions.

The primers used to amplify *Ac*Or11 (see Table III-11-1) were designed using the primer premier 5.0 software (Premier Biosoft International Co., Palo Alto, CA) with the input mRNA sequence of the *Am*Or11 gene (GenBank accession: NM_001242962.1) deposited in NCBI. The aforementioned primers were synthesized by Sangon Biotech (Sangon Biotech Shanghai, China Co., Ltd). The PCR thermocycling conditions were as follows: 94°C for 2 min, followed by 30 cycles of 94°C for 30 s, 58°C for 45 s, 72°C for 90 s, and a final extension at 72°C for 10 min. The PCR products were then electrophoretically resolved on a 1.5% agarose gel, and purified using a Gel Extraction Kit (Cwbiotech, www.cwbiotech.bioon.com.cn). Next, the purified products were ligated into an pEASY-T3 Clone Vector, and subsequently transformed into Trans5α Chemically Competent Cell (TransGen Biotech). Positive clones were screened and sequenced by Sangon Biotech.

Table III-11-1  Primers used to AcOr11gene clone and qRT-PCR

| Primer names | Primer sequences |
| --- | --- |
| AcOr11-F1 | 5'-TCACGAACAAGCTTTCATCGG-3' |
| AcOr11-R1 | 5'-GAAAGTGAACAAAGTGCTGTGTACA-3' |
| AcOr11-R2 | 5'-TCAATATCATTTTTGGCTAATCAGA-3' |
| AcOr11-QF | 5'-ATGTGCGGTTTGCTGAAGA-3' |
| AcOr11-QR | 5'-CGAGAAGGTGCCAATGACG-3' |
| AcOr2-QF | 5'-GGATCAGAGGAGGCCAAAAC-3' |
| AcOr2-QR | 5'-CCAACACCGAAGCAAAGAGA-3' |
| Ac-actin-QF | 5'-GGCTCCCGAAGAACATCC-3' |
| Ac-actin-QR | 5'-TGCGAAACACCGTCACCC-3' |

### *Sequence Analysis*

After sequencing, the cDNA sequence of *Ac*Or11 was obtained by assembling forward and reverse sequencing reads using SeqMan program in DNAstar 5.0 software (Lynnon Biosoft, Quebec, Canada). The amino acid sequence was translated by the Bioedit software. Similarity searches were conducted using BLAST programs on the NCBI website (http://www.ncbi.nlm.nih.gov). The isoelectric point (pI) and molecular weights (MW) were computed using Compute pI/MW (http://www.expasy.ch/tools/pi_tool.html). The post-translational modification sites were predicted using PROSITE SCAN (https://npsa-prabi.ibcp.fr/cgi-bin/npsa_automat. pl?page=/NPSA/npsa_server.html). The secondary structures were predicted using the SOPMA program (http://npsa-pbil.ibcp.fr/cgi-bin/npsa_automat.pl?page=/NPSA/npsa_sopma.html). The transmembrane helix (TMH) was predicted by using TMHMM Server v.2.0 (http://www.cbs.dtu.dk/services/TMHMM). Alignments of multiple sequences were carried out using ClustalW (Thompson *et al.*, 1994). The phylogenetic tree was constructed using MEGA4.0 (http://www.megasoftware.net/index.php), with a portion of sequences of known *A. mellifera* ORs obtained from GenBank.

### *Expression of the AcOr11 and AcOr2 genes*

Total RNA was isolated from the antennae and brains of drones to determine the expression levels of *Ac*Or11 and *Ac*Or2. Quantitative Real-Time PCR (qRT-PCR, ABI 7500 instrument) was performed using the SYBR Premix Ex Taq kit (Takara) in a total reaction volume of 10μl. The reaction mixture was prepared as follows: 4.2 μl cDNA (water for the negative control) and 0.4 μl of each primer. The primers (see Table 1) for the *Ac*Or11, *Ac*Or2 (GenBank accession: JN792581) and *A. cerana* actin genes (*Ac*-actin, GenBank accession: HM640276.1) were designed respectively to amplify 296, 118 and 195 bp fragments using the primer premier 5.0 software. qRT-PCR was performed with an initial denaturation step of 10 min at 95°C, followed by 40 cycles of 94°C for 15 s, (*Ac*Or11, 58.9°C; *Ac*Or2 and *Ac*-actin, 60°C) for 40 s, 72°C for 35 s, and a melting curve analysis was conducted to verify the specificity of the amplification. The *Ac*-actin gene was used as the internal control. The relative expression levels of *Ac*Or11 and *Ac*Or2 mRNAs were calculated using the $2^{-\Delta\Delta Ct}$ comparative CT method (Schmittgen and Livak, 2008).

***Statistical Analysis***

Differences in the relative expression of the *Ac*Or11 and *Ac*Or2 genes were determined using a t-test analysis in SPSS 17.0 (IBM, Armonk, NY). Values of $P < 0.05$ were considered significant in all treatments.

## Results

### *Cloning and sequence analysis of the AcOr11gene*

To explore the molecular functions of the *A. cerana* odorant receptor 11, the cDNA sequence of *Ac*Or11 containing the complete coding region was cloned, and the amino acid sequences were predicted by *in silico* translation. The *Ac*Or11 cDNA (GenBank accession: MG793195) contains a 5'-terminal untranslated region (UTR) of 29 bp, a 3'-terminal UTR of 93 bp, and an open reading frame (ORF) of 1185 bp encoding a polypeptide of 394 amino acids (Fig. Ⅲ-11-1). The molecular mass of the deduced *Ac*OR11 protein is predicted to be 45.14 kDa, and the calculated isoelectric point (pI) is 8.98. Moreover, the *Ac*OR11 protein belongs to the 7-transmembrane_6 receptor (7TM_6) superfamily, which consist of TM-I (13 - 32), TM-II (42 - 64), TM-III

```
-29                                                TCACGAACAAGCTTTCATCGGATTGAAAA
1     ATGGTCCAAATTAGAAACGCGAAAGAGGGGTTGAAGCATACCTTCTGGTTTGCATATCCGTTCTCAAGAACGCTC
1     M  V  Q  I  R  N  A  K  E  G  L  K  H  T  F  W  F  A  Y  P  F  S  R  T  L
76    GGTTATTGGCCACTGGTCTCTCCTTCCGCGGTTACCAAATTTTTTAATTCCTTCACAATTTTTATCCTATATTTA
26    G  Y  W  P  L  V  S  P  S  A  V  T  K  F  F  N  S  F  T  I  F  I  L  Y  L
151   TTGCAATTGATAGTCTTGATCCCAGGCCTTCTATACGTTCTGCAAGTGAAAAATGCCAGGACGAAGATTAAATTG
51    L  Q  L  I  V  L  I  P  G  L  L  Y  V  L  Q  V  K  N  A  R  T  K  I  K  L
226   TTGATGCCACATCTCAACAGCATCGCGCAAATGGCTAAATATACAATTATATTGCAACGGGCGAAGGAATTTAGC
76    L  M  P  H  L  N  S  I  A  Q  M  A  K  Y  T  I  I  L  Q  R  A  K  E  F  S
301   AAATTATTGGACGAGATAAAGAAGGATTGGTTGATAGCCACGGAAGAGAATCGACAAATTTTTAGCGAGAGGGCG
101   K  L  L  D  E  I  K  K  D  W  L  I  A  T  E  E  N  R  Q  I  F  S  E  R  A
376   TCCATCGAGCACAAATTGACAACGGTAATAGTGGTCACCATGTACGGTGGAGGTTTCTTTTACAGAACGATTCTT
126   S  I  E  H  K  L  T  T  V  I  V  V  T  M  Y  G  G  G  F  F  Y  R  T  I  L
451   CCACTTTCGAAAGGCAAAATCCTTCTACCGAATAACATGACAGTAAGATTATTACCTTGTCCGAGTTACTTTGGT
151   P  L  S  K  G  K  I  L  L  P  N  N  M  T  V  R  L  L  P  C  P  S  Y  F  G
526   TCCTTGAACGAGCAAGCTACGCCCAATTACGAGATAATTTTCACGCTACAAGTTCTCGGAGGATTTATCATTTAC
176   S  L  N  E  Q  A  T  P  N  Y  E  I  I  F  T  L  Q  V  L  G  G  F  I  I  Y
601   ACAGTTTTATGCGGCACCAAGAGTGCTTGTTTGATGCTGTGTTTGCATATGTGCGGTTTGCTGAAGATCTTGACG
201   T  V  L  C  G  T  K  S  A  C  L  M  L  C  L  H  M  C  G  L  L  K  I  L  T
676   AACAAGATAGTAGATCTTACGAATGACAGTGATGAACAAATTGTGCAGGAGAAGATCGTGCACATCGTTGAATAT
226   N  K  I  V  D  L  T  N  D  S  D  E  Q  I  V  Q  E  K  I  V  H  I  V  E  Y
751   CAGACGAGGATCAAAGAATTTTTGAATCAACTTGATCAATTCGTACCCGCTATTTATCTTATCGAAGTTGTTATC
251   Q  T  R  I  K  E  F  L  N  Q  L  D  Q  F  V  P  A  I  Y  L  I  E  V  V  I
826   CAAGTACTGATTATATGTATAATTGGTTATTGCATAATCATGGAATGGGAAGACAGCAATGCTATGGCTATGGTG
276   Q  V  L  I  I  C  I  I  G  Y  C  I  I  M  E  W  E  D  S  N  A  M  A  M  V
901   ATTTATGTTGTCTTCCAAGTAACTTGCGTCATTGGCACCTTCTCGGTATGCTACGTCGGTCAACTTCTACTTGAT
301   I  Y  V  V  F  Q  V  T  C  V  I  G  T  F  S  V  C  Y  V  G  Q  L  L  L  D
976   GAAAGCGAGAATATTAGACAGGCATATAACACATTGAACTGGTATCGATTACCTGTAAACAAGGCGCGCAGCTTG
326   E  S  E  N  I  R  Q  A  Y  N  T  L  N  W  Y  R  L  P  V  N  K  A  R  S  L
1051  ATATTATTGATCCTTATGTCACATTATCCAATAAAAGTAACTGCAGGAAGAATTATGGATTTATCTTTAGTCACC
351   I  L  L  I  L  M  S  H  Y  P  I  K  V  T  A  G  R  I  M  D  L  S  L  V  T
1126  TTCACTAGTATCATTAAAAGTGCAGTGGGATATATGAATATGTTACGTACAGTTACGTAAGATTTTATGTTTTTT
376   F  T  S  I  I  K  S  A  V  G  Y  M  N  M  L  R  T  V  T  *
1201  AGAAGATTCATGAAATCTACTATATCTGATTAGCCAAAAATGATATTGAATTATGTACACAGCACTTTGTTCACT
1276  TTC
```

**Figure Ⅲ-11-1** The nucleotide and deduced amino acid sequences of AcOr11. The positions of the nucleotides and amino acids are indicated in the left margin. The start codons used in cloning and sequencing are boxed, and the termination codon is marked with a star. The Shaded amino acid sequences indicate predicted 7-transmembrane (7TM- 6) domains, including TM-I (13 - 32), TM-II (42 - 64), TM-III (131 - 150), TM-IV (193 - 215), TM-V (265 - 287) and TM-VI (302 - 324).

(131 - 150), TM-IV (193 - 215), TM-V (265 - 287) and TM-VI (302 - 324) (Fig. III-11-1). The secondary structure was predicted using SOPMA, which consists of 54.31% alpha helices, 22.84% beta sheets, 4.06% turns and 18.78% is random coils. Comparison of the deduced amino acid sequence of AcOR11 (NCBI, BLASTP) to that of AmOR11(NP_001229891.1) reveals a high degree of identity 98%. Similarly, alignments with ORs of other Hymenopterans also exhibit high homology with A. dorsata OR85b-like (XP_006615208.1 identity, 99%) and A. florea OR4-like (XP_003691312.1, identity 97%) (Fig. III-11-2, see page 535). A phylogenetic tree was constructed using the MEGA 4.0 software using the deduced amino acid sequences of various A. mellifera ORs and AcOR11(Fig. III-11-3). Phylogenetic analysis showed that OR11s in A. mellfera and A. cerana belong to a single subfamily. Moreover, the phylogenetic tree demonstrated that the OR11 exhibits a relatively distant genetic relationship to the OR2 (ortholog OR83b family).

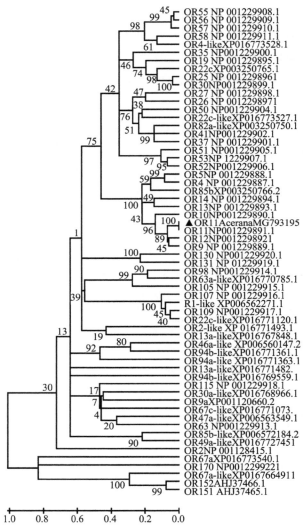

**Figure III-11-3** Phylogenetic analysis of amino acid sequences of various A. mellifera ORs and AcOr11. Clusters of tandem arrays of AcOr11 (triangle icon) and AmOrs on particular chromosomes are indicated by vertical lines on the right. The branch labels correspond to bootstrap values. Bootstrap supported values (%) are based on 1,000 replicates, as indicated by the scale bar. The protein accession numbers from NCBI are displayed respectively behind corresponding A. mellifera OR proteins.

### Analysis of AcOr11 and AcOr2 expression by qRT-PCR

The expression profiles of *Ac*Or11 were characterized across different developmental stages and physiological statuses of *A. cerena* using qRT-PCR. The expression profiles of *Ac*Or11 were determined in brains and antennae from immature and mature drones. It was observed that expression of *Ac*Or11 in drone antennae was significantly higher than that of brains ($t = -3.381$, df = 4, $P = 0.028$, Fig. III-11-4a. left; $t = -5.332$, df = 4, $P = 0.006$. Fig. III-11-4a. right). In brains, *Ac*Or11 in mature drones was more highly expressed than those of immature drones ($t = -2.883$, df = 4, $P = 0.045$, Fig. III-11-4b. left), whereas no differences in expression were observed in antennae ($t = -0.620$, df = 4, $P = 0.569$, Fig. III-11-4b. right). Moreover, *Ac*Or11 expression in mature drone brains in flying status was significantly higher than those in crawling status ($t = -2.790$, df = 4, $P = 0.049$, Fig. III-11-4c. right). However, there was no significant difference between the 2 physiological statuses of immature drones ($t = -0.773$, df = 4, $P = 0.483$, Fig. III-11-4c. left). In antenna, *Ac*Or11 expression of immature drones was significantly higher in flying status compared to crawling status

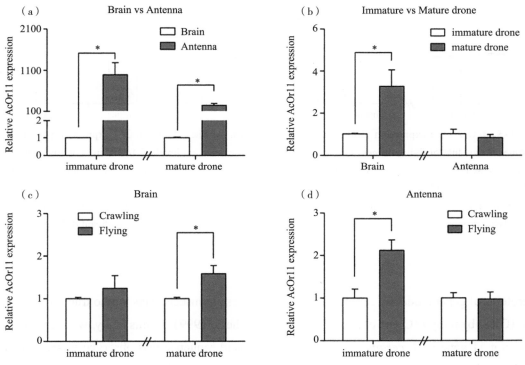

**Figure III-11-4** *Ac*Or11 gene expression profiles in *A. cerana* drones. (a) Comparison of *Ac*Or11 expression levels in drone brains versus antennae. The open bars represent the normalized expression level of *Ac*Or11 in antennae. The gray bars represent the relative expression in brains. Drones from two sexually developmental stages (immature and mature) were collected for this analysis respectively, same in (c) and (d). (b) Comparison of *Ac*Or11 expression levels in immature drones versus mature drones. The open bars represent expression level of *Ac*Or11 in immature drones, and the gray bars represent the relative expression in mature drones. Brains and antennae were collected from these immature and mature drones and were analyzed separately. (c) The relative *Ac*Or11 expression pattern in brains of crawling drones versus that of flying drones. The open bars represent expression level of *Ac*Or11 in crawling drone brains, and the gray bars represent the relative expression in flying ones. (d) The relative *Ac*Or11 expression pattern in antennae of crawling drones versus that of flying drones. The open bars represent expression level of *Ac*Or11 in crawling drone antennae, and the gray bars represent the relative expression in flying ones. The data was expressed as the mean ± SE, and the "*" indicates a significant difference ($P < 0.05$).

($t$ = -3.516, df = 4, $P$ = 0.025, Fig. III-11-4d. left), but no differences were observed for the mature stage ($t$ = -0.150, df = 4, $P$ = 0.888, Fig. III-11-4d. right). As is presented in Fig. III-11-5, the expression patterns of *Ac*Or11 and *Ac*Or2 were similar in brains and antennae between flying and crawling statuses. The expression level of *Ac*Or2 in mature drone brains in flying status was significantly higher than that in crawling status ($t$ = -6.350, df = 4, $P$ = 0.003, Fig. III-11-5a. right). In contrast, between the physiological statuses of immature drones, there were no significant differences ($t$ = -2.389, df = 4, $P$ = 0.075, Fig. III-11-5a. left). In antenna, *Ac*Or2 of immature drones was also expressed significantly higher in flying status compared to crawling status ($t$ = -7.286, df = 4, $P$ = 0.002, Fig. III-11-5b. left), but was not different in mature drones ($t$ = 1.994, df = 4, $P$ = 0.117, Fig. III-11-5b. right).

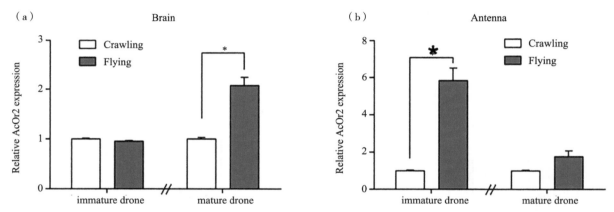

**Figure III-11-5** *Ac*Or2 gene expression profiles in *A. cerana* drones. (a) The relative *Ac*Or2 expression pattern in brains of crawling drones versus that of flying drones. The open bars represent expression level of *Ac*Or2 in crawling drone brains, and the gray bars represent the relative expression in flying ones. Immature and mature drones were collected for this analysis respectively, same in (b). (b) The relative *Ac*Or2 expression pattern in antennae of crawling drones versus that of flying drones. The open bars represent expression level of *Ac*Or2 in crawling drone antennae, and the gray bars represent the relative expression in flying drones. The data was expressed as the mean ± SE, and the "*" indicates significant difference ($P < 0.05$).

## Discussion

The proteins of general odorant receptors are highly diverse in insects. This is also true for the conservative co-receptor (OR83b) family (Clyne *et al.*, 1999; Gao and Chess, 1999). In this study, we identified a putative odorant receptor 11 gene in *A. cerana*. The *Ac*Or11 amino acid sequence was observed to share many similar characteristics with that of the *A. mellifera* Or11 orthologue. Interestingly, homologs of the *Ac*Or11 gene was not observed in other insects outside of *Apis*. In general, Ors exhibited a high sequence divergence among insects including classical model insects such as *Anopheles gambiae* and *Drosophila melanogaster* (Hill *et al.*, 2002; Robertson *et al.*, 2003). These observations are consistent with an ancient origin of the OR family. As demonstrated by the phylogenetic tree constructed here, the *Ac*Or11 was more closely related to *Am*Or11 than *Am*Or2 and *Am*Or170 of *A. mellifera*. This result served as a molecular confirmation for the traditional phylogenetic classes of honeybees (Hugh M. Robertson and Wanner, 2006), suggesting that *Ac*Or11 belongs to a typical odorant-receptor protein family in *A. cerana*.

Expression of *Ac*Or11 was assessed in both the antennae and brain of *A. cerana* drones. The data indicated that *Ac*Or11 was expressed at significantly higher levels in the antennae compared to the brain (Fig. III-11-4a). These observations were in agreement with our expected results, as antennae are the primary sensory organ in honeybees and contain numerous odorant receptors (Akers and Getz, 1992; Joerges *et al.*, 1997). Furthermore, this result is consistent with those published previously (Hugh M. Robertson and Wanner, 2006). In this study, we detected the expression pattern of *Ac*Or11 gene in drone brains under different sexual developmental stages and physiological statuses. Our results clearly indicated that the process of sexual development of honeybee drones strongly effects the expression of *Ac*Or11 in brains. Expression of *Ac*Or11 in brains of mature drones was significantly higher than those of immature drones, whereas there was no difference in antennae between immature and mature drones (Fig. III-11-4b). Both the antennal lobes (AL, the first synaptic processing station), and the mushroom bodies (MB, the multi-sensory integration centers) are the main olfactory brain regions of insects (Szyszka, 2005). The AL is enriched with the terminal axons of odorant receptor neurons (ORNs) and the glomeruli in olfactory sensory neurons (OSNs) (Gao and Chess, 1999; Mombaerts, 1999; Hill *et al.*, 2002; Forêt and Maleszka, 2006). It has been reported that the olfactory glomeruli in young honeybee brains are not well developed, and their volume increases significantly with age (Fahrbach and Robinson, 1995). Therefore, it was hypothesized that the immature drone brains (4 day old drones) were not fully developed, resulting in reduced expression of the *Ac*Or11 gene in their brains, therefore repressing their mating behavior. The flights of sexually immature drones are generally orientation flights rather mating flights (Graham B.M, 2015a). Moreover, it was observed that the expression of *Ac*Or11 in brains of flying mature drones was significantly higher than those of the crawling mature drones, but was not different in the antenna (Fig. III-11-4c). In many insects, a number of olfactory neurons express sex pheromone receptors in order to increase sensitivity to respond pheromone in the AL (Graham B.M, 2015b), Their function is transferring odor molecules to the sensory receptors distributed on the dendritic. This study indicated that physiological statuses of honeybee drones also correlated with *Ac*Or11 gene expression in brains, reflecting that daily mating flights of mature drones possibly perform as an important role in improving the mating behavior of honeybee drones by regulating the expression of *Ac*Or11 in drone brains. Furthermore, there was no significant difference between flying and crawling immature drone brains, which is likely main outside activity are orientation flights, rather mating flights in immature drones in nature.

Interestingly, expression of *Ac*Or11 was opposite in antenna compared to that of brains. Expression in the antennae of the flying drones was dramatically higher than those of the crawling drones in the immature stage (Fig. III-11-4d). It is unclear how orientation flying activity up-regulates *Ac*Or11 gene expression in antennae of immature drones. Perhaps the *Ac*Or11 is employed by immature drones to detect environmental scents for orientation flying, since the olfactory nervous system of young honeybees is most sensitive to environmental odors from inside and outside of their hive (Masson *et al.*, 1993; Sandoz and Menzel, 2001). This is consistent with other studies reporting that many insects utilize floral scents and other environmental odors for orientation (Phelan and Baker, 1987; Hern and Dorn, 1999; Anton *et al.*, 2007). For mature drones, Villar *et al.* (2015) reported that *Am*Or11 in drone antennae is correlated with mating behavior in response to 9-ODA stimulation. In Fig. 4d, the expression of *Ac*Or11 in mature drone antennae was not affected by physiological status. This could be explained by the fact that Ors expression in mature drone antennae tends to be stable and increases only when stimulated by QMPs, rather than in daily flights. Nevertheless, this phenomenon requires further investigation.

Furthermore, the results of *Ac*Or2 also showed a similar expression pattern in drone brains (Fig. III-11-5a) and antennae (Fig. III-11-5b) at different physiological statuses to that of *Ac*Or11. As the OR2 is the co-receptor of OR11 in honeybees (Wanner *et al.*, 2007), these data serve as additional confirmation of our observations *Ac*Or11.

In summary, *Ac*Or11 from the antennae of *A. cerana* was cloned, and expression patterns were analyzed in drones of different stages of sexual maturity and physiological statuses. The expression of *Ac*Or11 in drone brains was closely correlated with both sexual development and physiological status. This suggests that *Ac*Or11 in brain may have some biological functions involved in the progression to sexual maturity and mating behavior. This study provides an insight into the molecular basis underlying mating flights of *A. cerana* drones.

## Acknowledgments

Zhi Jiang Zeng, Xu Jiang He and Zi Long Wang conceived and designed the experiments. Jun Feng Liu, Le Yang and Mang Li performed the experiments. Jun Feng Liu, Xu Jiang He and Le Yang analyzed the data. Jun Feng Liu, Zhi Jiang Zeng and Xu Jiang He wrote the paper. All authors read and approved the final manuscript. This work is supported by the National Natural Science Foundation of China (31572469), and the Earmarked Fund for China Agriculture Research System (CARS-44-KXJ15).

## References

AKERS, R.P., GETZ, W.M., 1992. A test of identified response classes among olfactory receptor neurons in the honey-bee worker. *Chem. Senses* 17, 191-209. https://doi.org/10.1093/chemse/17.2.191.

ANTON, S., DUFOUR, M.C., GADENNE, C., 2007. Plasticity of olfactory-guided behaviour and its neurobiological basis: Lessons from moths and locusts. *Entomol. Exp. Appl.* 123, 1-11. https://doi.org/10.1111/j.1570-7458.2007.00516.x.

BOCH, R., SHEARER, D.A., YOUNG, J.C., 1975. Honey bee pheromones: Field tests of natural and artificial queen substance. *J. Chem. Ecol.* 1, 133-148. https://doi.org/10.1007/BF00987726.

BROCKMANN, A., BRÜCKNER, D., 2001. Structural differences in the drone olfactory system of two phylogenetically distant Apis species, A. florea and A. mellifera. *Naturwissenschaften* 88, 78-81. https://doi.org/10.1007/s001140000199.

BROCKMANN, A., BRÜCKNER, D., CREWE, R.M., 1998. The EAG Response Spectra of Workers and Drones to Queen Honeybee Mandibular Gland Components: The Evolution of a Social Signal. *Naturwissenschaften* 85, 283-285. https://doi.org/10.1007/s001140050500.

BROCKMANN, A., DIETZ, D., SPAETHE, J., TAUTZ, J., 2006. Beyond 9-ODA: SEX Pheromone Communication in the European Honey Bee Apis mellifera L. *J. Chem. Ecol.* 32, 657-667. https://doi.org/10.1007/s10886-005-9027-2.

BUTLER, C.G., 1971. The mating behaviour of the honeybee (Apis mellifera L.). *J. Entomol. Ser. A, Gen. Entomol.* 46, 1-11. https://doi.org/10.1111/j.1365-3032.1971.tb00103.x.

BUTLER, C.G., CALLOW, R.K., JOHNSTON, N.C., 1959. Extraction and Purification of "Queen Substance" from Queen Bees. *Nature* 184, 1871-1871. https://doi.org/10.1038/1841871a0.

CLAUDIANOS, C., LIM, J., YOUNG, M., YAN, S., CRISTINO, A.S., NEWCOMB, R.D., GUNASEKARAN, N., REINHARD, J., 2014. Odor memories regulate olfactory receptor expression in the sensory periphery. *Eur. J. Neurosci.* 39, 1642-1654. https://doi.org/10.1111/ejn.12539.

CLYNE, P.J., WARR, C.G., FREEMAN, M.R., LESSING, D., KIM, J., CARLSON, J.R., 1999. A Novel Family of Divergent Seven-Transmembrane Proteins. *Neuron* 22, 327-338. https://doi.org/10.1016/S0896-6273(00)81093-4.

DU, Y.L., PAN, J.F., WANG, S.J., YANG, S., ZHAO, H.T., JIANG, Y.S., 2017a. Cloning and expression analysis of odorant receptor gene AcerOr167 in the Chinese honeybee, Apis cerana cerana (*Hymenoptera:Apidae*). *J. Environ. Entomol.* 39, 39-47. https://doi.org/10.3969/j.issn.1674-0858.2017.01.3.

DU, Y.L., WANG, S.J., ZHAO, H.T., PAN, J.F., YANG, S., GUO, L.N., XU, K., JIANG, Y.S., 2017b. Cloning and temporal-spatial expression profiling of the odorant receptor gene AcerOR113 in the Chinese honeybee, Apis cerana cerana. *Acta Entomol. Sin.* 60, 533-543. https://doi.org/10.16380/j.kcxb.2017.05.005.

FAHRBACH, S.E., ROBINSON, G.E., 1995. Behavioral development in the honey bee: toward the study of learning under natural conditions. *Learn. Mem.* 2, 199-224. https://doi.org/10.1101/lm.2.5.199.

FORÊT, S., MALESZKA, R., 2006. Function and evolution of a gene family encoding odorant binding-like proteins in a social insect, the honey bee (*Apis mellifera*). *Genome Res.* 16, 1404-1413. https://doi.org/10.1101/gr.5075706.

GAO, Q., CHESS, A., 1999. Identification of candidate Drosophila olfactory receptors from genomic DNA sequence. *Genomics* 60, 31-9. https://doi.org/10.1006/geno.1999.5894.

GARY, N.E., MARSTON, J., 1971. Mating behaviour of drone honey bees with queen models (*Apis mellifera* L.). *Anim. Behav.* 19, 299-304. https://doi.org/10.1016/S0003-3472(71)80010-6.

GRAHAM B.M, 2015a. The hive and the honey bee. Dadant & Sons Press, Hamilton, Illnois, USA.

GRAHAM B.M, 2015b. The hive and the honey bee. Dadant & Sons Press, Hamilton, Illnois, USA.

HANSSON, B.S., ANTON, S., 2000. Function and Morphology of the Antennal Lobe: New Developments. *Annu. Rev. Entomol.* 45, 203-231. https://doi.org/10.1146/annurev.ento.45.1.203.

HERN, A., DORN, S., 1999. Sexual dimorphism in the olfactory orientation of adult Cydia pomonella in response to α-farnesene. *Entomol. Exp. Appl.* 92, 63-72. https://doi.org/10.1023/A:1003756428631.

HILL, C.A., FOX, A.N., PITTS, R.J., KENT, L.B., TAN, P.L., CHRYSTAL, M.A., CRAVCHIK, A., COLLINS, F.H., ROBERTSON, H.M., ZWIEBEL, L.J., 2002. G protein coupled receptors in Anopheles gambiae. *Science* (80-.). 298, 176-178. https://doi.org/10.1126/science.1076196.

JOERGES, J., KÜTTNER, A., GALIZIA, C.G., MENZEL, R., 1997. Representations of odours and odour mixtures visualized in the honeybee brain. *Nature* 387, 285-288. https://doi.org/10.1038/387285a0.

KEELING, C.I., SLESSOR, K.N., HIGO, H. A, WINSTON, M.L., 2003. New components of the honey bee (Apis mellifera L.) queen retinue pheromone. *Proc. Natl. Acad. Sci. U.S.A.* 100, 4486-4491. https://doi.org/10.1073/pnas.0836984100.

KURTOVIC, A., WIDMER, A., DICKSON, B.J., 2007. A single class of olfactory neurons mediates behavioural responses to a Drosophila sex pheromone. *Nature* 446, 542-546. https://doi.org/10.1038/nature05672.

LASKA, M., GALIZIA, C.G., GIURFA, M., MENZEL, R., 1999. Olfactory discrimination ability and odor-structure-activity relationships in honeybees. *Chem. Senses* 24, 429-438. https://doi.org/doi.org/10.1093/chemse/24.4.429.

LOPER, G.M., WOLF, W.W., TAYLOR, O.R., 1993. Radar detection of drones responding to honeybee queen pheromone. *J. Chem. Ecol.* 19, 1929-1938. https://doi.org/10.1007/BF00983797.

MASSON, C., PHAM DELEGUE, M.H., FONTA, C., GASCUEL, J., ARNOLD, G., NICOLAS, G., KERSZBERG, M., 1993. Recent advances in the concept of adaptation to natural odour signals in the honeybee, *Apis mellifera L*. Recent progress in the neurobiology of honeybees. *Apidologie* 24, 169-194. https://doi.org/10.1051/apido:19930302.

MOMBAERTS, P., 1999. Seven-Transmembrane Proteins as Odorant and Chemosensory Receptors. *Science* (80-.). 286, 707-711. https://doi.org/10.1126/science.286.5440.707.

PARK, D., JUNG, J.W., CHOI, B.-S., JAYAKODI, M., LEE, J., LIM, J., YU, Y., CHOI, Y.-S., LEE, M., PARK, Y., CHOI, I., YANG, T., EDWARDS, O.R., NAH, G., KWON, H.W., 2015. Uncovering the novel characteristics of Asian honey bee, Apis cerana, by whole genome sequencing. *BMC Genomics* 16, 1. https://doi.org/10.1186/1471-2164-16-1.

PHELAN, P.L., BAKER, T.C., 1987. An Attracticide for Control of Amyelois transitella (Lepidoptera: Pyralidae) in Almonds. *J. Econ. Entomol.* 80, 779-783. https://doi.org/10.1093/jee/80.4.779.

QIN, Q.H., HE, X.J., TIAN, L.Q., ZHANG, S.W., ZENG, Z.J., 2012. Comparison of learning and memory of Apis cerana and Apis mellifera. *J. Comp. Physiol.* A 198, 777-786. https://doi.org/10.1007/s00359-012-0747-9.

ROBERTSON, H.M., WANNER, K.W., 2006. The chemoreceptor superfamily in the honey bee, Apis mellifera: Expansion of the odorant, but not gustatory, receptor family. *Genome Res.* 16, 1395-1403. https://doi.org/10.1101/gr.5057506.

ROBERTSON, H.M., WANNER, K.W., 2006. The chemoreceptor superfamily in the honey bee, Apis mellifera: Expansion of the odorant, but not gustatory, receptor family. *Genome Res.* 16, 1395-1403. https://doi.org/10.1101/gr.5057506.

ROBERTSON, H.M., WARR, C.G., CARLSON, J.R., 2003. Molecular evolution of the insect chemoreceptor gene superfamily in Drosophila melanogaster. *Proc. Natl. Acad. Sci. U.S.A.* 100 Suppl, 14537-42. https://doi.org/10.1073/pnas.2335847100.

SANDOZ, J.-C., DEISIG, N., DE BRITO SANCHEZ, M.G., GIURFA, M., 2007. Understanding the logics of pheromone processing in the honeybee brain: from labeled-lines to across-fiber patterns. Front. Behav. *Neurosci.* 1, 5. https://doi.org/10.3389/neuro.08.005.2007.

SANDOZ, J.C., MENZEL, R., 2001. Side-Specificity of Olfactory Learning in the Honeybee:Generalization between Odors and Sides. *Learn. Mem.* 8, 286-294. https://doi.org/10.1101/lm.41401.an.

SCHMITTGEN, T.D., LIVAK, K.J., 2008. Analyzing real-time PCR data by the comparative CT method. *Nat. Protoc.* 3, 1101-1108. https://doi.org/10.1038/nprot.2008.73.

SLESSOR, K.N., KAMINSKI, L.-A., KING, G.G.S., BORDEN, J.H., WINSTON, M.L., 1988. Semiochemical basis of the retinue response to queen honey bees. *Nature*. https://doi.org/10.1038/332354a0.

SLESSOR, K.N., WINSTON, M.L., LE CONTE, Y., 2005. Pheromone communication in the honeybee (Apis mellifera L.). *J. Chem. Ecol.* 31, 2731-2745. https://doi.org/10.1007/s10886-005-7623-9.

SZYSZKA, P., 2005. Sparsening and Temporal Sharpening of Olfactory Representations in the Honeybee Mushroom Bodies. *J. Neurophysiol.* 94, 3303-3313. https://doi.org/10.1152/jn.00397.2005.

THOMPSON, J.D., HIGGINS, D.G., GIBSON, T.J., 1994. CLUSTAL W: improving the sensitivity of progressive multiple sequence alignment through sequence weighting, position-specific gap penalties and weight matrix choice. *Nucleic Acids Res.* 22, 4673-4680. https://doi.org/10.1093/nar/22.22.4673.

VETTER, R.S., VISSCHER, P.K., 1997. Influence of Age on Antennal Response of Male Honey Bees, *Apis mellifera*, to Queen Mandibular Pheromone and Alarm Pheromone Component. *J. Chem. Ecol. 23*, 1867-1880. https://doi.org/10.1023/B:JOEC.0000006456.90528.94.

VILLAR, G., BAKER, T.C., PATCH, H.M., GROZINGER, C.M., 2015. Neurophysiological mechanisms underlying sex- and maturation-related variation in pheromone responses in honey bees (*Apis mellifera*). *J. Comp. Physiol. A* 201, 731-739. https://doi.org/10.1007/s00359-015-1006-7.

VILLAR, G., GROZINGER, C.M., 2017. Primer effects of the honeybee, Apis mellifera, queen pheromone 9-ODA on drones. *Anim. Behav.* 127, 271-279. https://doi.org/10.1016/j.anbehav.2017.03.023.

WANNER, K.W., NICHOLS, A.S., WALDEN, K.K.O., BROCKMANN, A., LUETJE, C.W., ROBERTSON, H.M., 2007. A honey bee odorant receptor for the queen substance 9-oxo-2-decenoic acid. *Proc. Natl. Acad. Sci. U. S. A. 104*, 14383-14388. https://doi.org/10.1073/pnas.0705459104.

WU, X.B., WANG, Z.L., GAN, H.Y., LI, S.Y., ZENG, Z.J., 2016. Transcriptome comparison between newly emerged and sexually matured bees of *Apis mellifera*. *J. Asia. Pac. Entomol.* 19, 893-897. https://doi.org/10.1016/j.aspen.2016.08.002.

YANG, L., ZHANG, L.Z., ZENG, Z.J., 2017. Cloning and sequencing analysis of Odour receptor 170 in honeybees, *Apis cerana cerana*. *J. Environ. Entomol.* 39, 48-54. https://doi.org/10.3969/j.issn.1674-0858.2017.01.4.

ZENG, Z.-J., 2017. Apiculture. China Agriculture Press, Beijing.

ZHANG, L.Y., XIE, B.H., NI, C.X., ZHAO, L., LI, H.L., SHANG, H.W., 2012. cloning,expression and subcellar localization of the olfactory co-receptor Orco gen in the Chinese honeybee, Apis cerana cerana(*Hymentoptera: Apidae*). *Acta Entomol. Sin.* 55, 1246-1254. https://doi.org/10.16380/j.kcxb.2012.11.002.

ZHANG, L.Z., ZHANG, S.W., WANG, Z.L., YAN, W.Y., ZENG, Z.J., 2014. Cross-modal interaction between visual and olfactory learning in Apis cerana. *J. Comp. Physiol. A Neuroethol. Sensory, Neural, Behav. Physiol.* 200, 899-909. https://doi.org/10.1007/s00359-014-0934-y.

ZHANG, Z.Y., YANG, S.S., ZHAO, H.T., JIANG, Y.S., 2016. Expression and localization analysis of olfactory receptor AcerOrco in the Chinese honeybee,Apis cerana cerana(*Hymentoptera: Apidae*). *Acta Entomol. Sin.* 59, 185-191. https://doi.org/10.16380/j.kcxb.2016.02.008.

ZHAO, H.T., GAO, P.F., DU, H. YAN, MA, W.H., TIAN, S.H., JIANG, Y.S., 2014. Molecular characterization and differential expression of two duplicated odorant receptor genes, AcerOr1 and AcerOr3, in *Apis cerana cerana*. *J. Genet.* 93, 53-61. https://doi.org/10.1007/s12041-014-0332-9.

ZHAO, H.T., GAO, P.F., ZHANG, C.X., MA, W.H., JIANG, Y.S., 2013. Molecular Identification and Expressive Characterization of an Olfactory Co-Receptor Gene in the Asian Honeybee, *Apis cerana cerana*. *J. Insect Sci.* 13, 1-14. https://doi.org/10.1673/031.013.8001.

ZHAO, H.T., GAO, P.F., ZHANG, G.X., MA, W.H., JIANG, Y.S., 2012. Gene cloning and sequence analysis of Or1 and Or2in *Apis cerana*. *Chinese J. Appl. Entomol.* 49, 1117-1124.

ZHAO, H.T., GAO, P.F., ZHANG, G.X., TIAN, S. HAO, YANG, S.S., MENG, J., JIANG, Y.S., 2015. Expression and Localization Analysis of the Odorant Receptor Gene Orco in Drones Antennae *of Apis cerana cerana*. *Sci. Agric. Sin.* 48, 796-803. https://doi.org/10.3864/j.issn.0578-1752.2015.04.17.

# 12. Expression Patterns of Four Candidate Sex Pheromone Receptors in Honeybee Drones (*Apis mellifera*)

J.F. Liu[1,2], X.J. He[1], M. Li[1], Z.L. Wang[1], X.B. Wu[1], W.Y. Yan[1] & Z.J. Zeng[1*]

1. Honeybee Research Institute, Jiangxi Agricultural University, Nanchang, 330045, P.R. China.
2. Environment and Plant Protection Institute, Chinese Academy of Tropical Agricultural Sciences, Haikou, 571101, P. R. China.

**Abstract**

In *Apis mellifera* odorant receptors (Ors) play a crucial role in special recognition of sex pheromones in honeybee mating activities. Four candidate sex pheromone Ors (*Am*Or10, *Am*Or11, *Am*Or18 and *Am*Or170) have been identified and found to be preferentially expressed in drone antennae. However, few studies have investigated the regulation of these four drone Ors on drone mating behaviour. This study characterised the expression patterns of these Ors across different sexual developmental stages (immature and sexually mature) and different physiological statuses (flying and crawling), using both the antennae and brains of drones. qRT-PCR results indicated that the expression of four Ors were not significantly different in drone antennae between flying and crawling statuses at immature stage. However, all four Ors expression levels in brains of flying drones were significantly higher than those of crawling drones at mature stage. Moreover, only the expression level of Or170 was significantly higher in mature flying drones antennae than crawling ones. Therefore, this study demonstrated a link between four candidate sex pheromone Ors transcriptional expression in the brains of honeybee drones and behaviour associated with sexual maturity and mating flight. In addition, Or170 might be involved in the maturation of honeybee drones' olfactory system, and in the organisation of odour-mediated mating behaviours.

**Keywords:** *Apis mellifera*, odorant receptors, drones, mating flight, sexual development

## Introduction

In insects, odorant receptors (Ors) play important roles in environmental odours recognition and social communication, being critical for feeding, oviposition, predator avoidance and mate recognition (Leary *et al*. 2012; Lebreton *et al*. 2017; Sakurai *et al*. 2011). Olfaction is mediated by the interaction of volatile ligands with a set of specialised membrane proteins in olfactory sensory neurons (OSNs) of their antennae (Dweck *et al*. 2015; Smart *et al*. 2008; van der Goes van Naters & Carlson 2007). Odorant molecule bound by these

---

* Corresponding author: bees1965@sina.com (ZJZ).
注：此文发表在 *African Entomology* 2020 年第 2 期。

receptors results in OSNs depolarisation and produces a neuronal signal that is decoded by the insect brains, informing the decisions of behavioural responses (Carraher *et al.* 2015).

Honeybees (*Apis mellifera*) originated in Africa and expanded into Eurasia and New World, it mainly includes African honeybees and European honeybees (Whitfield *et al.* 2006). Honeybees display a striking copulate behaviour in the air. During the mating seasons, sexually mature males gather in the air and form drone congregations area (DCA), searching for the virgin queens. Virgin queens fly to this DCA and release queen mating pheromones to attract males (G. Koeniger *et al.* 1989; Koeniger 1986; Koeniger *et al.* 2005; N. Koeniger *et al.* 1989). Therefore, the drone's mating behaviour is triggered by queen sex pheromones such as queen mandibular pheromone (QMP), highly relying on their Ors in olfactory system (Brockmann & Brückner 2001; Gries & Koeniger 1996; Wanner *et al.* 2007). From the genome point of view, the honeybee *Apis mellifera* and *Apis cerana* encode 170 and 119 Ors respectively (Park *et al.* 2015; Robertson & Wanner 2006). Moreover, Wu *et al.* (2016) revealed that 16 ORs were up-regulated in the sexually matured drones of *A. mellifera* by using high-throughput RNA-Seq. Mounting evidence revealed that sex pheromone receptors play important roles in drones' physiological behaviours. Four candidate sex pheromone Ors (*Am*Or10, *Am*Or11, *Am*Or18 and *Am*Or170) have been identified from the honeybee genome based on their biased expression in male bee antennae. Particularly, *Am*Or11 can specifically bind 9-oxo-2-decenoic acid (9-ODA), which is one of the predominantly detected compounds of QMP (Wanner *et al.* 2007). A subsequent investigation reported that the expression level of *Am*Or11 was higher in the antennae of sexually mature drones than immature drones (Villar *et al.* 2015). Furthermore, our previous study showed that the *Ac*Or11 expression pattern in brains was dramatically higher in flying mature drones than crawling ones, suggesting that Or11 is associated with sexual maturity and mating behaviour (Liu *et al.* 2019). Despite of the importance of candidate sex pheromone Ors in honeybees, their biological functions have not been deeply studied especially those Ors in drones at different behavioural statuses and mature stages.

In this study, we detected and compared the expression patterns of four sex pheromone receptors (*Am*Or10, *Am*Or11, *Am*Or18 and *Am*Or170) during sexual developmental stages (immature and sexually mature) and different physiological statuses (flying and crawling) in antennae and brains of drones. The results allow further exploration of the biological functions of these Ors in drone copulation behaviour and eventually to provide the basis for elucidating the regulation mechanism of honeybee mating flights.

## Materials and Methods

### *Honeybee*

Three honeybee (*A. mellifera*) colonies from Honeybee Research Institute of Jiangxi Agricultural University (28.46°N 115.49°E) were used for all experiments. These colonies were maintained using standard keeping practices. Three healthy mated queens were caged to lay in the empty drone frames for 24 h, and then these frames with drone eggs were removed into queenless area (the top box of the hive) in colonies. The sealed frames were moved into a dark incubator at 34 °C and 50% relative humidity until the day prior to adult emergence. The sampling methods were modified from Villar *et al.* (2015). After drones emerged, they were paint-marked on their thorax and placed back into their natal colonies. We collected 10 crawling drones (inside the hive) and 10 flying drones (return back to the hive) from each colony at two age points of day 4 (sexually immature) and day 14 (sexually mature), respectively. All bees were snap-frozen in liquid $N_2$ until processed. Ten pairs of drone antennae or brains were collected from each group (total 4 groups: 4-day flying and crawling groups, and 14-day flying and crawling groups) for RNA extraction. Each group had three biological replicates from three honeybee colonies.

### Gene expression quantification and statistical analysis

Total RNA was isolated from the antennae and brains of drones using TransZol reagent (Transgen Biotech, www.transgen.com.cn) according to the manufacturer's instructions, and stored in a freezer at -80 °C until use. The cDNA was synthesised from the total RNA isolated from antennae using the Primer-Script RT reagent Kit (TaKaRa, www.takara-bio.com) according to the manufacturer's instructions. Then, the expression levels of these *Am*Ors (*Am*Or10, *Am*Or11, *Am*Or18 and *Am*Or170) were determined using quantitative Real-Time PCR and normalised to the β-actin gene. The primers for these *Am*Ors and *Am*-β-actin were designed by the primer premier 5.0 software (Table III-12-1). Finally, the relative expression levels of these *Am*Ors mRNAs were calculated using the $2^{-\Delta\Delta Ct}$ comparative CT method.

Table III-12-1  Primers used to *Am*Ors and β-actin gene for qRT-PCR

| Primer names | Primer sequences | GenBank accession |
| --- | --- | --- |
| *Am*Or10-F | 5'-CCGCATCTGAACAGTATCGTG-3' | NM_001242961.2 |
| *Am*Or10-R | 5'-ATTCTCCTCCGTGGCTATCG-3' | |
| *Am*Or11-F | 5'-ATGTGCGGTTTGCTGAAGA-3' | NM_001242962.1 |
| *Am*Or11-R | 5'-CGAGAAGGTGCCAATGACG-3' | |
| *Am*Or18F | 5'-TTTTATTACATCGCTTTGCC-3' | XM_003250678.4 |
| *Am*Or18-R | 5'-TCTTCCTTCCATCCACCA-3' | |
| *Am*Or170-F | 5'-CCAGTGTTGCCTCGCTC-3' | NM_001242993.1 |
| *Am*Or170-R | 5'-TTTCGTTATCTCACGCTCC-3' | |
| *Am*-β-actin-F | 5'- GGTATTGTATTGGATTCGGGTG -3' | NM_001185146.1 |
| *Am*-β-actin-R | 5'- TGCCATTTCCTGTTCAAAGTCA -3' | |

### Statistical Analysis

Differences in the relative expression of the *Am*Ors genes were determined using a t-test analysis in SPSS 17.0 (IBM, Armonk, NY). Values of $P < 0.05$ were considered significant in all treatments.

## Results

In antennae, the expression levels of four *Am*Ors were not significantly different between 4-day-old flying and crawling drones (Or10: $t = -1.832$, df = 4, $P = 0.208$; Or11: $t = -1.052$, df = 4, $P = 0.352$; Or18: $t = -2.044$, df = 4, $P = 0.110$; Or170: $t = -0.202$, df = 4, $P = 0.850$, Fig. III-12-1a). The expression levels of three genes (*Am*Or10, *Am*Or11 and *Am*Or18) also were not different in mature drones from flying/crawling comparisons (Or10: $t = 2.689$, df = 4, $P = 0.055$; Or11: $t = 1.155$, df = 4, $P = 0.313$; Or18: $t = 0.251$, df = 4, $P = 0.814$, Fig. III-12-1b). However, the expression level of *Am*Or170 in antennae was significantly higher in 4-day-old flying drones than those of crawling ones ($t = -4.211$, df = 4, $P = 0.014$, Fig. III-12-1b).

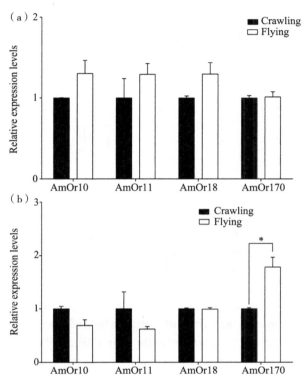

**Figure III-12-1** Four candidate sex pheromone receptors gene expression pattern in antennae of immature (a) and mature (b) drones (n = 3 pools). The grey bars represent the normalised expression level of crawling drone. The open bars represent the relative expression of flying drone. The data was expressed as the mean ± SE, and the "*" indicates a significant difference ($P < 0.05$).

In brains, the expression levels of three genes (*Am*Or11, *Am*Or18 and *Am*Or170) were not different in mature drones by compared flying with crawling drones (Or11: $t = -2.004$, df = 4, $P = 0.116$; Or18: $t = 0.860$, df = 4, $P = 0.479$; Or170: $t = 0.813$, df = 4, $P = 0.462$, Fig. III-12-2a). But, the expression level of *Am*Or10 in brains of 4-day-old flying drones was significantly higher than those of crawling drones ($t = -4.397$, df = 4, $P = 0.048$, Fig. III-12-2a). Moreover, the expression levels of four *Am*Ors in mature drone brains in flying status were significantly higher than those in crawling status (Or10: $t = -4.520$, df = 4, $P = 0.045$; Or11: $t = -3.433$, df = 4, $P = 0.026$; Or18: $t = -6.026$, df = 4, $P = 0.026$; Or170: $t = -4.833$, df = 4, $P = 0.008$, Fig. III-12-2b).

## Discussion

In honeybees, odorant receptors which are expressed in male neurons are critical for detecting the queen sex pheromones released during mating flights (Park *et al*. 2015; Robertson & Wanner 2006). In this study,

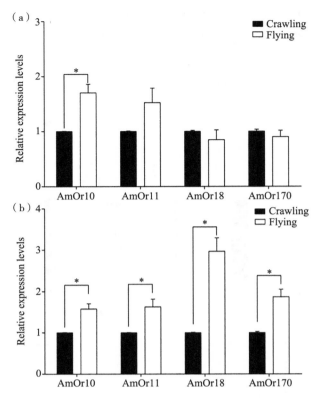

**Figure III-12-2** Four candidate sex pheromone receptors gene expression pattern in brains of immature (a) and mature (b) drones ($n = 3$ pools). The grey bars represent the normalised expression level of crawling drone. The open bars represent the relative expression of flying drone. The data was expressed as the mean ± SE, and the "*" indicates a significant difference ($P < 0.05$).

we documented the expression patterns of four candidate sex pheromone receptor genes (*Am*Or10, *Am*Or11, *Am*Or18 and *Am*Or170) in drone brains and antennae under different sexual developmental stages and physiological statuses.

It was observed that the expression patterns of four *Am*Ors in antennae have no difference between flying and crawling immature drones (Fig. III-12-1a). Many insects utilise floral scents and other chemical signals for orientation at immature stage (Phelan & Baker 1987; Hern & Dorn 1999; Anton *et al*. 2007). Therefore, the flights of sexually immature drones (less than 12 days) are generally orientation flights rather than mating flights, which is also consistent with a previous study (Graham 2015). Therefore, these results suggested that the biological foundation of these odorant receptors related to copulation may be underdeveloped in antennae of drones at the juvenile stage. The expression levels of three genes (*Am*Or10, *Am*Or11 and *Am*Or18) from all flying/crawling comparisons were not significantly different in antennae of mature drones (Fig. III-12-1b). Interestingly, only the expression of *Am*Or170 was dramatically higher in antennae of mature flying drones than the mature crawling ones, though there was no significant difference in the immature drone comparison. Presumably because *Am*Or170 gene is sensitive to physiological variation and employed by mature drones to detect environmental scents for mating flights. A similar result is reported in male *Drosophila melanogaster* that the expression level of sex pheromone receptor Or67d also is highly correlated the physiological variation

(Kurtovic *et al.* 2007; Zhou *et al.* 2009).

In addition, the physiological status of honeybee drones strongly effected their Ors expression in brains at the mature stage (Fig. III-12-2b). In many insects, a number of olfactory neurons expressed sex pheromone receptors in order to increase sensitivity to respond to pheromone in the antennae lobe (the first level of olfactory processing), which can effectively transfer odour molecules to the sensory receptors distributed on the dendritic of olfaction receptor neuron (Graham 2015). Particularly, *Am*Or11 was the most highly expressed of four candidate sex pheromone receptors and specially responded to 9-ODA, the major QMP component. When the antennae of drones were specifically stimulated with 9-ODA, the enlarged macroglomerulus 2 (MG2, the largest of the four macroglomeruli) in the drone antennae lobe was activated, suggesting that olfactory neurons projecting to MG2 express *Am*Or11 in drone brain (Sandoz 2006; Sandoz *et al.* 2007). Moreover, these results are similar to those of previous studies on Asian honeybees that *Ac*Or10 and *Ac*Or11 were significantly higher expressed in the brains of mature flying drones than those of crawling drones (Liu *et al.* 2019; Yang *et al.* 2018). This study indicated that there were not only Or10 and Or11 taking part in the daily mating flights of mature drones, but also Or18 and Or170 possibly perform as important roles in improving the mating behaviour of honeybee drones. Although we preliminary observed the characteristic of four Or genes, the molecular mechanism of how Ors play a role in the sophisticated mating behaviour of honeybees is still obscure. This requires further investigations in future work.

In summary, the expression patterns of candidate sex pheromone receptor genes (*Am*Or10, *Am*Or11, *Am*Or18 and *Am*Or170) in the antennae and brains of *A. mellifera* drones were analysed at different stages of sexual maturity and physiological statuses. These Ors expression patterns in drone brains were closely correlated with both sexual development and physiological status, suggesting that they were involved in drone sexual maturity and mating behaviour. Our results provide an insight into the molecular basis underlying mating flights of honeybee drones.

## Acknowledgements

All authors read and approved the final manuscript. This work is supported by the National Natural Science Foundation of China (31572469), and the Earmarked Fund for China Agriculture Research System (CARS-44-KXJ15).

## Compliance with Ethical Standards

The authors declare that they have no conflict of interest. All applicable international, national, and/or institutional guidelines for the care and use of animals were followed. All procedures performed in studies involving animals were in accordance with the ethical standards of the institution or practice at which the studies were conducted.

## Orcid iDs

X.J. He: orcid.org/0000-0001-7445-8944

M. Li : orcid.org/0000-0002-7664-3227

Z.L. Wang: orcid.org/0000-0002-9651-6129

X.B. Wu: orcid.org/0000-0002-9865-1623

W.Y. Yan : orcid.org/0000-0002-5183-7543

Z.J. Zeng: orcid.org/0000-0001-5778-4115

## References

ANTON, S., DUFOUR, M.C. & GANDENNE, C. 2007. Plasticity of olfactory-guided behaviour and its neurobiological basis: Lessons from moths and locusts. *Entomologia Experimentalis et Applicata* 123(1): 1-11.

BROCKMANN, A. & BRÜCKNER, D. 2001. Structural differences in the drone olfactory system of two phylogenetically distant *Apis* species, *A. florea* and *A. mellifera*. *Naturwissenschaften* 88(2): 78-81.

CARRAHER, C., DALZIEL, J., JORDAN, M.D., CHRISTIE, D.L., NEWCOMB, R.D. & KRALICEK, A.V. 2015. Towards an understanding of the structural basis for insect olfaction by odorant receptors. *Insect Biochemistry and Molecular Biology* 66: 31-41.

DWECK, H.K.M., EBRAHIM, S.A.M., THOMA, M., MOHAMED, A.A.M., KEESEY, I.W., TRONA, F., LAVISTA-LLANOS, S. SVATOŠ, A., SACHSE, S., KNADEN, M. & HANSSON, B.S. 2015. Pheromones mediating copulation and attraction in *Drosophila*. *Proceedings of the National Academy of Sciences* 112(21): E2829-E2835.

GRAHAM, B.M. 2015. *The Hive and the Honey Bee*. Dadant & Sons Press, Hamilton, IL, U.S.A.

GRIES, M. & KOENIGER, N. 1996. Straight forward to the queen: pursuing honeybee drones (*Apis mellifera* L.) adjust their body axis to the direction of the queen. *Journal of Comparative Physiology* 179(4): 539-544.

HERN, A. & DORN, S. 1999. Sexual dimorphism in the olfactory orientation of adult *Cydia pomonella* in response to α-farnesene. *Entomologia Experimentalis et Applicata* 92(1): 63-72.

KOENIGER, G. 1986. Mating sign and multiple mating in the honeybee. *Bee World* 67(4): 141-150.

KOENIGER, G., KOENIGER, N., PECHHACKER, H., RUTTNER, F. & BERG, S. 1989. Assortative mating in a mixed population of European honeybees, *Apis mellifera ligustica* and *Apis mellifera carnica*. *Insectes Sociaux* 36(2): 129-138.

KOENIGER, N., KOENIGER, G. & PECHHACKER, H. 2005. The nearer the better? Drones (*Apis mellifera*) prefer nearer drone congregation areas. *Insectes Sociaux* 52(1): 31-35.

KOENIGER, N., KOENIGER, G. & WONGSIRI, S. 1989. Mating and sperm transfer in *Apis florea*. *Apidologie* 20(5): 413-418.

KURTOVIC, A., WIDMER, A. & DICKSON, B.J. 2007. A single class of olfactory neurons mediates behavioural responses to a *Drosophila* sex pheromone. *Nature* 446: 542-546.

LEARY, G.P., ALLEN, J.E., BUNGER, P.L., LUGINBILL, J.B., LINN, C.E., MACALLISTER, I.E., KAVANAUGH, M.P. & WANNER, K.W. 2012. Single mutation to a sex pheromone receptor provides adaptive specificity between closely related moth species. *Proceedings of the National Academy of Sciences* 109(35): 14081-14086.

LEBRETON, S., BORRERO-ECHEVERRY, F., GONZALEZ, F., SOLUM, M., WALLIN, E.A., HEDENSTRÖM, E., HANSSON, B.S., GUSTAVSSON, A.L., BENGTSSON, M., BIRGERSSON, G., WALKER, W.B., DWECK, H.K.M., BECHER, P.G. & WITZGALL, P. 2017. A *Drosophila* female pheromone elicits species-specific long-range attraction via an olfactory channel with dual specificity for sex and food. *BMC Biology* 15(1): 1-14.

LIU, J.F., YANG, L., LI, M., HE, X.J., WANG, Z.L. & ZENG, Z.J. 2019. Cloning and expression pattern of odorant receptor 11 in Asian honeybee drones, *Apis cerana* (Hymenoptera, Apidae). *Journal of Asia-Pacific Entomology* 22(1): 110-116.

PARK, D., JUNG, J.W., CHOI, B-S., JAYAKODI, M., LEE, J., LIM, J., YU, Y., CHOI, Y-S., LEE, M., PARK, Y., CHOI, I., YANG, T., EDWARDS, O.R., NAH, G. & KWON, H.W. 2015. Uncovering the novel characteristics of Asian honey bee, *Apis cerana*, by whole genome sequencing. *BMC Genomics* 16(1): 1.

PHELAN, P.L. & BAKER, T.C. 1987. An attracticide for control of *Amyelois transitella* (Lepidoptera: Pyralidae) in almonds. *Journal of Economic Entomology* 80(4): 779-783.

ROBERTSON, H.M. & WANNER, K.W. 2006. The chemoreceptor superfamily in the honey bee, *Apis mellifera*: Expansion of the odorant, but not gustatory, receptor family. *Genome Research* 16(11): 1395-1403.

SAKURAI, T., MITSUNO, H., HAUPT, S.S., UCHINO, K., YOKOHARI, F., NISHIOKA, T., KOBAYASHI, I., SEZUTSU, H., TAMURA, T. & KANZAKI, R. 2011. A single sex pheromone receptor determines chemical response specificity of sexual behavior in the silkmoth *Bombyx mori*. *PLoS Genetics* 7(6): e1002115.

SANDOZ, J-C. 2006. Odour-evoked responses to queen pheromone components and to plant odours using optical imaging in the antennal lobe of the honey bee drone *Apis mellifera* L. *The Journal of Experimental Biology* 209: 3587-3598.

SANDOZ, J-C., DEISIG, N., DE BRITO SANCHEZ, M.G. & GIURFA, M. 2007. Understanding the logics of pheromone processing in the honeybee brain: from labeled-lines to across-fiber patterns. *Frontiers in Behavioral Neuroscience* 1: 5.

SMART, R., KIELY, A., BEALE, M., VARGAS, E., CARRAHER, C., KRALICEK, A.V., CHRISTIE, D.L., CHEN, C., NEWCOMB, R.D. & WARR, C.G. 2008. *Drosophila* odorant receptors are novel seven transmembrane domain proteins that can signal independently of heterotrimeric G proteins. *Insect Biochemistry and Molecular Biology* 38(8): 770-780.

VAN DER GOES VAN NATERS, W. & CARLSON, J.R. 2007. Receptors and neurons for fly odors in *Drosophila*. *Current Biology* 17(7): 606-612.

VILLAR, G., BAKER, T.C., PATCH, H.M. & GROZINGER, C.M. 2015. Neurophysiological mechanisms underlying sex- and maturation-related variation in pheromone responses in honey bees (*Apis mellifera*). *Journal of Comparative Physiology A* 201(7): 731-739.

WANNER, K.W., NICHOLS, A.S., WALDEN, K.K.O., BROCKMANN, A., LUETJE, C.W. & ROBERTSON, H.M. 2007. A honey bee odorant receptor for the queen substance 9-oxo-2-decenoic acid. *Proceedings of the National Academy of Sciences of the United States of America* 104(36): 14383-14388.

WHITFIELD C.W., BEHURA S.K., BERLOCHER S.H., CLARK A.G., JOHNSTON J. S., SHEPPARD W.S., SMITH D.R., SUAREZ A.V., WEAVER D., TSUTSUI N.D. 2006. Thrice out of Africa: ancient and recent expansions of the honey bee, *Apis mellifera*. *Science* 314(5799):642-645.

WU, X.B., WANG, Z.L., GAN, H.Y., LI, S.Y. & ZENG, Z.J., 2016. Transcriptome comparison between newly emerged and sexually matured bees of *Apis mellifera*. *Journal of Asia-Pacific Entomology* 19(3): 893-897.

YANG, L., JIANG, W., HE, X. & ZENG, Z.J. 2018. Cloning and expression analysis of olfactory gene AcOr10 in the Chinese honeybee, *Apis cerana cerana* (Hymenoptera: Apidae). *Acta Entomologica Sinica* 61(3): 292-299.

ZHOU, S., STONE, E.A., MACKAY, T.F.C. & ANHOLT, R.R.H. 2009. Plasticity of the chemoreceptor repertoire in *Drosophila melanogaster*. *PLoS Genetics* 5(10): e1000681.

# 13. Age Components of Queen Retinue Workers in Honey Bee Colony (*Apis mellifera*)

Y. Yi, Y. Li, Z.J. Zeng*

Honeybee Research Institute, Jiangxi Agricultural University, Nanchang, Jiangxi 330045, P. R. of China

**Abstract**

It's known that elaborate age is closely associated with polyethism in honeybee colonies, and the circle composed of queen retinue workers is a usual phenomenon in honeybee colonies. In this study, we showed that the age-bracket of retinue workers is 2-23 d, but mainly 6-18 d by marking newly hatched workers in two colonies.

**Keywords:** Queen, Retinue workers, Age

Honeybees are widely acknowledged as a model insect for the studies of social organization, social behaviour and neurology (Robinson *et al.*, 2005; He *et al.*, 2015; Liu *et al.*, 2015). It is known that the age-dependent division of labor in honeybee colonies is one of the most typical features of this social insect (Robinson *et al.*, 2005). Newly hatched workers undergo the task of cleaning during their first few days (Kryger *et al.*, 2000). Later, they begin to consume protein-rich pollen. During the age of 5-15 d, their hypopharyngeal glands fully develop and most of them play the role of nursing the queen and larvae (Pan *et al.*, 2013). Then, they begin to build comb by secreting wax, cleaning the hive and compacting pollen. After 15-20 days of age, some workers begin to undertake tasks that are out of the nest whilst other workers remain inside to hive to act as receivers and distributors (Kryger *et al.*, 2000). However, this division of labor is not immutable. Environmental changes can accelerate, delay, or even reverse the transition of nurse bee and forager bee (Robision 1992; Huang & Robision 1996). Furthermore, polyethism in honeybee colonies might be affected by genotypical variability (Page & Robinson 1991; Kryger *et al.*, 2000; Wang *et al.*, 2012).

The queen retinue pheromone (QRP) released by a honeybee queen attracts workers that we name queen retinue workers to surround beside her (Slessor *et al.*, 2005). Within the hive, it can be easily observed that there is a circle consisting of queen retinue workers, however, little is known about them as few studies have been performed on them. In this study, we intend to understand how the age-bracket of queen retinue workers is determined by marking newly hatched workers.

* Corresponding author: bees1965@sina.com(ZJZ).
注：此文发表在 *Sociobiology* 2016 年第 2 期。

Two honeybee (*Apis mellifera*) colonies (colony A and B) were maintained at the Honeybee Research Institute, Jiangxi Agricultural University, Nanchang, China (28.46 °N, 115.49 °E). Both colonies consisted of a new comb for the queen to lay eggs, and also a honey comb. Additionally, we placed extra brood combs, which contain sealed honeybee brood that will emergence during the night, into the queen spawn controller at 5 pm (all apertures of the queen bee spawn controller were too small for workers to pass through). Three hundred newly hatched workers were labeled with dyes at 9 am the next day. Following this, they were put into their natal colonies. In total 6000-8000 worker bees were present in each observation hive. We used colors to represent their date of birth and identify their age. Every two days, we marked workers with different color. In total we marked bees ten times and then began recording videos to monitor their behaviour.

We recorded the queen's behavior and the forming process of the queen retinue workers' circle in an open transparent awning at 9:00-11:30 am and 2:00-4:30 pm. In order to minimize the impact of frequent interruptions on colonies, we began recording 2 days after bees were placed into the hive. We then continued recording until there were no more marked bees around the queen, a process which took about 20 days. After recording, we counted the number of marked workers in each color by watching the videos. We were then able to come to a conclusion regarding the age-bracket of queen retinue workers. A frame from a video is shown in Fig. III-13-1 (see page 535). We analyzed the differences in the number of retinue workers of each age in the two colonies with the Z-test (SPSS Statistics 17.0).

It was easy for us to see a circle composed of queen retinue workers when the queens were laying eggs, standing still or being fed in their respective colonies during the recording period. The footage showed that there were many queen retinue workers within the age-bracket of 2-23 days. Additionally, there was only one worker at the age of 29 d in colony A, and three 28 d workers and one 30 d worker in colony B (Fig. III-13-2). However, we deduce that this data should be disregarded for the reason that the number of queen retinue workers aged over 23 d old was negligible.

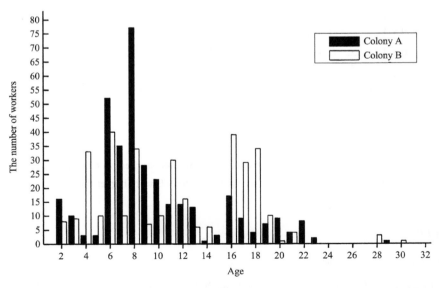

**Figure III-13-2** The number of queen retinue workers at each age. Most queen retinue workers were within the age-bracket of 2-23 d.

In addition, Fig. III-13-3 showed that two spots (ages 6 and 8) in colony A and four spots (ages 6, 8, 16 and 18) in colony B were higher than branch line (the number stands for the age of marked workers in colonies). It is known that some workers aged 5-15d play the role of nurse to feed the queen and small larvae (Pan *et al.*, 2013). After 15-20 days of age, some workers begin to undertake tasks that are outside of the nest (Kryger *et al.*, 2000). Thus, we indicate that the age-bracket of queen retinue workers is 2-23 day, but mainly 6-18 day. The honeybee queen releases QRP to encourage queen retinue workers to feed and groom her, and also to acquire and distribute her pheromone messages to other workers throughout the colony (Keeling *et al.*, 2003).

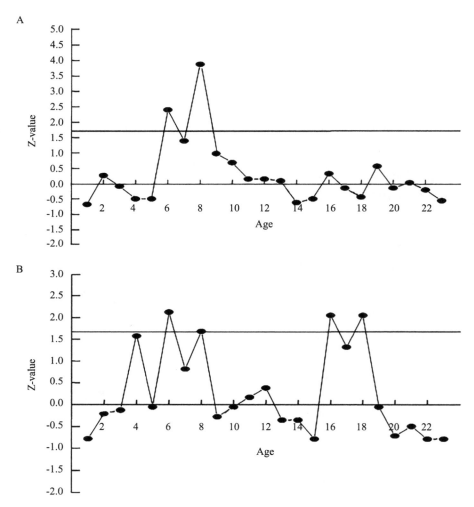

**Figure III-13-3** The Z values of same aged queen retinue workers in Colony A and B. The Z-values of two spots in colony A and four spots in colony B were higher than 1.65.

## Acknowledgements

We thank Dr. Qiang Huang and Ms. Lianne Meah for revising the paper. This work was supported by the Earmarked Fund for the China Agriculture Research System (No.CARS-45-KXJ12). All experimental procedures outlined in this work were performed in accordance with current Chinese laws on animal experimentation.

## References

HE, X.J., TIAN, L.Q., WU, X.B. & ZENG, Z.J. (2015). RFID monitoring indicates honeybees work harder before a rainy day. *Insect Science*, doi: 10.1111/1744-7917.12298.

HUANG, Z.Y. & ROBINSON, G.E. (1996). Regulation of honey bee division of labor by colony age demography. *Behavioral Ecology and Sociobiology*, 39: 147-158.

KEELING, C.I., SLESSOR, K.N., HIGO, H.A. & WINSTON, M.L. (2003). New components of the honey bee (*Apis mellifera L.*) queen retinue pheromone. *Biochemistry* 100: 4486-4491. doi: 10.1073/pnas.0836984100.

KRYGER, P., KRYGER, U. & MORITZ, R.F.A. (2000). Genotypical variability for the tasks of water collecting and scenting in a honey bee colony. *Ethology*, 106: 769-779.

LIU, H., WANG, Z.L., ZHOU, L.B. & ZENG, Z.J. (2015). Quantitative analysis of the genes affecting development of the hypopharyngeal gland in honey bees (*Apis mellifera L.*). *Sociobiology*, 62: 412-416. doi: 10.13102/sociobiology.v62i3.760.

PAGE, R.E. & ROBINSON, G.E. (1991). The genetics of division of labour in honey bee colonies. *Advance in Insect Physiology*, 23:117-169.

PAN, Q.Z., LIN, J.L., WU, X.B., ZHOU, L.B., ZHANG, F., YAN, W.Y. & ZENG, Z.J. (2013). Research and Application of Key Technique for Mechanized Production of Royal Jelly (IV)-Design and Application of A Machine for Gathering Royal Jelly. *Acta Agriculturae Universitatis Jiangxiensis*, 35: 1266-1271.

ROBINSON, G.E. (1992). Regulation of divtion of labor in insect societies. *Annual Review of Entomology*, 37: 637-665.

ROBINSON, G.E., GROZINGER C.M. & WHITFIELD, C.W. (2005). Sociogenomics: social life in molecular terms. *Nature Reviews Genetics*, 6: 257-270. doi:10.1038/nrg1575.

SLESSOR, K.N., WINSTON, M.L. & LE, C.Y. (2005). Pheromone communication in the honeybee (*Apis mellifera L.*). *Journal of Chemical Ecology*, 31: 2731-2745. doi:10.1007/s10886-005-7623-9.

WANG, H., ZHANG, S.W., ZHANG, F. & ZENG, Z.J. (2012). Analysis of lifespans of workers from different subfamilies in a honeybee colony. *Chinese Journal of Applied Entomology*, 49: 1172-1175.

# 14. Genotypic Variability of the Queen Retinue Workers in Honeybee Colony (*Apis mellifera*)

Y. Yi, W.Y. Yan, Y. Li, Y.H. Guo, L.Z. Zhang, Z.J. Zeng*

Honeybee Research Institute, Jiangxi Agricultural University, Nanchang 330045, P.R. China

**Abstract**

The polyandrous mating behaviour of the honeybee queen increases the genetic variability amongst her worker offsprings and the genetic variability within the honeybee colony can affect their polyethism. In this study, we are intended to understand whether there is a genetic variability in the task of queen retinue. Microsatellite DNA analyses revealed a total of 13 and 12 subfamilies in two colonies, respectively. It shows that the subfamily proportion of the queen retinue workers significantly deviated from random distribution, which suggests that there might have a genetic preference in the task of queen retinue.

**Keywords:** Honeybee, Queen retinue workers, Microsatellite, Subfamily

## Introduction

Honeybees are widely acknowledged as a model species for the studies of social behavior, social organization and even neurology (Robinson *et al.* 2005; Qin *et al.* 2014; He *et al.* 2016). Previous studies showed that the task choices of honeybee workers are closely associated with their ages. In other words, the tasks of honeybees vary with their age. The honeybee workers tend to clean the cells or keep the brood warm during 1-3 days old. When they reach the age of 3-6 days, they started undertaking the job of packing the pollen and honey. Later, the workers start to feed larvae or the queen with royal jelly secreted from their hypopharyngeal glands which are fully developed during their 6-12 days age. Then, the workers do the task of wax-making, hive-cleaning or pollen-handling during the age of 12-18 days. 18 days later, workers tend to collect water, pollen, nectar, propolis for the colony or guard their colony (Lindauer 1952, 1954; Zeng 2007).

In addition, for the reason that queens mate multiply, there are many subfamilies (patrilines) within a colony that are genetically different, and these genetic differences also have translated into different behaviors among them (Page and Robinson 1991). Page and Robinson (1991) found that workers of one subfamily were more likely to become pollen gatherers, whereas others were more likely to gather nectar. Genetic variability has been described for several behavioural traits: grooming behaviour (Frumhoff and Banker 1988), guarding and undertaking (Robinson and Page 1988), nectar and pollen foraging (Page and Robinson 1991) and nest site

---

\* Corresponding Author: bees1965@sina.com(ZJZ).
注：此文发表在 *African Entomology* 2018 年第 1 期。

scouting (Robinson and Page 1989), plant choice for pollen collection (Oldroyd et al. 1992), foraging distance (Oldroyd et al. 1993), queen rearing (Robinson et al. 1994), oophagy, oviposition and larval caring in queenless colonies (Robinson et al. 1990; Page and Robinson 1994), egg-laying in queen right colonies (Visscher 1996), water collecting and scenting (Kryger et al. 2000), fanning (Su et al. 2007), emergency queen-cell building (Xie et al. 2008), mite (Varroa destructor) parasitism rate (Liu et al. 2009) and survival differences (Wang et al. 2012) to name but a few. However, in all these studies only two or three patrilines could be distinguished, due to the low variability of allozymes in honeybees. Furthermore, the use of lines selected for specific allozyme markers may have an effect on the division of labour (Harrison et al. 1996).

The semiochemicals released by a honeybee queen have many effects for the colony (Winston and Slessor 1998). The most obvious effect is to attract workers around her, which are named as queen retinue workers (There is a circle consisting of queen retinue workers in Fig. S1). The age-bracket of honeybee queen retinue workers is 2-23 d, but mainly in 6-18 d (Yi et al. 2016). Queen retinue workers feed the queen or groom the queen with their antennae or proboscis, and then transmit the queen pheromone messages to their other nestmates throughout the colony (Keeling et al. 2003). It showed that there are three transmitting modes (queen bee, worker bee and air) of queen bee pheromones existing in honeybee colony. And only when three kind of modes exist in colony, queen bee can play a role of governing in the honeybee colony and make honeybee colony under a good discipline (Qi et al. 2008). Thus, it is important for us to learn about the task of queen retinue.

Previous studies indicate that the queen's mandibular gland pheromone (QMP) is the most important component of queen-produced compounds. And QMP is the main substance that inhibits the development of worker honeybee ovaries and attracts honeybee queen retinue workers (Hoover et al. 2003; Xuan and Chen 2005). Thus, all workers may have the willing to be a member of queen retinue workers in theory. That is, workers from all partrilines may be willing to become queen retinue workers. In this study, we are intended to understand whether there is a genetic variability in the task of queen retinue. 94 queen retinue workers and 94 workers that collected randomly (we named them as 'control workers') were sampled from two colonies respectively. After the procedure of DNA extraction, PCR amplification reaction and Microsatellite DNA analyses, Fisher exact test was used for assessing the genetic variability across these groups.

## Materials and Methods

Two honeybee (*Apis mellifera*) colonies (colony A and colony B) were used for this study and maintained at the Honeybee Research Institute of Jiangxi Agricultural University in Nanchang, China (28.46 °N, 115.49 °E).

Neilsen and Tarpy (2003) simulations suggest that 94 is sufficient sample size. We therefore sampled 94 queen retinue workers from each colony. Before sampling, we need to find the circle consist of queen retinue workers first. The circle consisting of queen retinue workers was usually formed when the queen was laying eggs, standing still or being fed (Yi et al. 2016). We then used a pair of tweezers to catch the retinue workers when the queen stops walking with the surrounded retinue workers. The queen retinue workers can be defined as: 1) following the queen to walk and become queen retinue workers soon when the queen stops walking, 2) tapping the queen's body fast with their antennae or 3) licking the queen with the proboscis. The samples were snap frozen in liquid nitrogen and preserved in a refrigerator of -80°C for DNA extraction. When sampling, catching all queen retinue workers was too difficult to be achieved. On the one hand, the sampling action may break the balance between the queen and the queen retinue workers. On the other hand, when we caught a queen retinue worker, the one we caught may release alarm pheromone to remind or warn the queen and other workers about escaping (Pankiw 2004). Hence, We could only catch 1-3 queen retinue workers from each circle of queen retinue workers at a time.

Additionally, we collected 94 workers (named as 'control workers') randomly from each colony to represent the background subfamily proportion. The total number of collected the 'control workers' are equal with the queen retinue workers each day. Furthermore, we used the tweezers to catch the 'control workers' randomly from the same hive of the queen retinue workers.

After the sampling, a genomic DNA extraction kit (TaKaRa MiniBEST Universal Genomic DNA Extraction Kit) was used to to extract DNA of each sample. Then, four microsatellite loci (Peter et al. 1999) were selected to determine the subfamily using Matesoft (Moilanen et al. 2004) (Table III-14-1). The four microsatellite loci (Ap289, Ap043, A113 and Ap226) was based on the reports of Tian et al. (2013) and Zhang et al. (2015). The allele of each microsatellite marker was analyzed with QIAxcel Advanced system (Shanghai Konecranes Co. Ltd. products). The statistical evaluation of the distribution of each patriline between the queen retinue workers and 'control workers' was carried out with a Fisher exact test (Raymond and Rousset 1995). $P$ values were calculated by comparing the distribution of each patriline against all other patrilines in each of the two honeybee colonies respectively with Fisher exact tests (Kryger et al. 2000).

## Results

Fig. III-14-1 shows that there are 13 patrilines in colony A and 12 patrilines in colony B (The genotypes of paternity of each drone tissue were shown in Table S1). In addition, there are 4 patrilines (nos 6, 8, 11 and 12) in colony A and 6 patrilines (nos 4, 8, 9, 10, 11 and 12) in colony B which could not find workers from the group of queen retinue workers (Nos number stands for the number of patrilines in each colony). By comparing the subfamily distribution between queen retinue workers and 'control workers', a significant difference was found in both colonies (colony A, $p < 0.05$; colony B, $p < 0.01$).

**Table III-14-1  Core sequences in cloned alleles and part of PCR condition for 4 used microsatellites**

| Locus | Sequence of primers | Size (bp) | Annealing Temperature (°C) | No. Cycle |
|---|---|---|---|---|
| AP289 | F 5'-AGCTAGGTCTTTCTAAGAGTGTTG-3'<br>R 5'-TTCGACCGCAATAACATTC-3' | 174 | 55 | 30 |
| AP043 | F 5'-GGCGTGCACAGCTTATTCC-3'<br>R 5'-CGAAGGTGGTTTCAGGCC-3' | 137 | 58 | 30 |
| A113 | F 5'-CTCGAATCGTGGCGTCC-3'<br>R 5'-CCTGTATTTTGCAACCT CGC-3' | 220 | 60 | 30 |
| AP226 | F 5'-AACGGTGTTCGCGAAACG-3'<br>R 5'-AGCCAACTCGTGCGGTCA-3' | 231 | 58 | 30 |

Furthermore, Fisher exact tests were carried out for the distribution of each patriline against all other patrilines on the two categorized groups in both colony A and B (Table III-14-2, last column). In colony A, two patrilines (nos 2 and 3) were significant at $p<0.05$. In colony B, two patrilines (nos 4 and 6) were significant at $p<0.01$ and one patriline (no 2) at $p<0.001$. For eleven patrilines in colony A and nine patrilines in colony B no significant differences were detected. However, some of these patrilines (colony A: nos 8, 11 and 12; colony B: nos 8. 9. 10. 11 and 12) had very low frequencies which reduces the power of the Fisher exact test. In multiple comparisons, some results are by chance expected to be significant, but clearly the number of significant results and the levels of significance found here exceed such stochastic

expectations. In summary, we find evidence in favour of the hypothesis that there is a genotypical component for the performance of the different tasks examined in a colony headed by a naturally mated queen.

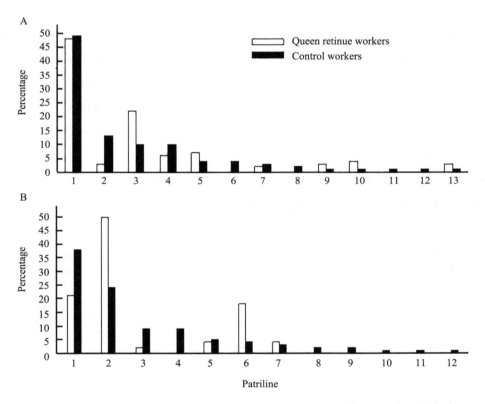

**Figure III-14-1** Percentages of each of the 13 patrilines in colony A and 12 patrilines in colony B in the group of queen retinue workers and the group of 'control workers'. The number of workers of each subfamilies was listed from the highest to the lowest in the group of 'control workers', and the group of queen retinue workers shared the same serial number of the subfamily in both colonies. In addition, it showed that 4 patrilines (nos 6, 8, 11 and 12) in colony A and 6 patrilines (nos 4, 8, 9, 10, 11 and 12) in colony B were not found bee in the group of queen retinue workers.

## Discussion

Queen retinue workers play an important role in feeding their queen and transferring queen pheromones for helping the queen to govern the whole honeybee colony under a good discipline (Qi *et al*. 2008). Our previous study showed that queen retinue workers have a age-bracket of day 6-18 (Yi *et al*. 2016). This study showed that the subfamily distribution of queen retinue workers was significantly different from the control workers (Fig. III-14-1). Furthermore, two (nos 3 and 4) and three patriline (nos 2, 4 and 6) were strongly overrepresented or underrepresented against control workers in colony A and B respectively (Table III-14-2). These findings suggest that there might be a genotypical component for the performance of queen retinue, not only being affected by worker ages. Accumulating evidences indicate that honeybee patrilines are strongly related to their division of labor and some special behavioural traits, such as guarding and undertaking (Robinson and Page 1988), nectar and pollen foraging (Page and Robinson 1991) and even queen rearing (Robinson et al.1994). Consequently, these studies potentially support our hypothesis that queen retinue task may also require a special worker caste in terms of genetic variability.

There are two transmitting modes for workers to obtain the information that there is a queen around her: by airborne transmission or toughing with queens' body (Huang 1994). Some queen retinue pheromones such as 9-ODA and 9-HDA are volatile chemicals (Zhang et al. 2008). Workers who around the queen might be able to get this message through the airborne transmission or by body tough and then follow to the queen. In our observation, only a few workers followed to their queens whereas most workers did not followed to their queens even toughing with their queen's body. Perhaps there is a selection of workers who are able to be a queen retinue worker after stimulated by queen pheromones, and some subfamilies might be more willing to do this task of becoming a queen retinue worker.

Table III-14-2 The number of workers between two groups in each patriline. The last column give the *p* values for Fisher exact tests for the distribution of each patriline against all other patrilines in the two groups

| Subfamilies | Colony A | | | Colony B | | |
| --- | --- | --- | --- | --- | --- | --- |
| | The number of queen retinue workers | The number of control workers | P | The number of queen retinue workers | The number of control workers | P |
| 1 | 45 | 46 | 1.000 | 20 | 36 | 0.160 |
| 2 | 3 | 12 | 0.028 | 47 | 23 | 0.000 |
| 3 | 21 | 9 | 0.027 | 2 | 8 | 0.100 |
| 4 | 6 | 9 | 0.592 | 0 | 8 | 0.007 |
| 5 | 7 | 4 | 0.536 | 4 | 5 | 0.744 |
| 6 | 0 | 4 | 0.121 | 17 | 4 | 0.004 |
| 7 | 2 | 3 | 1.000 | 4 | 3 | 1.000 |
| 8 | 0 | 2 | 0.497 | 0 | 2 | 0.497 |
| 9 | 3 | 1 | 0.621 | 0 | 2 | 0.497 |
| 10 | 5 | 1 | 0.368 | 0 | 1 | 1.000 |
| 11 | 0 | 1 | 1.000 | 0 | 1 | 1.000 |
| 12 | 0 | 1 | 1.000 | 0 | 1 | 1.000 |
| 13 | 7 | 1 | 0.621 | | | |

In addition, Robinson *et al.* (1994) showed that there were significant genotypic biases in the relative likelihood of rearing queens. Queen rearing workers are a part of queen retinue workers sometimes (Butler and Fairey 1964, Simpson 1966). Yi *et al.* (2016) showed that worker ages also affect the the queen retinue behaviour. These findings, and our results, are consistent with the interpretation that the ontogeny of queen retinue behavior might be affected by the stimulation of queen bee pheromones, , worker age and genotype, or by a combination of these three factors. Further investigations should focus on the mechanism of how workers response to the queen retinue demands and how their patriline affects their behaviour.

## Acknowledgments

We thank Dr. Qiang Huang, Dr. Cui Guan, Dr. XuJiang He and Ms. Susie hewlett for revising the paper, and thank Mr. Jing-Hua Hu, Wu-jun Jiang and Wan-Wan Hu for helping the experiment. This work was supported by the Earmarked Fund for the China Agriculture Research System (No.CARS-45-KXJ12) and the National Natural Science Foundation of China (No.31572469).

## Compliance with Ethical Standards

The authors declare that they have no conflict of interest. All procedures performed in studies involving human participants were in accordance with the ethical standards of the institutional and/or national research committee and with the 1964 Helsinki declaration and its later amendments or comparable ethical standards. All applicable international, national, and/or institutional guidelines for the care and use of animals were followed. All procedures performed in studies involving animals were in accordance with the ethical standards of the institution or practice at which the studies were conducted. Additional informed consent was obtained from all individual participants for whom identifying information is included in this article.

## References

BUTLER, C.G. & FAIREY, E.M. 1964. Pheromones of the honeybee: biological studies of the mandibular gland secredtion of the queen. *Journal of Apicultural Research* 5:65-76.

FRUMHOFF, P.C. & BANKER, J. 1988. A genetic component to division of labour within honey bee colonies. *Nature* 333: 359-361.

HARRISON, J.F., NIELSEN, D.I. & PAGE, R.E. 1996. Malate dehydrogenase phenotype, temperature and colony effects on fight metabolic rate in the honey bee *Apis mellifera*. *Functional Ecology* 10: 1-8.

HE, X.J., TIAN, L.Q., WU, X.B. & ZENG, Z.J. 2016. RFID monitoring indicates honeybees work harder before a rainy day. *Insect Science* 23:157-159.

HOOVER, S.E.R., KELING, C.I., WINSTON, M.L. & SLESSOR, K.N. 2003. The effect of queen pheromones on worker honey bee ovary development. *Naturwissenschaften* 90: 477-480.

HUANG, W.C. 1994. Honeybee pheromones. *Apiculture of China* 1:30-32.

KEELING, C.I., SLESSOR, K.N., HIGO, H.A. & WINSTON, M.L. 2003. New components of the honey bee (*Apis mellifera* L.) queen retinue pheromone. *Biochemistry* 100: 4486-4491.

KRYGER, P., KRYGER, U. & MORITZ, R.F.A. 2000. Genotypical variability for the tasks of water collecting and scenting in a honey bee colony. *Ethology* 106: 769-779.

LINDAUER, M. 1952. Ein Beitrag zur Frage der Arbeitsteilung im Bienenstaat. *Zeitschrift für vergleichende Physiologie* 34: 299-345.

LINDAUER, M. 1954. Temperaturregulierung und Wasserhaushalt im Bienenstaat. *Zeitschrift für vergleichende Physiologie* 36: 391-432.

LIU, Y.B., HUANG, Q. & ZENG, Z.J. 2009. Study on the Selective Parasitism of the Varroa Mite to the Different Subfamilies by ISSR. *Journal of Bee* 8: 6-9.

MOILANEN, A., SUNDSTROEM, L. & PEDERSEN, J.S. 2004. Matesoft: a program for deducing parental genotypes and estimating mating system statistics in haplodiploid species. *Molecular Ecology Notes* 4: 795-797.

NEILSEN, R., TARPY, D.R. & REEVE, H.K. 2003. Estimating effective paternity number in social insects and the effective number of alleles in a population. *Molecular Ecology* 12: 3157-3164.

OLDROYD, B.P., RINDERER, T.E. & BUCO S.M. 1992. Intra-colonial foraging specialism by honey bees (*Apis mellifera*) (Hymenoptera: apidae). *Behavioral Ecology Sociobiology* 30: 291-295

OLDROYD, B.P., RINDERER, T.E., BUCO, S.M. & BEAMAN, L. 1993. Genetic variance in honey bees for preferred foraging distance. *Animal Behaviour* 45: 323-332.

PAGE, R.E. & ROBINSON, G.E. 1991. The genetics of division of labour in honey bee colonies. *Advances in Insect Physiology* 23: 117-169.

PAGE, R.E. & ROBINSON, G.E. 1994. Reproductive competition in queenless honeybee colonies (*Apis mellifera* L.). *Behavioral Ecology and Sociobiology* 35: 99-107.

PANKIW, T. 2004. Cued in: honey bee pheromones as information flow and collective decision-making. *Apidologie* 35: 217-226.

PETER, N., ROBIN, F.A.M & JOB, V.P. 1999. Queen mating frequency in different types of honey bee mating apiaries. *Journal of Apicultural Research* 38:11-18.

QI, H.P., SHAO, Y.Q., GUO, Y., MA, W.H., LIU, Y.M. & ZHAO, J. 2008. Study on Queen Bee Pheromones transmitting mode in bee colony to worker bee ovarian development. *Apiculture of China* 59:11-12.

QIN, Q.H., WANG, Z.L., TIAN, L.Q., GAN, H.Y., ZHANG, S.W. & ZENG, Z.J. 2014. The integrative analysis of microRNA and mRNA expression in *Apis mellifera* following maze-based visual pattern learning. *Insect Science* 21: 619-636.

RAYMOND, M. & ROUSSET, F. 1995. An exact test for population differentiation. *Evolution* 38: 1280-1283.

ROBINSON, GE. & PAGE, R.E. 1988. Genetic determination of guarding and undertaking in honey-bee colonies. *Nature* 333: 356-358.

ROBINSON, G.E. & PAGE, R.E. 1989. Genetic determination of nectar foraging, pollen foraging, and nest-site scouting in honey bee colonies. *Behavioral Ecology and Sociobiology* 24: 317-323.

ROBINSON, GE., PAGE, R.E. & FONDRK, M.K. 1990. Intracolonial behavioural variation in worker oviposition, oophagy, and larval care in queenless honey bee colonies. *Behavioral Ecology and Sociobiology* 26: 315-323.

ROBINSON, G.E., PAGE, R.E. & ARENSEN, N. 1994. Genotypic differences in brood rearing in honey bee colonies:

context-specific? *Behavioral Ecology and Sociobiology* 34: 125-137.

ROBINSON, G.E., GROZINGER, C.M. & WHITFIELD, C.W. 2005. Sociogenomics: social life in molecular terms. *Nature Reviews Genetics* 6: 257-270.

SIMPSON, J. 1966. Repellency of the mandibular gland scent of worker honey bees. *Nature* 209: 531-532.

SU, S., ALBER, T.S., ZHANG, S., MAIER, S., CHEN, S., DU, H. & TAUTZ, J. 2007. Non-destructive genotyping and genetic variation of fanning in a honeybee colony. *Journal of Insect Physiology* 53: 411-417.

TIAN, L.Q., HE, X.J., WANG, H. & ZENG, Z.J. 2013. Analysis of Genetic Background between Two Types of nurses in Honeybee (*Apis mellifera*). *Journal of Bee* 10: 1-3.

VISSCHER, P.K. 1996. Reproductive conflict in honey bees: a stalemate of worker egg-laying and policing. *Behavioral Ecology and Sociobiology* 39: 237-244.

WANG, H., ZHANG, S.W., ZHANG, F. & ZENG, Z.J. 2012. Analysis of lifespans of workers from different subfamilies in a honeybee colony. *Entomological Knowledge* 49: 1172-1175.

WINSTON, M.L. & SLESSOR, K.N. 1998. Honey bee primer pheromones and colony organization: gaps in our knowledge. *Apidologie* 29: 81-95.

XIE, X.B., SUN, L.X., HUANG, K. & ZENG, Z.J. 2008. Worker nepotism during emergency queen rearing in Chinese honeybees Apis cerana cerana. *Acta Zoologica Sinica* 54: 695-700.

XUAN, H.Z. & CHEN, K. 2005. The Effect of Queen Pheromones on Worked Bee Ovary Development. *Special Wild Economic Animal and Plant Research* 1: 63-64.

YI, Y., LI, Y. & ZENG, Z.J. 2016. Age components of queen retinue workers in honeybee colony (*Apis mellifera*). *Sociobiology* 63 (2): 848-850.

ZENG, Z.J. 2007. The biology of the honeybee. BeiJing, China agriculture press.

ZHANG, G.S., Qiu, B., CHEN, Z.X. & LI, Y. 2008. Synthesis of Queen Honeybee Pheromones Ones. *Chinese Journal of Applied Chemistry* 25 (7): 871-873.

ZHANG, L.Z., YUAN, A., JIANG, W.J. & ZENG, Z.J. 2015. Analysis of the genetic background of worker bees with different olfactory learning ability. *Chinese Journal of Applied Entomology* 52: 1421-1428.

## 15. 蜂王上颚腺信息素对雄蜂候选性信息素受体基因表达的影响

刘俊峰[1,2]，李茫[1]，王子龙[1]，何旭江[1]，曾志将[1,*]

（1. 江西农业大学蜜蜂研究所，南昌 330045；2. 中国热带农业科学院环境与植物保护研究所，海口 571101）

**摘要**：性信息素受体（sex pheromone receptors, PRs）是雄蜂感受蜂王上颚腺信息素（queen mandibular pheromone, QMP）的重要受体。本研究分析中华蜜蜂 *Apis cerana cerana*（简称"中蜂"）和意大利蜜蜂 *Apis mellifera ligustica*（简称"意蜂"）雄蜂触角和大脑中候选性信息素受体基因 *Prs* 受 QMP 刺激下的表达特征，为探索蜜蜂气味受体（OR）基因的功能研究提供理论依据。利用 qRT-PCR 技术检测分析分别用 10 μL QMP[7.04 μg/μL 反式-9-氧代-2-癸烯酸（9-ODA）+1.26 μg/μL 9-羟基-2-癸烯酸（9-HDA）+0.03 μg/μL 对羟基苯甲酸甲酯（HOB）]和 10 μL 7.04 μg/μL 9-ODA 处理对飞行状态和爬行状态下的中蜂雄蜂和意蜂雄蜂触角和大脑中 4 个气味受体基因（*Or*10, *Or*11, *Or*18 和 *Or*170）的 mRNA 表达量的影响。与空白对照组相比，QMP 及 9-ODA 均能显著下调中蜂雄蜂和意蜂雄蜂触角与大脑中 *Or*11 的 mRNA 表达量；中蜂雄蜂触角中 *AcOr*18 和 *AcOr*170 基因受 QMP 及 9-ODA 刺激后 mRNA 表达量显著下调；QMP 与 9-ODA 均能显著降低中蜂雄蜂和意蜂雄蜂大脑中 *Or*170 的 mRNA 表达量。在 QMP 或 9-ODA 刺激下，飞行中蜂雄蜂大脑中 *AcOr*11 的 mRNA 表达量显著高于爬行中蜂雄蜂大脑中的，而飞行与爬行中蜂雄蜂触角中 *AcOr*11 的 mRNA 表达量没有显著差异（$P>0.05$）。中蜂和意蜂雄蜂触角与大脑中气味受体基因 *Or*11 均能应答蜂王上颚腺信息素 9-ODA，且受 9-ODA 刺激后 *Or*11 的 mRNA 表达下调。

**关键词**：中华蜜蜂；意大利蜜蜂；雄蜂；蜂王上颚腺信息素；气味受体

## Effects of Queen Mandibular Pheromone on the Expression of Candidate Sex Pheromone Receptor Genes in Honeybee Drones

Liu Junfeng[1,2], Li Mang[1], Wang Zilong[1], He Xujiang[1], Zeng Zhijiang[1,*]

(1. Honeybee Research Institute, Jiangxi Agricultural University, Nanchang 330045, China; 2. Environment and Plant Protection Institute, Chinese Academy of Tropical Agricultural Sciences, Haikou 571101, China)

**Abstract**

Sex pheromone receptors (PRs) are important receptors for the perception of queen mandibular pheromone

---

* 通讯作者：bees1965@sina.com（ZJZ）.
注：此文发表在《昆虫学报》2020 年第 6 期。

(QMP) in honeybee drones. The aim of this study is to analyze the expression characteristics of candidate sex pheromone receptor genes in the antennae and brains of *Apis cerana cerana* and *Apis mellifera ligustica* drones subjected to QMP stimulation, so as to provide the theoretical basis for further studies on the biological function of odorant receptor (OR) genes in honeybees. qRT-PCR was employed to detect and analyze the mRNA expression levels of four OR genes including *Or*10, *Or*11, *Or*18 and *Or*170 in the antennae and brains of flying and crawling drones of *A. c. cerana* and *A. m. ligustica* exposed to 10 μL QMP [7.04 μg/μL (*E*)-9-oxodec-2-enoic acid (9-ODA)+1.26 μg/μL 9-hydroxydec-2-enoic acid (9-HDA)+0.03 μg/μL methylp-hydroxybenzoate (HOB)] and 10 μL 7.04 μg/μL 9-ODA, respectively. The results showed that QMP and 9-ODA significantly inhibited the transcription of *Or*11 in the antennae and brains of *A. c. cerana* and *A. m. ligustica* drones compared to the blank control group. The mRNA expression levels of *AcOr*18 and *AcOr*170 in the antennae of *A. c. cerana* drones were down-regulated significantly by QMP and 9-ODA ($P<0.05$). Moreover, QMP and 9-ODA significantly decreased the mRNA expression levels of *Or*170 in the brains of drones of the two bees. After *A. c. cerana* drones were subjected to the stimulation of QMP or 9-ODA, the mRNA expression level of *AcOr*11 was significantly higher in the brain of flying drones than in the brain of crawling drones, whereas showed no significant difference in the antennae between flying drones and crawling drones. *Or*11 can respond to 9-ODA stimulation in the antennae and brains of *A. c. cerana* and *A. m. ligustica* drones, and its expression is down-regulated after 9-ODA stimulation.

**Keywords:** *Apis cerana cerana*, *Apis mellifera ligustica*, drone, queen mandibular pheromone, odorant receptor

蜜蜂是研究行为的良好模式生物，其具有丰富的行为特征和可塑性。自从2006年蜜蜂基因组成功测序后（Honey Bee Genome Sequencing Consortium, 2006），蜜蜂就被作为连接基因与复杂行为的动物模型。蜜蜂交尾行为是其复杂行为之一。雄蜂唯一的生物学职责是与处女蜂王交配（曾志将，2017）。在空中追逐处女蜂王的飞行雄蜂与正常有王蜂群内的爬行雄蜂同样受到蜂王上颚腺信息素（queen mandibular pheromone, QMP）影响，但是它们的行为表现不同。在户外飞行的性成熟雄蜂能被处女蜂王或QMP混合物所吸引（Loper *et al*., 1993; Brockmann *et al*., 2006），而在蜂群内的性成熟雄蜂却回避处女蜂王（Ohtani and Fukuda, 1977; Graham, 2015）。目前，雄蜂识别QMP的生理调控机理尚不清晰。

雄蜂拥有高度特化和极其灵敏的嗅觉识别系统。雄蜂触角的主要嗅觉感器——板形感器约有18 000个，数量约是工蜂的7倍（Sandoz *et al*., 2007）。嗅觉板形感器中识别气味的功能性外周蛋白主要由气味结合蛋白（odorant binding proteins, OBPs）、气味降解酶（odorant degrading enzymes, ODEs）、感觉神经元膜蛋白（sensory neuron membrane proteins, SNMPs）和气味受体（odorant receptors, ORs）等组成（Vosshall *et al*., 1999; Hallem *et al*., 2004）。其中，ORs以数量多、特异性强的特征受到了学者们广泛的关注与研究（Forêt and Maleszka, 2006; Robertson and Wanner, 2006; Li Z *et al*., 2012; Li H *et al*., 2013; 赵慧婷等，2015; 张中印等，2016; 杜亚丽等，2017; 杨乐等，2017, 2018）。从东方蜜蜂 *Apis cerana* 和西方蜜蜂 *Apis mellifera* 中已鉴定出的 ORs 分别为 119 和 170 个（Robertson and Wanner, 2006; Park *et al*., 2015），远远超过黑腹果蝇 *Drosophila melanogaster* 的62个ORs（Clyne *et al*., 1999; Vosshall *et al*., 1999）、冈比亚按蚊 *Anopheles gambiae* 的79个ORs（Hill *et al*., 2002）及家蚕 *Bombyx mori* 的48个

ORs（Wanner et al., 2007a），这也表明蜜蜂的嗅觉更为发达。Wanner 等（2007b）研究发现候选性信息素受体（sex pheromone receptors, PRs）基因 Or10, Or11, Or18 和 Or170 在西方蜜蜂雄蜂触角中高度表达，并证实 Or11 为反式-9-氧代-2-癸烯酸 [（E）-9-oxodec-2-enoic acid, 9-ODA] 的专一性受体。但蜜蜂 ORs 分别发挥何种机能的研究鲜有报道。Liu 等（2019）利用 RT-PCR 技术对东方蜜蜂 AcOr11 基因克隆基础上，使用 qRT-PCR 技术鉴定其在飞行回巢与蜂箱内爬行状态下的表达特性，提示了雄蜂大脑中 Or11 基因表达可能与婚飞行为有关。虽然目前 Or10, Or18 和 Or170 基因暂未鉴定出其功能，但我们推测这 3 个气味受体基因与性信息素受体基因 Or11 功能相似，并且可能在雄蜂识别 QMP 与婚飞过程中发挥作用。因此，以中华蜜蜂 A. c. cerana（简称"中蜂"）和意大利蜜蜂 A. m. ligustica（简称"意蜂"）性成熟的飞行与爬行雄蜂为研究材料，通过 qRT-PCR 技术检测分析 QMP 与 9-ODA 对中蜂、意蜂雄蜂候选性信息素受体基因 Or10, Or11, Or18 和 Or170 的 mRNA 表达量影响，以探索雄蜂对 QMP 的生理调控机制，从而为揭示雄蜂婚飞行为提供理论依据。

## 1 材料与方法

### 1.1 供试蜜蜂

试验蜂群是由江西农业大学蜜蜂研究所按活框饲养技术进行饲养的中华蜜蜂、意大利蜜蜂。试验选取蜂群遗传性状、群势基本相近的 4 群中蜂和 4 群意蜂。试验时间为 2019 年 6～8 月。

培育适龄性成熟雄蜂：将 4 张中蜂、意蜂空雄蜂巢脾各插入 4 群中蜂、意蜂群内，控制蜂王产卵 24 h 后继续留置原群培育孵化，待雄蜂出房前 24 h 置于恒温恒湿培养箱至羽化出房，使用 4 种不同颜色水性记号笔分别标记 4 蜂群各 300 头雄蜂，并放回原蜂群发育成熟。

收集适龄性成熟雄蜂：为了符合雄蜂生物学习性，行为学试验时间选在天气晴朗下午 13:00-16:00 时雄蜂飞行高峰期间，收集性成熟雄蜂（14 d）。使用 50 mL 离心管分别在蜂箱巢门口抓取已标记颜色、飞行回巢的性成熟雄蜂（简称"飞行雄蜂"），以及在蜂箱内壁上抓取已标记颜色、巢内爬行的性成熟雄蜂（简称"爬行雄蜂"）。

### 1.2 主要仪器与试剂

主要试剂：9-ODA, 9-HDA, HOB（Contech, 加拿大）；TransZol up 试剂盒、GelStain 荧光核酸染色试剂和 6×DNA Loading Buffer（北京全式金公司）；无水乙醇、异丙醇和氯仿（西陇化工，分析纯）；UltraPure™ 琼脂糖、TBE Buffer 及 DNase/RNase-Free Distilled Water（Invitrogen）；PrimeScript RT Reagent Kit 反转录试剂盒、DNA Marker DL2000 和荧光定量试剂盒 TB Green™ Premix Ex Taq™（TaKaRa）。

### 1.3 信息素刺激雄蜂

本试验使用信息素 QMP 与 9-ODA 分别刺激中蜂和意蜂性成熟雄蜂。将收集待测 1.1 节中的中蜂和意蜂性成熟雄蜂置于测试室适应环境 5 min 左右。测试室环境为红光暗室，行为学测试时间：13:00-16:00 时，测试室内温度 25 ℃±1 ℃，相对湿度 75%±5%。测试信息素 QMP[9-ODA（7.04 μg/μL）+9-羟基-2-癸烯酸（9-hydroxydec-2-enoic acid, 9-HDA）（1.26 μg/μL）+对羟基苯甲酸甲酯（methylp-hydroxybenzoate, HOB, 0.03 μg/μL）] 和 9-ODA（7.04 μg/μL）气流由 Y 型嗅觉仪（基管长 30 cm，两臂长 35 cm，内径 5 cm，两臂夹角 75°，南昌大学玻璃仪器厂）样品瓶流向雄蜂释放管一端，雄蜂由飞行通道逆风飞入样品瓶。在气流进入样品瓶之前，由空气泵推动空气先经过活性炭过滤和蒸馏水加湿，空气泵流速为 300 mL/min。测试信息素 QMP 与 9-ODA 成分含量参考 Plettner 等（1997），即每头性成熟西方蜜蜂处女蜂王含约 70.4 μg 9-ODA, 12.6 μg 9-HDA, 0.3 μg HOB。

处理组在 Y 型嗅觉仪左右 2 个测试室中均放入滴有 5 μL 信息素乙醇溶液的滤纸；空白对照组在 Y 型嗅觉仪左右 2 个测试室中均放入滴有 5 μL 乙醇溶液的滤纸。每次试验仅测试 1 头中蜂雄蜂或意蜂雄蜂，将离心管置于 Y 型管基臂端释放雄蜂，信息素刺激 5 min 后抓取雄蜂装入 EP 管，迅速放入液氮中速冻，转至 -80℃ 冰箱保存，用于提取 RNA。测定时每处理分 4 组，每组 8 头重复。测试完一组试验后清洗 Y 型管，先使用蒸馏水清洗，然后 75% 乙醇擦拭，再使用蒸馏水冲洗，最后用电吹风吹干后进行下一组试验。

### 1.4 RNA 提取与 cDNA 合成

将 1.3 节处理的雄蜂从液氮中取出，使用干净镊子及刀片剥离头部 1 对完整触角，装入 1.5 mL RNase-free EP 管，每组各收集 8 头雄蜂的 8 对触角为 1 个待测样品，迅速放入液氮中速冻，再转至 -80℃ 冰箱保存，用于提取 RNA。

将剥离触角后剩余的雄蜂头部放于置于磷酸缓冲盐溶液（phosphate buffer saline, PBS）（含 137 mmol/L NaCl, 2.7 mmol/L KCl, 10 mmol/L $Na_2HPO_4$, 2 mmol/L $KH_2PO_4$, pH 值 7.4）中。使用干净镊子及刀片去除头部外壳和复眼组织，迅速装入 1.5 mL RNase-free EP 管，放入液氮中速冻，每组各收集 8 头雄蜂的 8 对脑部组织为 1 个待测样品，再转至 -80℃ 冰箱保存，用于提取 RNA。

将采集得到的触角及脑部待测样品，分别按照 TransZol up 试剂盒的操作说明提取 RNA，并用分光光度计测量、琼脂糖凝胶电泳检测每个样品的 RNA 浓度及质量。每个 RNA 样品取 1 μg 按照反转录试剂盒说明进行反转录，将反转录的 cDNA 放置 -20℃ 冰箱保存，用于 qRT-PCR 检测。

### 1.5 qRT-PCR 检测基因表达

根据 NCBI 数据库中查询及比对出中蜂气味受体基因 *AcOr*10（GenBank 登录号：MF693365），*AcOr*11（GenBank 登录号：MG793195），*AcOr*18（中蜂转录组中比对 *AmOr*18 序列），*AcOr*170（GenBank 登录号：KX264359）和意蜂气味受体基因 *AmOr*10（GenBank 登录号：NM_001242961.1），*AmOr*11（GenBank 登录号：NM_001242962.1），*AmOr*18（GenBank 登录号：XM_003250678.2），*AmOr*170（GenBank 登录号：NM_001242993.1）。分别以 *Ac-β-actin*（GenBank 登录号：HM_640276.1）和 *Am-β-actin*（GenBank 登录号：NM_001185146.1）作为中蜂、意蜂气味受体基因表达量检测的内参基因，使用 Primer Premier 5.0 软件设计 qRT-PCR 引物序列（引物由上海生工公司合成），引物序列如表 Ⅲ-15-1 所示。

qRT-PCR 反应体系（10 μL）：TB Green® Premix Ex Taq™ II（2×）5 μL, ROX II（50×）0.2 μL, 上下游引物（10 μmol/L）各 0.4 μL, 超纯灭菌水 3 μL, cDNA 1 μL, 混匀、离心后放入 qRT-PCR 仪中进行扩增。空白对照模板用超纯灭菌水代替。每个样本重复 3 次。

qRT-PCR 扩增条件：95℃ 5 min, 94℃ 2 min; 95℃ 10 s, 退火（温度见表 Ⅲ-15-1）15 s, 72℃ 15 s, 40 个循环；72℃ 10 min。为了建立产物熔解曲线，上述引物扩增反应结束之后继续从 72℃ 缓慢加热到 99℃（每 5 s 升高 1℃）。最后通过 Applied Biosystems™ 7500 qRT-PCR 仪（Thermofisher）将试验结果直接导出，得到内参基因与目的基因的 $Ct$ 值。

表 III-15-1 qRT-PCR 引物

| 引物 | 引物序列（5′–3′） | Tm（℃） |
|---|---|---|
| AcOr10-QF | GACGGCATCCATTGAGCAT | 57.7 |
| AcOr10-QR | CCAAGGAACCAAAGTAACCC | |
| AcOr11-QF | ATGTGCGGTTTGCTGAAGA | 58.9 |
| AcOr11-QR | CGAGAAGGTGCCAATGACG | |
| AcOr18-QF | TACTGCACGGGATCAGAATG | 54 |
| AcOr18-QR | TTGAGAACAATCGGCACTTG | |
| AcOr170-QF | CGTTTGAGAAAGGTTGTTTGG | 58.9 |
| AcOr170-QR | CGAGAAGGTGCCAATGACG | |
| Ac-β-actin-QF | GGCTCCGAAGAACATCC | 60 |
| Ac-β-actin-QR | TGCGAACACCGTCACCC | |
| AmOr10-QF | CCGCATCTGAACAGTATCGTG | 57 |
| AmOr10-QR | ATTCTCCTCCGTGGCTATCG | |
| AmOr11-QF | CTTTTACCGAACAACATGACAG | 54 |
| AmOr11-QR | TTATCTCGTAATTAGGTGTGG | |
| AmOr18-QF | TTTTATTACATCGCTTTGCC | 58 |
| AmOr18-QR | TCTTCCTTCCATCCACCA | |
| AmOr170-QF | CCAGTGTTGCCTCGCTC | 58 |
| AmOr170-QR | TTTCGTTATCTCACGCTCC | |
| Am-β-actin-QF | GGCTCCCGAAGAACATCC | 60 |
| Am-β-actin-QR | TGCGAAACACCGTCACCC | |

## 1.6 数据分析

采用 $2^{-\Delta\Delta Ct}$ 方法统计中蜂、意蜂各基因的相对表达量，使用 SPSS 17.0 软件中 one-way ANOVA 与 $t$ 检验方法比较分析，$P<0.05$ 表示差异显著。通过 GraphPad Prism 5.2 软件进行作图。

## 2 结果

### 2.1 蜂王上颚腺信息素对中蜂雄蜂触角中气味受体基因 mRNA 表达量的影响

结果表明，QMP 和 9-ODA 处理组的飞行（图Ⅲ-15-1 A）与爬行（图Ⅲ-15-1 B）中蜂雄蜂触角中 4 个气味受体基因的 mRNA 表达量均显著低于对照组的（$P<0.05$）；9-ODA 处理组中飞行中蜂雄蜂触角中 $AcOr18$ 和 $AcOr170$ 的 mRNA 表达量显著低于 QMP 处理组中的（$P<0.05$，图Ⅲ-15-1A）；9-ODA 处理组爬行中蜂雄蜂触角中 $AcOr11$ 与 $AcOr170$ 的 mRNA 表达量显著低于 QMP 处理组的（$P<0.05$，图Ⅲ-15-1B）。

由图Ⅲ-15-2 结果可知，在 QMP 或 9-ODA 处理下，爬行中蜂雄蜂触角中 4 个 $AcOrs$ 基因的表达量均与飞行中蜂雄蜂触角中的没有显著差异（$P>0.05$）。

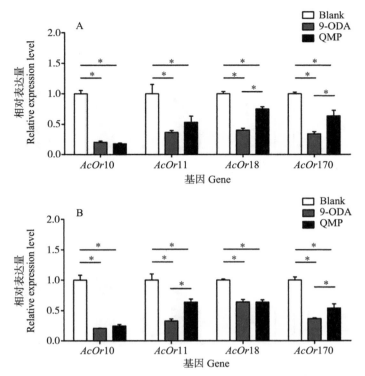

**图Ⅲ-15-1 QMP 和 9-ODA 处理后飞行（A）和爬行（B）中蜂雄蜂触角中 *AcOrs* 的表达量**

Blank: 空白对照（乙醇）；QMP: 蜂王上颚腺信息素，1 处女蜂王 QMP 当量包括反式 -9- 氧代 -2- 癸烯酸（9-ODA, 70.4 μg）+9- 羟基 -2- 癸烯酸（9-HDA, 12.6 μg）+ 对羟基苯甲酸甲酯（HOB, 0.3 μg），下图同。图中数据为平均值 ± 标准误；柱上星号表示组间基因表达量差异显著（$P<0.05$, one-way ANOVA）；图Ⅲ-15-3，图Ⅲ-15-5 和图Ⅲ-15-7 同。

图中数据为平均值 ± 标准误；柱上星号表示组间基因表达量差异显著（$P<0.05$, t 检验）；图Ⅲ-15-4，图Ⅲ-15-6 和图Ⅲ-15-8 同。

### 2.2 蜂王上颚腺信息素对中蜂雄蜂大脑中气味受体基因 mRNA 表达量的影响

结果表明，QMP 和 9-ODA 处理组的飞行中蜂雄蜂大脑中 *AcOr*11，*AcOr*18 和 *AcOr*170 基因的表达量均显著低于对照组的（$P<0.05$, 图Ⅲ-15-3A）；9-ODA 处理组飞行中蜂雄蜂大脑中 *AcOr*170 的表达量显著低于 QMP 处理组大脑中的（$P<0.05$, 图Ⅲ-15-3A）；而 QMP 和 9-ODA 处理组均与对照组飞行中蜂雄蜂大脑中的 *AcOr*10 基因表达量没有显著差异（$P>0.05$, 图Ⅲ-15-3A）。

由图Ⅲ-15-3B 结果可知，QMP 和 9-ODA 处理组爬行中蜂雄蜂大脑中 *AcOr*11 和 *AcOr*170 基因的表达量均显著低于对照组的（$P<0.05$）；QMP 与 9-ODA 处理组中蜂雄蜂大脑中 *AcOr*11 和 *AcOr*170 的表达量没有显著差异（$P>0.05$）；而 QMP 和 9-ODA 处理组与对照组的爬行中蜂雄蜂大脑中 *AcOr*10 及 *AcOr*18 基因表达量相比差异不显著（$P>0.05$）。

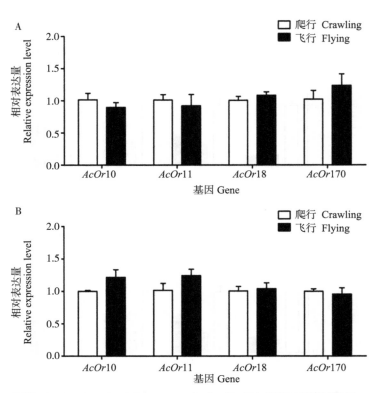

图Ⅲ-15-2 QMP（A）和 9-ODA（B）对飞行与爬行中蜂雄蜂触角中 *AcOr*s 表达量的影响

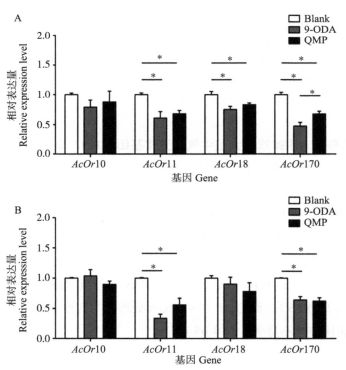

图Ⅲ-15-3 QMP 和 9-ODA 处理后飞行（A）和爬行（B）中蜂雄蜂大脑中 *AcOr*s 的表达量

由图Ⅲ-15-4可知，在QMP处理下，爬行中蜂雄蜂大脑中$AcOr11$和$AcOr170$基因的表达量均显著低于飞行中蜂雄蜂大脑中的（$P<0.05$，图Ⅲ-15-4A），而爬行与飞行中蜂雄蜂大脑中$AcOr10$和$AcOr18$基因的表达量没有显著差异（$P>0.05$，图Ⅲ-15-4A）；在9-ODA处理下，爬行中蜂雄蜂大脑中$AcOr11$基因的表达量均显著低于飞行中蜂雄蜂大脑中的（$P<0.05$，图Ⅲ-15-4B），爬行中蜂雄蜂大脑中$AcOr18$基因的表达量均显著高于飞行中蜂雄蜂大脑中的（$P<0.05$，图Ⅲ-15-4B），而爬行与飞行中蜂雄蜂大脑中$AcOr10$和$AcOr170$基因的表达量没有显著差异（$P>0.05$，图Ⅲ-15-4B）。

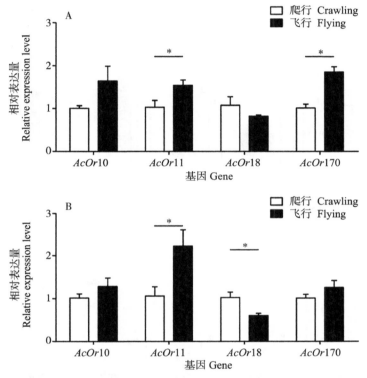

**图Ⅲ-15-4 QMP（A）和9-ODA（B）对飞行与爬行中蜂雄蜂大脑中 AcOrs 表达量的影响**

### 2.3 蜂王上颚腺信息素对意蜂雄蜂触角中气味受体基因 mRNA 表达量的影响

结果表明，QMP和9-ODA处理组飞行（图Ⅲ-15-5A）与爬行（图Ⅲ-15-5B）意蜂雄蜂触角中$AmOr10$和$AmOr11$基因的表达量均显著低于对照组的（$P<0.05$）；而处理组与对照组相比，其飞行（图Ⅲ-15-5A）与爬行（图Ⅲ-15-5B）意蜂雄蜂触角中$AmOr18$及$AmOr170$基因表达量均没有显著差异（$P>0.05$）。

由图6可知，在QMP或9-ODA处理下，爬行意蜂雄蜂触角中$AmOr170$基因的表达量均显著高于飞行意蜂雄蜂触角中的（$P<0.05$），而爬行与飞行意蜂雄蜂触角中$AmOr10$、$AmOr11$和$AmOr18$基因的表达量没有显著差异（$P>0.05$）。

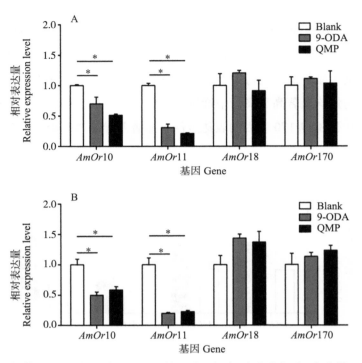

图Ⅲ-15-5 QMP 和 9-ODA 处理后飞行（A）和爬行（B）意蜂雄蜂触角中 *AmOrs* 的表达量

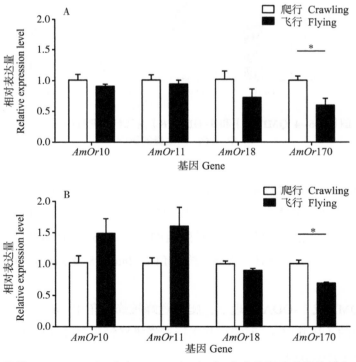

图Ⅲ-15-6 QMP（A）和 9-ODA（B）对飞行与爬行意蜂雄蜂触角中 *AmOrs* 表达量的影响

## 2.4 蜂王上颚腺信息素对意蜂雄蜂大脑中气味受体基因 mRNA 表达量的影响

由图Ⅲ-15-7A 可知，QMP 与 9-ODA 处理组飞行意蜂雄蜂大脑中 *AmOr*11, *AmOr*18 及 *AmOr*170 基因的表达量均显著低于对照组的（$P<0.05$），而处理组与对照组相比，其飞行意蜂雄蜂大脑中 *AmOr*10 基因的表达量没有显著差异（$P>0.05$）；QMP 处理组意蜂雄蜂大脑中 *AmOr*18 的表达量显著低于 9-ODA 处理组意蜂雄蜂大脑中的（$P<0.05$）。由图Ⅲ-15-7B 可知，QMP 与 9-ODA 处理组爬行意蜂雄蜂大脑中 *AmOr*11 及 *AmOr*170 基因的表达量均显著低于对照组雄蜂大脑中的（$P<0.05$），而处理组与对照组相比，其爬行意蜂雄蜂大脑中 *AmOr*10 和 *AmOr*18 基因的表达量没有显著差异（$P>0.05$）。

由图Ⅲ-15-8 可知，在 QMP 处理下，爬行意蜂雄蜂大脑中 *AmOr*10 和 *AmOr*18 基因的表达量均显著高于飞行意蜂雄蜂大脑中的（$P<0.05$，图Ⅲ-15-8A），而爬行与飞行意蜂雄蜂大脑中 *AmOr*11 和 *AmOr*170 基因的表达量没有显著差异（$P>0.05$，图Ⅲ-15-8A）；在 9-ODA 处理下，爬行与飞行意蜂雄蜂大脑中 4 个 *AmOr*s 基因的表达量没有显著差异（$P>0.05$，图Ⅲ-15-8B）。

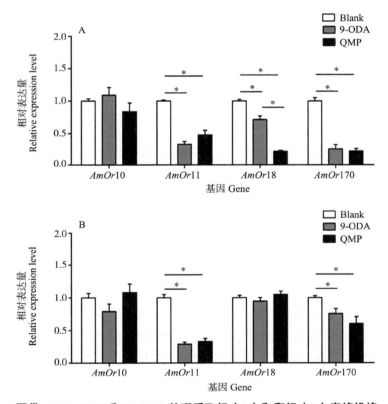

图Ⅲ-15-7 QMP 和 9-ODA 处理后飞行（A）和爬行（B）意蜂雄蜂大脑中 *AmOr*s 的表达量

## 3 讨论

性信息素受体在昆虫两性交流中发挥重要作用，雄虫能长距离地识别性成熟雌虫释放的性信息素，来寻找配偶并与之交配（Brockmann *et al.*, 2006; Becher *et al.*, 2010; Andersson *et al.*, 2014, 2016）。候选性信息素受体基因 *Or*10, *Or*11, *Or*18 和 *Or*170 在蜜蜂雄蜂触角中高表达，其中 *Or*11 特异性识别 9-ODA（Wanner *et al.*, 2007b）。

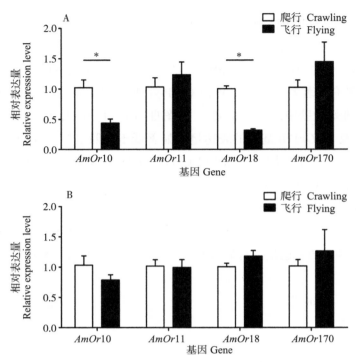

图Ⅲ-15-8 QMP（A）和9-ODA（B）对飞行与爬行意蜂雄蜂大脑中 *AmOrs* 表达量的影响

本研究发现 QMP 和 9-ODA 均能抑制 *Or*11 基因在中蜂雄蜂与意蜂雄蜂触角中的转录表达（图Ⅲ-15-1，图Ⅲ-15-3，图Ⅲ-15-5 和图Ⅲ-15-7）。这与斜纹夜蛾 *Spodoptera litura* 受性信息素处理后与未处理相比，其气味受体基因 *Or*3，*Or*13 和 *Or*54 表达下调结果（林欣大等，2015）相似。蜜蜂气味受体与学习记忆有着密切的关系。训练工蜂对芳樟醇（linalool）和 9-ODA 进行喙伸反应（proboscis extension reflex, PER）嗅觉学习后，其对应的气味受体基因 *Or*151 与 *Or*11 转录水平均下调，且触角电位（electroantennography, EAG）反应也随之降低（Claudianos et al., 2014）。这一研究报道与本试验结果一致。蜜蜂对食物气味的联想记忆能力非常突出（Menzel and Muller, 1996; Giurfa, 2007），微弱熟悉的气味吹入蜂巢内足以触发蜜蜂相关的视觉和嗅觉刺激的跨模式回忆，引诱蜜蜂搜寻食物来源（Reinhard et al., 2004a, 2004b, 2006）。蜜蜂为了适应外界不断变化的气味环境，在成功学习了某种气味后，会相应地减少该气味受体表达，并上调其他气味受体表达以确保能继续灵敏地察觉、识别新的气味；在食物源消失或不再加强气味记忆时，蜜蜂气味受体的表达水平也将波动（Menzel and Muller, 1996）。本研究发现，中蜂雄蜂与意蜂雄蜂脑部组织中 *Or*11 基因的表达量也在 9-ODA 刺激后降低，这可能是 Or11 识别 9-ODA 后在蜜蜂大脑神经传导中继续发挥重要作用。昆虫触角中气味受体特异性结合在气味受体神经元（olfaction receptor neurons, ORNs）的细胞膜上，ORNs 的功能取决于 ORs 的功能，相似的 ORNs 的轴突会聚集到昆虫中脑部嗅觉神经中枢——触角叶（antennal lobe, AL）内一个或几个定型的嗅神经球（glomeruli）上（Strausfeld, 2009），其中雄蜂 AL 中宏嗅神经球（macroglomeruli 2, MG2）特异性应答 9-ODA（Sandoz, 2006），最终 MG2 通过投射神经元（projection neurons, PNs）将嗅觉信息传递至中枢神经系统——蘑菇体（mushroom body, MB）产生指令后做出相应的行为反应（Hansson and Stensmyr, 2011）。*Or*11 具体在雄蜂脑部哪些神经组织中表达分布仍不清楚，需要进一步研究探索。

值得注意的是，在 QMP 或 9-ODA 刺激下，飞行中蜂雄蜂大脑中 *AcOr*11 基因的表达量显著高于爬行中蜂雄蜂大脑中的（图Ⅲ-15-4），而飞行与爬行中蜂雄蜂触角中 *AcOr*11 基因的表达没有显著差异（图Ⅲ-15-2）。这与性成熟时期中蜂雄蜂触角与大脑中 *AcOr*11 基因表达特性的结果（Liu et al., 2019）相似。我们推测，中蜂雄蜂脑组织中的气味受体 AcOr11 参与交配行为的调节。雄蜂在蜂群内、外同样能感知 QMP，而其反应却截然不同。性成熟雄蜂在蜂群内不能与处女蜂王交配，雄蜂更多地集中在蜜粉脾、蜂箱内壁及底部等远离蜂王的位置（Ohtani and Fukuda, 1977），而户外飞行中的雄蜂能被处女蜂王或人工合成的 QMP 与 9-ODA 引诱追逐（Brockmann et al., 2006）。黑腹果蝇 *D. melanogaster* 通过调整嗅觉神经元性信息素受体的表达，以提高其对性信息素反应的灵敏性（Graham, 2015）。交配行为能抑制雄性小地老虎 *Agrotis ipsilon* 对性信息素的嗅觉神经生理反应（Gadenne et al., 2001）。

本研究还发现，中蜂雄蜂触角 *AcOr*18 和 *AcOr*170 基因受 QMP 或 9-ODA 刺激后表达显著下调（图Ⅲ-15-1），而意蜂雄蜂触角 *AmOr*18 和 *AmOr*170 基因的 mRNA 表达量不受 QMP 影响（图Ⅲ-15-5）。这可能是由于 *Or*18 与 *Or*170 在中蜂、意蜂雄蜂触角中分布水平存在差异，致使该受体基因表达量也不同。蜜蜂感受蜂王信息素的嗅觉感器主要为板形感受器（赵慧婷等，2019），其在意蜂雄蜂触角上的分布数量约为中蜂雄蜂的 2.5 倍（宋飞飞，2011）。因此，我们推测中蜂雄蜂通过板形感受器表达出更多气味受体以应答性信息素，而意蜂雄蜂进化出更多的板形感受器来应对其对蜂王信息素的识别。此外 QMP 刺激后中蜂、意蜂在大脑中 *Or*170 表达特性与 *Or*11 一致，可能提示 Or170 也参与大脑组织对 QMP 识别过程。对 Or170 如何识别具体的蜂王性信息素成分或参与调控哪些气味受体，还需要进一步研究。

本研究通过对中蜂、意蜂雄蜂候选性信息素受体基因表达特性的研究，为探究雄蜂嗅觉识别机制以及婚飞行为研究提供了理论基础。本研究原计划取在空中追逐处女蜂王的飞行的性成熟雄蜂，由于很难寻找到处女蜂王交配的雄蜂聚集区，也尝试过用遥控无人机在空中取飞行雄蜂样品，都没有成功。后来是在巢门中取标记飞行回巢的 14 日龄性成熟雄蜂，但这种性成熟雄蜂与空中追逐处女蜂王的飞行雄蜂的生理以及信息素受体基因表达差异，目前没有报道，还有待于以后深入对比分析。

## 参考文献

ANDERSSON MN, CORCORAN JA, ZHANG DD, HILLBUR Y, NEWCOMB RD, LÖFSTEDT C, 2016. A sex pheromone receptor in the Hessian fly Mayetiola destructor (*Diptera, Cecidomyiidae*). *Front. Cell. Neurosci.*, 10: 212.

ANDERSSON MN, VIDEVALL E, WALDEN KKO, HARRIS MO, ROBERTSON HM, LÖFSTEDT C, 2014. Sex- and tissue-specific profiles of chemosensory gene expression in a herbivorous gall-inducing fly (*Diptera: Cecidomyiidae*). *BMC Genomics*, 15(1): 501.

BECHER PG, BENGTSSON M, HANSSON BS, WITZGALL P, 2010. Flying the fly: long-range flight behavior of Drosophila melanogaster to attractive odors. *J. Chem. Ecol.*, 36(6): 599-607.

BROCKMANN A, DIETZ D, SPAETHE J, TAUTZ J, 2006. Beyond 9-ODA: sex pheromone communication in the European honey bee Apis mellifera L. *J. Chem. Ecol.*, 32(3): 657-667.

CLAUDIANOS C, LIM J, YOUNG M, YAN S, CRISTINO AS, NEWCOMB RD, GUNASEKARAN N, REINHARD J, 2014. Odor memories regulate olfactory receptor expression in the sensory periphery. *Eur. J. Neurosci.*, 39(10): 1642-1654.

CLYNE PJ, WARR CG, FREEMAN MR, LESSING D, KIM J, CARLSON JR, 1999. A novel family of divergent seven-transmembrane proteins. *Neuron*, 22(2): 327–338.

杜亚丽, 王树杰, 赵慧婷, 潘建芳, 杨爽, 郭丽娜, 徐凯, 姜玉锁, 2017. 中华蜜蜂气味受体基因 AcerOr113 的克隆与时空表达分析. 昆虫学报, 60(5): 533–543.

FORÊT S, MALESZKA R, 2006. Function and evolution of a gene family encoding odorant binding-like proteins in a social insect, the honey bee (*Apis mellifera*). *Genome Res.*, 16(11): 1404–1413.

GADENNE C, DUFOUR MC, ANTON S, 2001. Transient post-mating inhibition of behavioural and central nervous responses to sex pheromone in an insect. *Proc. R. Soc. B Biol. Sci.*, 268(1476): 1631–1635.

GIURFA M, 2007. Behavioral and neural analysis of associative learning in the honeybee: a taste from the magic well. *J. Comp. Physiol. A Neuroethol. Sens. Neural Behav. Physiol.*, 193(8): 801–824.

GRAHAM BM, 2015. The Hive and The Honey Bee. Dadant & Sons Press, Hamilton, Illnois, USA.

HALLEM EA, NICOLE FOX A, ZWIEBEL LJ, CARLSON JR, 2004. Mosquito receptor for human-sweat odorant. *Nature*, 427(6971): 212–213.

HANSSON BS, STENSMYR MC, 2011. Evolution of insect olfaction. *Neuron*, 72(5): 698–711.

HILL CA, FOX AN, PITTS RJ, KENT LB, TAN PL, CHRYSTAL MA, CRAVCHIK A, COLLINS FH, ROBERTSON HM, ZWIEBEL LJ, 2002. G protein-coupled receptors in Anopheles gambiae. *Science*, 298(5591): 176–178.

HONEY BEE GENOME SEQUENCING CONSORTIUM, 2006. Insights into social insects from the genome of the honeybee Apis mellifera. *Nature*, 443(7114): 931–949.

LI H, ZHANG L, NI C, SHANG H, ZHUANG S, LI J, 2013. Molecular recognition of floral volatile with two olfactory related proteins in the Eastern honeybee (*Apis cerana*). *Int. J. Biol. Macromol.*, 56: 114–121.

LI Z, LIU F, LI W, ZHANG S, NIU D, XU H, HONG Q, CHEN S, SU S, 2012. Differential transcriptome profiles of heads from foragers: comparison between Apis mellifera ligustica and *Apis cerana cerana*. *Apidologie*, 43(5): 487–500.

林欣大, 劳冲, 姚云, 杜永均, 2015. 性信息素对斜纹夜蛾雄蛾嗅觉相关基因 abp, pbp 和 or 表达的影响. 昆虫学报, 58(3): 237–243.

LIU JF, YANG L, LI M, HE XJ, WANG ZL, ZENG ZJ, 2019. Cloning and expression pattern of odorant receptor 11 in Asian honeybee drones, Apis cerana (*Hymenoptera, Apidae*). *J. Asia-Pac. Entomol.*, 22(1): 110–116

LOPER GM, WOLF WW, TAYLOR OR, 1993. Radar detection of drones responding to honeybee queen pheromone. *J. Chem. Ecol.*, 19(9): 1929–1938.

MENZEL R, MULLER U, 1996. Learning and memory in honeybees: from behavior to neural substrates. *Annu. Rev. Neurosci.*, 19: 379–404.

OHTANI T, FUKUDA H, 1977. Factors governing the spatial distribution of adult drone honeybees in the hive. *J. Apic. Res.*, 16: 14–26.

PARK D, JUNG JW, CHOI BS, JAYAKODI M, LEE J, LIM J, YU Y, CHOI YS, LEE M, PARK Y, CHOI I, YANG T, EDWARDS OR, NAH G, KWON HW, 2015. Uncovering the novel characteristics of Asian honey bee, *Apis cerana*, by whole genome sequencing. *BMC Genomics*, 16(1): 1.

PLETTNER E, OTIS GW, WIMALARATNE PDC, WINSTON ML, SLESSOR KN, PANKIW T, PUNCHIHEWA PWK, 1997. Species- and caste-determined mandibular gland signals in honeybees (Apis). *J. Chem. Ecol.*, 23(2): 363–377.

REINHARD J, SRINIVASAN MV, GUEZ D, ZHANG SW, 2004a. Floral scents induce recall of navigational and visual memories in honeybees. *J. Exp. Biol.*, 207(25): 4371–4381.

REINHARD J, SRINIVASAN MV, ZHANG SW, 2006. Complex memories in honeybees: can there be more than two? *J. Comp. Physiol. A Neuroethol. Sens. Neural Behav. Physiol.*, 192(4): 409–416.

REINHARD J, SRINIVASAN MV, ZHANG SW, 2004b. Scent-triggered navigation in honeybees. *Nature*, 427(6973): 411.

ROBERTSON HM, WANNER KW, 2006. The chemoreceptor superfamily in the honey bee, *Apis mellifera*: expansion of the odorant, but not gustatory, receptor family. *Genome Res.*, 16(11): 1395–1403.

SANDOZ JC, DEISIG N, DE BRITO SANCHEZ MG, GIURFA M, 2007. Understanding the logics of pheromone processing in the honeybee brain: from labeled-lines to across-fiber patterns. Front. Behav. *Neurosci.*, 1: 5.

SANDOZ JC, 2006. Odour-evoked responses to queen pheromone components and to plant odours using optical imaging in the antennal lobe of the honey bee drone *Apis mellifera L. J. Exp. Biol.*, 209(18): 3587–3598.

STRAUSFELD NJ, 2009. Brain organization and the origin of insects: an assessment. *Proc. R. Soc. B Biol. Sci.*, 276(1664): 1929–1937.

宋飞飞, 2011. 意大利蜜蜂和中华蜜蜂雄蜂与工蜂触角差异蛋白质组分析. 郑州: 郑州大学.

VOSSHALL LB, AMREIN H, MOROZOV PS, RZHETSKY A, AXEL R, 1999. A spatial map of olfactory receptor expression in the *Drosophila antenna. Cell*, 96(5): 725–736.

WANNER KW, ANDERSON AR, TROWELL SC, THEILMANN DA, ROBERTSON HM, NEWCOMB RD, 2007a. Female-biased expression of odourant receptor genes in the adult antennae of the silkworm, Bombyx mori. *Insect Mol. Biol.*, 16(1): 107–119.

WANNER KW, NICHOLS AS, WALDEN KKO, BROCKMANN A, LUETJE CW, ROBERTSON HM, 2007b. A honey bee odorant receptor for the queen substance 9-oxo-2-decenoic acid. *Proc. Natl. Acad. Sci. USA*, 104(36): 14383–14388.

杨乐, 曾志将, 张丽珍, 2017. 中华蜜蜂气味受体基因 AcOr170 的克隆与序列分析. 环境昆虫学报, 39(1): 48–54.

杨乐, 江武军, 何旭江, 曾志将, 2018. 中华蜜蜂气味受体基因 AcOr10 的克隆及表达分析. 昆虫学报, 61(3): 292–299.

曾志将, 2017. 养蜂学第 3 版. 北京: 中国农业出版社.

张中印, 杨珊珊, 孟娇, 赵慧婷, 姜玉锁, 2016. 中华蜜蜂嗅觉受体 AcerOrco 的表达及定位分析. 昆虫学报, 59(2): 185–191.

赵慧婷, 高鹏飞, 张桂贤, 田嵩浩, 杨珊珊, 孟娇, 姜玉锁, 2015. 气味受体基因 Orco 在中华蜜蜂雄蜂触角中的表达及定位分析. 中国农业科学, 48(4): 796–803.

赵慧婷, 彭竹, 杜亚丽, 姜玉锁, 2019. 中华蜜蜂触角感器扫描电镜观察. 中国蜂业, 70(3): 68–72.

## 16. 蜜蜂蜂王与雄蜂幼虫饥饿信息素鉴定及其生物合成通路

何旭江[†]，江武军[†]，颜伟玉，曾志将[*]

（江西农业大学蜜蜂研究所，南昌 330045）

**摘要**：蜜蜂幼虫在饥饿状态下会通过释放特定信息素来向外界传递饥饿信号，称为蜜蜂幼虫饥饿信息素（hunger pheromone）。本文主要研究西方蜜蜂（*Apis mellifera*）蜂王与雄蜂幼虫饥饿信息素化学成分及在幼虫体内的生物合成通路，明确蜜蜂幼虫与成年蜂之间的化学信息素交流机制。采集 2 日龄与 4 日龄 60487 西方蜜蜂蜂王与雄蜂幼虫及其食物，将其分为饥饿组、饲喂组与纯食物组。饲喂组幼虫平躺在预先准备好的食物表面，饥饿处理组则不提供食物。所有样品分别放入 20 mL 密封采样瓶中并在 35℃条件下放置 45 min。利用 Needle trap 技术从样品瓶中采集 10 mL 气体，富集气体中的挥发性化学物质。在气质联用进样口以 250℃高温将富集的化学物质解离，并通过气质联用技术分析鉴定蜂王与雄蜂幼虫饥饿信息素。同时，采用 RNA-Seq 技术分析饥饿信息素在蜂王与雄蜂幼虫体内的生物合成通路及相关基因表达。从蜂王与雄蜂幼虫中分别分析鉴定出 10 种与 9 种信息素，其中蜂王幼虫含有一种特有的幼虫信息素——2-庚酮，且在蜂王食物中含量最高。E-β-罗勒烯在蜂王与雄蜂幼虫各饥饿组含量均显著高于其饲喂幼虫组与食物组，表明蜂王与雄蜂幼虫均以 E-β-罗勒烯为其饥饿信息素。2 日龄蜂王与雄蜂幼虫饥饿组 E-β-罗勒烯含量均差异不显著，但 4 日龄蜂王幼虫饥饿组 E-β-罗勒烯含量显著低于雄蜂幼虫组。其余 8 种信息素为肉豆蔻酸、棕榈酸、甲基棕榈酸脂、硬脂酸、棕榈油酸、十五烷酸、乙酸与乙酸乙酯，均未出现饥饿组高于其他两组的规律。其中乙酸乙酯与乙酸在食物组与饲喂组含量高于饥饿组，推测其可能来自于幼虫食物。RNA-Seq 结果表明蜂王与雄蜂幼虫通过甲羟戊酸途径由乙酰辅酶 A 从头合成 E-β-罗勒烯。发现 *Geranylgeranyl pyrophosphate synthase-like* 与 *Farnesyl pyrophosphate synthase* 等 9 个基因参与该通路，但在饥饿组与其饲喂组幼虫中表达差异均不显著。蜜蜂蜂王与雄蜂幼虫利用 E-β-罗勒烯作为其饥饿信息素乞求食物，并且在体内从头合成 E-β-罗勒烯。工蜂利用 2-庚酮标记蜂王幼虫，但未对雄蜂幼虫进行标记。

**关键词**：西方蜜蜂；饥饿信息素；E-β-罗勒烯；生物合成；表达分析

---

[†] These two authors contributed equally to this paper.
[*] 通讯作者：bees1965@sina.com (ZJZ).
注：此文发表在《中国农业科学》2016 年第 23 期。

# Identification and Biosynthetic Pathway of a Hunger Pheromone in Honeybee Queen and Drone Larvae

He Xujiang, Jiang Wujun, Yan Weiyu, Zeng Zhijiang

(Institute of Honeybee, Jiangxi Agricultural University, Nanchang 330045)

**Abstract**: The objective of this study is to identify the hunger pheromone of honeybee queen and drone larvae and its biosynthetic pathway, which would enormously contribute to understanding of the mechanism of communication between adults and larvae in honey bees *Apis mellifera*. Two- and four-day-old queen and drone larvae and their food were collected and divided into three groups: fed larvae, starving larvae and food. For the fed larvae they were lain on their relative foods prepared in advance, and for the starving larvae their foods were totally deprived. All samples were immediately put into 20 mL sealed glass bottles and were kept in an incubator under 35 °C for 45 min. Afterward, a needle trap system was employed to extract 10 mL gas from those bottles and the volatile chemicals were enriched in needles. The needles were injected into a gas chromatography-mass spectrometry system and the chemicals were dissociated by a high temperature of 250 °C for identifying the hunger pheromone of honeybee queen and drone larvae. RNA-Seq was used for identifying the biosynthetic pathway of their hunger pheromone and the expression of related genes. Nine and ten chemicals were identified in drone larvae and queen larvae, respectively, in which queen larvae had one more chemical (2-heptanone) that was the highest royal jelly. E-β-ocimene was identified as the hunger-signal pheromone of queen and drone larvae, since their starving larvae had significantly more E-β-ocimene than related fed larvae. The E-β-ocimene released from queen and drone starving larvae was not significantly different at 2-day-old, but queen larvae released significantly less E-β-ocimene than drone larvae at 4-day-old. Other eight chemicals were myristic acid, palmitic acid, methyl palmitic ester, stearic acid, palmitoleic acid, pentadecanoic acid, acetic acid and ethyl acetate, but did not show a clear pattern that significantly more amount of these chemicals were detected in starving larvae released than fed larvae and food groups. Acetic acid and ethyl acetate were detected higher in food and fed larvae groups compared to starving larvae groups, indicating that these two chemicals may be from their food rather themselves. RNA-Seq analysis showed that there was a *de novo* E-β-ocimene biosynthetic pathway in queen and drone larvae via a mevalonate pathway. Nine genes such as *Geranylgeranyl pyrophosphate synthase-like* and *Farnesyl pyrophosphate synthase* were involved in this biosynthetic pathway, but all these genes were not significantly differentially expressed between starving larvae and their relative fed larvae. Honeybee queen and drone larvae both use E-*β*-ocimene as their hunger pheromone for food begging, and they have a *de novo* E-β-ocimene biosynthetic pathway. Further, nurse honeybees specifically use 2-heptanone to mark queen larvae for guiding other nurses.

This study will enrich our understanding of the biological characteristics of honeybees.

**Keywords:** *Apis mellifera*, hunger pheromone, E-β-ocimene, biosynthetic pathway, gene expression

## 引言

蜜蜂是一种真社会性昆虫，其内部具有一套十分完整的信息交流系统，保证群体与个体能够正常生长、发育与繁衍[1]。研究蜜蜂蜂群内部的信息交流机制是深入探索其社会性的重要内容，也可为人类及其他动物的社会性研究提供科学借鉴。蜜蜂的信息交流主要分为舞蹈语言系统与化学语言系统。其中蜜蜂信息素是蜂群内社会管理、婚飞交尾、个体交流以及报警防卫等主要交流媒介[2-6]。例如，蜂王利用反式-9-氧代-2-癸烯酸（（2E）-9-oxodecenoic acid）等9种蜂王信息素来引起工蜂的侍从行为、雄蜂的追逐行为以及参与蜂群的社会管理[7-9]；工蜂利用牻牛儿醇（geraniol）、橙花醛（neral）和金合欢醇（farnesol）信息素招引处女王与工蜂回巢，招引本群的工蜂结团，以及招引其他工蜂前来采集[10-11]；而蜜蜂幼虫利用甲基棕榈酸酯（methyl palmitate）与甲基油酸酯（methyl oleate）等10种表皮信息素可以诱导工蜂封盖幼虫巢房行为与刺激工蜂出巢采集等[12]。在正常蜂群中，哺育蜂每天需要饲喂大量幼虫以保证整个蜂群的正常繁衍。He 等[13]研究表明工蜂幼虫利用信息素 E-β-罗勒烯（E-β-ocimene，$C_{10}H_{16}$）向其哺育蜂乞求食物，并发现了该饥饿信息素（hunger pheromone）的生物合成通路。然而，蜂群中含有蜂王、工蜂与雄蜂，称为三型蜂，它们在发育过程中所需的食物存在明显差异。蜂王幼虫在 5 d 饲喂期均食用蜂王浆，而工蜂与雄蜂幼虫在 1～3 日龄时采食工蜂浆与雄蜂浆，4～5 日龄采食蜂蜜与花粉混合而成的蜂粮[14]。有研究表明蜂王浆与工蜂浆及雄蜂浆成分不同，且即使是同一类型的蜜蜂幼虫，每天采食的王浆在成分与数量上也存在显著差异[14-17]。蜂王与雄蜂幼虫如何进行乞食，其乞食信息素成分是否有所不同？哺育蜂又是如何区分三型蜂幼虫以及其日龄，并饲喂不同成分的食物，针对上述问题，利用 Needle trap 与气质联用技术，分析鉴定蜂王与雄蜂幼虫的饥饿信息素，并利用 RNA-Seq 技术分析鉴定蜂王与雄蜂幼虫饥饿信息素的生物合成通路。通过分析鉴定蜂王与雄蜂幼虫的饥饿信息素，并对比两者之间的饥饿信息素差异，从而深入探索蜂群内蜜蜂幼虫的乞食行为及哺育蜂的饲喂行为机制。

## 1 材料与方法

信息素鉴定试验于 2014 年 4～6 月在南京中医药大学完成，转录组测序试验于 2014 年 4～7 月在北京百迈克生物科技有限公司完成。

### 1.1 供试昆虫

信息素鉴定试验使用 3 个西方蜜蜂（*Apis mellifera*）种群，饲养于南京中医药大学（32.6°N，118.56°E）。RNA-Seq 试验使用西方蜜蜂 3 个种群，饲养于江西农业大学蜜蜂研究所（28.46°N，115.49°E）。试验蜜蜂蜂群参照西方蜜蜂的营养标准进行饲养[18]。

### 1.2 气质联用标准曲线的建立

利用捕集针（needle trap）系统（德国 PAS 科技公司）与气质联用仪（安捷伦公司，7890A/5975C）结合来鉴定并精确定量蜜蜂饥饿信息素，在试验前首先建立了标准曲线。从 Sigma 公司购得内标物 1-壬烯（CAS：124-11-8，$C_9H_{18}$，纯度 > 99.5%）与标准物 E-β-罗勒烯（CAS：13877-91-3，$C_{10}H_{16}$，纯度 > 90%）。将 2 μL 的 1-壬烯配入 20 mL 的无水乙醇（南京化学试剂股份有限公司，分析纯）中，并分为 6 组，每组分别加入 0、1、2、4、8 和 16 μL 的 E-β-罗勒烯，振荡搅匀。用移液枪从各组中选取 2 μL 溶液，并迅速移入 20 mL 采样瓶中，封盖后立即放入 35℃ 培养箱（扬州三发电子有限公司，SHP-250）中放置 45 min。然后利用捕集针从采样瓶中吸取 10 mL 气体，注入气质联用仪进行分析。结果表明 DB-5 与 VOC 的标准曲线 $R^2$ 值为 0.9916 与 0.9926，均大于 0.99，符合气质联用一般的曲线标准[19]。

### 1.3 幼虫样品的采集

利用免移虫产卵技术[20-21]，控王产卵 12 h，并将卵移到王台中进行培育蜂王，在 2 日龄与 4 日龄幼虫期取蜂王幼虫，做进一步试验分析；利用同群已婚飞交尾的成年蜂王，在预先准备好的雄蜂巢脾上控王产卵 12 h。控王产卵时，将蜂王用隔王栅隔离在单一的雄蜂巢脾上，并放在蜜蜂平箱群所有巢脾的最内侧，使其产卵。在雄蜂幼虫发育到 2 日龄与 4 日龄幼虫期，采集样品进行试验。

信息素鉴定试验中，利用移虫针在 30℃ 室温条件下采集 20 只 2 日龄蜂王或雄蜂幼虫，立即放入 20 mL 采样瓶中，并迅速加入 2μL 1-壬烯内标溶液（$V:V$=1:10 000）。在 35℃ 的培养箱中放置 45 min，作为饥饿幼虫组；另外采集 20 只幼虫，放入之前已经铺好 2 日龄蜂王浆或雄蜂浆的采样瓶中，使每只幼虫横躺在王浆表面，加入内标后放置到培养箱 45 min，作为饲喂组；最后采集 2 日龄蜂王浆或雄蜂浆到采样瓶中，作为对照组。饥饿 45 min 后，利用捕集针从采样瓶中采集 10 mL 气体，并注入气质联用仪进行分析。每个样品采集时间严格控制在 5 min 内。每组样品采集 12 个平行样，其中 6 个样品由 DVB、PDMS 与 CAR1000 捕集针采集，剩余 6 个样品由脂肪酸型捕集针采集。由于 4 日龄工蜂个体较大，每个样品只采集 10 只幼虫。

分别取 2 日龄与 4 日龄蜂王、雄蜂幼虫，进行饥饿与饲喂处理 45 min，然后立即液氮冷冻，用于 RNA-Seq 试验。

### 1.4 Needle Trap 与 GC-MS 分析

利用捕集针（needle trap）技术与气质联用系统分析鉴定蜂王与雄蜂幼虫饥饿信息素。将捕集针一头插入密闭的 20 mL 采样瓶，一头连接采集器，从采样瓶中抽取气体。本试验利用 20 MPa 的压强以 5 mL/min 的速度从采样瓶中采集 10 mL 气体。采集时将采样瓶放入 35℃ 的温控装置中恒温采集。捕集针为 DVB、PDMS 与 CAR1000 与脂肪酸填料两种，分别配对 DB-5（长度 30 m，直径 25 μm）与 VOC 柱子（长度 30 m，直径 25 μm）分离柱。

采集好的捕集针立即注入安捷伦公司生产的 7890A/5975C 型气质联用仪。通过摸索，按如下最佳程序进样分析：进样口温度为 250℃；柱温程序为 35℃ 保持 2 min，然后以 8℃/min 的速度从 35℃ 上升至 250℃，最后 250℃ 保持 5 min。离子源为电子轰击源（electron impact ionization，EI），流动相为氦气。氦气气压保持 6.7776 psi。获得的化学物质离子图谱采用 NIST32I 数据库进行比对，最终鉴定出捕集针中所吸附的挥发性化学物质。在进样时取一个 1 mL 容量的无菌注射器，从气质联用仪中抽取 1 mL 氦气，然后与采集完样品的捕集针相连，并手动插入气质联用仪的进样口，30 s 内将 1 mL 氦气缓缓注入气质联用仪。该处理可使捕集针中的化学物质完全洗脱，进入气质联用仪分析。

### 1.5 RNA-Seq

采集蜂王与雄蜂 2 日龄与 4 日龄幼虫的饥饿组与饲喂组幼虫，进行 RNA-Seq 测序，获得饥饿信息素的生物合成通路与调控基因。按标准 TRIzol Reagent 方法提取 RNA（Life technologies，California，

USA），用于 RNA-Seq 测序分析。利用 3 群西方蜂群，每个试验组含有 3 个生物学重复。

首先，利用 TRlzol 法提取样品的总 RNA，再通过 NEBNext Poly（A）mRNA 磁性分离技术（NEB，E7490）分离出所需的 mRNA。将 mRNA 反转录为 cDNA，并连接到 Illumina 上构建上机文库。制备好的文库用 1.8% 琼脂糖凝胶电泳进行检测文库插入片段大小，然后用 Library Quantification Kit-Illumina GA Universal（Kapa，KK4824）进行 qPCR 定量。检测合格的文库在 illuminacbot 上进行簇的生成，最后用 Illumina HiSeq™ 2500 进行测序。

为了保证测序结果的高度准确性，去除了未知碱基比率 > 5% 的 reads，只保留测序错误率 < 1% 的 reads（Q20 > 98%）。然后利用 Tophat2 软件[22]将所获得的所有 clean reads 与蜜蜂基因组（OGSv3.2）进行比对。比对后的各个基因对应的 reads 在 cufflinks 软件[23]中对其表达量水平进行估计，并以 FPKM 值来记录每个基因的表达量。最后，利用 DESeq 差异分析与 Q-value 统计方法来比对饥饿组与饲喂组之间基因的 FPKM 值，最终获得差异表达基因。当 FDR 值 < 0.05 且 lg2 ≥ 1.5 时，即为基因表达存在显著差异[24]。使用 BLAST 软件对提取的基因序列与 nr、SwissProt、GO、COG 与 KEGG 数据库进行比对，获得各基因的注释信息。

### 1.6 数据统计方法

气质联用实验所获得的数据全部利用"ANOVA or ANCOVA"在 Statview 5.01（SAS Institute，Cary，NC，USA）软件中进行比较分析。

## 2 结果

### 2.1 蜂王与雄蜂幼虫饥饿信息素鉴定

通过 Needle trap 与气质联用系统分析，从 2 日龄与 4 日龄的雄蜂幼虫中获得了 9 个化学物质：E-β-罗勒烯（E-β-ocimene）、肉豆蔻酸（myristic acid）、棕榈酸（palmitic acid）、甲基棕榈酸脂（methyl palmitic ester）、硬脂酸（stearic acid）、棕榈油酸（palmitoleic acid）、十五烷酸（pentadecanoic acid）、乙酸（acetic acid）、乙酸乙酯（ethyl acetate），与其对应标准物的离子图谱匹配率均高于 78%（表Ⅲ-16-1、图Ⅲ-16-1、图Ⅲ-16-2）。在蜂王幼虫中，除了发现这 9 种化学物质，还发现了一种特殊的化学物质，即 2-庚酮（2-haptanone）（表Ⅲ-16-1）。同时，图Ⅲ-16-3 表明，E-β-罗勒烯是雄蜂与蜂王幼虫中唯一在饥饿组中的释放量显著高于饲喂组的信息素，且在蜂王浆中未测到任何 E-β-罗勒烯，表明蜂王与雄蜂幼虫饥饿信息素为 E-β-罗勒烯。在饥饿幼虫组对比中，2 日龄的蜂王与雄蜂幼虫 E-β-罗勒烯含量没有差异，而 4 日龄组对比中，蜂王幼虫的 E-β-罗勒烯含量显著低于雄蜂幼虫组。值得注意的是，2 日龄饥饿幼虫组 E-β-罗勒烯释放量显著高于 4 日龄饥饿幼虫组（图Ⅲ-16-4）。

## III. 蜜蜂行为生物学及其分子机制
### Behavioral Biology of Honeybee and its Molecular Mechanism

表 III-16-1 气质联用所测单只西方蜜蜂蜂王与雄蜂幼虫释放的化学物质及含量

| 样品名 | 组别 | E-β-罗勒烯 | 棕榈酸 | 肉豆蔻酸 | 甲基棕榈酸酯 | 硬脂酸 | 棕榈油酸 | 十五烷酸 | 乙酸 | 乙酸乙酯 | 2-庚酮 |
|---|---|---|---|---|---|---|---|---|---|---|---|
| 与标准物匹配率 | | 97% | 98% | 99% | 99% | 78% | 95% | 99% | 86% | 86% | 95% |
| 2日龄蜂王幼虫 | 食物组 | 0.00±0.00a | 0.15±0.08a | 0.06±0.03a | 0.3±0.00a | 0.02±0.01a | 0.05±0.02 | 0.02±0.01a | 0.12±0.04a | 0.08±0.03a | 0.06±0.02a |
| | 饲喂组 | 3.30±0.28b | 0.22±0.05a | 0.06±0.01a | 0.02±0.01a | 0.02±0.00a | 0.10±0.05a | 0.02±0.00a | 0.11±0.05a | 0.14±0.05a | 0.03±0.01a |
| | 饥饿组 | 5.61±0.64c | 0.23±0.18a | 0.10±0.04a | 0.01±0.00a | 0.03±0.01a | 0.07±0.02a | 0.02±0.01a | 0.18±0.07a | 0.05±0.02a | 0.00±0.00b |
| 4日龄蜂王幼虫 | 食物组 | 0.00±0.00a | 0.07±0.01a | 0.02±0.00a | 0.01±0.00a | 0.01±0.00a | 0.02±0.01a | 0.01±0.00a | 0.05±0.02a | 0.16±0.04a | 0.11±0.04a |
| | 饲喂组 | 0.38±0.03b | 0.16±0.06a | 0.05±0.01a | 0.01±0.00a | 0.01±0.00a | 0.04±0.01a | 0.01±0.00a | 0.10±0.02ab | 0.21±0.09a | 0.10±0.02a |
| | 饥饿组 | 1.12±0.27c | 0.26±0.11a | 0.08±0.03a | 0.01±0.00a | 0.05±0.03a | 0.07±0.02a | 0.02±0.01a | 0.13±0.03b | 0.08±0.02a | 0.00±0.00b |
| 2日龄雄蜂幼虫 | 食物组 | 0.18±0.02a | 0.02±0.00a | 0.03±0.01a | 0.02±0.01a | 0.02±0.01a | 0.01±0.00a | 0.00±0.00a | 0.45±0.06a | 0.51±0.04a | 0.00±0.00a |
| | 饲喂组 | 1.50±0.13b | 0.10±0.04a | 0.07±0.02a | 0.07±0.01b | 0.10±0.03a | 0.05±0.02a | 0.01±0.00b | 0.14±0.04b | 0.45±0.11a | 0.00±0.00a |
| | 饥饿组 | 4.57±0.66c | 0.16±0.06a | 0.12±0.45a | 0.08±0.02b | 0.10±0.02a | 0.04±0.02a | 0.01±0.00b | 0.80±0.10c | 0.01±0.02b | 0.00±0.00a |
| 4日龄雄蜂幼虫 | 食物组 | 0.25±0.01a | 0.05±0.02a | 0.03±0.01a | 0.05±0.02a | 0.02±0.01a | 0.01±0.00a | 0.02±0.01a | 0.09±0.02a | 0.16±0.04a | 0.00±0.00a |
| | 饲喂组 | 0.59±0.07b | 0.10±0.05a | 0.07±0.02a | 0.13±0.06a | 0.10±0.03a | 0.05±0.02a | 0.07±0.01b | 0.07±0.01ab | 0.07±0.01a | 0.00±0.00a |
| | 饥饿组 | 1.89±0.27c | 0.23±0.14a | 0.12±0.05a | 0.23±0.14a | 0.10±0.02a | 0.04±0.02a | 0.07±0.02b | 0.04±0.01b | 0.07±0.03a | 0.00±0.00a |

注：表中数据为平均数±标准误，是各化学物质对1-王烯的相对含量，E-β-罗勒烯为所测得的质量（ng）。各幼虫组内食物组、饲喂组与饥饿组进行方差分析比较，同列不同小写字母表示在5%水平上差异显著性。

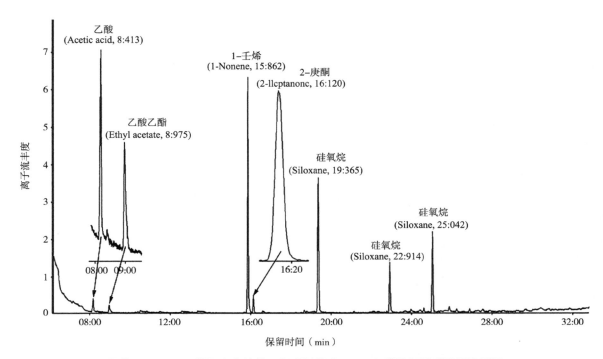

**图Ⅲ-16-1　2日龄西方蜜蜂蜂王饲喂组幼虫VOC色谱柱气质联用总离子图**
各个化学物质的离子峰及出峰时间已标注。其中1-壬烯为内标物，硅氧烷为色谱柱的柱流失

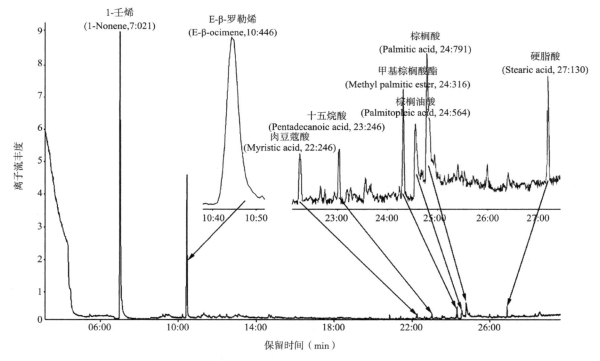

**图Ⅲ-16-2　2日龄西方蜜蜂蜂王饥饿组幼虫DB-5色谱柱气质联用总离子图**
各个化学物质的离子峰及出峰时间已标注。1-壬烯为内标物

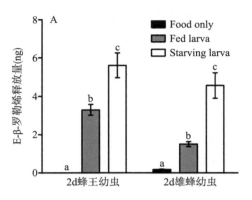
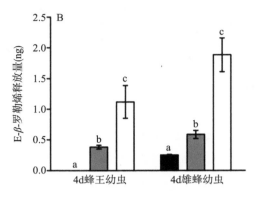

图Ⅲ-16-3 2日龄（A）与4日龄（B）西方蜜蜂蜂王与雄蜂幼虫E-β-罗勒烯单只幼虫释放量

不同字母表示同日龄内对比差异显著（$P < 0.05$）。下同

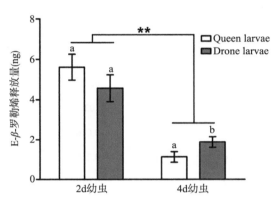

图Ⅲ-16-4 西方蜜蜂蜂王与雄蜂饥饿组幼虫E-β-罗勒烯含量对比

"**"表示两个日龄间差异显著（$P < 0.05$）

### 2.2 蜂王与雄蜂幼虫饥饿组与饲喂组差异表达基因

2日龄蜂王幼虫饥饿组与饲喂组差异基因为6个，4日龄对比组为287个；2日龄雄蜂幼虫饥饿组与饲喂组差异基因为8个，4日龄对比组为0个（表Ⅲ-16-2）。这些差异基因中，未发现任何基因与幼虫的信息素合成相关。

表Ⅲ-16-2 西方蜜蜂蜂王与雄蜂幼虫饥饿组与饲喂组差异基因统计

| 对比组 | 2日龄蜂王饥饿与饲喂组 | 4日龄蜂王饥饿与饲喂组 | 2日龄雄蜂饥饿与饲喂组 | 4日龄雄蜂饥饿与饲喂组 |
| --- | --- | --- | --- | --- |
| 差异基因数量 | 6 | 287 | 8 | 0 |

### 2.3 E-β-罗勒烯合成通路与相关基因

蜂王幼虫与雄蜂幼虫具有一条从头合成E-β-罗勒烯的生物合成通路（图Ⅲ-16-5），表明E-β-罗勒烯在幼虫体内由乙酰辅酶A经甲羟戊酸途径合成，这与工蜂幼虫E-β-罗勒烯的生物合成通路相同[13]。参与该通路的9个基因在饥饿组与饲喂组中均表达差异不显著（表Ⅲ-16-3）。

表 III-16-3 蜂王与雄蜂幼虫 E-β-罗勒烯合成通路基因表达

| 基因编号 | 饲喂幼虫组 vs 饥饿幼虫组 | | FDR 值 | lg2 FC 值 | 基因注释 |
|---|---|---|---|---|---|
| | 幼虫组别 | 日龄 | | | |
| GB47436 | 蜂王幼虫对比组 | 2 | 1 | −0.088 | Acetyl−CoA−C−acetyl transferase 1 [A. mellifera] |
| | | 4 | 0.067 | 0.470 | |
| | 雄蜂幼虫对比组 | 2 | 1 | −0.009 | |
| | | 4 | 1 | −0.129 | |
| GB49784 | 蜂王幼虫对比组 | 2 | 1 | −0.107 | Acetyl−CoA acetyl transferase, cytosolic−like [A. mellifera] |
| | | 4 | 0.954 | −0.026 | |
| | 雄蜂幼虫对比组 | 2 | 1 | −0.054 | |
| | | 4 | 1 | −0.205 | |
| GB44420 | 蜂王幼虫对比组 | 2 | 1 | −0.185 | PREDICTED: Hydroxymethylglutaryl−CoA synthase 1[A. mellifera] |
| | | 4 | 0.422 | −0.253 | |
| | 雄蜂幼虫对比组 | 2 | 1 | −0.105 | |
| | | 4 | 1 | 0.039 | |
| GB51591 | 蜂王幼虫对比组 | 2 | 1 | −0.006 | PREDICTED: 3−hydroxy−3−methylglutaryl−coenzyme A reductase[A. mellifera] |
| | | 4 | 0.459 | −0.210 | |
| | 雄蜂幼虫对比组 | 2 | 1 | −0.010 | |
| | | 4 | 1 | 0.031 | |
| GB47666 | 蜂王幼虫对比组 | 2 | 1 | −0.116 | PREDICTED: Mevalonate kinase−like [A. mellifera] |
| | | 4 | 0.137 | 0.428 | |
| | 雄蜂幼虫对比组 | 2 | 1 | 0.050 | |
| | | 4 | 1 | −0.259 | |
| GB49221 | 蜂王幼虫对比组 | 2 | 1 | −0.071 | PREDICTED: phosphomevalonate kinase−like [A. mellifera] |
| | | 4 | 0.192 | −0.506 | |
| | 雄蜂幼虫对比组 | 2 | 1 | −0.191 | |
| | | 4 | 1 | −0.028 | |
| GB44718 | 蜂王幼虫对比组 | 2 | 1 | −0.042 | Similar to mevalonate (diphospho) decarboxylase [A. mellifera] |
| | | 4 | 0.998 | 0.010 | |
| | 雄蜂幼虫对比组 | 2 | 1 | −0.129 | |
| | | 4 | 1 | −0.028 | |

（续表）

| 基因编号 | 饲喂幼虫组 vs 饥饿幼虫组 | | FDR 值 | lg2 FC 值 | 基因注释 |
| --- | --- | --- | --- | --- | --- |
| | 幼虫组别 | 日龄 | | | |
| GB48898 | 蜂王幼虫对比组 | 2 | 1 | −0.062 | PREDICTED: Farnesyl pyrophosphate synthase[A. mellifera] |
| | | 4 | 0.447 | 0.201 | |
| | 雄蜂幼虫对比组 | 2 | 1 | 0.083 | |
| | | 4 | 1 | −0.001 | |
| GB43964 | 蜂王幼虫对比组 | 2 | 1 | −0.037 | PREDICTED: Geranylgeranyl pyrophosphate synthase-like [A. mellifera] |
| | | 4 | 0.268 | 0.397 | |
| | 雄蜂幼虫对比组 | 2 | 1 | −0.150 | |
| | | 4 | 1 | −0.097 | |

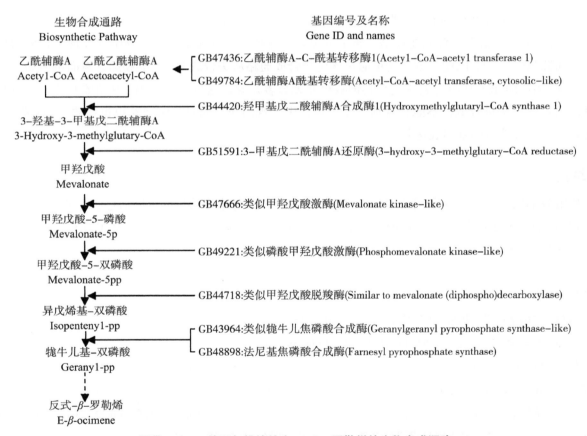

图Ⅲ-16-5　蜂王与雄蜂幼虫 E-β-罗勒烯的生物合成通路

## 3　讨论

信息素是蜜蜂幼虫与成年蜂之间的重要交流方式[12,25]。本研究鉴定分析出蜂王与雄蜂幼虫饥饿信息素的主要成分为 E-β-罗勒烯，这与工蜂幼虫饥饿信息素成分相同[13]。前人研究表明，蜂王幼虫采食 5 d 蜂王浆，而雄蜂幼虫采食 3 d 雄蜂浆，后 2 d 采食蜂粮。即使在 1~3 日龄，蜂王浆与雄蜂浆成分也不同[14]。图 4 表明 2 日龄时蜂王与雄蜂幼虫释放的罗勒烯并无显著差异。那么哺育蜂是如何在黑暗的蜂箱中分辨出蜂王与雄蜂幼虫，并饲喂其所需的各种食物？有研究表明，蜂王在产卵过程中，先利用其触角与前足丈量巢房的大小，再进行产受精卵或未受精卵[26]。因此，推测哺育蜂也可能利用相似的测量来分辨巢房大小，进而饲喂不同食物。有趣的是，蜂王幼虫含有独特的 2-庚酮这类物质。该信息素是工蜂上颚腺分泌的标记信息素，主要用于标记蜂巢的巢门方向与采集的花朵上，同时也可作为标记敌人的报警信息素[27-29]。因此，哺育蜂也可能利用 2-庚酮标记需要饲喂的蜂王幼虫，从而引导其他哺育蜂对蜂王幼虫进行准确定位与饲喂。

本试验表明 2 日龄蜂王与雄蜂幼虫释放 E-β-罗勒烯量均显著高于与其 4 日龄幼虫，这与前人在不同日龄工蜂幼虫中测得的 E-β-罗勒烯量相一致[13,30]。哺育蜂是否可通过感知 E-β-罗勒烯不同释放量来区分幼虫的日龄，仍需进一步研究。蜂王 4 日龄幼虫 E-β-罗勒烯释放量显著低于 4 日龄雄蜂幼虫，这可能是蜂王幼虫采食蜂王浆且个体较大，其耐饿性较强，从而在饥饿处理状态下 E-β-罗勒烯释放量较低。

蜂王幼虫与雄蜂幼虫的 E-β-罗勒烯由乙酰辅酶 A 与乙酰乙酰辅酶 A 通过甲羟戊酸途径合成（图Ⅲ-16-5）。这一单萜生物合成途径是一条十分保守的途径，在其他昆虫与植物中均发现了相似的合成途径与同源基因[31-32]。因此，蜜蜂幼虫通过该生物合成通路从头合成 E-β-罗勒烯，然而饥饿幼虫组的 E-β-罗勒烯通路相关基因的表达量与其饲喂组并无显著差异。这可能是 E-β-罗勒烯的合成量只有几纳克，其基因表达量较低，不能直观地从基因表达中表现出来。同时，饲喂幼虫也释放 E-β-罗勒烯，从而减小了与饥饿组的基因表达差异。

## 4 结论

利用 Needle trap 与气质联用技术分析鉴定出了蜜蜂蜂王与雄蜂幼虫的饥饿信息素为 E-β-罗勒烯，并且利用 RNA-Seq 技术证实在幼虫体内通过甲羟戊酸途径从头合成该信息素。蜂王具有一种特有的信息素——2-庚酮。

## References

[1] 陈盛禄. 中国蜜蜂学. 北京：中国农业出版社, 2001: 158-175.

[2] 曾志将. 养蜂学. 2 版. 北京：中国农业出版社, 2009: 39-56.

[3] 颜伟玉, Le Conte Y, Beslay D, 曾志将. 中华蜜蜂幼虫信息素鉴定. 中国农业科学, 2009, 42(6): 2250-2254.

[4] 张含, 颜伟玉, 曾志将. 蜜蜂幼虫信息素对工蜂发育和采集行为的影响. 蜜蜂杂志, 2009(9): 4.

[5] 赵红霞, 梁勤, 张学锋, 黄文忠, 陈华生, 罗岳雄. 蜜蜂化学通讯的研究进展. 中国农学通报, 2014, 30(2): 1-6.

[6] 胡福良, 玄红专. 蜜蜂蜂王信息素研究进展. 昆虫知识, 2004, 41(3): 208-211.

[7] Callow RK, Johnston NC. The chemical constitution and synthesis of queen substance of honeybees (Apis mellifera) preliminary note. *Bee World*, 1960, 6(41): 152-153.

[8] Keeling C I, Slessor K N, Higo H A, Winston M L. New components of the honey bee (*Apis mellifera* L.) queen retinue pheromone. *Proceedings of the National Academy of Sciences of the United States of America*, 2003, 100(8): 4486-4491.

[9] Pankiw T, Winston M L, Fondrk M K, Slessor K N. Selection on worker honeybee responses to queen pheromone (Apis mellifera L.). *Naturwissenschaften*, 2000, 87(11): 487-490.

[10] Butler C G, Calam D H. Pheromones of the honey bee- The secretion of the Nassanoff gland of the worker. *Journal of Insect Physiology*, 1969, 2(15): 237-244.

[11] Ferguson A W, Free J B, Pickett J A. Techniques for studying honeybee pheromones involved in clustering, and experiments on the effect of Nasonov and queen pheromones. *Physiological Entomology*, 1979, 4(4): 339-344.

[12] Le Conte Y, Arnold G, Trouiller J, Masson C, Chappe B. Identification of a brood pheromone in honeybees. *Naturwissenschaften*, 1990, 7(77): 334-336.

[13] He X J, Zhang X C, Jiang W J, Barron A B, Zhang J H, Zeng Z J. Starving honey bee (Apis mellifera) larvae signal pheromonally to worker bees. *Scientific Reports*, 2016, 6: 22359.

[14] Haydak M H. Honey bee nutrition. *Annual review of entomology*, 1970, 1(15): 143-156.

[15] Brouwers E V M. Glucose/fructose ratio in the food of honeybee larvae during caste differentiation. *Journal of Apicultural Research*, 1984, 2(23): 94-101.

[16] Asencot M, Lensky Y. Juvenile hormone induction of 'queenliness' on female honey bee (Apis mellifera L.) larvae reared on worker jelly and on stored royal jelly. *Comparative Biochemistry and Physiology Part B: Comparative Biochemistry*, 1984, 1(78): 109-117.

[17] Kamakura M. Royalactin induces queen differentiation in honeybees. *Nature*, 2011, 473(7348): 478-483.

[18] 董文滨, 马兰婷, 王颖, 李成成, 郑本乐, 冯倩倩, 王改英, 李迎军, 焦震, 刘锋, 杨维仁, 胥保华. 意大利蜜蜂春繁, 产浆, 越冬和发育阶段营养需要建议标准. 动物营养学报, 2014, 26(2): 342-347.

[19] Trefz P, Kischkel S, Hein D, James E S, Schubert J K, Miekisch W. Needle trap micro-extraction for VOC analysis: effects of packing materials and desorption parameters. *Journal of Chromatography* A, 2012, 1219: 29-38.

[20] Pan Q Z, Wu X B, Guan C, Zeng Z J. A new method of queen rearing without grafting larvae. *American Bee Journal*, 2013, 12: 1279-1280.

[21] 张飞, 吴小波, 颜伟玉, 王子龙, 曾志将. 蜂王浆机械化生产关键技术研究与应用（Ⅱ）——仿生免移虫蜂王浆生产技术. 江西农业大学学报, 2013, 35(5): 1036-1041.

[22] Kim D, Pertea G, Trapnell C, Pimentel H, Kelley R, Salzberg S L. TopHat2: accurate alignment of transcriptomes in the presence of insertions, deletions and gene fusions. *Genome Biology*, 2013, 14(4): R36.

[23] Trapnell C, Williams B A, Pertea G, Mortazavi A, Kwan G, Van Baren M J, Salzberg S L, Wold B J, Pachter L. Transcript assembly and quantification by RNA-Seq reveals unannotated transcripts and isoform switching during cell differentiation. *Nature Biotechnology*, 2010, 28(5): 511-515.

[24] Anders S, Huber W. Differential expression analysis for sequence count data. *Genome Biology*, 2010, 11(10): R106.

[25] 曾云峰, 曾志将, 颜伟玉, 吴小波. 幼虫信息索中三种酯类对中华蜜蜂和意大利蜜蜂工蜂哺育和封盖行为以及蜂王发育影响. 昆虫学报, 2010, 53(2): 154-159.

[26] Koeniger N. Uber die Fahigkeit der Bienenkonigin (Apis mellifica L.) zwischen Arbeiterinnen-und Drohnenzellen zu unterscheiden. *Apidologie*, 1970, 1: 115-142.

[27] Ferguson A W, Free J B. Production of a forage-marking pheromone by the honeybee. Journal of *Apicultural Research*, 1979, 2(18): 128-135.

[28] Shearer D A, Boch R. 2-Heptanone in the mandibular gland secretion of the honey-bee. Nature, 1965, 206: 530.

[29] Kerr W E, Blum M S, Pisani J F, Stort A C. Correlation between amounts of 2-heptanone and iso-amyl acetate in honeybees and their aggressive behaviour. *Journal of Apicultural Research*, 1974, 3(13): 173–176.

[30] Maisonnasse A, Lenoir J C, Beslay D, Crauser D, Le Conte Y. E-beta-ocimene, a volatile brood pheromone involved in social regulation in the honey bee colony (*Apis mellifera*). *PLoS One*, 2010, 5(10): e13531.

[31] Seybold S J, Bohlmann J, Raffa K F. Biosynthesis of coniferophagous bark beetle pheromones and conifer isoprenoids: evolutionary perspective and synthesis. *The Canadian Entomologist*, 2000, 6(132): 697–753.

[32] Dudareva N, Martin D, Kish CM, KolosovaN, Gorenstein N, Fäldt J, Miller B, Bohlmann J. (E)-beta-ocimene and myrcene synthase genes of floral scent biosynthesis in snapdragon: function and expression of three terpene synthase genes of a new terpene synthase subfamily. *The Plant Cell*, 2003, 15(5): 1227–1241.

# 17. 东方蜜蜂幼虫封盖信息素含量及生物合成通路

秦秋红 [1,2]，何旭江 [1]，江武军 [3]，王子龙 [1]，曾志将 [1*]

（1. 江西农业大学蜜蜂研究所/江西省蜜蜂生物学与饲养重点实验室，南昌 330045；
2. 广西科技大学医学部，柳州 545005；3 江西省养蜂研究所，南昌 330052）

**摘要：** 在蜜蜂中，甲基棕榈酸酯（MP）、甲基油酸酯（MO）、甲基亚油酸酯（ML）和甲基亚麻酸酯（MLN）是重要的封盖信息素成分，它们触发成年工蜂对幼虫的封盖行为。本研究旨在比较封盖信息素化学成分在不同封盖时期的东方蜜蜂（*Apis cernana*）工蜂与雄蜂幼虫体内的含量，并分析其在幼虫体内的生物合成通路，进一步探索蜜蜂幼虫与成年工蜂之间的信息素交流机制。以中华蜜蜂（*Apis cernana cernana*）为实验材料，分别取未封盖、正在封盖和已封盖的工蜂与雄蜂幼虫，利用 GC/MS 分析技术，比较 4 种封盖信息素成分在不同封盖时期的工蜂与雄蜂幼虫体内的含量；同时利用 RNA-seq 技术对不同封盖时期的工蜂与雄蜂幼虫进行转录组测序，分析其基因表达差异，并根据差异表达基因 KEGG 富集分析推测封盖信息素的生物合成通路。在工蜂幼虫中，4 种封盖信息素成分在正在封盖和已封盖幼虫体内的含量均显著高于未封盖幼虫，其中 MP 和 MO 的含量均随幼虫日龄增长而显著增加，而 ML 和 MLN 的含量在正在封盖和已封盖幼虫体内差异不显著；在雄蜂幼虫中，4 种信息素成分含量均随日龄增长而增加，且已封盖幼虫的信息素含量显著高于未封盖和正在封盖的幼虫。对工蜂与雄蜂 3 个封盖时期幼虫的基因表达量进行组间比较分析，分别从 3 个比较组中获得 4 299 和 3 926 个差异表达基因，并且在差异表达基因 KEGG 注释分析中分别获得 152 和 130 个 KEGG 通路。根据差异表达基因 KEGG 富集结果，推测出东方蜜蜂工蜂与雄蜂幼虫可能利用乙酰辅酶 A 合成 MP、MO、ML 和 MLN 的生物合成通路以及 11 个调控候选基因，并发现该生物合成通路与西方蜜蜂相同。东方蜜蜂工蜂幼虫与雄蜂幼虫在被封蜡盖的关键阶段增加了 MP、MO、ML 和 MLN 的释放量，进一步验证了它们是与蜜蜂封盖行为相关的信息素，并且推测这些信息素可能是在相关基因的调控下由乙酰辅酶 A 从头合成，而东方蜜蜂幼虫与西方蜜蜂幼虫可能利用相同的生物合成通路进行信息素的生物合成。

**关键词：** 东方蜜蜂；封盖信息素；含量；转录组；生物合成通路

---

\* 通讯作者：bees1965@sina.com (ZJZ).
注：此文发表在《中国农业科学》2021 年第 11 期。

# The Capping Pheromone Contents and Putative Biosynthetic Pathways in Larvae of Honeybees *Apis cernana*

Qin Qiuhong[1,2], He Xujiang[1], Jiang Wujun[3], Wang Zilong[1], Zeng Zhijiang[1✉]

1.Honeybee Research Institute, Jiangxi Agricultural University/Jiangxi Province Key Laboratory of Honeybee Biology and Beekeeping, Nanchang 330045; 2.School of Medicine, Guangxi University of Science and Technology, Liuzhou 545005, Guangxi; 3Apicultural Research Institute of Jiangxi Province, Nanchang 330052

**Abstract**: In honeybees, methyl palmitate (MP), methyl oleate (MO), methyl linoleate (ML) and methyl linolenate (MLN) are important capping pheromone components, which trigger the capping behavior of adult workers. The objective of this study is to compare the contents of these four pheromone components in the larvae of workers and drones of *Apis cernana* at different capping stages, analyze their biosynthetic pathways, and to further explore the mechanism of pheromone communication between larvae and adult workers. Using *A. c. cernana* as the experimental material, the larvae of workers and drones of prior to be capped, in the process of being capped and had been capped were collected for comparing the contents of these four pheromone components by using GC/MS. Simultaneously, RNA-seq was used for gene expression analysis, and the biosynthetic pathways were speculated based on KEGG enrichment of differential expressed genes. In worker larvae, the contents of the four capping pheromone components were significantly higher at the capping and capped stage than those of the prior to be capped larvae, and the contents of MP and MO significantly increased with aging of the larvae, while the contents of ML and MLN were not significantly different between the capping and capped stage. Whereas in drone larvae, the contents of the four pheromone components were higher overall and increased with aging, and the content at capped stage was significantly higher than that at prior to be capped and capping. RNA-seq results showed that there were 4 299 and 3 926 differential expressed genes among the larvae groups of three stages of workers and drones, respectively. In addition, 152 and 130 KEGG pathways were obtained from the KEGG annotation analysis of the differential expressed genes, respectively. Furthermore, the possible *de novo* biosynthetic pathways were proposed for MP, MO, ML and MLN from acetyl-CoA, regulating under 11 related candidate genes, and these biosynthesis pathways were found to be similar to those of *Apis mellifera*. The release contents of MP, MO, ML and MLN were increased during the critical stage of capping in worker and drone larvae of *A. cernana*, which further verified that these four pheromones were related to capping behavior of honeybees, and it was speculated that they were possibly *de novo* biosynthesized from acetyl-CoA under the control of related genes. *A. cernana* larvae and *A. mellifera* larvae may use the same biosynthesis pathway for pheromone biosynthesis.

**Keywords:** *Apis cernana*, capping pheromone, content, transcriptome, biosynthetic pathway

# 引言

蜜蜂是一种高度社会化的昆虫，蜂群成员间需要进行信息交流来协调个体活动，以确保群体能够健康生存和持续繁衍。作为完全依赖群体的成员，蜜蜂幼虫通过各种化学信号向成年工蜂表明自己的生物状态和需求，以便得到必需的喂养和照料[1-4]。从分子水平研究东方蜜蜂（*Apis cernana*）幼虫封盖信息素的生物合成通路，进一步探索蜜蜂幼虫诱导成年工蜂封盖幼虫巢房行为的分子机理，可为深入了解东方蜜蜂信息素交流的内在机制提供新的启示。Le Conte 等首次从西方蜜蜂（*Apis mellifera*）雄蜂幼虫中鉴定出 10 种幼虫酯类信息素（the brood ester pheromone，BEP）：甲基棕榈酸酯（methyl palmitate，MP）、甲基油酸酯（methyl oleate，MO）、甲基亚油酸酯（methyl linoleate，ML）、甲基硬脂酸酯（methyl stearate，MS）、甲基亚麻酸酯（methyl linolenate，MLN）、乙基棕榈酸酯（ethyl palmitate，EP）、乙基油酸酯（ethyl oleate，EO）、乙基亚油酸酯（ethyl linoleate，EL）、乙基硬脂酸酯（ethyl stearate，ES）和乙基亚麻酸酯（ethyl linolenate，ELN）[5]。近 30 年来科研工作者发现，蜜蜂幼虫信息素中单一或组合的脂肪酸酯能够引起工蜂不同的行为与发育变化，这些信息素成分随着幼虫日龄和性别的不同而发生变化（工蜂或雄蜂），成年工蜂也相应地调整它们对幼虫的行为反应[1,6-7]。然而目前大多数的研究仅涉及信息素的化学成分鉴定和功能描述，对蜜蜂信息素的合成途径研究较少。He 等研究发现 E-β-罗勒烯可能是工蜂幼虫的饥饿信息素，并利用 RNA-seq 技术分析推测出 E-β-罗勒烯在工蜂幼虫体内通过乙酰辅酶 A 和乙酰乙酰辅酶 A 从头合成的生物合成通路及其调控候选基因[8]。进一步研究表明，蜜蜂蜂王与雄蜂幼虫通过同样的通路在体内从头合成 E-β-罗勒烯作为其饥饿信息素来乞求食物[9]。蜜蜂是一种变态发育的昆虫，其生活史要经过卵、幼虫、蛹和成虫 4 个阶段。在正常的蜂巢环境中（巢房温度约 35℃），卵孵化成幼虫并发育到一定时期（蜂王幼虫 5 d、工蜂幼虫 6 d、雄蜂幼虫 7 d），成年工蜂就会从蜡腺分泌蜂蜡对其巢房进行封盖，以利于幼虫在一个干净、稳定的封闭环境中化蛹[10]。Le Conte 等研究发现，蜜蜂幼虫信息素中 MP、MO、ML 和 MLN 4 种成分中的单一或混合成分都可以诱导成年工蜂对幼虫巢房的封盖行为[11]。Qin 等利用 RNA-seq 技术分析推测出西方蜜蜂工蜂与雄蜂幼虫利用乙酰辅酶 A 合成 MP、MO、ML 和 MLN 的生物合成通路以及 12 个调控候选基因，并利用稳定同位素示踪剂证实了这些封盖信息素成分是由幼虫合成的，而不是从它们的食物中获得[12]。蜂螨是危害蜜蜂最严重的寄生性病害之一，导致全球蜜蜂很高的死亡率[13]。研究表明，大蜂螨能利用蜜蜂幼虫信息素找到即将封盖的工蜂或雄蜂幼虫[5,11]，在幼虫被封盖前潜入其巢房内，吸食幼虫血淋巴，繁殖后代，而蜂螨更偏好于寄生在雄蜂的幼虫巢房内[10]。东方蜜蜂和西方蜜蜂是目前世界上广泛饲养的两个蜂种，它们在外部形态、个体发育、生活习性和学习记忆等生物学特性方面存在较大差异[14-16]。目前对蜜蜂信息素的研究大多集中在以意大利蜜蜂（*Apis mellifera ligustica*）为代表的西方蜜蜂上。虽然近年来我国学者逐渐对东方蜜蜂信息素展开研究[17-18]，获得了一些东方蜜蜂化学通讯的知识，但相对于西方蜜蜂而言，人们对东方蜜蜂信息素仍知之甚少，尤其是信息素分子水平层面的研究较为缺乏。为了丰富对东方蜜蜂信息素的了解，分别利用 GC/MS 技术和 RNA-seq 技术，分析封盖信息素化学成分在不同封盖时期的东方蜜蜂工蜂与雄蜂幼虫体内的含量变化及转录组差异，验证 MP、MO、ML 和 MLN 是与东方蜜蜂封盖行为相关的信息素并探究其生物合成分子机理，进一步揭示东方蜜蜂幼虫与成年工蜂之间的信息素交流内在机制。

## 1 材料与方法

信息素鉴定试验于 2017 年 4~5 月在江西中烟工业有限责任公司完成，转录组测序分析于 2017 年 4~8 月在北京百迈客生物科技有限公司完成，荧光定量 PCR 验证于 2018 年 4~5 月在江西农业大学蜜蜂研究所完成。

### 1.1 供试昆虫

供试样品均采自中华蜜蜂（Apis cernana cernana）蜂群，由江西农业大学蜜蜂研究所（28.46°N，115.49°E）根据标准的养蜂技术饲养。

幼虫样品均分为以下 3 个不同封盖时期：4 日龄未封盖的工蜂与雄蜂幼虫（4-day-old WL/DL）；正在封盖的工蜂与雄蜂幼虫（capping WL/DL），即蜂房上的蜡盖正在被筑建的幼虫；已封盖的工蜂与雄蜂幼虫（capped WL/DL），样品取自完全封盖的蜂房，蜡盖下有一层薄薄的茧衣，工蜂约 8 日龄，雄蜂约 9 日龄。各组幼虫均为从蜂箱中直接取出巢脾新鲜采集，每个生物学重复分别来自于群势相同的不同健康蜂群。

### 1.2 气相色谱—质谱（GC/MS）分析

3 个不同封盖时期的幼虫样品数量分别为 4 日龄未封盖的工蜂与雄蜂幼虫各 40 只，正在封盖和已封盖的工蜂幼虫各 20 只，正在封盖和已封盖的雄蜂幼虫各 10 只，每组样品进行 4 个生物学重复。将采集的幼虫样品（未经清洗）分别放入装有 3 mL 二氯甲烷（分析纯，南京化学试剂股份有限公司）的玻璃瓶中，并添加 20 μL 甲基十九烷酸酯（≥98%，标准品，10 μg/mL，Sigma）作为内标。随后，将装有幼虫的玻璃小瓶放置在水平脱色摇床（ZD-9560，江苏盛蓝仪器制造有限公司）上轻轻摇晃 30 min（120 r/min），最后将上清液转移到干净的玻璃瓶中并用氮气浓缩至约 20 μL。分别取 1 μL 浓缩后的样品注入气相色谱—质谱（GC/MS）系统（7890A/5975C，Agilent）进行检测，具体参数参照 Qin 等[12]的报道。

在 MP、MO、ML 和 MLN 的基峰重建了单个离子监测色谱图，内部标准分别为质荷比（m/z）270、264、294、292 和 312。对于感兴趣的化合物，通过参考每种化合物的校准曲线，将峰面积转换为数量。外部标准品为 MP、MO、ML 和 MLN（≥99.5%，Sigma），7 个浓度的外标（0.25、0.5、1、2、5、10 和 25 μg/mL）被用来构建标准曲线。所构建的 4 条标准曲线的相关系数均大于 99.96%，说明 GC/MS 系统足够稳定[19]，可用于后期对蜜蜂幼虫 MP、MO、ML 和 MLN 的定量分析。

### 1.3 RNA-seq 分析

采集未封盖、正在封盖和已封盖的工蜂与雄蜂幼虫各 6 只，每组样品进行 3 个生物学重复。取样时将幼虫放入 5 mL EP 管中，随后迅速放入液氮速冻，并于 -80℃ 保存。按标准 TRIzol Reagent 法提取 RNA（Life technologies，California，USA），并使用琼脂糖凝胶电泳和核酸蛋白分析仪（IMPLEN）对 RNA 进行质量检测，将检测合格的 RNA 样品用于 cDNA 文库构建。cDNA 文库的构建和测序由北京百迈客生物科技有限公司完成，具体操作方法参见余爱丽等[20]的报道。

对测序获得的 18 个样品的 Raw Data 进行去接头、去引物序列、过滤低质量 Reads 等质量控制，最终获得高质量 Clean Reads。采用中华蜜蜂的全基因组序列（ftp://ftp.ncbi.nlm.nih.gov/genomes/all/GCA/002/290/385/GCA_002290385.1_ApisCC1.0/）作为参考基因进行比对。以 FPKM（Fragments per Kilobase of transcript per Million fragments mapped）作为衡量指标，对测序样品的基因表达量进行统计。

用斯皮尔曼相关系数 $r^2$ 评估各组样品 3 个生物学重复的相关性[21]。以 fold change ≥ 2 且 FDR < 0.05 作为筛选差异表达基因（differential expressed gene，DEG）的标准检测各幼虫样品组间的基因表达差异。

为了进一步解读差异表达基因的功能，探究东方蜜蜂幼虫封盖信息素的生物合成途径，利用 KEGG（Kyoto Encyclopedia of Genes and Genomes）数据库对差异表达基因进行通路（Pathway）注释和富集分析。首先将差异表达基因最长的转录本转化为蛋白序列，然后利用 BLAST 软件与 KEGG 蛋白数据库进行比较，界定标准为 e < $10^{-5}$。为了识别 RNA-seq 数据中假定的代谢通路富集，使用基于 KEGG Orthology 的注释系统 2.0（KOBAS 2.0）分析 KEGG 映射的结果。KOBAS 2.0 将提供的基因数据库与代谢和信号通路的多个现有数据库进行对照，以识别 RNA-seq 样本中富集的通路[22]。

### 1.4 差异表达基因 qRT-PCR 验证

从本研究的转录组测序结果中推测出东方蜜蜂工蜂与雄蜂幼虫封盖信息素的生物合成通路及参与该通路的 11 个基因。为了验证转录组测序结果的可靠性，以中华蜜蜂工蜂幼虫为实验材料，随机选取 5 个参与该通路的候选基因进行 qRT-PCR 定量分析：*Kat*（gene-APICC_00421：3-ketoacyl-CoA thiolase）、*Hadha*（gene-APICC_06057：Trifunctional enzyme subunit alpha）、*Mecr*（gene-APICC_00389：PREDICTED: probable trans-2-enoyl-CoA reductase, mitochondrial）、*Hacd*（gene-APICC_02088：PREDICTED: very-long-chain（3R）-3-hydroxyacyl-CoA dehydratase hpo-8）和 Δ11 desaturase（gene-APICC_01118：Acyl-CoA Delta（11）desaturase）。

采集未封盖、正在封盖和已封盖的工蜂幼虫各 3 只，每组样品进行 3 个生物学重复。将采集的样品迅速放入液氮中速冻，并于 -80℃保存。按标准 TRlzol Reagent 法提取 RNA 并检测 RNA 样品的质量，随后用反转录试剂盒（TaKaRa）将检测合格的总 RNA 反转录合成第一链 cDNA。用 Primer 5.0 软件进行基因引物设计，并由上海生物工程有限公司进行引物合成（引物序列详见表Ⅲ-17-1）。选用 β-actin 为"管家基因"。qRT-PCR 反应体系（10 μL）：cDNA 模板 1 μL，上、下游引物各 0.4 μL（10 μmol/L），SYBR® Premix Ex Taq™ Ⅱ 5 μL，无 RNA 酶水 3.2 μL。qRT-PCR 反应条件：95℃预变性 30 s；95℃ 10 s，60℃ 1 min，40 个循环。扩增反应结束后继续从 50℃缓慢加热到 90℃，建立 PCR 产物的熔解曲线。对于每个基因，每个样品进行 3 个技术重复。用公式 $2^{-\Delta\Delta Ct}$ 计算各个基因在各样品中的相对表达水平[23-24]，然后做平方根转换后进行方差分析，检验各基因在工蜂幼虫不同封盖时期的表达量差异情况。

表 Ⅲ-17-1　实时荧光定量 PCR 引物

| 基因编号 | 基因名称 | 正向引物 | 反向引物 |
| --- | --- | --- | --- |
| gene-APICC_00421 | *Kat* | CCACGGCCTCAAACGACTC | CGGCACCATCAGAAATACCAG |
| gene-APICC_06057 | *Hadha* | GGAGGAGGCTTGGAGATGG | GCACCAGGCAAGATACCTAACA |
| gene-APICC_00389 | *Mecr* | TGCCTTCACCAAAATTAGCCC | CCATGCTGTCATCCAAAATCC |
| gene-APICC_01118 | Δ11 desaturase | GCTTGTCAATTCCGCTGCTC | AACCTTCGCCAAGTGCTCCTA |
| gene-APICC_02088 | *Hacd* | GCAGCTTCATCTTTTGCTCTTC | CTATAACTCCATGCTTCTGGGTG |
|  | *β-actin* | TCCTGGAATCGCAGATAGAATG | GGAAGGTGGACAAAGAAGCAAG |

## 1.5 数据统计方法

使用 StatView 5.01（SAS Institute，Cary，NC，USA）分析 GC/MS 和 qRT-PCR 的数据。数据采用"ANOVA and $t$-test"中的"ANOVA or ANCOVA"进行统计分析，各组间用 Fisher's PLSD 进行差异显著性比较分析。组间差异被认为是显著的概率水平为 0.05。

## 2 结果

### 2.1 不同封盖时期幼虫封盖信息素含量

在工蜂中，4 个封盖信息素成分在正在封盖和已封盖幼虫体内的含量均显著高于 4 日龄未封盖的幼虫。其中，MP 和 MO 的含量均随幼虫年龄的增长而增加，且在 3 个封盖时期间差异显著（$P < 0.05$），而 ML 和 MLN 的含量在正在封盖和已封盖的幼虫中差异不显著（$P > 0.05$）（图Ⅲ-17-1）。

在雄蜂中，从 4 日龄未封盖幼虫到已封盖幼虫，4 种信息素成分含量均随年龄增长而增加，且已封盖幼虫的信息素含量显著高于 4 日龄和正在封盖的幼虫（$P < 0.05$）。此外，除了 ML 在正在封盖幼虫体内的含量显著高于 4 日龄未封盖幼虫外（$P < 0.05$），其余 3 种信息素成分在这两个封盖时期的幼虫中均差异不显著（$P > 0.05$）（图Ⅲ-17-2）。

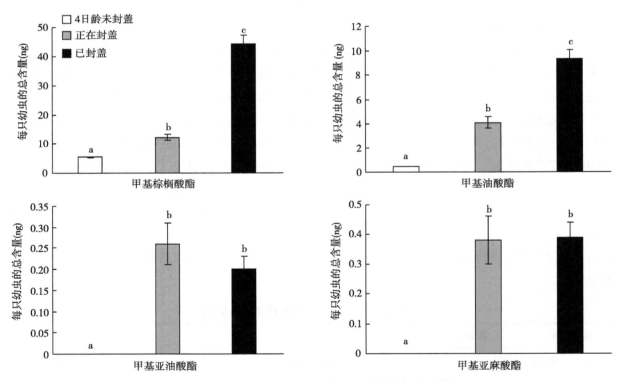

**图Ⅲ-17-1　工蜂不同封盖时期幼虫封盖信息素含量**

图中数据为平均值 ± 标准误，柱上标不同字母表示差异显著（$P < 0.05$）。图Ⅲ-17-2 同。

### 2.2 RNA-seq 分析

完成了 18 个样品 RNA-seq 测序，经过质量控制，共获得 57.26 Gb Clean Data，各样品 Clean Data 均达到 2.65 Gb，且 Q30 碱基百分比均 ≥ 94.74%。分别将各样品的 Clean Reads 与中华蜜蜂参考基因组进行序列比对，比对效率均在 88.89% ~ 95.37%，且 Unique Reads 的比对率 ≥ 87.99%，数据利用率正常。各试验组3个生物学重复的斯皮尔曼相关系数 $r^2$ 均大于0.75，说明本研究中工蜂与雄蜂各封盖时期幼虫3个重复样品的相关性较强。

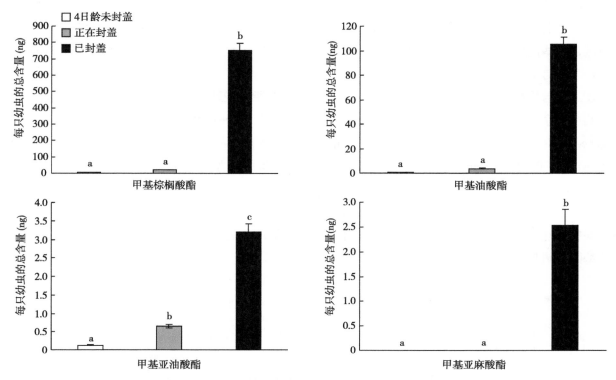

图Ⅲ-17-2　雄蜂不同封盖时期幼虫封盖信息素含量

对工蜂与雄蜂3个不同封盖时期幼虫的基因表达量进行组间比较分析，分别获得4 299和3 926个差异表达基因。在未封盖与正在封盖、正在封盖与已封盖以及未封盖与已封盖3个比较组中，工蜂幼虫分别有62、3 288和3 701个基因存在表达差异，其中表达上调的基因数分别为20、1 812和2 022个，表达下调的基因数分别为42、1 476和1 679个（图Ⅲ-17-3A，见536页）；雄蜂幼虫分别有355、2 343和3 489个基因存在表达差异，其中表达上调的基因数分别为204、1 517和1 936个，表达下调的基因数分别为151、826和1 553个（图Ⅲ-17-3C）。分别对工蜂与雄蜂幼虫的差异表达基因集进行集合分析（http://bioinformatics.psb.ugent.be/webtools/ Venn/），在未封盖与正在封盖、正在封盖与已封盖以及未封盖与已封盖3个比较组中，工蜂幼虫有21个共有的差异表达基因，4、587和983个独有的差异表达基因（图Ⅲ-17-3B）；雄蜂幼虫有116个共有的差异表达基因，58、341和1 382个独有的差异表达基因（图Ⅲ-17-3D）。

在KEGG注释分析中，工蜂不同封盖时期幼虫间的所有差异表达基因共注释到152个KEGG通路，其中未封盖与正在封盖、正在封盖与已封盖以及未封盖与已封盖3个比较组中的差异表达基因分别注释到31、99和22个通路。雄蜂不同封盖时期幼虫间的所有差异表达基因共注释到130个KEGG通路，其中上述3个比较组中的差异表达基因分别注释到65、128和125个通路。KEGG分类注释结果表明，上述KEGG通路涉及新陈代谢、细胞过程、遗传信息处理、环境信息处理和有机系统5个方面，其中与代谢有关的通路数量最多。

进一步对注释到的所有 KEGG 通路进行分析，发现两个脂肪酸延伸通路（ko00062：Fatty acid elongation）和不饱和脂肪酸生物合成通路（ko01040：Biosynthesis of unsaturated fatty acids），它们可能涉及长碳链的生物合成途径：首先由乙酰辅酶 A 通过脂肪酸延伸通路从头合成十六碳酸（C16），然后以 C16 为前体通过不饱和脂肪酸生物合成通路形成一个或多个双键的不饱和脂肪酸，如油酸和亚油酸。据报道，一些昆虫通过这一途径进行信息素的生物合成[25-27]。因此推测，这两个 KEGG 通路可能与东方蜜蜂工蜂与雄蜂幼虫封盖信息素的生物合成有关。

在工蜂与雄蜂幼虫中，共有 11 个差异表达基因富集在脂肪酸延伸通路和不饱和脂肪酸生物合成通路。其中 5 个基因 gene-APICC_00389、gene-APICC_00421、gene-APICC_05477、gene-APICC_06057 和 gene-APICC_10032 富集在脂肪酸延伸通路，4 个基因 gene-APICC_01118、gene-APICC_01590、gene-APICC_03413 和 gene-APICC_06057 富集在不饱和脂肪酸生物合成通路，此外还有 2 个基因 gene-APICC_05451 和 gene-APICC_02088 同时富集到上述两个通路中。推测这 11 个基因可能参与调控东方蜜蜂工蜂与雄蜂幼虫封盖信息素 MP、MO、ML 和 MLN 的生物合成过程（表Ⅲ-17-2）。

根据上述结果，提出东方蜜蜂工蜂与雄蜂幼虫封盖信息素的生物合成通路，以及调控这些通路的相关基因（图Ⅲ-17-4）。

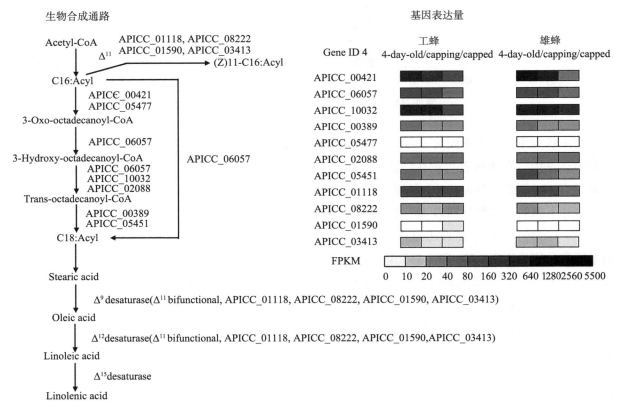

**图Ⅲ-17-4 东方蜜蜂工蜂与雄蜂幼虫封盖信息素生物合成通路及相关调控基因**

网格表示工蜂与雄蜂不同封盖时期幼虫的基因表达量。每个格子的灰度等级代表各个基因在幼虫中的绝对表达量，各灰度等级对应的 FPKM 值分别为 0～10、10～20、20～40、40～80、80～160、160～320、320～640、640～1280、1280～2560 和 2560～5500。

## 2.3 qRT-PCR

调控东方蜜蜂工蜂与雄蜂幼虫封盖信息素生物合成的 5 个候选基因 *Kat*、*Hadha*、*Mecr*、*Hacd* 和 Δ11 desaturase 在封盖前后 3 个时期的表达量上、下调趋势和差异情况均与转录组测序结果相一致（图Ⅲ-17-5）。qRT-PCR 结果验证了转录组测序结果的可靠性。

表 Ⅲ-17-2　工蜂与雄蜂幼虫封盖信息素生物合成相关候选基因注释

| 基因编号 | 基因注释 |
| --- | --- |
| gene-APICC_00421 | 3-ketoacyl-CoA thiolase (*Apis cerana cerana*) |
| gene-APICC_06057 | Trifunctional enzyme subunit alpha (*Apis cerana cerana*) |
| gene-APICC_10032 | enoyl-CoA hydratase (*Apis cerana cerana*) |
| gene-APICC_00389 | PREDICTED: probable trans-2-enoyl-CoA reductase, mitochondrial (*Apis cerana*) |
| gene-APICC_05477 | PREDICTED: elongation of very long chain fatty acids protein 4-like isoform X1 (*Apis dorsata*) |
| gene-APICC_02088 | PREDICTED: very-long-chain (3R)-3-hydroxyacyl-CoA dehydratase hpo-8 (*Apis cerana*) |
| gene-APICC_05451 | Trans-2,3-enoyl-CoA reductase (*Apis cerana cerana*) |
| gene-APICC_01118 | Acyl-CoA Delta (11) desaturase (*Apis cerana cerana*) |
| gene-APICC_08222 | Acyl-CoA desaturase (*Apis cerana cerana*) |
| gene-APICC_01590 | PREDICTED: acyl-CoA Delta (11) desaturase-like (*Apis cerana*) |
| gene-APICC_03413 | Acyl-CoA desaturase (*Apis cerana cerana*) |

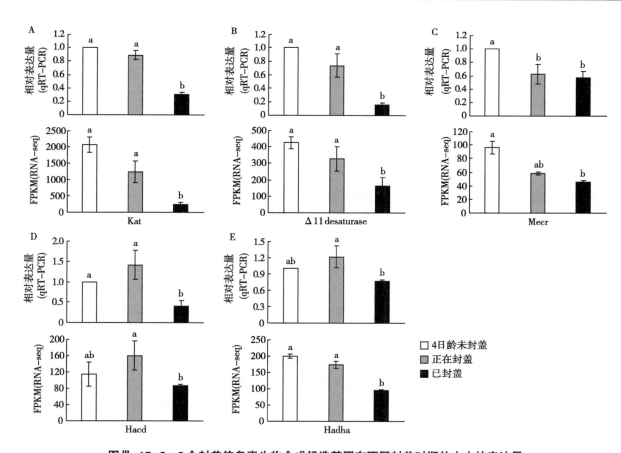

图 Ⅲ-17-5　5 个封盖信息素生物合成候选基因在不同封盖时期幼虫中的表达量

图中数据为平均值 ± 标准误，柱上标不同字母表示表达量差异显著（qRT-PCR：$P < 0.05$；RNA-seq：FDR≤0.05）

## 3 讨论

Le Conte 等提出 MP、MO、ML 和 MLN 是诱导蜜蜂封盖行为的重要信号[11]。本研究表明，工蜂与雄蜂幼虫在被封蜡盖的关键阶段均增加了这 4 种信息素成分的释放量，这与前人报道的西方蜜蜂结果相一致[12,28]，进一步验证了 MP、MO、ML 和 MLN 是与东方蜜蜂封盖行为相关的信息素。

在蜂螨的生活周期里，可分为体外寄生期和蜂房繁殖期两个不同的时期。体外寄生期的雌性成螨寄生在巢房外的成年蜂体上，依靠吸取成年蜂的血淋巴生活，当巢房内的蜜蜂幼虫将要被工蜂封盖时，便潜入到幼虫巢房中，并于巢房封盖后转移到幼虫或预蛹体上，靠吮吸其血淋巴补充营养，随后产卵繁殖。蜂螨更偏好于寄生在雄蜂的幼虫巢房内[10]。据报道，在蜂群中有 3 种 BEP 成分对蜂螨有吸引作用，其中 MP 对蜂螨的吸引力最强，而 EP 和 ML 的作用相对较弱[5]。在本研究中，MP 和 ML 在正在封盖的雄蜂幼虫中的含量均显著高于同发育时期的工蜂幼虫，试验数据支持早期的研究结果，即这些 BEP 成分可以被蜂螨用来从正在封盖的工蜂幼虫中区分和挑选出雄蜂幼虫。

颜伟玉等[17]测定了 10 种幼虫信息素成分在中华蜜蜂不同日龄幼虫和成年蜂个体中的含量分布情况，张含等[18]和曾云峰等[29]先后研究了幼虫信息素 3 种酯类成分 MP、EP 和 EO 对中华蜜蜂工蜂和蜂王个体发育以及工蜂采集、哺育和封盖等行为的影响，并与意大利蜜蜂作比较，均发现中华蜜蜂幼虫与意大利蜜蜂幼虫所分泌的幼虫信息素的成分和含量不同，且这两个蜂种的工蜂对幼虫信息素的反应灵敏度也不同。本研究中，东方蜜蜂工蜂与雄蜂幼虫的 MO、ML 和 MLN 含量整体来说均低于已报道的相同实验条件下的西方蜜蜂[12]，笔者推测这可能是由于东方蜜蜂嗅觉更灵敏[16]，较低的信息素水平即可引起工蜂对幼虫的封盖行为。然而，MP 在东方蜜蜂工蜂与雄蜂幼虫中的含量却高于西方蜜蜂。除了能引发工蜂的封盖行为外，MP 还被报道与工蜂的发育和采集等行为有关[18,29]，但东方蜜蜂幼虫 MP 含量高的具体原因还有待进一步研究。

信息素的前体主要有 3 个来源：从头合成、宿主植物或基质提供的前体的转化以及宿主分子的直接结合，但大部分的信息素还是通过昆虫自身的从头合成而来[30]。脂类信息素广泛存在于鳞翅目昆虫中且种类繁多，其生物合成通常是在一系列酶的作用下，利用软脂酸和硬脂酸等脂肪酸中间产物作为前体，经过碳链缩短、去饱和化和官能团修饰等步骤逐步形成信息素。本研究利用转录组测序技术分析了 3 个不同封盖时期的东方蜜蜂工蜂与雄蜂幼虫各组间的差异表达基因，并对差异表达基因进行 KEGG 注释和富集分析，发现脂肪酸延伸和不饱和脂肪酸生物合成通路有差异表达基因显著富集，由此笔者推测出东方蜜蜂工蜂与雄蜂幼虫封盖信息素可能的生物合成通路及相关的调控基因。这些生物合成通路是由乙酰辅酶 A 从头合成 C16，然后以 C16 为前体通过去饱和作用形成一个或多个双键的不饱和脂肪酸。在许多昆虫中，乙酰辅酶 A 是主导脂肪酸和酯类生物合成的几个不同代谢途径的前体[31-33]，昆虫能利用上述途径进行信息素的生物合成[26-28]，如在家蚕、钩端螺旋体和大豆中均发现了类似的甲酯合成途径[26-27,34-35]。这些研究支持关于东方蜜蜂幼虫 4 种封盖信息素成分生物合成通路的推测。本研究中提出的东方蜜蜂工蜂与雄蜂幼虫封盖信息素生物合成通路与 Qin 等[12]报道的西方蜜蜂工蜂与雄蜂幼虫封盖信息素生物合成通路相同。东方蜜蜂与西方蜜蜂生存的自然环境、气候条件和植物区系等因素不同，其生物学特性存在较大的差异，如外部形态、个体发育、生活习性和学习记忆等[14-16]。东方蜜蜂和西方蜜蜂是同属于蜜蜂属的两个蜂种，其遗传背景相似，都具有蜜蜂属的特性，如高度的社会性生活、采蜜和贮蜜以及筑造双面六边形巢房等。已有研究表明，东方蜜蜂幼虫含有与西方蜜蜂幼虫相同的10种脂肪酸酯类信息素[17],其中一些信息素成分已被证实在影响工蜂发育和行为方面具有与西方蜜蜂相似的效应。因此,笔者推测东方蜜蜂与西方蜜蜂有可能利用相同的生物合成途径进行信息素的生物合成。

作用于信息素合成过程的脱氢酶基因已在许多昆虫中被发现。在本研究中,2个 *Δ11-desaturase* 基因(gene-APICC_01118 和 gene-APICC_01590)以及2个 *Acyl-CoA desaturase* 基因(gene-APICC_08222 和 gene-APICC_03413)富集在笔者提出的东方蜜蜂幼虫封盖信息素的生物合成通路,且GO注释表明 gene-APICC_08222 和 gene-APICC_03413 具有 Δ11-desaturase 的生物功能。Δ11 脱氢酶是蛾类和蝴蝶类在C11上形成双键最主要的酶[36-41]。Δ11-desaturase 是一种生物功能酶,具有类似于 Δ11 和 Δ12 的去饱和活性,在昆虫信息素生物合成过程中发挥着重要作用[42-46]。Roelofs 等研究发现,红带卷叶蛾、大白菜尺蛾和家蚕中的油酸酯、亚油酸酯和亚麻酸酯是从头合成的,并由 Δ11-desaturase 去饱和[47]。因此,推测这4个基因可能参与了东方蜜蜂幼虫封盖信息素成分的生物合成途径。将参与调控东方蜜蜂封盖信息素生物合成通路的11个候选基因与已报道的西方蜜蜂的12个候选基因进行比对,发现东方蜜蜂中的10个候选基因(gene-APICC_05477 除外)与西方蜜蜂中的10个候选基因有相同的注释或基因功能,这些基因分别在东方蜜蜂和西方蜜蜂幼虫封盖信息素生物合成通路中发挥着相同的调控作用。进一步比较这些候选基因在未封盖、正在封盖和已封盖工蜂幼虫与雄蜂幼虫中的表达量可知,这些基因在同一封盖时期的工蜂与雄蜂幼虫中的表达量相近。上述比较结果进一步为这些基因作为参与蜜蜂幼虫封盖信息素生物合成通路的调控基因提供了证据,但确切的调控机制还有待进一步研究。

## 4　结论

东方蜜蜂工蜂幼虫与雄蜂幼虫在被封蜡盖的关键阶段增加了甲基棕榈酸酯(MP)、甲基油酸酯(MO)、甲基亚油酸酯(ML)和甲基亚麻酸酯(MLN)的释放量,进一步验证了它们是与蜜蜂封盖行为相关的信息素,推测这些信息素可能是在相关基因的调控下由乙酰辅酶A从头合成,而东方蜜蜂幼虫与西方蜜蜂幼虫可能利用相同的生物合成通路进行信息素的生物合成。

**致谢**:本论文得到了江西中烟工业有限责任公司郭磊和江西农业大学麻俊武老师的帮助,在此表示感谢!本研究得到国家自然科学基金(31872432)和国家蜂产业技术体系(CARS-44-kxj15)资助。

## 参考文献

[1] FREE J B, WINDER M E. Brood recognition by honey bee (Apis mellifera) workers. *Animal Behavior*, 1983, 31: 539-545.

[2] HAYDAK M H. Honey bee nutrition. *Annual review of entomology*, 1970, 15: 143-156.

[3] HUANG Z Y, OTIS G W. Inspection and feeding of larvae by worker honey bees (Hymenoptera: Apidae): effect of starvation and food quantity. *Journal of insect behavior*, 1991, 4(3): 305-317.

[4] HUANG Z Y, OTIS G W. Nonrandom visitation of brood cells by worker honey bees (Hymenoptera: Apidae). *Journal of Insect Behavior*, 1991, 4(2): 177-184.

[5] LE CONTE Y, ARNOLD G, TROUILLER J, MASSON C, CHAPPE B, OURISSON G. Attraction of the parasitic mite

Varroa to the drone larvae of honey bees by simple aliphatic esters. *Science*, 1989, 245(4918): 638-639.

[6] SLESSOR K N, WINSTON M L, LE CONTE Y. Pheromone communication in the honeybee (*Apis mellifera* L.). *Journal of Chemical Ecology*, 2005, 31(11): 2731-2745.

[7] LE CONTE Y. The recognition of larvae by worker honeybees. *Naturwissenschaften*, 1994, 81: 462-465.

[8] HE X J, ZHANG X C, JIANG W J, BARRON A B, ZHANG J H, ZENG Z J. Starving honey bee (*Apis mellifera*) larvae signal pheromone ally to worker bees. *Scientific Reports*, 2016, 6: 22359.

[9] 何旭江, 江武军, 颜伟玉, 曾志将. 蜜蜂蜂王与雄蜂幼虫饥饿信息素鉴定及其生物合成通路. 中国农业科学, 2016, 49(23): 4646-4655.

[10] 曾志将. 蜜蜂生物学. 北京: 中国农业出版社, 2007.

[11] LE CONTE Y, ARNOLD G, TROUILLER J, MASSON C, CHAPPE B. Identification of a brood pheromone in honeybees. *Naturwissenschaften*, 1990, 77: 334-336.

[12] QIN Q H, HE X J, BARRON A B, GUO L, JIANG W J, ZENG Z J. The capping pheromones and putative biosynthetic pathways in worker and drone larvae of honey bees Apis mellifera. *Apidologie*, 2019, 50(6): 793-803.

[13] TROUILLER J, ARNOLD G, CHAPPE B, LE CONTE Y, MASSON C. Semiochemical basis of infestation of honey bee brood by Varroa jacobsoni. *Journal of Chemical Ecology*, 1992, 18(11): 2041-2053.

[14] 曾志将. 养蜂学. 3版. 北京: 中国农业出版社, 2017.

[15] 陈盛禄. 中国蜜蜂学. 北京: 中国农业出版社, 2001.

[16] QIN Q H, HE X J, TIAN L Q, ZHANG S W, ZENG Z J. Comparison of learning and memory of Apis cerana and Apis mellifera. *Journal of Comparative Physiology* A, 2012, 198(10): 777-786.

[17] 颜伟玉, LE CONTE Y, BESLAY D, 曾志将. 中华蜜蜂幼虫信息素鉴定. 中国农业科学, 2009, 42(6): 2250-2254.

[18] 张含, 曾志将, 颜伟玉, 吴小波, 郑云林. 信息素中三种酯类对中华蜜蜂工蜂发育和采集行为的影响. 昆虫学报, 2010, 53(1): 55-60.

[19] TREFZ P, KISCHKEL S, HEIN D, JAMES E S, SCHUBERTA J K, MIEKISCH W. Needle trap micro-extraction for VOC analysis: effects of packing materials and desorption parameters. *Journal of Chromatography* A, 2012, 1219: 29-38.

[20] 余爱丽, 赵晋锋, 成锴, 王振华, 张鹏, 刘鑫, 田岗, 赵太存, 王玉文. 谷子萌发吸水期关键代谢途径的筛选与分析. 中国农业科学, 2020, 53(15): 3005-3019.

[21] LOVE M I, HUBER W, ANDERS S. Moderated estimation of fold change and dispersion for RNA-seq data with DESeq2. *Genome Biology*, 2014, 15: 550.

[22] XIE C, MAO X, HUANG J, DING Y, WU J M, DONG S, KONG L, GAO G, LI C Y, WEI L P. KOBAS 2.0: a web server for annotation and identification of enriched pathways and diseases. *Nucleic Acids Research*, 2010, 39: W316-W322.

[23] LIU W, SAINT D A. A new quantitative method of real time reverse transcription polymerase chain reaction assay based on simulation of polymerase chain reaction kinetics. *Analytical Biochemistry*, 2002, 302(1): 52-59.

[24] DVINGE H, BERTONE P. HTqPCR: high-throughput analysis and visualization of quantitative real-time PCR data in R. *Bioinformatics*, 2009, 25(24): 3325-3326.

[25] ROELOFS W, BJOSTAD L. Biosynthesis of lepidopteran pheromones. *Bioorganic Chemistry*, 1984, 12(4): 279-298.

[26] ANDO T, HASE T, ARIMA R, UCHIYAMA M. Biosynthetic pathway of bombykol, the sex pheromone of the female silkworm moth. *Agricultural and Biological Chemistry*, 1998, 52(2): 473-478.

[27] TANG J D, CHARLTON R E, JURENKA R A, WOLF W A, PHELAN P L, SRENG L, ROELOFS W L. Regulation of pheromone biosynthesis by a brain hormone in two moth species. *Proceedings of the National Academy of Sciences of the United States of America*, 1989, 86(6): 1806-1810.

[28] LE CONTE Y, ELLIS M, RITTER W. Varroa mites and honey bee health: can Varroa explain part of the colony losses? *Apidologie*, 2010, 41(3): 353-363.

[29] 曾云峰，曾志将，颜伟玉，吴小波. 幼虫信息素中三种酯类对中华蜜蜂和意大利蜜蜂工蜂哺育和封盖行为以及蜂王发育影响. 昆虫学报, 2010, 53(2): 154-159.

[30] TILLMAN J A, SEYBOLD S J, JURENKA R A, BLOMQUIST G J. Insect pheromones—An overview of biosynthesis and endocrine regulation. *Insect Biochemistry and Molecular Biology*, 1999, 29(6): 481-514.

[31] MCGEE R, SPECTOR A A. Fatty acid biosynthesis in Erlich cells. The mechanism of short term control by exogenous free fatty acids. *The Journal of Biological Chemistry*, 1975, 250(14): 5419-5425.

[32] CRIPPS C, BLOMQUIST G J, DE RENOBALES M. De novo biosynthesis of linoleic acid in insects. *Biochimica et Biophysica Acta (BBA) - Lipids and Lipid Metabolism*, 1986, 876(3): 572-580.

[33] MOSHITZKY P, MILOSLAVSKI I, AIZENSHTAT Z, Applebaum S W. Methyl palmitate: a novel product of the medfly (ceratitis capitata) corpus allatum. *Insect Biochemistry and Molecular Biology*, 2003, 33(12): 1299-1306.

[34] STERN N, SHENBERG E, TIETZ A. Studies on the metabolism of fatty acids in leptospira: the biosynthesis of $\Delta 9$- and $\Delta 11$- monounsaturated acids. *European Journal of Biochemistry*, 1969, 8(1): 101-108.

[35] BACHLAVA E, DEWEY R E, BURTON J W, CARDINAL A J. Mapping candidate genes for oleate biosynthesis and their association with unsaturated fatty acid seed content in soybean. *Molecular Breeding*, 2009, 23(2): 337-347.

[36] KNIPPLE D C, ROSENFIELD C L, MILLER S J, LIU W, TANG J, MA P W K, ROELOFS W L. Cloning and functional expression of a cDNA encoding a pheromone gland-specific acyl-CoA $\Delta^{11}$-desaturase of the cabbage looper moth, Trichoplusia ni. *Proceedings of the National Academy of Sciences of the United States of America*, 1998, 95: 15287-15292.

[37] LIENARD M A, STRANDH M, HEDENSTROM E, JOHANSSON T, LÖFSTEDT C. Key biosynthetic gene subfamily recruited for pheromone production prior to the extensive radiation of Lepidoptera. *BMC Evolutionary Biology*, 2008, 8: 270.

[38] DING B J, LIENARD M A, WANG H L, ZHAO C H, LÖFSTEDT C. Terminal fatty-acyl-CoA desaturase involved in sex pheromone biosynthesis in the winter moth (*Operophtera brumata*). *Insect Biochemistry and Molecular Biology*, 2011, 41(9): 715-722.

[39] HAGSTROM A K, LIENARD M A, GROOT A T, HEDENSTROM E, LÖFSTEDT C. Semi-selective fatty acyl reductases from four heliothine moths influence the specific pheromone composition. *PLoS OnE*, 2012, 7(5): e37230.

[40] LIENARD M A, WANG H L, LASSANCE J M, LÖFSTEDT C. Sex pheromone biosynthetic pathways are conserved between moths and the butterfly Bicyclus anynana. *Nature Communications*, 2014, 5: 3957.

[41] Wang H L, Brattström O, Brakefield P M, Francke W, LÖFSTEDT C. Identification and biosynthesis of novel male specific esters in the wings of the tropical butterfly, Bicyclus martius sanaos. *Journal of Chemical Ecology*, 2014, 40(6): 549−559.

[42] 王镜岩, 朱圣庚, 徐长法. 生物化学. 3 版. 北京: 高等教育出版社, 2002.

[43] SERRA M, PIÑA B, BUJONS J, CAMPS F. FABRIAS G. Biosynthesis of 10, 12−dienoic fatty acids by a bifunctional $\Delta$11 desaturase in Spodoptera littoralis. *Insect Biochemistry and Molecular Biology*, 2006, 36(8): 634−641.

[44] LOFSTEDT C, ELMFORS A, SJÖGREN M, WIJK E. Confirmation of sex pheromone biosynthesis from (16−$d_3$) palmitic acid in the turnip moth using capillary gas chromatography. *Experientia*, 1986, 42(9): 1059−1061.

[45] RODRIGUEZ F, HALLAHAN D L, PICKETT J A, CAMPS F. Characterization of the $\Delta^{11}$−palmitoyl−coa−desaturase from spodoptera littoralis (lepidoptera: noctuidae). *Insect Biochemistry and Molecular Biology*, 1992, 22(2): 143−148.

[46] FOSTER S P. Sex pheromone biosynthesis in the tortricid moth planotortrix excessana (walker) involves chain−shortening of palmitoleate and oleate. *Archives of Insect Biochemistry and Physiology*, 1998, 37(2): 158−167.

[47] ROELOFS W, BJOSTAD L. Biosynthesis of lepidopteran pheromones. *Bioorganic Chemistry*, 1984, 12(4): 279−298.

# 蜜蜂基因组生物学及 CRISPR/Cas9 技术
## Honeybee Genome Biology and CRISPR/Cas9

西方蜜蜂基因组序列 2006 年 10 月在 *Nature* 上发表，随后东方蜜蜂基因组及其他蜜蜂基因组测序也相继完成。随着基因组测序技术创新，蜜蜂基因版本也不断更新。蜜蜂生物学也进入"后基因组"时代。利用蜜蜂基因组相关信息，为解析蜜蜂发育及其复杂行为生物学机理提供了更有效的工具。2007 年 3 月 *Science* 发表一篇题为 *CRISPR Provides Acquired Resistance Against Viruses in Prokaryotes* 的研究论文，他们发现 CRISPR/Cas 是细菌和古细菌的一种获得性免疫防御机制，具体原理是细菌或者古细菌可将入侵噬菌体和质粒 DNA 片段整合到 CRISPR 中。CRISPR/Cas9 技术在蜜蜂基因功能验证领域中展现出极大应用前景。

利用单分子实时（SMRT）PacBio 测序技术和高通量染色质构象捕获（Hi-C）技术，获得了中蜂染色体水平的基因组序列。新的中蜂基因组大小为 215.67 Mb，contig N50 为 4.49 Mb，比以前基于 Illumina 的版本提高了 212 倍。该基因组由 126 条 scaffold 组成，包含 16 条拟染色体和 110 个未定位到染色体上的 scaffold。16 条拟染色体总长度为 211.03Mb，占全基因组序列的 97.85%。预测出了 10741 个蛋白质编码基因，其中 9627 个基因得到注释，共鉴定出 314 个新基因。新的高质量中蜂参考基因组为中蜂生物学研究提供精确序列信息。

通过结合长片段测序技术和三维基因组染色体构想捕获技术，将卡尼鄂拉蜜蜂（*Apis mellifera carnica*）基因组组装到染色体水平。通过比较基因组学，我们发现参与社会分工的基因在社会性蜜蜂与独居蜂之间有显著的趋向性。在卡尼鄂拉蜜蜂基因组中，耐寒性基因受到正向选择，这可促进卡尼鄂拉蜜蜂适应北半球气候。卡尼鄂拉蜜蜂基因组对后续比较基因组、亚种形成和人工育种有重要促进作用。

研究了中蜂、意蜂胚胎发育图谱，揭示了蜜蜂胚胎早期合子形成和卵裂细胞迁移规律，为显微注射进行 CRISPR/Cas9 基因敲除注射位点的选择、关王取卵时间确定奠定了科学依据。采用转录组技术分别对 24h、48h 和 72h 卵期进行测序，结果表明：共鉴定到 18 284 个基因，包括 8065 个核心基因和 8819 个新注释基因；通过中蜂、意蜂胚胎发育早期转录组学比较，筛选出大量两者在胚胎早期发育过程中的差异表达基因，为利用 CRISPR/Cas9 技术进行蜜蜂胚胎发育期关键基因功能研究提供了候选靶基因库。之后将 Mrjp1 基因作为靶基因，分别在蜂卵头腹侧和尾背侧注射 sgRNA 和 Cas9 蛋白到 0~2h 蜂卵。结果表明：头腹侧注射的编辑率（93.3%）明显高于尾背侧注射的编辑率（11.8%）。通过 0~2h 蜂卵头腹侧注射 sgRNA 和 Cas9 蛋白实现了蜜蜂靶基因的高效编辑。为了验证该编辑系统的高效性，用同样方法测试了另一基因 Pax6，获得了 100% 的编辑率。此外 TA 克隆结果显示，头腹侧注射的两批蜂卵分别获得了 73.3% 和 76.9% 的双等位基因敲除胚胎。建立的 CRISPR–Cas9 编辑系统实现了蜜蜂胚胎靶基因的一步双等位高效敲除，从而减少了嵌合体蜂王逃逸所导致的生态风险，提高了蜜蜂基因编辑科研工作的生态安全性。

利用 RNA 干扰技术比较分析了人工饲喂 siRNA 和蜜蜂头部注射 siRNA 干扰效果，结果发现：这两种方法都能降低蜜蜂脑部基因表达，但人工饲喂方法需要更大剂量 siRNA。

# 1. A Chromosome-scale Assembly of the Asian Honeybee *Apis cerana* Genome

Zilong Wang[1], Yongqiang Zhu[2], Qing Yan[2], Weiyu Yan[1], Huajun Zheng[2*], Zhijiang Zeng[1*]

1. Honeybee Research Institute, Jiangxi Agricultural University, Nanchang, Jiangxi, China.
2. Shanghai-MOST Key Laboratory of Health and Disease Genomics, Chinese National Human Genome Center at Shanghai, Shanghai, China.

**Abstract**

*Apis cerana* is one of the main honeybee species in artificial farming, which is widely distributed in Asian countries. The genome of *A. cerana* has been sequenced by several different research groups using second generation sequencing technologies. However, it is still necessary to obtain more complete and accurate genome sequences. Herewe present a chromosome-scale assembly of the *A. cerana* genome using single-molecule real-time (SMRT) Pacific Biosciences sequencing and high-throughput chromatin conformation capture (Hi-C) genome scaffolding. The updated assembly is 215.67 Mb in size with a contig N50 of 4.49 Mb, representing an ~212-fold improvement over the previous Illumina-based version. Hi-C scaffolding resulted in 16 pseudochromosomes occupying 97.85% of the assembled genome sequences. A total of 10,741 protein-coding genes were predicted and 9,627 genes were annotated. Besides, 314 new genes were identified compared to the previous version. The improved high-quality *A. cerana* reference genome will provide precise sequence information for biological research of *A. cerana*.

**Keywords:** *Apis cerana*, genome assembly, Hi-C, SMRT, chromosome-scale

**Introduction**

*Apis cerana* is one of the main honeybee species that can be systematically raised by human beings and has been widely cultivated in the eastern countries such as China, Japan and India for a long time, bringing considerable economic benefits to beekeepers. Compared with western honeybees, *A. cerana* is more sensitive to smell, better at using sporadic honey sources, and more suitable for collecting a variety of honey plants, while western honeybees usually prefer to collect large single honey source. Moreover, the *A. cerana* has stronger disease and stress resistance, stronger cold tolerance, lower feed consumption and longer collection period compared to western honeybees (Chen, 2001).

---

\* Corresponding authors: bees1965@sina.com (ZJZ); zhenghj@chgc.sh.cn (HJZ).
注：此文发表在 *Frontiers in Genetics* 2020, 11:279。

Genome sequences are of great significance to the basic biological research of a species. The first genome assembly of *A. cerana* of 228.32 Mb was reported by a Korean research group in 2015 (Park *et al.*, 2015). Then, in 2018 our research group reported the genome assembly of the Chinese native species *A. cerana cerana*, with genome size of 228.79 Mb (Diao *et al.*, 2018). Also, the genome assembly of another important eastern honeybee subspecies *A. cerana* japonica of 211.20 Mb is available in Genbank database (NCBI GCA_002217905.1).

The third generation sequencing (TGS) technologies do not need PCR amplification, which can effectively avoid systematic errors caused by bias of PCR amplification. At the same time, the lengths of DNA sequence fragments sequenced from a single run are so long that the average length of the reads can reach 10,000 bp, which is very helpful for assembling repetitive sequences. Compared with the second generation sequencing technologies, the TGS technologies also maintain the advantages of high throughput and low cost. Typical TGS technologies are single-molecule real-time (SMRT) sequencing technology developed by Pacific Biosciences company and Nanopore sequencing technology launched by Oxford Nanopore company, which are now widely used in genome sequencing of animals and plants (Jiang *et al.*, 2014; Jiao *et al.*, 2017; Gong *et al.*, 2018; Kronenberg *et al.*, 2018; Ghurye *et al.*, 2019; Low *et al.*, 2019; Wallberg *et al.*, 2019; Zhang *et al.*, 2019).

Here we used PacBio and Hi-C technologies to generate a highly contiguous de novo assembly of the eastern honeybee, *A. cerana*. The PacBio long reads from haploid drones were assembled into contigs, then, they were further assembled into scaffolds using Hi-C proximity ligation data. The completeness and contiguity of this new assembly was greatly improved compared to previous genome assemblies.

## Materials and Methods

### *Library construction and sequencing*

The experimental bees used in this study were 6-day old drone pupae sampled from a single colony in a *A. cerana* apiary raised by a local bee keeper in Fulong Township, Baisha Li Nationality Autonomous County, Hainan province (19°22'26"N, 109°28'20"E), China, in October 2018. The intestines of the pupae were removed to avoid contamination of gut microbes before construction of SMRTbell and Hi-C libraries. Two SMRTbell libraries were constructed and sequenced. For each library, DNA was extracted from a singleintestine removed drone pupa by AxyPrep™ Multisource Genomic DNA Miniprep Kit (Axygen, USA). Genomic DNA concentration was measured using the Qubitfluorimetry system with the High Sensitivity kit for detection of double-stranded DNA (Thermo Fisher, USA). Fragment size distribution of the genomic DNA was assessed using the Agilent 2100 Bioanalyzer with the 12000 DNA kit (Agilent, USA). Then, 5 μg of high molecular weight genomic DNA was sheared using g-Tube (Covaris, USA) to 10 kb, and the sheared DNA was used as input into the SMRTbell library preparation. SMRTbell library was prepared using PacBio 10 kb library preparation protocol.Once library was completed, it was size selected from 10 kb using the Blue Pippin instrument (Sage Science, USA) to enrich for the longest insert size. The library was sequenced on a PacBio Sequel system using Sequencing Kit 3.0, 1200 min movies with 120 min pre-extension and software v6.0 (PacBio).

One Hi-C library was created from a single intestine removed drone pupa of *A. cerana*. Briefly, the drone pupa was fixed with formaldehyde for 10 mins and then was lysed in lysing solution (500 μL 10 mmol/L Tris-HCl pH 8.0, 10 mmol/L NaCl, 0.2% Igepal CA-630 and 50 μL protease inhibitors). Then, the cross-linked genomic DNA was digested with Hind III overnight. Sticky ends of the genomic DNA were biotinylated and proximity-ligated to form chimeric junctions that were enriched for and then physically sheared to a size of 300-700 bp. Chimeric fragments representing the original cross-linked long-distance physical interactions were then processed into paired-end sequencing library, and sequenced on the Illumina HiSeq X Ten platform.

*Genome assembly*

After Pacbio sequencing and removal of low quality reads, the remaining Pacbio clean reads were assembled into contigs using HGAP4 of SMRT Link (version 6.0) (Chin et al., 2013). To assemble contigs into scaffolds, long-range contact reads were generated by Hi-C sequencing and adapter sequences of raw reads for Hi-C sequencing were trimmed and low quality PE reads were removed. Then the raw contigs were split into segments of 50 kb on average. The Hi-C clean data were mapped to these segments using BWA (version 0.7.10-r789) (Li and Durbin, 2009). The uniquely mapped pair-end reads were retained to perform assembly by using LACHESIS software (Burton et al., 2013). Any two segments which showed inconsistent connection with information from the raw contigs were checked manually. These corrected contigs were then assembled with LACHESIS.

*Repetitive sequences and genome annotation*

The RepeatModeler (version 1.73; RepeatModeler, RRID: SCR 015027) (Smit and Hubley, 2008) and RepeatMasker (version 3.3.0; RepeatMasker, RRID: SCR 012954) (Chen, 2004) were used to detect repeat sequences in the assembled genome. Gene prediction was performed using Augustus (version 3.3.2) (Stanke et al., 2006)with an *A. mellifera* model and known *A. cerana* genes as input. First, known *A. cerana* genes were mapped to the new genome assembly by BLAT v35 (Kent, 2002), then the Augustus scripts were used to create a hints file. De novo gene structure identification was performed using Augustus. The functional annotation of predicted genes was performed through BlastP against the NCBI non-redundant peptide database (NR), with parameters setting at E-value 1e-5. Gene ontology analysis was performed using Blast2GO (Conesa et al., 2005) through BlastP against the Swiss-Prot database (Bairoch and Apweiler, 2000) with a parameter of E-value 1e-5. Protein motif and KOG assignment was predicted through RPS-BLAST with the Conserved Domain Database (CDD) (Marchler-Bauer et al., 2011) with $E$-value $1e^{-5}$. The metabolic pathway was constructed based on the KEGG database (Kanehisa and Goto, 2000) by BBH method.

*Evaluation of the genome quality*

The quality of the assembled genome and annotated gene sets were assessed first using the Bench marking Universal Single-Copy Orthologs (version 3.1.0; BUSCO, RRID: SCR_015008) (Simão et al., 2015) with the hymenoptera_odb9 dataset. To further assess the completeness of the predicted genes, we used three Illumina RNA-seq data from our previous study of the genome v2.0 (Diao et al., 2018) to map the predicted gene sets of genome assembly v2.0 and v3.0 respectively using bowtie2 software (Langmead and Salzberg, 2012) with default parameters. In this study, we named the previous genome assembly of *A. cerana* reported by us as genome version 2.0, and this new assembly as version 3.0.

### *Genome comparison*

Comparison of the predicted genes between genome v2.0 and v3.0 was performed using BlastP program with default parameters. The transcripts of the identified new genes were analyzed by mapping RNA-seq data from *A. cerana* in our previous research (Diao *et al.*, 2018) to the predicted genes of genome v3.0 using bowtie2 software. Gene fusion and splitting between genome v2.0 and v3.0 were detected by sequence alignment of all the predicted CDSs of these two genomes using BlastN program with E-value $1e^{-100}$.

Collinearity analysis between *A. cerana* genome v3.0 and the latest genome sequences of *A. mellifera* (version Amel_HAv3.1, ftp://ftp.ncbi.nlm.nih.gov/genomes/Apis_mellifera/) was performed using Mummer (Delcher *et al.*, 2003) software with default parameters. Comparison of the domains of all the predicted genes between *A. cerana* genome v3.0 and the *A. mellifera* genome Amel_HAv3.1 was conducted in the Pfam database (El-Gebali *et al.*, 2019) by RPS-BLAST with *E*-value $1e^{-3}$. Unique genes in *A. cerana* and *A. mellifera* were analyzed using BlastP program with *E*-value $1e^{-3}$.

## Results and discussion

### *Genome sequencing and assembly*

After Pacbio sequencing, 1,379,634 clean reads with total size of 28,919,033,693 bp, N50 of 38,250 bp and mean length of 20,961 bp were obtained (Table IV-1-1). The 28.92 Gb Pacbio clean reads were assembled into 200 contigs with size ranging from 3,587 bp to 11,106,448 bp using HGAP4 of SMRT Link (version 6.0) (Chin *et al.*, 2013) (Table IV-1-2). The total size of the assembled contigs is 215,661,233 bp with average length of 20,961 bp and N50 of 4,485,954 bp, which represents an about 212-fold improvement in completeness contiguity compared to genome v2.0 of *A. cerana*.

Table IV-1-1  Summary of sequencing data for the new assembly of *A. cerana* genome

| | PacBio | Hi-C |
|---|---|---|
| Number of Bases | 28,919,033,693 | 42,964,780,942 |
| Number of Reads | 1,379,634 | 143,436,579 |
| N50 Read Length(bp) | 38,250 | 150 |
| Mean Reads Length(bp) | 20,961 | 150 |

In order to assemble contigs into scaffolds further, 143,436,579 Hi-C reads with total size of 42.96 Gb were generated through Illumina HiSeq X Ten platform (Table IV-1-1), and they were used to link the 200 contigs into 126 scaffolds with an N50 length of 13.42 Mb (Table IV-1-2, Table S1). The final scaffolds contain 16 pseudomolecules representing the 16 chromosomes of *A. cerana* (Fig. IV-1-1, see page 537) and 110 unplaced scaffolds. The total length of the 16 pseudochromosomes with length range from 7.10 Mb to 26.80 Mb is 211.03 Mb, consisting of 97.85% of the whole genome sequences (Table S1). The mean length of the remaining 110 unplaced scaffolds is 0.04 Mb which is far below 13.19 Mb of mean length of the 16 pseudochromosomes, suggesting these unplaced scaffolds may consist mainly of repetitive sequences.

Table IV-1-2  Statistics of the Pacbio and Hi-C assembly

|  | Contig | Scaffold |
| --- | --- | --- |
| Assembly length (bp) | 215,661,233 | 215,670,033 |
| Number | 200 | 126 |
| N50 (bp) | 4,485,954 | 13,422,783 |
| Largest (bp) | 11106448 | 26,804,424 |

*Genome annotation and evaluation*

Repeat sequences are widely distributed in higher organisms and play an important role in regulation of gene transcription (Fotsing *et al*., 2019; Wu *et al*., 2019), DNA replication (Madireddy and Gerhardt, 2017), and so on. We analyzed categories and proportions of repeat sequences in this new assembly. Finally, 19,731,355 bp of repeat sequences were detected accounting for 9.15% of the genome, which is significant higher than the 14.79 Mb (6.48%) in the assembly reported by Korean scientists (Park *et al*., 2015) and the 9.61 Mb (4.2%) in the assembly v2.0 reported by us (Diao *et al*., 2018),indicating that third generation sequencing technologies are beneficial to identification of repeat sequences. These repeat sequences includelong interspersed nuclear elements (LINEs) (0.04%), long terminal repeats (LTRs) (0.20%), DNA elements (0.65%), small RNAs (0.01%), simple repeats (4.03%), low complexity sequences (1.06%) and unclassified repeat sequences (3.16%). Of them, simple repeats with total length of 8,693,708 bp are the richest type (Table IV-1-3), which is consistent with the previous reports (Park *et al*., 2015; Diao *et al*., 2018) and is the same as in *A. mellifera* (Wallberg *et al*., 2019), suggesting that simple repeats are the major type of repeat sequences in honeybees.

Table IV-1-3  Transposable elements and repeat sequence statistics

| Classification | Number of elements | Length (bp) | percentage (%) |
| --- | --- | --- | --- |
| LINEs | 786 | 84,104 | 0.04 |
| LTR elements | 983 | 441,743 | 0.20 |
| DNA elements | 5,809 | 1,393,124 | 0.65 |
| Unclassified | 24,299 | 6,823,093 | 3.16 |
| Small RNA | 45 | 31,168 | 0.01 |
| Simple repeats | 191,534 | 8,693,708 | 4.03 |
| Low complexity | 43,237 | 2,285,374 | 1.06 |

A total of 10,741 protein-encoding genes were predicted, with an average gene size of 5,069 bp and an average coding DNA sequence (CDS) size of 1,519 bp. The average exon and intron sizes were 247 bp and 688 bp, respectively. 9,627 genes occupying 89.63% of the whole set of predicted genes were annotated through

KEGG, KOG, Pfam and uniprot databases (Table IV-1-4).

Table IV-1-4  General statistics of the functional annotation

| Database | Number | Percentage (%) |
|---|---|---|
| Total | 10741 | 100 |
| KEGG | 9303 | 86.61 |
| KOG | 7629 | 71.03 |
| Pfam | 7776 | 72.40 |
| uniprot | 4884 | 45.47 |
| unannotated | 1114 | 10.37 |

The quality of the assembled genome and annotated gene sets were assessed using BUSCO. Among the 4,415 BUSCO groups searched, 4,116 BUSCO groups (including complete and fragmented BUSCOs) were identified, occupying 93.23% of the total BUSCO groups. Among them, 3,801 groups were complete BUSCO groups, occupying 86.09% of the total BUSCO groups supporting the high quality of the genome assembly (Table IV-1-5).

The completeness of the predicted genes was further assessed by mapping Illumina RNA-seq data from our previous study (Diao et al., 2018) to the predicted gene sets of genome v2.0 and v3.0 respectively, which results in 31.27%-41.59% of the reads mapping to gene sets of genome v3.0, while the ratio is just 20.66%-24.83% when gene sets of genome v2.0 was used as mapping reference (Table S2). It suggests that the completeness of the predicted genes is significantly improved in the novel assembly of A. cerana genome.

Table IV-1-5  Statistics of the BUSCO assessment

| | BUSCO groups | Percentage (%) |
|---|---|---|
| Complete BUSCOs | 3,801 | 86.09 |
| Complete and single-copy BUSCOs | 3,792 | 85.89 |
| Complete and duplicated BUSCOs | 9 | 0.20 |
| Fragmented BUSCOs | 315 | 7.13 |
| Missing BUSCOs | 299 | 6.77 |
| Total BUSCO groups searched | 4,415 | 100 |

*Comparison between genome v2.0 and v3.0*

We compared this new genome assembly with the previous version of A. cerana genome reported by us (Diao et al., 2018). Compared to genome v2.0, the size of this improved genome was significantly reduced, which mainly due to the elimination of a large number of "N" representing sequencing gaps that exist in genome v2.0 (there are a total of 19.55 Mb "N" in the genome v2.0) (Table S3).

We compared the predicted genes between genome v2.0 and v3.0, a total of 314 new genes were identified in genome v3.0 (Table S4). Of them, 154 genes are newly assembled in genome v3.0 that their nucleotide sequences do not exist in genome v2.0, and 160 genes are newly predicted in genome v3.0 while not predicted in v2.0 because of sequencing gaps, assembling errors and so on. The mean length of the CDS region of all the newly predicted genes is just 321 bp which is significantly shorter than that of all the predicted genes, suggesting that most of the identified new genes encode proteins with small molecular weight, but the mean length of the introns of the 314 new genes is significantly longer than that of all the predicted genes (Fig. S1). 268 of the 314 new genes (85.35%) are annotated as "hypothetical protein" or "uncharacterized conserved protein", suggesting that most of the new genes are functionally unknown. Of the genes with definite functions, genes annotated as "Predicted ubiquitin-protein ligase/hyperplastic discs protein, HECT superfamily", "HIV-1 Vpr-binding protein" and "Dosage compensation complex, subunit MLE" have more than 10 copies in the genome, suggesting that third generation sequencing technologies have greatly improved success rate of identifying multi copy genes during whole genome sequencing which is very difficult using second generation sequencing technologies.Besides, RNA polymerase I large subunit (g2476.t1), ATP-dependent RNA helicase (g4187.t1), nuclear protein Ataxin-7 (g7042.t1), and so on, are important functional genes in the new gene list. We found that 164 new genes have transcription evidence when mapping RNA-seq data from *A. cerana* to the predicted genes of genome v3.0, implying that most of the new genes identified in genome v3.0 are really existed and have transcriptional activity in *A. cerana*.

We detected gene fusion and splitting between genome v2.0 and v3.0. The results showed that 78 genes in genome v2.0 were fused into 38 genes in genome v3.0; on the other hand, 135 genes in genome v2.0 each was spitted into more than two genes in genome v3.0 (Table S5).

### *Genome comparison between A. cerana and A. mellifera*

Within the genus *Apis* species, *A. cerana* and *A. mellifera* have the closest relationship. Their niche overlaps and competition between these two species is fierce, especially the drones of *A. mellifera* can interfere with the mating behavior of *A. cerana* queens and drones resulting in decline of the *A. cerana* colonies year by year. We compared the collinearity between *A. cerana* genome v3.0 and the *A. mellifera* genome Amel_HAv3.1 (Fig. IV-1-2, see page 537, Fig. S2). The 16 pseudochromosomes we identified in the new assembly of *A. cerana* genome aligned exactly against the 16 chromosomes of the *A. mellifera* genome Amel_HAv3.1 with high similarity. For the 16 chromosomes, chromosomes 1, 2, 5, 6, 9, 15 show almost completely collinear between the *A. cerana* genome v3.0 and *A. mellifera* genome Amel_HAv3.1; and chromosome 3, 4, 10, 13, 14, 16 just have small stretches of non-collinear region; for chromosome 7, 8, 11, 12, the collinear lines are split into several fragments because of large chromosome inversion, but in each fragment it is highly collinear between *A. cerana* and *A. mellifera*. These results suggest that our assembly is of high continuity compared with the *A. mellifera* genome.

We compared the domains of all the predicted genes between *A. cerana* genome v3.0 and the *A. mellifera* genome Amel_HAv3.1. A total of 4,026 Pfam family domains existed in both *A. cerana* and *A. mellifera* (Table S6). The total number of genes containing Pfam family domains showed no obvious difference between *A. cerana* and *A. mellifera*. Among the identified Pfam family domain containing genes, the number of "7tm

Odorant receptor" (PF02949) domain containing genes showed great difference between *A. cerana* and *A. mellifera*, 63 genes vs 137 genes, which is consistent with the results reported in the previous two genome assemblies of *A. cerana* (Park *et al.*, 2015; Diao *et al.*, 2018), suggesting that there are significant differences in olfactory ability between these two species. Besides, 43 Pfam family domains are unique to *A. cerana*, and 314 Pfam family domains are unique to *A. mellifera*.

In addition, we analyzed unique genes in *A. cerana* and *A. mellifera*. Finally, 111 genes unique to *A. mellifera* and 71 genes unique to *A. cerana* were identified (Table S7). Of the *A. mellifera* unique genes, 55 genes have KOG annotation, and "Signal transduction mechanisms" "General function Translation" are the major KOG categories. For *A. cerana* unique genes, most of them are annotated as "hypothetical protein", just 30 genes have KOG annotation and half of them belong to "transcription" category, suggesting that these genes might promote the formation of unique characters of *A. cerana* by regulating gene transcription.

## Conclusion

We reported the newly assembled genome of *A. cerana* using an integrated strategy of PacBio and Hi-C technologies. This new genome assembly has an about 212-fold improvement in contiguity over the previous version based on second generation sequencing technologies, and many new genes were identified. This genome assembly will be an important genome reference for studies of functional genes in *A. cerana*. Also, it will provide important data resource for evolutionary and comparative genomic studies.

## Data Availability Statement

The genome assembly 3.0 of *A. cerana* data has been submitted to the NCBI under Bio-Project number PRJNA579740. The Hi-C and PacBio sequencing data are available from NCBI via accession numbers SRR10377217, SRR10377218 and SRR10377219.

## Author Contributions

Z. Z. and H. Z. conceived and designed the research. Z. W. and W. Y. prepared the experimental materials. Y. Z. conducted the experiments. Y. Z., H. Z., Q. Y. and Z. W. analyzed the data. Z. W. and Y. Z. wrote the manuscript. All authors read and approved the publication of the manuscript. This work was supported by the China Agriculture Research System (No. CARS-45-KXJ12) and the National natural science foundation (No. 31860686).

## Competing Interests

The authors declare that they have no competing interests.

## References

BAIROCH, A., AND APWEILER, R. (2000). The SWISS-PROT protein sequence database and its supplement TrEMBL in 2000. *Nucleic Acids Res*. 28, 45-48. doi: 10.1093/nar/28.1.45.

Burton, J. N., Adey, A., Patwardhan, R. P., Qiu, R., Kitzman, J. O., and Shendure, J. (2013). Chromosome-scale scaffolding of de novo genome assemblies based on chromatin interactions. *Nat. Biotechnol.* 31, 1119-1125.doi: 10.1038/nbt.2727.

CHEN, N. (2004). Using RepeatMasker to identify repetitive elements in genomic sequences. *Curr. Protoc. Bioinformatics* Chapter 4:Unit 4.10. doi: 10.1002/0471250953.bi0410s05.

CHEN, S. (2001). The Apicultural Science in China. Beijing: China Agricultural Press.

CHIN, C. S., ALEXANDER, D. H., MARKS, P., KLAMMER, A. A., DRAKE, J., HEINER, C., et al. (2013). Nonhybrid, finished microbial genome assemblies from long-read SMRT sequencing data. *Nat. Methods* 10, 563-569. doi: 10.1038/nmeth.2474.

CONESA, A., GÖTZ, S., GARCÍA-GÓMEZ, J. M., TEROL, J., TALÓN, M., and ROBLES, M. (2005). Blast2GO: a universal tool for annotation, visualization and analysis in functional genomics research. *Bioinformatics* 21, 3674-3676. doi: 10.1093/bioinformatics/bti610.

DELCHER, A. L., SALZBERG, S. L., and PHILLIPPY, A. M. (2003). Using MUMmer to identify similar regions in large sequence sets. *Curr. Protoc. Bioinformatics* 00,10.3.1-10.3.18. doi: 10.1002/0471250953.bi1003s00.

DIAO, Q., SUN, L., ZHENG, H., ZENG, Z., WANG, S., XU, S., et al. (2018). Genomic and transcriptomic analysis of the Asian honeybee *Apis cerana* provides novel insights into honeybee biology. *Sci. Rep.* 8, 822. doi: 10.1038/s41598-017-17338-6.

EL-GEBALI, S., MISTRY, J., BATEMAN, A., EDDY, S. R., LUCIANI, A., POTTER, S. C., et al. (2019). The Pfam protein families database in 2019. *Nucleic Acids Res.* 47, D427-D432. doi: 10.1093/nar/gky995.

Fotsing, S. F., Margoliash, J., Wang, C., Saini, S., Yanicky, R., Shleizer-Burko, S., et al. (2019). The impact of short tandem repeat variation on gene expression. *Nat. Genet.* 51, 1652-1659. doi: 10.1038/s41588-019-0521-9.

GHURYE, J., KOREN, S., SMALL, S. T., REDMOND, S., HOWELL, P., PHILLIPPY, A. M., et al. (2019). A chromosome-scale assembly of the major African malaria vector *Anopheles funestus*. *Gigascience* 8, pii: giz063. doi: 10.1093/gigascience/giz063.

GONG, G., DAN, C., XIAO, S., GUO, W., HUANG, P., XIONG, Y., et al. (2018). Chromosomal-level assembly of yellow catfish genome using third-generation DNA sequencing and Hi-C analysis. *Gigascience* 7, giy120. doi: 10.1093/gigascience/giy120.

JIANG, X., PEERY, A., HALL, A. B., SHARMA, A., CHEN, X. G., WATERHOUSE, R. M., et al. (2014). Genome analysis of a major urban malaria vector mosquito, *Anopheles stephensi*. *Genome Biol.* 15, 459. doi: 10.1186/s13059-014-0459-2.

JIAO, W. B., ACCINELLI, G. G., HARTWIG, B., KIEFER, C., BAKER, D., SEVERING, E., et al. (2017). Improving and correcting the contiguity of long-read genome assemblies of three plant species using optical mapping and chromosome conformation capture data. *Genome Res.* 27, 778-786. doi: 10.1101/gr.213652.116.

KANEHISA, M., AND GOTO, S. (2000). KEGG: kyoto encyclopedia of genes and genomes. *Nucleic Acids Res.* 28, 27-30. doi: 10.1093/nar/28.1.27.

KENT, W. J. (2002). BLAT--the BLAST-like alignment tool. *Genome Res.* 12, 656-664. doi: 10.1101/gr.229202.

KRONENBERG, Z. N., FIDDES, I. T., GORDON, D., MURALI, S., CANTSILIERIS, S., MEYERSON, O. S., et al. (2018). High-resolution comparative analysis of great ape genomes. *Science* 360, pii: eaar6343. doi: 10.1126/science.

aar6343.

LANGMEAD, B., and SALZBERG, S. L. (2012). Fast gapped-read alignment with Bowtie 2. *Nat. Methods* 9, 357-359. doi: 10.1038/nmeth.1923.

LI, H., and DURBIN, R. (2009). Fast and accurate short read alignment with Burrows-Wheeler transform. *Bioinformatics* 25, 1754-1760. doi: 10.1093/bioinformatics/btp324.

LOW, W. Y., TEARLE, R., BICKHART, D. M., ROSEN, B. D., KINGAN, S. B., SWALE, T., et al. (2019). Chromosome-level assembly of the water buffalo genome surpasses human and goat genomes in sequence contiguity. *Nat. Commun.* 10, 260. doi: 10.1038/s41467-018-08260-0.

MADIREDDY, A., and GERHARDT, J. (2017). Replication Through Repetitive DNA Elements and Their Role in Human Diseases. *Adv. Exp. Med. Biol.* 1042, 549-581.doi: 10.1007/978-981-10-6955-0_23.

MARCHLER-BAUER, A., LU, S., ANDERSON, J. B., CHITSAZ, F., DERBYSHIRE, M. K., DEWEESE-SCOTT, C., et al. (2011). CDD: a Conserved Domain Database for the functional annotation of proteins. *Nucleic Acids Res.* 39, D225-229. doi: 10.1093/nar/gkq1189.

PARK, D., JUNG, J. W., CHOI, B. S., JAYAKODI, M., LEE, J., LIM, J., et al. (2015). Uncovering the novel characteristics of Asian honey bee, *Apis cerana*, by whole genome sequencing. *BMC Genomics* 16:1. doi: 10.1186/1471-2164-16-1.

SIMÃO, F. A., WATERHOUSE, R. M., IOANNIDIS, P., KRIVENTSEVA, E. V., and ZDOBNOV, E. M. (2015). BUSCO: assessing genome assembly and annotation completeness with single-copy orthologs. *Bioinformatics* 31, 3210-3212. doi: 10.1093/bioinformatics/btv351.

SMIT, A. F. A., and HUBLEY, R.(2008). RepeatModeler Open-1.0. http://www.repeatmasker.org.

STANKE, M., KELLER, O., GUNDUZ, I., HAYES, A., WAACK, S., and MORGENSTERN, B. (2006). AUGUSTUS: ab initio prediction of alternative transcripts. *Nucleic Acids Res.* 34, W435-439. doi: 10.1093/nar/gkl200.

WALLBERG, A., BUNIKIS, I., PETTERSSON, O. V., MOSBECH, M. B., CHILDERS, A. K., EVANS, J. D., et al. (2019). A hybrid de novo genome assembly of the honeybee, *Apis mellifera*, with chromosome-length scaffolds. *BMC Genomics* 20, 275. doi: 10.1186/s12864-019-5642-0.

WU, Z. C., XIA, X. J., LI, H. R., JIANG, S. J., MA, Z. Y., WANG, X. (2019). Tandem repeat sequence of duck circovirus serves as downstream sequence element to regulate viral gene expression. *Vet. Microbiol.* 239, 108496. doi: 10.1016/j.vetmic.2019.108496.

ZHANG, L., HU, J., HAN, X., LI, J., GAO, Y., RICHARDS, C. M., et al. (2019). A high-quality apple genome assembly reveals the association of a retrotransposon and red fruit colour. *Nat. Commun.* 10, 1494. doi: 10.1038/s41467-019-09518-x.

# 2. Signatures of Positive Selection in the Genome of *Apis mellifera carnica*: A subspecies of European Honeybees

Qiang Huang [1,2], Yongqiang Zhu [3], Bertrand Fouks [4], Xujiang He [1,2],
Qingsheng Niu [5], Huajun Zheng [3]* and Zhijiang Zeng [1,2]*

1. Honeybee Research Institute, Jiangxi Agricultural University, Nanchang, Jiangxi, 330045, China
2. Jiangxi Province Key Laboratory of Honeybee Biology and Beekeeping, Nanchang 330045, China
3. Shanghai-MOST Key Laboratory of Health and Disease Genomics, Chinese National Human Genome Center at Shanghai and Shanghai Institute for Biomedical
4. Institute for Evolution and Biodiversity, Molecular Evolution and Bioinformatics, Westfälische Wilhelms Universität, 48149 Münster, Germany
5. Apiculture Science Institute of Jinlin Province, Yuanlin Rd., Jinlin 132108, China

**Abstract**

The technology of long reads substantially improved the contingency of the genome assembly, particularly resolving contiguity of the repetitive regions. By integrating the interactive fragment using Hi-C, and the HiFi technique, a solid genome of the honeybee *Apis mellifera carnica* was assembled at the chromosomal level. A distinctive pattern of genes involved in social evolution was found by comparing it with social and solitary bees. A positive selection was identified in genes involved with cold tolerance, which likely underlies the adaptation of this European honeybee subspecies in the north hemisphere. The availability of this new high-quality genome will foster further studies and advances on genome variation during subspeciation, honeybee breeding and comparative genomics.

**Keywords:** honeybee, subspecies, selection, sociality, cold tolerance

**Introduction**

Insect pollination is essential for maintaining the balance of the ecosystem, which contributes approximately 35% to crop pollination [1]. The honeybee *Apis mellifera* is a key managed pollinator, with an estimated annual global economic value of $195 billion [2]. The recent decline of honeybee colonies has provoked serious concerns regarding the biodiversity, as well as food security [3]. A number of stressors have been proven to cause the honeybee collapses, including parasites, pesticide, climate change and habitat loss [4-8].

---

\* Corresponding author: zhenghj@chgc.sh.cn (HJZ.); bees1965@sina.com (ZJZ).
注：此文发表在 *Life* 2022,12(10):1642。

Among those stressors, parasites are a major cause leading to colony losses, with a synergistic effect with pesticide. It is known that honeybee strains showed variation in tolerance towards parasites and climates [9-13]. However, the genetic mechanism underlying that tolerance is yet not fully understood. In this study, we aim to reveal the genomic variation within *A. mellifera* species, which is essential to understand their evolution, adaptation to different climates, as well as to refine breeding strategies for disease-tolerant strains.

In this study, we provided a highly contiguous genome assembly of a valuable subspecies of the European honeybee, *Apis mellifera carnica*. Two complementary sequencing technologies of HIFI and Hi-C were used to generate a high-quality chromosome genome assembly. Comparative genomic analysis revealed gene families selected during the social evolution and climate adaptation.

## Material and Methods

### DNA extraction and sequencing library preparation

The honeybee *A. mellifera carnica* were collected from the national honeybee breeding center in Apicultural Science Institute of Jilin, China. Six *A. mellifera* carnica drone pupae were collected from a single colony, which were pooled for DNA extraction by AxyPrep TM Multisource Genomic DNA Miniprep Kit (Axygen, Irvine, CA, USA). The Qubitfluorimetry system was used to define the DNA concentration (Thermo Fisher, Waltham, MA, USA). Fragment size distribution was assessed using the Agilent 2100 Bioanalyzer with the 12,000 DNA kit (Agilent, Santa Clara, CA, USA). Then, 5 μg of high molecular weight genomic DNA was used to prepare the library. The DNA which uses g-Tube (Covaris, Woburn, MA, USA) to shear into 10 kb was used as input into the SMRTcell library preparation according to PacBio 10 kb library preparation protocol. The library was sequenced on a PacBio Sequel system using Sequencing Kit 3.0.

Additionally, five drone pupae from the same colony were fixed with formaldehyde and lysed for Hi-C library. The cross-linked genomic DNA was digested with Hind III overnight. Sticky ends of the genomic DNA were biotinylated and proximity-ligated to form chimeric junctions that were enriched for and then physically sheared to a size of 300-700 bp. Chimeric fragments representing the original cross-linked long-distance physical interactions were then processed into paired-end sequencing libraries and sequenced on the Illumina HiSeq X Ten platform.

### Genome assembly and gene annotation

The reads were filtered through Fastp with default parameters, and 2,602,139 clean Pacbio reads with a total size of 22,818,161,032 bp were obtained [14]. The reads were assembled by HGAP4 of SMRT Link (version 6.0) with default parameters. Additionally, 280,437,292 Hi-C reads with a total size of 84.13 Gb was obtained. The raw contigs were split into segments of 50 Kb on average. The Hi-C reads were aligned to the segments using BWA (version 0.7.10) with default parameters [15]. The uniquely mapped reads were retained to assemble the genome using LACHESIS package [16].

Augustus (v3.3.2) was used to train and predict the gene features with Amel_HAv3.1 gene set [17,18]. Based on the deduced amino acid sequences, the annotation was performed through BLASTP against the non-redundant peptide database with the cut-off Evalue at 10-5 and RPS-BLAST against the Conserved Domain

Database at *E*-value at 10⁻³ [19]. Gene ontology analysis was performed using BLASTP against the InterProScan [20]. The Pathway was constructed based on the KEGG database [19].

### *The genome completeness and honeybee evolution*

The protein sequences were used to query the BUSCO arthropod ortholog set to evaluate the genome completeness. The genomes of honeybee subspecies A. *mellifera mellifera* and *A. mellifera* DH4 were further queried to *A. mellifera carnica* using minimap2 [21]. The alignment files were viewed using package pafr (https://github.com/dwinter/pafr, accessed on 7 January 2021). The microsatellite was identified using MISA (microsatellite identification tool) with default parameters [22]. The microsatellite markers primers were blasted against the three genomes and the congruent was analyzed using Chi-squared test, R [23].

### *Phylogenetic analysis of social genes*

To investigate the molecular mechanisms underlying sociality between social and solitary bees, the protein sequences of four insulin family proteins insulin receptor substrate 1 (IRS1), insulin receptor substrate 4 (IRS4), insulin receptor like (IR-like) and insulin-like peptides (IPR) were retrieved from Eastern honeybees (*Apis cerana*), dwarf honeybees (*Apis florea*), giant honeybees (*Apis dorsata*), bumble bees (*Bombus terrestris*), digger bees (*Habropoda laboriosa*), red mason bees (*Osmia bicornis bicornis*) and small carpenter bees (*Ceratina calcarata*) by NCBI blast with E-value cutoff $\leq 1 \times 10^{-5}$. The protein sequences were aligned using Muscle with default parameters by MEGA X package (Version 10.0.2) [24]. The phylogenetic tree was constructed using the Neighbor joining model with 1000 bootstraps and the small carpenter bees were used to root the tree.

### *Identifying genome selection pressures*

Orthologs between 7 species (*Nasonia vitripennis, Bombus terrestris, Bombus impatiens, Apis florea, Apis dorsata, Apis laboriosa, Apis cerana*) and 4 subspecies (*Apis mellifera carnica, Apis mellifera caucasica, Apis mellifera ligustica, Apis mellifera mellifera*) were discovered using Orthofinder (v2.5.2) [25]. To optimize the number of single-copy orthologs, we categorized them as such if at least 3 *Apis mellifera* subspecies and *Apis cerana* had a single copy gene for the given orthogroup, culminating at 6328 single-copy ortholog families. Phylogenetic tree reconstruction, including all species described above, was undertaken by OrthoFinder. For each single-copy ortholog family, the longest protein isoforms for each of the species' gene were used in multiple sequence alignment with MAFFT (using local-pair algorithm and 1000 iterations) [26] and unreliably aligned residues and sequences were masked with GUIDANCE (v2.02) [27]. To optimize alignment length without gaps, we ran a maxalign script and removed subsequent sequences leading to more than 30% of gapped alignment as long as it did not result in the removal of any *A. mellifera* subspecies and *A. cerana* [28]. The protein sequences were replaced with coding sequences in the multiple alignments using the pal2nal script [29]. Furthermore, sequences containing a stop codon or having a length inconsistency between protein and DNA coding sequences (after removal of undefined bases) were filtered out. Alignments regions, where gapped positions were present, were removed with a custom python script, as these are the most problematic for positive selection inference [30,31]. Finally, CDS shorter than 100 nucleotides were eliminated [32].

Phylogenetic tests of positive selection in protein-coding genes usually contrast substitution rates at non-synonymous sites to substitution rates at synonymous sites taken as a proxy to neutral rates of evolution. The adaptive branch-site random effects model (aB-SREL) from Hyphy software package was used to detect positive selection experienced by a gene family in a subset of sites in a specific branch of its phylogenetic tree [33]. The test for positive selection was run only on the branches leading to the origin of *A. mellifera* and on each *A. mellifera* subspecies. Results from the adaptive branch-site random effects model were corrected for multiple testing as one series using False Discovery Rate (FDR) and set up our significant threshold at 10% [34].

*Test for functional category enrichment*

Gene Ontology (GO) annotations for our gene families were taken from Hymenoptera Genome database [35]. The enrichment of functional categories was evaluated with the package topGO version 2.4 of Bioconductor [36,37]. To identify functional categories enriched for genes under positive selection, strengthened, and relaxed selection pressure, the SUMSTAT test was used [38,39]. The SUMSTAT test is more sensitive than other methods and minimizes the rate of false positives [40-43]. To be able to use the distribution of log-likelihood ratios of the aBSREL and RELAX tests as scores in the SUMSTAT test, a fourth root transformation was used [39]. This transformation conserves the ranks of gene families [44]. Gene Ontology categories mapped to less than 10 genes were discarded. The list of significant gene sets resulting from enrichment tests is usually highly redundant. We therefore implemented the "elim" algorithm from the Bioconductor package topGO, to decorrelate the graph structure of the Gene Ontology. To account for multiple testing, the final list of *p*-values resulting from this test was corrected with the FDR and set up our significant threshold at 20%. To cluster the long list of significant functional categories, we used REVIGO with the SimRel semantic similarity algorithm and medium size (0.7) result list [45,46].

## Results and Discussion

*The genome assembly statistics of A. mellifera carnica*

A robust genome of 226.02 Mbp comprised of 313 contigs was assembled, which were further collapsed into 169 scaffolds (GCA_013841245.2) (Table IV-2-1). By aligning the predicted protein sequences to 1066 core arthropod Benchmarking Universal Single-copy orthologs (BUSCOs) [47], 93.53% of complete BUSCOs were identified. The results suggest that the assembled genome and predicted gene set were complete. Phylogenetic tree reconstruction revealed the subspeciation events of the European honeybees and the topology agreed with the evolution from solitary to social living (Fig. IV-2-1) [48,49].

Table IV-2-1 Statistics of the studied *A. mellifera carnica* genome and other recently assembled honeybee genomes. NA indicates that the genome was assembled at contig level

| Technology | *A. mellifera carnica* | *A. mellifera caucasica* | *A. mellifera* DH4 | *A. mellifera mellifera* | *A. cerana cerana* |
|---|---|---|---|---|---|
| | Pacbio, Hi–C | Pacbio | Pacbio, Hi–C | Pacbio, Hi–C | Pacbio, Hi–C |
| Coverage | 101 | 112 | 192 | 100 | 134 |
| Assembly size (Mbp) | 226.02 | 224.7 | 225.25 | 227.03 | 215.6 |

|                      | A. mellifera carnica | A. mellifera caucasica | A. mellifera DH4 | A. mellifera mellifera | A. cerana cerana |
|----------------------|----------------------|------------------------|------------------|------------------------|------------------|
| Technology           | Pacbio, Hi-C         | Pacbio                 | Pacbio, Hi-C     | Pacbio, Hi-C           | Pacbio, Hi-C     |
| Number of Contigs    | 313                  | 224                    | 227              | 199                    | 214              |
| N50 Scaffold size (Mbp) | 13.4              | NA                     | 13.6             | 13.5                   | 13.7             |
| Releasing date       | 2020                 | 2020                   | 2018             | 2018                   | 2020             |

(continued)

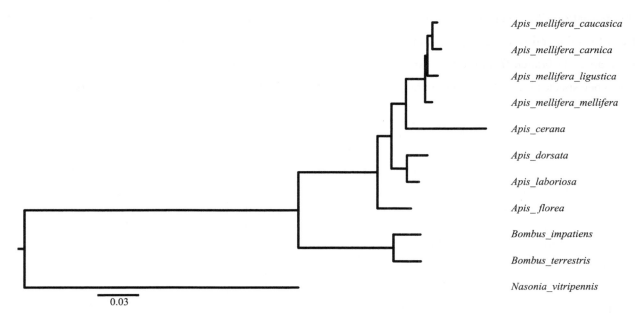

**Figure IV-2-1** Phylogenetic tree of the studies insects for the genome selection. All nodes were 100% bootstrap supported. N. vitripennis was used to root the tree. For *A. mellifera carnica*, 93.53% of complete BUSCOs were found, which suggests the assembly is complete.

### *Genome alignment among honeybee subspecies*

By pair-wised alignment, 83% and 89% of *A. mellifera carnica* genome can be perfectly aligned to *A. mellifera mellifera* and A. mellifera DH4 genome, respectively (Fig. IV-2-2, see page 538, and S1). The average length of the aligned region was 103 Kbp and 108 Kbp for *A. mellifera mellifera* and *A. mellifera* DH4, respectively (Fig. S2). The inverted fragment may reflect natural structural variation among the genomes, reflecting local adaption [50,51]. Overall, 88,380 microsatellites with dinucleotides motif were identified in *A. mellifera carnica*. Comparatively, 89,099 and 89,569 were identified in A. mellifera DH4 and *A. mellifera mellifera*, respectively. The relative abundance of microsatellites along the motifs were not significantly different among the three genomes (Pearson's Chi-squared test, df = 24, $p = 0.26$). However, the number of microsatellites decreased with the increasing number of repeats for all three genomes (Pearson's correlation coefficient, df = 9, $p < 0.001$, Fig. IV-2-3A, see page 538). A set of linkage map makers were further compared among the three genomes [52-54]. Out of 1081 paired microsatellite primers, 839 (77%) could be aligned to all three genomes (Fig. S3), with an average density of 4.8 cM per locus (Fig. IV-2-3B). For the remaining 242 markers, 162 were aligned to at least one genome. *A. mellifera* DH4 shared a higher number of markers with *A. mellifera carnica* compared with *A. mellifera mellifera*, which significantly deviated from random (Pearson's Chi-squared test, df = 2, $p < 0.01$) and is congruent with the genome phylogenetic tree in general (Fig. S4).

### *Phylogenetic analysis of sociality-related proteins*

The evolutionary process of bee sociality is fascinating, and highlights how genomes evolved to give rise to new and complex behaviors [49,55,56]. Insulin is an essential gene family regulating honeybee caste determination [57,58]. Hexamerin regulates the reproductive tissue development after honeybee caste differentiation [59-61]. The two gene families were selected to indicate the social evolution. The phylogenetic tree of insulin and hexamerin gene families clearly showed that solitary bees (digger bees and red mason bees) were an early branch from the root (small carpenter bees), followed by bumble bees (Fig. IV-2-4). The four honeybee species were clustered together, indicating that the a distinctive gene selection of sociality [62-64].

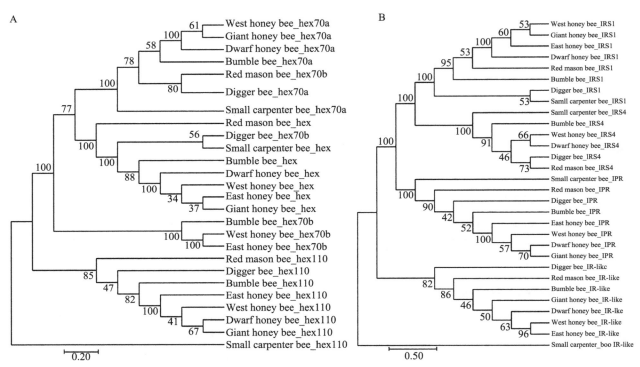

**Figure IV-2-4** Phylogenetic tree of social genes. (A): phylogenetic tree of hexamerin family proteins. B: phylogenetic tree of insulin family. The orthologs were retrieved by BLAST in NCBI manually with E-value cutoff $\leq 1 \times 10^{-5}$. The sequences were aligned with Muscle and the phylogenetic tree was constructed using Neighbor joining model with 1000 bootstrap.

### *Signature of positive selection*

Overall, 4897 single-copy orthologous groups were identified, out of which 245 orthologous groups displayed signs of positive selection in at least one branch test. The number of orthologous groups under positive selection within each branch varied significantly (Chi-squared test, $p < 2.2 \times 10^{-16}$) and ranged from 10 to 114 with 10, 27, 45, 78, and 114 for *A. mellifera caucasica*, *A. mellifera mellifera*, *A. mellifera spp.*, *A. mellifera carnica*, and *A. mellifera ligustica*, respectively (Table S1). Such a variation in the number of genes under positive selection among the different *A. mellifera* subspecies may highlight a fast pace of adaptation or directed domestication in *A. mellifera carnica* and *A. mellifera ligustica* compared with others [65,66]. However, this result could also be the consequence of different population structures among *A. mellifera* subspecies with differential gene flow and intro gression levels. Only one orthologous group, encoding for the protein obscurin involved in myogenesis and Hippo signaling pathway [67,68], was found to be under positive selection in three branches (*A. m. carnica, A. m. ligustica,* and *A. m. spp.*, Table IV-2-2). Moreover, a few orthologous groups were found under positive selection in more than one branch (Table IV-2-2). Additionally, Hippo signaling pathway was involved in cold temperature adaptation in bees [13].

Table IV-2-2  Orthologous found to be under positive selection in more than one branch tested

| Gene/Protein Name | Putative Function | Branches |
| --- | --- | --- |
| obscurin | Involved in myogenesis and the Hippo signaling pathway | *A. mellifera carnica* <br> *A. mellifera ligustica* <br> *A. mellifera* spp. |
| ATP-dependent RNA helicase abstrakt | axonal growth (visual system) | *A. mellifera carnica* <br> *A. mellifera ligustica* |
| bifunctional heparan sulfate N-deacetylase/N-sulfotransferase | involved in Wnt signaling | *A. mellifera carnica* <br> *A. mellifera ligustica* |
| foxP protein | Central Nervous System/learning-memory | *A. mellifera carnica* <br> *A. mellifera ligustica* |
| serine/threonine-protein kinase tricorner | Post-transcriptional regulation, development (dendrite morphogenesis) | *A. mellifera carnica* <br> *A. mellifera ligustica* |
| protein turtle homolog B | Establishing coordinated motor control, axonal targeting of the R7 photoreceptor | *A. mellifera carnica* <br> *A. mellifera ligustica* |
| GTPase-activating Rap/Ran-GAP domain-like protein 3 | signal transduction | *A. mellifera carnica* <br> *A. mellifera ligustica* |
| zinc finger protein 512B | involved in transcriptional regulation | *A. mellifera carnica* <br> *A. mellifera mellifera* |
| prohormone-2 precursor | neuropeptide/social behavior regulation | *A. mellifera carnica* <br> *A. mellifera mellifera* |
| protein lingerer | copulation/short-term memory | *A. mellifera carnica* <br> *A. mellifera mellifera* |
| retinoid-inducible serine carboxypeptidase | involved in vascular wall | *A. mellifera carnica* <br> *A. mellifera mellifera* |
| cAMP-specific 3',5'-cyclic phosphodiesterase | signaling, phsysiology, female fertility, learning/memory | *A. mellifera carnica* <br> *A. mellifera mellifera* |
| 7SK snRNA methylphosphate capping enzyme | methylation, development | *A. mellifera carnica* <br> *A. mellifera mellifera* |
| RB1-inducible coiled-coil protein 1 | involved in autophagy | *A. mellifera carnica* <br> *A. mellifera caucasica* |

(continued)

| Gene/Protein Name | Putative Function | Branches |
|---|---|---|
| muscle LIM protein Mlp84B | cell differentiation late in myogenesis | A. mellifera carnica<br>A. mellifera spp. |
| junctophilin-1 | formation of junction membrane in sarcomere | A. mellifera carnica<br>A. mellifera spp. |
| RNA binding protein fox-1 homolog 2 | alternative splicing | A. mellifera ligustica<br>A. mellifera mellifera |
| chitin synthase chs-2 | chitin biosynthetic synthesis | A. mellifera ligustica<br>A. mellifera spp. |
| ell-associated factor Eaf | regulation of transcription elongation | A. mellifera ligustica<br>A. mellifera spp. |
| nuclear protein localization protein 4 homolog | ubiquitination | A. mellifera ligustica<br>A. mellifera spp. |
| kinesin-like protein unc-104 | synaptic vesicle transport/locomotion | A. mellifera ligustica<br>A. mellifera spp. |
| actin-interacting protein 1 | sarcomere organization/development/locomotion | A. mellifera ligustica<br>A. mellifera spp. |
| ubiquitin carboxyl-terminal hydrolase 36 | stem cell line maintenance | A. mellifera ligustica<br>A. mellifera spp. |
| teneurin-a | neural development | A. mellifera ligustica<br>A. mellifera spp. |
| troponin I | muscle contraction | A. mellifera ligustica<br>A. mellifera spp. |

### *Functional categories enriched of positively selected genes*

We identified 11 significant functional categories in *A. m. carnica*, 28 in A. m. ligustica, 5 in *A. m. mellifera* and *caucasica*, and 6 in *A. mellifera* branch, which were enriched in genes under positive selection at 20% FDR. The long list of significant GO-terms found to be significantly enriched of positively selected genes in A. m. ligustica were mainly related to larval development (Fig. IV-2-5, see page 539), as demonstrated with clustering from REVIGO [45]. Interestingly, the two most significant enriched functional genes under positive selection in *A. mellifera* branch, chitin metabolism and mitochondrial translation (Fig. IV-2-6, see page 539), matched functional genes previously found in *A. mellifera* [34]. While functions found to be enriched of positively selected genes in *A. m. mellifera* were mainly related to the nervous system, in *A. m. caucasica*. They were mainly related to autophagy/cell death (Fig. IV-2-6). In *A. m. carnica*, most of significant GO terms seems to be linked with stress tolerance, nota bly response to wounding, reactive oxygen species (ROS) metabolism, and larval midgut programmed cell death (Fig. IV-2-6). More precisely, several significant GO terms are likely involved in cold resistance in honeybees, such as developmental growth [69], ROS metab olism [70], and hippo signaling [13]. Interestingly, the gene encoding the protein Dachsous, which was under positive selection in *A. m. carnica* and found in several significant GO terms, plays a key role in the adaptation to temperate climate in the *A. m. sinisxinyuan* [13]. Furthermore, the genes Amel_mTOR and Amel_Imp, encoding serine/threonine-protein kinase mTOR and insulin-like growth factor 2 mRNA-binding protein 1, respectively, might also play a role in cold tolerance as they both are involved in the insulin pathway regulating food intake, essential for cold tolerance [71,72]. Moreover, cold tolerance in *A. cerana* involved serine/threonine-protein kinases like mTOR [73]. The protein RB1-inducible coiled-coil protein 1, part of the GO term 'larval midgut cell programmed cell death', was found to be under positive selection in *A. m. caucasica* and involved in cold adaptation in amphipods [74].

## Supplementary Materials

The following supporting information can be downloaded at: https://www.mdpi.com/article/10.3390/life12101642/s1, Fig. S1. Dot plot of among three honeybee subspecies. The honey bee *Apis mellifera carnica* showed frame shifts and inversion compared with *A. mellifera* DH4 and *A. mellifera mellifera*. Fig. S2. Aligned sequences between the genomes. The average length of the aligned region was 103 Kbp and 108 Kbp for *A. mellifera mellifera* and *A. mellifera* DH4 respectively. Fig. S3. Linkage map markers using to align among the three honeybee genomes. Fig. S4. Phylogenetic tree and estimated completeness of the studies 7 genomes. (a) The phylogenetic tree was constructed on protein sequences of 874 single-copy orthologs shared among all 7 genomes. All nodes were 100% bootstrap supported. D. melanogaster was used to root the tree. (b) Completeness of predicted protein sets of each genome was assessed by aligning to the arthropod BUSCOs. For *A. mellifera carnica*, 93.53% of complete BUSCOs were found, which suggest the assembly is complete. Table S1. Number of positively selected genes in each *A. mellifera* subspecies tested and as well as the *A. mellifera* branch and number of overlapping genes under positive selection in more than one branch.

## Author Contributions

Z.-J.Z. conceived the experiment, Q.-S.N. collected the sample, Y.-Q.Z. and H.-J.Z. assembled the genome, Q.H. analyzed genome completeness, B.F. analyzed genome selection, X.-J.H. analyzed the phylogeny of sociality-related proteins, Q.H., Y.-Q.Z., B.F., X.-J.H., Q.-S.N.,H.-J.Z. and Z.-J.Z. organized the manuscript. All authors have read and agreed to the published version of the manuscript.

## Funding

This work was supported by the Initiation Package of Jiangxi Agricultural University (050014/923230722), National Natural Science Foundation of China (32172790), the Earmarked Fund for China Agriculture Research System (CARS-44-KXJ15) and EU-H2020 MSCA-IF-2020 fellowship(101024100, TEEPI).

**Institutional Review Board Statement**: Not applicable.

**Informed Consent Statement**: Not applicable.

**Data Availability Statement**: The raw sequencing reads have been deposited in BioProject

PRJNA 644991. The genome and gene annotation file are at: https://www.ncbi.nlm.nih.gov/assem-bly/GCA_013841245.2 (accessed date 31 August 2022).

**Conflicts of Interest**: The Authors declare no conflict of interest.

# References

1. IPBES. *The Assessment Report of the Intergovernmental Science-Policy Platform on Biodiversity and Ecosystem Services on Pollinators, Pollination and Food Production*; Potts, S.G., Imperatriz-Fonseca, V.L., Ngo, H.T., Eds.; Secretariat of the Intergovernmen: 2016.
2. BAUER, D.M.; SUE WING, I. The macroeconomic cost of catastrophic pollinator declines. *Ecol. Econ.* 2016, *126*, 1-13.
3. PARREÑO, M.A.; ALAUX, C.; BRUNET, J.-L.; BUYDENS, L.; FILIPIAK, M.; HENRY, M.; KELLER, A.; KLEIN, A.-M.; KUHLMANN, M.; LEROY, C.; *et al.* Critical links between biodiversity and health in wild bee conservation. *Trends Ecol. Evol.* 2022, *37*, 309-321.
4. FLORES, J.M.; GÁMIZ, V.; JIMÉNEZ-MARÍN, Á.; FLORES-CORTÉS, A.; GIL-LEBRERO, S.; GARRIDO, J.J.; HERNANDO, M.D. Impact of Varroa destructor and associated pathologies on the colony collapse disorder affecting honey bees. *Res. Vet. Sci.* 2021, *135*, 85-95.
5. ANDERSON, D.; EAST, I.J. The latest buzz about colony collapse disorder. *Science* 2008, *319*, 724-725.
6. VAN ENGELSDORP, D.; HAYES, J.; UNDERWOOD, R.M.; PETTIS, J. A Survey of Honey Bee Colony Losses in the U.S., Fall 2007 to Spring 2008. *PLoS ONE* 2008, *3*, e4071.
7. COX-FOSTER, D.L.; CONLAN, S.; HOLMES, E.; PALACIOS, G.; EVANS, J.D.; MORAN, N.A.; QUAN, P.L.; BRIESE, T.; HORNIG, M.; GEISER, D.M. A metagenomic survey of microbes in honey bee colony collapse disorder. *Science* 2007, *318*, 283-287.
8. FLORES, J.M.; GIL-LEBRERO, S.; GÁMIZ, V.; RODRÍGUEZ, M.I.; ORTIZ, M.A.; QUILES, F.J. Effect of the climate change on honey bee colonies in a temperate Mediterranean zone assessed through remote hive weight monitoring system in conjunction with exhaustive colonies assessment. *Sci. Total Environ.* 2019, *653*, 1111-1119.
9. KURZE, C.; LE CONTE, Y.; DUSSAUBAT, C.; ERLER, S.; KRYGER, P.; LEWKOWSKI, O.; MULLER, T.; WIDDER, M.; MORITZ, R.F.A. *Nosema* Tolerant Honeybees (*Apis mellifera*) Escape Parasitic Manipulation of Apoptosis. *PLoS ONE* 2015, *10*, e0140174.
10. KURZE, C.; ROUTTU, J.; MORITZ, R.F.A. Parasite resistance and tolerance in honeybees at the individual and social level. *Zoology* 2016, *119*, 290-297.
11. THADURI, S.; STEPHAN, J.G.; DE MIRANDA, J.R.; LOCKE, B. Disentangling host-parasite-pathogen interactions in a varroa-resistant honeybee population reveals virus tolerance as an independent, naturally adapted survival mechanism. *Sci. Rep.* 2019, *9*, 6221.
12. LI, X.; MA, W.; SHEN, J.; LONG, D.; FENG, Y.; SU, W.; XU, K.; DU, Y.; JIANG, Y. Tolerance and response of two honeybee species *Apis cerana* and *Apis mellifera* to high temperature and relative humidity. *PLoS ONE* 2019, *14*, e0217921.
13. CHEN, C.; LIU, Z.; PAN, Q.; CHEN, X.; WANG, H.; GUO, H.; LIU, S.; LU, H.; TIAN, S.; LI, R.; *et al.* Genomic

Analyses Reveal Demographic History and Temperate Adaptation of the Newly Discovered Honey Bee Subspecies *Apis mellifera sinisxinyuan* n. ssp. *Mol. Biol. Evol.* 2016, *33*, 1337-1348.

14. CHEN, S.; ZHOU, Y.; CHEN, Y.; GU, J. Fastp: An ultra-fast all-in-one FASTQ preprocessor. *Bioinformatics* 2018, *34*, i884-i890.

15. LI, H.; DURBIN, R. Fast and accurate short read alignment with Burrows-Wheeler transform. *Bioinformatics* 2009, *25*, 1754-1760.

16. BURTON, J.N.; ADEY, A.; PATWARDHAN, R.P.; QIU, R.; KITZMAN, J.O.; SHENDURE, J. Chromosome-scale scaffolding of de novo genome assemblies based on chromatin interactions. *Nat. Biotechnol.* 2013, *31*, 1119-1125.

17. STANKE, M.; KELLER, O.; GUNDUZ, I.; HAYES, A.; WAACK, S.; MORGENSTERN, B. AUGUSTUS: Ab initio prediction of alternative transcripts. *Nucleic Acids Res.* 2006, *34*, 435-439.

18. WALLBERG, A.; BUNIKIS, I.; PETTERSSON, O.V.; MOSBECH, M.-B.; CHILDERS, A.K.; EVANS, J.D.; MIKHEYEV, A.S.; ROBERTSON, H.M.; ROBINSON, G.E.; WEBSTER, M.T. A hybrid de novo genome assembly of the honeybee, *Apis mellifera*, with chromosome-length scaffolds. *BMC Genom.* 2019, *20*, 275.

19. MARCHLER-BAUER, A.; LU, S.; ANDERSON, J.B.; CHITSAZ, F.; DERBYSHIRE, M.K.; DEWEESE-SCOTT, C.; FONG, J.H.; GEER, L.Y.; GEER, R.C.; GONZALES, N.R.; *et al.* CDD: A Conserved Domain Database for the functional annotation of proteins. *Nucleic Acids Res.* 2011, *39*, D225-D229.

20. MULDER, N.J.; APWEILER, R. The InterPro Database and Tools for Protein Domain Analysis. *Curr. Protoc. Bioinforma.* 2008, *21*, 2-7.

21. LI, H. Minimap2: Pairwise alignment for nucleotide sequences. *Bioinformatics* 2018, *34*, 3094-3100.

22. BEIER, S.; THIEL, T.; MÜNCH, T.; SCHOLZ, U.; MASCHER, M. MISA-WEB: A web server for microsatellite prediction. *Bioinformatics* 2017, *33*, 2583-2585.

23. R CORE TEAM. R: A language and environment for statistical computing. *R Found. Stat. Comput. Vienna Austria* 2013. Available online: http://www.R-project.org/ (accessed on 4 June 2021).

24. EDGAR, R.C. MUSCLE: Multiple sequence alignment with high accuracy and high throughput. *Nucleic Acids Res.* 2004, *32*, 1792-1797.

25. EMMS, D.M.; KELLY, S. OrthoFinder: Phylogenetic orthology inference for comparative genomics. *Genome Biol.* 2019, *20*, 238.

26. ROZEWICKI, J.; LI, S.; AMADA, K.M.; STANDLEY, D.M.; KATOH, K. MAFFT-DASH: Integrated protein sequence and structural alignment. *Nucleic Acids Res.* 2019, *47*, W5-W10.

27. SELA, I.; ASHKENAZY, H.; KATOH, K.; PUPKO, T. GUIDANCE2: Accurate detection of unreliable alignment regions accounting for the uncertainty of multiple parameters. *Nucleic Acids Res.* 2015, *43*, W7-W14.

28. GOUVEIA-OLIVEIRA, R.; SACKETT, P.W.; PEDERSEN, A.G. MaxAlign: Maximizing usable data in an alignment. *BMC Bioinform.* 2007, *8*, 312.

29. SUYAMA, M.; TORRENTS, D.; BORK, P. PAL2NAL: Robust conversion of protein sequence alignments into the corresponding codon alignments. *Nucleic Acids Res.* 2006, *34*, W609-W612.

30. FLETCHER, W.; YANG, Z. The effect of insertions, deletions, and alignment errors on the branch-site test of positive selection. *Mol. Biol. Evol.* 2010, *27*, 2257-2267.

31. MARKOVA-RAINA, P.; PETROV, D. High sensitivity to aligner and high rate of false positives in the estimates of positive selection in the 12 *Drosophila genomes. Genome Res.* 2011, *21*, 863-874.

32. HAMBUCH, T.M.; PARSCH, J. Patterns of synonymous codon usage in *Drosophila melanogaster* genes with sex-biased expression. *Genetics* 2005, *170*, 1691-1700.

33. POND, S.L.K.; FROST, S.D.W.; MUSE, S. V HyPhy: Hypothesis testing using phylogenies. *Bioinformatics* 2005, *21*, 676-679.

34. FOUKS, B.; BRAND, P.; NGUYEN, H.N.; HERMAN, J.; CAMARA, F.; ENCE, D.; HAGEN, D.E.; HOFF, K.J.; NACHWEIDE, S.; ROMOTH, L.; *et al.* The genomic basis of evolutionary differentiation among honey bees. *Genome Res.* 2021, *31*, 1203-1215.

35. WALSH, A.T.; TRIANT, D.A.; LE TOURNEAU, J.J.; SHAMIMUZZAMAN, M.; ELSIK, C.G. Hymenoptera Genome Database: New genomes and annotation datasets for improved go enrichment and orthologue analyses. *Nucleic Acids Res.* 2022, *50*, D1032-D1039.

36. MI, H.; MURUGANUJAN, A.; EBERT, D.; HUANG, X.; THOMAS, P.D. PANTHER version 14: More genomes, a new PANTHER GO-slim and improvements in enrichment analysis tools. *Nucleic Acids Res.* 2019, *47*, D419-D426.

37. ALEXA, A.; RAHNENFUHRER, J. TopGO: Enrichment Analysis for Gene Ontology. R Package Version 2.46.0 2021.

38. EFRON, B.; TIBSHIRANI, R. On testing the significance of sets of genes. *Ann. Appl. Stat.* 2007, *1*, 107-129.

39. ROUX, J.; PRIVMAN, E.; MORETTI, S.; DAUB, J.T.; ROBINSON-RECHAVI, M.; Keller, L. Patterns of positive selection in seven ant genomes. *Mol. Biol. Evol.* 2014, *31*, 1661-1685.

40. ACKERMANN, M.; STRIMMEr, K. A general modular framework for gene set enrichment analysis. *BMC Bioinform.* 2009, *10*, 47.

41. DAUB, J.T.; HOFER, T.; CUTIVET, E.; DUPANLOUP, I.; QUINTANA-MURCI, L.; ROBINSON-RECHAVI, M.; EXCOFFIEr, L. Evidence for polygenic adaptation to pathogens in the human genome. *Mol. Biol. Evol.* 2013, *30*, 1544-1558.

42. FEHRINGER, G.; LIU, G.; BRIOLLAIS, L.; BRENNAN, P.; AMOS, C.I.; SPITZ, M.R.; BICKEBÖLLER, H.; WICHMANN, H.E.; RISCH, A.; HUNG, R.J. Comparison of pathway analysis approaches using lung cancer GWAS data sets. *PLoS ONE* 2012, *7*, e31816.

43. TINTLE, N.L.; BORCHERS, B.; BROWN, M.; BEKMETJEV, A. Comparing gene set analysis methods on single-nucleotide polymorphism data from Genetic Analysis Workshop 16. *BMC Proc.* 2009, *3*, S96.

44. CANAL, L. A normal approximation for the chi-square distribution. *Comput. Stat. Data Anal.* 2005, *48*, 803-808.

45. SUPEK, F.; BOŠNJAK, M.; ŠKUNCA, N.; ŠMUC, T. REVIGO Summarizes and Visualizes Long Lists of Gene Ontology Terms. *PLoS ONE* 2011, *6*, e21800.

46. RIMAL, R.; ALMØY, T.; SÆBØ, S. A tool for simulating multi-response linear model data. *Chemom. Intell. Lab. Syst.* 2018, *176*, 1-10.

47. SEPPEY, M.; MANNI, M.; ZDOBNOV, E.M. *BUSCO: Assessing Genome Assembly and Annotation Completeness BT—Gene Prediction: Methods and Protocols*; Kollmar, M., Ed.; Springer: New York, NY, USA, 2019; pp. 227-245; ISBN 978-1-4939-9173-0.

48. PETERS, R.S.; KROGMANN, L.; MAYER, C.; DONATH, A.; GUNKEL, S.; MEUSEMANN, K.; KOZLOV, A.;

PODSIADLOWSKI, L.; PETERSEN, M.; LANFEAR, R.; et al. Evolutionary History of the Hymenoptera. *Curr. Biol.* 2017, *27*, 1013-1018.

49. KAPHEIM, K.M.; PAN, H.; LI, C.; SALZBERG, S.L.; PUIU, D.; MAGOC, T.; ROBERTSON, H.M.; HUDSON, M.E.; VENKAT, A.; FISCHMAN, B.J.; et al. Social evolution. Genomic signatures of evolutionary transitions from solitary to group living. *Science* 2015, *348*, 1139-1143.

50. TICHKULE, S.; CACCIÒ, S.M.; ROBINSON, G.; CHALMERS, R.M.; MUELLER, I.; EMERY-CORBIN, S.J.; EIBACH, D.; TYLER, K.M.; VAN OOSTERHOUT, C.; JEX, A.R. Global Population Genomics of Two Subspecies of *Cryptosporidium hominis* during 500 Years of Evolution. *Mol. Biol. Evol.* 2022, *39*, msac056.

51. WALLBERG, A.; SCHÖNING, C.; WEBSTER, M.T.; HASSELMANN, M. Two extended haplotype blocks are associated with adaptation to high altitude habitats in East African honey bees. *PLOS Genet.* 2017, *13*, e1006792.

52. SOLIGNAC, M.; MOUGEL, F.; VAUTRIN, D.; MONNEROT, M.; CORNUET, J.-M. A third-generation microsatellite-based linkage map of the honey bee, *Apis mellifera*, and its comparison with the sequence-based physical map. *Genome Biol.* 2007, *8*, R66.

53. BEHRENS, D.; HUANG, Q.; GESSNER, C.; ROSENKRANZ, P.; FREY, E.; LOCKE, B.; MORITZ, R.F.A.; KRAUS, F.B. Three QTL in the honey bee *Apis mellifera* L. suppress reproduction of the parasitic mite *Varroa destructor*. *Ecol. Evol.* 2011, *1*, 451-458.

54. HUANG, Q.; KRYGER, P.; LE CONTE, Y.; LATTORFF, H.M.G.; KRAUS, F.B.; MORITZ, R.F.A. Four quantitative trait loci associated with low *Nosema ceranae* (Microsporidia) spore load in the honeybee *Apis mellifera*. *Apidologie* 2014, *45*, 248-256.

55. SALEH, N.W.; RAMÍREZ, S.R. Sociality emerges from solitary behaviours and reproductive plasticity in the orchid bee *Euglossa dilemma*. *Proc. R. Soc. B Biol. Sci.* 2019, *286*, 20190588.

56. SHELL, W.A.; STEFFEN, M.A.; PARE, H.K.; SEETHARAM, A.S.; SEVERIN, A.J.; TOTH, A.L.; REHAN, S.M. Sociality sculpts similar patterns of molecular evolution in two independently evolved lineages of eusocial bees. *Commun. Biol.* 2021, *4*, 253.

57. DE AZEVEDO, S.V.; HARTFELDER, K. The insulin signaling pathway in honey bee (*Apis mellifera*) caste development—Differential expression of insulin-like peptides and insulin receptors in queen and worker larvae. *J. Insect Physiol.* 2008, *54*, 1064-1071.

58. WANG, Y.; AZEVEDO, S.V.; HARTFELDER, K.; AMDAM, G. V Insulin-like peptides (AmILP1 and AmILP2) differentially affect female caste development in the honey bee (*Apis mellifera* L.). *J. Exp. Biol.* 2013, *216*, 4347-4357.

59. LAGO, D.C.; HUMANN, F.C.; BARCHUK, A.R.; ABRAHAM, K.J.; HARTFELDER, K. Differential gene expression underlying ovarian phenotype determination in honey bee, *Apis mellifera* L., caste development. *Insect Biochem. Mol. Biol.* 2016, *79*, 1-12.

60. MARTINS, J.R.; NUNES, F.M.F.; CRISTINO, A.S.; SIMÕES, Z.L.P.; BITONDI, M.M.G. The four hexamerin genes in the honey bee: Structure, molecular evolution and function deduced from expression patterns in queens, workers and drones. *BMC Mol. Biol.* 2010, *11*, 23.

61. MARTINS, J.R.; MORAIS FRANCO NUNES, F.; LUZ PAULINO SIMÕES, Z.; MARIA GENTILE BITONDI, M. A honeybee storage protein gene, hex 70a, expressed in developing gonads and nutritionally regulated in adult fat body. *J.*

*Insect Physiol.* 2008, *54*, 867-877.

62. PAGE, R.E., JR.; SCHEINER, R.; ERBER, J.; AMDAM, G.V. The development and evolution of division of labor and foraging specialization in a social insect (*Apis mellifera* L.). *Curr. Top. Dev. Biol.* 2006, *74*, 253-286.

63. WITTWER, B.; HEFETZ, A.; SIMON, T.; MURPHY, L.E.; ELGAR, M.A.; PIERCE, N.E.; KOCHER, S.D. Solitary bees reduce investment in communication compared with their social relatives. *Proc. Natl. Acad. Sci. USA* 2017, *114*, 6569-6574.

64. SMITH, A.; SIMONS, M.; BAZARKO, V.; SEID, M. The influence of sociality, caste, and size on behavior in a facultatively eusocial bee. *Insectes Soc.* 2019, *66*, 153-163.

65. OLDROYD, B.P. Domestication of honey bees was associated with expansion of genetic diversity. *Mol. Ecol.* 2012, *21*, 4409-4411.

66. WALLBERG, A.; HAN, F.; WELLHAGEN, G.; DAHLE, B.; KAWATA, M.; HADDAD, N.; SIMÕES, Z.L.P.; ALLSOPP, M.H.; KANDEMIR, I.; DE LA RÚA, P.; *et al.* A worldwide survey of genome sequence variation provides insight into the evolutionary history of the honeybee *Apis mellifera*. *Nat. Genet.* 2014, *46*, 1081-1088.

67. YAMAUCHI, T.; MOROISHI, T. Hippo Pathway in Mammalian Adaptive Immune System. *Cells* 2019, *8*, 398.

68. MA, S.; MENG, Z.; CHEN, R.; GUAN, K.-L. The Hippo Pathway: Biology and Pathophysiology. *Annu. Rev. Biochem.* 2019, *88*, 577-604.

69. RAMIREZ, L.; LUNA, F.; MUCCI, C.A.; LAMATTINA, L. Fast weight recovery, metabolic rate adjustment and gene-expression regulation define responses of cold-stressed honey bee brood. *J. Insect Physiol.* 2021, *128*, 104178.

70. MUCCI, C.A.; RAMIREZ, L.; GIFFONI, R.S.; LAMATTINA, L. Cold stress induces specific antioxidant responses in honey bee brood. *Apidologie* 2021, *52*, 596-607.

71. AZEVEDO, S.V.; CARANTON, O.A.M.; DE OLIVEIRA, T.L.; HARTFELDER, K. Differential expression of hypoxia pathway genes in honey bee (*Apis mellifera L.*) caste development. *J. Insect Physiol.* 2011, *57*, 38-45.

72. LIU, N.; REN, Z.; REN, Q.; CHANG, Z.; LI, J.; LI, X.; SUN, Z.; HE, J.; NIU, Q.; XING, X. Full length transcriptomes analysis of cold-resistance of *Apis cerana* in Changbai Mountain during overwintering period. *Gene* 2022, *830*, 146503.

73. XU, K.; NIU, Q.; ZHAO, H.; DU, Y.; JIANG, Y. Transcriptomic analysis to uncover genes affecting cold resistance in the Chinese honey bee (*Apis cerana cerana*). *PLoS ONE* 2017, *12*, e0179922.

74. LIPAEVA, P.; VERESHCHAGINA, K.; DROZDOVA, P.; JAKOB, L.; KONDRATEVA, E.; LUCASSEN, M.; BEDULINA, D.; TIMOFEYEV, M.; STADLER, P.; LUCKENBACH, T. Different ways to play it cool: Transcriptomic analysis sheds light on different activity patterns of three amphipod species under long-term cold exposure. *Mol. Ecol.* 2021, *30*, 5735-5751.

# 3. Histomorphological Study on Embryogenesis of the Honeybee *Apis cerana*

Xiaofen Hu[1], Li Ke[1], Zhijiang Zeng[1*]

1. Honeybee Research Institute, Jiangxi Agricultural University, Nanchang, Jiangxi 330045, China.

**Abstract**

As important pollination species, honeybees play substantial impacts on the balance of global ecosystem, including two best-known honeybees *Apis mellifera* and *Apis cerana*. Embryogenesis is a fundamental stage of honeybee development and plays important roles in supporting the whole-life developmental process. However, few studies were reported on honeybee embryonic morphology using egg section, possibly due to the fragility of honeybee eggs and the difficulty of making embryonic sections. In this study, we reported a simply equipped method of frozen sectioning and PI (propidium iodide) staining to show the inner structure and cell distribution of *A. cerana* embryos at the different embryonic developmental stages. We found that the stages of *A. cerana* embryogenesis could also be typically classified into ten developmental stages, which are similar with the sister honeybee species, *A. mellifera*. To be noted, besides the cell distribution in the whole egg, we clearly observed the migration route of embryonic cells during the early embryonic development in *A. cerana*. This study provides a new insight into the whole process of honeybee embryogenesis from the perspective of egg sectioning, a histological basis for genetic manipulation using *A. cerana* eggs, and a reference method for egg sectioning for other insect species.

**Keywords:** *Apis cerana*, histomorphological observation, whole course of embryonic development, PI staining

**Introduction**

Honeybees are an important group of Hymenoptera insects with complete metamorphosis, which have great economic and ecological values and are closely related to human. The well-known honeybees are *Apis mellifera* and *Apis cerana*. Both honeybees are domesticated and used to produce honeys or other bee products; meanwhile, both honeybees are raised for pollination of crops (Winston, 1991; Aizen and Harder, 2009). The whole development process of honeybees includes four stages: eggs, larvae, pupae and adult. It was widely accepted that honeybee embryogenesis was an important process and of great significance to the late-stage

---

\* Corresponding author: bees1965@sina.com (ZJZ).
注：此文发表在 *Journal of Asia-Pacific Entomology* 2019 年第 1 期。

development of honeybees and formation of their social behavior.

The morphological studies of embryonic development of honeybees (*A. mellifera*) began in the late nineteenth and early twentieth century. Optical microscopy was mainly used for morphological observation of honeybee embryos during those days. Through observation under optical microscopy, Nelson (1915) and Schnetter (1934) roughly divided the development of honeybee embryos into eight stages: (1) fertilized eggs, (2) cleavage, (3) blastoderm formation and completion, (4) germ band formation, (5) mesoderm formation, (6) gastrointestinal tract and amniotic membrane formation, (7) basic completion of internal system, and (8) mature embryos. Later, DuPraw (1963) improved the observation method by continuously culturing an *A. mellifera* embryo at a constant temperature of 35.5 °C and observing its complete embryonic development. DuPraw (1967) clearly divided the process of honeybee embryogenesis into ten stages. Each of the stages was further divided into early, middle and late intervals. Overall, 10 developmental stages were defined with a detailed morphology of *A. mellifera* embryos. From then on, many studies on morphology and molecular biology of bee embryos referred to DuPraw's embryonic staging. For examples, Milne et al. (1988) used optical imaging system to take a series of pictures of honeybee embryos and depicted the development of honeybee embryos sorted by DuPraw's embryonic staging times. Dearden et al. (2006, 2014) staged the experimental honeybee embryos according to the scheme of DuPraw. Cridge et al. (2017) also referred to DuPraw's hand-drawing and their staging times to give a dark optical field view of the developing honeybee embryos.

With the development of observation equipment, electron microscopy has been employed to study the morphology of honeybee embryos. Based on the observations in the scanning electron microscope, Fleig and Sander (1985) described the successive changes at the cell surface of fertilized honeybee eggs, the very long process of blastoderm formation and the morphological change and arrangement of peripheral nuclei during this process. They further observed and described the whole process of honeybee embryogenesis based on embryos fixed at one hour intervals from oviposition to hatching using the scanning electron microscope (Fleig and Sander, 1986). Fleig and Sander (1988) continued to employ the scanning electron microscope to observe the migration of epithelial cell populations (epithelial expansion or translocation) in the honeybee embryogenesis, which appears to result from various combinations of ameboid movement, cell flattening, cell rearrangement (intercalation) and cell contraction along the free epithelial edge.

Since the late 1980s, many molecular studies of honeybee embryogenesis have been conducted and provided a strong base for understanding honeybee embryonic development (Cridge et al., 2017). And since the 21th century, with the wide use of fluorescence microscopy, many important techniques for visualizing gene expression in honeybee embryos have been developed, such as immunohistochemistry (Dearden, 2006; Dearden et al., 2009) or *in situ* hybridization (Osborne and Dearden, 2005a; Osborne and Dearden, 2005b; Wilson et al., 2014). For example, Dearden used the blue-fluorescent dye 4′,6-diamidino-2-phenylindole (DAPI) to counterstain *in situ* hybridized *empty-spiracles* gene, which revealed the expression of *empty-spiracles* gene at different embryonic stages and also showed the morphologic changing of honeybee embryos (Dearden, 2014). Additionally, the images of honeybee embryos have been usually taken using a stereo microscope with dark-field illumination at different embryonic stages (Cridge et al., 2017).

Most of the above morphological studies on honeybee embryos were focused on *A. mellifera*. There were few studies on the histomorphology of *A. cerana* embryos. It is generally accepted that making sections of honeybee embryos was difficult due to the fragility of honeybee eggs, especially at the early stages. When we looked up and read the relevant literature about honeybee embryogenesis, we seldom saw real images of complete embryonic section, instead hand-drawn images of inner embryonic structure in early literature. Here in this study, we employed a simply equipped method of frozen sectioning and PI (propidium iodide) staining in the section making of *A. cerana* embryos. We successfully made a series of *A. cerana* embryonic sections to show the inner structure and cell distribution of *A. cerena* embryos at ten embryonic developmental stages. These histological developmental atlas of *A. cerana* embryos provides a basic histological basis for the behavioral and neurological studies in the later stages of Asian honeybees.

## Materials and methods

### *Experimental honeybees and ethics statement*

The experimental Asian honeybees (*A. cerena*) were reared in the apiary of Honeybee Research Institute, Jiangxi Agricultural University, China (28.46 °N, 115.49 °E), same as our previous reports (Yan *et al*., 2016, Liu *et al*., 2019). All animal procedures were performed in accordance with guidelines developed by the China Council on Animal Care; and protocols were approved by the animal care and use committee of Jiangxi Agricultural University, China.

### *Collection of honeybee embryos*

The *A. cerana* embryos were collected from honeybee colonies in the breeding season of *A. cerana* (the temperature usually ranges from 15 to 30 °C). A 100 μL plastic tip was made as a thin and soft embryo-transferring picker (as shown in Fig. S1). We used the end of the modified picker to softly contact the adhered posterior end of the embryos and gently move the embryos from the bottom of the cell.

To collect age-controlled embryos, mated egg-laying queens were first held in queen cages for 2-3 h, and then placed in the egg- and brood-free areas of a pre-cleaned comb; before the queen was placed, the comb was cleaned overnight by worker bees. During queen ovulation, the queen was confined to the free space of half a comb by a wooden or bamboo fence that only allows worker bees to come in and out but does not allow the queen out. In the process of limiting the queen, the operation of the queen and the colony was as gentle as possible, so as not to frighten the queen and the worker bees. The whole process was carried out in quiet conditions. After four hours, the queen was released from the egg-containing honeycombs. Generally, the queen needs to adapt to the new honeycomb for a period of time (more than 1h), so the age of honeybee embryos should be about 0-3 h. To gain the appropriate honeybee eggs with specific developmental ages, the cultivation time of embryos in incubator were determined according to this developmental interval. If no embryo was not taken, the queen would be limited for another 4 h until the embryos were collected. Then the honeycomb with embryos was put in a thermostat (35 °C, 85% humidity) and the embryos were cultured to the corresponding age of each developmental stage. Then the embryos were taken for section making.

*Observation of fixed embryos*

Honeybee embryos were taken carefully and gently without any damage from the honeycomb using a self-made egg-taken picker (Fig. S1), and then fixed in 4% polyformaldehyde solution for 12 - 24 h. The fixed time varied according to the developmental stages of embryos; usually the shorter developmental time the shorter fixing hour. The fixed embryos at the different developmental stages were transferred to a slide and immersed in the polyformaldehyde solution. Then the immersed embryos were observed and photographed under a microscope.

*Frozen section making*

The fixed embryos were transferred into 20% sucrose solution for 24 - 48 h dehydration at 4 °C. The dehydration time varied according to the developmental stages of embryos; usually the shorter developmental time the longer dehydration hour. Due to more protoplasm and more water, the sections of early-stage eggs were much easier to deform and rupture. The dehydrated eggs were moved into a plastic box for embedding. Before egg embedding, the sucrose solution around the eggs was carefully sucked up with filter paper. Then the eggs were filled with optimal cutting temperature (OCT) embedding agent, and no bubbles should be generated. Then the box containing embedded eggs were put into the isopentane pre-frozen on liquid nitrogen surface.

After frozen, the embedded embryos were transferred to slice-making machine (LEICA CM1950) to make frozen sections. The frozen sections with thickness of 5-7 μm were cut out. The sections were placed in a cool place with room temperature below 20 °C for natural drying.

*PI staining and microscopy observation*

The 50 μg/mL PI staining solution was gently dropped to the dried embryonic section to keep the whole embryo immersed in a thin layer of PI staining solution. When dropping PI staining solution to the section, the action should be gentle. The dropping site was directly above the embryonic section. The dosage of one dropping was within 5 μl just covering the whole embryo and avoiding egg tissue floating. Then, the embryonic section was placed and observed under a fluorescence microscope (Leica, Germany) using green fluorescence as the excitation light. Photographs of honeybee embryonic sections were taken with red fluorescence.

## Results

To study the development of *A. cerana* embryos at different stages more clearly, we not only observed the sections of *A. cerana* embryos at each stage, but also collected the pictures of polyformaldehyde-fixed *A. cerana* embryos at each stage under a microscope for comparison with the corresponding sections of *A. cerana* embryos. In the present study, the pictures were displayed from a lateral view with the downward abdomen and the upward back.

*Stage one*

The age of the collected embryos at this stage was 0-4 h. The *A. cerana* embryos looked more like sausages: the interior was homogeneous in texture and there were no gaps at both ends of the eggshell (Fig. IV-3-1A1, see page 540). From the embryonic section, uniform and light-colored flocculent structures were found (Fig. IV-3-1B1), which were the protoplasm and yolk fluid of the first-stage eggs. At the first stage, the cleavage cells were very few; it was difficult to detect the energids on a section. Because the peak period of cleavage did not reach, the inner nutrients were abundant and the embryonic texture was thick.

**Stage two**

The age of our collected embryos at the second stage was 5-8 hours after oviposition. As shown in Fig. IV-3-1A2, significant protoplast contractions occurred at the cephalic and posterior ends of the embryo, resulting in obvious gaps at both embryonic ends, which was a typical morphological feature of honeybee embryos at the second stage. Meanwhile, a columnar void was observed in the middle axis of the embryo, and the shadows of the gray clumps were seen around the voids, which were possibly the migrating energids. This chart showed that the cleavage cells had completed the migration to the posterior end and were migrating outward along the middle axis of the embryo (Fig. IV-3-1A2). The inner structures of *A. cerana* embryos at the early, middle, late and end intervals of the second stage were shown in Fig. IV-3-1B2, C2, D2 and E2. During the early interval of stage 2, a small number of distinct energids were detected around the region less than one third of the body length near the cephalic side of the embryo (Fig. IV-3-1B2). During the middle interval of stage 2, cells proliferated quickly and the number of energids increased significantly. Most energids were around the middle axis of the embryo; there were more cells in the anterior half of the embryo than in the posterior half (Fig. IV-3-1C2). Several energids have been distributed on the posterior region of the embryo, which indicated that some energids were moving to the posterior end of the embryo. As the developmental time went on, the energids at the later interval of stage 2 (Fig. IV-3-1D2) were closer to the surface of embryo than the middle interval. At the end of stage 2, most energids had migrated into the embryonic periplasm and reached the inner surface of the embryo, only a few of energids remained in the protoplasm (Fig. IV-3-1E2). Taken together, looking at these four images, we could clearly reveal the migration route of energids during the whole developmental time of stage 2.

**Stage three**

The age of our collected embryos at stage three was about 8-11 h. The typical morphology of honeybee embryo at the early third stage was shown in Fig. IV-3-1A3. Many gemlike protuberances were observed at the cephalic side of the embryo. There were still gaps on both cephalic and posterior ends of the embryo; the gap at the posterior end was smaller than that at the cephalic end; but both gaps were less than those at stage two, which would finally disappear at the end of stage three. The histological sections of *A. cerana* embryos at the early and late intervals of stage three were shown in Fig. IV-3-1B3 and C3, respectively. At the early third stage, most cleavage nuclei were just approaching the embryonic surface, the periplasm around each cleavage nucleus begun to indent. The gemlike protuberances on the embryonic surface were clearly observed at the anterior pole (Fig. IV-3-1B3). While at the late interval of stage three, both gaps at the anterior and posterior ends disappeared, cell boundaries around the cleavage nuclei formed; monolayered embryonic blastoderm cells were visible on the surface of the egg, but there was still no arrangement of monolayer cells at the posterior side of embryo (Fig. IV-3-1C3).

In Fig. IV-3-1-B3 and C3, a certain number of cleavage nuclei (about 10%), visible as bright red granules, were found in the yolk fluid of eggs. These cleavage nuclei remained in the interior of the protoplast, did not migrate to the periplasm and would become vitellophages. These nuclei were called as "yolk nuclei" or primary vitellophages. At the later stage of embryonic development, they would play important roles in transformation of the yolk fluid (DuPraw, 1967).

### Stage four

At this stage, the age of embryos we collected was 16-22 h. The fixed embryo at this stage resembled the initial morphology at the first stage in appearance.

The liquid-filled gaps at both ends disappeared; the whole surface of the embryo could be seen surrounded by the blastoderm cells; and a dense line or layer was visible between the blastoderm and the yolk mass (Fig. IV-3-1A4). This layer of inner periplasm first appeared in the anterior ventral region of the embryo, then gradually extended to both anterior and posterior ends, and finally surrounded the entire dorsal side.

In Fig. IV-3-1-B4, we found that the whole surface of embryo was surrounded by the blastoderm cells. On the ventral side, the structure of interlaced two-layer blastoderm cells was formed; while on the dorsal side, a single layer of blastoderm cells was visible. There was a foggy periplasm surrounding the inner side of the whole blastoderm where the yolk mass intersected; the periplasm on the ventral side was obviously thicker than that on the dorsal side (Fig. IV-3-1B4). The blastoderm cells on the dorsal side do not rearrange and remain single layer structure during the whole stage of embryonic blastoderm; they do not form part of the embryo proper but may contribute to the formation of embryo membrane (amnion-serosa) in the later stage (DuPraw, 1967). Fig. IV-3-1C4 showed a section of *A. cerana* embryo at late stage four. It was visible that the ventral cells of the embryo begun to transit to the regular single cell morphology. The periplasm and cytoplasm were gradually fusing, and the periplasm at this time became thinner than that at the early stage four.

### Stage five

The age of the collected embryos at this stage was about 26-29 h. There were obvious fluid-filled spaces in the anterior and posterior ends of the egg again (Fig. IV-3-1A5). The recurrent spaces were first detected at the anterior end and then observed at the posterior end. This embryo was in the late interval of stage five, and the dorsal single-layer cell area was more transparent than the ventral region (Fig. IV-3-1A5). The embryo in Fig. IV-3-1B5 also belonged to the later stage five. The large vacuoles in the blastoderm cells could be clearly visible, which was a typical developmental feature of the blastoderm cells in the later interval of stage five; the whole periplasm of the embryo was relatively uniform (Fig. IV-3-1B5).

### Stage six

The age of our collected embryos in this period was about 32-38 h. Fig. IV-3-2A6(see page 540) showed external morphology of a fixed embryo at the early stage six. There were still fluid-filled spaces at both anterior and posterior ends of the egg. The anterior side of the embryo became a little pointed; and the serosa structure of the single cell layer covering the anterior end was visible, which looks like a transparent cap. A conspicuous, plate-like thickening appeared in the ventroanterior blastoderm, which is the anterior midgut rudiment. The brighter band on the ventral side of the anterior end was the mesoderm which had not been covered by the ectoderm. Another image of fixed embryo in the a-little-later interval of stage six was showed in Fig. IV-3-2 B6. It was seen that the cap-like single-cell layer was more obvious and moving down to the ventral region at the anterior end. In the ventral side of the embryo head, a large number of ectoderm cells could be seen to form long strips of shadow, and the shadow at the anterior ventral end near to the cap was much thicker.

As shown in the embryonic section of early stage six (Fig. IV-3-2C6), there was a distinctly thicken dish-like structure at the anterior ventral side of the embryo, which was caused by the rapid ventral migration of the blastoderm cells at the anterior end of the protoplast. At the same time, a hood-like single-cell layer was visible between the chorion and the protoplast at the anterior end of the egg, which is the anterior part of the developing embryonic amnion-serosa. In addition, the embryonic amniotic chorion, a single layer of cells similar to that at the anterior end, was forming around the dorsal side of the embryo. By the end of stage 6, the amniotic serosa would surround the entire dorsal side, and the ventral margin of the serosa would coincide with the dorsal margin of the ectoderm. This was the developing embryonic amniotic chorion. Fig. IV-3-2D6 showed the embryonic sections at the later interval of stage six. The ventrolateral ectoderm was gradually covering the abdominal mesoderm from both sides and extending toward the posterior end at the junction of the ventrolateral midline. This extension would terminate after reaching the posterior end and eventually converge with the gradually separated single layer cells (chorion) on the dorsal side. The posterior ends of the embryos in Fig. IV-3-2C6 had some distortion, which was caused by the process of making sections. Its posterior end should be consistent with the posterior ends of the eggs of Fig. IV-3-2A6. It would be blunt and round. The shape of the posterior ends was basically unchanged throughout the sixth stage.

### *Stage seven*

The embryos we collected in stage seven were at the age of 38-44 h. The main developmental events in the seventh stage were the formation of the whole amniotic chorion and its complete separation from the embryo. The whole fixed embryo looked like a canoe, and there were dents in the middle ventral part between the anterior and posterior ends (Fig. IV-3-2A7). A small arc protrusion was seen at the anterior end, which was the shadow of the separated amnion-serosa structure. The embryo of Asian honeybee in Fig. IV-3-2B7 was in the late stage seven. The whole embryo was surrounded by the single cell layer of the amnion-serosa. The ventral part between the anterior and posterior ends had obvious depression. The shape at the anterior end was very similar to that at the posterior end. Based on the embryonic section, the whole embryo was obviously like a canoe. Compared to the sixth stage, the embryonic cells of the progenitor midgut at the anterior and posterior ends proliferated more frequently and the cell layer was obviously thicker.

### *Stage eight*

The age of the collected embryos in this stage was 44-47 h. In the eighth stage, one of the most typical developmental events was that the progenitor of the midgut develops into the midgut. As shown in Fig. IV-3-2A8, the embryonic protoplast contracted sharply during this period, the inner length of the embryo was the shortest in the whole embryonic development stage. The anterior and posterior ends resembled as the water tank, and there were no obvious head and other symmetrical appendages. It was difficult to distinguish the embryonic head and tail, but the space at the anterior end was smaller than that at the posterior end. At the anterior end there was a precursor of the lip formed by the thickening of the ectoderm.

In Fig. IV-3-2B8, a conspicuous protuberance at the anterior end of the embryo seemed as the precursor of the lip. At the dorsal side of the egg, there was no longer a sparse single-celled serous chorionic structure, but a well-arranged single-celled structure. The midgut precursor completed the development of the midgut, and the whole yolk fluid was surrounded by a single layer of cells, forming a complete midgut cavity. And the indentation on the ventral surface from the anterior part to the posterior part disappeared.

***Stage nine***

The age of our collected embryos at this stage is 47-67 h. The main developmental events in the ninth stage are the formation of body segments and internal tissues. In Fig. IV-3-2A9, a fixed image of Asian honeybee embryo at the early ninth stage was shown. Thirteen body segments and the front lip of the head were clearly visible; while the posterior tail was blunt and round. Fig. IV-3-2B9 showed another fixed embryo at the middle or late ninth stage. It was seen that most of the external morphology of the larvae head was formed, the development of the ventral side was close to that of the larvae, and the tail was changed. The apex begun to form the tail organ, but the dorsal part was not well developed and no obvious thickness of the back was seen. Fig. IV-3-2C9, D9, E9 and F9 showed the early, middle, late and end intervals of stage nine, respectively. The basic structure of the anterior lip just formed was seen in Fig. IV-3-2C9. The body segments were not very clear yet. The dorsal side of the midgut cavity had just formed, so there was only a very thin layer of cells. Fig. IV-3-2D9 showed 13 distinct body segments in the ventral region, which begun to form the embryonic stomatal prototype, the cell layer on the dorsal side of the embryo begun to thicken, the tail begun to change from a circular shape to a slightly protruding shape, and the shape of the tail begun to change. In Fig. IV-3-2E9, we could see clearly the structure of the large-capacity brain, the outline of the mouth was clearer, the rudiments of three pairs of appendages (the mandibles and two maxillae) of the head were basically formed, the tail organs were sharpened, the basic shape had been formed, the back continues to thicken, and the pouch-like midgut cavity was very clear. Fig. IV-3-2F9 showed the complete intracranial structure, clear oral structure and tubular structure of the whole digestive system, the dorsal morphology of the larvae, the bronchial structure from the second to the eleventh ganglia and the clear nerve cord in the abdomen.

***Stage ten***

The embryos were collected at 67-70 h. The main developmental event in the tenth stage was larvae hatching. A fixed-in-vivo view of Asian honeybee embryo at this stage was shown in Fig. IV-3-2A10. Clear blood vessels were visible. The embryo had not yet ruptured the amnion-serosa, but it was ready to hatch. The tissue development of the larvae was basically completed at the end of ninth stage; the hatching processes were mainly carried out in this stage, including muscle activity, head and tail shaking and body rotation. The embryonic section at the tenth stage was same to the one at the end of the ninth stage, so it was not shown.

## Discussion

In the morphological studies on honeybee embryos, optical, electron and fluorescence microscopies have been widely used to observe and describe honeybee eggs. It has provided us an abundant morphological data of honeybee embryogenesis and detailed honeybee embryonic stages. However, many of the internal structures of honeybee eggs and the processes of energid migration and blastoderm formation were presented in the form of textual descriptions, photos of the whole eggs and/or hand-drawn drawings. There were few reports on honeybee embryonic sections, especially few *A. cerana* embryonic sections. The reasons might be that the presence of a large number of egg fluids makes it difficult to make sections, especially for early embryos. Recently, we have provided some images of *A. cerana* embryonic sections in the related studies of transcriptome (Hu *et al.*, 2018) and gene editing (Hu *et al.*, 2019), but it was only a part of the whole embryogenesis. In this paper, applying a simple method of making sections, as far as possible to reduce the intermediate steps in the making process, we clearly showed the images of *A. cerana* embryos at ten stages one by one. The whole migration route of energids during early developmental stages of bee eggs were clearly shown through real images of embryonic sections. Meanwhile, in the whole embryo picture, the formation and migration path of the blastoderm cells into the endoderm, mesoderm and ectoderm, the formation of the intestinal cavity and the formation of the internal organs of the embryo were well reflected. The display of these morphological changes through embryonic sections was not only a supplement but also a good verification to previous morphological studies of honeybee embryos.

Generally, two methods were widely used in making tissue sections in divergent species: paraffin section and frozen section. Previously, we applied a common paraffin sectioning method, an improved paraffin sectioning method of pre-embedding the eggs with swine fat and a frozen sectioning method to make embryonic section and observe their internal structure in *A. mellifera* (Hu *et al.*, 2016). Our results showed that it was very hard to obtain the sections of honeybee eggs directly by the common paraffin sectioning method, mainly because that the process of making paraffin sections contained many immersing and cleaning steps and honeybee embryos were very small and easy to lose. We improved the paraffin sectioning method by pre-encapsulating honeybee eggs with swine fat to avoid egg losing; however, this improved method easily resulted in the whole contraction or collapse of honeybee eggs. We found that this improved method was suitable for local morphological observation of honeybee embryos at the late stages but still difficult to make sections for the early-stage eggs. Instead, the frozen sectioning method would be better to make sections for honeybee embryos, especially for early-stage embryos. It was relatively simple using frozen section method to make honeybee egg sections, which only needs polyformaldehyde fixation and sucrose dehydration processes to prepare honeybee eggs for slicing. The Sucrose dehydration process would also lead to a small contraction of the egg, but it would not change the overall shape and structure. So the completement of the embryonic sections was relatively better compared to the paraffin sections.

In the choice of dyeing method, we did not choose the traditional hematoxylin and eosin (HE) staining method, because this method needs many cleaning steps. Cleaning steps easily lead to the loss of the inner materials (such as yolk mass and vitellophages) in eggs and the incompleteness of embryonic external structure. The PI staining method used in this study almost omitted most of the cleaning processes before microscope observation (Nicoletti *et al.*, 1991). A very small amount of PI staining solution was gently dropped and covered on the embryonic sections and then the sections were observed directly. This treatment avoided huge loss of the inner materials in the eggs and reduced the chance to deform the whole embryonic structure. In our present study, this floating phenomenon of embryonic single layer cells was very hard to avoid. Especially, it was easy to occur in the early *A. cerana* embryo before stage three, the monolayer of blastoderm at the fifth stage and the dorsal amniotic serosa at the seventh stage. Because PI staining was mainly aimed at the specific staining of nuclei, under fluorescence microscope, green fluorescence will stimulate the PI-specific staining of nuclei to emit red fluorescence, but other tissues will also have a relatively light staining, so that the structure of the whole egg could be clearly seen and the overall distribution of embryonic cells with a brighter red fluorescence could be seen. In this study, we gained and showed good section results of *A. cerana* embryos at ten developmental stages. This dyeing method could also be used as a reference in making sections for other similar tissues or insect eggs in the future.

From our fixed pictures and histological pictures, we could see that the key events of embryonic development and the development of *A. cerana* embryo was basically similar to those of *A. mellifera*, and the same treatment of *A. cerana* according to the embryonic development zoning schedule of *A. mellifera* will find that the embryonic development zoning table of *A. mellifera* is basically applicable to the embryonic development zoning of *A. cerana*. In all the morphological and histological sections, it was found that the typical morphological and histological characteristics of each developmental stage of *A. cerana* were similar to those of *A. mellifera*. From our histological study, we could conclude that the time range and histological characteristics of embryonic development of *A. cerana* are basically the same as those of *A. mellifera*. Therefore, the Histomorphology of *A. cerana* embryo in this paper could also be used as a reference for embryonic development section of *A. mellifera*.

## Conflict of Interest

No conflict of interest exits in the submission of this manuscript.

## Author Contributions

Xiao Fen Hu and Zhi Jiang Zeng provided the experimental design and developed the methodology. Xiao Fen Hu and Li Ke completed all the experiments, including sample collection and making embryo slices. Zhi Jiang Zeng provided the financial support for publishing the manuscript. Xiao Fen Hu and Zhi Jiang Zeng wrote the manuscript. All authors read and approved the final manuscript and agreed to be accountable for all aspects of the work.

## Acknowledgements

This work was supported by the National Natural Science Foundation of China (31572469,31872432), the Earmarked Fund for China Agriculture Research System (CARS-44-KXJ15) and the Jiangxi Provincial Education Department Research Project (GJJ150415). We thank Dr. Qiang Huang for improving the English of this manuscript.

## References

AIZEN, M.A. AND HARDER, L.D., 2009. The global stock of domesticated honey bees is growing slower than agricultural demand for pollination. *Curr. Biol.* 19(11), 915-918.

CRIDGE, A.G., LOVEGROVE, M.R., SKELLY, J.G., TAYLOR, S.E., PETERSEN, G.E.L., CAMERON, R.C. AND DEARDEN, P.K., 2017. The honeybee as a model insect for developmental genetics. *Genesis* 55(5), e23019.

DEARDEN, P.K., WILSON, M.J., SABLAN, L., OSBORNE, P.W., HAVLER, M., MCNAUGHTON, E., KIMURA, K., MILSHINA, N.V., HASSELMANN, M., GEMPE, T., SCHIOETT, M., BROWN, S.J., ELSIK, C.G., HOLLAND, P.W.H., KADOWAKI, T. AND BEYE M., 2006. Patterns of conservation and change in honey bee developmental genes. *Genome Res.* 16(11), 1376-1384.

DEARDEN, P.K., DUNCAN, E.J. AND WILSON, M.J., 2009. RNA interference (RNAi) in honeybee (*Apis mellifera*) embryos. *Cold Spring Harb Protoc.* 6(6), prot5228.

DEARDEN, P.K., 2014. Expression pattern of empty-spiracles, a conserved head-patterning gene, in honeybee (*Apis mellifera*) embryos. *Gene Expr. Patterns* 15(2), 142-148.

DEARDEN, PK., 2006. Germ cell development in the Honeybee (*Apis mellifera*); Vasa and Nanos expression. *BMC Dev. Biol.* 6(1), 6.

DUPRAW, E.J., 1963. Techniques for the analysis of cell function and differentiation, using eggs of the honeybee. XVI Int. *Congress* Zool. 2, 238.

DUPRAW, E.J., 1967. The honeybee embryo. In: Wilt, F.H. and Wessells, N.K. (Eds), Methods in Developmental Biology. Thomas Y. Crowell, New York, pp. 183-217.

FLEIG, R. and SANDER, K., 1985. Blastoderm development in honeybee embryogenesis as seen in the scanning electron microscope. *Int. J. Insect Morphol. Embryol.* 8(4-5), 279-286.

FLEIG, R. and SANDER, K., 1986. Embryogenesis of the honeybee *Apis mellifera L.* (Hymenoptera: Apidae): an SEM study. *Int. J. Insect Morphol. Embryol.* 15(5-6), 449-462.

FLEIG, R. and SANDER, K., 1988. Honeybee morphogenesis: embryonic cell movements that shape the larval body. *Development* 103(3), 525-534.

HU, X., KE, L. and ZENG, Z., 2016. Comparison of three methods for honeybee egg histological section. Acta *Agriculturae Universitatis Jiangxiensis*, 38(4), 729-733.

HU, X., KE, L., WANG, Z. and ZENG, Z., 2018. Dynamic transcriptome landscape of Asian domestic honeybee (*Apis cerana*) embryonic development revealed by high-quality RNA sequencing. *BMC Dev. Biol.* 18(1), 11.

HU, X.F., ZHANG, B., LIAO, C.H. and ZENG, Z.J., 2019. High-efficiency CRISPR/Cas9- mediated gene editing in honeybee (*Apis mellifera*) embryos. *G3: Genes, Genomes, Genetics*, g3-400130.

LIU, J.F., YANG, L., LI, M., HE, X.J., WANG, Z.L. and ZENG, Z.J., 2019. Cloning and expression pattern of odorant receptor 11 in Asian honeybeedrones, *Apis cerana* (Hymenoptera, Apidae). *J. Asia Pac. Entom.* 22(1), 110-116.

MILNE, JR. C.P., PHILLIPS, J.P. and KRELL, P.J., 1988. A photomicrographic study of worker honeybee embryogenesis. *J. Apicult. Res.* 27(2), 69-83.

NELSON, J.A., 1915. The embryology of the honey bee. Princeton Univ. Press, Princeton, NJ, USA.

NICOLETTI, I., MIGLIORATI, G., PAGLIACCI, M.C., GRIGNANI, F. and RICCARDI, C., 1991. A rapid and simple method for measuring thymocyte apoptosis by propidium iodide staining and flow cytometry. *J. Immun. Methods*, 139(2), 271-279.

OSBORNE P.W. and DEARDEN P.K., 2005a. Expression of Pax group III genes in the honeybee (*Apis mellifera*). *Dev. Genes Evol.* 215(10), 499-508.

OSBORNE, P.W. and DEARDEN, P.K., 2005b. Non-radioactive in-situ hybridisation to honeybee embryos and ovaries. *Apidologie* 36(1), 113-118.

SCHNETTER, M., 1934. Morphologische untersuchungen über das differenzierungszen- trum in der embryonal- entwicklung der honigbiene. *Zeitschrift für Morphologie und Ökologie der Tiere* 29(1), 114-195.

WINSTON, M.L., 1991. The biology of the honey bee. Harvard Univ. Press, Cambridge, MA, USA.

WILSON, M.J., KENNY, N.J. and DEARDEN, P.K. 2014. Components of the dorsal-ventral pathway also contribute to anterior-posterior patterning in honeybee embryos (*Apis mellifera*). *EvoDevo* 5(1), 11.

YAN, W.Y., GAN, H.Y., LI, S.Y., HU, J.H., WANG, Z.L., WU, X.B. and ZENG, Z.J. 2016. Morphology and transcriptome differences between the haploid and diploid drones of *Apis cerana*. *J. Asia Pac. Entom.* 19(4), 1167-1173.

# 4. Dynamic Transcriptome Landscape of Asian Domestic Honeybee (*Apis cerana*) Embryonic Development Revealed by High-quality RNA-seq

Xiaofen Hu[1], Li Ke[1], Zilong Wang[1] and Zhijiang Zeng[1*]

1. Honeybee Research Institute, Jiangxi Agricultural University, Nanchang, Jiangxi 330045, China

**Abstract**

**Background**: Honeybee development consists of four distinct stages: egg, larva, pupa and adult. Embryogenesis is a key process of cell division and cellular differentiation in the egg stage, which takes three days in honeybee. However, the overview of embryonic transcriptome and the dynamic regulation of embryonic transcription are still largely uncharacterized in honeybee, especially in the Asian honeybee (*Apis cerana*). Here we employed high-quality RNA-Seq to explore the transcriptome over the complete time course of honeybee embryogenesis at ages of 24 h (also referred as Day1), 48 h (Day2), and 72 h (Day3).

**Results**: Nine embryo samples at the three ages were collected and an average of 57.3 million clean reads per sample were generated. Hierarchical clustering and principal component analysis showed that embryo samples from same day were well grouped into same clusters and the Day2 samples were clustered more closely with the Day3 ones than the Day1 ones. Finally, a total of 18,284 genes harboring 55,646 transcripts were detected and expressed in the *Apis cerana* embryos, of which 44.5% consisted of the core transcriptome shared by all the three ages of embryos. 4,088 up-regulated and 3,046 down-regulated differential expression genes were identified among the three ages of embryo samples, of which 2,010, 3,177 and 1,528 genes were up-regulated and 2,088, 2,294 and 303 genes were down-regulated from Day 1 to Day 2, from Day 1 to Day 3 and from Day 2 to Day 3, respectively.

**Conclusions**: The developmental status between Day2 and Day3 embryos was closer than the one between the first day and the second or third day in *Apis cerana*. Based on the dynamic expression of core transcriptome, we speculated that the developmental stages of 24, 48 and 72 hours in honeybee embryogenesis were roughly correspondent to the stages 7-9, 11-13, and 15-17 of *Drosophila* embryo, respectively. Notably, eye development of *Apis cerana* might be earlier than previous thought, because several biological processes related to eye development were detected in the first-day embryogenesis, including detect of light stimulus and eye-antennal disc development. Our transcriptomic data substantially expand the number of known transcribed elements in the *Apis cerana* genome and provide a high-quality view of transcriptome dynamics throughout

* Corresponding author: bees1965@sina.com (ZJZ).
注：此文发表在 *BMC Developmental Biology* 2018 年第 11 期。

embryonic development in honeybee.

**Keywords:** *Apis cerana*, Embryonic transcriptome, Dynamic regulation of transcripts, Embryonic development

**Introduction**

Bees are flying insects playing important roles in pollination and keystone species supporting critical ecosystem services [1]; honey bees represent a small fraction of known bee species [2]. The best-known honey bees are European honey bee (*Apis mellifera*) and Asian honey bee (*Apis cerana*). Both European and Asian honey bees are eusocial flying insects playing great impacts on global ecological environment and can be domesticated for honey production and crop pollination [3, 4]. Additionally, honey bees could serve as an excellent model animal to elucidate the molecular and neural mechanisms underlying their social behaviors [5, 6].

European honey bees are more popularly imported to around the globe and can produce more honey products than Asian honey bees. On the contrary, Asian honey bees have several advantages compared to European honey bees: (1) Asian honey bees has a strong cold resistance to work in low temperature that could be lethal to European honey bees [7]; (2) Asian honeybees have strong collection ability of collecting honey in the sporadic nectar sources and longer period of honey collection [8]; (3) Asian honeybees are more active or more suitable to raise in the mountain Area; and (4) Asian honeybees possess strong resilience to the ectoparasitic mite [9], which has a catastrophic effect on the population of European honey bees [10].

The development of honey bee consists of four life cycle phases: egg, larva, pupa and adult [3, 11]. Among these phases, embryogenesis is the first and the most important stage, which is precisely controlled and modulated by both environmental and intracellular signals. And during the period of embryogenesis, the rudimentary organs of adult honey bees are gradually formed. Therefore, gene expression in the embryonic stage contributes greatly to adult morphology, appearance and behavior.

Currently, several genetic manipulations of the honey bee embryo have been developed, such as RNA interference [12], transgenesis using the transposon piggyBac [13], and genome editing by CRISPR/Cas9 method [14]. And short-term [15], long-term [16], and immortalized cell lines [17] have been successfully developed and non-apis genes could be expressed in cultured embryonic cells [18]. Meanwhile, high-throughput sequencing, also referred to as next-generation sequencing (NGS), has become essential for modern biological researches, and it has been widely applied to honey bee genome [19, 20], transcriptome [21, 22] and metagenome analyses [23]. Moreover, MS-based proteomics has been developed to be a powerful technology that enables a system wide view of honey bee proteomes and their changes [24, 25]. Altogether, these current technologies can bring the study on honey bee embryogenesis into a new era, compared to their previous morphological observation using light/scan microscope [11, 26].

The genome of *Apis mellifera* has been assembled in 2006 [19]. More recently, the genome and gene annotation of *Apis* mellifera are upgraded and appear nearly complete [27]. And several transcriptomic and proteomic studies of *Apis mellifera* have been performed to reveal the embryogenesis of honey bee [21, 24, 25]. However, the genome of Apis cerana was roughly completed in 2015 [20]; molecular biology research and sequence information for *Apis cerana* are extremely lacking and need further exploration, especially for *Apis cerana* embryogenesis.

Recently, transcriptome analyses were employed to investigate the different gene expressions between workers and Queens in Apis cerena [22]. And transcriptome analyses of haploid and diploid embryos have been also applied to reveal early zygotic transcription during cleavage in *Apis mellifera* [21]. But the developmental strategies of Asian honey bees during the embryogenesis of egg stage are still largely unknown. Here we collected nine samples over the complete time course of Asian honeybee embryogenesis at 24 (Day1), 48 (Day2), and 72 h (Day3) of age, and employed high coverage RNA-Seq to explore their transcriptomes. Our study would provide a high-quality view of transcriptomic dynamics throughout embryonic development in Asian honeybee and offer novel embryologic insights for other social insects.

## Methods and Materials

### Sample collection and next generation sequencing

The embryo samples of *Apis cerena* were obtained from 3 honey bee colonies maintained in the apiary of Honeybee Research Institute, Jiangxi Agricultural University, China. To collect age-controlled embryos, mated egg-laying queens were first limited in queen cages for 2-3 h, and then placed into egg- and brood-free areas of their own hives. After nine hours, the queens were released from the egg-laid honeycombs. After 24 h, 40 embryos per colony were collected and stored together in liquid nitrogen as a first-day embryo sample, the remaining embryos were continuously placed back into their hives; after 48 hours, another 40 embryos per colony were collected and stored in liquid nitrogen as a second-day embryo sample; after 72 h, 40 embryos per colony from the rest were collected and put into liquid nitrogen as a third-day embryo sample (Fig. IV-4-1A, see page 540). Finally, we collected nine embryo samples for three days, in other words, three biological replicates per day.

Total RNA was extracted from the above-mentioned embryo samples using the TRIzol reagent according to the manufacture's protocol (Invitrogen). The concentration, RIN, the 28S/18S rRNA ratio and size of total RNA samples were determined using the RNA 6000 Nano Kit on the Agilent 2100 Bioanalyzer (Agilent Technologies, CA, US). The complementary DNA was prepared for each embryo sample using the Illumina mRNA sequencing kit (Illumina, CA, US) and the Clontech SMART cDNA Library Construction Kit (Invitrogen). Libraries were sequenced using the Illumina HiSeq XTen platform (Illumina, CA, US).

### Mapping and quality control of RNA-seq data

Before mapping, raw data was filtered as the following steps: 1) Remove reads with adapters; 2) Remove reads in which unknown bases are more than 10%; 3) Remove low quality reads, in which the percentage of low quality bases (the quality is no more than 10) is over 50%. The remaining reads were defined as "clean reads" and used for next bioinformatics analyses. Quality control on clean data was performed through drawing base composition chart and quality distribution chart using FastQC (http://www.bioinformatics.babraham.ac.uk/projects/fastqc/). We used Hisat2 [28] to map clean reads to the *A. cerana* genome assembly version 2.0 (ACSNU-2.0 GCA_001442555.1). Prior to this, exons and splice sites were extracted from the genome annotation GTF files and used to index the genomes. The mapping statistics and the coverage over gene body of RNA-seq reads were calculated using bam_stat.py and geneBody_coverage2.py implemented in the RSeQC package [29], respectively.

### *Transcriptome assembly and annotation*

We applied a reference-based transcript assembler, StringTie [30], to assemble the transcripts for each tested embryo sample. The gffcompare program (https://github.com/gpertea/gffcompare) was employed to evaluate the transcript assembles (such as summary of gene and transcript counts, novel vs. known etc.) and to perform basic tracking of assembled isoforms across multiple RNA-Seq experiments. Hmmer2go (https://github.com/sestaton/HMMER2GO) was used to identify the possible open reading frame (ORF) of the assembled transcripts. We identified putative homologs to Drosophila melanogaster database using BLAST with an E-value threshold of $1e^{-5}$, which was implemented in the online annotation tools of KOBAS 3.0 [31].

### *Differential expression analysis*

Transcript abundances were estimated by running StringTie with the -eB options. Read count information was extracted from the above files of transcript abundances by prepDE.py for the next differential expression analysis. We used the R statistical packages DESeq2 [32] from the Bioconductor repository to test for differential expression between the different-day embryo samples.

### *GO and KEGG analysis*

To perform functional enrichment of the candidate genes, the fly Gene Ontology database (http://www.geneontology.org) was then queried by the ClueGO plugin [33] of Cytoscape using Uniprot ID or EntrezID as input parameters. The enriched GO terms were characterized according to the default setting. The KEGG pathways were characterized on the online annotation tools of KOBAS 3.0 (http://kobas.cbi.pku.edu.cn/annotate.php).

### *Quantitative Real-time PCR*

cDNAs were generated using Reverse Transcriptase kit reagents (Transgen, Beijing, China), according to the manufacturer's instructions. 14 differentially expressed genes were selected for qRT-PCR analysis and actin 3 was used as an internal control to normalize the data. The primer pairs are provided in Table S1. Real-time PCR was conducted using an ABI 7500 real time PCR detection system (ABI, CA, USA). The cycling conditions were as follows: preliminary 95°C for 30 sec, 40 cycles including 95°C for 10 sec, 60°C for 1 min, and 72°C for 30 sec. The specificity of the PCR products was verified by melting curve analysis for each sample. For each unigene, three biological replicates (with three technical replicates for each biological replicate) were performed. The control and target unigene for each sample were run in the same plate to eliminate interplate variation. The Ct value for each biological replicate was obtained by calculating the arithmetic mean of three technical replicate values. Relative mRNA expression was determined using the comparative $2^{-\Delta\Delta Ct}$ method. The statistical analysis of gene expression was performed by analysis of variance (ANOVA) using R version 3.3.3.

## Results

### *Illumina sequencing and mapping statistics*

Illumina RNA sequencing generated an average of 65.7 million (M) raw reads per sample. After filtration, 57.3 M average clean reads with accumulated length of 8.6 million base (Mb) per sample remained for the next transcriptomic analyses. The average Q20 percentage (sequencing error rate, 1%) per sample was 95.3%, and the average GC percentage was 38.3% (Table S2). With the embryo age increasing, the GC percentage of clean reads significantly increased from 36.7% (Day1) to 37.6% (Day2, $P = 9.63 \times 10^{-3}$), and to 40.6% (Day3, $P = 5.37 \times 10^{-5}$). The clean reads were mapped to the *Apis cerana* genome reference (Assembly ACSNU-2.0) assembled and submitted by Park et al.[20]. The average mapping rate was 74.2%, Still 25.8% of clean reads cannot map into the reference genome, which indicated that the current *Apis cerana* reference genome should be further improved. However, when we mapped these *Apis cerana* RNA-seq reads to the *Apis mellifera* genome reference (Assembly Amel_4.5), the average mapping rate was much smaller of 25.6% (Fig. S1). This significant reduction of mapping rate across species suggested that the genetic basis between *Apis cerana* and *Apis mellifera* is largely divergent and the reference genome of *Apis mellifera* was not suitable for mapping the RNA-seq reads of *Apis cerana*.

In the present study, 74.5%, 73.1% and 75.0% of clean reads at the first-, second- and third- day were mapped to the Apis cerana genome reference, respectively. Of these mapped read data, the percentage (77.1%) of properly-paired read at the first day is less than the second day (80.0%, $P = 7.10 \times 10^{-3}$) and the third day (82.9%, $P = 4.41 \times 10^{-4}$) (Table S3). Meanwhile with the age of embryo increasing, the percentage of singleton were decreasing from 16.1% (Day1) to 13.9% (Day2, $P = 1.05 \times 10^{-2}$), and to 10.6% (Day3, $P = 6.08 \times 10^{-4}$). It is also interesting that with the embryo age increasing, splice reads increased from 26.7% at the first day to 28.3% ($P = 8.11 \times 10^{-3}$) at the second day, and to 29.9% ($P = 2.76 \times 10^{-3}$) at the third day (Table S3). These findings suggested that: (1) the transcripts of the first day were more likely located on the regions which were harder to be resequenced possibly due to high GC or AT content, or high extensive of repeat sequence; (2) with the development of Apis cerana embryo, more splice events happened.

To check if reads coverage was uniform and if there was any 5'/3' bias, we estimated the gene body coverage for each sample. We found that the gene body coverage was nearly symmetry between 5' end and 3' end of gene and reads were evenly distributed on reference genes in each embryo sample (Fig. S2). The results indicated that the RNA quality was high and the randomness of reads was good, and suggested that the sequencing data of the embryo samples was suitable for the next bioinformatic analysis.

### *Hierarchical clustering and principal component analysis for samples*

To evaluate the consistency of sample collection and investigate the transcriptomic relationship among the first-, second- and third- day embryos, we performed hierarchical clustering and principal component analysis (PCA) for these samples using the whole gene expression data. Both analyses could give us an overview of similarities and dissimilarities between samples. Before these two analyses, we first normalized the gene expression levels among the tested samples using the method of regularized log transformation implemented in the R package of DESeq2 (Fig. S3). We found that the embryo samples collected at the same day were well grouped into the same clusters. On the other words, the embryo samples of the first-day clustered together, the samples of the second-, third- day clustered into their own subtrees together, too (Fig. IV-4-1B). Notably, the samples of the second day and the third day clustered closer than the relationship between the second/third day and the first day. This result suggested that the developmental status between the second day and the third day was closer than the one between the first day and the second or third day in *Apis cerana*. In the PCA plot, PC1 and PC2 together explained more than 96% variance for the Apis cerana samples; and embryo samples from the same day were grouped together (Fig. IV-4-1C). The result of PCA was in perfect agreement with the one of hierarchical clustering analysis.

*Overview of the Apis cerana embryonic transcriptome*

We employed StringTie software to reconstruct the embryonic transcriptome for each *Apis cerana* embryo sample and to estimate the relative abundance of genes/transcripts in these embryonic transcriptomes. In the present study, we detected an average of 28,561 transcripts per first-day embryo, 32,155 transcripts per second-day embryo and 34,406 transcripts per third-day embryo, which located on an average of 20,128, 21,441 and 23,876 transcribed loci per sample, respectively (Fig. IV-4-2A, see page 541). If filtered the low expression transcripts with the criterion of fragments per kilobase of exon per million fragments mapped (FPKM) > 1, an average of 13,830 transcripts representing 9,298 genes for the first-day embryo, 17463 transcripts representing 11,311 genes for the second-day, and 18012 transcripts representing 12,238 genes for the third-day were remained (Fig. IV-4-2A). These results showed that the number of transcripts and transcribed loci generally increased with the increasing embryonic age and indicated that the transcription activity of embryonic genes became brisker along with the developmental time during the egg stage. These dynamics of gene expression during the embryonic development are generally consistent with the one of Drosophila melanogaster [34].

In the reconstructed transcriptome, 51.5%, 45.7% and 47.7% of transcripts at the first-, second- and third- days showed low expression level of FPKM ≤ 1. With the expression abundance increasing, the percentage of the transcripts decreased. And the expression abundance (FPKM) of ~95% transcripts at all these three days was no more than 35 (Fig. IV-4-2B). After filtering the low expressed transcripts with FPKM ≤ 1, the transcriptomes at all three days of the embryonic stage contained ~20% single-exon transcripts and ~80% multi-exon transcripts. Except for single-exon transcripts, in the embryonic transcriptomes the number of three-exon transcripts reached maximum value, and gently the number of multi-exon transcripts decreased with the increased exon units (Fig. IV-4-2C). While in the reference genes (GCA_001442555.1) of ACSNU-2.0, there are fewer single-exon transcripts than embryo samples, but more multi-exon transcripts except for the two-exon transcripts; also, three-exon transcripts had maximum volumes, then the number of transcripts gently decreased with the increased exon units. We further found that the numbers of single-exon and multi-exon transcripts at the first day were almost all significantly smaller than the corresponding number at the second day or at the third day. In addition, we found that the single-isoform genes were much more than the genes with multiple isoforms in the embryonic transcriptomes, and the number of single-isoform genes generally increased along with the embryonic development time (Fig. IV-4-2D).

***Discovery of new transcribed regions in the embryonic stage***

The transcriptomes (FPKM >1) of nine tested embryo samples were compared with the annotated *Apis cerana* genes deposited in the NCBI Genbank database; the statistic summary of comparison was shown in Table S4. Only less than half of transcripts in the embryonic transcriptomes (47.3% for the first day, 46.2% for the second day, 44.7% for the third day) were completely matched with the annotated genes, which suggested that our transcriptomic data identified in this study largely expanded the number of known transcribed elements in the Apis cerana genome and could make an important complement to the current reference genes.

A total of 55,646 consensus transcripts located in 18,284 transcribed regions was detected in the above embryo samples. Of these transcribed regions, 8,819 were novel compared to the annotated *Apis cerana* genes deposited in the NCBI database (Fig. IV-4-2E). These novel transcript regions harbored 15,292 consensus transcripts. Of the novel transcripts 55.5% had a predicted open reading frame (ORF) greater than 100 amino acids; the remaining novel transcripts could encode small peptides but many were likely to be non-coding RNAs. We annotated these novel genes with Drosophila Melanogaster protein database by online KOBAS 3.0 annotation tools [31], only 364 genes harboring 727 transcripts were annotated. The biological function of other unannotated novel genes or novel transcripts need to be further validated by other experimental methods.

GO terms and KEGG pathways were enriched using the annotated novel genes. We found 7 KEGG pathways (Table S5) and 112 GO terms were enriched with corrected $P$ value < 0.05 (Table S6) in these new genes. The enriched KEGG pathways included phototransduction - fly (dme04745), hippo signaling pathway - fly (dme04391), phagosome (dme04145), ECM-receptor interaction (dme04512), oxidative phosphorylation (dme00190), metabolic pathways (dme01100) and endocytosis (dme04144). GO analysis showed that some key tissue or organ development were enriched during the embryogenesis, like tube, epithelium, compound eye, nervous system, muscle structure, wing disc and imaginal disc (Table S6).

***Core or specific transcriptomes in the development of different day embryos***

There were 8,139 common genes shared with embryo samples of all three days (Fig. IV-4-2E). These genes harbored 37,434 transcripts, of which 33,578 (89.7%) were possible coding RNAs with predicted ORF greater than 100 amino acids. Of these core transcripts 15% was single-exon transcripts, 85% was multiple-exon transcripts and 46% was multiple-exon multiple isoform transcripts. 68.9% of these core transcripts were successfully annotated. We performed GO and KEGG analysis for these annotated core transcripts in the embryonic development, and 1098 GO terms (Table S7) and 50 KEGG pathways (Table S8) were enriched with corrected $P$-value less than 0.05. The Top GO biological processes (Table IV-4-1) included cellular macromolecule metabolic process (GO:0044260), system development (GO:0048731), gene expression (GO:0010467), and cell differentiation (GO:0030154), *et al.*; while the most significant KEGG pathways (Table IV-4-1) were metabolic pathways (dme01100), RNA transport (dme03013), Endocytosis (dme04144), Spliceosome (dme03040) and Protein processing in endoplasmic reticulum (dme04141). These results inferred that the genes involved in the above biological processed and pathways play key roles in the embryonic development of *Apis cerana* during the egg stage.

1,986, 1,799 and 2,903 genes were specifically expression in the first-, second- and third-day embryos

(Fig. IV-4-2E). We also performed GO (Table IV-4-2) and KEGG (Table S9) enrichment analysis for these specific transcripts in the different stages of embryogenesis. At the first day, the top GO term was neuropeptide signaling pathway (GO:0007218), which was reported to regulate synaptic growth in Drosophila [35] and suggested that the nervous system of Apis cerana had been initiated to develop on the first day of embryo. The second top GO term was response to water (GO:0009415), which indicated that water utilization or water response was important in early embryonic development. To our surprise, eye-antennal and leg disc development and morphogenesis have start at the first day of embryo development. Possibly, pigment metabolic process (GO:0042440) was related to the development of the eye and eye pigment during the embryonic development of *Apis cerana*. At Day 2, only seven GO terms were found, which mainly included organic acid biosynthetic process, olfactory behavior, photoreceptor cell axon guidance, tRNA processing, and ion transport. At Day 3, the biological process of ion transport and homeostasis were found frequently, indicating ion transport and homeostasis was an important event during the third day development of *Apis cerana* embryo.

Table IV-4-1 The top 10 Go terms and KEGG pathways enriched with the common genes in the core transcriptome

| ID | GO Term or KEGG pathway | P Value | Corrected P value[1] | Nr. Genes |
|---|---|---|---|---|
| | **GO term** | | | |
| GO:0044260 | cellular macromolecule metabolic process | 3.50E−221 | 5.50E−218 | 2499 |
| GO:0034641 | cellular nitrogen compound metabolic process | 2.30E−171 | 3.60E−168 | 1961 |
| GO:0010467 | gene expression | 2.00E−158 | 3.20E−155 | 1554 |
| GO:0048731 | system development | 4.00E−157 | 6.30E−154 | 1737 |
| GO:0006139 | nucleobase-containing compound metabolic process | 6.10E−141 | 9.60E−138 | 1654 |
| GO:0048869 | cellular developmental process | 1.20E−140 | 2.00E−137 | 1739 |
| GO:0046483 | heterocycle metabolic process | 1.00E−139 | 1.50E−136 | 1689 |
| GO:0006725 | cellular aromatic compound metabolic process | 1.30E−139 | 2.10E−136 | 1727 |
| GO:1901360 | organic cyclic compound metabolic process | 5.40E−138 | 8.40E−135 | 1752 |
| GO:0030154 | cell differentiation | 7.90E−138 | 1.20E−134 | 1673 |
| | **KEGG pathway** | | | |
| dme01100 | Metabolic pathways | 6.94E−24 | 1.42E−21 | 558 |
| dme03013 | RNA transport | 1.74E−10 | 1.78E−08 | 112 |
| dme04144 | Endocytosis | 3.79E−10 | 2.57E−08 | 102 |
| dme03040 | Spliceosome | 2.10E−09 | 1.07E−07 | 100 |
| dme04141 | Protein processing in endoplasmic reticulum | 7.75E−09 | 3.16E−07 | 96 |
| dme00240 | Pyrimidine metabolism | 1.70E−07 | 4.96E−06 | 69 |
| dme04120 | Ubiquitin mediated proteolysis | 1.70E−07 | 4.96E−06 | 76 |

| ID | GO Term or KEGG pathway | P Value | Corrected P value[1] | Nr. Genes |
|---|---|---|---|---|
| dme04013 | MAPK signaling pathway – fly | 6.29E–07 | 1.60E–05 | 72 |
| dme03018 | RNA degradation | 1.29E–06 | 2.93E–05 | 50 |
| dme00230 | Purine metabolism | 3.05E–06 | 6.23E–05 | 89 |

1, GO Term *P*-values were corrected with the method of Bonferroni step down; KEGG pathway p-values were corrected with the method of Benjamini and Hochberg.

## *Differential expression genes and their expression patterns*

In the present study, we identified 4088 up-regulated and 3,046 down-regulated differential expression genes (DEGs) among three ages of embryo samples. Of these DEGs, 2,010, 3,177 and 1,528 genes were up regulated, and 2,088, 2,294 and 303 genes were down regulated, from Day 1 to Day 2, from Day 1 to Day 3 and from Day 2 to Day 3, respectively (Fig. IV-4-3A, see page 542). Generally, the total number of up-regulated genes was larger than the one of down-regulated genes, especially for the DEGs from Day 2 to Day 3. The number of DEGs from Day 2 to Day 3 is much less than the one of DEGs from Day 1 to Day 2, which indicated that the development change between Day 1 and Day 2 was much larger than the one between Day 2 and Day 3 (Fig. IV-4-3A and B). This result is consistent with the sample cluster using full gene set (Fig. IV-4-1B) or using DEG set (Fig. IV-4-3C).

We further classified these DEGs into 6 groups based on the expression patterns (Fig. S4): (1) Class one, continuous up regulated; (2) Class two, up regulated from Day 1 to Day 2, but not changed significantly between Day 2 and Day 3; (3) Class three, up regulated from Day 2 to Day 3, but not changed significantly between Day 1 and Day 2; (4) Class four, continuous down regulated; (5) Class five, down regulated from Day 1 to Day 2, but not changed significantly between Day 2 and Day 3; (6) Class six, down regulated from Day 2 to Day 3, but not changed significantly between Day 1 and Day 2. We found that the DEG number of Class 2 (up regulated from Day 1 to Day 2) is largest of 1781, the second largest gene expression pattern is Class 5 (down regulated from Day 1 to Day 2), which also suggested the expression change from Day 1 to Day 2 was very large, and the development between Day 1 and Day 2 was of great dynamic activity. We selected 14 DEGs from different classes of the above expression patterns, most of which are with known biological function. We employed RT-qPCR to validate their gene expression. The trends of mRNA expression showed that gene expression of tested genes identified by RT-qPCR were consistent with those detected with RNA-Seq (Fig. S5).

**Table IV-4-2 Go terms enriched with the specific genes expressed in the first-, second- and third- day embryos**

| Day | GOID | GOTerm | Term P Value | Term P Value Corrected[1] | Nr. Genes | Associated Genes Found |
|---|---|---|---|---|---|---|
| 1 | GO:0007218 | neuropeptide signaling pathway | 5.40E–06 | 1.00E–04 | 6 | [CCAP-R, CG33639, CapaR, ETHR, FMRFaR, Lkr] |
| 1 | GO:0009415 | response to water | 1.50E–05 | 2.80E–04 | 3 | [CapaR, Poxn, ppk28] |

(continued)

| Day | GOID | GOTerm | Term P Value | Term P Value Corrected[1] | Nr. Genes | Associated Genes Found |
|---|---|---|---|---|---|---|
| 1 | GO:0042440 | pigment metabolic process | 1.90E−04 | 3.30E−03 | 6 | [CG1885, Cdk5alpha, Itgbn, cd, dl, yellow−d2] |
| 1 | GO:0035214 | eye-antennal disc development | 8.20E−03 | 8.20E−03 | 3 | [Poxn, bab1, dl] |
| 1 | GO:0006582 | melanin metabolic process | 6.90E−04 | 1.10E−02 | 4 | [Cdk5alpha, Itgbn, dl, yellow−d2] |
| 1 | GO:0001101 | response to acid chemical | 1.00E−03 | 1.50E−02 | 3 | [CapaR, Poxn, ppk28] |
| 1 | GO:0040012 | regulation of locomotion | 7.90E−03 | 1.50E−02 | 3 | [Hr51, Hrs, Ten−m] |
| 1 | GO:0072507 | divalent inorganic cation homeostasis | 5.10E−03 | 1.50E−02 | 3 | [FMRFaR, Lkr, trpl] |
| 1 | GO:0010035 | response to inorganic substance | 4.60E−03 | 1.80E−02 | 3 | [CapaR, Poxn, ppk28] |
| 1 | GO:0018958 | phenol-containing compound metabolic process | 1.50E−03 | 1.90E−02 | 4 | [Cdk5alpha, Itgbn, dl, yellow−d2] |
| 1 | GO:0071214 | cellular response to abiotic stimulus | 1.30E−03 | 1.90E−02 | 4 | [Arr1, CapaR, Gat, ppk28] |
| 1 | GO:0007478 | leg disc morphogenesis | 1.80E−03 | 2.20E−02 | 4 | [Poxn, bab1, dl, trio] |
| 1 | GO:0042551 | neuron maturation | 4.40E−03 | 2.20E−02 | 3 | [Drice, Hr51, Hrs] |
| 1 | GO:0006874 | cellular calcium ion homeostasis | 2.10E−03 | 2.30E−02 | 3 | [FMRFaR, Lkr, trpl] |
| 1 | GO:0007455 | eye-antennal disc morphogenesis | 2.10E−03 | 2.30E−02 | 3 | [Poxn, bab1, dl] |
| 1 | GO:0016322 | neuron remodeling | 4.00E−03 | 2.40E−02 | 3 | [Drice, Hr51, Hrs] |
| 1 | GO:0055074 | calcium ion homeostasis | 2.40E−03 | 2.40E−02 | 3 | [FMRFaR, Lkr, trpl] |
| 1 | GO:0072503 | cellular divalent inorganic cation homeostasis | 3.80E−03 | 2.60E−02 | 3 | [FMRFaR, Lkr, trpl] |
| 1 | GO:0035006 | melanization defense response | 3.40E−03 | 2.70E−02 | 3 | [Cdk5alpha, Itgbn, dl] |
| 1 | GO:0009755 | hormone-mediated signaling pathway | 3.20E−03 | 2.90E−02 | 3 | [ETHR, Hr51, Lkr] |
| 2 | GO:0016053 | organic acid biosynthetic process | 7.90E−04 | 5.50E−03 | 5 | [Baldspot, CG5278, Dh44−R2, Oat, b] |
| 2 | GO:0042048 | olfactory behavior | 2.00E−03 | 1.00E−02 | 5 | [CG14509, Calr, NaCP60E, TkR99D, prt] |
| 2 | GO:0072499 | photoreceptor cell axon guidance | 1.70E−03 | 1.00E−02 | 3 | [Nrk, gogo, jbug] |
| 2 | GO:0008033 | tRNA processing | 1.60E−02 | 1.60E−02 | 3 | [CG10495, CG15618, CG4611] |
| 2 | GO:0046394 | carboxylic acid biosynthetic process | 4.50E−03 | 1.80E−02 | 4 | [Baldspot, CG5278, Dh44−R2, Oat] |
| 2 | GO:0098656 | anion transmembrane transport | 1.50E−02 | 3.10E−02 | 3 | [Best1, Indy, NaPi−T] |
| 2 | GO:0043269 | regulation of ion transport | 1.00E−02 | 3.20E−02 | 3 | [Best1, NaCP60E, Rgk1] |
| 3 | GO:0006811 | ion transport | 7.00E−07 | 5.30E−05 | 24 | [CG10960, CG14507, CG31028, CG31547, CG3690, CG42269, CG5002, CG5621, CG6125, CG6356, CG8249, Ca−alpha1T, Ctr1A, Gat, HisCl1, Mco1, Nckx30C, Nmdar1, RyR, Trpgamma, Zip42C.1, Zip89B, nAChRbeta1, para] |

(continued)

| Day | GOID | GOTerm | Term *P* Value | Term *P* Value Corrected[1] | Nr. Genes | Associated Genes Found |
|---|---|---|---|---|---|---|
| 3 | GO:0055080 | cation homeostasis | 1.30E−06 | 9.90E−05 | 11 | [CG5002, CG6125, Ctr1A, Dat, Mco1, Nckx30C, Nmdar1, RyR, Trpgamma, Zip42C.1, norpA] |
| 3 | GO:0098771 | inorganic ion homeostasis | 1.60E−06 | 1.20E−04 | 11 | [CG5002, CG6125, Ctr1A, Dat, Mco1, Nckx30C, Nmdar1, RyR, Trpgamma, Zip42C.1, norpA] |
| 3 | GO:0050801 | ion homeostasis | 2.30E−06 | 1.70E−04 | 11 | [CG5002, CG6125, Ctr1A, Dat, Mco1, Nckx30C, Nmdar1, RyR, Trpgamma, Zip42C.1, norpA] |
| 3 | GO:0055065 | metal ion homeostasis | 3.80E−06 | 2.70E−04 | 9 | [Ctr1A, Dat, Mco1, Nckx30C, Nmdar1, RyR, Trpgamma, Zip42C.1, norpA] |
| 3 | GO:0009072 | aromatic amino acid family metabolic process | 7.70E−06 | 5.50E−04 | 5 | [CG1461, Trh, hgo, v, y] |
| 3 | GO:0072507 | divalent inorganic cation homeostasis | 1.20E−05 | 8.40E−04 | 7 | [Dat, Nckx30C, Nmdar1, RyR, Trpgamma, Zip42C.1, norpA] |
| 3 | GO:0030003 | cellular cation homeostasis | 1.40E−05 | 1.00E−03 | 9 | [CG5002, CG6125, Ctr1A, Dat, Mco1, Nckx30C, RyR, Trpgamma, norpA] |
| 3 | GO:0006873 | cellular ion homeostasis | 2.10E−05 | 1.40E−03 | 9 | [CG5002, CG6125, Ctr1A, Dat, Mco1, Nckx30C, RyR, Trpgamma, norpA] |
| 3 | GO:0055074 | calcium ion homeostasis | 2.70E−05 | 1.80E−03 | 6 | [Dat, Nckx30C, Nmdar1, RyR, Trpgamma, norpA] |
| 3 | GO:0098660 | inorganic ion transmembrane transport | 3.30E−05 | 2.20E−03 | 14 | [CG31028, CG31547, CG5002, CG6125, Ca-alpha1T, Ctr1A, HisCl1, Mco1, Nckx30C, RyR, Trpgamma, Zip42C.1, Zip89B, para] |
| 3 | GO:0055082 | cellular chemical homeostasis | 4.90E−05 | 3.20E−03 | 9 | [CG5002, CG6125, Ctr1A, Dat, Mco1, Nckx30C, RyR, Trpgamma, norpA] |
| 3 | GO:0006875 | cellular metal ion homeostasis | 5.70E−05 | 3.60E−03 | 7 | [Ctr1A, Dat, Mco1, Nckx30C, RyR, Trpgamma, norpA] |
| 3 | GO:0000041 | transition metal ion transport | 1.20E−04 | 7.80E−03 | 5 | [Ctr1A, Mco1, Trpgamma, Zip42C.1, Zip89B] |
| 3 | GO:1901565 | organonitrogen compound catabolic process | 1.30E−04 | 8.30E−03 | 9 | [CG5418, CG8129, CG9380, Dat, GLS, PGRP-LB, hgo, v, verm] |
| 3 | GO:0048878 | chemical homeostasis | 1.40E−04 | 8.90E−03 | 11 | [CG5002, CG6125, Ctr1A, Dat, Mco1, Nckx30C, Nmdar1, RyR, Trpgamma, Zip42C.1, norpA] |
| 3 | GO:0019725 | cellular homeostasis | 1.90E−04 | 1.10E−02 | 11 | [CG5002, CG6125, CG6888, Ctr1A, Dat, Mco1, Nckx30C, RyR, Trpgamma, norpA, tn] |
| 3 | GO:0006874 | cellular calcium ion homeostasis | 2.50E−04 | 1.50E−02 | 5 | [Dat, Nckx30C, RyR, Trpgamma, norpA] |
| 3 | GO:0042430 | indole-containing compound metabolic process | 3.30E−04 | 1.90E−02 | 3 | [Dat, Trh, v] |

| Day | GOID | GOTerm | Term $P$ Value | Term $P$ Value Corrected[1] | Nr. Genes | Associated Genes Found |
|---|---|---|---|---|---|---|
| 3 | GO:0044550 | secondary metabolite biosynthetic process | 3.80E-04 | 2.10E-02 | 5 | [Elo68alpha, v, y, yellow-c, yellow-e2] |
| 3 | GO:0046189 | phenol-containing compound biosynthetic process | 5.10E-04 | 2.80E-02 | 4 | [Trh, y, yellow-c, yellow-e2] |
| 3 | GO:1901606 | alpha-amino acid catabolic process | 5.10E-04 | 2.80E-02 | 4 | [CG8129, GLS, hgo, v] |
| 3 | GO:0070838 | divalent metal ion transport | 6.40E-04 | 3.50E-02 | 6 | [Ca-alpha1T, Nckx30C, RyR, Trpgamma, Zip42C.1, Zip89B] |
| 3 | GO:0072511 | divalent inorganic cation transport | 6.40E-04 | 3.50E-02 | 6 | [Ca-alpha1T, Nckx30C, RyR, Trpgamma, Zip42C.1, Zip89B] |
| 3 | GO:0072503 | cellular divalent inorganic cation homeostasis | 6.60E-04 | 3.60E-02 | 5 | [Dat, Nckx30C, RyR, Trpgamma, norpA] |
| 3 | GO:0006820 | anion transport | 7.90E-04 | 4.20E-02 | 8 | [CG14507, CG31547, CG3690, CG5002, CG6125, CG6356, Gat, HisCl1] |
| 3 | GO:0015711 | organic anion transport | 8.30E-04 | 4.30E-02 | 6 | [CG14507, CG3690, CG5002, CG6125, CG6356, Gat] |

1, Term P Value Corrected with Bonferroni step down.

GO analysis (Supplemental Table S10) showed that more than hundred of continuous-up-regulated genes participated in the biological processes of multicellular organism development, system development, cell differentiation and development, and anatomical structure morphogenesis, and hundreds of continuous-down-regulated genes took part in the biological processes of regulation of cellular process and biosynthetic process, macromolecule modification and metabolic process, and so on. In the patterns of Class 2, 3, 5 and 6, the top GO terms again referred to system development, animal organ development and multicellular organism development, which indicated that during the whole embryonic development many differentially expressed genes involved the key biological processes of system, organ or multicellular organism development. Notably in Class 2, the related GO terms to epithelium, tube and neuron development were enriched, while in Class 3, GO terms related to muscle development were repeatedly observed. These findings indicated that neuron, epithelium development was earlier than muscle development during the honeybee embryogenesis.

In addition, many core genes differentially expressed in the three ages of *Apis cerana* embryo, which indicated that these core genes were regulated during the embryonic development. We found that 203 GO terms specifically overlapped with the ones enriched by down-regulated genes, 43 GO terms specifically overlapped with the ones enriched by up-regulated genes and 110 GO terms overlapped simultaneously with the terms enriched by up- and down- regulated genes (Supplemental Table S7).

### *Comparison with previous studies on honeybee embryonic development*

Previously, early zygotic transcription of European honeybee (*Apis mellifera*) at 0-2h, 0-6h and 18-24h were investigated [21]. Among their embryonic samples, the collection time of 18-24h age sample was very close to our first-day embryo samples. Therefore, we downloaded the raw RNA-seq data of 18-24h age embryo sample, and made a comparison with our transcriptome of the first-day Apis cerana embryos. The descriptive statistics of RNA-seq, mapping and annotation were listed on Supplemental Table S11. In general, the quality of our RNA-seq data was higher, the data size was about six times larger than the downloaded data, and more transcripts were identified in our *Apis cerana* embryos. We annotated both transcripts of *Apis mellifera* and *Apis cerana* against the protein database of Drosophila melanogaster using online annotation tool KOBAS 3.0 [31]. 5,961 *Apis mellifera* and 6,002 *Apis cerana* genes were annotated. We found 86.9% (5,215) of *Apis cerana* annotated genes were shared in *Apis mellifera* first-day embryo, and 787 genes were specifically expressed in *Apis cerana* while 746 genes in *Apis mellifera* (Fig. IV-4-4, see page 543). Among the common genes, 98% (5,116 genes) were core genes expressed in all *Apis cerana* embryos during the whole embryonic development.

GO and KEGG analyses were performed to identify the biological processes and pathways for the specific genes expressed in *Apis cerana* and *Apis mellifera*. 79 and 28 GO terms for the first-day *Apis cerana* and *Apis mellifera* embryos have corrected enrichment $P$-value less than 0.05 (Supplemental Table S12), while 3 and 2 KEGG pathways passed the same filtering criterion, respectively (Fig. IV-4-4). Notably in *Apis cerana*, the top significant KEGG pathway was phtotransduction - fly (dme04745), and 6 of 10 top GO terms were related to the biological process of light detection, which including detection of light stimulus (GO:0009583), detection of external stimulus (GO:0009581), detection of abiotic stimulus (GO:0009582), response to light stimulus (GO:0009416), phototransduction (GO:0007602) and detection of visible light (GO:0009584). These findings suggested that the ability of light detection of *Apis cerana* was different with *Apis mellifera* or possibly stronger than *Apis Mellifera*. On the other hand, in *Apis mellifera*, the top KEGG pathway (Fig. IV-4-4) was Glycosaminoglycan biosynthesis - chondroitin sulfate / dermatan sulfate (dme00532). Top 10 GO terms were associated with glycosylation and glycoprotein metabolic process (Fig. IV-4-4). Chondroitin sulfate proteoglycan is an important component of the extracellular matrix in the central nervous system, and it takes part in cell migration of the central nervous system [36]. Therefore, the development of central nervous system in *Apis mellifera* might take place ahead compared to *Apis cerana*.

Another work on embryo development of European honeybee (*Apis mellifera*) was performed by Fang, et al. [24] recently. They employed mass spectrometry-based proteomics to investigate the proteomic alterations of honeybee embryogenesis at 24, 48 and 72 h of age. In our *Apis cerana* transcriptome analyses, the sample collection and analysis strategies were similar as those of Fang, et al. We found that the clustering of the embryo samples was consistent with their clustering result of *Apis mellifera*, in which the samples of the second- and third- days were close to each other and clustered together, and the first- day sample was grouped into another cluster. Fang, et al. successfully annotated 845, 1,078 and 1,116 proteins to the KEGG database at 24, 48 and 72h. While in the present study, we identified 6,002, 6,677 and 6,831 annotated proteins at 24 (Day1), 48 (Day2) and 72h (Day3). Far more common metabolic pathways were significantly enriched in the embryos during the three stages, and more genes were included in each enriched metabolic pathway in our transcriptomic data. For example, in the proteomic analysis, 16 proteins were enriched in the glycolysis/gluconeogenesis pathway, while in our transcriptome, 27 genes were detected and enriched in the same pathway, which involved all the above proteins (Fig. S6).

## Discussion

Embryogenesis is an elementary and fundamental stage of insect development, and plays an important role in supporting the developmental process of whole life cycle. Our knowledge about insect embryogenesis mainly comes from Drosophila. Drosophila embryogenesis includes 17 different stages along with five major embryogenetic events: cleavage divisions (1-4 stages), gastrulation (5-7 stages), germ band elongation (8-11 stages) and shortening (12-14 stages), dorsal closure and head involution (15-17 stages) [37]. During the embryonic development, the first larval instar and the imaginal disks of various tissues are gradually formed, and sex differentiation and cell or tissue functional differentiation are generated.

Before the omics era, honeybee embryogenesis, exactly *Apis mellifera*, has been characterized through morphological observation using light microscope [38, 39] and/or scanning electron microscope [11, 26]. These previous works have provided us lots of important developmental knowledges about honeybee embryogenesis. However, Morphological observation could not reveal the intrinsic developmental mechanism during honeybee embryogenesis. Recent studies on honey bee embryogenesis also mainly focused on *Apis mellifera*, including the proteomics characterization of *Apis mellifera* embryogenesis [24, 25] and early zygotic transcriptome analysis for haploid and diploid honeybee embryonic cleavage [21]. But nowadays the proteomics technology could not yield developmental information as much as the transcriptomics technology for tissue functional research; in other words, there might be a lot of information lost in proteomics. Hitherto, there was still no report for Asian honeybee (*Apis cerana*) embryogenesis. It was the first time to employ transcriptomics to reveal *Apis cerana* embryonic development in the present study.

In this study, we analyzed 9 deep-sequenced RNA libraries from age-controlled embryos of Asian honey bee (*Apis cerana*). In total, more than 500 million clean reads were mapped to the *Apis cerana* database (ACSNU-2.0). The results of hierarchical clustering and principal component analysis for samples showed that the sample homogeneity was good and these time-collected samples could well represent for the development nodes of honey bee embryo. Also, the analyses of gene body coverage showed that the quality of extracted RNA and RNA-seq data were high and the RNA-seq data was suitable to the next transcriptomic analyses. In addition, the RT-qPCR results validated the reliability of transcritome data and the analyses of DEGs. In the present study, we identified lots of new elements, including thousands of genes, coding and non-coding transcripts, exons and splicing events. These data substantially expand the number of known transcribed elements in the *Apis cerana* genome and provide a high-quality view of transcriptome dynamics throughout embryonic development in honeybee.

Both embryonic developmental period of *Apis cerana* and *Apis mellifera* are 3 days (72 h). We sampled the first-, second- and third- day embryos of *Apis cerana* to perform the transcriptomics research. The overall mRNA expression generally increased along with the developmental time of *Apis cerana* embryos, which was comparable to that of Drosophila [34]. Meanwhile the gene expression difference between the first and third day was most significant and the expression pattern of the second day was much closer to that of the third day than the first day. On the other words, the mRNA expression change from the first day to the second day was larger than the one from the second day to the third day. This trancriptomic pattern was consistent with the proteomic changing of *Apis mellifera* revealed by proteomics [24], which suggested that the developmental patterns of both *Apis mellifera* and *Apis cerana* were similar from the overall perspective of embryogenesis.

In this study, we detected a total of 18,284 genes expressed in the *Apis cerana* embryos, of which 44.5% were core genes commonly expressed in the whole period of embryonic development. These core genes participated in a series of important biological processes and pathways, and many core genes differentially expressed in the three ages (24, 48 and 72h) of Apis cerana embryo. The down-regulated GO terms involved many basic biological processes, such as RNA processing and splicing, macromolecule modification, cellular component assembly, oogenesis and oocyte development, mitotic cell cycle process, axis specification, and so on. Most of the down-regulated GO terms were expressed in early embryogenesis of *Apis cerana*. We also found morphogenesis of embryonic epithelium, embryonic axis specification and neuroblast proliferation were down-regulated indicating these biological processes occur in early period of Apis cerana embryonic development. These development events occurred in the early period (stages 7-9) of Drosophila embryonic development [37]. The up-regulated GO terms involved in cell fate determination and many tissue or system developments, including respiratory system, circulatory system, muscle, gland, brain, head, and leg disc developments. These development events were consistent with the ones in the late period (stages 15-17) of Drosophila embryonic development [37]. The GO terms overlapped simultaneously with the terms enriched by up- and down- regulated genes mainly participated in segmentation, nervous system development, epithelial tube, wing disc and eye morphogenesis, which usually develop during the stages 11-13 of Drosophila embryogenesis [37, 40].

Previously, morphological studies on honey bee embryogenesis [11, 26] suggested that the whole honey bee embryogenesis could be divided into five stages as follows: (1) cleavage (0-6h), (2) blastoderm formation and development (7-33h), (3) gastrulation (33-40h), (4) the germ band stage (40-55h) and (5) completion of the larval body (from 55h onward). Based on the previous morphological studies [11, 26] and the dynamic changing of these core genes in our present transcriptomic studies, we generally figured out the corresponding periods of honeybee embryogenesis to Drosophila: the development stages of 24, 48 and 72 hours in honeybee embryogenesis were roughly consistent with stages 7-9, 11-13, and 15-17 of Drosophila embryo, respectively.

A total of 10,943 genes were expressed on the first-day of *Apis cerana* embryonic development, of which 1,986 were specifically expressed in the first day compared to the expressed genes in the second and third day. These specific expressed genes involved in neuropeptide signaling pathway, neuronal maturation, eye-antennal disc development, leg disc morphogenesis, pigment metabolic process, and response to water and so on. Neuropeptide signaling is integral to many aspects of neural communication, particularly modulation of membrane excitability and synaptic transmission and growth [35]. In Drosophila embryo, the first reported appearance of specific neuropeptides is at stage 15 [41, 42]. However, we found neuropeptide signaling pathway appeared in the first day of the development of Apis cerana which was equivalent to the stage 7-8 of Drosophila embryo. Similarly, the development of eye-antennal disc in Drosophila began in the stage 17, while expression of related genes occurred in the first day of the *Apis cerana* embryonic development. These findings indicated that the basic components of the nervous system and eye-antennal disc of *Apis cerana* developed earlier than Drosophila. During the embryonic development of Drosophila melanogaster, pigments were deposited mainly on wing disc [43] and eye-antennal disc [44]. Therefore, the biological process of pigment deposition in *Apis cerana* embryo in the first day might prepare for their early development of wing disc.

On the second day of *Apis cerana* embryo, a total of 12,907 genes were expressed. Among these expressed genes, 1,799 were specifically expressed in the second day. These genes were involved in the organic acid biosynthetic process, olfactory behavior, photoreceptor cell axon guidance, anion transmembrane transport and ion transport regulation. These results indicated that eye morphogenesis continued to and olfactory tissue might be developed on the mid-stage of *Apis cerana* embryonic development. Meanwhile organic acids might play important role in this mid-stage. And a total of 13,947 genes were expressed on the third day of embryonic development of *Apis cerana*. Among them, of which 2,903 were specifically expressed in the third day compared to the expressed genes in the first and second day, which are involved in ion transport and ion homeostasis. Ion channels, pumps and exchangers are also useful markers of cell fate determination, as electrical stimulation modulates fate determination of differentiating embryonic stem cells [45]. Therefore, Ion transport and homeostasis play important roles in cell differentiation and cell fate determination during *Apis cerana* embryogenesis.

We made a comparison between *Apis cerana* and *Apis mellifera* embryogenesis, the overall trends of their embryonic developments were similar. The developmental status between the second day and the third day was closer than the one between the first day and the second or third day in both *Apis cerana* (Fig. IV-4-1B) and *Apis mellifera* [24]. In addition, we compared the trascriptome of the first-day embryo between *Apis cerana* and *Apis mellifera*. We found more than 80% of the expressed genes are shared in both honeybees. Notably, the biological processes related to eye development, like response to and detection of light stimulus, and phototransduction, were repeatedly enriched using the specifically expressed genes in *Apis cerana* first-day embryogenesis. On the other side of *Apis mellifera*, glycosylation and glycoprotein metabolic processes were enriched, and glycosaminoglycan biosynthesis - chondroitin sulfate / dermatan sulfate were detected using the specifically expressed genes in *Apis mellifera* first-day embryogenesis. Chondroitin sulfate proteoglycan was reported as an important component of the extracellular matrix in the central nervous system and played roles in cell migration of the central nervous system [36]. These finding suggested that: (1) eye development of *Apis cerana* was different with *Apis mellifera*, and (2) the development of central nervous system in *Apis mellifera* was earlier than *Apis cerana*. In the future, it should be compared in different time periods of embryo between *Apis cerana* and *Apis mellifera*, so that we can more completely understand the honeybee embryogenesis and their difference, which provide the developmental basis for us to study adult honeybee morphogenesis and behavior.

## Conclusion

We applied high-quality RNAseq to analyze the dynamic expression of the transcriptome during the embryogenesis of *Apis cerana*. A total of 18,284 genes were identified, including 8,139 core genes and 8,819 new genes, which extend the existing known gene database of *Apis cerana*. We speculate that the first day (referred as the age of 24h) of the honeybee's development is equivalent to stages 7-8 of *Drosophila*, the second day (48 h) is equivalent to stages 11-13, and the third day (72 h) is equivalent to stages 15-17. We found that the eye-antennal disc development of *Apis cerana* was developed earlier than that of *Drosophila*. There are two notable differences in the first-day embryonic development between *Apis cerana* and *Apis mellifera*: (1) eye development in the embryonic development is obviously different between both honeybees, and *Apis cerana* possibly can sense light earlier than *Apis mellifera*; and (2) the central nervous development of *Apis mellifera* is earlier than that of *Apis cerana*.

## Acknowledgements

This work was supported by the National Natural Science Foundation of China (31572469), the Earmarked Fund for China Agriculture Research System (CARS-44-KXJ15) and Jiangxi Provincial Education Department Research Project (GJJ150415). We have declared that no conflict of interests exists.

## References

1. BROMENSHENK JJ, HENDERSON CB, SECCOMB RA, WELCH PM, DEBNAM SE, FIRTH DR: Bees as Biosensors: Chemosensory Ability, Honey Bee Monitoring Systems, and Emergent Sensor Technologies Derived from the Pollinator Syndrome. *Biosensors* 2015, 5(4):678-711.
2. MICHENER CD: The bees of the world (2nd Edition). The Johns Hopkins University Press 2007:66-75.
3. WINSTON ML: The biology of the honey bee. Harvard university press 1991.
4. RUTTNER F: Biogeography and taxonomy of honeybees. Berlin New York: Springer-Verlag 2013.
5. GIURFA M, ZHANG S, JENETT A, MENZEL R, SRINIVASAN MV: The concepts of 'sameness' and 'difference' in an insect. *Nature* 2001, 410(6831):930-933.
6. MENZEL R, MULLER U: Learning and memory in honeybees: from behavior to neural substrates. *Annual review of neuroscience* 1996, 19:379-404.
7. LI J, QIN H, WU J, SADD BM, WANG X, EVANS JD, PENG W, CHEN Y: The prevalence of parasites and pathogens in Asian honeybees *Apis cerana* in China. *PloS One* 2012, 7(11):e47955.
8. YANG GH: Harm of introducing the western honeybee Apis mellifera L. to the Chinese honeybee *Apis cerana* F. and its ecological impact. *Acta Entomologica Sinica* 2005, 48(3):401-406.
9. PENG YS, FANG YZ, XU SY, GE LS: The resistance mechanism of the Asian honey bee, *Apis cerana* Fabr., to an ectoparasitic mite, Varroa jacobsoni Oudemans. *Journal of invertebrate pathology* 1987, 49(1):54-60.
10. NAZZI F, LE CONTE Y: Ecology of Varroa destructor, the Major Ectoparasite of the Western Honey Bee, Apis mellifera. *Annual Review of Entomology* 2016, 61:417-432.
11. FLEIG R, SANDER K: Embryogenesis of the honeybee apis mellifera l. (hymenoptera : apidae): An sem study. *International Journal of Insect Morphology & Embryology* 1986, 15:449-462.
12. MALESZKA J, FORET S, SAINT R, MALESZKA R: RNAi-induced phenotypes suggest a novel role for a chemosensory protein CSP5 in the development of embryonic integument in the honeybee (*Apis mellifera*).

*Development Genes and Evolution* 2007, 217(3):189-196.

13. SCHULTE C, THEILENBERG E, MULLER-BORG M, GEMPE T, BEYE M: Highly efficient integration and expression of piggyBac-derived cassettes in the honeybee (*Apis mellifera*). *Proceedings of the National Academy of Sciences of the United States of America* 2014, 111(24):9003-9008.

14. KOHNO H, SUENAMI S, TAKEUCHI H, SASAKI T, KUBO T: Production of Knockout Mutants by CRISPR/Cas9 in the European Honeybee, *Apis mellifera* L. *Zoological Science* 2016, 33(5):505-512.

15. GASCUEL J, MASSON C, BERMUDEZ I, BEADLE DJ: Morphological analysis of honeybee antennal cells growing in primary cultures. *Tissue & Cell* 1994, 26(4):551-558.

16. BERGEM M, NORBERG K, AAMODT RM: Long-term maintenance of in vitro cultured honeybee (*Apis mellifera*) embryonic cells. *BMC Developmental Biology* 2006, 6:17.

17. KITAGISHI Y, OKUMURA N, YOSHIDA H, NISHIMURA Y, TAKAHASHI J, MATSUDA S: Long-term cultivation of in vitro Apis mellifera cells by gene transfer of human c-myc proto-oncogene. *In vitro Cellular & Developmental Biology Animal* 2011, 47(7):451-453.

18. CHAN MM, CHOI SY, CHAN QW, LI P, GUARNA MM, FOSTER LJ: Proteome profile and lentiviral transduction of cultured honey bee (Apis mellifera L.) cells. *Insect Molecular Biology* 2010, 19(5):653-658.

19. HONEYBEE GENOME SEQUENCING C: Insights into social insects from the genome of the honeybee *Apis mellifera*. *Nature* 2006, 443(7114):931-949.

20. PARK D, JUNG JW, CHOI BS, JAYAKODI M, LEE J, LIM J, YU Y, CHOI YS, LEE ML, PARK Y *et al*: Uncovering the novel characteristics of Asian honey bee, *Apis cerana*, by whole genome sequencing. *BMC Genomics* 2015, 16:1.

21. PIRES CV, FREITAS FC, CRISTINO AS, DEARDEN PK, SIMOES ZL: Transcriptome Analysis of Honeybee (*Apis Mellifera*) Haploid and Diploid Embryos Reveals Early Zygotic Transcription during Cleavage. *PLoS One* 2016, 11(1):e0146447.

22. WANG ZL, LIU TT, HUANG ZY, WU XB, YAN WY, ZENG ZJ: Transcriptome analysis of the Asian honey bee Apis cerana cerana. *PloS One* 2012, 7(10):e47954.

23. COX-FOSTER DL, CONLAN S, HOLMES EC, PALACIOS G, EVANS JD, MORAN NA, QUAN PL, BRIESE T, HORNIG M, GEISER DM *et al*: A metagenomic survey of microbes in honey bee colony collapse disorder. *Science* 2007, 318(5848):283-287.

24. FANG Y, FENG M, HAN B, LU X, RAMADAN H, LI J: In-depth proteomics characterization of embryogenesis of the honey bee worker (*Apis mellifera ligustica*). *Molecular & cellular proteomics : MCP* 2014, 13(9):2306-2320.

25. FANG Y, FENG M, HAN B, QI Y, HU H, FAN P, HUO X, MENG L, LI J: Proteome analysis unravels mechanism underling the embryogenesis of the honeybee drone and its divergence with the worker (*Apis mellifera lingustica*). *Journal of Proteome Research* 2015, 14(9):4059-4071.

26. FLEIG R, SANDER K: Blastoderm development in honey bee embryogenesis as seen in the scanning electron microscope. *International Journal of Invertebrate Reproduction and Development* 1985, 8:279-286.

27. ELSIK CG, WORLEY KC, BENNETT AK, BEYE M, CAMARA F, CHILDERS CP, DE GRAAF DC, DEBYSER G, DENG J, DEVREESE B *et al*: Finding the missing honey bee genes: lessons learned from a genome upgrade. *BMC Genomics* 2014, 15:86.

28. KIM D, LANGMEAD B, SALZBERG SL: HISAT: a fast spliced aligner with low memory requirements. *Nature Methods* 2015, 12(4):357-360.

29. WANG L, WANG S, LI W: RSeQC: quality control of RNA-seq experiments. *Bioinformatics* 2012, 28(16):2184-2185.

30. PERTEA M, PERTEA GM, ANTONESCU CM, CHANG TC, MENDELL JT, SALZBERG SL: StringTie enables improved reconstruction of a transcriptome from RNA-seq reads. *Nature Biotechnology* 2015, 33(3):290-295.

31. XIE C, MAO X, HUANG J, DING Y, WU J, DONG S, KONG L, GAO G, LI CY, WEI L: KOBAS 2.0: a web server for annotation and identification of enriched pathways and diseases. *Nucleic Acids Research* 2011, 39(Web Server issue):W316-322.

32. LOVE MI, HUBER W, ANDERS S: Moderated estimation of fold change and dispersion for RNA-seq data with DESeq2. *Genome Biology* 2014, 15(12):550.

33. BINDEA G, MLECNIK B, HACKL H, CHAROENTONG P, TOSOLINI M, KIRILOVSKY A, FRIDMAN WH, PAGES F, TRAJANOSKI Z, GALON J: ClueGO: a Cytoscape plug-in to decipher functionally grouped gene ontology and pathway annotation networks. *Bioinformatics* 2009, 25(8):1091-1093.

34. GRAVELEY BR, BROOKS AN, CARLSON JW, DUFF MO, LANDOLIN JM, YANG L, ARTIERI CG, VAN BAREN MJ, BOLEY N, BOOTH BW et al: The developmental transcriptome of Drosophila melanogaster. *Nature* 2011, 471(7339):473-479.

35. CHEN X, GANETZKY B: A neuropeptide signaling pathway regulates synaptic growth in Drosophila. *The Journal of Cell Biology* 2012, 196(4):529-543.

36. GALTREY CM, FAWCETT JW: The role of chondroitin sulfate proteoglycans in regeneration and plasticity in the central nervous system. *Brain Research Reviews* 2007, 54(1):1-18.

37. CAMPOS-ORTEGA JA, HARTENSTEIN V: The embryonic development of Drosophila melanogaster. *Springer Science & Business Media* 2013.

38. NELSON JA: The Embryology of the Honeybee. Princeton University Press, Princeton 1915.

39. SCHNETTER M: Morphologische Untersuchungen über das Differenzierungszentrum in der Embryonalentwicklung der Honigbiene. *Z Morphol Ökol Tiere* 1935, 29:114-195.

40. BEJSOVEC A, WIESCHAUS E: Segment polarity gene interactions modulate epidermal patterning in Drosophila embryos. *Development* 1993, 119(2):501-517.

41. ISSHIKI T, PEARSON B, HOLBROOK S, DOE CQ: Drosophila neuroblasts sequentially express transcription factors which specify the temporal identity of their neuronal progeny. *Cell* 2001, 106(4):511-521.

42. TAGHERT PH, VEENSTRA JA: Drosophila neuropeptide signaling. *Advances in Genetics* 2003, 49:1-65.

43. TRUE JR, EDWARDS KA, YAMAMOTO D, CARROLL SB: Drosophila wing melanin patterns form by vein-dependent elaboration of enzymatic prepatterns. *Current Biology : CB* 1999, 9(23):1382-1391.

44. HANLY EW, FULLER CW, STANLEY MS: The morphology and development of Drosophila eye. I. In vivo and in vitro pigment deposition. *Journal of Embryology and Experimental Morphology* 1967, 17(3):491-499.

45. YAMADA M, TANEMURA K, OKADA S, IWANAMI A, NAKAMURA M, MIZUNO H, OZAWA M, OHYAMA-GOTO R, KITAMURA N, KAWANO M et al: Electrical stimulation modulates fate determination of differentiating embryonic stem cells. *Stem Cells* 2007, 25(3):562-570.

# 5. High-efficiency CRISPR/Cas9-mediated Gene Editing in Honeybee (*Apis mellifera*) Embryos

Xiaofen Hu[1], Bo Zhang[1], Chunhua Liao[1], Zhijiang Zeng[1*]

1 Honeybee Research Institute, Jiangxi Agricultural University, Nanchang, Jiangxi 330045, China.

**Abstract**

The honeybee (*Apis mellifera*) is an important insect pollinator of wild flowers and crops, playing critical roles in the global ecosystem. Additionally, the honeybee serves as an ideal social insect model. Therefore, functional studies on honeybee genes are of great interest. However, until now, effective gene manipulation methods have not been available in honeybees. Here, we reported an improved CRISPR/Cas9 gene-editing method by microinjecting sgRNA and Cas9 protein into the region of zygote formation within 2 h after queen oviposition, which allows one-step generation of biallelic knockout mutants in honeybee with high efficiency. We first targeted the *Mrjp1* gene. Two batches of honeybee embryos were collected and injected with *Mrjp1* sgRNA and Cas9 protein at the ventral cephalic side and the dorsal posterior side of the embryos, respectively. The gene-editing rate at the ventral cephalic side was 93.3%, which was much higher than that (11.8%) of the dorsal-posterior-side injection. To validate the high efficiency of our honeybee gene-editing system, we targeted another gene, *Pax6*, and injected *Pax6* sgRNA and Cas9 protein at the ventral cephalic side in the third batch. A 100% editing rate was obtained. Sanger sequencing of the TA clones showed that 73.3% (for *Mrjp1*) and 76.9% (for *Pax6*) of the edited current-generation embryos were biallelic knockout mutants. These results suggest that the CRISPR/Cas9 method we established permits one-step biallelic knockout of target genes in honeybee embryos, thereby demonstrating an efficient application to functional studies of honeybee genes. It also provides a useful reference to gene editing in other insects with elongated eggs.

**Keywords:** honeybee, CRISPR/Cas9, gene editing, biallelic knockout

## Introduction

The honeybee Apis mellifera is an important pollinator of wild flowers and crops, playing a great impact on plant diversity in the ecological environment (Aizen and Harder 2009; Klein *et al.* 2007). Honeybees have complex social behaviors including communication of food source locations via waggle dances, task specialization of colony members and group responding to environmental perturbations (Bonabeau *et al.* 1997;

---

\* Corresponding author: bees1965@sina.com (ZJZ).

注：此文发表在 *G3: Genes|Genomes|Genetics* 2019 年第 5 期。

Franks et al. 2002; Williams et al. 2010). Therefore, the honeybee is often used as a model organism for the studies on insect social organization and behavior, physiology and development, molecular nerve mechanism, and insect genetics (Winston 1991; Menzel and Muller 1996; Caron and Connor 2013; Cridge et al. 2017). To understand these interesting features of honeybees, many functional genes have been studied using different molecular biological techniques, including RNA interference (Beye et al. 2002; Dearden et al. 2009), in situ hybridization (Fleig et al. 1988; Osborne and Dearden 2005; Walldorf et al. 1989), immunohistochemistry (Fleig 1990), transgenesis with the transposon piggyBac (Schulte et al. 2014), and genome editing by the clustered regular interspaced palindromic repeats (CRISPR)-associated protein (Cas9) (Kohno et al. 2016; Kohno and Kubo 2018).

Currently, CRISPR/Cas9-mediated gene editing has been widely used in many different species (Belhaj et al. 2013; Gratz et al. 2013; Hwang et al. 2013; Jiang et al. 2013) to understand the function of target genes due to its easy operation and high efficiency. The first application of the CRISPR/Cas9 system in arthropods was in the model insect fruit fly Drosophila melanogaster (Gratz et al. 2013) in 2013, then followed by silk worm Bombyx mori (Wang et al. 2013) in the same year; afterwards, this system was used in other diverse insects including mosquito (Dong et al. 2015), jewel wasp (Li et al. 2017), butterfly (Perry et al. 2016) and honeybee (Kohno et al. 2016; Kohno and Kubo 2018). However, the CRISPR/Cas9 editing efficiency in insects was generally lower than that in mammals (Sun et al. 2017). There have been almost no reports of biallelic knockout mutants in insect gene-editing; usually mosaic mutants were produced, and multiple-generation breeding procedure should be conducted to produce homozygous knockout mutants, especially in the case of social insects with elongated eggs.

We speculated previous gene editing in most insects did not directly target the zygotes, as it did in mammals, instead targeting the embryonic primordial germ cells. We believed that editing efficiency would be greatly improved if the insect zygotes were exactly targeted and gene editing were performed at the correct time. Therefore, how to determine the exact region of zygote formation and suitable injection time is the key to achieve a high efficiency of gene editing in insects.

Since the latter half of the 19th Century, the studies on honeybee embryonic morphology have been carried out (Cridge et al. 2017). A plenty of morphological data of honeybee embryos was provided. Most importantly, ten standard stages of honeybee embryos were defined (DuPraw 1967) and have been widely used to sort random and untimed embryos from any colony. Stage one (about 4.5 h) was defined as the period begins with oviposition and ends when the cleavage nuclei initiate their migration toward the posterior pole. In the begin of stage one, newly laid honeybee egg was in the course of first meiotic division (Nachtsheim 1913; DuPraw 1967), and the center of cell division is located near the cephalic pole toward the ventral side of the egg. After about 2 h, the ovum complete its maturation and is fertilized with sperm to form zygote. This physiological process provides the possibility of gene editing at the course of zygote formation using honeybee embryos.

Hitherto, three Cas9 deliveries of plasmid, mRNA and protein were developed and used in the arthropod organisms (Sun et al. 2017). Injecting the complex of sgRNA and Cas9 protein at the region of zygote formation allows Cas9 nuclease to edit genes immediately after injection. Both of Cas9 plasmid and mRNA could not function immediately (Kouranova et al. 2016); they need a certain time to experience a protein synthesis process in the embryo (Kouranova et al. 2016) and may miss the most suitable editing time. In this present study, we injected sgRNA + Cas9 protein complex into the cleavage center of honeybee embryos during the early zygote formation, we finally obtained one-step biallelic knockout honeybee embryos. This improved CRISPR/Cas9 editing system is efficient in honeybee gene-editing and can provide a solid foundation for gene functional studies in the honeybees.

## Materials and Methods

### Sample collection and Embryo slice making

The embryo samples of Apis mellifera were obtained from honeybee colonies maintained in the apiary of the Honeybee Research Institute, Jiangxi Agricultural University, China (28.46 °N, 115.49 °E). To collect age-controlled embryos, a mated egg-laying queen was first held in queen cages with the size of 5 cm × 3 cm × 1.5 cm for 2-3 h, and then placed in the egg- and brood-free areas of a movable built-up comb; before the queen was placed, the built-up comb was cleaned overnight by worker bees. During queen ovulation, the queen was confined to the free space of half a comb (containing a total of 1024 cells) by a wooden or bamboo fence that only allows worker bees come in and out but does not allow the queen out. In the process of limiting the queen, the operation of the queen and the colony was as gentle as possible, so as not to frighten the queen and the worker bees. The whole process was carried out in quiet conditions. After two hours, the queen was released.

We collected all fresh Apis mellifera embryos (usually 30-40 embryos) within 2 h after queen oviposition and used ten embryos to make sections. Ten remained embryos each time were taken and incubated in the incubator at 35 °C and 85% humidity for 4 and 6 hours respectively, and were used for making sections. All collected 10 eggs were fixed in 4% paraformaldehyde for 24 hours and then transferred into 20% sucrose solution for 48-hour dehydration at 4 °C ready for making sections. The dehydrated embryos were embedded in optimal cutting temperature (OCT) compound and then cut into 7-μm sections and placed on glass slides. Each embryo section was stained by propidium iodide (PI) with a concentration of 50 μg/mL and was immediately observed using a fluorescence microscope (Leica, Germany) as described in our previous report (Hu et al. 2018).

### Preparation of single-guide RNA and Cas9 protein

In this study, Mrjp1 and Pax6 genes were selected as target genes for gene editing. The sequence of Mrjp1 sgRNA target site was 5'-TTGTTTATGCTGGTATGCCTTGG-3' (Underlined is the PAM sequence), which was designed according to the previous report and located on the exon 2 of Mrjp1 gene (Kohno et al. 2016). For Pax6 gene, we identified an sgRNA target site (5'-GACCATTACCAGACTCTACAAGG-3'; form +1106 to +1128 on the reference sequence of XM_006565377.2) in exon 2 of Pax6 using the CCTop online tool (Stemmer et al. 2015). A PCR-based approach was used to produce sgRNAs of Mrjp1 and Pax6. A specific oligonucleotide encoding a T7 polymerase-binding site and the sgRNA target sequences of Mrjp1 or Pax6 was designed as the forward primers (F-sgRNAMrjp1: 5'-TAATACGACTCACTATAGTTG TTTATGCTGGTAT

GCCTgttttagagctagaaatagc-3' for Mrjp1; F-sgRNAPax6: 5'-TAATACGA CTCACTATAGACCATTACCAG ACTCTACAgttttagagctagaaatagc-3' for Pax6) and a common oligonucleotide encoding the remaining sgRNA sequences was designed as the reverse primer (R-Common: 5'-AA AAAAAGCACCGACTCGGTGCCAC-3'). The two pairs of primers for Mrjp1 and Pax6 were annealed by PCR to synthesize template DNA. The PCR reaction mixture (40 μL) contained 20 μL of 2 × Pfu Mastermix (Transgene), 14 μL of $H_2O$, 2 μL of 5 μmol/L forward and reverse primers (F-sgRNAMrjp1 or F-sgRNAPax6 and R-Common) and 2 μL of 20 ng/μL pYSY-sgRNA plasmid (YaoShunYu, China). PCR was performed at 95 °C 3 min, 30 cycles of (95 °C 30 s, 56 °C 30 s, 72 °C 30 s), 72 °C 10 min and 12 °C ∞. PCR products were purified by a PCR clean-up (Axygen) kit. In vitro transcription was performed with the HighMAXIscript SP6/T7 RNA in vitro transcription kit (Ambion, USA) according to the manufacturer's instruction.

The Cas9 protein (TrueCut™ Cas9 Protein v2) was purchased from Thermo Fisher Scientific (Shanghai, China).

### *Microinjection and rearing*

A honeybee queen was placed in a quiet, clean and egg-free comb, and allowed to lay eggs. After two hours, the queen was released. The plastic plugs together with the laid eggs were taken off from the built-up comb, and the attached eggs were ready for injection.

We used a microinjection device (Eppendorf FemtoJet) and an Oxford micromanipulator (Eppendorf TransferMan NK2) to inject the embryos under an inverted microscope. One delivery of 200 ng/μL Mrjp1 sgRNA and 200 ng/μL Cas9 protein was injected into honeybee embryos from the m1 ($n$=24) and m2 ($n$=26) batches at the sites of the ventral side near to embryonic cephalic pole (the region of zygote formation) and the dorsal posterior side of the embryos, respectively. Another delivery of 200 ng/μL Pax6 sgRNA and 200 ng/μl Cas9 protein was injected into honeybee embryos from the p3 ($n$=22) batch at the sites of the ventral side near to the embryonic cephalic pole. Before injection, the solution of sgRNA and Cas9 protein was mixed well and then placed on ice for half an hour. The tips of the injection needles were rigid and the internal diameter of the glass needle tip was 4 μm. When we performed the injection operation, the angle between the needle and the side of the egg was adjusted to less than 30 degrees and the needle was inserted following the direction from the anterior side to the posterior side. Once the tip of the needle entered the egg, injection took place, avoiding deep insertion. The injection time was 0.1 s, the injection pressure was 600 hPa, and the balance pressure was 50 hPa. The injection parameters were similar to the previous description (Schulte *et al*. 2014). The integrity of the egg should be ensured after injection. Eggs with protoplasm overflowing were removed, as they would die because of damage. A transparent spot was found at the injection site of the egg, which then disappeared slowly.

The embryos were incubated in plastic boxes at 35 °C (relative humidity 85%) with a small amount of 16% (vol/vol) sulfuric acid to prevent mold formation (Schulte *et al*. 2014).

### *Pre-sequencing process of mutants*

After microinjection, the eggs were incubated for 48-60 h. Then the embryonic morphology was observed under the microscope. At this time, the normally developed egg is in the ninth stage of embryonic development, and there is a very obvious fluid-filled gap in the anterior pole and a protuberance at the head of the amnion-serosa (Fig. S1), which can be used as a criterion for selection. Embryos with normal developmental morphogenesis were used as the materials for PCR analysis of knockout target genes. PCRs for target genes were performed following the instructions of the TransDirect Animal Tissue PCR Kit purchased from Transgen Biotech (Beijing, China).

The fragment (403 bp) flanking the editing target site in Mrjp1 was amplified using a pair of specific primers (forward: 5'-ATATTCCATTGCTTCGTTACTCG-3', reverse: 5'-TGGATATGAAGA ATTTTGGACAAG-3'). The fragment (446 bp) flanking the editing target site in Pax6 was amplified using another pair of specific primers (forward: 5'-GCCGGTGTGTGTTTATTCAA-3', reverse: 5'-TGCAAAAGTGACATCCTTGC T-3'). The PCR reaction mixture (20 μL) contained 10 μL of 2 × TransDirect PCR supermix (+dye), 0.4 μL of forward primers, 0.4 μL of reverse primers, 5.2 μL of ddH$_2$O, 4 μL of embryonic lysis fluid. PCR was performed at 94°C 10min, 35 cycles of (94°C 30 s, 52°C 30 s, 72°C 1min), 72°C 10 min and 12°C until the PCR products were taken out. The PCR products were ready for Sanger sequencing or inserted into TA vectors for Sanger sequencing.

### *PCR of negative controls*

Honeybees used in this experiment came from a common Apis mellifera colony without special breeding. To better determine the type of knockout mutants, we investigated the flanking sequences around the PAM sites of Mrjp1 and Pax6 genes in the unedited embryos laid by the experimental queens. For each gene, ten unedited embryos were collected and PCR amplification was performed same as the corresponding candidate genes. The PCR products were used as negative controls for Sanger sequencing or TA cloning.

### *Sequencing of PCR products and TA clones*

Sanger sequencing was conducted using the forward primers of Mrjp1 or Pax6 as the sequencing primer. Different single sequencing peak from wild-type samples and double sequencing peaks present as a cluster of bases indicated a mutation event. All PCR products, except the samples with obvious clean single peaks, were TA-cloned, and 20 colonies were collected for Sanger sequencing by Tsingke Biological Technology (Changsha, China) to determine the exact indel types.

### *Data and reagent availability*

The authors state that all data necessary for confirming the conclusions presented in the article are represented fully within the figures and the tables. All honeybee strains and reagents are available upon request. Supplemental material available at Figshare.

## Results

### *Morphological slice analysis of early honeybee embryos*

To determine reasonable injection site and time for the gene editing, we first analyzed morphological slices of early honeybee embryos at the ages of approximate 0-2, 4-6 and 6-8 hours (h). From the slice of 0-2 h honeybee embryo (Fig. IV-5-1A, see page 543), we observed that the homogeneous protoplasm occupied the whole embryo and there was relatively little structure. We could not find any obvious cleavage nuclei with deeper staining. According to the description of DuPraw (1967), the egg of this period is still in the course of zygote formation which is near the cephalic pole toward the ventral side of the egg, so the energid with deep staining could not be observed on a slice. As shown in the slice of 4-6 h honeybee embryo (Fig. IV-5-1B), several energids, formed by nuclei recruiting their own plasm islands, were found together with deeper staining near the anterior pole of embryo. It was a typical structure of late stage one or early stage two embryo, which was consistent with the morphology descripted by DuPraw (1967): "At the beginning of stage 2 a cluster of eight energids lies near the 10% level (measured from the anterior pole), about equidistant from the dorsal, ventral, and laternal surfaces". The 6-8 hours honeybee embryos (Fig. IV-5-1C) were in the middle or late stage two of embryonic development. Many energids were found but not clustered together in the cleavage center. All energids were migrating dispersedly towards the surface of embryo and some had reached the surface of anterior pole.

These embryonic slices convinced us that zygote formation happens in the region near the cephalic pole toward the ventral side of the egg around 2 hours after queen oviposition.

### *Mutagenesis of Mrjp1 gene by CRISPR/Cas9*

In the $m_1$ and $m_2$ batches, 17 and 15 normal embryos were obtained; the normal embryonic development rates were 65.4% and 62.5%, respectively (Table IV-5-1). The PCR products of all these embryos were obtained. The editing types of these two batches of target genes are shown in Fig. S2 and S3. We sequenced the flanking region around the PAM site of *Mrjp1* gene in 10 blank samples and found a T/C mutation at the sgRNA site of *Mrjp1* gene (5'-TTGTTTA TGCTGTA(T/C)GCCTTGG-3', Fig. IV-5-2A), which was not in the core recognition region of *Mrjp1* sgRNA.

**Table IV-5-1 Differences in gene editing efficiency of *Mrjp*1 or *Pax6* at the different injection sites**

| Batch | Target gene | Injection site | Total | Survival[1] (%) | Mutated (%) | Biallelic knockout | |
|---|---|---|---|---|---|---|---|
| | | | | | | Non-mosaic (%) | Mosaic (%) |
| $m_1$ | *Mrjp1* | Dorsal posterior side | 26 | 17 (65.4%) | 2 (11.8%) | 0 | 0 |
| $m_2$ | *Mrjp1* | Ventral cephalic side | 24 | 15 (62.5%) | 14 (93.3%) | 5 (33.3%) | 6 (40.0%) |
| $p_3$ | *Pax6* | Ventral cephalic side | 22 | 13 (59.1%) | 13 (100.0%) | 3 (23.1%) | 7 (53.8%) |

[1], the number of injected embryos that have developed into stage 9.

In the $m_1$ batch, only two chimeric embryos, 8C and 6C, were detected among 17 injected samples (Fig. S2). The results in the $m_2$ batch showed that there were three types of results in the samples at the injection site of the ventral side near to embryonic cephalic pole: (1) the first type is chimera with obvious double peaks shown as the 1A sample in Figure S3, which included 8 other samples of 2A, 3A, 6A, 9A, 10A, 12A, 14A and 15A; (2) the second type is complete knockout mutant with clean single peaks different from wild-type shown as the 11A sample in Figure S3. Four other samples of 4A, 5A, 8A and 13A were included; and (3) the third type is an entirely unedited sample like 7A, only one sample belongs to this type (Fig. S3).

The PCR products of 11 chimera samples (9 samples injected at the ventral-cephalic-side and 2 samples injected at the dorsal-posterior-side) from the $m_1$ and $m_2$ batches were inserted into TA-clones, and 20 clones were randomly selected and sequenced. The sequencing results for different types of gene editing were shown in Fig. IV-5-2. We found that the editing efficiency of samples injected in the ventral side near cephalic pole (93.3%) was significantly higher than that of samples injected at the dorsal posterior side (11.3%). Overall, 73.3% of the individuals injected near the ventral side of the cephalic pole were biallelic knockout mutants (5 biallelic homozygous mutants and 6 biallelic heterozygous mutants), while only two low-editing-rate chimeras (Table IV-5-2) were obtained when injecting at the dorsal posterior side of embryos. The results showed that the editing efficiency of an injection site near the ventral side of the cephalic pole was significantly higher than that of injection site at the dorsal posterior side.

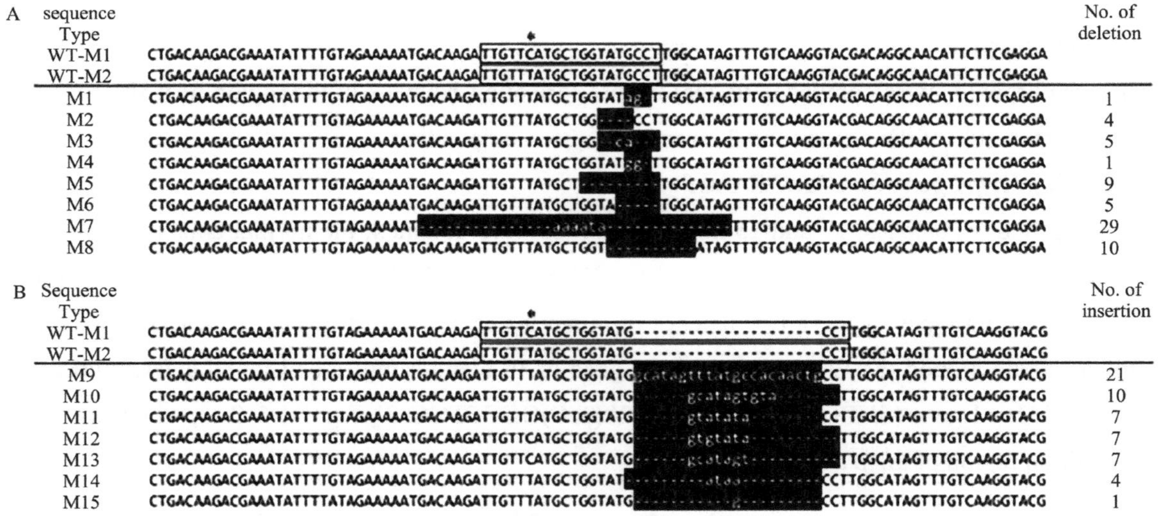

**Figure IV-5-2** Gene editing patterns of *Mrjp1* gene. A deletion; B insertion. The two sequences shown at the top are wild-type sequences of WT-M1 and WT-M2. Letters in gray boxes and in blank boxes with a black frame indicate PAM and sgRNA target site, respectively. Star indicates SNP of T/C in wild-type sequence. White letters and dashes in the black boxes indicate inserted and deleted nucleotide sequences, respectively. Sequence type from which each sequence type was detected are shown on the left. The numbers of nucleotide deletions or insertion that differed between the genome-edited embryos and wild-type (WT) sequences are shown on the right.

### *Mutagenesis of Pax6 gene by CRISPR/Cas9*

In the $p_3$ batch, 13 normal embryos were obtained; the normal embryonic development rate was 59% (Table IV-5-1). The PCR products of all these embryos were obtained. The editing types of *Pax6* gene are shown in Fig. S4. For *Pax6* gene, we also sequenced the target gene sequences of 10 wild-type samples, and found a G/A variant 5'-GACCATTACCAG ACTACAAG(G/A)-3' on the sgRNA site, which happened to be in the PAM site (see the WT-P1 and WT-P2 sequences in Figure 3), but AGA is also a non-canonical PAM site. The cutting efficiency was reported lower than AGG (Kleinstiver *et al*. 2015; Zhang *et al*. 2014), which may affect the editing efficiency.

There were three editing results for the $p_3$ samples injected into the ventral site near the cephalic pole: (1) the first type is chimeras with obvious double peaks, including eight samples of 1D, 3D, 4D, 5D, 6D, 7D, 8D and 11D; (2) the second type is complete-like knockout with little weak peaks, including three samples of 10D, 12D and 13D; and (3) the third type is complete knockout with clean single peaks shown in Fig. S4, including two samples of 2D and 9D. The PCR products of 11 samples from the first and second types were inserted into TA-clones, and then 20 clones for each PCR product were selected for sequencing. The sequencing results for different types of gene editing were shown in Fig. IV-5-3. The results showed that the efficiency of target gene *Pax6* editing was very high, reaching 100% editing rate (Table IV-5-2). 76.9% of the individuals injected near the ventral side of the cephalic pole were biallelic knockout mutants (3 biallelic homozygous mutants and 7 biallelic heterozygous mutants).

**Figure IV-5-3** Gene editing patterns of *Pax6* gene. A deletion; B insertion. The two sequences shown at the top are wild-type sequences of WT-P1 and WT-P2. Letters in gray boxes and in blank boxes with a black frame indicate PAM and sgRNA target site, respectively. Star indicates SNP of G/A in wild-type sequence. White letters and dashes in the black boxes indicate inserted and deleted nucleotide sequences, respectively. Sequence type from which each sequence type was detected are shown on the left. The numbers of nucleotide deletions or insertion that differed between the genome-edited embryos and wild-type (WT) sequences are shown on the right.

**Table IV-5-2 The editing types of all edited individuals in three batches of experiments**

| Injection site | ID | Target gene | Editing type (number) | Editing Rate |
| --- | --- | --- | --- | --- |
| Ventral cephalic side | 1A | *Mrjp1* | Wild-type(12) M15(8) | 8/20=40% |
|  | 2A | *Mrjp1* | Wild-type (14) M12(1) M15(5) | 6/20=30% |
|  | 3A | *Mrjp1* | M1(14) M2(1) M3(1) M6(3) M15(1) | 20/20=100% |
|  | 4A | *Mrjp1* | M8 | 100% |
|  | 5A | *Mrjp1* | M6 | 100% |
|  | 6A | *Mrjp1* | Wild-type (8) M8(12) | 12/20=60% |
|  | 8A | *Mrjp1* | M6 | 100% |
|  | 9A | *Mrjp1* | M5(12) M8(8) | 20/20=100% |

| | | | | (continued) |
|---|---|---|---|---|
| Injection site | ID | Target gene | Editing type (number) | Editing Rate |
| Ventral cephalic side | 10A | *Mrjp1* | M4(14) M13(6) | 20/20=100% |
| | 11A | *Mrjp1* | M8 | 100% |
| | 12A | *Mrjp1* | M11(2) M15(18) | 20/20=100% |
| | 13A | *Mrjp1* | M6 | 100% |
| | 14A | *Mrjp1* | M8(14) M9(5) M14(1) | 20/20=100% |
| | 15A | *Mrjp1* | M8(10) M10(10) | 20/20=100% |
| | 1D | *Pax6* | Wild-type (1) P6(19) | 19/20=95% |
| | 2D | *Pax6* | P11 | 100% |
| | 3D | *Pax6* | Wild-type (1) P7(16) P14(3) | 19/20=95% |
| | 4D | *Pax6* | P9(2) P16(18) | 20/20=100% |
| | 5D | *Pax6* | P7(16) P15(4) | 20/20=100% |
| | 6D | *Pax6* | P1(3) P7(17) | 20/20=100% |
| | 7D | *Pax6* | P4(17) P5(1) P7(1) P10(1) | 20/20=100% |
| | 8D | *Pax6* | P7(1) P3(19) | 20/20=100% |
| | 9D | *Pax6* | P7 | 100% |
| | 10D | *Pax6* | P7(1) P8(1) P13(18) | 20/20=100% |
| | 11D | *Pax6* | P2(10) P7(1) P13(9) | 20/20=100% |
| | 12D | *Pax6* | P12(20) | 20/20=100% |
| | 13D | *Pax6* | Wild-type(1) P7(19) | 19/20=95% |
| Dorsal posterior side | 6C | *Mrjp1* | Wild-type (19) M7(1) | 1/20=5% |
| | 8C | *Mrjp1* | Wild-type (14) M6(6) | 6/20=30% |

Note: 4A, 5A, 8A, 11A, 13A, 2D, 9D and 12D were biallelic homozygous mutants; 3A, 9A, 10A, 12A, 14A, 15A, 4D, 5D, 6D, 7D, 8D, 10D and 11D were biallelic heterozygous mutants.

## Discussion

CRISPR/Cas9-mediated gene-editing has achieved tremendous success and impact in many species. However, the editing efficiency in insects was not as high as that in mammals, especially for those with elongated eggs, such as honeybee (Kohno et al. 2016; Kohno and Kubo 2018; the rates of edited offspring were below than 12.5%) and mosquito (Dong et al. 2015; the knockout efficiency was 5.5%). It led to the fact that CRISPR/Cas9 technology has not been widely used in honeybees. In this study, we reported an improved CRISPR/Cas9 gene-editing method by microinjecting sgRNA and Cas9 protein into the region of zygote formation within 2 hours after queen oviposition, which allows one-step generation of biallelic knockout mutants in honeybee with high efficiency.

We first targeted *Mrjp*1 gene. The design of the *Mrjp*1 sgRNA was same as previous report by Kohno *et al.* (2016). The main aims of editing honeybee Mrjp1 gene were: (1) determining injection time, (2) determining injection site, and (3) determining the feasibility of the sgRNA and Cas9 protein delivery. We wanted to know whether the ventral cephalic side (near the position of zygote formation) was more suitable for the injection site than the dorsal posterior site as previously described (Kohno *et al.* 2016), and whether the eggs at the developmental stage of zygote formation collected by our method were suitable for gene editing. Through our exploration, the results showed that the improvement in the above three aspects could help us get a high gene-editing rate in honeybees. To further demonstrate the efficiency of the improved Crispr/Cas9 editing system, we designed an sgRNA for another target gene Pax6. The results showed that the gene editing efficiency for Pax6 gene was also high.

### *Effect of mutation in sgRNA of target genes*

We examined the chimeras (1A, 2A, 6A, 1D, 3D and 10D) in the two batches of Mrjp1 and Pax6 gene editing. We found that the TA clones derived from the Mrjp1 chimera embryos share a common feature: the fifteenth base at the left flank to PAM site is the C sequence (Fig. IV-5-2). The Mrjp1 sgRNA was same to previous report originally designed based on the reference sequence of Mrjp1 gene. However, our sequencing result of wild type embryos showed that there was a mutation of T>C at the fifteenth base at the left flank to PAM site. Although this mutation is not located in the core sequence of the 6-12 base near PAM (Jiang *et al.* 2013), the results show that the heterozygosity of this site has a negative impact on editing efficiency.

We designed the Pax6 sgRNA according to the NCBI reference sequence of Pax6 gene. However, based on the sequencing results of wild-type embryos, we found that there was a mutation of G>A at the PAM site (Fig. S3) on the guide RNA recognition sequence. Previous study showed that both AGG and AGA are recognizable PAM sites, but the editing efficiency of AGG was higher than that of AGA (Kleinstiver *et al.* 2015; Zhang *et al.* 2014). In our case, based on the sequencing results of TA clones, all the unedited sequences ($n=3$) contained the AGA PAM, which was found in only a low proportion of all tested embryos; most of sequences ($n=40$) containing AGA were edited in the mutagenesis of Pax6. The lowest editing rate of target gene Pax6 was 95%. So it suggested that the Pax6 sgRNA used in our study was relatively specific.

### *Improvements in our honeybee gene editing system*

In this study, we have adjusted injection time, injection site, and injection delivery to honeybee embryos, with the aim of directly targeting the honeybee zygote and determining the optimum time point to perform gene editing. We have made significant improvements in the methodology of gene editing in honeybees from the technical perspective. Fortunately, we finally achieved high efficiency of gene editing in honeybee and one-step production of honeybee biallelic mutants. Therefore, we have provided a method for efficient gene-editing in honeybee, which was an obvious technological improvement compared to the previously established methodology (Kohno *et al.* 2016).

Here, we summarized three major improvements in our honeybee gene editing system as follows:

First, for honeybee, an insect with elongated embryos, gene editing at right embryonic position and at accurate embryonic developmental time is an important technological improvement and is the key to gain high efficient gene-editing results. In previous studies on honeybee genetic manipulation (Schulte *et al.* 2014; Kohno *et al.* 2016; Kohno and Kubo 2018), the position of primitive gonad cells (dorsal posterior region of embryos) was chosen for injection, which resulted in generation of mosaic queens or low efficient rate of transgenesis (27% and 20%) or gene-editing (< 12.5%). And it needed a complex breeding process to obtain the offspring mutants. However in mammals, such as mice (Wang *et al.* 2013), rats (Li *et al.* 2013) and pigs (Hai *et al.* 2014), zygote was directly targeted as the injection object of gene editing, and the biallelic knockdown mutants could be obtained at the current generation (G0 generation) by one-step injection. Therefore, we considered that targeting the honeybee zygote instead of the primitive gonad cells should be the key to achieving highly efficient gene-editing results in honeybee. Before our embryonic injection, we paid more attention to the development of honeybee embryos. The elongated egg of honeybee is greatly different from the round or nearly round egg of fruit fly, silkworm or other insects in the morphological and anatomical structure. We carefully read and understood previous literature about the development of honeybee embryo and observed histological structure of early honeybee embryos. We were sure that the course of zygote formation in honeybee embryos occurred about two hours after queen ovulation and the position of honeybee zygote formation was the site of the ventral side near to embryonic cephalic pole. In addition, to allow Cas9 nuclease to edit genes immediately after injection, we chose the complex of sgRNA and Cas9 protein as injection delivery and injected the complex into the region of honeybee zygote formation. While both Cas9 plasmid and mRNA need a certain time for protein synthesis process in the embryo (Kouranova *et al.* 2016) and may miss the suitable editing time at the 1- or 2- cell stage of honeybee zygote. Finally, we successfully achieved highly efficient gene editing in honeybee embryos. We could make a conclusion that the improvement of injection site, time and delivery resulted in high efficiency of honeybee gene editing. This method we established also provided a reference for gene editing in other insects with elongated embryos.

Second, compared to previous honeybee gene-editing methodology, the generation efficiency of mutants has been greatly improved. To our knowledge, two papers have reported CRISPR/Cas9 gene editing in honeybees from Kubo's group (Kohno *et al.* 2016; Kohno and Kubo 2018). Their generation efficiency of honeybee mutants were 12.4%, 5.1% and 10%, respectively. While in our study, the efficiency of bi-allelic mutants for two candidate genes were 73.3% and 76.9%, respectively, which were much higher than those in the studies of Kubo's group.

Third, one-step high-efficiency gene-editing method established in our present study could greatly accelerate the production of biallelic knockout honeybee mutants. In previous application of CRISPR/Cas9 in honeybees, the biallelic knockout honeybee mutants could only be achieved through complex breeding procedures over several generations. Usually, three generations of queens with mutated gene needed to be cultivated (Kohno *et al.* 2016; Kohno and Kubo 2018). Honeybee is a typical social insect. Honeybees have three castes: drones, workers, and queens. Drones are male, while workers and queens are female. Haploid embryos develop into drones; diploid embryos develop into worker bees or queens. Honeybee development consists of four stages: embryo, larva, pupa and adult. In the stage of larvae, there is a great difference in bee diet between the larvae developing into queens and that developing to worker bees, which is controlled by worker bees with the duty of feeding brood. In other words, the production of queen needs the feeding of nurse bees. To our knowledge, there were no reports that queens could be artificially bred without nurse bees' feeding. It was difficult to achieve artificial breeding of queen with reproductive ability. The larvae ready to develop into queen needed to be transferred to the natural colony and fed by nurse bees. In addition, nurse bees would strictly supervise the hatching eggs and larvae in the natural colony. If there were weak embryos or damaged embryos, they would be cleaned out. The injected embryos were unavoidably damaged to a certain extent compared to the normal embryos. So there was a large probability that the injected embryos would be cleaned out by nurse bees. For these above reasons, in the process of obtaining mutant worker bees by microinjection CRISPR/Cas9 gene editing method, the developmental process from injected embryos to queens became a bottleneck problem, which easily led to the failure of the whole experiment or the great reduction of the experimental efficiency.

In our present study, we established one-step high-efficiency gene-editing method. We could obtain biallelic mutants through one generation cultivation of worker bees. Our method avoids the complex process of queen breeding and greatly reduces the difficulty of the whole experiment of honeybee gene editing. Its application in honeybees could also rapidly reveal the phenotypes of gene knockout mutants. In addition, this efficient gene editing approach provides the possibility of functional studies of genes critical for development from embryo into adult. All together, we considered that the improvement of honeybee gene-editing method was obvious from the technological perspective, and its effect and significance were remarkable.

Finally, it should be noted that we have not reared any of the edited embryos into larvae or adults. The efficacy of our method awaits testing in a full egg-to-adult system to ensure that there are no unexpected problems associated with the injection innovations during the rearing process.

## Conclusions

In the present study, we targeted the region of zygote formation at the 1-2 cell stage, injected the complex formed by sgRNA and Cas9 protein as the delivery into honeybee embryos, and carried out gene editing experiments on *Mrjp*1 and *Pax*6 genes. The results showed that the CRISPR/Cas9 editing system we established could produce G0 knockout mutant embryos with high efficiency in honeybees. This efficient CRISPR/Cas9 editing system paves the road for the gene functional studies in honeybees and provides a useful reference to the gene editing in other insects.

## Acknowledgements

XFH, BZ and CHL completed all the experiments, including sample collection, making embryo slices, sgRNA designation and generation, PCR analysis and microinjection. XFH and ZJZ provided the experimental design and developed the methodology. ZJZ provided the financial support for all sample sequencing and gene quantification. ZJZ provided the financial support for publishing the manuscript. XFH and ZJZ wrote the manuscript. All authors read and approved the final manuscript and agreed to be accountable for all aspects of the work. We thank Dr. Qiang Huang and Dr. Frederick Partridge for improving the English of this manuscript.

This work was supported by the National Natural Science Foundation of China (31572469,31872432), the Earmarked Fund for China Agriculture Research System (CARS-44-KXJ15) and the Jiangxi Provincial Education Department Research Project (GJJ150415).

## References

AIZEN, M. A., and L. D. HARDER, 2009 The global stock of domesticated honey bees is growing slower than agricultural demand for pollination. *Curr Biol* 19 (11):915-918.

BELHAJ, K., A. CHAPARRO-GARCIA, S. KAMOUN, and V. NEKRASOV, 2013 Plant genome editing made easy: targeted mutagenesis in model and crop plants using the CRISPR/Cas system. *Plant Methods* 9 (1):39.

BEYE, M., S. HARTEL, A. HAGEN, M. HASSELMANN, and S. W. OMHOLT, 2002 Specific developmental gene silencing in the honey bee using a homeobox motif. *Insect Mol Biol* 11 (6):527-532.

BONABEAU, E., G. THERAULAZ, J. L. DENEUBOURG, S. ARON, and S. CAMAZINE, 1997 Self-organization in social insects. *Trends Ecol Evol* 12 (5):188-193.

CARON, D. M., and L. J. CONNOr, 2013 *Honey bee biology and beekeeping*. Kalamazoo, Michigan, USA: Wicwas Press.

CRIDGE, A. G., M. R. LOVEGROVE, J. G. SKELLY, S. E. TAYLOR, G. E. L. PETERSEN, R. C. CAMERON, and P. K. DEARDEN, 2017 The honeybee as a model insect for developmental genetics. *Genesis* 55 (5).

DEARDEN, P. K., E. J. DUNCAN, and M. J. WILSON, 2009 RNA interference (RNAi) in honeybee (Apis mellifera) embryos. *Cold Spring Harb Protoc* 2009 (6):pdb prot5228.

DONG, S., J. LIN, N. L. HELD, R. J. CLEM, A. L. PASSARELLI, and A. W. E. FRANZ, 2015 Heritable CRISPR/Cas9-mediated genome editing in the yellow fever mosquito, Aedes aegypti. *PLoS One* 10 (3):e0122353.

DUPRAW, E. J., 1967 The honeybee embryo. *Methods in developmental biology*:183-217.

FLEIG, R., 1990 Engrailed expression and body segmentation in the honeybee Apis mellifera. *Roux Arch Dev Biol* 198 (8):467-473.

FLEIG, R., U. WALLDORF, W. J. GEHRING, and K. SANDER, 1988 In situ localization of the transcripts of a homeobox gene in the honeybee Apis mellifera L. (Hymenoptera). *Roux Arch Dev Biol* 197 (5):269-274.

FRANKS, N. R., S. C. PRATT, E. B. MALLON, N. F. BRITTON, and D. J. SUMPTER, 2002 Information flow, opinion polling and collective intelligence in house-hunting social insects. *Philos Trans R Soc Lond B Biol Sci* 357 (1427):1567-1583.

GRATZ, S. J., A. M. CUMMINGS, J. N. NGUYEN, D. C. HAMM, L. K. DONOHUE, M. M. HARRISON, J. WILDONGER, and K. M. O'CONNOR-GILES, 2013 Genome engineering of Drosophila with the CRISPR RNA-guided Cas9 nuclease. *Genetics* 194 (4):1029-1035.

HAI, T., F. TENG, R. GUO, W. LI, and Q. ZHOU, 2014 One-step generation of knockout pigs by zygote injection of CRISPR/Cas system. *Cell Res* 24 (3):372-375.

HU, X., L. KE, Z. WANG, and Z. ZENG, 2018 Dynamic transcriptome landscape of Asian domestic honeybee (Apis cerana) embryonic development revealed by high-quality RNA sequencing. *BMC Dev Biol* 18 (1):11.

HWANG, W. Y., Y. FU, D. REYON, M. L. MAEDER, S. Q. TSAI, J. D. SANDER, R. T. PETERSON, J. R. YEH, and J. K. JOUNG, 2013 Efficient genome editing in zebrafish using a CRISPR-Cas system. *Nat Biotechnol* 31 (3):227-229.

JIANG, W., D. BIKARD, D. COX, F. ZHANG, and L. A. MARRAFFINI, 2013 RNA-guided editing of bacterial genomes using CRISPR-Cas systems. *Nat Biotechnol* 31 (3):233-239.

KLEIN, A. M., B. E. VAISSIERE, J. H. CANE, I. STEFFAN-DEWENTER, S. A. CUNNINGHAM, C. KREMEN, and T. TSCHARNTKE, 2007 Importance of pollinators in changing landscapes for world crops. *Proc Biol Sci* 274 (1608):303-313.

KLEINSTIVER, B. P., M. S. PREW, S. Q. TSAI, V. V. TOPKAR, N. T. NGUYEN, Z. ZHENG, A. P. GONZALES, Z. LI, R. T. PETERSON, J. R. YEH, M. J. ARYEE, and J. K. JOUNG, 2015 Engineered CRISPR-Cas9 nucleases with altered PAM specificities. *Nature* 523 (7561):481-485.

KOHNO, H., and T. KUBO, 2018 mKast is dispensable for normal development and sexual maturation of the male European honeybee. *Sci Rep* 8 (1):11877.

KOHNO, H., S. SUENAMI, H. TAKEUCHI, T. SASAKI, and T. KUBO, 2016 Production of Knockout Mutants by CRISPR/Cas9 in the European Honeybee, Apis mellifera L. *Zoolog Sci* 33 (5):505-512.

KOURANOVA, E., K. FORBES, G. ZHAO, J. WARREN, A. BARTELS, Y. WU, and X. CUI, 2016 CRISPRs for Optimal Targeting: Delivery of CRISPR Components as DNA, RNA, and Protein into Cultured Cells and Single-Cell Embryos. *Hum Gene Ther* 27 (6):464-475.

LI, M., L. Y. C. AU, D. DOUGLAH, A. CHONG, B. J. WHITE, P. M. FERREE, and O. S. AKBARI, 2017 Generation of heritable germline mutations in the jewel wasp Nasonia vitripennis using CRISPR/Cas9. *Sci Rep* 7 (1):901.

LI, W., F. TENG, T. LI, and Q. ZHOU, 2013 Simultaneous generation and germline transmission of multiple gene mutations in rat using CRISPR-Cas systems. *Nat Biotechnol* 31 (8):684-686.

MENZEL, R., and U. MULLER, 1996 Learning and memory in honeybees: from behavior to neural substrates. *Annu Rev Neurosci* 19:379-404.

NACHTSHEIM, H., 1913 Cytologische Studien über die Geschlechtsbestimmung bei der Honigbiene Apis Mellifica L. *Wilhelm Engelmann*.

OSBORNE, P., and P. K. DEARDEN, 2005 Non-radioactive in-situ hybridisation to honeybee embryos and ovaries. *Apidologie* 36 (1):113-118.

PERRY, M., M. KINOSHITA, G. SALDI, L. HUO, K. ARIKAWA, and C. DESPLAN, 2016 Molecular logic behind the three-way stochastic choices that expand butterfly colour vision. *Nature* 535 (7611):280-284.

SCHULTE, C., E. THEILENBERG, M. MULLER-BORG, T. GEMPE, and M. BEYE, 2014 Highly efficient integration and expression of piggyBac-derived cassettes in the honeybee (Apis mellifera). *Proc Natl Acad Sci U S A* 111 (24):9003-9008.

STEMMER, M., T. THUMBERGER, M. DEL SOL KEYER, J. WITTBRODT, and J. L. MATEO, 2015 CCTop: An Intuitive, Flexible and Reliable CRISPR/Cas9 Target Prediction Tool. *PLoS One* 10 (4):e0124633.

SUN, D., Z. GUO, Y. LIU, and Y. ZHANG, 2017 Progress and Prospects of CRISPR/Cas Systems in Insects and Other Arthropods. *Front Physiol* 8:608.

WALLDORF, U., R. FLEIG, and W. J. GEHRING, 1989 Comparison of homeobox-containing genes of the honeybee and Drosophila. *Proc Natl Acad Sci U S A* 86 (24):9971-9975.

WANG, Y., Z. LI, J. XU, B. ZENG, L. LING, L. YOU, Y. CHEN, Y. HUANG, and A. TAN, 2013 The CRISPR/Cas system mediates efficient genome engineering in Bombyx mori. *Cell Res* 23 (12):1414-1416.

WILLIAMS, N. M., E. E. CRONE, H. R. T'AI, R. L. MINCKLEY, L. PACKER, and S. G. POTTS, 2010 Ecological and life-history traits predict bee species responses to environmental disturbances. *Biological Conservation* 143 (10):2280-2291.

WINSTON, M. L., 1991 *The biology of the honey bee*. the United States of America: Harvard University Press.

ZHANG, Y., X. GE, F. YANG, L. ZHANG, J. ZHENG, X. TAN, Z. JIN, J. QU, and F. GU, 2014 Comparison of non-canonical PAMs for CRISPR/Cas9-mediated DNA cleavage in human cells. *Sci Rep* 4:5405.

# 6. A Comparison of RNA Interference via Injection and Feeding in Honey Bees

Yong Zhang [1,2,†], Zhen Li [1,2,†], Zilong Wang [1,2], Lizhen Zhang [1,2] and Zhijiang Zeng [1,2,*]

1. Honeybee Research Institute, Jiangxi Agricultural University, Nanchang, Jiangxi, 330045, P. R. of China;
2. Jiangxi Province Key Laboratory of Honeybee Biology and Beekeeping, Nanchang 330045, P. R. of China

**Abstract**

RNA interference (RNAi) has been used successfully to reduce target gene expression and induce specific phenotypes in several species. It has proved useful as a tool to investigate gene function and has the potential to manage pest populations and reduce disease pathogens. However, it is not known whether different administration methods are equally effective at interfering with genes in bees. Therefore, we compared the effects of feeding and injection of small interfering RNA (siRNA) on the messenger RNA (mRNA) levels of alpha-aminoadipic semialdehyde dehydrogenase (*ALDH7A1*), 4-coumarate-CoA ligase (*4CL*), and heat shock protein 70 (*HSP70*). Both feeding and injection of siRNA successfully knocked down the gene but feeding required more siRNA than the injection. Our results suggest that both feeding and injection of siRNA effectively interfere with brain genes in bees. The appropriateness of each method would depend on the situation.

**Keywords:** RNAi, siRNA, honey bee, brain

## Introduction

RNA interference (RNAi) was first discovered in transgenic plants [1], followed by the discovery of its use to analyze gene function, and its important applications in several insect species [2-4]. In honey bees, RNAi has contributed to the understanding of gene functions in caste differentiation [5,6], sex determination [7,8], lifespan [9,10], social behavior [11], learning, and memory [12,13].

The efficiency and stability of RNAi are of great importance when studying gene functions. RNAi application and efficacy remain variable between insect species, life stages, and genes [14]. Small interfering RNA (siRNA) can induce degradation of the complementary messenger RNA (mRNA) of the target gene, reducing the expression levels of the target gene [15]. Quantifying mRNA levels by quantitative real-time polymerase chain reaction (qRT-PCR) has been widely used to characterize the efficiency of RNAi [12-14]. Gene

---

† These two authors contributed equally to this paper.
* Corresponding author: bees1965@sina.com (ZJZ).
注：此文发表在 *Insects* 2022 年第 10 期。

knockdown efficacy depends on the transcript level of the target gene, protein turnover rates, and the efficiency of siRNA uptake by organs or cells [15]. Different interference methods have different locations and times of action on target genes. Two common methods used to administer RNAi in honey bees for gene function studies are via feeding or injection [15]. The responses of cells to these two administration methods are considerably different and lead to significant differences in the effectiveness of RNAi treatments [15,16]. For example, the injection of double-stranded RNA (dsRNA) into the body cavity of a locust caused a higher sensitivity than that induced by the feeding of dsRNA [17]. Although the feeding of dsRNA often requires more dsRNA than injection, this method is both less invasive and has a longer-lasting silencing effect in honey bees [18,19]. Both methods have their merits, but, as far as we know, feeding of RNAi has never been used to suppress gene expression in the brains of honey bees. It is thought that gene expression in the bees' brains can only be knocked down by local injections [20]. However, chemically modified siRNA has been developed that could successfully interfere with brain genes by intravenous injection [21]. We selected *ALDH7A1*, *4CL* and *HSP70* as the target genes, which have been reported to be expressed in honey bees [22]. We also compared the efficiency of feeding and injection of chemically modified siRNA and unmodified siRNA on gene expression in the brains of bees.

*ALDH7A1* is a member of the *ALDH* family and is mainly involved in aldehyde oxidation and aldehyde detoxification [23-25]. It affects a large number of neurotransmitters and neurohormones involved in learning, memory, behavior, and energy metabolism [26]. Additionally, *ALDH7A1* may be involved in the regulation of honey bee caste differentiation [22]. The *4CL* gene is involved in p-coumaric acid synthesis in honey bee larval diets and may be involved in honey bee caste differentiation [27-30]. *HSP70* is a member of the *HSP* family, which protects cells from both biotic and abiotic stress stimuli [31]. It is involved in the regulation of natural bee metabolism, flight behavior, learning, and memory [32-35]. Therefore, the ability to interfere with the expression of these genes in the honey bee brain is important for future experiments.

Here, we used three genes to study the effect of different siRNA delivery methods on honey bee mortality and gene expression. These data can contribute to better understanding of the importance of siRNA in honey bee RNAi, and inform us about the RNAi methods to be used in different experimental conditions.

## Materials and Methods

### Insects

We obtained the honey bees (*Apis mellifera*) for this study from Jiangxi Agricultural University (28.46 °N, 115.49 °E), Nanchang, China, in 2021. A frame of capped brood was removed from a colony and placed in a cage within a 34 °C humidified incubator overnight. Honey bees were collected within eight hours of emergence to ensure that they were the same age. Newly emerged honey bees were kept in a humidified incubator for six days before the experimental treatment began. The honey bees were starved for 3 h prior to injection and feeding of siRNA.

### siRNA Preparation and Injection

ALDH7A1-specific siRNA (forward: GCAUGGAUUCAAUGGG-CAUTT, reverse:

AUGCCCAUUGAAUCCAUGCTT), HSP70-specific siRNA (forward: GCUCGAUGCAACCAAUUATT, reverse: UAAUUGGUUAGCAUCGAGCTT) and 4CL-specific siRNA (forward: GGUGAAAGAUAUGCUAAUATT, reverse: UAUUAGCAUAUCUUUCACCTT) sequences were designed by siDirect (http://sidirect2.rnai.jp/; accessed on 13 September 2022) and DSIR (http://biodev.extra.cea.fr/DSIR/DSIR.html; accessed on 13 September 2022). GenePharma (Shanghai, China; Shanghai Jima Pharmaceutical Technology) helped us synthesize 2'-O-methyl (2'-Ome) modified and unmodified siRNAs. Negative control siRNA is widely used as a control and has no effect on gene expression in bees [36]. Therefore, in this experiment, siRNA-NC (forward: UUCUCC GAACGUGUCACGUTT, reverse: *ACGUGACACGUUCGGAGAATT)* was used in the control group.

During siRNA injecting, honey bees were tied inside a copper tube and then placed under a stereomicroscope with 1.5 mm sponge double-sided tape under the bee's brain (to ensure that the bee's brain did not move around; the sides of the proboscis could also be fastened with a pin). The fluff was scraped from the bee's brain with a 5 mL syringe needle. The tip of a 5 mL syringe was used to make a crack of about 1 mm in front of the median ocellus. (The tip should not be too deep in the bee's brain to ensure survival). The siRNA was injected into the bee's brain through the fissure in the bee's brain using a microinjector (FemtoJet 4i, Eppendorf; Fig. S1a). Honey bees in the experimental group were injected with 1 μL of siRNA solution (siRNA-ALDH7A1, siRNA-4CL, siRNA-HSP70) by microinjector. siRNA solutions were diluted in ddH$_2$O to six different concentrations (0.5 μg/μL, 1 μg/μL, 2 μg/μL, 5 μg/μL, 10 μg/μL, and 15 μg/μL). The honey bees in the control group were treated the same as the honey bees in the experimental group, except that the honey bees in the control group were injected siRNA-NC. After the injection, Vaseline was applied to the fissure in the bees' brains to avoid infection. One hundred honey bees were injected for each group. Fifty of the honey bees were used for survival analysis and the remaining honey bees were used for qRT-PCR.

During siRNA feeding, honey bees in the experimental group were fed 5 μL of siRNA solution (siRNA-ALDH7A1, siRNA-4CL, siRNA-HSP70) by a pipettor (Figure S1b). siRNA solutions were diluted in ddH$_2$O to six different concentrations (0.1 μg/μL, 0.2 μg/μL, 0.4 μg/μL, 1 μg/μL, 2 μg/μL, and 3 μg/μL). The honey bees in the control group were treated the same as the honey bees in the experimental group, except that the honey bees in the control group were fed siRNA-NC. If a honey bee could not completely eat all 5 μL of siRNA solution, this honey bee was abandoned. One hundred honey bees were fed for each group. Fifty of the honey bees were used for survival analysis and the remaining honey bees were used for qRT-PCR.

At 8 h, 16 h, 24 h, 48 h, and 72 h after injection and feeding, the brains of the honey bees were dissected, and knockdowns were verified using qRT-PCR.

### *RNA Preparation and qRT-PCR Assay*

Total RNA was extracted from the pooled brains with Trizol (Transgen; Beijin China), and reverse transcribed to obtain cDNA using the PrimeScript™RT reagent kit (Takara; Tokyo Japan). The obtained cDNA was used for qRT-PCR analysis. The qRT-PCR analysis was performed using the ABI 7500 real-time quantitative PCR system (ABI; Massachusetts USA) to detect the expression levels of genes, with GAPDH as an internal control. Two bee brains were pooled as a sample. Four biological replicates were performed for each sample, and each biological replicate included three technical replicates. Primers were designed by Premier

5.0 software based on the sequences. The primers for the qRT-PCR assay are provided in Table IV-6-1. The cycle threshold value for each sample was obtained by calculating the mean of technical replicates. The data were analyzed by $2^{-\Delta\Delta CT}$. When the $p$ value is less than 0.05 on an ANOVA test, it is considered as a significant difference.

Table IV-6-1  Primer sequences for quantitative qRT-PCR

| Genes | Forward Primer | Reverse Primer | Length |
| --- | --- | --- | --- |
| ALDH7A1 | GATGGGTCCTCTTGGTTCAG | TATAGTGGCACGTCGCATGT | 157 |
| HSP70 | GATTCGCAAAGGCAAGCTAC | CCGCTGTTGACTTCACTTCA | 217 |
| 4CL | CAAGTGGACCTTTCGTGGTT | TCTTGTGCGTCAACATGACA | 198 |
| GAPDH | GCTGGTTTCATCGATGGTTT | ACGATTTCGACCACCGTAAC | 180 |

***Effects of Different Modes of siRNA Delivery on the Survival of Honey Bees***

To confirm whether the different ways of delivering siRNA influence the survival of honey bees, honey bees were collected and fed using the previous experimental method. At six days old, the experimental group of honey bees was administered RNAi by the method mentioned in Section "siRNA Dreparation and Injection". They were then re-placed in an incubator for rearing ($n$ = 50 per cage). The control group received no treatment. The number of dead honey bees was recorded at 12 noon each day and the dead bees were removed. When all the honey bees were dead, the data were counted and analyzed.

***Data Analysis and Statistics***

An ANOVA test was conducted to analyze the difference between mRNA levels. The results were expressed as mean ± SE. The Kaplan-Meier method was used to analyze the differences between the control and treatment groups. A value of $p < 0.05$ was considered statistically significant. All statistical data were analyzed with SPSS 25.0 (IBM, USA).

## Results

***The Effects of RNA Interference Methods on the Survival of Honey Bees***

Recent studies have shown that both injection-induced damage and high doses of the reagent can cause a rapid increase in honey bee mortality. The high mortality rate of bees may have an impact on subsequent experiments. Therefore, we compared the effects of feeding and injecting high concentrations of siRNA on honey bee mortality.

The different delivery methods, and whether the siRNA was modified or not, had no effect on the survival rate of honey bees (log-rank, chi-square = 2.99, df = 4, $p$ = 0.56, Fig. IV-6-1a, see page 544; log-rank, chi-square = 0.77, df = 4, $p$ = 0.94, Figure 1b; log-rank, chi-square = 0.85, df = 4, $p$ = 0.93, Fig. IV-6-1c, see page 544). The pairwise comparison between samples is shown in supplementary Table S1.

***The Effects of ALDH7A1 RNAi Knockdown on mRNA Levels***

We quantified the mRNA transcripts of ALDH7A1 in the honey bee brains 8 h, 16 h, 24 h, 48 h, and

72 h after administering RNAi using qRT-PCR. At 16 h after the injection, the qRT-PCR results showed that different doses of siRNA reduced the expression of ALDH7A1. The effect lasted up to 72 h (Fig. IV-6-2a-f, see page 545; Table S2). The injection of 2′Ome modified siRNA (siRNA-ALDH7A1-2′Ome) and unmodified siRNA (siRNA-ALDH7A1-un) had the same effect on the expression of the ALDH71 gene. Feeding low doses of siRNA-ALDH7A1-un or siRNA-ALDH7A1-2′Ome had no effect on *ALDH7A1* expression (Fig. IV-6-1g-i). The expression of *ALDH7A1* was affected by feeding high doses of siRNA-ALDH7A1-2′Ome but was not affected by feeding high doses of siRNA-ALDH7A1-un (Fig. IV-6-2j-l). Both the feeding and injection of 2′Ome modified siRNA successfully reduced *ALDHA71* mRNA (Fig. IV-6-2; Table S2). However, the knockdown required different siRNA dosages. When injecting siRNA-ALDH7A1-2′Ome, only 1 μg RNA was required to produce the best knockout effect, while 10 μg siRNA-ALDH7A1-2′Ome was required when feeding.

### *The Effects of 4CL RNAi Knockdown on mRNA Levels*

We quantified the mRNA transcripts of *4CL* in the honey bee brains 8 h, 16 h, 24 h, 48 h, and 72 h after administering RNAi using qRT-PCR. At 16 h after the injection, the qRT-PCR results showed that all doses of siRNA except 0.5 μg reduced the expression of *4CL*. The effect lasted up to 72 h (Fig. IV-6-3b-f, see page 546; Table S3). The injection of 2′Ome modified siRNA (siRNA-4CL-2′Ome) and unmodified siRNA (siRNA-4CL-un) had the same effect on the expression of the *4CL* gene. Feeding low doses of siRNA-4CL or siRNA-4CL-2′Ome had no effect on the gene expression of *4CL* (Fig. IV-6-1g-i). The expression of *4CL* was affected by feeding high doses of siRNA-4CL-2′Ome but was not affected by feeding high doses of siRNA-4CL-un (Fig. IV-6-3j-l). Both the feeding and injection of siRNA successfully reduced *4CL* mRNA (Fig. IV-6-3; Table S3). However, the knockdown required different siRNA dosages. When injecting siRNA-4CL-2′Ome, only 2 μg RNA was required to produce the best knockout effect, while 10 μg siRNA-4CL-2′Ome was required when feeding.

### *The Effects of HSP70 RNAi Knockdown on mRNA Levels*

We quantified the mRNA transcripts of *HSP70* in the honey bee brains 8 h, 16 h, 24 h, 48 h, and 72 h after administering RNAi using qRT-PCR. At 8 h after the injection, the qRT-PCR results showed that different doses of siRNA reduced the expression of *HSP70*. The effect lasted up to 72 h (Fig. IV-6-4a-f, see page 547; Table S4). The injection of 2′Ome modified siRNA (siRNA-HSP70-2′Ome) and unmodified siRNA (siRNA-HSP70-un) had the same effect on the expression of the *HSP70* gene. Feeding low doses of siRNA had no effect on the gene expression of *HSP70* (Fig. IV-6-4g-i). The expression of *HSP70* was affected by feeding high doses of siRNA-HSP70-2′Ome but was not affected by feeding high doses of siRNA-HSP70-un (Fig. IV-6-4j-l). Both the feeding and injection of siRNA successfully reduced *HSP70* mRNA (Fig. IV-6-4; Table S4). However, the knockdown required different siRNA dosages. When injecting siRNA-HSP70-2′Ome, only 1 μg RNA was required to produce the best knockout effect, while 10 μg siRNA-HSP70-2′Ome was required when feeding.

## Discussion

In this study, we investigated the efficiency of gene knockdown using different dosages and different administration methods (via injection and feeding) in honey bee brains at mRNA levels. We determined the optimal time-window and dosage for studying the functions of the honey bee brain using RNAi.

In the experiment using the siRNA injection, we compared the efficacy of different RNAi molecules. Although the initial time of effect is different, they can effectively reduce the expression of target genes, and the effect can last for at least 48 h. In general, the knockdown effect is also different in different target genes of the same organism, depending on the specific structure of the siRNA and the molecular dose of RNAi [37]. Several researches have indicated that there is also great variability in the knockdown effect of siRNA due to the different regions of the targeted gene [20,38]. This would explain why *HSP70* expression began to decline as early as 8 h after siRNA injection. However, the expression of *ALDH7A1* and *4CL* did not begin to decrease until 16 h later.

Moreover, the expression of *HSP70* was decreased by injection of 0.5 μg siRNA, while *ALDH7A1* and *4CL* needed an injection of 1 μg siRNA.

Although modified siRNA has been shown to interfere with brain genes by injection in mammals [39,40], the feeding of siRNA has not been previously reported to inhibit gene expression in honey bee brains [41]. This may be due to the low accumulation and poor stability of siRNA in the brain. Researchers are working on ways to deliver siRNA systematically, efficiently, and safely to the brain. Presently, there are two main methods. Firstly, siRNA can be encapsulated in nanoparticles to avoid degradation. It interacts with cell-surface receptors expressed in the brain to provide cell uptake of siRNA. Alternatively, siRNA can be chemically modified so that it can enter specific tissues [21,42-44]. Our results clearly show that feeding 2'Ome-modified siRNA can reduce gene expression in the honey bee brain, whereas unmodified siRNA cannot.

We found no difference in honey bee mortality between the siRNA-treated bees and the control group. This indicated that neither of the siRNA administration methods affected the survival rate of the honey bees. When siRNA was injected, only 1 μg was required to achieve the highest level of interference, while 10 μg was required when feeding siRNA. This suggests that the effect of gene knockdown depends on the delivery mode and dose of siRNA, which is consistent with Mittal's view [37].

Injection and feeding each have their advantages and disadvantages. Injection has an important advantage in that it allows researchers to deliver the siRNA immediately to the tissue or into the hemolymph and hence avoiding possible barriers such as the blood-brain barrier or the gut epithelium which could be a problem in feeding. Another advantage is that the exact amount of dsRNA brought into an organism is known, in contrast to delivery by soaking or in some cases by feeding. However, this method has some disadvantages. The work itself is more delicate than other methods. Factors such as the choice of needle, the angle of the injection, and the volume and position of the injection are all very important and vary greatly between organisms. For example, in *Acyrthosiphon pisum*, the volume of the injection has been reported to be critical to the survival of the aphid after injection [45]. Damage to the cuticle caused by the injection may stimulate immune function, which could further complicate the interpretation of the results [46,47].

Feeding also has many advantages. It is easy to manipulate, convenient, and causes less damage to the insect [48,49]. It also has advantages in small insects, which are harder to manipulate using microinjections [49,50].

This method is also very suitable for the screening of pest control genes [47]. However, feeding is not suitable for all species. For example, the dsRNA designed for *Spodoptera litura* did not succeed in disrupting the target [51]. Sometimes feeding is less effective than injection, such as in *Caenorhabditis elegans* [52] and *Rhodnius prolixus* [50]. In addition, the RNAi efficiency of siRNA ingestion in different species may vary depending on the intestinal environment. Another limitation of siRNA feeding is the difficulty of determining the amount of siRNA that enters the insect through ingestion, which may affect many investigations. In addition, from the results of this article, to interfere with gene expression in the honey bees' brain by feeding, modified siRNA must be used. Therefore, the appropriateness of the injection compared to the feeding of siRNA needs to be decided based on the requirements of each experiment.

## Conclusions

The results showed that the injection of unmodified or 2'Ome-modified siRNA could reduce the expression of honey bee brain genes. However only feeding 2'Ome-modified siRNA could reduce the expression of bee brain genes. Feeding unmodified siRNA did not reduce gene expression in the bee brain. siRNA feeding and siRNA injection had no significant effect on honey bee mortality, but less siRNA was required for siRNA injection.

*Supplementary Materials*

The following supporting information can be downloaded at: www.mdpi.com/xxx/s1, Fig. S1. (a) Schematic diagram of honey bee injection. The green arrow shows the microinjector used to inject the bees with siRNA. (b) Schematic diagram of honey bee feeding. The green arrow shows the pipettor used to inject the bees with siRNA. Table S1: Pairwise comparison of the effects of different delivery methods and different modified siRNA methods on honey bee mortality. Table S2: Effects of different delivery methods and different modified siRNA methods on the expression of *ALDH7A1* in honey bees. Table S3: Effects of different delivery methods and different modified siRNA methods on the expression of *4CL* in honey bees. Table S4: Effects of different delivery methods and different modified siRNA methods on the expression of *HSP70* in honey bees.

*Author Contributions*

Conceptualization, Y.Z., Z.L., Z.L.W., L.Z.Z. and Z.J.Z.; methodology, Y.Z.; software, Y.Z.; validation, Z.L.W., Z.J.Z.; formal analysis, Y.Z., Z.L.; investigation, Z.J.Z.; resources, Z.J.Z.; data curation, Y.Z., Z.L.; writing—original draft preparation, Y.Z.; writing—review and editing, Y.Z., Z.L.; visualization, Y.Z., Z.L.; supervision, Z.J.Z.; project administration, Z.J.Z.; funding acquisition, Z.J.Z. All authors have read and agreed to the published version of the manuscript.

*Funding*

This work was supported by the National Natural Science Foundation of China (32172790, 31872432), and the Earmarked Fund for China Agriculture Research System (CARS-44-KXJ15).

Institutional Review Board Statement: Not applicable.

Data Availability Statement: The data presented in this study are available in this article.

*Acknowledgments*

The authors are thankful to the National Natural Science Foundation of China (32172790, 31872432), and the Earmarked Fund for China Agriculture Research System (CARS-44-KXJ15).

*Conflicts of Interest*

The authors declare that there is no conflict of interests.

## References

1. MATZKE, M.; PRIMIG, M.; TRNOVSKY, J.; MATZKE, A. Reversible methylation and inactivation of marker genes in sequentially transformed tobacco plants. *EMBO J.* 1989, *8*, 643-649.

2. BELLÉS, X. Beyond Drosophila: RNAi in vivo and functional genomics in insects. *Annu. Rev. Entomol.* 2010, *55*, 111-128.

3. HUVENNE, H.; SMAGGHE, G. Mechanisms of dsRNA uptake in insects and potential of RNAi for pest control: A review. *J. Insect Physiol.* 2010, *56*, 227-235.

4. LIU, C.; PITTS, R.J.; BOHBOT, J.D.; JONES, P.L.; WANG, G.; ZWIEBEL, L.J. Distinct olfactory signaling mechanisms in the malaria vector mosquito Anopheles gambiae. *PLoS Biol.* 2010, *8*, e1000467.

5. AMENT, S.A.; CORONA, M.; POLLOCK, H.S.; ROBINSON, G.E. Insulin signaling is involved in the regulation of worker division of labor in honey bee colonies. *Proc. Natl. Acad. Sci. USA* 2008, *105*, 4226-4231.

6. WANG, Y.; AZEVEDO, S.V.; HARTFELDER, K.; AMDAM, G.V. Insulin-like peptides (AmILP1 and AmILP2) differentially affect female caste development in the honey bee (Apis mellifera L.). *J. Exp. Biol.* 2013, *216*, 4347-4357.

7. BEYE, M.; HASSELMANN, M.; FONDRK, M.K.; PAGE, R.E., JR.; OMHOLT, S.W. The gene csd is the primary signal for sexual development in the honeybee and encodes an SR-type protein. *Cell* 2003, *114*, 419-429.

8. KOHNO, H.; KUBO, T. mKast is dispensable for normal development and sexual maturation of the male European honeybee. *Sci. Rep.* 2018, *8*, 1-10.

9. PAOLI, P.P.; WAKELING, L.A.; WRIGHT, G.A.; FORD, D. The dietary proportion of essential amino acids and Sir2 influence lifespan in the honeybee. *Age* 2014, *36*, 1239-1247.

10. SEEHUUS, S.-C.; NORBERG, K.; GIMSA, U.; KREKLING, T.; AMDAM, G.V. Reproductive protein protects functionally sterile honey bee workers from oxidative stress. *Proc. Natl. Acad. Sci. USA* 2006, *103*, 962-967.

11. ANTONIO, D.S.M.; GUIDUGLI-LAZZARINI, K.R.; DO NASCIMENTO, A.M.; SIMÕES, Z.L.P.; HARTFELDER, K. RNAi-mediated silencing of vitellogenin gene function turns honeybee (Apis mellifera) workers into extremely precocious foragers. *Sci. Nat.* 2008, *95*, 953-961.

12. AWATA, H.; WAKUDA, R.; ISHIMARU, Y.; MATSUOKA, Y.; TERAO, K.; KATATA, S.; MATSUMOTO, Y.; HAMANAKA, Y.; NOJI, S.; MITO, T. Roles of OA1 octopamine receptor and Dop1 dopamine receptor in mediating appetitive and aversive reinforcement revealed by RNAi studies. *Sci. Rep.* 2016, *6*, 1-10.

13. CRISTINO, A.S.; BARCHUK, A.R.; FREITAS, F.C.; NARAYANAN, R.K.; BIERGANS, S.D.; ZHAO, Z.; SIMOES, Z.L.; REINHARD, J.; CLAUDIANOS, C. Neuroligin-associated microRNA-932 targets actin and regulates memory in the honeybee. *Nat. Commun.* 2014, *5*, 1-11.

14. SCOTT, J.G.; MICHEL, K.; BARTHOLOMAY, L.C.; SIEGFRIED, B.D.; HUNTER, W.B.; SMAGGHE, G.; ZHU,

K.Y.; DOUGLAS, A.E. Towards the elements of successful insect RNAi. *J. Insect Physiol.* 2013, *59*, 1212-1221.

15. YANG, D.; XU, X.; ZHAO, H.; YANG, S.; WANG, X.; ZHAO, D.; DIAO, Q.; HOU, C. Diverse factors affecting efficiency of RNAi in honey bee viruses. *Front. Genet.* 2018, *9*, 384.

16. WHANGBO, J.S.; HUNTER, C.P. Environmental RNA interference. *Trends Genet.* 2008, *24*, 297-305.

17. WYNANT, N.; DURESSA, T.F.; SANTOS, D.; VAN DUPPEN, J.; PROOST, P.; HUYBRECHTS, R.; BROECK, J.V. Lipophorins can adhere to dsRNA, bacteria and fungi present in the hemolymph of the desert locust: A role as general scavenger for pathogens in the open body cavity. *J. Insect Physiol.* 2014, *64*, 7-13.

18. LI, W.; EVANS, J.D.; HUANG, Q.; RODRÍGUEZ-GARCÍA, C.; LIU, J.; HAMILTON, M.; GROZINGER, C.M.; WEBSTER, T.C.; SU, S.; CHEN, Y.P. Silencing the honey bee (Apis mellifera) naked cuticle gene (nkd) improves host immune function and reduces Nosema ceranae infections. *Appl. Environ. Microbiol.* 2016, *82*, 6779-6787.

19. MAORI, E.; PALDI, N.; SHAFIR, S.; KALEV, H.; TSUR, E.; GLICK, E.; SELA, I. IAPV, a bee-affecting virus associated with colony collapse disorder can be silenced by dsRNA ingestion. *Insect Mol. Biol.* 2009, *18*, 55-60.

20. GUO, X.; WANG, Y.; SINAKEVITCH, I.; LEI, H.; SMITH, B.H. Comparison of RNAi knockdown effect of tyramine receptor 1 induced by dsRNA and siRNA in brains of the honey bee, Apis mellifera. *J. Insect Physiol.* 2018, *111*, 47-52.

21. SHUKLA, S.; SUMARIA, C.S.; PRADEEPKUMAR, P. Exploring chemical modifications for siRNA therapeutics: A structural and functional outlook. *Chemmedchem* 2010, *5*, 328-349.

22. HASEGAWA, M.; ASANUMA, S.; FUJIYUKI, T.; KIYA, T.; SASAKI, T.; ENDO, D.; MORIOKA, M.; KUBO, T. Differential gene expression in the mandibular glands of queen and worker honeybees, Apis mellifera L.: Implications for caste-selective aldehyde and fatty acid metabolism. *Insect Biochem. Mol. Biol.* 2009, *39*, 661-667.

23. FONG, W.-P.; CHENG, C.; TANG, W.-K. Antiquitin, a relatively unexplored member in the superfamily of aldehyde dehydrogenases with diversified physiological functions. *Cell. Mol. Life Sci.* 2006, *63*, 2881-2885.

24. TANG, W.-K.; CHENG, C.H.; FONG, W.-P. First purification of the antiquitin protein and demonstration of its enzymatic activity. *FEBS Lett.* 2002, *516*, 183-186.

25. DEMOZAY, D.; ROCCHI, S.; MAS, J.-C.; GRILLO, S.; PIROLA, L.; CHAVEY, C.; VAN OBBERGHEN, E. Fatty aldehyde dehydrogenase: Potential role in oxidative stress protection and regulation of its gene expression by insulin. *J. Biol. Chem.* 2004, *279*, 6261-6270.

26. PENA, I.A.; ROUSSEL, Y.; DANIEL, K.; MONGEON, K.; JOHNSTONE, D.; WEINSCHUTZ MENDES, H.; BOSMA, M.; SAXENA, V.; LEPAGE, N.; CHAKRABORTY, P. Pyridoxine-dependent epilepsy in zebrafish caused by Aldh7a1 deficiency. *Genetics* 2017, *207*, 1501-1518.

27. MAO, W.; SCHULER, M.A.; BERENBAUM, M.R. A dietary phytochemical alters caste-associated gene expression in honey bees. *Sci. Adv.* 2015, *1*, e1500795.

28. ISLAM, M.T.; LEE, B.-R.; LEE, H.; JUNG, W.-J.; BAE, D.-W.; KIM, T.-H. *p*-Coumaric acid induces jasmonic acid-mediated phenolic accumulation and resistance to black rot disease in Brassica napus. *Physiol. Mol. Plant Pathol.* 2019, *106*, 270-275.

29. STUIBLE, H.-P.; BÜTTNER, D.; EHLTING, J.; HAHLBROCK, K.; KOMBRINK, E. Mutational analysis of 4-coumarate: CoA ligase identifies functionally important amino acids and verifies its close relationship to other

adenylate-forming enzymes. *FEBS Lett.* 2000, *467*, 117-122.

30. CUKOVICA, D.; EHLTING, J.; ZIFFLE, J.A.V.; DOUGLAS, C.J. Structure and evolution of 4-coumarate: Coenzyme A ligase (4CL) gene families. *Biol. Chem.* 2001, *382*, 645-654.

31. MOSSER, D.D.; CARON, A.W.; BOURGET, L.; DENIS-LAROSE, C.; MASSIE, B. Role of the human heat shock protein hsp70 in protection against stress-induced apoptosis. *Mol. Cell. Biol.* 1997, *17*, 5317-5327.

32. WILLIAMS, J.B.; ROBERTS, S.P.; ELEKONICH, M.M. Age and natural metabolically-intensive behavior affect oxidative stress and antioxidant mechanisms. *Exp. Gerontol.* 2008, *43*, 538-549.

33. MORAMMAZI, S.; SHOKROLLAHI, B. The pattern of HSP70 gene expression, flight activity and temperature in Apis mellifera meda Colonies. *J. Therm. Biol.* 2020, *91*, 102647.

34. REITMAYER, C.M.; RYALLS, J.M.; FARTHING, E.; JACKSON, C.W.; GIRLING, R.D.; NEWMAN, T.A. Acute exposure to diesel exhaust induces central nervous system stress and altered learning and memory in honey bees. *Sci. Rep.* 2019, *9*, 1-9.

35. ROBERTS, S.P.; ELEKONICH, M.M. Muscle biochemistry and the ontogeny of flight capacity during behavioral development in the honey bee, Apis mellifera. *J. Exp. Biol.* 2005, *208*, 4193-4198.

36. NUNES, F.M.F.; SIMÕES, Z.L.P. A non-invasive method for silencing gene transcription in honeybees maintained under natural conditions. *Insect Biochem. Mol. Biol.* 2009, *39*, 157-160.

37. MITTAL, V. Improving the efficiency of RNA interference in mammals. *Nat. Rev. Genet.* 2004, *5*, 355-365.

38. KRAUTZ-PETERSON, G.; RADWANSKA, M.; NDEGWA, D.; SHOEMAKER, C.B.; SKELLY, P.J. Optimizing gene suppression in schistosomes using RNA interference. *Mol. Biochem. Parasitol.* 2007, *153*, 194-202.

39. ZHOU, Y.; ZHU, F.; LIU, Y.; ZHENG, M.; WANG, Y.; ZHANG, D.; ANRAKU, Y.; ZOU, Y.; LI, J.; WU, H. Blood-brain barrier-penetrating siRNA nanomedicine for Alzheimer's disease therapy. *Sci. Adv.* 2020, *6*, eabc7031.

40. PARDRIDGE, W.M. shRNA and siRNA delivery to the brain. *Adv. Drug Del. Rev.* 2007, *59*, 141-152.

41. KOHNO, H.; KUBO, T. Genetics in the honey bee: Achievements and prospects toward the functional analysis of molecular and neural mechanisms underlying social behaviors. *Insects* 2019, *10*, 348.

42. PÉREZ-MARTÍNEZ, F.C.; GUERRA, J.; POSADAS, I.; CEÑA, V. Barriers to non-viral vector-mediated gene delivery in the nervous system. *Pharm. Res.* 2011, *28*, 1843-1858.

43. HUANG, R.; MA, H.; GUO, Y.; LIU, S.; KUANG, Y.; SHAO, K.; LI, J.; LIU, Y.; HAN, L.; HUANG, S. Angiopep-conjugated nanoparticles for targeted long-term gene therapy of Parkinson's disease. *Pharm. Res.* 2013, *30*, 2549-2559.

44. YIN, Y.; HU, B.; YUAN, X.; CAI, L.; GAO, H.; YANG, Q. Nanogel: A versatile nano-delivery system for biomedical applications. *Pharmaceutics* 2020, *12*, 290.

45. JAUBERT-POSSAMAI, S.; LE TRIONNAIRE, G.; BONHOMME, J.; CHRISTOPHIDES, G.K.; RISPE, C.; TAGU, D. Gene knockdown by RNAi in the pea aphid Acyrthosiphon pisum. *BMC Biotechnol.* 2007, *7*, 1-8.

46. HAN, Y.S.; CHUN, J.; SCHWARTZ, A.; NELSON, S.; PASKEWITZ, S.M. Induction of mosquito hemolymph proteins in response to immune challenge and wounding. *Dev. Comp. Immunol.* 1999, *23*, 553-562.

47. YU, N.; CHRISTIAENS, O.; LIU, J.; NIU, J.; CAPPELLE, K.; CACCIA, S.; HUVENNE, H.; SMAGGHE, G. Delivery of dsRNA for RNAi in insects: An overview and future directions. *Insect Sci.* 2013, *20*, 4-14.

48. CHEN, J.; ZHANG, D.; YAO, Q.; ZHANG, J.; DONG, X.; TIAN, H.; CHEN, J.; ZHANG, W. Feeding-based RNA

interference of a trehalose phosphate synthase gene in the brown planthopper, Nilaparvata lugens. *Insect Mol. Biol.* 2010, *19*, 777-786.

49. TIAN, H.; PENG, H.; YAO, Q.; CHEN, H.; XIE, Q.; TANG, B.; ZHANG, W. Developmental control of a lepidopteran pest Spodoptera exigua by ingestion of bacteria expressing dsRNA of a non-midgut gene. *PLoS ONE* 2009, *4*, e6225.

50. ARAUJO, R.; SANTOS, A.; PINTO, F.; GONTIJO, N.; LEHANE, M.; PEREIRA, M. RNA interference of the salivary gland nitrophorin 2 in the triatomine bug Rhodnius prolixus (Hemiptera: Reduviidae) by dsRNA ingestion or injection. *Insect Biochem. Mol. Biol.* 2006, *36*, 683-693.

51. RAJAGOPAL, R.; SIVAKUMAR, S.; AGRAWAL, N.; MALHOTRA, P.; BHATNAGAR, R.K. Silencing of Midgut Aminopeptidase N of Spodoptera litura by Double-stranded RNA Establishes Its Role asBacillus thuringiensis Toxin Receptor. *J. Biol. Chem.* 2002, *277*, 46849-46851.

52. HUNTER, C.P. Genetics: A touch of elegance with RNAi. *Curr. Biol.* 1999, *9*, R440-R442.

# 附录 A 彩图

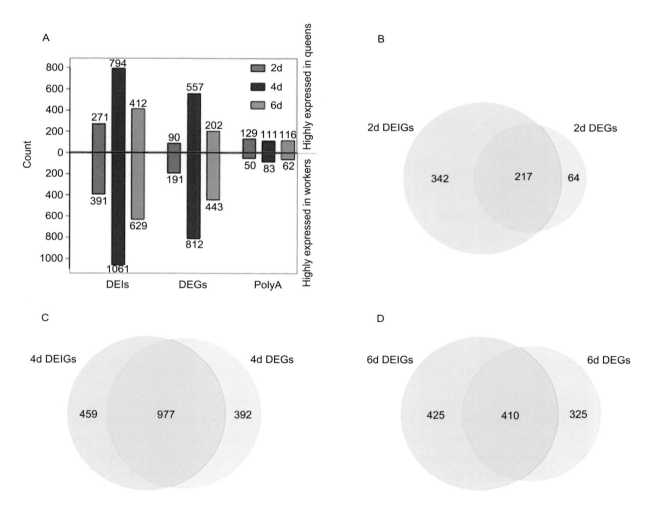

**Figure I-1-1** Comparison of isoform and gene expression, poly(A) length between queen and worker larvae. See also Table S1-S9 and S13-S17. (A) Numbers of differentially expressed isoforms (DEIs) and genes (DEGs), different poly(A)-length related isoforms between queens and workers at three larval stages (day 2, day 4 and day 6). The up bars are DEIs and DEGs length highly expressed in 2d, 4d and 6d queen larvae as well as isoforms with longer poly(A) tails in queen larvae, and the down bars are that highly expressed or with longer poly(A) tails in worker larvae. (B) The venn diagram of DEIGs (DEI mapped genes) and DEGs from 2d queen-worker comparison. (C) The venn diagram of DEIGs and DEGs from 4d queen-worker comparison. (D) The venn diagram of DEIGs and DEGs from 6d queen-worker comparison.

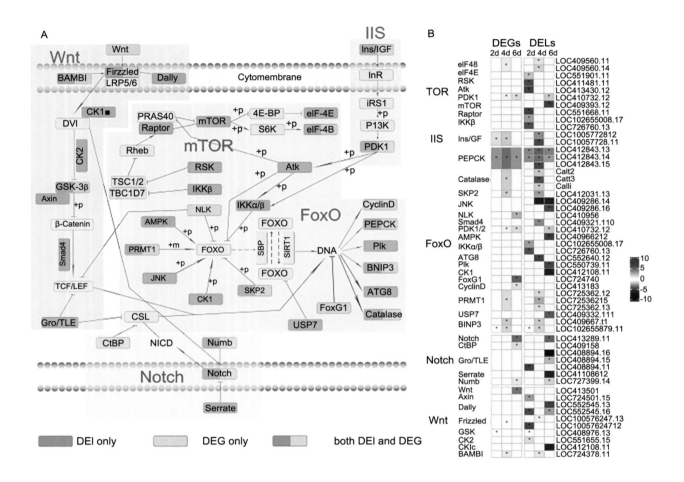

**Figure I-1-2** Expression and enrichment of DEIs and DEGs in five KEGG pathways. See also Table S1-S6, S16 and S17. (A) DEIs and DEGs enriched in key steps of five KEGG pathways. Five KEGG pathways were shown with transparent boxes and the names of these pathways were red marked. Green boxes inside of transparent boxes are DEI enriched proteins; yellow boxes are DEG enriched proteins; green-yellow mixed boxes are both DEI and DEG enriched proteins; blank boxes are non DEI and DEG enriched proteins. (B) Expression of DEIs and DEGs in the five key pathways. The left are pathway and protein names, and the right are isoform expression and names. Expression of isoforms and genes are presented with their log2 TPM values and shown with color scales. DEIs or DEGs are marked with "*" in the middle of the boxes.

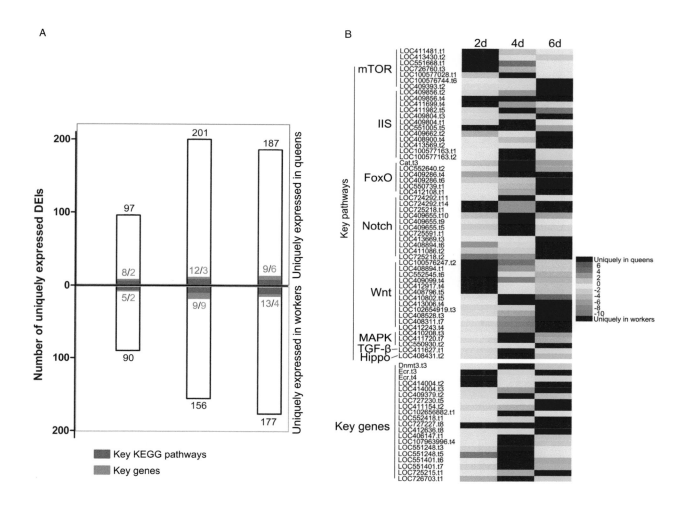

**Figure I-1-3** Uniquely expressed isoforms in queen and worker larvae. See also Table S10-S12, S16 and S17. (A) Numbers of uniquely expressed isoforms in queen larvae (up bars) or worker larvae (down bars) at 2d, 4d and 6d stages. The red bars are numbers of uniquely expressed isoforms enriched in eight key KEGG signaling pathways (mTOR, IIS, FoxO, Notch, Wnt, MAPK, Hippo and TGF-β). The purple bars are numbers of uniquely expressed isoforms that are key genes reported in previous studies (details see Table S10-12). (B) The heat map of expression of uniquely expressed isoforms which are shown in red bars and green bars in A. The red color blocks in heat map are uniquely expressed isoforms in queen larvae at 2d, 4d and 6d age, whereas the dark green blocks are uniquely expressed isoforms in worker larvae. Other color scales are the log2 fold change values of isoform expression between queen and worker larvae at 2d, 4d and 6d stages.

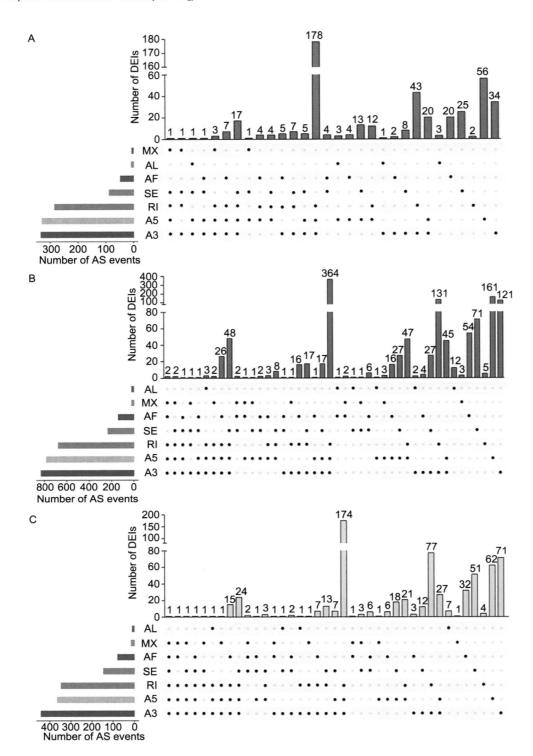

**Figure I-1-4** The alternative splicing events (AS) in DEIs of three comparisons. See also Figure S6. (A) DEIs of 2d comparison containing AS events. The colorful bars in the left bottom diagram are the different AS patterns and the number of AS events in DEIs. Seven different AS patterns are 3'splice site (A3), 5'splice site (A5), First exon (AF), Last exon (AL), Retained intron (RI), Skipping exon (SE) and Mutually exclusive exon (MX). The pink vertical bars in the top of the right diagram are the numbers of DEIs that were spliced by a particular combination of AS forms. The black spots in a column mean DEIs that were spliced by a combination of related AS forms. Same in B and C. (B) DEIs of 4d comparison containing AS events. (C) DEIs of 6d comparison containing AS events.

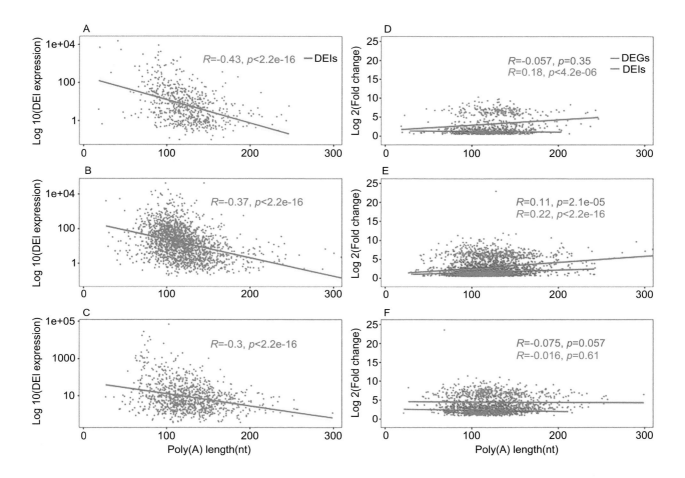

**Figure I-1-5** Correlation between poly(A) length and expression of DEIs and DEGs. See also Fig. S7, S8 and Table S7-S9. (A) Correlation between poly(A) lengths and the expression of DEIs of 2d queen-worker comparison. Expression of each DEI was the log10 TPM value. Same in B and C. (B) Correlation between poly(A) lengths and DEIs of 4d comparison. (C) Correlation between poly(A) lengths and DEIs of 6d comparison. (D) Correlation between poly(A) lengths and $\log_2$ fold change values of DEIs and DEGs of 2d comparison. Same in E and F. (E) Correlation between poly(A) lengths and $\log_2$ fold change values of DEIs and DEGs of 4d comparison. (F) Correlation between poly(A) lengths and log2 fold change values of DEIs and DEGs of 6d comparison.

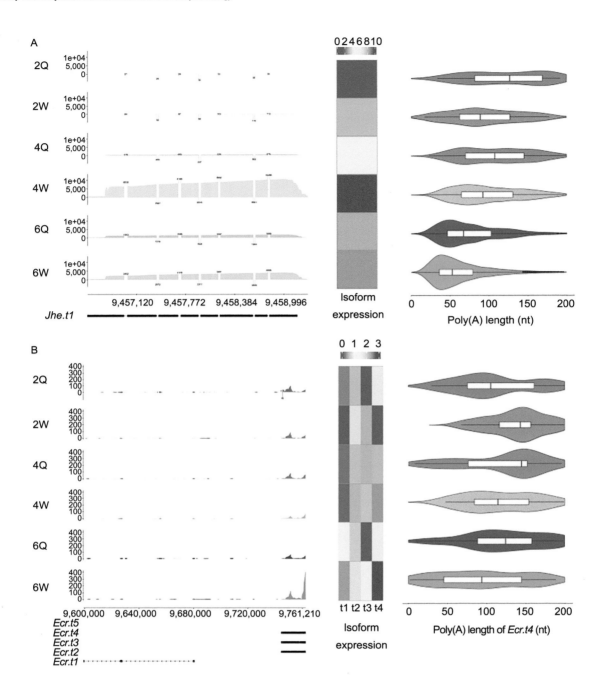

**Figure I-1-6** The poly(A) lengths and isoform expression of two key genes. See also Table S1-S3, S16 and S17. (A) Expression and related poly(A) lengths of *Jhe* isoform. The left is the exon coverage of clean reads (namely gene expression) of *Jhe* gene of 6 larval groups. The *Jhe* has only one isoform (*Jhe.t1*) and its structure is present (see its exons below marked with black color). The middle heat map is the isoform expression of *Jhe.t1* ($\log_2$ TPM values) in 6 groups and presented with color scales. The right is the poly(A) length distribution of Jhe.t1 in 6 groups. The mean poly(A) length is presented in the middle of each violinplot. (B) Expression and related poly(A) lengths of Ecr isoforms. The left is the exon coverage of clean reads (namely gene expression) of *Ecr* gene The Ecr gene has 4 isoforms and their expression is presented using heat map as above. The poly(A) length distribution of *Ecr.t4* in 6 groups is presented on the right side, it was the most significantly differentially expressed isoform of gene *Ecr*. The poly(A) of *Ecr.t4* in 2dW was presented with a blank box which means few clean data of poly(A) was detected, since 2dW did not express *Ecr.t4*.

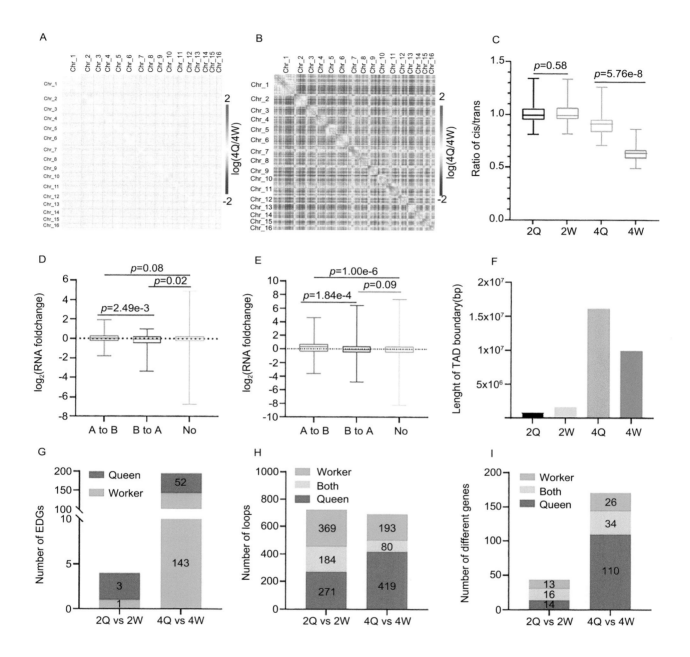

**Figure I-2-1** Hi-C analysis of queen and worker larvae. (A) Interaction map of 2Q vs 2W (observed/control). (B) Interaction map of 4Q vs 4W (observed/control). In A and B, the sixteen *Apis mellifera* chromosomes (chr) are shown from left to right and top to bottom. Chromosomes are separated by thin black bars. (C) The cis and trans ratios in queen and worker larvae. Test by Mann-Whitney U test. (D and E) Box plot comparing gene expression fold changes between genes in switch regions (A-B and B-A) and no switch regions (A-A and B-B) in 2Q/2W and 4Q/4W, respectively. Test by Kruskal-Wallis H test. (F) Length of TAD boundaries in larval samples. (G) Number of DEGs located at TAD boundary, (H) Number of loops and (I) Number of DEGs located at loop in comparisons of 2Q and 2W larvae and 4Q and 4W larvae.

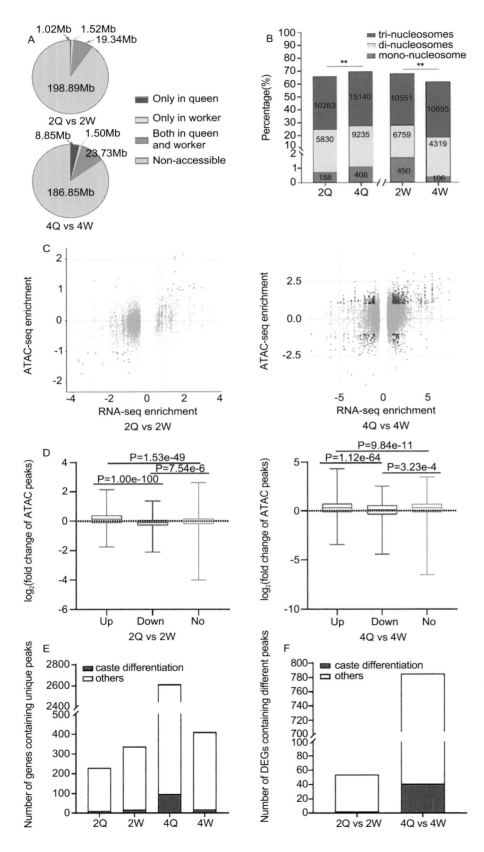

**Figure I-2-2** ATAC-seq analysis of queen and worker larvae. (A) Pie chart of comparison of accessible regions in 2Q and 2W, and 4Q and 4W. (B) Histogram of the proportions of different nucleosomes between queen and worker larvae. The bars are the ratios of mononucleosome, di-nucleosomes and tri-nucleosomes in 2Q, 2W, 4Q and 4W. The numbers inside the bars are the accurate number of three nucleosome types in each group. "**" indicates a significant difference ($p<0.01$, chi-square test) (C) Scatter plot of the difference in significant ATAC-seq enrichment between queen and worker ($y$-axis) against the $\text{Log}_2\text{FC}$ of transcript expression between queen and worker ($x$-axis). (D) Box plots comparing fold changes of ATAC peaks between protein coding genes in comparisons of 2Q and 2W, and 4Q and 4W, respectively. Test by Kruskal-Wallis H test. (E) Number of genes containing unique ATAC-seq peaks in queen or worker larvae, and (F) Number of DEGs containing significantly different ATAC-seq peaks between queen and worker larvae. The red part in each bar represents the number of genes that are involved in caste differentiation.

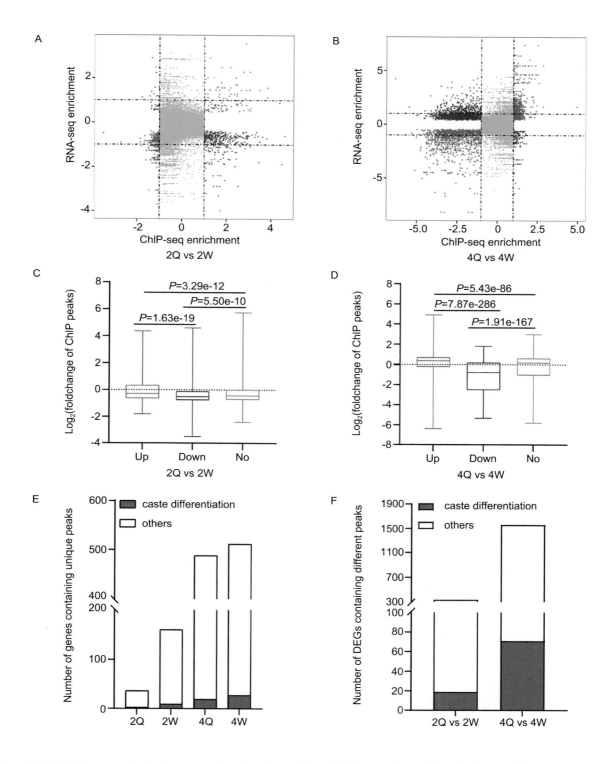

**Figure I-2-3** ChIP-seq analysis of queen and worker larvae. (A and B) Scatter plots of the significant differences in ChIP-seq enrichment between queen worker larvae samples (*x*-axis) against the $Log_2FC$ of transcript expression between queen worker larvae samples (*y*-axis). (C and D) Box plot comparing fold change of ChIP peaks between protein coding gene in 2Q/2W and 4Q/4W, respectively. Test by Kruskal-Wallis H test. (E) Number of genes containing unique ChIP-seq peaks in queen or worker larvae. (F) Number of DEGs containing significantly different ChIP-seq peaks between queen and worker larvae. The red part in each bar represents the number of genes that are involved in caste differentiation.

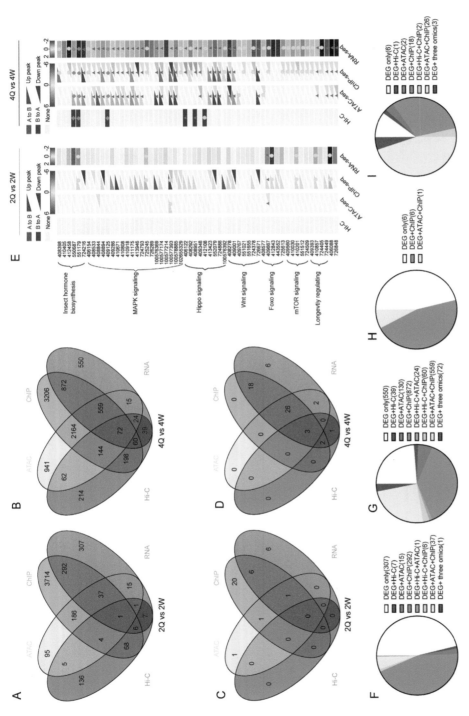

**Figure I-2-4** Multiomics analysis between queen and worker larvae. Venn diagram showing the overlap in DEGs identified by different omics methods in 2Q/2W comparisons (A) and 4Q/4W comparisons (B). (C) We selected 58 genes that have been reported to be associated with the caste differentiation in the honey bee. The figure shows which epigenomic modifications of these 58 DEGs differed in 2Q/2W comparisons. (D) The 58 DEGs and their epigenetic modofications differed in 4Q/4W comparisons. More of these genes were differentially regulated by at least one epigenomic system in 4 day comparisons compared to 2 day comparison. (E) Summary of differences in epigenetic regulation of genes associated with caste differentiation in queen and worker larvae comparisons. Yellow circle means there were significant differences observed in two omic analyses. Green triangle means there were significant differences observed in three omic comparisons. Pink square means there were significant differences observed in four omic comparisons. The locations of the yellow circle, green triangle, and pink square indicate differences in these omics between queen and worker larvae. (F, G) Pie chart of number of genes that differ in epigenetic regulation in comparisons of 2Q/2W (F) and 4Q/4W (G). (H, I) Pie chart of number of 58 key genes that differ in epigenetic regulation in comparisons of 2Q/2W (H) and 4Q/4W (I).

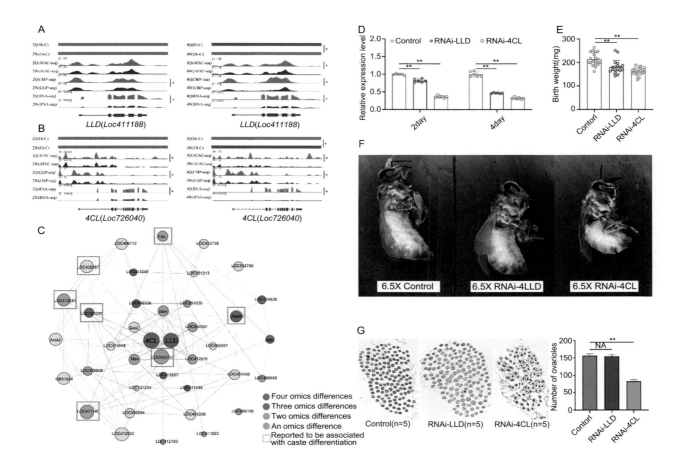

**Figure I-2-5** RNAi verification of the *LLD* and *4CL* gene. (A) Example showing the correlation of Hi-C, ATAC signal, ChIP signal and RNA-seq reads in 2Q/2W (left) and 4Q/4W (right) for the *LLD* gene. In Hi-C data, red bars represent A compartments, whereas blue ones represent B compartments. The red waves are read coverage of queens in ATAC signal, ChIP signal and RNA-seq, while the blue ones are that of workers. Data in "[]" are the scales of read counts. "*" indicates that there is a difference between two groups. Same to B. (B) Example showing the correlation of Hi-C, ATAC signal, ChIP signal and RNA-seq reads in 2Q/2W (left) and 4Q/4W (right) for the *4CL* gene. (C) Network analysis of two key genes, *4CL* and *LLD*. Red circles indicate that genes were different in all four omics; Pink circles indicate that genes were different in three omics; Purple circles indicate that genes were different in two omics, and grey indicate that genes were different in an omic. Green squares indicate genes that have been reported to be associated with honeybee caste differentiation. (D) Gene expression in siRNA fed and control larvae. "**" indicate a significant difference ($p<0.01$ t test). (E) Weight of newly-emerged queens, from control (feeding the siRNA-NC), RNAi-*LLD* (feeding the siRNA-*LLD*) and RNAi-*4CL* groups (feeding the siRNA-*4CL*). Bars are the mean ± SE. "**" indicate a significant difference ($p<0.01$ Mann-Whitney U test). (F) Photo of queen after feeding siRNA reagent, left is the control group, middle is the RNAi-*LLD* group and right is the RNAi-*4CL* group. (G) Section of the newly-emerged queen's left ovaries, left is the control group, middle is the RNAi-*LLD* group and right is the RNAi-*4CL* group. "n" is the number of samples. Bar graphs represent statistical plots of ovarian number in the three groups. Bars are the mean ± SEM. "**" indicate a significant difference ($p<0.01$ t test).

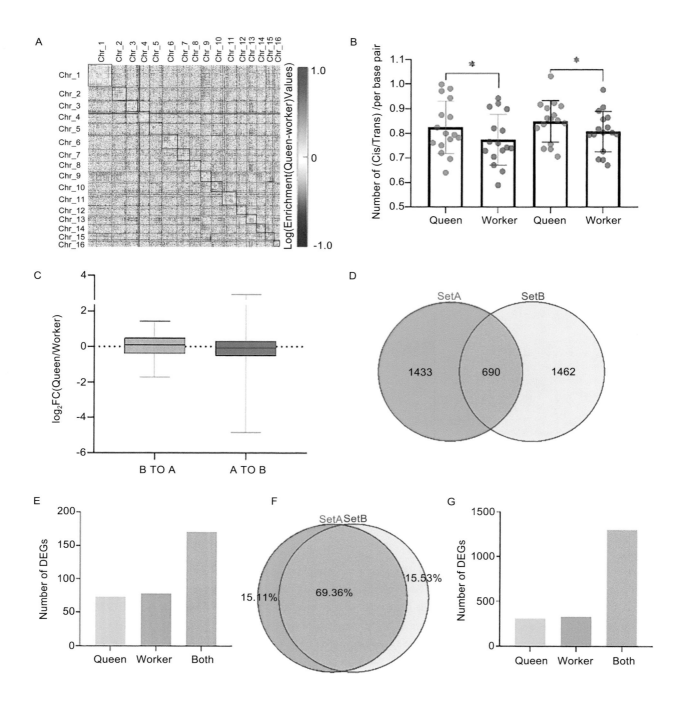

**Figure I-3-1 Hi-C analysis of queen and worker.** (A) Interaction map of queen vs worker (observed/control). The sixteen chromosomes (chr) o *A. mellifera* are presented from left to right and top to bottom. Chromosomes are separated by black lines. (B) The numbers of *cis* and *trans* interactions in queen and worker. Data were presented as mean ± SE. $P<0.01$ by t test. (C) The expression fold changes (queen/worker) of genes associated with A/B compartment switches (Unpaired t test with Welch's correction). Data were presented as mean ± SE. $P<0.05$. (D) The numbers of chromatin loops in queen and worker. (E) The numbers of DEGs (queen vs worker) paired with chromatin loops. (F) The percentages of unique TAD regions in queen and worker. (G) The numbers of DEGs in the unique and shared TAD regions.

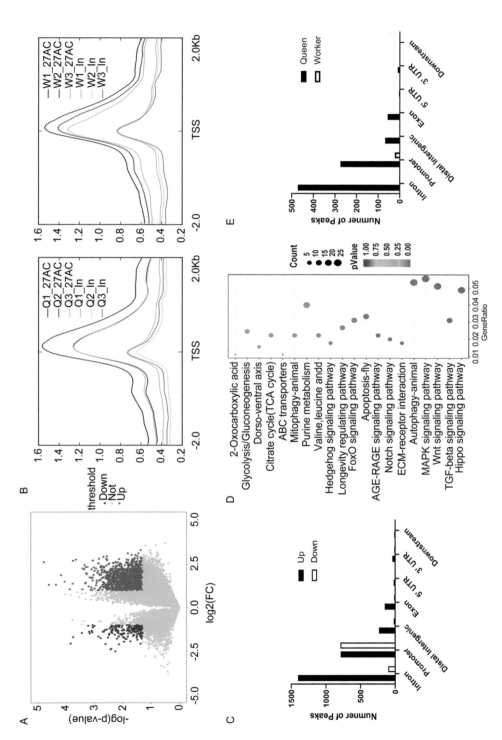

**Figure 1-3-2 The queen and worker show caste-specific differences in the enrichment of H3K27ac that correlate with differential gene expression.** (A) A volcano plot of the difference in enrichment between queen and worker against the negative log p-value for H3K27ac. Regions in gray fall below the genome-wide threshold of significance ($P > 0.05$). Regions in red (queen) and blue (worker) are those that reach genome-wide significance ($P \leq 0.05$) and have a greater than twofold difference in enrichment between queen and worker. (B) Plots of the ChIP-seq enrichment above input around the TSS (±2 kbp) of genes profiled across queen (left) and worker (right). There are three replicates performed for each ChIP-seq experiment. (C) The distribution of differential peaks between queen and worker in the genome region. The black bars indicate peaks up-regulated in the queen compared with the worker; the white bars indicate peaks down-regulated in the queen. (D) The top 20 significantly enriched KEGG pathways of differential H3K27ac peak-related genes between queen and worker. (E) The distribution of unique peaks in queen and worker in the genome region.

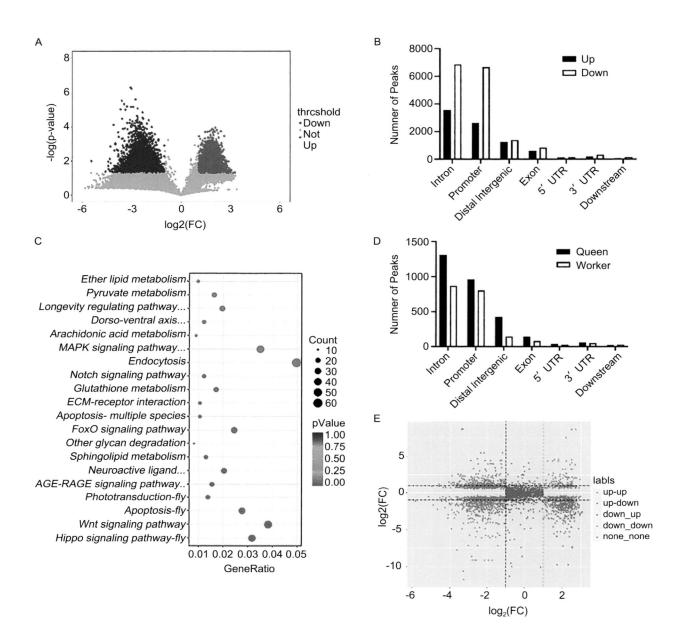

**Figure I-3-3 The queen and worker show caste-specific differences in the enrichment of H3K4me1 that correlate with differential gene expression.** (A) A volcano plot of the difference in enrichment between queen and worker against the negative log p-value for H3K4me1. Regions in gray fall below the genome-wide threshold of significance ($P > 0.05$). Regions in red (queen) and blue (worker) are those that reach genome-wide significance ($P \leq 0.05$) and have a greater than twofold difference in enrichment between queen and worker. (B) The distribution of differential peaks between queen and worker in the genome region. The black bars indicate peaks up-regulated in the queen compared with the worker; the white bars indicate peaks down-regulated in the queen. (C) The top 20 significant enriched KEGG pathways of differential H3K4me1 peak-related genes between queen and worker. (D) The distribution of unique peaks in queen and worker in the genome region. (E) A scatter plot of the promoter ChIP-seq H3K4me1 $\log_2$ (fold change) for peaks related genes (x-axis) and $\log_2$ (fold change) (y-axis) for RNA-seq FPKM between queen and worker.

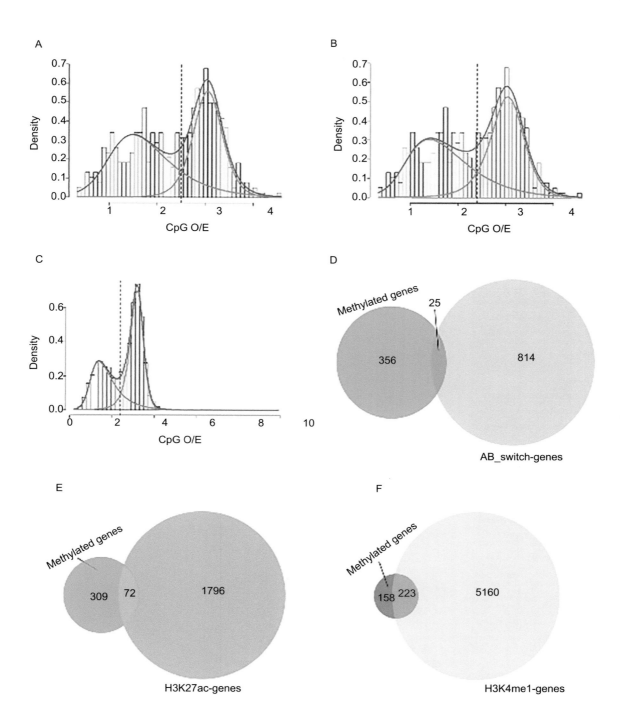

**Figure I-3-4 Association analysis of Hi-C, H3K27ac and H3K4me1 with DNA methylation.** The upper part indicates the distribution of CpG O/E in genes associated with A/B compartment switches (A) differential H3K27ac peaks (B) and differential H3K4me1 peaks (C) The lower part indicates overlapped genes between the DMEs reported by Frank Lyko et al. and the A/B compartment switch related genes (D), differential H3K27ac peak related genes (E) and differential H3K4me1 peak related genes (F).

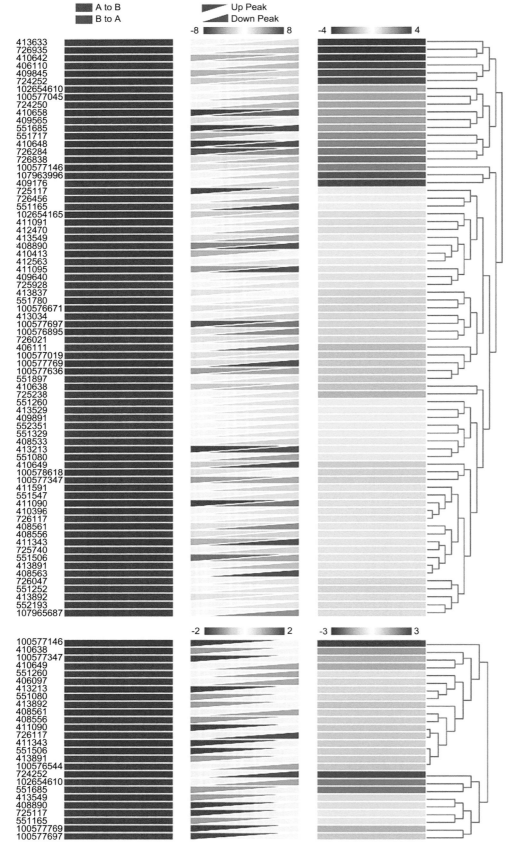

Figure I-3-5 Heat map of three omics. The genes in the upper part are differential genes in group Hi-C, H3K4me1 and RNA-seq. The genes in the lower part are differential genes in group Hi-C, H3K27ac and RNA-seq.

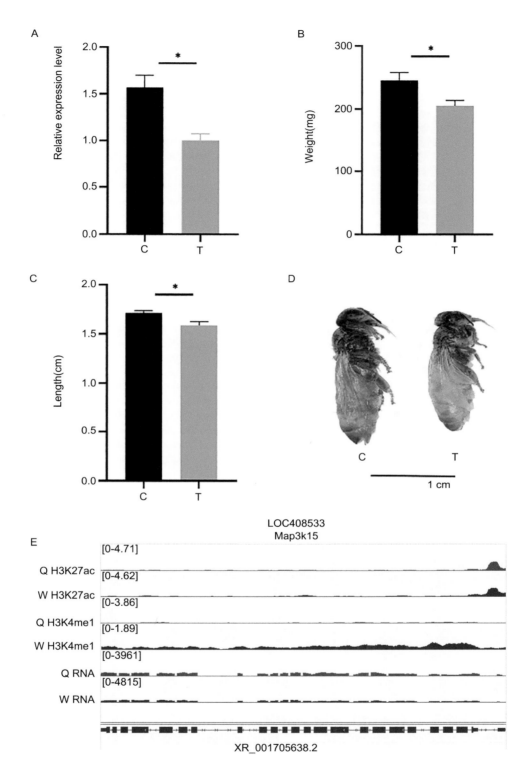

**Figure I-3-6 Map3k15 has a significant effect on caste differentiation of honeybees.** (A) The expression change of *Map3k115* after RNAi. Data were presented as mean ± SE. *P*<0.05 by t test. (B) Birth weight of bees between siRNA group and control group. Data were presented as mean ± SE. *P*<0.01 by t test. (C) Body length of bees between siRNA group and control group. Data were presented as mean ± SE. *P*<0.05 by t test. (D) The morphology of bees in the RNAi group and control group. (E) Expression and histone modification of *Map3k115*.

**Figure I-4-5** Gene expression of 33 genes among 6 larval groups. DEGs were measured as their read counts under a statistic value of FDR<0.01 and the absolute value of log2 FC>1. Genes significantly upregulated in worker or queen larvae compared to drone larvae are coloured red. Genes significantly downregulated in the same comparisons are coloured green. Black represents no significant difference in expression ratio. WL and QL represent worker larvae and queen larvae respectively.

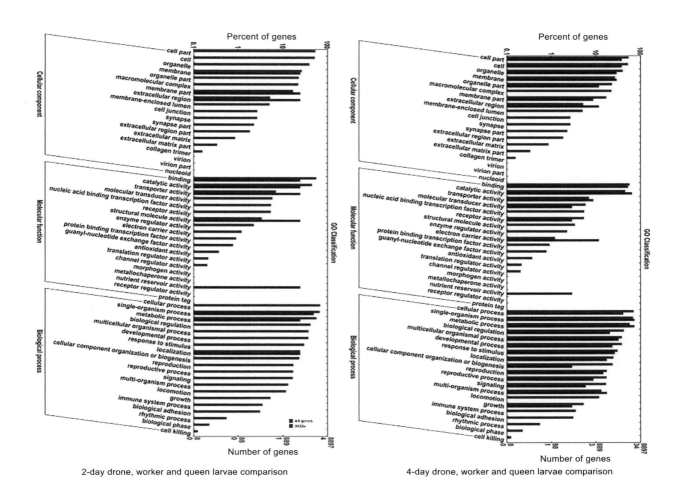

**Figure I-4-6 Gene ontology classification of same DEGs from queen vs drone larvae and worker vs drone larvae comparisons**. The results are summarized in three main categories: biological process, cellular component and molecular function. *Y*-axis indicates category, *X*-axis indicates the percentage of DEGs. Red bars are all genes, blue bars are DEGs.

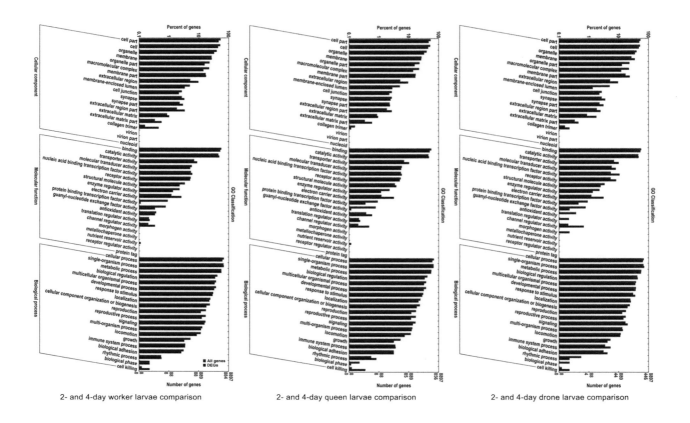

**Figure I-4-7 Gene ontology classification of DEGs between 2- and 4-day larvae in three larval castes.** The results are summarized in three main categories: biological process, cellular component and molecular function. Y-axis indicates category, X-axis indicates the percentage of DEGs. Red bars are all genes, blue bars are DEGs.

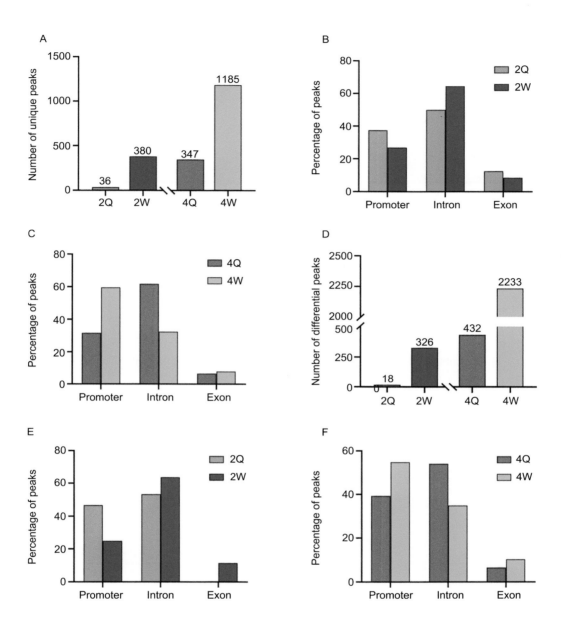

**Figure I-6-1** (A) Bar plot showing the number of unique H3K4me1 peaks in queen and worker larvae. (B,C) Bar plot showing the percentage of unique H3K4me1 ChIP-seq peaks within promoters, introns, and exons in 2Q vs. 2W and 4Q vs. 4W. (D) Bar plot showing the number of differential H3K4me1 peaks in queen and worker larvae. (E,F) Bar plot showing the percentage of differential H3K4me1 ChIP-seq peaks within promoters, introns, and exons in 2Q vs. 2W and 4Q vs. 4W.

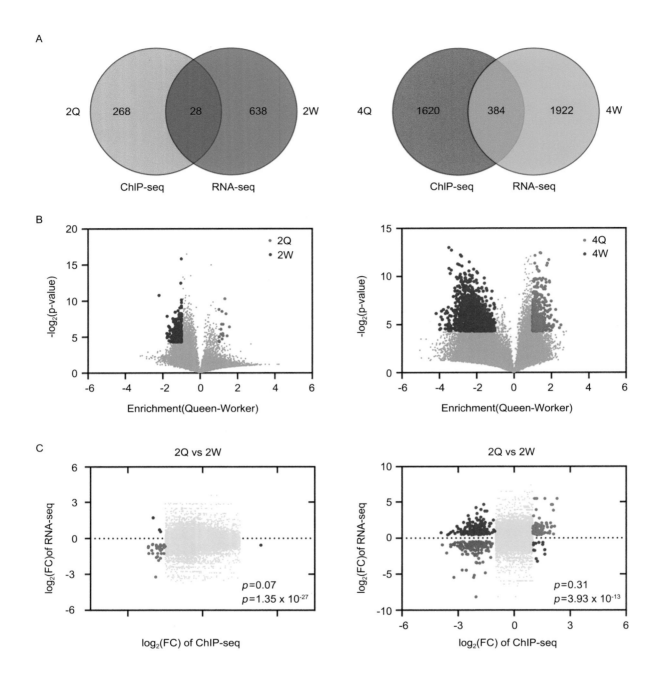

**Figure I-6-2** (A) Venn diagram showing overlap of differentially H3K4me1-modified genes and DEGs between queens and workers. (B) Volcano plot of the difference in enrichment between queens and workers against the negative log $p$-value for the H3K4me1 signal. Red areas indicate that the H3K4me1 modification is up-regulated in queens. Blue areas indicate that the H3K4me1 modification is up-regulated in workers. Gray indicates a lack of a significant difference ($p > 0.05$). (C) Scatter plots of the significant differences in expression, determined by ChIP-seq, between queen and worker larvae ($x$-axis) against the Log$_2$FC of transcript expression between queen and worker larvae ($y$-axis). Red indicates that the differences determined by RNA-seq and ChIP-seq are in agreement. Blue indicates that the opposite patterns were obtained by RNA-seq and ChIP-seq. Gray indicates that there is no significant difference ($p > 0.05$).

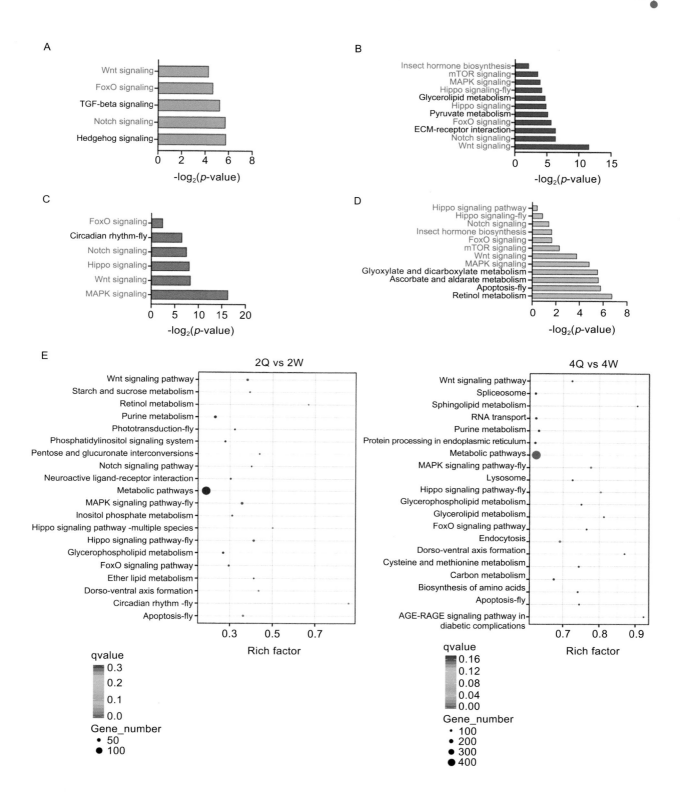

**Figure I-6-3** Negative $\log_2$ $p$-values for the honey bee caste differentiation-related KEGG pathways that are enriched with unique H3K4me1-related genes in 2Q (A), 2W (B), 4Q (C), and 4W (D). Pathways marked in red font are pathways associated with caste differentiation, while black font is not. (E) KEGG pathway enrichment analysis of differentially H3K4me1 peak-related genes in 2Q vs. 2W. (F) KEGG pathway enrichment of differentially H3K4me1 peak-related genes in 4Q vs. 4W.

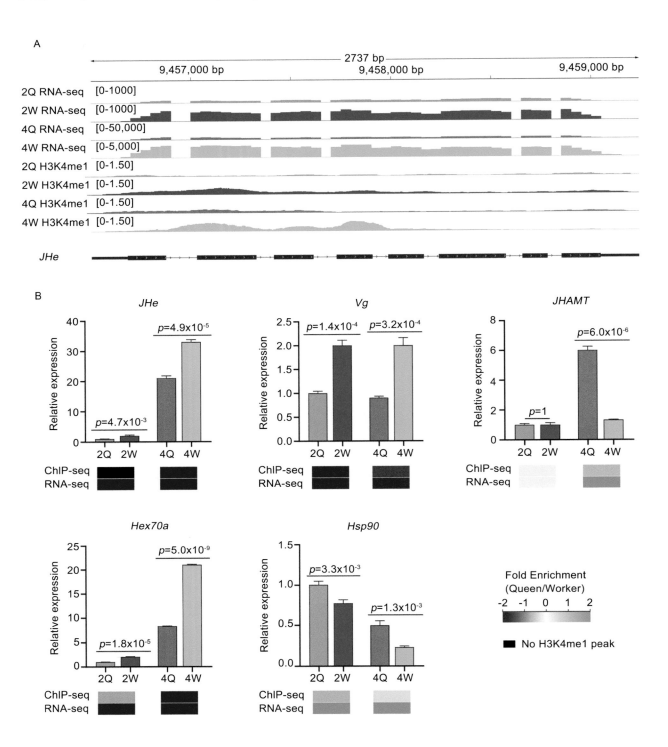

**Figure I-6-4** Function of H3K4me1 modification in caste differentiation by modulating the expression levels and modification abundance of genes. (A) Juvenile hormone esterase (*JHe*; LOC406066). The gene in this region reached genome-wide significance ($p \leq 0.01$) at both the 2nd and 4th instars, and had a greater than two-fold difference in H3K4me1 modification. (B) RNA expression levels and H3K4me1 abundance of differentially expressed candidate caste-differentiation-related transcripts are shown. Expression levels are expressed as mean ± SEM relative to a reference gene (*GAPDH*; LOC726445) in three replications. Fold enrichment is the difference in abundance between queen larvae and worker larvae; green denotes a higher fold enrichment, while purple denotes a lower fold enrichment. Differences in relative expression levels were analyzed using *t*-tests.

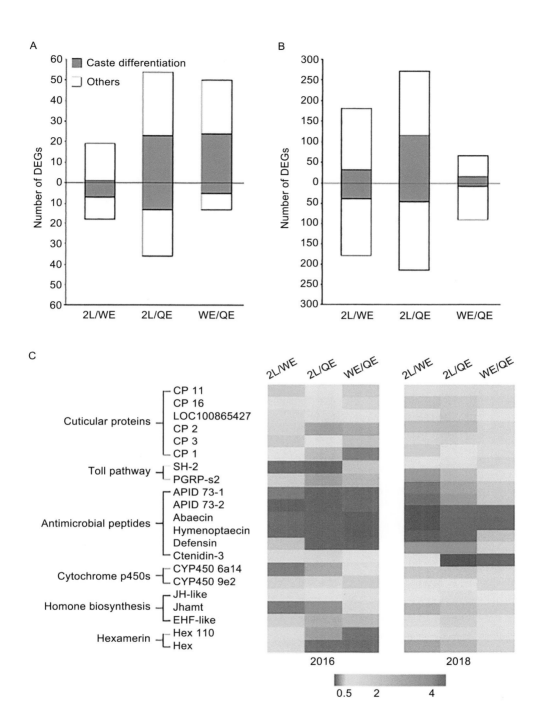

**Figure II-1-4** Summary of gene expression differences in pairwise comparisons between QE, WE and 2L in 2016 (A) and 2018 (B). A. total numbers of DEGs detected as significantly upregulated (above line) and downregulated (below line) in each comparison from 2016 RNA-Seq. Grey areas mark the numbers of DEGs that have previously been identified as differing between either queens and workers, or between queens of different quality (details and references in Table S1). B. total numbers of DEGs detected as significantly upregulated (above line) and downregulated (below line) in each comparison from 2018 RNA-Seq (details and references in Table S2). C. gene expression ratios (colour coded by scale bar) of selected DEGs with proposed functional roles in hormone synthesis, caste differentiation, immune function and detoxification. See also Table S1 and S2.

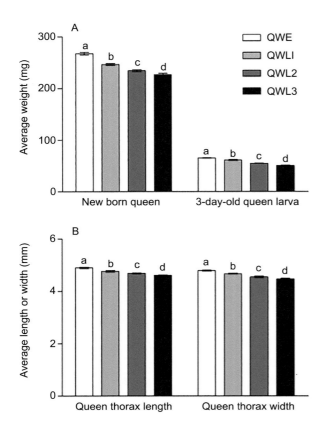

**Figure II-2-1** A. Mean (+SE) weight of new born queens and 3-day-old queen larvae, B. Mean (+SE) thorax width and length of new born queens, from QWE (open), QWL1 (grey), QWL2 (black) and QWL3 (diagonal stripes). Different letters above each bar indicate significant differences ($P < 0.05$, ANOVA test followed with Fisher's PLSD test).

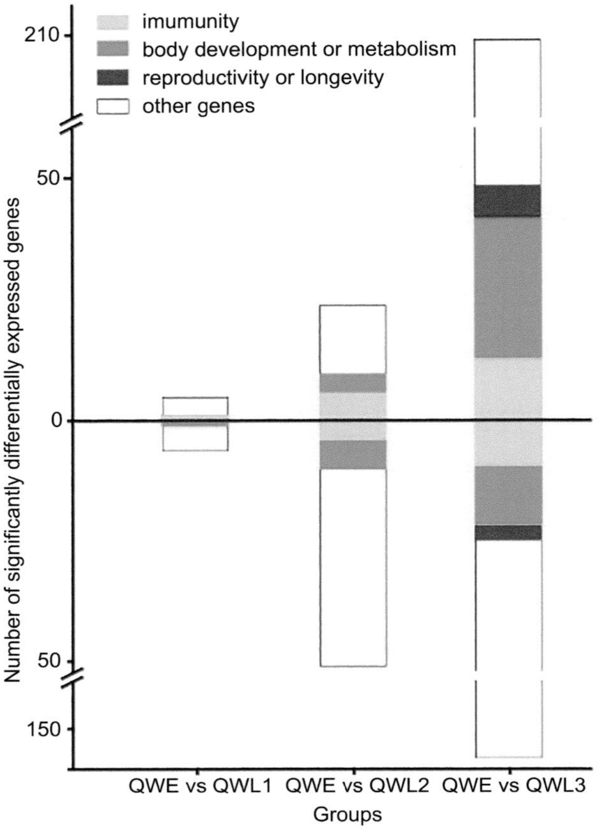

**Figure II-2-2** Significantly differentially expressed genes in three comparisons. Genes were identified as differentially expressed if both the FDR < 0.05, and the absolute value of the log2 fold change was ≥ 1. Genes involved in immunity (yellow), body development (green), reproduction or longevity (purple) and other functions (open bars) are shown. Top bars represent up-regulated genes in QWE compared to other groups. Lower bars are down-regulated genes.

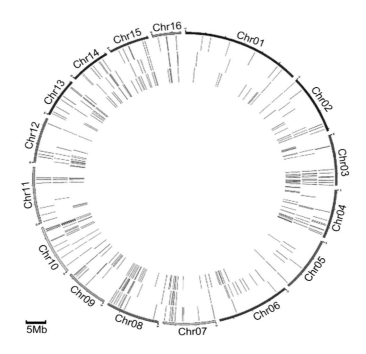

**Figure II-2-3** Distribution of significantly differentially methylated regions (DMRs) from three comparisons for the 16 honeybee chromosomes. DMRs were deemed significantly differentially methylated across QWE, QWL1, QWL2 and QWL3 with a false discovery rate (FDR) <0.05 and log2 fold change ≥ 1.5 in sequence counts. The DMRs of QWE/QWL1, QWE/QWL2 and QWE/QWL3 comparisons are presented from outer to inner respectively. Red plots are up-regulated DMRs in QWE compared to other three groups, whereas green plots are down-regulated ones. Chromosome name and scale are indicated on the outer rim.

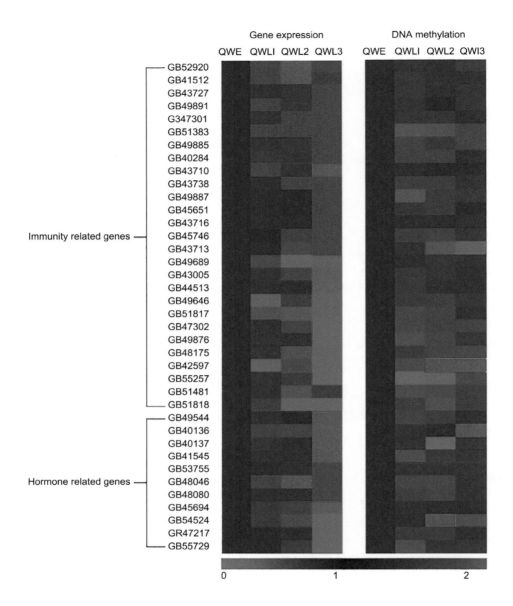

**Figure II-2-4** Expression and DNA methylation of 38 immunity and hormone related genes among QWE, QWL1, QWL2 and QWL3. These genes were identified by their functions in immunity or hormone biosynthesis, and were at least significantly differentially expressed between one comparison of QWE and QWLs. The ratio of gene expression in QWLs against QWE were used for presenting the expression level of each gene. Green indicates down-regulation in QWLs compared to QWE, red indicates up-regulation and black indicates no difference. Left side is the gene ID and gene function, middle is the ratio of gene expression of each gene and right is the ratio of DNA methylation for each gene. More detailed information of these 38 refers to Table S6.

**Figure II-3-1** Experimental design. (A) Lineage of queens and measurements. (B) Timing of the grafting and brood transfers. In the first generation (G1) a queen artificially inseminated with the semen of a single drone was caged on a plastic frame for 6 hours to obtain eggs of a known age. Some of the eggs (E) were transferred to queen cells after 48 h, while other eggs were left for 96 h and 120h and transferred after they reached the first (L1) or second (L2) larval instar respectively. The queens obtained were in turn caged on a plastic frame for 6 h and their brood was again grafted onto queen cells after 48 h, 96 h and 120 h and so on for the next three consecutive generations (generations G2 - G4). At each generation queens that were not used to raise the next generation of eggs were killed, the number of ovarioles was counted and the DNA genome-wide level of methylation from the brain, thorax and ovarioles was measured.

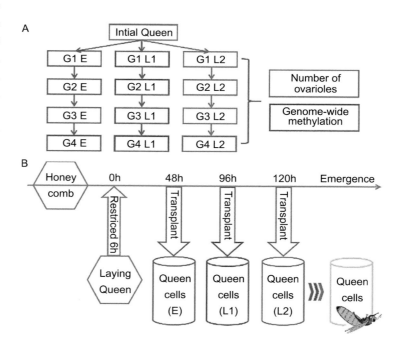

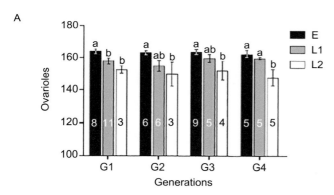

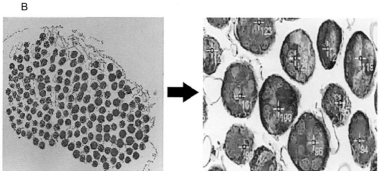

**Figure II-3-2** (A) Number of ovarioles in E, L1 and L2 queens in generations G1 - G4. Bars show mean ± SEM. Sample size was shown in each bar. One-way ANOVA was performed on each generation. Different letters above bars indicated significantly difference within each generation. (B) Specimen ovariole section illustrating ovariole counting methods (see Methods).

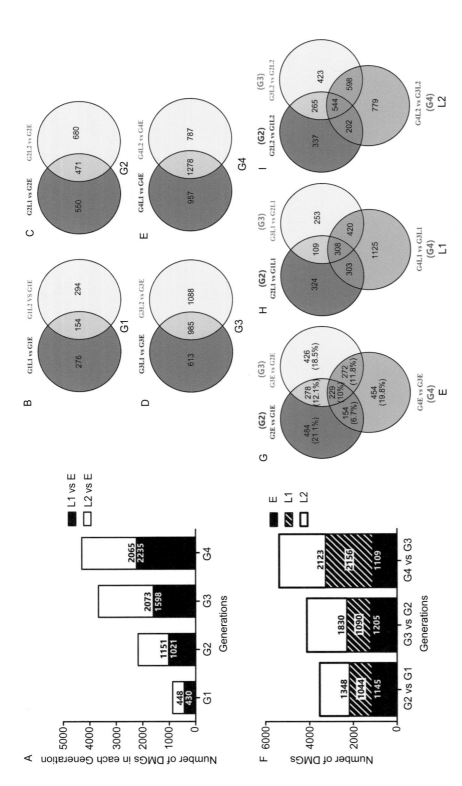

**Figure II-3-3** (A) Summary of numbers of differentially methylated genes (DMGs) comparing L1 with E (black), and L2 with E (white) in generations G1 - G4. The number of DMGs in each comparison was written into each bar (Summary of DMG IDs in Table S4). (B-E) Venn diagrams of the numbers of DMGs comparing L1 with E, and L2 with E in generations G1 - G4. (F) Numbers of DMGs comparing each queen rearing type across successive generations. The three bars showed DMGs in comparisons of G2 with G1, G3 with G2, and G4 with G3. Within each stacked bar we showed the number of DMGs for each of the three different queen rearing groups (for example, comparing G2E with G1E, G2L1 with G1L1, and G2L2 with G1L2). The number of DMGs was shown in each bar. Summary of DMG IDs in Table S5. (G-I) Venn diagrams of the numbers of DMGs comparing GnE with Gn-1E (G), GnL1 with Gn-1L1 (H), and GnL2 with Gn-1L2 (I) (n = 2, 3, or 4).

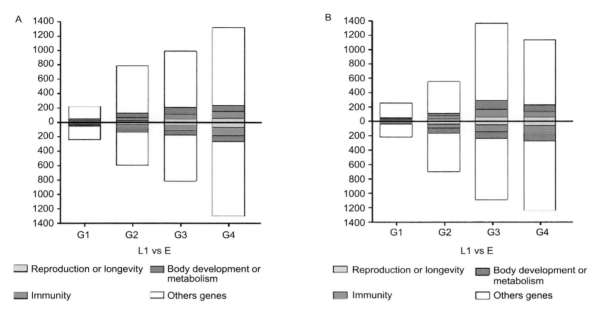

**Figure II-3-4** Number of differentially methylated genes in different functional classes within each generation, when comparing L1 with E (A) and L2 with E (B).

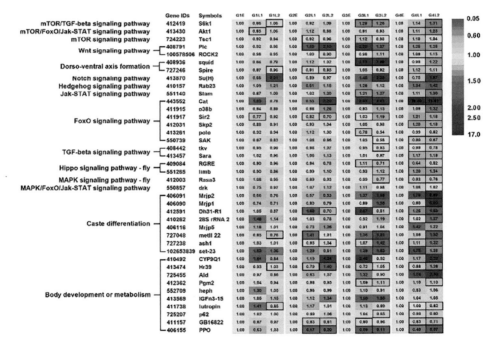

**Figure II-3-5** 40 focal genes were selected for their known functions in caste differentiation, body development or metabolism of honey bee. For these 40 genes we compared within each generation the relative methylation level of L1 and L2 queens with E queens for each gene. Methylation level was calculated by comparing the proportions of methylated reads at each site in all exons of a gene for each sample group. Relative methylation level (shown by color of each box) was then calculated as the ratio of methylation levels for each comparison (within a generation, L1 with E or L2 with E). Green and blue indicated hypo-methylated genes in groups compared with E. Red and purple indicated hyper-methylated genes, and yellow indicates no difference. The deeper the color, the greater the difference. Black borders indicated that there was at least one exon in this gene that was significantly differentially methylated between the compared groups. For each gene we showed gene functions, gene IDs and gene symbols from left to right. More detailed information of these 40 genes have been provided in Table S8.

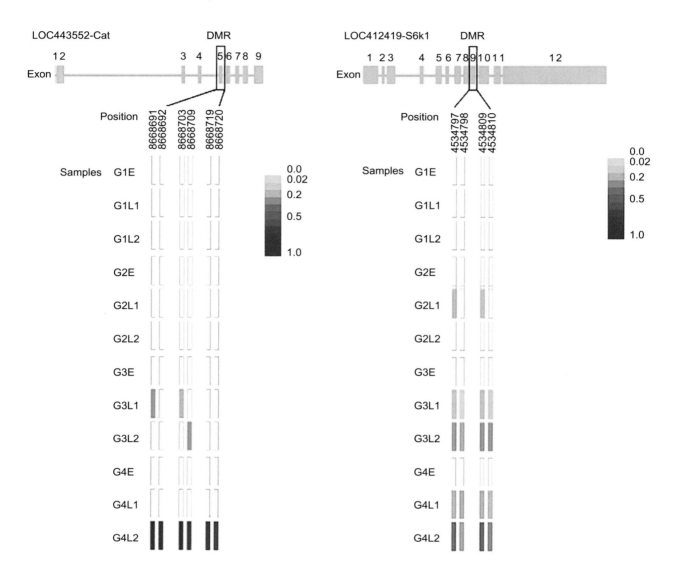

**Figure II-3-6** Two genes (*Cat* and *S6k1*) associated with caste differentiation were analyzed in detail. For each gene we showed the methylation sites within focal exons. The color indicated the level of methylation at each site in all rearing groups across all generations. Yellow, red and purple indicated low, medium and high methylation levels respectively. White indicated no detected methylation. From the top to the bottom and left to right are gene IDs, gene symbols, DMRs, number of all exons, number of sites and samples with methylation level indicated by colors. More detailed information of these two genes has been provided in Table S9.

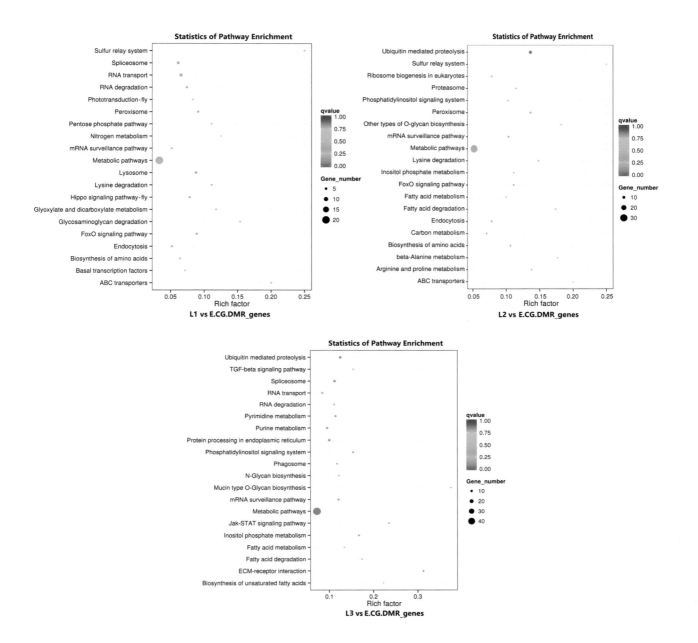

**Figure II-4-4** Scatter plot of KEGG pathways enriched in DEGs in comparisons of L1 vs E, L2 vs E, and L3 vs E (Details shown in Table S5). The size of the dots indicates the number of DEGs in that functional pathway. The color of the dots indicates $p$-value. The rich factor indicates the relative enrichment of DEGs in the KEGG pathway (number of DEGs in the KEGG pathway / total number of annotated genes in the same KEGG pathway). The larger the Rich factor, the greater the degree of enrichment.

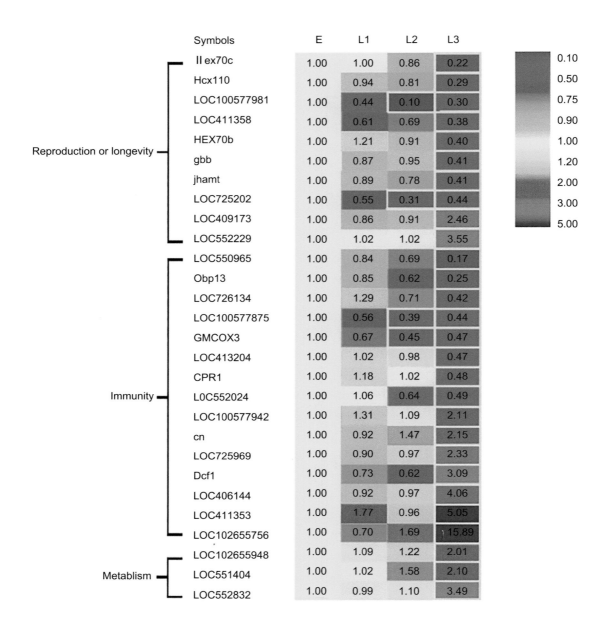

**Figure II-4-5** 28 focal genes related to functions in reproduction, longevity, or immunity were selected for further analysis. The value in the box indicates the ratio of the gene expression levels of all queen groups (E, L1, L2 and L3) with E group of each gene. Red, purple and black indicates up-regulated genes in groups compared with E, green and blue indicates down-regulated genes in groups compared with E. Yellow indicates no difference. Grey border indicates that there is a significant difference in gene expression between the group with E group for that gene. More detailed information shown in Table S6.

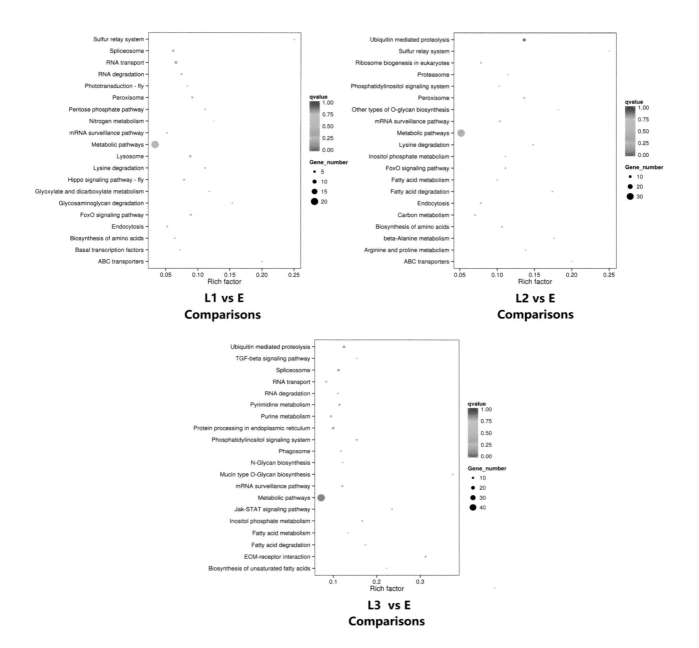

**Figure II-5-2** Scatter plot of the 20 most enriched KEGG pathways enriched in DMGs in comparisons of L1 vs E, L2 vs E, and L3 vs E. The size of the dots indicates the number of DMGs in that functional pathway. The color of the dots indicates $q$-value (corrected $p$-value). The rich factor indicates the relative enrichment of DMGs in the KEGG pathway (number of DMGs in the KEGG pathway / total number of annotated genes in the same KEGG pathway). The larger the Rich factor, the greater the degree of enrichment.

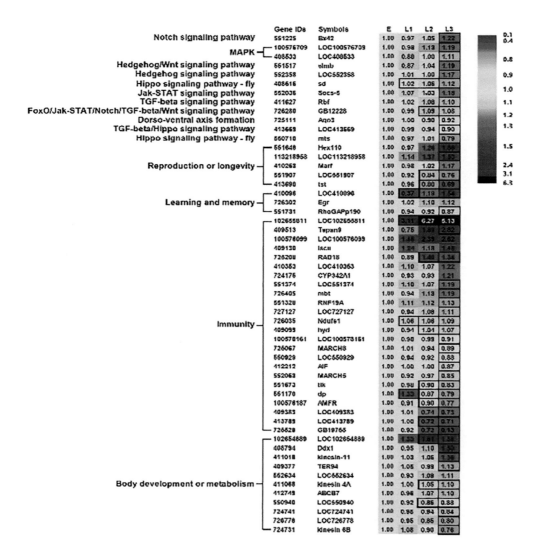

**Figure II-5-3** 55 focal genes, related to functions in reproduction, longevity, immunity, learning and memory, body development or metabolism, were selected for further analysis. Relative methylation level (given as the value in the box and indicated by the colour of each cell) was calculated by comparing the proportions of methylated reads at each CG site in all exons of the gene between larval reared queen groups (L1-L3) compared with E. Green and blue indicates hypo-methylated genes in groups compared with E. For the scare bar, red, purple and black indicates hyper-methylated genes, and yellow indicates no difference. Black borders indicate that there was at least one exon in this gene that was significantly differentially methylated in the comparisons. Boxes with no black border indicates that no exon in this gene was significantly differentially methylated in the comparisons. A differentially methylated exon is part of the differentially methylated region (DMR). More detailed information is given in Table SVI.

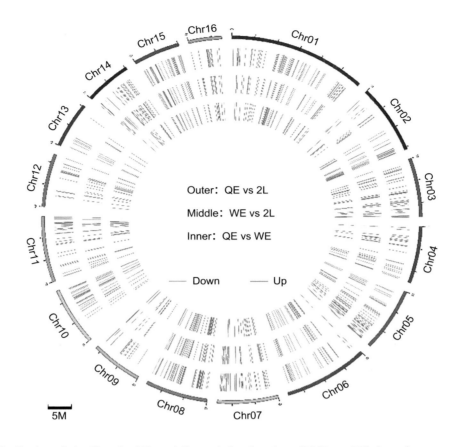

**Figure II-6-2** Distribution of significantly differentially methylated regions (DMRs, mCG) from three comparisons for the 16 honeybee chromosomes. The DMRs of QE/2L, WE/2L and QE/WE comparisons are presented from outer to inner, respectively. Red plots are upregulated DMRs and green plots are downregulated ones. Chromosome name and scale are indicated on the outer rim.

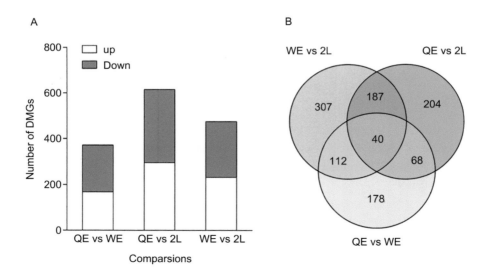

**Figure II-6-3** (A) Significantly differentially methylated genes (DMGs) in three comparisons. White bars represent upregulated DMGs in each comparison (former compares to latter) and grey bars represent downregulated ones. Detailed information of each DMG refers to Table S2 to 4. (B) The venn diagram of DMGs among QE, WE and 2L.

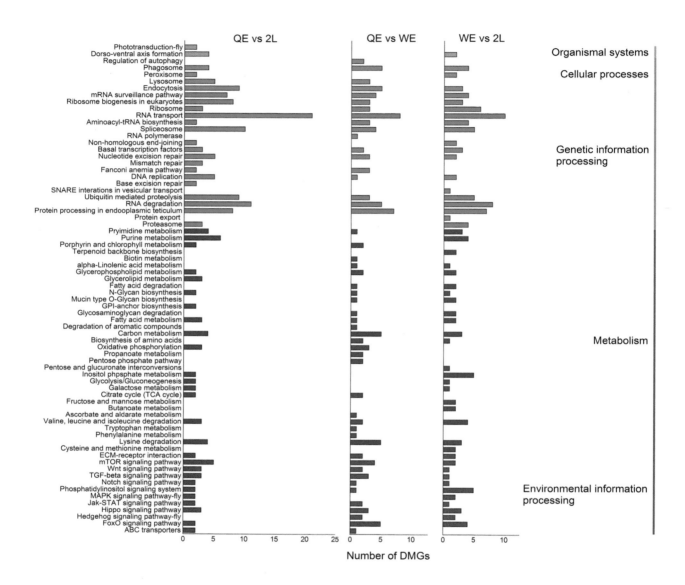

**Figure II-6-4** Enrichment of DMGs to KEGG categories and pathways. All DMGs from QE/2L, WE/2L and QE/WE comparisons were mapped to 72 KEGG pathways which belong to four categories (color marked, right side). Bars indicate number of DMGs in each pathway.

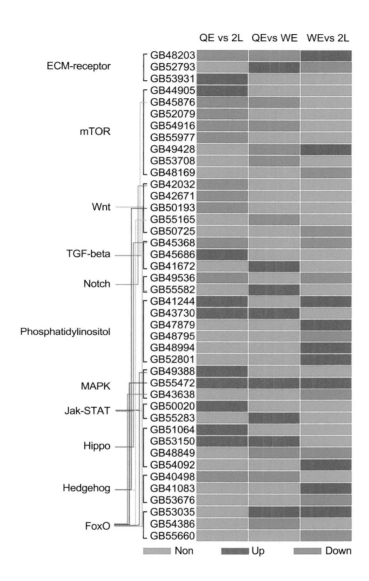

**Figure II-6-5** DNA methylation of 42 DMGs enriched in 11 caste-differentiation related KEGG pathways among QE, WE and 2L. Blue indicates significant downregulation in each comparison, red indicates upregulation and green indicates no difference. Left side is the gene ID and KEGG pathways. Some genes were involved in two or more pathways; therefore they were marked with different color lines.

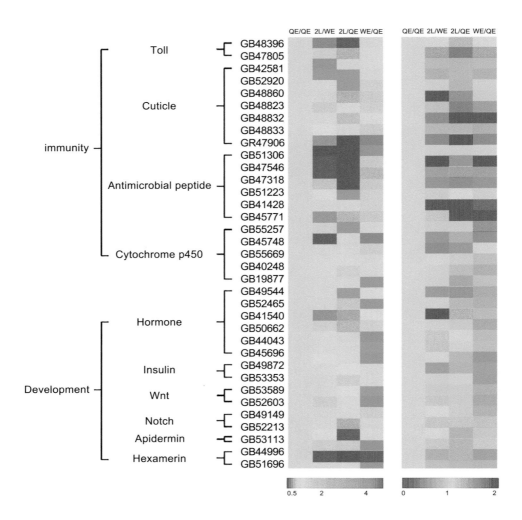

**Figure II-6-6** Gene expression and related DNA methylation levels of 35 differentially expressed genes. Gene expression ratios and DNA methylation ratios (colour coded by scale bars) of selected DEGs with proposed functional roles in immunity and development. The left heat map is gene expression and the right is DNA methylation.

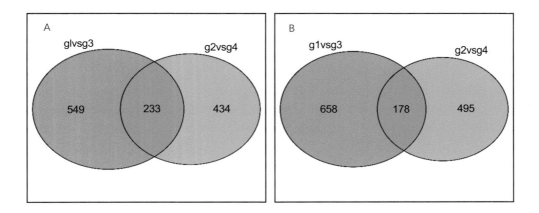

**Figure II-7-1** Venn analysis of the sexual maturity related DEGs in queens and drones. (A) up-regulated genes during sexual maturity both in queens and drones. (B) down-regulated genes during sexual maturity both in queens and drones.

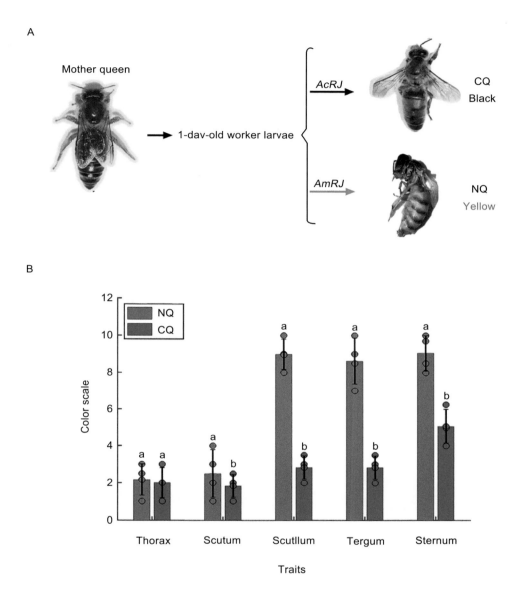

**Figure II-9-1** (A)The body color alternation in *Apis cerana-Apis mellifera* nutritional crossbreed. The 1-day worker larvae from their *Apis cerana* mother queen were fed with *Apis mellifera* royal jelly (AmRJ) based diet or *Apis cerana* royal jelly (AcRJ), resulted in yellow color nutritional crossbreed queens (NQ) or black color control queens (CQ) respectively. (B) The body color quantification had significant differences of (NQs, $n=4$, Scutum's $M=2.5$, $SD=1.29$, $t=0.85$, Scutellum's $M=8.9$, $SD=0.81$, $t=5.84$, Tergum's $M=8.5$, $SD=1.25$, $t=63.81$, and Sternum's $M=9.5$, $SD=0.95$, $t=15.4$) compared to (CQs, $n=4$, Scutum's $M=1.83$, $SD=0.62$ $t=5.11$, Scutellum's $M=0.81$, $SD=0.62$, $t=-16.12$, Tergum's $M=2.8$, $SD=0.62$ $t=63.81$, and Sternum's $M=5.8$, $SD=0.92$, $t=15.14$) all $P$ were<0.05) based on Ruttner's color scales. Bars present as values of Mean ± SD, the black dots on the top of each bar represent replicates. Different letters on the top of bars indicate significant difference ($p<0.05$, Independent-sample t-test).

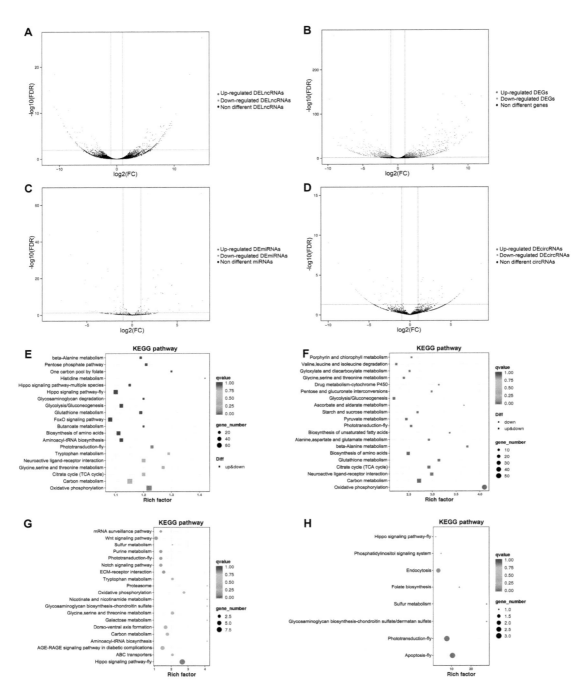

**Figure II-9-2** (A) The volcano diagram of lncRNAs between NC and NQ. The red spots represent up-regulated DElncRNAs in NQ compared to CQ, whereas the green spots represent down-regulated DElncRNAs; The black spots are non-different lncRNAs. (B) The volcano diagram of DEGs between NC and NQ. The red spots represent up-regulated DEGs in NQ compared to CQ, whereas the green spots represent down-regulated DEGs. lncRNAs with FDR <0.05, |log2 (Fold change) | ≥ 1 were identified as DElncRNAs. Same to DEGs, DEmiRNAs and DEcircRNAs. (C) The volcano diagram of miRNAs between NC and NQ. The red spots represent up-regulated DEmiRNAs in NQ, whereas green spots represent down-regulated DEmiRNAs. (D) The volcano diagram of circRNAs between NC and NQ. The red spots represent up-regulated DEcircRNAs in NQ, whereas green spots represent down-regulated DEcircRNAs. (E), (F), (G) and (H) are the top pathways of KEGG enrichment of DEGs, DElncRNAs, DEmiRNAs and DEcircRNAs respectively. The sizes of circles represent the number of DEGs, DElncRNAs, DEmiRNAs and DEcircRNAs, and the colors of circles represent the p-values of enrichment.

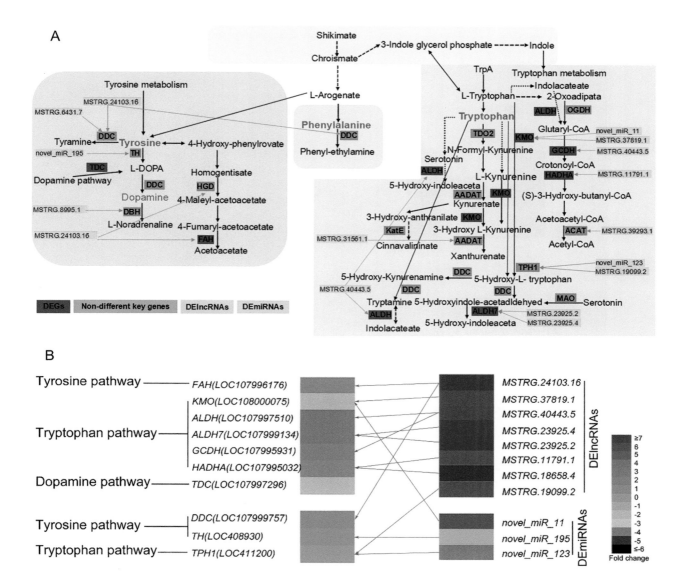

**Figure II-9-3** (A) The predicted KEGG network of honeybee body color alternation in nutritional crossbreed. This network is based on four key KEGG pathways including tyrosine, tryptophan, dopamine, and phenylalanine pathways. The key DEGs, DElncRNAs and DEmiRNAs involved in this network are also presented. Yellow bars represent DElncRNAs, sky blue bars represent DEmiRNAs, and purple bars represent DEGs and green bars represent key genes but not DEGs. Blue arrows mean DElncRNAs or DEmiRNAs involved into the regulation of related genes. (B) The heatmap of key DEGs, DElncRNAs and DEmiRNAs are involved into three key KEGG pathways. Different colors represent significantly differential expression of key DEGs, DElncRNAs and DEmiRNAs using their log2(Fold change) values. Olive green arrows represent the regulatory relationship between key DEGs and non-coding RNAs (DElncRNAs and DEmiRNAs).

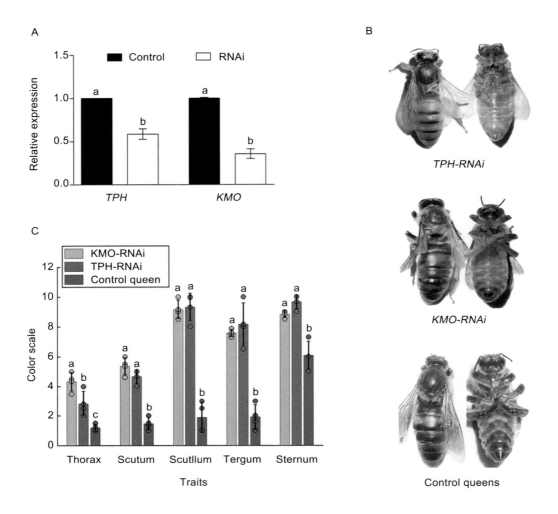

**Figure II-9-4** (A) The expression of TPH1 and KMO genes in 4-day *Apis cerana* queen larvae by RNAi. Bars represent mean ± SD values of relative gene expression. Different letters on the top of each bar represent the significant difference ($p<0.05$, independent-sample t-test). (B) The body color alternation of *Apis cerana* queens in RNAi experiment. The upper queens with a yellow body color are TPH1-RNAi group, the middle ones are KMO RNAi group with slight body color changes compared to control bees. The lower ones are control queens fed with negative siRNA. (C)The body color quantification of KMO-iRNA ($n=4$), TPH1-iRNA ($n=4$), and control queens($n=4$) based on Ruttner's color scales. Bars are present as values of Mean ± SD. Different letters on the top of the bars indicate significant differences ($p<0.0001$ $F=25.41$ $F$ critical $= 4.25$, One-Way ANOVA), the same letter indicates no significant difference, and the black dots on the top of each bar represent replicates.

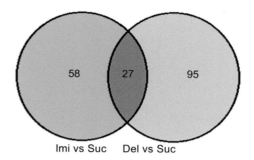

**Figure III-7-2**  Venn diagram of DEGs between groups. The light green part indicated that 58 DEGs specific to the imidacloprid - treatment group versus the control group. The bluish violet part indicated that 95 DEGs were specific to the deltamethrin - treatment group versus the control group. The brown part where two circles cross indicated that 27 DEGs were shared in both treatment groups.

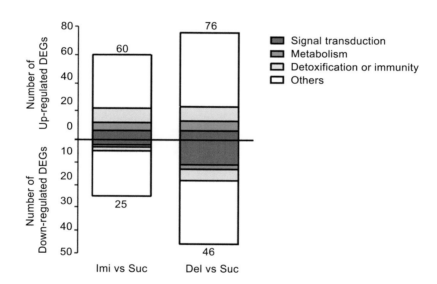

**Figure III-7-3**  Up and downregulated DEGs. In the imidacloprid - treatment group versus the control group, 60 genes were upregulated, and 25 genes were downregulated. In the deltamethrin - treatment group versus the control group, 76 genes were upregulated, and 46 genes were downregulated. The red colour represents sensory and signalling genes, the green colour represents metabolically related genes, the yellow colour represents detoxification or immunity genes, the white colour represents other genes.

Interestingly, 35 genes of the DEGs of the insecticide-treated versus control bees were involved in detoxification, immunity, sensation and signal transduction. There were 17 genes with significant differences in the imidacloprid-treatment group and 25 of such genes in the deltamethrin-treatment group (Table S3, Fig. III-7-4). There were eight commonly upregulated genes in both groups, including cytochrome P450 6a14, peroxisomal multifunctional enzyme type 2-like, apidaecins type 73, hymenoptaecin preproprotein, troponin C type I, troponin C type IIb and calmodulin-like protein 4. The significantly downregulated gene was major royal jelly protein 1 in both groups.

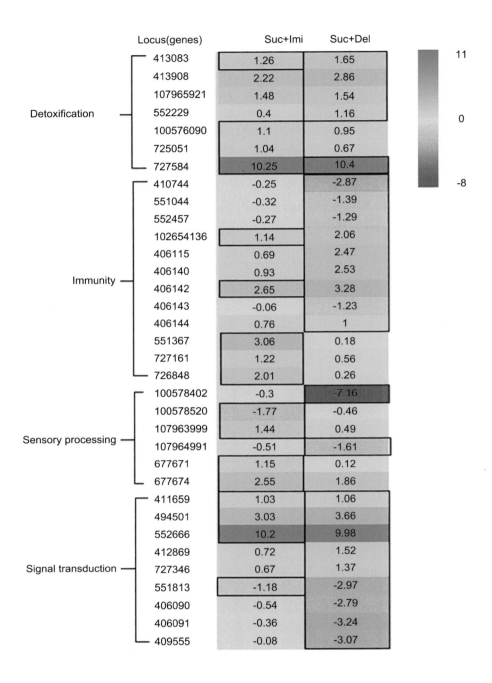

**Figure III-7-4** Heat map of DEGs between groups. Left column represents the imidacloprid - treated group compared with the control group and right column represents deltamethrin - treated group compared with the control group. Each row represents a gene, and the log2 ratio of the normalized transcript content is relative to the control, which is shown using different colours: Red to green represent the gradation of gene expression abundance from low to high. The data with a border represent a significant difference (the FDR ≤0.05 threshold and an absolute value of the log2 ratio≥1).

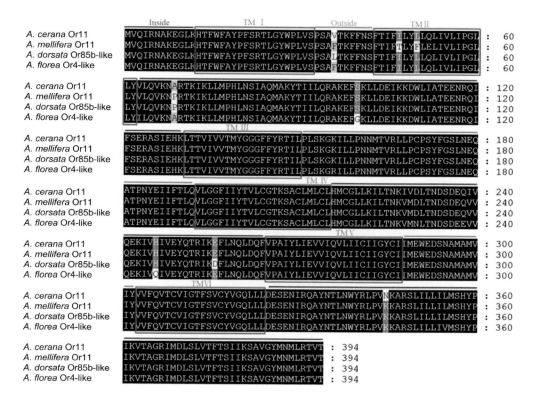

**Figure III-11-2** Alignments of AcOr11 with other honeybee OR sequences. Black shade: identity of sequences =100%; Gray shade: identity of sequences >75%. The six red boxes of AcOr11 amino acids sequences in respectively represent conserved transmembrane domains from TM Ⅰ to Ⅵ. The blue and green lines respectively reveal inside and outside membrane of AcOr11 protein positions.

**Figure III-13-1** A circle composed of queen retinue workers surrounding a feeding queen. Different colors on worker bee thoraxes stands represent different ages.

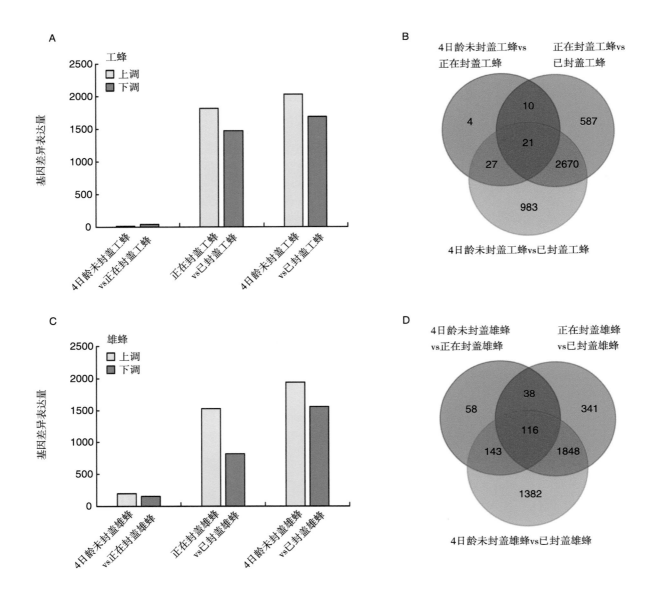

**图 III-17-3 工蜂与雄蜂不同封盖时期幼虫差异表达基因**

附录 A 彩图

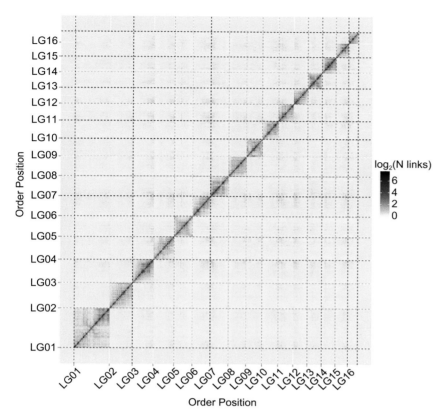

**Figure IV-1-1** Heat map of Hi-C contact information of the 16 chromosomes. Pixel colors represent different normalized counts of Hi-C links between 50-kb non-overlapping windows for all 16 chromosomes (chr) on a logarithmic scale.

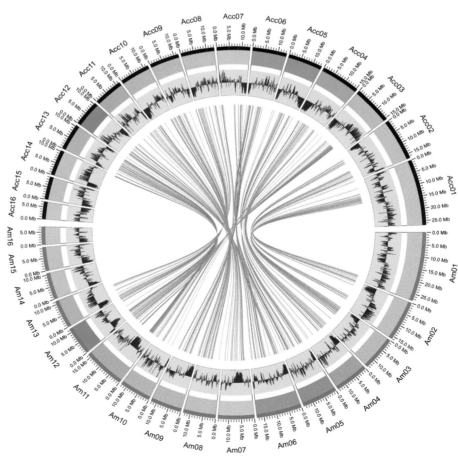

**Figure IV-1-2** The atlas of the *A. cerana* chromosomes.

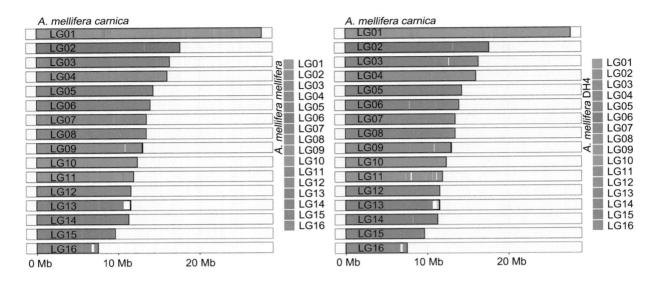

**Figure IV-2-2** Coverage map of the honeybee subspecies. The genomes of *A. mellifera mellifera* and *A. mellifera* DH4 were compared with *A. mellifera carnica*. Overall, 80% of the nucleotides were aligned.

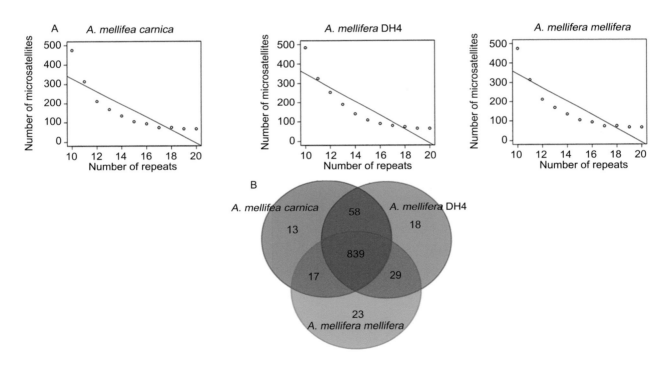

**Figure IV-2-3** Microsatellite analysis of the three honeybee subspecies. (A) the distribution of microsatellites of the dinucleotide motif. The number of microsatellites decreased with the increasing number of repeats for all three genomes. Overall, the variation of the microsatellite distribution among the three genomes was not significant. (B) Venn diagram of the linkage map makers among the three honeybee genomes. *A. mellifera* DH4 shared significantly higher number of markers with *A. mellifera carnica* compared with *A. mellifera mellifera*.

附录 A 彩图

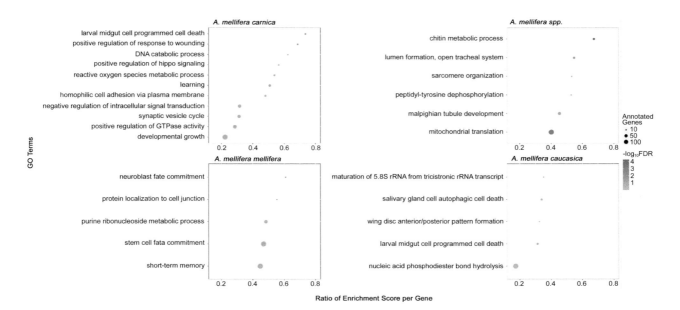

**Figure IV-2-5** Functional categories significantly enriched of genes under positive selection in the branch *A. m. ligustica*, using REVIGO to cluster the long list of significant GO terms. The medium cluster size was 0.7 and the semantic similarity measure SimRel.

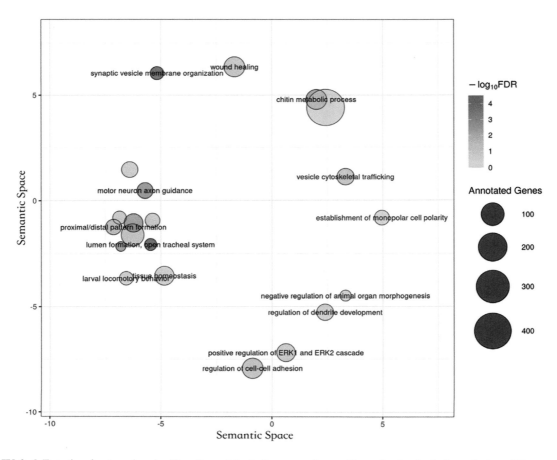

**Figure IV-2-6** Functional categories significantly enriched of genes under positive selection in the branch *A. mellifera* spp.

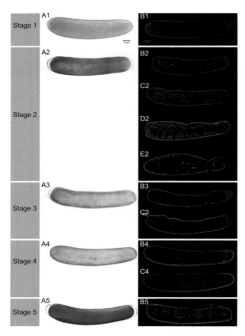

**Figure IV-3-1** Morphological observation of *Apis cerana* embryos at stages 1-5 after fixing (A1, A2, A3, A4 and A5) and PI staining (B1, B2, C2, D2, E2, B3, C3, B4, C4 and B5).

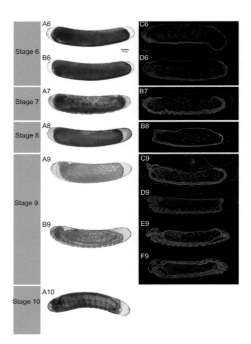

**Figure IV-3-2** Morphological observation of *Apis cerana* embryos at stages 6-10 after fixing (A6, B6, A7, A8, A9, B9 and A10) and PI staining (C6, D6, B7, B8, C9, D9, E9 and F9).

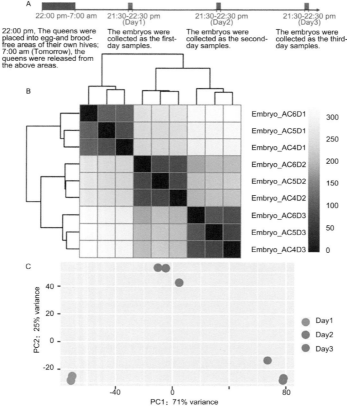

**Figure IV-4-1** Sample collection and clustering for Apis cerana in the egg stage. (A) The embryo samples were collected at 24 h (day 1), 48 h (day 2) and 72 h (day 3) after egg laying and rapidly stored in liquid nitrogen for next step of RNA extraction and RNA sequencing. (B) Heat map and hierarchical clustering of the embryo samples using the whole transcripts data. (C) Principle component analysis for the embryo samples using the whole transcripts data.

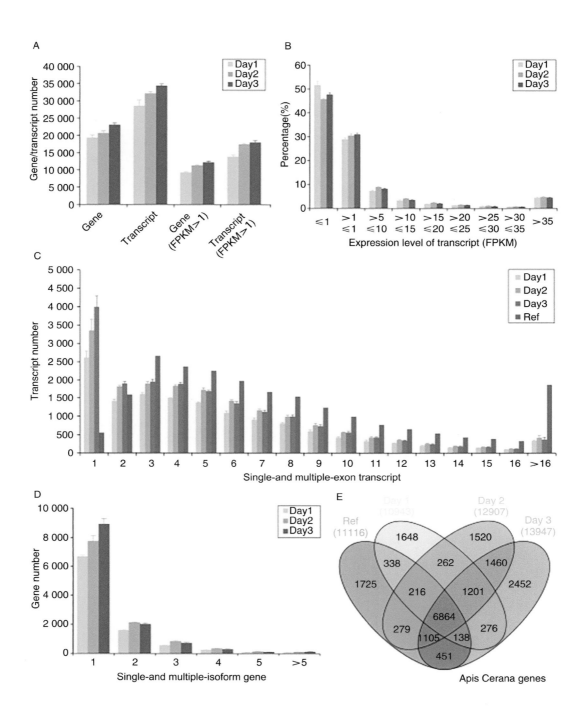

**Figure IV-4-2** The overview of embryonic transcriptome. (A) the genes or transcripts detected in the first-, second- and third-day embryos. (B) the expression level of transcript in the first-, second- and third- day embryos. (C) the distribution of single- and multi- exon transcripts. (D) the distribution of single- and multi- isoform genes. (E) Comparison for the expressed genes among the first-, second- and third- day embryos and reference genes.

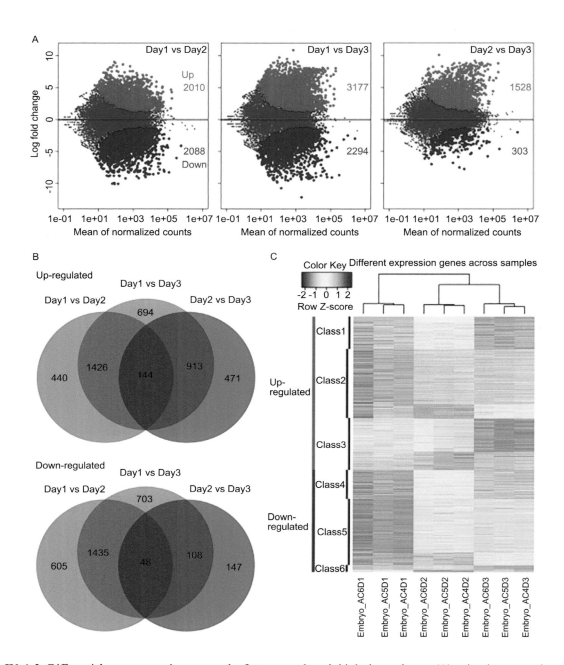

**Figure IV-4-3** Differential gene expression among the first-, second- and third- day embryos. (A) pair-wise comparison for the differential expressed genes among the first-, second- and third- day embryos. (B) up- and down- regulated genes among the first-, second- and third- day embryos. (C) Hierarchical clustering and heat map for all embryo samples using differential expressed gene data.

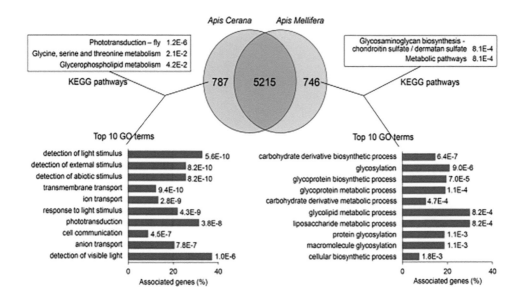

**Figure IV-4-4** Comparison of annotated genes in the first day embryo transcriptome between Apis cerana and Apis mellifera. Venn diagram showed the number of specific and common genes expressed in *Apis cerana* and *Apis mellifera*. All KEGG pathways listed have enrichment $P$ values < 0.05, and the top 10 GO terms were highlighted (for the complete GO terms and KEGG pathways, refer to Table S7). Corrected $P$-values were shown on the right flanking sites.

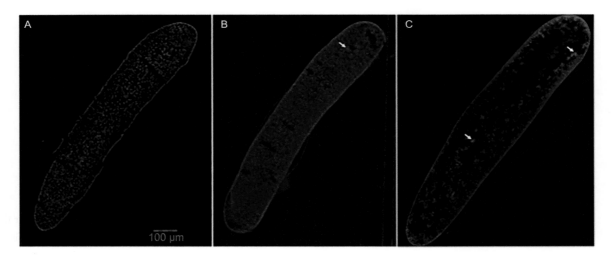

**Figure IV-5-1** Embryo slices for the 0-2, 4-6 and 6-8 h samples. The white arrow denotes the energid. Scale bar = 100 μm.

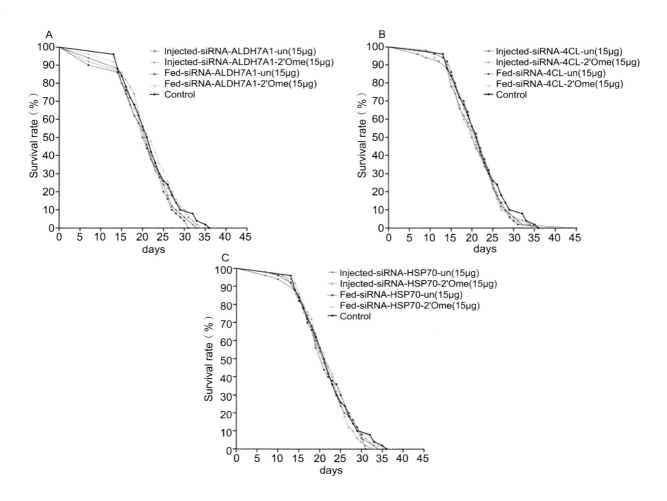

**Figure IV-6-1** Survival curves of honey bees after the injection and feeding of siRNA. (a) Effects of different delivery methods and different modified siRNAs on the survival rate of honey bees. Red solid line represents the injection of unmodified siRNA-ALDH7A1; Orange solid line represents the injection of 2'Ome modified siRNA-ALDH7A1; Blue dashed line represents the feeding of unmodified siRNA-ALDH7A1; Pink dashed line represents the feeding of 2'Ome modified siRNA-ALDH7A1; Black solid line represents honey bees without any treatment. (b) siRNA-4CL and (c) siRNA-HSP70 were used, otherwise consistent with (a). Each group has 50 honey bees.

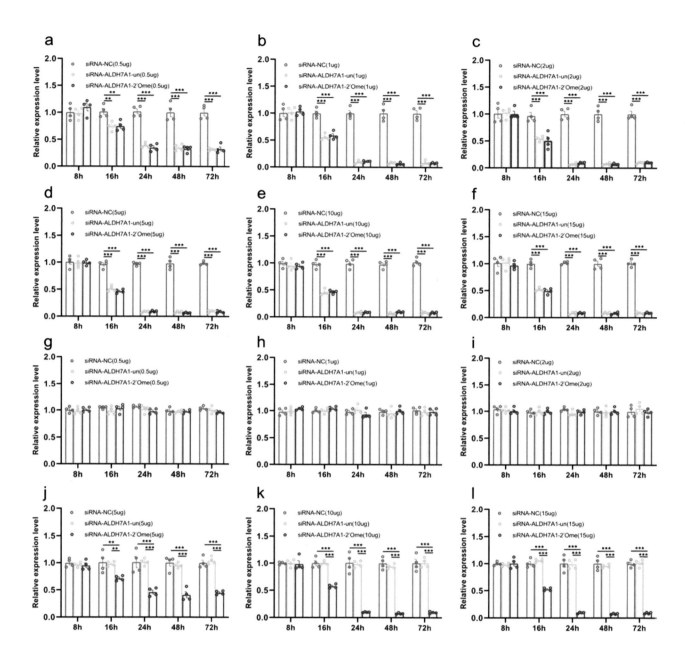

**Figure IV-6-2** The effects of siRNA on *ALDH7A1* mRNA transcripts in honey bee brains. The mRNA levels of *ALDH7A1* were analyzed at 8 h, 16 h, 24 h, 48 h, and 72 h after injection of 0.5 μg (a), 1 μg (b), 2 μg (c), 5 μg (d), 10 μg (e), and 15 μg (f) siRNA, or after feeding of 0.5 μg (g), 1 μg (h), 2 μg (i), 5 μg (j), 10 μg (k), and 15 μg (l) siRNA. The red column shows the relative expression of siRNA-NC injected. The orange column shows the relative expression of siRNA-ALDH7A1-un injected. The blue column shows the relative expression of siRNA-ALDH7A1-2′Ome injected. Each circle represents a biological repeat. ** indicates $p < 0.01$; *** indicates $p < 0.001$, tested by ANOVA test. Detailed data are shown in Table S2.

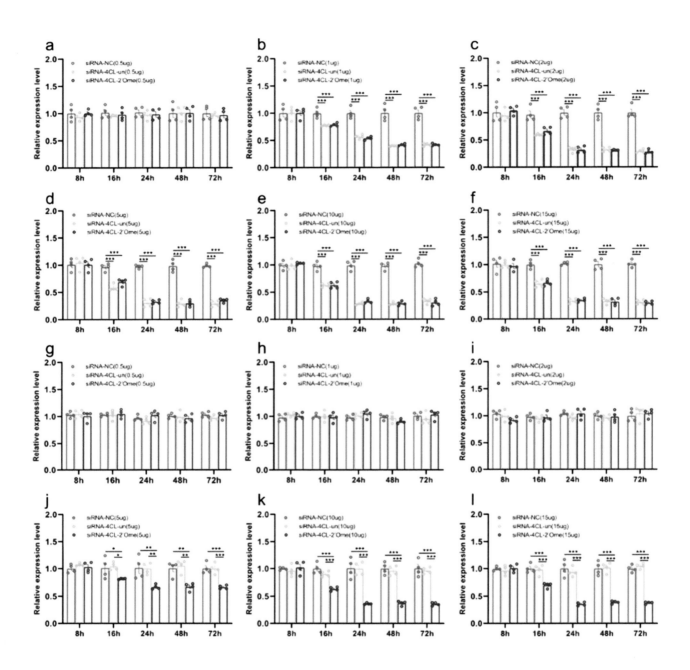

**Figure IV-6-3** The effects of siRNA on 4CL mRNA transcripts in honey bee brains. The mRNA levels of *4CL* were analyzed at 8 h, 16 h, 24 h, 48 h, and 72 h after injection of 0.5 μg (a), 1 μg (b), 2 μg (c), 5 μg (d), 10 μg (e), and 15 μg (f) siRNA, or after feeding of 0.5 μg (g), 1 μg (h), 2 μg (i), 5 μg (j), 10 μg (k), and 15 μg (l) siRNA. The red column shows the relative expression of siRNA-NC injected. The orange column shows the relative expression of siRNA-4CL-un injected. The blue column shows the relative expression of siRNA-4CL-2′Ome injected. Each circle represents a biological repeat. * Indicates $p < 0.05$, ** indicates $p < 0.01$, *** indicates $p < 0.001$, tested by ANOVA test. Detailed data are shown in Table S3.

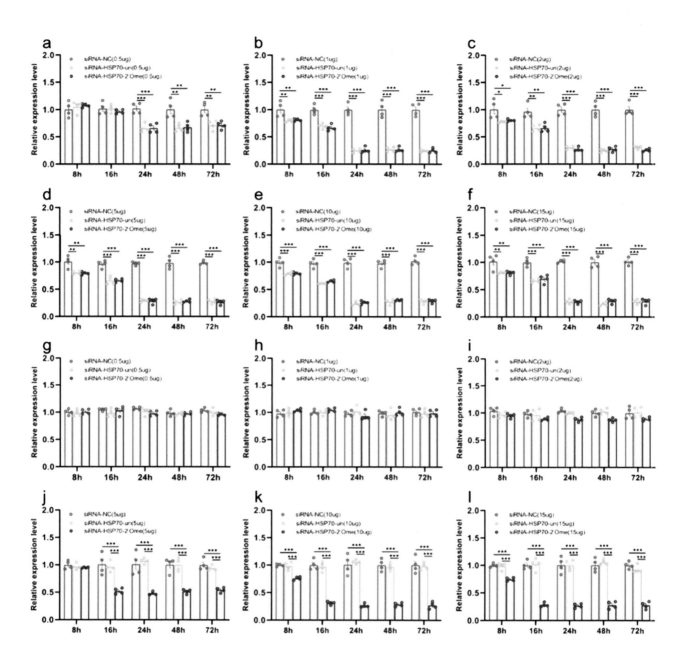

**Figure IV-6-4** The effects of siRNA on *HSP 70* mRNA transcripts in honey bee brains. The mRNA levels of *HSP 70* were analyzed at 8 h, 16 h, 24 h, 48 h, and 72 h after injection of 0.5 μg (a), 1 μg (b), 2 μg (c), 5 μg (d), 10 μg (e), and 15 μg (f) siRNA, or after feeding of 0.5 μg (g), 1 μg (h), 2 μg (i), 5 μg (j), 10 μg (k), and 15 μg (l) siRNA. The red column shows the relative expression of siRNA-NC injected. The orange column shows the relative expression of siRNA-HSP70-un injected. The blue column shows the relative expression of siRNA-HSP70-2′Ome injected. Each circle represents a biological repeat. * Indicates $p < 0.05$, ** indicates $p < 0.01$, *** indicates $p < 0.001$, tested by ANOVA test. Detailed data are shown in Table S4.

## 作者简介

曾志将，1965年7月生，江西吉水人。1987年毕业于福建农学院养蜂专业，获农学学士学位。同年分配至江西农业大学从事养蜂教学与研究工作，1992年破格晋升为讲师，1993年破格晋升为副教授，1995年聘任为硕士研究生导师，1998年破格晋升为教授，2002年获同济大学理学博士学位，2003年聘任为博士研究生导师。现任江西农业大学副校长/蜜蜂研究所所长、二级教授、博士研究生导师、特种经济动物饲养博士点学科带头人、江西省蜜蜂生物学与饲养重点实验室主任、国家畜禽遗传资源委员会委员兼蜂专业委员会副组长、农业农村部岗位科学家。是国家有突出贡献中青年专家、国家百千万人才工程入选者、享受国务院政府特殊津贴专家、全国蜂业突出贡献奖获得者、中国养蜂学会副理事长、农业农村部授粉昆虫生物学重点实验室学术委员会主任、江西省示范研究生导师创新团队带头人、江西省高等学校动物生产类专业教学指导委员会主任、江西省赣鄱英才555工程领军人才、江西省主要学科学术和技术带头人。

先后主持10项国家自然科学基金、1项国家重点研发计划课题、1项国家现代蜂产业技术体系岗位科学家项目（2008年连续资助至今）以及10多项省部级项目。以第一完成人获得江西省科学技术进步二等奖2项、江西省自然科学二等奖1项、江西省技术发明二等奖1项、江西省自然科学三等奖1项。参与完成科研成果获得省部级二等奖2项、三等奖1项。获得国家教学成果二等奖1项、江西省教学成果特等奖1项、一等奖5项、二等奖3项。在 Current Biology、Molecular Ecology、Insect Biochemistry and Molecular Biology、Insect Science、iScience、Journal of Agricultural and Food Chemistry、《中国农业科学》《昆虫学报》等国内外学术期刊发表学术论文200多篇，其中SCI收录论文90多篇，并且有多篇论文获《中国科学报》、"科学网"、New Scientist、Science Daily、PNAS-Front Matter 等国内外权威科学媒体专题报道。主编出版了全国高等农林院校统编教材《养蜂学》。制订了1项国家标准和3项地方标准。获授权国家发明专利12项。先后指导培养了2名博士后、27名博士研究生、76名硕士研究生，其中5名研究生获江西省优秀博士论文奖，8名研究生获江西省优秀硕士论文奖。

先后被评选为江西省优秀青年教师、江西省模范教师、江西省优秀共产党员、江西省师德标兵、江西省教学名师、全国优秀教师。曾到法国、美国、澳大利亚留学和合作研究。